W9-DGX-810

Critical Reviews of

OXIDATIVE STRESS AND AGING

Advances in Basic Science, Diagnostics and Intervention

Volume II

Critical Reviews of

OXIDATIVE STRESS
AND AGING

Advances in Basic Science, Diagnostics and Intervention

Volume II

Editors

Richard G. Cutler

Kronos Longevity Research Institute, Arizona, USA

Henry Rodriguez

National Institute of Standards and Technology, Gaithersburg, Maryland, USA

World Scientific
New Jersey • London • Singapore • Hong Kong

Published by

World Scientific Publishing Co. Pte. Ltd.

5 Toh Tuck Link, Singapore 596224

USA office: Suite 202, 1060 Main Street, River Edge, NJ 07661

UK office: 57 Shelton Street, Covent Garden, London WC2H 9HE

British Library Cataloguing-in-Publication Data
A catalogue record for this book is available from the British Library.

ISBN 981-02-4636-6 (Set)
ISBN 981-238-996-2 (Vol. 1)
ISBN 981-238-997-0 (Vol. 2)

This book is printed on acid-free paper.

Printed in Singapore by Uto-Print

I dedicate this book to my wife Jessica and sons Ronald and Roman that have not only been so encouraging, but also understanding of why science is so important to me particularly when it deals with the possibility of significantly extending the healthy and productive years of human life span.

Richard G. Cutler

This book is dedicated to my parents Ricardo and Micaela Rodriguez, who taught me to believe in myself and to my wife Joy and daughter Dana who gives me the strength to take on and conquer the many challenges that life has to offer.

Henry Rodriguez

Acknowledgments

The editors of *Critical Reviews of Oxidative Stress and Aging: Advances in Basic Science, Diagnostics, and Intervention* and The Oxidative Stress and Aging Association (O2SA) would like to thank the following associations for their outstanding patronage and support of the Second International Conference on Oxidative Stress and Aging and their continuing support towards advancing the field of Oxidative Stress and Aging in research, diagnostics and therapeutics worldwide.

American Aging Association
American Federation For Aging Research
Cayman Research
Ciphergen
Galileo Laboratories
Great Smokies Diagnostic Laboratories
HealthSpan Sciences
InterHealth
Karger
Kronos
Medinox
MitoKor
National Institute of Standards and Technology
NitroMed
Oxford Biomedical Research
Polyphenols Laboratory
The Kronos Longevity Research Institute
Zenith Technology

Preface

The fastest-growing segment of the United States population is age 65 years and older. At current rates, demographers forecast that by the year 2025, 65 year-olds will outnumber teenagers by almost 2:1. Fifty percent or more of the "baby boomer" population are projected to live to be Centenarians. Outside the United States, demographic trends are similar. As longer life span becomes a reality, there is a new emphasis on diagnostic and therapeutic applications that not only prolong life, but also increase the quality of life and productivity of the aging population.

Aging is associated with an increased incidence and prevalence of numerous diseases, many of them chronic. It is now known that free radicals cause extensive damage to cellular components that can lead to serious dysfunctions and death. Recently, free radicals have been associated with the aging process, several clinical disorders and a range of age-related diseases including atherosclerosis, cancer, and neurodegenerative diseases such as Parkinson's and Alzheimer's.

As a result, a new market is emerging: Oxidative Stress Diagnostics and Therapeutics. This new discipline in the field of Clinical Age Management and Longevity Medicine is revolutionizing the economics, science and practice of health care in new ways. Virtually every medical discipline will be reshaped by longevity research and its social implications.

The purpose of *Critical Reviews of Oxidative Stress and Aging: Advances in Basic Science, Diagnostics and Intervention* is to provide a comprehensive review of the most up-to-date information pertaining to the translational research field of oxidative stress and aging. **The book focuses on understanding the molecular basis of oxidative stress and its associated age-related diseases with the goal being the development of new and novel methods in treating human aging processes.** Over 100 of the leading experts in this field whose specialty includes biogerontology, geriatric medicine, free radical chemistry and biology, oncology, cardiology, neurobiology, dermatology, pharmacology, nutrition, and molecular medicine, have contributed information to this book. Accordingly, this reference book is essential reading to a broad range of individuals including researchers, physicians, corporate industry leaders, graduate and medical school students, as well as the many health conscious individuals who wish to learn more about the emerging field of oxidative stress and aging with an emphasis on diagnostics and intervention.

This is an exciting time to be involved in oxidative stress and longevity research. Answers to some of the most intractable problems in the field now appear to be within reach.

Richard G. Cutler   Henry Rodriguez

List of Contributors

In writing this book, the following authors have provided invaluable information and data. Special thanks are due to the following contributors (in alphabetical order), but the responsibility for any errors or omissions in the text is entirely that of the authors.

Acworth, Ian N.
Adinolfi, Christy
Akman, Steven A.
Alessio, Helaine M.
Allen, R.G.
Alleva, Renata
Ames, Bruce N.
Arik, Tali
Arking, Robert
Arteel, Gavin E.
Ayala, Antonio
Azzi, Angelo
Bagchi, Debasis
Balin, Arthur K.
Banks, Dwayne A.
Baudry, Michel
Beckman, Joseph S.
Block, Gladys
Blumberg, Jeffrey B.
Bogdanov, Mikhail B.
Bohr, Vilhelm
Bonkovsky, Herbert L.
Bourdat, A.-G.
Bray, Tammy M.
Bronkovsky, Herbert L.
Brown-Borg, Holly M.
Buettner, Garry R.
Cadet, Jean
Cai, Jiyang

Cao, Guohua
Carrillo, Maria-Cristina
Carroll, A.K.
Chen, Hsiu-Hua
Chen, XinLian
Christen, Stephan
Clarke, Charlotte H.
Clarke, Mark S.F.
Coles, L. Stephen
Collins, Andrew R.
Cooke, Marcus S.
Cutler, Richard G.
Cutler, Roy G.
Dai, Shu-Mei
Davies, Kelvin J. A.
de Grey, Aubrey D. N. J.
Dietrich, Marion
Dizdaroglu, Miral
Doctrow, Susan
Douki, T.
Duthie, Susan J.
Dykens, James A.
Elbirt, Kimberly K.
Flanagan, Steven D.
Fleck, B.
Fossel, Michael
Frelon, S.
Giorgio, Marco
Goto, Sataro

Griffiths, Helen R.
Harman, S. Mitchell
Hattori, Itaro
Heward, Christopher B.
Hiai, Hiroshi
Hirano, Takeshi
Hoberg, A-M.
Holmgren, Arne
Holmquist, Gerald P.
Huffman, Karl
Ivy, Gwen O.
Janero, David R.
Jaruga, Pawel
Jazwinski, S. Michal
Jensen, B.R.
Jiang, Qing
Jones, Dean P.
Jones, George D.D.
Joseph, James A.
Kanai, Setsuko
Kang, Y. James
Kasai, Hiroshi
Kelley, Eric E.
Khanna, Savita
Kirkwood, Thomas B.L.
Kitani, Kenichi
Kregel, Kevin C.
Krinsky, Norman I.
Lai, Ching-San (Monte)
Li, Jun
Loft, Stephen
Luccy, Britt
Lunec, Joseph
Machado, Alberto
Malfroy, Bernard
Marcotte, Richard
Marcus, Catherine Bucay
Maruyama, Wakako
Mason, Ronald P.
Masutani, Hiroshi
Matson, Wayne R.
Mattson, Mark P.

McCord, Joe M.
Mele, James
Melov, Simon
Meydani, Mohsen
Milbury, Paul
Minami, Chiyoko
Mitsui, Akira
Monnier, Vincent M.
Montine, Thomas J.
Morrow, Jason D.
Nakamura, Hajime
Niki, Etsuo
Nishinaka, Yumiko
Noguchi, Noriko
Norkus, Edward
O'Connor, Timothy R.
Okamoto, Takashi
Orida, Norman K.
Osawa, Toshihiko
Ozeki, Munetaka
Packer, Lester
Pelicci, Pier G.
Pong, Kevin
Pouget, J.-P.
Poulsen, Henrik E.
Preuss, Harry G.
Radák, Zsolt
Ravanat, J.-L.
Rayment, S. J.
Reaven, Peter
Reich, Erin E.
Remmen, Holly Van
Rice-Evans, Catherine
Richardson, Arlan
Riggs, Arthur D.
Roberts, L. Jackson
Rodell, Timothy C.
Rodriguez, Henry
Rong, Yongqi
Routledge, Michael N.
Roy, Sashwati
Saxena, Amit

Saxena, Poonam
Schafer, Freya Q.
Schroeter, Hagen
Seidel, Chris
Sell, David R.
Sen, Chandan K.
Sevy, Alexander
Shigenaga, Mark K.
Shmookler Reis, Robert J.
Shukitt-Hale, Barbara
Sies, Helmut
Singh, Keshav K.
Sinibaldi, Ralph M.
Smart, Janet L.
Sorenson, M.
Souza-Pinto, Nadja C.
Spencer, Jeremy P. E.
Spickett, Corinne M.
Sternberg, Jr., Paul
Subramaniam, Ram
Takahashi, Ryoya
Tanaka, Tomoyuki
Teoh, Cheryl Y.

Terai, Hidetomi
Termini, John
Tessier, Frederic
Tetsuka, Toshifumi
Tocco, Georges
Toyokuni, Shinya
Vacanti, Joseph P.
Villeponteau, Bryant
Visarius, Theresa
Wang, Eugenia
Warner, Huber R.
Weimann, A.
Weinberger, Scot R.
Weiss, Miriam F.
Wilson, Rhoda
Wise, Bradley
Wright, A.C.
Yarosh, Daniel
Yodoi, Junji
Youdim, Kuresh A.
Zubik, Ligia
Zullo, Steven J.

Contents

Volume 1

VOLUME 2

SECTION 7

Epidemiology and Intervention Studies

Chapter 48

Epidemiology of Antioxidants in Cancer and Cardiovascular Disease

L. Stephen Coles

L. Stephen Coles • Kronos Longevity Research Institute; 4455 East Camelback Road; Phoenix, AZ 85018
E-mail: scoles@grg.org

1. Introduction

Cancer is a disease that has afflicted our race since the beginning of recorded history. Egyptian mummies have exhibited bone tumors dating back to 3000 BC. In ancient Greece, Hippocrates described cancers of the breast, stomach, skin cervix, and rectum. Hydatidiform mole was recognized in ancient Rome. Ayurvedic medical books from India (circa 2000 years ago) talk about the treatment of tumors of the throat and esophagus. Chinese folklore suggests oral cancers go way back. Galen attributed cancer to an excess of black bile, consistent with the Greek balance of "the four humours" model of health and disease — blood, phlegm, yellow bile, and black bile. These are symmetrically related to the "four elements" (earth, air, water, and fire) out of which everything else is composed. (Ironically, not even one of these four "elements" turned out to be chemical elements in the sense of the Periodic Table.)

The relationship of cancer to occupational exposure was first documented by Percival Pott in 1775 in chimney sweeps and their high rate of scrotal cancer. Others noted that the distribution of cancer varied with race and with geographical location. Those who moved to a new continent and adopted the diet and life style of their new homes soon acquired the epidemiological distribution of cancer characteristic of their neighbors.

So cancer is an old disease. It is endemic to many species as well as all human societies, although in different forms and incidence rates. But the recent rise of cancer and heart disease as major killers is not a paradox. It is simply that we are living longer because we have reduced many of the competing causes of death through public health interventions, leaving cancer and heart disease more conspicuously on the playing field.[8]

As the conventional interventions for cancer treatment: surgery, chemotherapy, and radiation (slash, choke, and burn) are less than ideal, proposing means to *prevent* cancer by nutritional or other means is extremely important. For this, we require a brief review of the pathophysiology of this complex disease.

2. Understanding Cancer

First, cancer is not a single disease but a multiplicity of diseases. Furthermore, it is an intrinsic part of our evolutionary legacy. It is not subject to cure like the pinpoint mutation of an inheritable disease or an infection by a single micro-organism with a unique set of vulnerabilities that can be exploited by designers of pharmaceutical agents. Cancer is an essential ingredient in the programming of our biological world — from the DNA, genes, cells, tissues, organs, and bodies that inhabit this world — part of the way things work. Therefore, it follows, unhappily, that we should not expect any "magic bullet" for a cure that comes from the advertising hype of a drug company or a politically motivated "war on cancer" to be waged by government scientists at NIH or elsewhere anytime soon.

The problem of cancer is steeped in a complexity that is commensurate with the problem of understanding — at the molecular level — the nature of life itself. Yet, from this complexity a coherent pattern is emerging that is beginning to make sense. And, if cancer is the "enemy" to be fought at all costs, we will need to understand it on its own terms — from the point of view of a cancer cell itself.

Cancer cells are merely obeying a fundamental commandment of life — a Prime Directive, so to speak — to "go forth and multiply." The programming for immortalization is well embedded within its subroutines, if only it could escape from the constraints of "contact inhibition" and other inhibitors of its low-level programming. The multiple layers of constraints that seek to maintain a cell in a state of "good citizenship," as it were — incarcerated to participate in a chain gang as a unit in a larger tissue with a well-defined architecture (and specific function) — are not really consistent with its more primitive prime directive. These programmatic constraints were accumulated later in Darwinian evolution as "addons" or "plugins" and would be quickly abandoned, given the right opportunity.

The programmers of bacterial genomes did a lot of work to get to where they had to be over 3–4 billion years. This is our "legacy" code, so to speak. (The recent expense associated with solving the "Y2K problem" would not have existed were it not for legacy code that preceded it.) The modern code that produces mammals is really a cosmetic paste-on. The more modern code that produces primates among mammals is an even more superficial layer. The phenotypic differences that give bats wings, giraffes long necks, elephants trunks, zebras stripes, and whales spouts are minor tweaks on the legacy code. If ever the modern genetic software developers were tempted to look upon their work with pride, the ancient developers might regard them with contempt, given their greater contribution to the overall effort. Of course, all of this "program design" metaphor is anthropomorphic circumlocution. Everything, after all is, as Richard Dawkins would say, "a blind, digital river out of Eden".[29] It is all a stochastic process without concern for consequences.

Again speaking metaphorically, tissues with their monotonously-regular histology are really "prisons". All of the prisoners (somatic cells) are eagerly awaiting (if not actively planning for) their chance to escape into a clonal expansion, into what any worthy parasite would consider *nirvana* — an environment rich in nutrition, oxygen, and nearly perfect control of temperature, pressure, and pH. The "host" provides all of these resources in abundance. The observation that the internal parasite's mission versus the host's mission may be incompatible is irrelevant to the newly-escaped prisoner. Obliteration of normal tissue architecture is merely a side effect.

Indeed, one wonders how it could be otherwise. It makes one pause and ask how much work has really gone into keeping these prisoners in check, at least for the duration that really matters — the interval of reproductive competence? The answer is — a lot of work indeed. But when the period of reproductive competence is over with the onset of menopause in women (or the more gradual onset of andropause in men) there are no further incentives to confine the prisoners in the jail house. In fact, the rate of cancer development in multicellular organisms post reproductive competence can be expected to rise exponentially with age after menopause.

Why does the age of the host result in a paradigm shift? After all, there are childhood cancers, like leukemia, aren't there? And the incidence of certain cancers is more uniformly distributed over the life history of the organism, not just at the end. The answer as to why there is an exponential rise in the incidence of cancer after age 50 for humans lies in the fact that the "overlay" or "addon" programming that constrains this prime directive at the cellular level becomes inherently defective with time, exposing the more primitive code to the probability of execution without constraint.

There has been a misconception, particularly in the popular press, that the origin of cancer is due either to one or the other of two forces: (1) nature (genetic) or (2) nurture (environmental exposure to carcinogens). But it turns out that these two explanations are neither mutually exclusive nor jointly exhaustive. Cancer sheds the constraints of a somatic cell whenever it can and takes on the characteristics of an active embryonic stem cell or a free-swimming bacterium. (Note that the mathematical functions that govern that response of cancer cells to chemotherapy (log-kill per dose) and the response of bacteria to antibiotic therapy are essentially identical, so cancer cells and bacteria have a lot in common.)

One side of the coin is genetic susceptibility (e.g., BRCA-1/BRCA-2 genes in women with a family history of breast cancer) while the flip side of the coin is exposure to environmental carcinogens (e.g., smokers and lung cancer). Of course, it is argued that interactions amongst these causes will facilitate the onset of cancer and explains why identical twins reared apart will not normally come down with the identical type of cancer at the same time. But our point is that this popular model of cancer is too simplistic. Nature (inheritance) and nuture (carcinogenesis) are not jointly exhaustive alternatives. At least one new parameter needs to be taken into consideration: the rate of accumulated natural mutations (or nucleotide copy errors) with every cell division.

A more rigorous model would need to compare cancer to playing *Russian Roulette* for a very long time. Some players will die sooner, and some will die later. But in the end, all players will die, providing that each one plays long enough. The counterpart to spinning the barrel of the revolver is asking the cell's chromosomes to duplicate themselves during the process of cellular division ("S-phase"). The synthesis of DNA by inherently imperfect (unfaithful) enzymes leads to random mutations in each pair of daughter chromosomes. The net systematic error rate is measured empirically as 1 false nucleotide per billion copies (after post-editing by specific DNA Polymerase proofreading enzymes to repair obvious spontaneous missence copy-errors). Before editing, there is actually 1 incorrect base per 10 000 replications due to the natural infidelity of DNA Polymerase III; however, repair enzymes excise incorrect bases and replace them with the correct ones, giving us this net error rate of 1 per billion. Further evidence for the effectiveness of post editing is that if a mutation in the gene encoding the epsilon-portion (a subunit) of the enzyme were to occur, the proof-reading function of the enzyme is abolished, and this leads to a 1000-fold increase in the rate of spontaneous mutations. There are still other mechanisms for identifying and repairing mispaired bases in newly replicated DNA. Obviously, mismatch repair machinery (clean-up enzymes) must first determine which of the two strands is the one to be repaired by distinguishing the *newly-replicated* strand from the *template* strand before undertaking a DNA lesion repair [pp. 472 and 1081 of Refs. 17 and 18].

These initial errors are not significant at the beginning of one's life. Out of our three billion base pairs in each nucleus, 97 percent are non-coding and only 3 percent are involved in genes. (It has been recently estimated that there are somewhere in the range of 30 000–40 000 genes to be identified in the complete human genome. However, this estimation task should be completed in the next six months or so, with the completion of the Human Genome Project and its final publication in both the journals *Science* and *Nature* (February 2001).) But even within an actual gene, a random point mutation will be invisible approximately one-third of the time, owing to the inherent redundancy of the codons in the genetic code. And even for the remaining 2/3 of the nucleotides where a true substitution of an amino acid takes place, it may be a useless distinction without a functional difference in the active binding site of the protein or a transparent alteration in the tertiary structure (folding) of the protein simply replacing one hydrophilic amino acid with another or conversely a hydrophobic amino acid with another. It takes a very particular mutation (nonsense or frame-shift) to completely damage the function of a protein.

If we speculate that 0.03×0.3 mutations will actually result in a change in the primary amino acid sequence of a protein, and that 1 percent of those will actually make a protein defective, then there are 10^5 (= 100 000) defective cells arising *de novo* in our body every day. Sooner or later (75 years average life expectancy $\times$ 365 days/year) some of those cells are going to win the lottery and become cancerous. One really needs only 6–10 cumulative hits, each in critical regions of the genome. If one randomly rotates the dial of a safe back and forth, how long will it take to open it? Figuring this out is simple mathematics. Similarly, how long does it take to get three cherries or whatever in a slot machine and hit the jackpot? In an acre of slot machines, if enough people are playing simultaneously for long enough, sooner or later someone will hit a jackpot on one of the slot machines. Regular players of *Russian Roulette* will not survive more than a few iterations. Thus, the real surprise is that cancer is not more common in the geriatric population, not less. Probably potential cancer cells are arising all the time, except that they are "snuffed out" (apoptosis-triggered PCD or immune-surveillance detection followed by death by natural killer T-cells). Player cells that are snuffed out are removed from the "casino".

Therefore, cancer is a statistical inevitability, especially when the immune system weakens with age. The theory that cancer is a failure of immune system surveillance, as was suggested by MacFarlane Burnett of Australia a few decades ago, is not the whole story, however. If that were true then AIDS patients would experience a broad increase in all sorts of cancer. However, we have learned empirically that patients with AIDS mostly get opportunistic infections as their primary clinical manifestation, not cancer, although they frequently get an unusual form of cancer called "Kaposi's Sarcoma".

Genes, while tightly wound on chromosomes, are moderately well-protected against the natural world's radioactivity, solar, and even cosmic radiation. (Commercial pilots, who spend a lot of time at 35 000 feet have less atmospheric shielding against cosmic radiation than the rest of us living at or near sea level. And indeed they do suffer a somewhat higher incidence of cancer as an occupational hazard.) But whenever a cell falls out of one of its quiescent growth phases in the cell cycle, and moves into its "S" phase, when it must necessarily shed its local histone protection, the naked DNA becomes especially vulnerable to the ubiquitous exposure of radiation and free radicals in the neighborhood. Whenever the right collection of mutations occurs there is a well-orchestrated clonal expansion of cells that have a genetic million-year-old memory (subroutines) of how to clone themselves, survive stress, and expand their territory, particularly in the presence of superhighways like blood vessels or lymphatic channels for metastatic migration. Tissues have chemical controls to limit the prime directive of cells to multiply. Without the constraints of tissue architecture, the penalty for the host can be severe — ultimately death.

Note that if there were a zero mutation rate (perfect DNA copying), there would be hardly any cancer at all. But of course, there would be no humans either, since mutations are the very foundation for evolution of one species into a new species. So cancer is an intrinsic side effect of the machinery of life on Earth. Random mutations in the DNA at just the right places allows cancer cells to ignore the regulatory constraints of the body's own tissues, and ultimately allows them through this complex chain of events to follow their own Darwinian law of Natural Selection — to survive and multiply surreptitiously. According to this model, a consecutive series of single-point mutations is how cancers arise. See Table 1. If all 6–10 mutations accumulate simultaneously in a single cell, we have a "jail break", so to speak. The cell escapes from the constraints of its

Table 1. Series of Mutations that Collectively Cause Cancer (Clonal Expansion)

	Necessary Mutations	Possible Mechanism
1.	Immortalization	*Telomerase* is the enzyme that re-lengthens telomeres (active in 90% of all cancers)
2.	Loss of contact inhibition	Adherence (defect in *Fibronectin*)
3.	"Cloaking"	Immune surveillance transparency
		No detection by killer T-cells
4.	Inhibition of apoptosis or Programmed Cell Death (PCD)	Defective p53 (50% of all cancers), NF-kappa B, Tumor Necrosis Factor (TNF), proto-oncogenes converted to true oncogenes
5.	Vascularization	Angiogenesis factor(s) needed for tumor nutrition
6.	Capacity to metastasize — motility	"Rho C" gene, particularly for breast, liver, skin, and pancreatic cancers [Kenneth van Golen, University of Michigan Health Sciences in Ann Arbor]
7.	Dysdifferentiation	To an embryonic stem cell
8.–10.	Other reasons — necessary but not sufficient	To cause cancer, but *not* yet identified

tissue and goes on to expose its primitive subroutines of autonomous survival and replication, even though such behavior may be counterproductive to the long-term interests of the host.

Clonal emancipation will not occur if all conditions are not right. A tumor of modest size will remain subclinical if it cannot figure out how to feed its interior cells. A vascularized tumor, even if it impinges on surrounding tissue, will never become malignant until it figures out how to metastasize. Competition for nutrients with a fast-growing cancer will result in cachexia and death of the host. The synthesis of free radicals may lead to a persistent or chronic inflammatory lesion.

Simple point mutations, however, are not the only way for cancer to arise. There are other sorts of DNA lesions, including missing or extra bases (causing kinks or bulges), linked pyrimidines, cross-linked strands, and fragmented nucleotides (nucleotides disrupted by free radicals). Even worse are double-stranded breaks (which may lead to chromosomal fusion, inversion, or translocation). Hyper-methylation of bases can chemically alter the DNA in a way that silences inhibitory genes inappropriately, and this mechanism has been documented in nonhereditary breast and lung cancers. Aneuploidy (or duplication of entire chromosomes secondary to a failure to segregate on the spindle apparatus during mitosis) can increase cancer rates. Evolution itself has no malign intent, even though the cancer may be malignant — it is just the way things work; enzymes are like a blind watchmaker totally ignorant of consequences.

The fate map of all the cells that lead from a fertilized egg to 10–15 trillion somatic cells in an adult human, most of which are replicative (except for nerve and muscle cells), reveals cells that have undergone a dozen divisions from embryogenesis to adulthood over the 75-year life expectancy of a hypothetical human. Each day, we synthesize 2×10^{11} new cells in our bodies; each gene copied 10^{11} times every day means that there may be 10^{10} mutations taking place in our bodies every day, an alarming rate. Therefore, we begin to see that the error rate of ~1 per billion becomes significant in the "end game". But this is the nature of *Russian Roulette*. Therefore, we see that cancer is not just a process of genetics versus environmental exposure to carcinogens.

3. Epidemiology of Twin Studies

As recently reported in a series of Swedish/Danish/Finish twin studies,[19, 20] the most important causation for cancer was stated to be "environmental" by default (80 percent) given that the "inherited" contribution (20 percent) could be reasonable measured. But the association with the age-of-onset of this disease (which is probably available in the data base) was not stated. If there were a uniform distribution of cancer with age-of-onset — from adolescence to adulthood — then one could reasonably argue that the environment makes a linearly-independent contribution to cancer (cosmic rays, something in our water, some unidentified environmental toxin or pesticide, etc.). But, we already know that this assumption is false — environment and genetics are not independent variables, but play off against each other in subtle ways ("variable penetrance" is the term used by epidemiologists). And, since we know that risk is not uniform and that there is an exponential increase in the incidence of cancer after the age of 50, one has to go back to the "whole genome" as the principal contributor, even if a particular *heritable* gene cannot be identified as causal for a particular type of cancer. Indeed, this is the sort of relationship that could be teased out of an *identical* versus *fraternal* zygotic twin study.

What if, as we explained, the principal cause of cancer were random DNA mutations that are the result of "copy errors" in somatic cells undergoing mitosis (something over which one has no control)? Although there are known *heritable* causes (like susceptibilities to certain types of cancer that run in families) and certain known *lifestyle* causes (like exposure to certain carcinogens, occupational asbestos exposure, radiation, or smoking), after a certain age, these might not be the principal ones. We know a great deal about cancer in *congenic* mice (whole colonies of identical twins are raised as a single cohort) caged under ideal environmental conditions (same food, same water, same opportunity for exercise, same room temperature, no predators of course, etc.). The mice do not all die of the same cause on the same day; however, they nearly all do get fatal cancers at different times over a six month period in a seemingly random fashion sooner or later.

But what about the germ lines of older fathers and/or mothers? The rate of congenital anomalies is very small considering everything that can go wrong. Are copy enzymes more accurate during spermatogenesis (when telomerase is present)? Recall that sperm have to jump through a variety of hoops (swimming through the cervix, for example) before they are privileged to fertilize an egg, all of which demonstrate their competitive fitness. Or, a fertilized egg that must demonstrate fitness as it reaches the blastocyst stage of embryogenesis in its travels down the fallopian tube before implantation into the lining of the uterus, which initiates the formation of a placenta? A non-trivial percentage of unfit embryos (now shown to be as much as 50 percent) never implant and are unceremoniously expelled (spontaneously aborted) along with the menstrual period. Undoubtedly, somatic cells do not undergo such a rigorous fitness-testing process before they divide and potentially form a cancer capable of destroying the architecture of the tissue in which they reside or into which they happen to metastasize.

As another limitation of the methodology in the twin study, we must ask, what about identical twins-reared-apart during youth versus "virtual" twins (adopted siblings of identical age and gender raised together in the same household but with no common parents)? This would make any distinctions for heritable causality more dramatic if they were present. The problem is that this sort of "gold standard" for human demographic data is extremely hard to come by, even in countries that scrupulously maintain medical records for all their citizens by enrolling them in a national health insurance program at birth (the US is certainly not one of them).

The bottom line is that spending more tax-payer money on the Environmental Protection Agency (to reduce exposure to dangerous water pollution, air pollution, or agricultural pesticides, as much as possible) may not necessarily be cost/effective as a prevention for cancer, providing that epidemiologists are correct that cancer rates increase exponentially with age independently of environmental exposure. The increasing incompetence of T-cell immune surveillance may not fully explain such an exponential rise in random cancers with age, but rejuvenation of thymic function in the elderly may play a greater role in preventing cancer than reducing familial and lifestyle contributions combined.

Besides, how many women are willing to undergo a prophylactic mastectomy if they test positive for either the BRCA-1 or BRCA-2 genes (less than 0.25 percent of all women)? By the way, less than 5 percent of all breast cancers derive from this particular genotype. The solution to cancer will be a better understanding the molecular biology of the cancer cell itself, how its genotype can be identified, and how such a cell can be forced to undergo apoptosis before it does irreversible damage.

Cancer involves a stochastic process of DNA copy errors at the molecular level that has nothing to do with one's inherited genome or one's exposure to environmental toxins, except that they can make a bad situation worse. The natural situation is already bad to start with. Without a deeper understanding, one might be tempted to throw more money into a combination of better diagnostics of genetic susceptibility (catching cancer early is always better than waiting until it's too late) and spending more public tax money on the EPA (Environmental Protection Agency) to keep our air, water, and food pure, and teach people not to smoke through an effective educational awareness program. But it's not just nature versus nurture. There's something else going on at a deeper molecular level that is responsible for this disease. Throwing all our resources at genetic diagnoses and/or the EPA would be like "rearranging the deck chairs on the Titanic while the ship sinks".

4. Rate of Knowledge Expansion in Medicine

When I went to medical school, one of my standard textbooks in histology [published in 1956] stated that there were "48" human chromosomes (instead of 46 chromosomes).[16] A cytologist in Surgical Pathology should not be rendering diagnostic/prognostic opinions based on amniocentesis or Chorionic Villus Sampling (CVS) with that book as an exclusive reference, should they? But 1956 was not that long ago. The same argument could be made for all important discoveries about doing cardiovascular surgery before William Harvey told us that the blood circulates throughout the body or curing childhood infectious disease before Louis Pasteur taught us about his empirical observation regarding cow pox vaccination. Think about learning anatomy before dissection, or learning pathology before autopsy was common. Think of treating infection before penicillin. Discoveries of blood-typing (for transfusion) and anesthesia — as prerequisites for surgery — were not developed that long ago. Think of molecular biology before Watson and Crick. Embryonic and adult stem cell activation and reprogramming in culture was uncovered as a therapy only in the last year. The next stage of stem cell technology will teach us how to activate adult stem cells *in vivo* with the use of highly specialized growth factors. Stem cells from a different (non-HLA compatible) source will, in fact, trigger an NK T-cell immune response sooner or later. Only autologous stem-cell transfer will not require the use of immunosuppressive drugs. More knowledge will result in better interventions.

One of my colleagues at the University of California at Berkeley estimates that the world now produces 1–2 EB (Exabytes) of non-redundant digital information per year. Imagine 1000–2000 miles of CD-ROMS stacked in a row across the country. Remember that a Megabyte is 10^6 Bytes; a Gigabyte is 10^9 B, a Terabyte is 10^{12} B, a Petabyte is 10^{15} B, and an Exabyte is 10^{18} B. (For the cognoscenti, a Zettabyte is 10^{21} B and a Yottabyte is 10^{24} B; to my knowledge the ISO has not defined any nomenclature past the "Yotta".) Much of this data is medical data, and, according to informatics specialists, the time for medical knowledge to double is on the order of 18 months.

Although the interplay between theoretical knowledge and practical intervention in medicine is fraught with peril, 100 million consumers (compared with 70 million in 1999) sought health care information on the Internet over the past year. Physicians are still wary of the quality of the information found on the web. Only 35 percent said they were confident of the data on health-related Internet sites and just 7 percent said that they had any confidence in information patients find in on-line in chat rooms.[25] As one example, Internet search engines identify more than 2000 websites featuring *laetrile* (amygdalin or "Vitamin B17"), the vast majority of which promote this bogus product as a cure for cancer, despite the fact that this compound has been discredited by the FDA since 1982.[28]

What new medical interventions for cancer and heart disease are on the horizon can only be imagined with a cloudy crystal ball. But in order to take advantage of them when they arrive, you have to be here! Therefore, preventing cancer and heart disease should be high on our agenda if we are going to "bridge" ourselves from the unhappy state of our ignorant past to the luxurious state of the erudite future that should certainly benefit our grandchildren if not ourselves or our children. But could it benefit us in our lifetimes? Say, in the next 20–30 years? Maybe. That depends to some extent on what we do while we are waiting (life style [exercise, reduced stress levels, sleep pattern, reduced exposure to carcinogens within our control], nutrition [antioxidant vitamins and mineral supplements], and hormonal manipulations [sex steroids, melatonin, DHEA, pregnenolone, hGH, etc.]. But it also depends on what the medical research community can get done during this interval, given the trivial level of investment compared with, say, the investment that we have made in our collective national defense establishments or in entertainment. All I can say is that it is still a horse race.

So what we do in our personal lives could still make a difference. Well, what can we do to tilt the odds in our favor? Is there any scientific evidence that one could look toward in formulating a plan? That is, beyond the obvious — like "do not get hit by a truck" (i.e., buckle your seat belt and do not indulge in "sky diving" as a hobby).

Living centenarians as a group have very little to offer in the way of valid advice as to the secret of their longevity. Having interviewed a number of centenarians myself, I have found no pattern emerging that suggests a set of recommended interventions. Indeed, some common sense rules about how to live a long life are occasionally violated by Supercentenarians (those greater than 110 years) who regularly contradict conventional wisdom about not smoking or drinking only moderately. Indeed, some living Supercentenarians never saw a doctor before the age of 90, so the medical establishment can hardly take credit for its interventional skills, as far as this group is concerned. In other words, although public health measures in the last century have made a dramatic improvement in average life expectancy in the developed countries, no cause-and-effect rules have emerged for altering the maximum lifespan of our species.

Comparing maximum life span of different species, however, does lead one to believe that the key lies in our genes. But choosing one's parents wisely is no longer an option available to those of us now living. That decision has already been made, and so it would seem that our fate is sealed. Our modern proclivity to smoke, sunbathe for hours, indulge in sex for recreation, eat empty calories, coupled with our propensity to sit still for hours watching television, are at odds with our basic hunter/gatherer physiology. Evolution is too slow a process to compensate for the unhealthy habits cultivated over the last 100 years. But the odds in this lottery can be shifted somewhat, once you know the rules of the game.

There do appear to be a few "rules of thumb" regarding exercise (and avoiding obesity) and nutrition that are suggestive of how to cut down on the likelihood of both cancer and heart disease. For one thing, it is important to eat at least several servings of fruits and vegetables every day. But if you do not, will vitamin tablets provide the right sort of back up? And, if so, how many, how often, and at what dose?

5. Vitamin Supplements and Cancer Prevention

The US Government recommended dosages of essential nutrients derived recently from the Institute of Medicine (part of the National Academy of Sciences)[1] have evolved over the years to increasing doses modestly, but nowhere near the amounts recommended by the health food store community.[2, 3] Table 2 contains my personal recommendations that lie somewhere in between.

Yet, how can one explain this wide discrepancy in recommendations? Why are the authors of the Federal Government guidelines who determine the recommended dosages (MDRs) for the American public so far off the standards recommended by the health food industry (aside from their obvious financial self-interest)? The answer, in part, derives from the professional perspective that the government scientists bring to the process. They are largely biochemists or epidemiologists by training. They are generally quite familiar with the scientific literature of how certain vitamins in isolation can demonstrate toxicity (dose-response curves) *in vitro* at higher doses in DNA. Therefore, being naturally conservative, this group of individuals is unlikely to achieve a consensus about recommending really high dosages of anything, even though they are well aware that *in vivo* and *in vitro* experiments may lead to very different results. [The minimum dose needed to eliminate a vitamin deficiency disease should be quite adequate, thank you.] On the other hand, those scientists concerned with

Table 2. Target Dosages for Optimal Dietary Supplementation

	Nutrient	USRDA	USRMA	ORD	ORMA
1.	Vitamin A	5000 IU		5000 IU	
2.	Mixed carotenoids (alpha and beta carotene, etc.)	N/A		25 000 IU	
3.	Vitamin B1 (Thiamine HCl)	1.5 mg		[100–500] mg	
4.	Vitamin B2 (Riboflavin)	1.7 mg		[50–100] mg	
5.	Vitamin B3 (Niacin; Nicotinamide NADH)	20.0 mg		[100–250] mg^2	
6.	Vitamin B5 (Pantothenic acid)	10.0 mg		[250–500] mg	
7.	Vitamin B6 (Pyridoxine HCl)	2.0 mg		[25–50] mg	
8.	Vitamin B12	6.0 mcg		[50–1000] mcg	1500 mcg
9.	PABA (Para amino benzoic acid)	N/A		[0–50] mg	
10.	Biotin	300.0 mcg		300 mcg	
11.	Folic acid	400.0 mcg		[800–1600] mcg	
12.	Vitamin C (Ascorbic acid)	60.0 mg		[500–2000] mg	
	Women (new)	75.0 mg		1000 mg	
	Men (new)	90.0 mg (diarrhea)		1000 mg 1000 mg	
	Smokers (new)	add +35.0 mg			
13.	Vitamin D (Ergocalciferol)	400.0 IU		[200–600] IU	2000 IU
14.	Vitamin E (d-alpha tocopherol; mixed — tocopherols preferred)	30.0 IU		[200–800] IU	2000 IU
	New (d-alpha-tocopherol)	22.0 IU	1500 IU		
	New (dl-alpha-tocopherol/synthetic)		1100 IU (hemorrhaging)		

Table 2　(*Continued*)

Nutrient	USRDA	USRMA	ORD	ORMA
15. Vitamin K1 (Phytonadione)	80.0 mcg		[40–80] mcg	
16. Calcium (Pantothenate 100 mg)	1.0 g		[1000–1500] mg	
17. Iron	18.0 mg		0 mg^3	2.5 g^3
18. Phosphorus	1000 mg		1000 mg	
19. Iodine	150.0 mcg		150 mcg	
20. Magnesium	400.0 mg		1000 mg	
21. Copper	2.0 mg		2 mg	
22. Zinc (Nicotinate)	15.0 mg		[25–50] mg	
23. Chromium (Picolinate)	N/A		[100–800] mcg	
24. Selenium (old)	N/A		[100-300] mcg	750 mcg
(new)	55 mcg	400 mcg (Selenosis)		
25. Molybdenum (Gluconate)	N/A		200 mcg	
26. Manganese (Gluconate)	N/A		20 mg	
27. N-Acetyl Cysteine (NAC)	N/A		[500–2000] mg	
28. N-Acetyl-L-Carnitine	N/A		[500–2000] mg	
29. Alpha-Lipoic Acid (ALA)	N/A		[100–600] mg	
30. Co-Enzyme Q-10 (Ubiquinone)	N/A		[30–300] mg	
31. Pycnogenol (Pine bark extract; or grape seed extract)	N/A		[60–100] mg	
32. Betaine free base (Trimethyl glycine)	N/A		25.0 mg	
33. Inositol (IP-6 Hexaphosphate)	N/A		[250–5000] mg	
34. Boron (Calcium borate)	N/A		[0–5] mg	
35. Choline bitartrate	N/A		500.0 mg	
36. Phosphatidyl choline (Lecithin)	N/A		[250–1200] mg	
37. Phosphatidyl serine	N/A		[250–500] mg	
38. Garlic extract (Odor free)	N/A		[100–1200] mg	
39. Ginseng (Standardized)	N/A		[250–500] mg	
40. Ginkgo biloba	N/A		[150–350] mg	
41. Flaxseed and fish oils (DHA, EPA, and GLA)	N/A		[1000–3000] mg	

Table 2 (*Continued*)

Nutrient	USRDA	USRMA	ORD	ORMA
42. Lycopene	N/A		300 mg	
43. ASA (Aspirin) (Persons. > 50 yo with cholesterol > 280)	N/A		[81–325] mg [qd or qod]	
44. DMAE (Dimethylaminoethanol)	N/A		100 mg	
45. Flax seed oil (Refrigerated and kept in a dark bottle)	N/A		[1–2] tsp	

USRDA = (US Recommended Daily Dietary Allowance)
   [Intended for all adults and children older than five years]
USRMA = (US Recommended *Maximum* Allowance)
  ORDI = (Optimal Range of Recommended Daily Intake)
 ORMA = (Optimal Recommended *Maximum* Allowance)[1]

1. A dose beyond which there could be noticeable adverse side effects (not necessarily the same as the toxicity that is secondary to hypervitaminosis).
2. For the particular indication of elevated cholesterol, Niacin [*Niaspan* is the timed release version] may be taken 500–3000 mg every day; the "No Flush" formulation [Inositol Hexaniacinate] may be better tolerated by some individuals.
3. Pregnant females may wish to supplement with $FeSO_4$ tablets as recommended by their OB/GYN. However, iron supplements should only be taken by pre-menopausal women; conversely, adult men, girls, and post-menopausal women should never take supplemental iron (or risk subclinical hemachromatosis). On the contrary, certain men should donate one pint of blood to the Red Cross — or equivalent blood bank — on a regular basis (every 3–12 months). This should be done, not so much for the sake of the therapeutic benefits to a potential *recipient* badly in need of blood, but for the purely selfish, genuine health benefits to be obtained by having the *donor* reduce his iron load. [Remember that the mammalian cascade of clotting factors, iron hoarding, and fat storage mechanisms evolved to counteract the iron-deficiency anemia of periodic bleeding associated with trauma and the lack of cooperative prey, something that was much more prevalent in prehistoric times than it is today at modern fast food outlets.]
4. Many commercial multivitamins with minerals also contain Chloride, Potassium, Magnesium, Nickel, Silicon, Tin, and Vanadium, but I have no particular MDR for these minerals.
5. Common herbal supplements that may be used for various indications: Bilberry, Dong Quai, Ginger, Feverfew, St. John's Wort, Goldenseal, Horse Chestnut Extract, Milk Thistle.
6. Commonly supplemented hormones: Melatonin, DHEA, and Pregnenolone.
7. For children under age 12, a standard chewable multivitamin is sufficient. All children should also be exposed to fluoridated water for strong teeth. If your household drinks only bottled water known not to contain fluoride then a supplemental tablet may be indicated.

oxidative stress and aging as well as the health food industry are interested in chemical reactions whose time constants may be measured in years to decades not seconds to minutes. Biochemists are primarily interested in reactions whose time constants are measured at most in minutes to hours. Remember that experiments must lead to publishable results within a reasonable time span, essentially that of an Assistant Professor who is seeking tenure. Perhaps these temporal discrepancies can explain the substantial differences among the government bias and the long-term hypothetical bias of those concerned with oxidative stress.

Millions of Americans take antioxidant vitamin supplements as a preventative against the debilitating chronic diseases of cancer and CVD (Cardiovascular Disease). If you care about your health, you are probably one of them. However, this recent report prepared by a 14-member Committee on Dietary Antioxidants and Related Compounds issued by the prestigious Institute of Medicine (IOM), a part of the National Academy of Sciences which is charged by the US Government with responsibility for setting the RDAs (Recommended Daily Allowances) for all supplements, provided a presumably rigorous review of the scientific literature about what is known about vitamins C, E, Beta-Carotene, and the mineral Selenium, as they relate to chronic disease. Their conclusion essentially was "There is no proof that these supplements do what is claimed for them". (A free copy of the report is available on their website[1]).

Have Americans been misled all these years? "Perhaps they have", said Dr. Norman Krinsky, Chairman of the Panel and Professor of Biochemistry at Tufts University School of Medicine in Boston, Massachusetts.[2] Not all of the positive reports were credible. Some were based on unpublished and/or poorly substantiated research. By contrast, the IOM panel chose to focus only on data published in reputable peer-reviewed journals with the goal of finding out what had been *proven* about the relationship between antioxidants and health. They also wanted to find out whether new research warranted a change in the existing RDAs for each antioxidant. In a few cases they found that the RDA did need revision [mostly upward, but only by relatively small amounts].

The report provided solid evidence linking tissue-damaging oxidizing free radicals to a heightened risk for cancer, heart disease, and cataracts. Indeed, the higher the level of free radicals in the body, the greater the risk. And, it has been shown that antioxidants do prevent free-radical damage *in vitro*. The problem was in finding proof that these effects observed in the laboratory also occurred in human clinical trials. It may seem logical that boosting antioxidant levels via supplements will help guard against chronic disease, but this has never been proven. One of the difficulties, as we shall see, is that the healthy human body controls the capacity to neutralize free radicals within tight tolerances using a finely-tuned control system, in the same way that temperature, pH, and other physiological parameters are kept within exquisitely tight limits, regardless of variation in the environment. Therefore, it should come as no surprise that dietary supplementation, under conditions of vigorous health, should make little observable difference or the sort that can be measured in the laboratory without the benefit of a control system just for that purpose.

But the use of vitamin supplements in the treatment of cancer and CVD still remains a highly controversial topic. One of the basic reasons has to do with the natural tension that exists in the public mind between the ultra-conservative custodians of our medical establishment and the multi-billion dollar health-food/ vitamin-supplement industry, which is driven to a large extent by the profit motive rather than the expensive burden of rigorous scientific proof (the US FDA and the pharmaceutical industry play a similarly adversarial role, but their ground rules for confrontation are much more gentlemanly). Nevertheless, the health-food industry asks, "How much *proof* do you need?" And what if we are right that you should be supplementing with more vitamins than you could possibly get from a typical American diet? And what if there is little or no down-side (toxicity) to our higher-dose vitamin recommendations? And what if the cost of following our advice is reasonably affordable? Do you want to risk the possibility that we may be proven to be correct later on, and, by then, it could be too late because of irreversible chronic damage for medical science to intervene? So the argument continues.

Now, a vitamin-products company has come along to challenge the establishment on its own turf — on the scientific merit of the IOM's argument that "There is no [scientific] proof". Mr. William Faloon, Vice President of the Life Extension Foundation of Wilton Manors, Florida supported by a distinguished Scientific Advisory Board of 12 members — many of whom are pioneers in antioxidant biochemistry — and a 30-member Medical Advisory Board argues that the 512-page Committee Report *Dietary Reference Intakes for Vitamin C, Vitamin E, Selenium, and the Carotenoids* is filled with serious contradictions and omissions. How could reasonable men coming at the same issue with an open mind arrive at such diametrically opposed conclusions?

The Press Release issued by the National Academy of Sciences used to promote their book concluded, "Insufficient evidence exists to support claims that taking megadoses of dietary antioxidants, such as Selenium and vitamins C and E, or carotenoids, including Beta Carotene, can prevent chronic diseases". On the other hand Faloon says, "This statement was contradicted by published findings presented within the book itself showing disease risk-reduction benefits for vitamin C, vitamin E, Selenium, and the Carotenoids. Furthermore, the report omitted many other positive published studies about these supplements. When these omissions are added to the favorable findings reported in the book, it is difficult to ascertain how anyone concerned about protecting their health would not supplement with vitamin C, vitamin E, Selenium, and Carotenoids".[3] By the way, Ref. 3 contains 555 references to the medical literature of its own.

Table 3. ORAC (Oxygen Radical Absorbance Capacity) Units per 100 g of Selected Fruits and Vegetables[26]

Fruits	ORAC	Vegetables	ORAC
1. Prunes	5800	1. Kale	1800
2. Raisins	2800	2. Spinach	1300
3. Dates	2700	3. Brussels sprouts	1000
4. Blue berries	2400	4. Alfalfa sprouts	900
5. Strawberries	1500	5. Broccoli flowers	900
6. Raspberries	1200	6. Beets	800
7. Plums	900	7. Red bell peppers	700
8. Oranges	800	8. Onion	500
9. Red grapes	700	9. Corn	400
10. Kiwi	600	10. Egg plant	400

The media's response to the press release promoting the book was to question the value of all dietary supplements sold in health food stores. Yet, the book did not review folic acid, CoEnzyme Q-10, B-vitamins, zinc, or a host of other dietary supplements commonly consumed. These nutrients have shown considerable disease risk-reduction benefits in several thousand published studies.

Here are three fundamental questions for the vitamin supplement studies of the future:

(1) Natural foods are highly complex botanical combinations of vitamins and other medicinally active compounds. When is taking a pill as good as (or even better than) a serving of fruits or vegetables? For example, lycopenes are in fruits but may not be part of a pill. Table 3 lists the antioxidant capabilities of a few common fruits and vegetables. Without a proper controlled study, one cannot say whether eating the first five fruits and vegetables on this list every day would be fully equivalent to taking antioxidant vitamins like C, E, and others.

(2) The problem of vitamin-supplement interactions has rarely been studied scientifically. In a new survey by the American Institute for Cancer Research older Americans are abandoning healthy diets in exchange for supplements in the hopes of lowering their risk of cancer.[30] Furthermore, formal studies are hardly ever done comparing a control group taking no supplements to an experimental group taking a complete protocol of 30–40 supplements on a daily basis, which is what many people do in real life.

(3) And what is the proper dose (in mg/kg of body weight)? For example, the amount of Ascorbic acid necessary to cure scurvy is really very little compared with the 1000–10 000 mg promoted by Dr. Linus Pauling to prevent cancer. Do you want to have really expensive urine? The answer for some is "Yes, I do. That way, I will know I am getting the amount I need".

The controversy over whether vitamins should be obtained solely from food or from a combination of food plus supplements is not likely to be resolved any time soon. A great deal of indirect scientific evidence is accumulating to suggest that extra doses of anti-oxidants may be beneficial to our health by acting as free-radical scavengers. Other data, however, indicate little, if any, benefit to be derived from taking other supplements and, in certain cases like herbals, they may do harm.

Finally, with respect to the relationship between cancer and smoking, we can report some good news for California. That state's tough anti-smoking education measures and public health campaigns have resulted in a 14 percent decrease in lung cancer over the past 10 years while other regions of the country reported only a 2.7 percent decrease over the same period, according to a CDC report just released.[7] Lung cancer develops slowly and the full benefits of quitting smoking can take up to 15 years to be realized. However, epidemiologists are expected start seeing results within five years. The effect of the anti-tobacco efforts has been fewer smokers and fewer deadly cases of cancer related to smoking, but researchers said they expect the trend to continue. In its report, the CDC compared cancer registries in California, Connecticut, Hawaii, Iowa, New Mexico, and Utah, as well as the cities of Seattle, Atlanta, and Detroit. Unfortunately, cancer rates in many foreign countries can be expected to increase dramatically in the next few decades.

6. Cardiovascular Disease

Speaking simplistically, most cardiovascular disease is a disease of clogged plumbing. Young flexible elastic vessels slowly and silently transform themselves into stiff pipes with the deposition of calcium along the inner lining. This is exacerbated by increased levels of cholesterol and/or any chronic inflammatory agent, whether viral, bacterial, or other traumatic injury. The deposits that result in *hard* plaque formation are not as serious as those that result in *soft* plaque, with foam cells predominating. Soft plaque is the type that can readily break off leading to a fatal heart attack or stroke. Frequently (30–50 percent of the time), the first symptom of ruptured soft plaque is sudden death. Therefore, prevention of damage to vessels is an essential strategy for long life. What are the risk factors that we need to look for? Table 4 provides a tentative list. Unfortunately, these risk factors accurately predict high-risk CVD only 25 percent of the time.

Our current hypothesis for the most common form of heart disease is as follows: bacteria and/or viruses enter the circulation through any convenient portal where they are taken up by macrophages for sequestration and destruction. However, some white blood cells may enter arterial walls with early atherosclerotic damage. Once they enter the inner surface lining, bacteria/virus particles may promote further inflammation, adding to the damage and weakening the calcified

Table 4. Risk Factors for CVD and Atherosclerosis

Risk Factor	Possible Mechanism
1. Male gender	Estrogen
2. Family history	Genetics of LDL/HDL
3. Age	Calcification
4. Smoking	Free radicals in smoke
5. Hypertension	Work load
6. Abnormal blood lipids	LDL, HDL, Triglycerides, Lp(a), Apo A-I, A-II
7. Homocysteine	
8. Diabetes	Syndrome-X
9. Obesity	Type-II diabetes
10. Sedentary life style	Lack of exercise
11. Chronic stress	ACTH, cortisol
12. Depression	Serotonin, dopamine
13. Chronic infection [chlamydia pneumoniae, helicobacter pylori, cytomegalovirus (CMV)]	cytokines, interlucans-1,6
14. Periodontal disease	Bacterial infection of the gums
15. Chronic inflammation	C-reactive protein marker
16. Excess clotting factors	Factor VIII, von Willebrand
17. Sticky platelets	Clotting

plaque, so that it is more likely to rupture. Such a catastrophic event may trigger clot formation and the resulting embolic stenosis and blockage of distal vessels causes ischemia leading to angina pectoris and possible death secondary to a massive failure of contractility in a large portion of cardiac muscle.

7. Conclusion

First published in 1917, *On Growth and Form*[9] was a revolutionary work by a Scottish embryologist D'Arcy Wentworth Thompson (1860–1948) who grew up during the time when *Darwinism* first became popular. An iconoclast, he took issue with the central dogma of his day — "not because it was necessarily wrong", he said, "but because it violated the spirit of *Occam's Razor*, in which simpler explanations are preferable to complex ones". English biologist Sir Peter Medawar went so far as to say, "Thompson's tome is beyond comparison the finest work of literature in all the annals of biological science that has ever been recorded in the English language".

Thompson's comprehensive treatise is filled with drawings from cover to cover of botanical plants, fishes, and shells that demonstrate how seemingly sharp differences in outward morphology could be explained by relatively small mathematical (genetic) tweaks at the developmental (embryogenic) level. He did not live to see the mathematics of Fractal Geometry, Post Production Systems, or Chomsky Transformational Grammars, all of which came after his time, but he certainly would have appreciated them if he had lived into the present century. The feeling that you get by just browsing through the drawings (whether the feathers of a peacock or the petals of a rose) is that all of the superficial differences in complex organisms are really based on a common underlying embryonic machine whose capacity to express differences in tissue architecture is really quite simple. Furthermore, the grammatical structure that makes this possible is nearly ready to be exposed — in, say, the next ten years, when we will know exactly why giraffes have long necks or zebras have stripes. What is the evidence for such a bold assertion?

Scientists on three continents have just deciphered the entire genetic makeup of a flowering plant — a breakthrough in basic science that not only unlocks a secret of nature, but may help to feed a hungry world, reduce pollution, and identify potential medicines of the future. The new poster plant for the genetics revolution is not a towering sequoia or a fragrant rose. Instead, it is a spindly weed that grows along highways everywhere. *Arabidopsis thaliana* [or as gardeners know it — *thale cress*] joins the fruit fly, the nematode worm, various viruses, two-dozen bacteria, and yeast as organisms whose entire DNA blueprints have been deciphered thus far. The plant's genetic code (25 498 genes distributed over 5 chromosomes) was published in *Nature* (Refs. 11–13; December 14, 2000) based on work done in the USA, UK, France, Germany, and Japan. *Arabidopsis*, a member of the mustard family and a cousin to cauliflower, was chosen over 250 000 other plant species because it was biologically simple and grows quickly — as many as eight generations reproduce in a single year. The function of two-thirds of its genes has been identified so far. Its genes contain about 117 MBP (million base pairs). By way of comparison, corn has more than 3 GBP (billion base pairs), nearly 30 times as much. Like animal models that have been genetically sequenced, it is easily manipulated in laboratory experiments and is widely used as a stand-in for more complex organisms.

But even more interesting are the implications of a follow-on project. Rita Calwell, Director of the National Science Foundation, which helped to fund the plant research, announced that "the NSF will spend as much as $25 million over the next decade to fund a new effort — to be called *The 2010 Project* — whose goal is to catalog the functions of all 25 498 *Arabidopsis* genes". As a start, botanists have analyzed about 8000 genes with oligonucleotide mocroarrays and found that about 6 percent showed daily cycles of expression. Entire metabolic pathways exhibit coordinated circadian rhythms of their components in order to protect the plant against the direct rays of the sun and the cold of the night.[15]

In a recent literary essay in *Harper's* Magazine, Arthur B. Cody correctly points out that, although we have made a lot of progress in gene sequencing, we are still a long way from translating the "Book of Life" into a usable instruction manual.[14] Cody rhetorically asks, "… how do genes make a leg or an eye? …" He responds, "The answer is no one knows. Not only does no one know, no one has even the slightest idea of how to look for an answer". Perhaps, if all goes well with *The 2010 Project*, we will have an answer both for Mr. Cody and for Mr. Thompson.

Humans are not delivered with a "user's guide" when we are born — like a new VCR or a new television set with instructions in multiple languages (just in case we speak something other than English). Well, aside from simply imitating our siblings, our parents, and our grandparents, God *did* provide such a book. The problem is that it is not written in a standard natural language. (After all, how could He know what language we might be speaking when we grew up?) The user's guide in question is our human genome. But we cannot comprehend it yet? We need some sort of *Rosetta Stone*. But God provided the stone as well; it is the collection of all the other genomes surrounding us. If we compare our book with other books of all the other animals and plants with whom we share this planet, we will begin make sense of our own genome. After a while, by identifying all the homologies across species boundaries and building a complete phylogenetic tree of all organisms as they evolved throughout the biological history of the Earth, we will succeed in understanding our own genome. In the process, cancer and heart disease will finally succumb to a true understanding of these pathological processes with interventions to match and cure them.

References

1. Krinsky, N. I., Young, V., Garza, C. *et al.* (2000). *DRI (Dietary Reference Intakes) for Vitamin C, Vitamin E, Selenium, and Carotenoids: A Report on Dietary Antioxidants and Related Compounds, Subcommittees on Upper Reference Levels of Nutrients and Interpretation and Uses of Dietary Reference Intakes, and the Standing Committee on the Scientific Evaluation of Dietary Reference Intakes,* Food and Nutrition Board; Institute of Medicine (National Academy Press, Washington, D.C.). URL: http://books.nap.edu/catalog/9810.html (April 10, 2000).
2. Krinsky, N. I. (2000). Antioxidants: good and bad, surprising findings. *Bottom Line Health* **14**(8): 1–3.
3. Faloon, W. (2000). A critical analysis of the National Academy of Sciences attack on dietary supplements. *Live Extension Magazine* **6**(8): 18–42.

4. Coles, L. S. and Steinman, D. (1999). *Nature's Ultimate Anti-Cancer Pill: The IP-6 with Inositol Question and Answer Book*, Freedom Press, Topanga, CA.
5. Weiss, D. (2000). Cardiovascular disease: risk factors and fundamental nutrition. *Int. J. Integ. Med.* **2**(4): 6–13.
6. Ewald, P. W. (2000). *Plague Time: How Stealth Infections are Causing Cancers, Heart Disease, and Other Deadly Ailments*, Free Press, New York.
7. Cimons, M. (2000). State programs credited for dip in lung cancer, p. A1, 16, *The Los Angeles Times* (December 1, 2000) and California campaign trimmed cancer rates, p. A1, B5, *The Wall Street Journal* (December 1, 2000).
8. Greaves, M. (2000). *Cancer: The Evolutionary Legacy*, Oxford University Press, UK.
9. Thompson, D'Arcy W. (1942). *On Growth and Form*, Dover, New York; 1992; 1st Edition 1917; Cambridge University Press Edition.
10. Mesel, R. (2000). Scientists decode a plant's genome for first time, pp. A1, 3, 26, *The Los Angeles Times* (December 14, 2000); Regalado, A. (2000). DNA — decoded list now includes a plant, p. A1, B13, *The Wall Street Journal* (December 14, 2000).
11. Lin, X. T., Kaul, S., Rounsley, S., Shea, T. P., Benito, M.-I., Town, C. D. *et al.* (1999). Sequence and analysis of chromosome 2 of the plant *Arabidopsis thaliana*. *Nature* **402**(6763): 761–768.
12. Mayer, K., Schuller, C., Wambutt, R., Murphy, G., Volckaert, G., Pohl, T., Dusterhoft, A., Stiekema, W. *et al.* (1999). Sequence and analysis of chromosome 4 of the plant *Arabidopsis thaliana*. *Nature* **402**(6763): 769–777.
13. Theologis, A., Ecker, J. R., Palm, C. J., Federspiel, N. A., Kaul, S., White, O. *et al.* (2000). Sequence and analysis of chromosome 1 of the plant *Arabidopsis thaliana*. *Nature* **408**(6814): 816–820.
14. Cody, A. B. (December 2000). Messages from the genome. *Harpers Magazine* **302**(1807): 15–22.
15. Harmer, S. L., Hogenesch, J. B., Straume, M., Chang, H.-S., Han, B., Zhu, T., Wang, X., Kreps, J. A. and Kay, S. A. (2000). Orchestrated transcription of key pathways in *Arabidopsis* by the circadian clock. *Science* **290**(5499): 2110–2113.
16. Hoskins, M. M. and Bevelander, G. (1956). *Essentials of Histology*, 3rd Edition, p. 187, The C. V. Mosby Company, St. Louis, MO.
17. Friedberg, E. C., Walker, G. C. and Siede, W. (1995). *DNA Repair and Mutagenesis*, ASM Press, Washington, D.C.
18. Lodish, H., Berk, A., Zipursky, S. L., Matsudaira, P., Baltimore, D. and Darnell, J. (2000). *Molecular Cell Biology*, 4th Edition, W. H. Freeman and Company, New York.
19. Lichtenstein, P., Holm, N. V., Verkasalo, P. K., Iliadou, A., Kaprio, J., Koskenvuo, M., Pukkala, E., Skytthe, A. and Hemminki, K. (2000). Environmental and heritable factors in the causation of cancer: analyses of cohorts of twins from Sweden, Denmark, and Finland. *New England J. Med.* **343**(2): 78–85.

20. Hoover, R. N. (2000). Cancer — nature, nuture, or both, Editorial. *New England J. Med.* **343**(2): 135–136.
21. Mestel, R. (2000). Study ties most cancer to lifestyle, not genetics, pp. A1, A28, *The Los Angeles Times* (July 13, 2000).
22. Cancers caused by genes make up more than 25 percent of cases … , p. A1, *The Wall Street Journal* (July 13, 2000).
23. Genetic code is not enough to cure cancer, *USA Today*, p. D11 (July 13, 2000).
24. Varmus, H. and Collins, F. (2000). Joint interview on the Charlie Rose show (PBS-TV; KCET, Channel 28 in Los Angeles, 11:00 PM PDT; July 13, 2000; TRT = 30 minutes).
25. Landro, L. (2000). More people are using internet health sites, but fewer are satisfied, p. A9, *The Wall Street Journal* (December 29, 2000).
26. Woolley, C. *et al.* Breakthroughs in cascading anti-oxidants, New Ways International, Inc. of Salem, UT; Jean Mayer United States Department of Agriculture Human Nutrition Research Center on Aging; Tufts University, Boston, MA. http://www.hnrc.tufts.edu/researchprograms/USDALabResProgDes/Oracchrt.html
27. Ames, B. N., Shigenaga, M. K. and Hagen, T. M. (1993). Oxidants, antioxidants, and the degenerative diseases of aging. *PNAS* **90**: 7915–7922.
28. Cassileth, B. (2001). Laetrile by any other name is Still Bogus, pp. S1, 6, *The Los Angeles Times* (January 1, 2001).
29. Dawkins, R. (1995). *River Out of Eden: A Darwinian View of Life*, Basic Books, New York.
30. American Institute for Cancer Research (AICR) recommendations can be found on their website at http://www.aicr.org/r083100a.htm

Chapter 49

Oxidative Stress and Antioxidant Intervention

Freya Q. Schafer, Eric E. Kelley and Garry R. Buettner*

Keywords: Antioxidants, free radicals.

Freya Q. Schafer, Eric E. Kelley and **Garry R. Buettner** • Free Radical and Radiation Biology Program, EMRB 68, The University of Iowa, Iowa City, IA 52242-1101

*Corresponding Author.
Tel: 319-335-6749, Email: garry-buettner@uiowa.edu

1. Introduction

The term antioxidant is now a part of the American vocabulary. Antioxidant formulations are widely available in health food stores, pharmacies, supermarkets and through mail-order sources. These formulations range from the traditional antioxidant vitamins C and E, to phenolic antioxidants isolated from various plant sources, e.g. pycnogenol, xanthohumol, green tea extracts and other polyphenols. Producers and vendors suggest that their products lead to better health. Only a few of the claims have been supported by sound research. Other than for antioxidant vitamins, many claims of health benefits of antioxidants are extrapolations from experiments that simply demonstrate that a substance can serve as an antioxidant.

An antioxidant is a substance that, when present in small amounts, prevents oxidation of the bulk. Unfortunately, many compounds are inappropriately labeled as antioxidants. Many experiments have been reported to demonstrate a purported antioxidant reaction of a substance, but in reality the data only suggest a slight retarding reaction.[1] There is still much to learn about phytochemicals and antioxidant action.

A good deal is known about traditional antioxidants such as vitamin E and vitamin C.[2-4] Many investigators have examined the antioxidant action of these two small molecules using cell culture models as well as animal models. The results clearly demonstrate that these antioxidants protect cells and tissues from oxidative stress. They suggest that in stress situations, such as disease or exercise, intake of higher amounts of these antioxidants would be beneficial. But direct demonstration that higher than normal intakes of these antioxidants is beneficial is experimentally difficult. In this work we discuss the mechanisms of how antioxidants work together and look at examples that demonstrate this cooperativity.

2. Lipid Peroxidation and Antioxidants

There are many types of antioxidants; they can be classified by their mechanism of action. Preventative antioxidants include peroxide decomposers and metal ion decomposers, while chain-breaking antioxidants intercept chain-carrying radicals. Many chain-breaking antioxidants donate a hydrogen atom to the chain-carrying radical thereby stopping the oxidation process. This results in an antioxidant radical. However, this radical is much less reactive than the original chain-carrying radical. But even this much more domesticated radical must be removed. Tocopherol (TOH) is a typical donor antioxidant in this class; it protects against lipid peroxidation, Scheme 1. Lipid peroxidation is a chain reaction, reaction 1–3.

$$\text{L-H} + \text{oxidant}^{\bullet} \xrightarrow{\text{R}} \text{L}^{\bullet} + \text{oxidant-H} \qquad \qquad (initiation) \ (1)$$

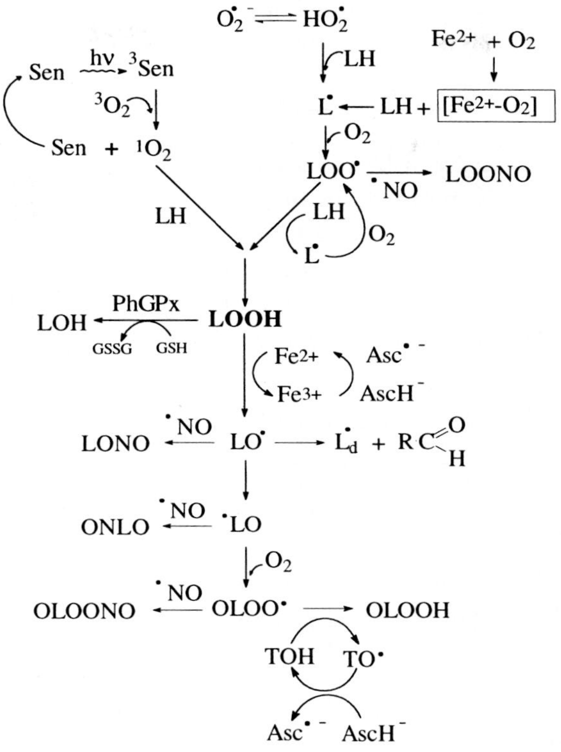

$$O_2^{\cdot -} \rightleftharpoons HO_2^{\cdot}$$

Scheme 1. **Lipid peroxidation and antioxidants.**

$$L^{\cdot} + O_2 \xrightarrow{\;k_o = 3 \times 10^8 \text{ M}^{-1}\text{s}^{-1}\;} LOO^{\cdot} \qquad \text{(propagation cycle) (2)}$$

$$LOO^{\cdot} + LH \xrightarrow{\;k_p = 10\text{--}50 \text{ M}^{-1}\text{s}^{-1}\;} L^{\cdot} + LOOH \qquad \text{(propagation cycle) (3)}$$

$$L^{\cdot} + L^{\cdot} \longrightarrow \text{non-radical products} \qquad \text{(termination) (4)}$$

$$L^{\cdot} + LOO^{\cdot} \longrightarrow \text{non-radical products.} \qquad \text{(termination) (5)}$$

Tocopherol breaks the propagation cycle by donating a hydrogen atom to the chain-carrying peroxyl radical, LOO·, thereby stopping the oxidation process, reaction 6.

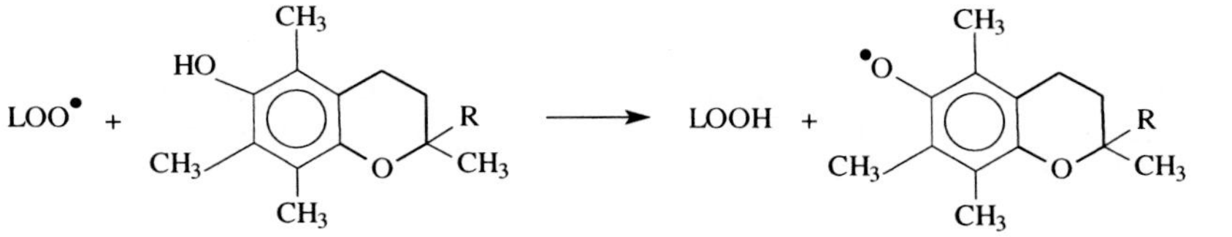

The tocopheroxyl radical (TO$^\bullet$) is much less reactive than LOO$^\bullet$, but it too can be a chain-carrying radical, albeit very poorly, reaction 7.

$$TO^\bullet + LH \xrightarrow{\quad k_p \approx 0.05\,M^{-1}s^{-1}\,[5,6,7] \quad} L^\bullet + TOH. \qquad \text{(propagation) (7)}$$

The tocopheroxyl radical can be repaired by additional donor compounds that are thermodynamically more reducing than TOH.[8,9] Examples are ascorbate (AscH$^-$) or ubiquinol (CoQH$_2$), see Table 1.

The reaction of either AscH$^-$ or CoQH$_2$ with TO$^\bullet$ naturally results in formation of a radical, Asc$^{\bullet-}$ or CoQ$^{\bullet-}$, respectively. These radicals are in turn removed by enzyme systems; thus AscH$^-$ and CoQH$_2$ are recycled.

Table 1. One-Electron Reduction Potentials

Redox Couple	E$^{o\prime}$/mV
HO$^\bullet$, H$^+$/H$_2$O	+2310
RO$^\bullet$, H$^+$/ROH (aliphatic alkoxyl radical)	+1600
ROO$^\bullet$, H$^+$/ROOH (alkyl peroxyl radical)	+1000
GS$^\bullet$/GS$^-$ (glutathione)	+920
PUFA$^\bullet$, H$^+$/PUFA-H (bis-allylic-H)	+600
TO$^\bullet$, H$^+$/TOH	+480
H$_2$O$_2$, H$^+$/H$_2$O, HO$^\bullet$	+320
Ascorbate$^{\bullet-}$, H$^+$/Ascorbate monoanion	+282
semiubiquinone, H$^+$/ubiquinol (CoQ$^{\bullet-}$, 2H$^+$/CoQH$_2$)	+200
Fe(III) EDTA/Fe(II) EDTA	+120
Ubiquinone, H$^+$/Semiubiquinone (CoQ/CoQ$^{\bullet-}$)	−36
O$_2$/O$_2^{\bullet-}$	−160[a]
Paraquat/Paraquat$^{\bullet+}$	−448
Fe(III)DFO/Fe(II)DFO	−450
RSSR/RSSR$^{\bullet-}$ (GSH)	−1500
H$_2$O/e$^-_{aq}$	−2870

[a]Two different thermodynamic reference states are used for O$_2$. Here we have chosen to use the aqueous concentration of oxygen, thus the appropriate reference state is a solution that is 1 molal ($\approx$ 1 M) in O$_2$ and E$^{o\prime}$ = −160 mV. The second reference state often used is 1 atmosphere of O$_2$; E$^{o\prime}$ is then −330 mV. A pressure of 1 atmosphere of O$_2$ will result in [O$_2$] $\cong$ 1.25 mM in room temperature aqueous solutions. If this reference state is used, then in the Nernst equation P$_{O_2}$ replaces [O$_2$] in all equations. The same value for E will result. Table derived from Ref. 31.

3. Vitamin E

3.1. Vitamin E Levels in Cells can be Manipulated

We have found that most cultured cells are deficient in vitamin E.[10] L1210 cells, cultured in standard growth media, contain only 2.3 ± 0.03 µg of tocopherol/ 10^8 cells, Table 2. When these cells are transplanted and grown for the same time in the ascites fluid of mice fed standard diets, the vitamin E content of the cells increases to 5.8 ± 0.6 µg of α-tocopherol/10^8 cells, a level lower than would be expected if the cells were grown in fully vascularized compartments. This apparent tocopherol deficiency in cultured cells is likely due to the low concentrations of tocopherol contained in most tissue culture media, even with the addition of serum.

To study this apparent deficiency and the relationship of cellular tocopherol to the cellular lipid composition, we supplemented the growth media of murine leukemia cells (L1210) with α-tocopherol and compared the resultant cellular tocopherol content to the degree of unsaturation of cellular lipids (membrane lipid *bis*-allylic hydrogen positions). The α-tocopherol was incorporated by cells in a time- and concentration-dependent manner with plateaus at 24 h and 100 µM, respectively. A maximum 400% increase in cellular tocopherol was easily achieved.

By experimentally modifying the fatty acid content of cellular lipids, we were able to determine that cellular tocopherol uptake and content is not a function of cellular lipid composition. Cells enriched with polyunsaturated lipids incorporated tocopherol to the same extent as those enriched with more saturated lipids. Thus, as the cellular polyunsaturated fatty acid content increases, the tocopherol:

Table 2. α-Tocopherol and *bis*-Allylic Positions in L1210 Cells

Cell Lipid Modification[a]	TOH Molecules per Cell ($\times 10^7$)[b]	*bis*-Allylic Positions per Cell ($\times 10^{11}$)[c]	*bis*-Allylic per TOH[d,e]
Unmodified	3.28	0.68	2075
18:1	2.78	0.25	899
18:3	3.08	2.07	6710
20:4	2.68	2.93	10 990
22:6	3.00	2.97	9900

[a]The lipid profile of cells was modified by supplementing the culture media with the indicated fatty acid. This manipulation allows us to vary the oxidizability of the cell. No supplemental tocopherol was provided to these cells.
[b]This is the number of tocopherol molecules per L1210 cell (SE = approximately 1).
[c]This is the number of *bis*-allylic positions in the lipids in an L1210 cell. It is the *bis*-allylic positions of the lipid that are susceptible to oxidation. (SE = approximately 0.1).
[d]This ratio indicates the average number of lipid *bis*-allylic positions that each TOH must protect from oxidation. The greater the number the less protection afforded by TOH.
[e]Data are derived from Ref. 10.

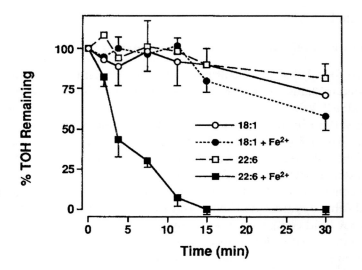

Fig. 1. **Effect of oxidative stress on cellular α-tocopherol.** L1210 cells (5×10^7) were enriched with 22:6 or 18:1 by adding 32 µM of the fatty acid to the growth media for 48 hours. During the final 24 h, 12.5 µM tocopherol acetate was added. The enriched cells were washed and then incubated in 0.9% NaCl with or without 100 µM Fe^{2+} and the cellular tocopherol determined at the time points shown. Shown are the levels as a percentage of time zero and values are the mean and SE of three determinations. The values at time zero were: 18:1, 1.84 ± 0.6 µg/10^8 cells; 18:1 + Fe^{2+}, 1.48 ± 0.4 µg/10^8 cells; 22:6, 1.70 ± 0.1 µg/10^8 cells; and for 22:6 + Fe^{2+}, 2.0 ± 0.5 µg/10^8 cells.

bis-allylic position ratio in the cells decreases, resulting in less antioxidant protection for each lipid double bond. It is the *bis*-allylic moieties in lipids that are susceptible to oxidation. When polyunsaturated fatty acid-enriched cells are exposed to an oxidative stress, the rate of oxidation is high and the tocopherol levels decline much faster than in cells enriched with saturated fatty acids. In effect, they have much less antioxidant protection. This loss of tocopherol correlates with the *bis*-allylic:tocopherol ratio, Fig. 1. Modifying the lipid content of cells, as well as their α-tocopherol content, affects their susceptibility to free radical-mediated lipid peroxidation.

To further investigate the effect of vitamin E supplementation on lipid per-oxidation in L1210 murine leukemia cells,[11] cells were exposed to an oxidative stress induced by 20 µM Fe^{2+} and 100 µM ascorbate. The kinetics of the generation of lipid-derived free radicals, as measured by EPR spin trapping (a product) and O_2 consumption (a reactant) were measured. Cells grown for 24 h with supple-mental vitamin E (5–100 µM) in their media had a slower rate of lipid radical generation compared to cells grown without vitamin E supplementation. This inhibition in the rate of oxidation was generally dependent upon the amount of vitamin E supplementation, Fig. 2.[11] In complementary studies measuring O_2

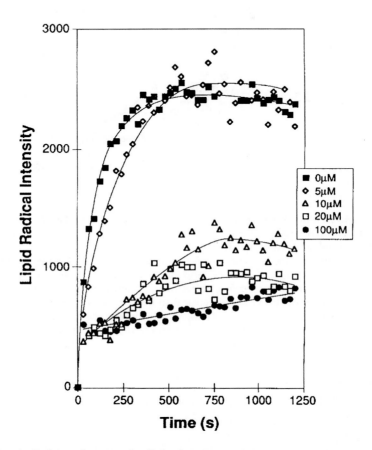

Fig. 2. **Vitamin E slows the rate of cellular lipid radical formation.** Cellular incorporation of α-tocopherol slows the rate, but induces no apparent lag phase in lipid radical formation during lipid peroxidation as shown by EPR-detectable POBN/L· adducts. L1210 cells, 5 × 10^6 cells/mL enriched with 22:6ω3 and supplemented with various concentrations of α-tocopherol acetate were subjected to the oxidative stress presented by 20 μM Fe^{2+} and 100 μM ascorbate in the presence of 50 mM of the spin trap POBN. Each data point is the mean of 5–7 experiments and represents the EPR peak height of the first peak of the low field doublet. The lipid radical intensity units of the ordinate are arbitrary values; 1000 corresponds to 0.24 μM POBN/L·.

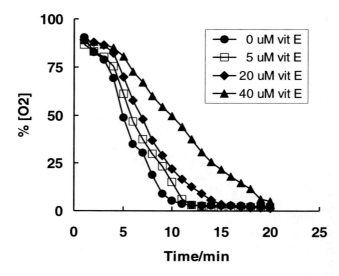

Fig. 3. **Vitamin E slows the rate of oxygen consumption during cellular lipid peroxidation.** Cellular incorporation of vitamin E slows the rate of oxygen consumption during cellular lipid peroxidation. L1210 cells (5×10^6 cells/mL) enriched with 22:6ω3 and supplemented with various concentrations of α-tocopherol acetate were subjected to the oxidative stress presented by 20 μM Fe^{2+} and 100 μM ascorbate in the presence of 50 mM POBN. Each point is the mean of 5–7 experiments from oxygen probe recordings. The conditions for these incubations were identical to those described in the legend of Fig. 2. The incubations were initially air-saturated, which implies the dissolved $[O_2] \approx 250$ μM. The POBN was included to maintain identical conditions compared to the spin trapping experiments of Fig. 2. However, there was little if any difference between oxygen-uptake experiments with or without POBN.

consumption, 5–100 μM vitamin E slowed the rate of oxidation (10-fold with 100 μM supplemental vitamin E) consistent with the EPR studies, Fig. 3.[1]

Figure 4 clearly shows the relationship of cell tocopherol content, rate of oxygen consumption, rate of lipid radical formation, and trypan blue dye exclusion.[11] Note that cellular membranes appear to "saturate" with vitamin E. Only about 10 μM of vitamin E in the media is required; the antioxidant protection provided by this vitamin E also "saturates".

These results are the first to actually demonstrate, by monitoring free radical production, that vitamin E inhibits lipid peroxidation in cells by slowing the rate of lipid peroxidation.

3.2. *In Vivo* Photoprotection by Vitamin E

If an organism is undergoing an oxidative stress, then tocopherol may provide some protection from this stress. We investigated the photoprotective effect of

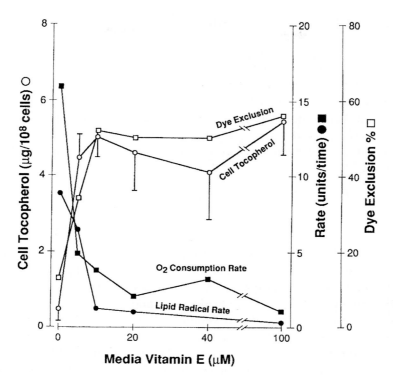

Fig. 4. **Effect of vitamin E concentration on inhibition of cellular lipid peroxidation.** The ordinate shows the initial rate of generation of POBN/L˙ (derived from Fig. 2), (negative) initial rate of oxygen consumption (Fig. 3), trypan blue dye exclusion, and cellular uptake of vitamin E. The abscissa is the concentration of α-tocopherol in the culture media.

topically applied tocopherols against chronic UVB radiation-induced skin damage, using a mouse model of photoaging.[12, 13] In previous testing using this model,[14] tocopherol acetate was poorly photoprotective against chronic UVB radiation-induced skin wrinkling, whereas α-tocopherol provided significant protection. We compared the efficacy of these two materials against tocopherol sorbate.[15] The results demonstrated that tocopherol sorbate is significantly more protective than the other two forms of vitamin E against skin wrinkling. In addition we found that the mice in the tocopherol sorbate group (1.8 tumors/mouse) and the α-tocopherol group (2.0 tumors/mouse) had fewer tumors compared to the vehicle control group (3.6 tumors/mouse) at the end of the study (week 23). Tocopherol sorbate and α-tocopherol reduced the average number of tumors per mouse, but they did not delay the onset of appearance of the first tumor, relative to vehicle. Tocopherol if delivered appropriately, can provide significant protection from UV-induced skin damage and tumor formation.

4. Nitric Oxide as an Antioxidant

Vitamin E (TOH) is an important lipid-soluble antioxidant and has been traditionally thought to be the principal chain-breaking antioxidant in blood and in lipid structures of cells.[16, 17] However, the experiments done that led to this conclusion were done in the absence of ·NO.

Nitric oxide can protect cells against the detrimental effects of reactive oxygen species.[18, 19] Using low density lipoprotein as well as model systems, it has been demonstrated that ·NO can serve as a chain-breaking antioxidant to blunt lipid peroxidation.[20–23]

$$LOO^{\bullet} + {}^{\bullet}NO \longrightarrow LOON{=}O \qquad\qquad \text{(chain-termination) (8)}$$

$$LO^{\bullet} + {}^{\bullet}NO \longrightarrow LON{=}O. \qquad\qquad \text{(chain-termination) (9)}$$

To test the hypothesis that ·NO can serve as a chain-breaking antioxidant in cell membranes, we examined the effect of ·NO on iron-induced lipid peroxidation in human leukemia cells.[24] We exposed HL-60 cells (enriched with docosahexaenoic acid, DHA) to an oxidative stress (20 μM Fe^{2+}) and monitored the consumption of oxygen as a measure of lipid peroxidation. While oxygen consumption in our cellular vitamin E experiments slowed with increasing vitamin E levels (Fig. 3), oxygen consumption was completely arrested by the addition of ·NO, Figs. 5 and 6. The duration of inhibition of oxygen consumption by ·NO was concentration-dependent in the 0.4–1.8 μM range. The inhibition ends upon depletion of ·NO. But note that once lipid peroxidation is under control only nM levels of ·NO are required to stop subsequent free radical chain reactions; these levels are in the physiological range.

4.1. A Theoretical Comparison of ·NO and Vitamin E as Cell Membrane-Antioxidants[24]

From the data above and kinetic information from the literature, it is possible to make an estimate of the importance of TOH versus ·NO as a chain-breaking antioxidant. A typical TOH level in cells is 1000–2000 PUFA/1 TOH.[26] If PUFA constitutes about 22% of the lipids in a cell membrane,[27] then there are about 5000–10 000 total fatty acid chains/1 TOH. As a first approximation, the mole fraction of TOH in lipid regions of membranes will be:

$$\text{Mole fraction (TOH)} \approx 1/(5000 \text{ to } 10\,000)$$
$$\approx 1/7500$$
$$\approx 1.3 \times 10^{-4}.$$

Because TOH and the various fatty acids have similar properties, we assume for simplicity, that they have the same partial molar volume in the lipid regions of

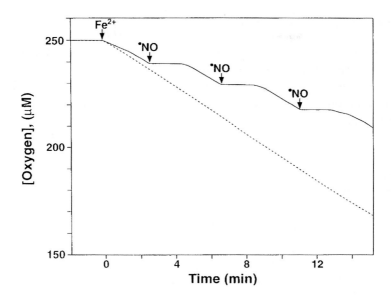

Fig. 5. Nitric oxide inhibits iron-induced lipid peroxidation. The rate of O_2 consumption of HL-60 cells (5×10^6/mL) enriched with docosahexaenoic acid was determined using a YSI oxygen monitor. Fe^{2+} (20 μM) was added at the first arrow and subsequently ·NO (1.8 μM) was added (other arrows). When ·NO was added, the O_2 consumption was inhibited for a period of a few minutes, then it resumed at near its initial rate until the reintroduction of additional ·NO. Also shown (lower dashed line) is a typical control of HL-60 cells subjected to Fe^{2+}-induced oxidative stress in the absence of ·NO addition.

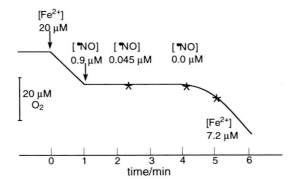

Fig. 6. Changes in concentration of ·NO and Fe^{2+} during cellular lipid peroxidation and its inhibition by ·NO. Shown are the concentrations of ·NO and Fe^{2+} at key time points. Peroxidation was initiated with 20 μM Fe^{2+}. At 1 min after the addition of Fe^{2+}, 0.9 μM ·NO was introduced. Nitric oxide was rapidly depleted and is below the limit of detection at about 4 minutes. At the time of ·NO depletion, rapid O_2 uptake resumes. This re-initiation of O_2 consumption is due to Fe^{2+} that is still present at 7.2 μM or about 36% of its original value.

cell membranes. Using a density of ≈ 0.9 g lipid/mL and an average molecular weight of 300 g/mole for the fatty acyl chains of the lipids in a cell membrane, the effective molarity of the fatty acyl chains in the lipid regions of membranes will be ≈ 3 M. The molarity of TOH will then be about $3 \text{ M} \times 1.3 \times 10^{-4} \approx 400$ mM.

The rate at which $^{\bullet}$NO would terminate the chain propagation reactions in lipid peroxidation by reacting with the chain-carrying peroxyl radical, LOO$^{\bullet}$, reaction 8, will be:

$$\text{rate inhibition by } ^{\bullet}\text{NO} = k_{NO} \, [^{\bullet}\text{NO}] \, [\text{LOO}^{\bullet}] \tag{10}$$

where $k_{NO} = 2 \times 10^9$ M^{-1}s^{-1} [28] and $[^{\bullet}\text{NO}] = 8 \times 45$ nM $= 360$ nM. Here 8 is the estimated membrane/water partition coefficient for $^{\bullet}$NO;[29, 30] 45 nM is the measured $^{\bullet}$NO concentration in the aqueous phase of the cell suspension during the inhibition phase of lipid peroxidation, Fig. 6. So,

$$\text{rate inhibition } (^{\bullet}\text{NO}) = 2 \times 10^9 \text{ M}^{-1}\text{s}^{-1} \times 360 \times 10^{-9} \text{ M } [\text{LOO}^{\bullet}]. \tag{11}$$

The rate at which TOH would terminate these reactions, reaction 6, would be:

$$\text{rate inhibition (TOH)} = k_{TOH} \, [\text{TOH}] \, [\text{LOO}^{\bullet}] \tag{12}$$

where $k_{TOH} = 8 \times 10^4$ M^{-1}s^{-1} [31] and $[\text{TOH}] = 400$ mM. Thus, rate inhibition (TOH) $= 8 \times 10^4$ M^{-1}s$^{-1} \times 400 \times 10^{-6}$ M $[\text{LOO}^{\bullet}]$. The ratio of these rates is:

$$\text{rate } (^{\bullet}\text{NO})/\text{rate (TOH)} = 20/1. \tag{13}$$

If the concentrations and kinetic parameters used to make this estimate accurately represent these processes in cells, then an aqueous concentration of $^{\bullet}$NO of only 2 nM would provide antioxidant protection for membrane lipids equal to that typical of vitamin E. This is a remarkably low concentration of $^{\bullet}$NO and is easily achievable in many cells and tissues. That these hypotheses have merit is the recently published observation that $^{\bullet}$NO spares vitamin E in liposomes.[32] In addition these investigators demonstrated that $^{\bullet}$NO does not react with tocopherol or its radical. Nitric oxide has been shown to protect against the cytotoxicity of oxidative stress in endothelial cells, fibroblasts, hepatocytes, intestinal epithelium, and cardiomyocytes.[33–38] However, there is much yet to be learned about the molecular mechanisms of this protection and the importance of $^{\bullet}$NO as an antioxidant in cells and tissues. $^{\bullet}$NO is a remarkable antioxidant.

5. Vitamin C

5.1. Ascorbate, the Terminal Small-Molecule Antioxidant

Ascorbate, an excellent reducing agent, undergoes two consecutive, reversible, one-electron oxidation processes forming the ascorbate radical (Asc$^{\bullet-}$) as an intermediate. $Asc^{\bullet-}$ has its unpaired electron in a highly delocalized π-system making it a relatively unreactive free radical. These properties make ascorbate a superior biological, donor antioxidant.[39] $Asc^{\bullet-}$ can be detected in biological fluids or tissues using electron paramagnetic resonance spectroscopy (EPR). This is consistent with ascorbate's role as the terminal small-molecule antioxidant, Table 1.[40] As seen in Table 1, ascorbate has a low reduction potential. Thus, it can repair many free radicals that are produced during oxidative stress, such as $HO^{\bullet}$, $RO^{\bullet}$, $LOO^{\bullet}$, $GS^{\bullet}$, urate, and even the tocopheroxyl radical ($TO^{\bullet}$). The electron transfer reaction:

$$AscH^- + X^{\bullet} \longrightarrow Asc^{\bullet-} + XH \tag{14}$$

is relatively rapid.[41] Consequently, both thermodynamically and kinetically, ascorbate can be considered to be an excellent aqueous antioxidant.

The ascorbate radical formed in these reactions is relatively stable and does not react with O_2 to form dangerous peroxyl radicals. Ascorbate (most likely Asc^{2-}, and/or $Asc^{\bullet-}$) appears to produce very low levels of superoxide.[42, 43] Superoxide dismutase provides protection from this possibility.[44] In addition, $Asc^{\bullet-}$ as well as the dehydroascorbic formed can be reduced back to ascorbate by enzyme systems. Hence, this antioxidant is recycled.

5.2. Ascorbate can be a Pro-Oxidant

Ascorbate is known to be and is widely used as a pro-oxidant.[45–53] This behavior results because it is an excellent reducing agent. As a reducing agent it is able to reduce catalytic metals such as Fe^{3+} and Cu^{2+} to Fe^{2+} and Cu^+. The redox cycling of these metals is essential to the oxidation of the vast majority of singlet state organic molecules.[54] In nearly all the experimental systems where ascorbate has pro-oxidant properties, there is the simultaneous presence of redox active metals. Thus, in living organisms catalytic metals and ascorbate are to be avoided.

On the other hand, ascorbate in combination with catalytic metals are used to induce a controlled, well defined oxidative stress, such as that used in the experiments of Figs. 3 and 4. The reduced metal can react with peroxides (Fenton reaction) forming hydroxyl or alkoxyl radicals, which are very oxidizing and will initiate biological damage.

$$Fe^{2+} + H_2O_2 \longrightarrow Fe^{3+} + OH^- + HO^{\bullet} \qquad \text{Fenton Reaction} \tag{15}$$

$$Fe^{2+} + ROOH \longrightarrow Fe^{3+} + OH^- + RO^{\bullet}. \qquad \text{Organic Fenton Reaction} \tag{16}$$

However, peroxides are not necessary for these reduced metals to be dangerous. For example, in the biological pH range, all that is required for Fe^{2+} to initiate detrimental oxidations is the presence of dioxygen.[55] It is thought that $Fe^{2+} + O_2$

results in the formation of iron-oxygen complexes that are oxidizing species; the reactivity of the iron in this chemistry depends upon its coordination environment. The reactivity of iron is also influenced by pH. Using a tissue culture model, it has been observed that the lower the extracellular pH, the higher the flux of free radicals induced by the addition of ferrous iron and the greater the cellular membrane damage, as determined by trypan blue exclusion.[56] The presence of reducing agents such as ascorbate will, in general, recycle the catalytic iron, ensuring that it will be dominantly in the ferrous state. Thus, ascorbate in the presence of iron can be detrimental rather than beneficial.

6. Antioxidant Network

Oxidative stress has many faces. There are a variety of reactive oxygen species producing unwanted oxidations in different locations. Each antioxidant has its niche in that each provides protection in distinct environments and against certain types of reactive species. To provide optimal protection from unwanted oxidations, a network of antioxidants is needed. Many of these antioxidants work together, i.e. they are co-antioxidants. They recycle each other or work in concert.

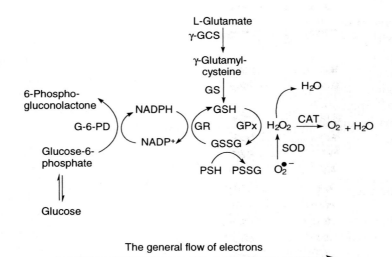

Scheme 2. **A subset of the antioxidant network.** This schematic shows the relationships between antioxidant enzymes and glutathione. Abbreviations: CAT = catalase; G-6-PD = glucose-6-phosphate dehydrogenase; γ-GCS = γ-glutamylcysteine synthetase; GS = glutathione synthetase; GPx = glutathione peroxidase; GR = glutathione reductase; PSH = protein thiol; PSSG = protein mixed disulfide with GSH; SOD = superoxide dismutase.

For example, glutathione recycles not only vitamin $C^{57, 58}$ but also glutathione peroxidase (GPx) an antioxidant enzyme known to reduce H_2O_2. GPx itself works in concert with another antioxidant enzyme, superoxide dismutase (SOD). SOD reacts with superoxide radical producing hydrogen peroxide, Scheme 2. These few examples show clearly the network that exits between various antioxidants *in vivo*.

Gey re-examined epidemiological data to see if the nutritional antioxidants vitamins E and C, are indeed co-antioxidants.[59] Vitamin E can protect LDL, cell membranes, and other lipid structures from oxidation. However, the resulting tocopheroxyl radical must be removed, preferably recycled. Ascorbate can do this. If this cooperativity is important, then there might be an ideal level of these co-antioxidants for optimal health. Gey found that, assuming sufficiency of both E and C, the data suggested that a ratio of C to E in plasma (i.e. [vitamin C]/ [vitamin E]) of 0.6–0.8 associates with increased risk of coronary heart disease, whereas a ratio of > 1.3–1.5 may be desirable to minimize risk of this disease. This is consistent with these antioxidant vitamins serving as co-antioxidants, providing conditions for optimal health.

7. Aging

The free radical theory of aging would suggest that keeping a full complement of antioxidants may be of benefit for better health and increase the probability of reaching one's maximum lifespan potential. Nutritional antioxidants have an important role in health. They work together with the endogenous antioxidants such as glutathione. Measurements of glutathione in plasma show a general decline with age.[60] Specifically, the half-cell reduction potential of the GSSG/2GSH couple in plasma becomes more positive with age, which suggests that as we age free radical oxidation processes increase.[61, 62]

8. Summary

Antioxidants are necessary for life. We have a network of small-molecule and enzyme antioxidants that work together to defend against unwanted, detrimental oxidations. New, quantitative viewpoints suggest that optimal health is associated with a "healthy" redox state. A shift in the redox state to more positive potentials is detrimental to cells, tissues and whole organisms.[63] The role of antioxidants may be to protect the structure and function of the machinery that maintains a healthy redox status. Thus, we propose that studies on antioxidants and health should include quantitative measures of redox state as an indicator of overall organism health.

9. Abbreviations

AscH$^-$, ascorbate monoanion; Asc$^{•-}$, ascorbate radical; Asc^{2-}, ascorbate dianion; CoQH$_2$, ubiquinol; CoQ$^{•-}$, semiubiquinone; DFO, deferrioximine; DHA, docosahexaenoic acid; EPR, electron paramagnetic resonance; Fe^{2+}, ferrous iron; Fe^{3+}, ferric iron; GSH, glutathione; GSSG, glutathione disulfide; GPx, glutathione peroxidase; HL-60, human leukemia cells; H$_2$O$_2$, hydrogen peroxide; $^•$OH, hydroxyl radical; L$^•$, lipid radical, LDL, low densitiy lipoprotein; LH, unsaturated lipid, LOO$^•$, lipid peroxyl radical, LOOH, lipid hydroperoxide, L1210, murine leukemia cells; $^•$NO, nitric oxide; O$_2$, dioxygen; O$_2^{•-}$, superoxide radical; PhGPx, phospholipid hydroperoxide glutathione peroxidase; POBN, α-(4-pyridyl-1-oxide)-*N-tert*-butylnitrone; PUFA, polyunsaturated fatty acid; RO$^•$, aliphatic alkoxyl radical; SOD, superoxide dismutase; TOH, tocopherol; TO$^•$, tocopheroxyl radical; XH, substrate; X$^•$, substrate radical.

References

1. Barclay, L. R. and Vinqvist, M. R. (2000). Do spin traps also act as classical chain-breaking antioxidants? A quantitative kinetic study of phenyl *tert*-butylnitrone (PBN) in solution and in liposomes. *Free Radic. Biol. Med.* **28**: 1079–1090.
2. Williams, G. M., Sies, H., Baker, G. T., Erdman, J. W. and Henry, C. J. (1993). *Antioxidants: Chemical, Physiological Nutritional and Toxicological Aspects*, Princeton Scientific Publishing Co. Inc., Princeton, New Jersey.
3. Packer, L. and Ong, A. S. H. (1998). *Biological Oxidants and Antioxidants: Molecular Mechanisms and Health Effects*, AOCS Press, Champaign, IL.
4. Cadenas, E. and Packer, L. (1996). *Handbook of Antioxidants*, Marcel Dekker Inc., New York.
5. Bowry, V. W. and Stocker, R. (1993). Tocopherol-mediated peroxidation. The prooxidant effect of vitamin E on the radical-initiated oxidation of human low-density lipoprotein. *J. Am. Chem. Soc.* **115**: 6029–6044.
6. Waldeck, A. R. and Stocker, R. (1996). Radical-initiated lipid peroxidation in low density lipoproteins: insights obtained from kinetic modeling. *Chem. Res. Toxicol.* **9**: 954–964.
7. Mukai, K., Morimoto, H., Okauchi, Y. and Nagaoka, S. (1993). Kinetic study of reactions between tocopheroxyl radicals and fatty acids. *Lipids* **28**: 753–755.
8. Sharma, M. K. and Buettner, G. R. (1993). Interaction of vitamin C and vitamin E during free radical stress in plasma: an ESR study. *Free Radic. Biol. Med.* **14**: 649–653.
9. Bowry, V. W., Mohr, D., Cleary, J. and Stocker, R. (1995). Prevention of tocopherol-mediated peroxidation in ubiquinol-10-free human low density lipoprotein. *J. Biol. Chem.* **270**: 5756–5763.

10. Kelley, E. E., Buettner, G. R. and Burns, C. P. (1995). Relative α-tocopherol deficiency in cultured tumor cells: free radical-mediated lipid peroxidation, lipid oxidizability, and cellular polyunsaturated fatty acid content. *Arch. Biochem. Biophys.* **319**: 102–109.

11. Wagner, B. A., Buettner, G. R. and Burns, C. P. (1996). Vitamin E slows the rate of free radical-mediated lipid peroxidation in cells. *Arch. Biochem. Biophys.* **334**: 261–267.

12. Bissett, D. L., Hannon, D. P. and Orr, T. V. (1990). Photoprotective effect of topical anti-inflammatory agents against ultraviolet radiation-induced chronic skin damage in the hairless mouse. *Photodermatol. Photoimmunol. Photomed.* **7**: 153–158.

13. Bissett, D. L., Hannon, D. P. and Orr, T. V. (1987). An animal model of solar-aged skin: histological, physical, and visible changes in UV-irradiated hairless mouse skin. *Photochem. Photobiol.* **46**: 367–378.

14. Bissett, D. L., Chatterjee, R. and Hannon, D. P. (1990). Photoprotective effect of superoxide-scavenging antioxidants against ultraviolet radiation-induced chronic skin damage in hairless mouse. *Photodermatol. Photoimmunol. Photomed.* **7**: 56–62.

15. Jurkiewicz, B. A., Bissett, D. L. and Buettner, G. R. (1995). The effect of topically applied tocopherols on ultraviolet light-mediated free radical damage in skin. *J. Invest. Derm.* **104**: 484–488.

16. Burton, G. W., Joyce, A. and Ingold, K. U. (1983). Is vitamin E the only lipid soluble, chain breaking antioxidant in human blood plasma and erythrocyte membranes? *Arch. Biochem. Biophys.* **221**: 281–290.

17. Burton, G. W. and Ingold, K. U. (1986). Vitamin E: application of the principles of physical organic chemistry to the exploration of its structure and function. *Acc. Chem. Res.* **19**: 194–201.

18. Wink, D. A., Cook, J. A., Pacelli, R., DeGraff, W., Gamson, J., Liebmann, J., Krishna, M. C. and Mitchell, J. B. (1996). The effect of various nitric oxide-donor agents on hydrogen peroxide-mediated toxicity: a direct correlation between nitric oxide formation and protection. *Arch. Bioch. Biophys.* **331**: 241–248.

19. Wink, D. A., Cook, J. A., Krishna, M. C., Hanbauer, I., DeGraff, W., Gamson, J. and Mitchell, J. B. (1995). Nitric oxide protects against alkyl peroxide-mediated cytotoxicity: further insights into the role nitric oxide plays in oxidative stress. *Arch. Biochem. Biophys.* **319**: 402–407.

20. Hogg, N., Kalyanaraman, B., Joseph, J., Struck, A. and Parthasarathy, S. (1993). Inhibition of low-density lipoprotein oxidation by nitric oxide. Potential role in atherogenesis. *FEBS Lett.* **334**: 170–174.

21. Rubbo, H., Parthasarathy, S., Barnes, S., Kirk, M., Kalyanaraman, B. and Freeman, B. A. (1995). Nitric oxide inhibition of lipoxygenase-dependent liposome and low-density lipoprotein oxidation: termination of radical chain

propagation reactions and formation of nitrogen-containing oxidized lipid derivatives. *Arch. Biochem. Biophys.* **324**: 15–25.

22. Yamanaka, N., Oda, O. and Nagao, S. (1996). Nitric oxide released from zwitterionic polyamine/NO adducts inhibits Cu^{2+}-induced low density lipoprotein oxidation. *FEBS Lett.* **398**: 53–56.

23. O'Donnell, V. B., Chumley, P. H., Hogg, N., Bloodsworth, A., Darley-Usmar, V. M. and Freeman, B. A. (1997). Nitric oxide inhibition of lipid peroxidation: kinetics of reaction with lipid peroxyl radicals and comparison with α-tocopherol. *Biochemistry* **36**: 15 216–15 223.

24. Kelley, E. E., Wagner, B. A., Buettner, G. R. and Burns, C. P. (1999). Nitric oxide inhibits iron-induced lipid peroxidation in HL-60 cells. *Arch. Biochem. Biophys.* **370**: 97–104.

25. Burns, C. P., Kelley, E. E., Wagner, B. A. and Buettner, G. R. (2002). Role of nitric oxide and membrane phospholipid polyunsaturation in oxidative cell death. *In* "Subcellular Biochemistry Vol 36, Phospholipid Metabolism in Apoptosis" (P. Quinn, and V. Kagan, Eds.), pp. 97–121, Kluwer Academic/Plenum Press, New York.

26. Kelley, E. E., Buettner, G. R. and Burns, C. P. (1995). Relative α-tocopherol deficiency in cultured tumor cells: free radical-mediated lipid peroxidation, lipid oxidizability, and cellular polyunsaturated fatty acid content. *Arch. Biochem. Biophys.* **319**: 102–109.

27. Wagner, B. A., Buettner, G. R., Oberley, L. W. and Burns, C. P. (1998). Sensitivity of K562 and HL-60 cells to edelfosin, an ether lipid drug, correlates with production of active oxygen species. *Cancer Res.* **58**: 2809–2816.

28. Padmaja, S. and Huie, R. E. (1993). The reaction of nitric oxide with organic peroxyl radicals. *Biochem. Biophys. Res. Commun.* **195**: 539–544.

29. Rubbo, H., Radi, R., Trujillo, M., Telleri, R., Kalyanaraman, B., Barnes, S., Kirk. M. and Freeman, B. A. (1994). Nitric oxide regulation of superoxide and peroxynitrite-dependent lipid peroxidation. Formation of novel nitrogen-containing oxidized lipid derivatives. *J. Biol. Chem.* **269**: 26 066–26 075.

30. Liu, X., Miller, M. J. S., Joshi, M. S., Thomas, D. D. and Lancaster, J. R., Jr. (1998). Accelerated reaction of nitric oxide with O_2 within the hydrophobic interior of biological membranes. *Proc. Natl. Acad. Sci. USA* **95**: 2175–2179.

31. Buettner, G. R. (1993). The pecking order of free radicals and antioxidants: lipid peroxidation, α-tocopherol, and ascorbate. *Arch. Biochem. Biophys.* **300**: 535–543.

32. Rubbo, H., Radi, R., Anselmi, D., Kirk, M., Barnes, S., Butler, J., Eiserich, J. P. and Freeman, B. A. (2000). Nitric oxide reaction with lipid peroxyl radicals spares alpha-tocopherol during lipid peroxidation. Greater oxidant protection from the pair nitric oxide/alpha-tocopherol than alpha-tocopherol/ascorbate. *J. Biol. Chem.* **275**: 10 812–10 818.

33. Wink, D. A., Hanbauer, I., Krishna, M. C., DeGraff, W., Gamson, J. and Mitchell, J. B. (1993). Nitric oxide protects against cellular damage and cytotoxicity from reactive oxygen species. *Proc. Natl. Acad. Sci. USA* **90**: 9813–9817.

34. Struck, A. T., Hogg, N., Thomas, J. P. and Kalyanaraman, B. (1995). Nitric oxide donor compounds inhibit the toxicity of oxidized low-density lipoprotein to endothelial cells. *FEBS Lett.* **361**: 291–294.

35. Gupta, M. P., Evanoff, V. and Hart, C. M. (1997). Nitric oxide attenuates hydrogen peroxide-mediated injury to porcine pulmonary artery endothelial cells. *Am. J. Physiol.* **272**: L1133–L1141.

36. Sergent, O., Griffon, B., Morel, I., Chevanne, M., Dubos, M. P., Cillard, P. and Cillard, J. (1997). Effect of nitric oxide on iron-mediated oxidative stress in primary rat hepatocyte culture. *Hepatology* **25**: 122–127.

37. Gorbunov, N. V., Tyurina, Y. Y., Salama, G., Day, B. W., Claycamp, H. G., Argyros, G., Elsayed, N. M. and Kagan, V. E. (1998). Nitric oxide protects cardiomyocytes against *tert*-butyl hydroperoxide-induced formation of alkoxyl and peroxyl radicals and peroxidation of phosphatidylserine. *Biochem. Biophys. Res. Commun.* **244**: 647–651.

38. Chamulitrat, W. (1998). Nitric oxide prevented peroxyl and alkoxyl radical formation with concomitant protection against oxidant injury in intestinal epithelial cells. *Arch. Biochem. Biophys.* **355**: 206–214.

39. Niki, E. (1991). Vitamin C as an antioxidant. *World Rev. Nutri. Diet.* **64**: 1–30.

40. Buettner, G. R. and Jurkiewicz, B. A. (1993). Ascorbate free radical as a marker of oxidative stress: an EPR study. *Free Radic. Biol. Med.* **14**: 49–55.

41. Buettner, G. R. and Jurkiewicz, B. A. (1996). Catalytic metals, ascorbate and free radicals: combinations to avoid. *Rad. Res.* **145**: 532–541.

42. Scarpa, M., Stevanato, R., Viglino, P. and Rigo, A. (1983). Superoxide ion as active intermediate in the autoxidation of ascorbate by molecular oxygen. *J. Biol. Chem.* **258**: 6695–6697.

43. Williams, N. H. and Yandell, J. K. (1982). Outer-sphere electron-transfer reaction of ascorbate anions. *Aust. J. Chem.* **35**: 1133–1144.

44. Winterbourn, C. C. (1993). Superoxide as an intracellular radical sink. *Free Radic. Biol. Med.* **14**: 85–90.

45. Wills, E. D. (1969). Lipid peroxide formation in microsomes. Relationship of hydroxylation to lipid peroxide formation. *Biochem. J.* **113**: 315–324.

46. Wills, E. D. (1969). Lipid peroxide formation in microsomes. The role of non-haem iron. *Biochem. J.* **113**: 325–332.

47. Wills, E. D. (1966). Mechanisms of lipid peroxide formation in animal tissues. *Biochem. J.* **99**: 667–675.

48. Girotti, A. W., Bachowski, G. J. and Jordan, J. E. (1985). Lipid photooxidation in erythrocyte ghosts: sensitization of the membranes toward ascorbate- and

superoxide-induced peroxidation and lysis. *Arch. Biochem. Biophys.* **236**: 238–251.

49. Girotti, A. W., Thomas, J. P. and Jordan, J. E. (1985). Prooxidant and antioxidant effects of ascorbate on photosensitized peroxidation of lipids in erythrocyte membranes. *Photochem. Photobiol.* **41**: 267–276.

50. Burkitt, M. J. and Gilbert, B. C. (1990). Model studies of the iron-catalyzed Haber-Weiss cycle and the ascorbate-driven Fenton reaction. *Free Radic. Res. Commun.* **10**: 265–280.

51. Buettner, G. R., Kelley, E. E. and Burns, C. P. (1993). Membrane lipid free radicals produced from L1210 murine leukemia cells by photofrin photosensitization: an EPR spin trapping study. *Cancer Res.* **53**: 3670–3673.

52. Lin, F. and Girotti, A. W. (1993). Photodynamic action of merocyanine 540 on leukemia cells: iron-stimulated peroxidation and cell killing. *Arch. Biochem. Biophys.* **300**: 714–723.

53. Wagner, B. A., Buettner, G. R. and Burns, C. P. (1993). Free radical-mediated lipid peroxidation in cells: oxidizability is a function of cell lipid *bis*-allylic hydrogen content. *Biochem.* **33**: 4449–4453.

54. Miller, D. M., Buettner, G. R. and Aust, S. D. (1990). Transition metals as catalysts of "autoxidation" reactions. *Free Radic. Biol. Med.* **8**: 95–108.

55. Qian, S. Y. and Buettner, G. R. (1999). Iron and dioxygen chemistry is an important route to initiation of biological free radical oxidations: an electron paramagnetic resonance spin trapping study. *Free Radic. Biol. Med.* **26**: 1447–1456.

56. Schafer, F. Q. and Buettner, G. R. (2000). Acidic pH amplifies iron-mediated lipid peroxidation in cells. *Free Radic. Biol. Med.* **28**: 1175–1181.

57. Vethanayagam, J. G., Green, E. H., Rose, R. C. and Bode, A. M. (1999). Glutathione-dependent ascorbate recycling activity of rat serum albumin. *Free Radic. Biol. Med.* **26**: 1591–1598.

58. Winkler, B. S., Orselli, S. M. and Rex, T. S. (1994). The redox couple between glutathione and ascorbic acid: a chemical and physiological perspective. *Free Radic. Biol. Med.* **17**: 333–349.

59. Gey, K. F. (1998). Vitamins E plus C and interacting conutrients required for optimal health. A critical and constructive review of epidemiology and supplementation data regarding cardiovascular disease and cancer. *Biofactors* **7**: 113–174.

60. Jones, D. P., Kagan, V. E., Aust, S. D., Reed, D. J. and Omaye, S. T. (1995). Impact of nutrients on cellular lipid peroxidation and antioxidant defense system. *Fundamental Appl. Toxicol.* **26**: 1–7.

61. Samiec, P. S., Drews-Botsch, C., Flagg, E. W., Kurtz, J. C., Sternberg, P., Reed, R. L. and Jones, D. P. (1998). Glutathione in human plasma: decline in association with aging, age-related macular degeneration, and diabetes. *Free Radic. Biol. Med.* **24**: 699–704.

62. Jones, D. P., Carlson, J. L., Mody, V. C., Cai, J., Lynn, M. J. and Sternberg, P. (2000). Redox state of glutathione in human plasma. *Free Radic. Biol. Med.* **28**: 625–635.
63. Schafer, F. Q. and Buettner, G. R. (2001). Redox state as viewed through the glutathione disulfide/glutathione couple. *Free Radic. Biol. Med.* **30**: 1191–1212.

Chapter 50

Oxidative Stress in Human Populations

Gladys Block*, Marion Dietrich, Edward Norkus and Lester Packer

Gladys Block, Marion Dietrich and **Lester Packer** • University of California, Berkeley, CA 94720

Edward Norkus • Department of Biomedical Research, Our Lady of Mercy Medical Center, Bronx, NY 10466-2697

*Corresponding Author, 426 Warren Hall, University of California, Berkeley, CA 94720
Tel: 510-643-7896, E-mail: gblock@uclink4.berkeley.edu

1. Summary

While data about antioxidant intake or antioxidant blood levels are available, little large-scale survey data exist about the oxidative stress status (OSS) of human populations and the causes and correlates of such oxidative stress. Such information is essential for the design of studies on the role of oxidative stress in health and disease. Data are presented on lipid peroxidation levels, one measure of oxidative stress, in 306 healthy adults ranging in age from 19 to 78 years.

In addition to current smoking, gender is found to be a strong predictor of OSS. Other significant predictors include plasma cholesterol level and inflammation (as measured by C-reactive protein). Plasma ascorbic acid was found to have a strong and statistically significant inverse relationship with OSS. Other plasma antioxidants, including alpha- and beta-carotene, other carotenoids, and alpha-tocopherol, had no significant inverse relationships with OSS, once ascorbic acid was included in the model.

Future surveys and epidemiologic studies should include one or more markers of OSS, and plasma ascorbic acid. Understanding of the role of OSS and antioxidants in the health of human populations cannot advance materially without these data.

2. Introduction

There is increasing evidence that the oxidation of biomolecules (DNA, proteins, lipids) may play a role in susceptibility to disease, especially the diseases of aging such as cancer and heart disease.[1,2] Animal data have shown that DNA damage accumulates with age,[3] and Harman suggested almost a half a century ago that oxidative damage is related to the debilities associated with aging.[4] Consequently, it would seem that the oxidative stress status (OSS) of human populations should be of considerable interest and importance.

However, while surveys and studies on dietary and plasma antioxidant levels have been published, little is available on oxidative stress itself in human populations. Numerous large-scale, representative national and international surveys have provided information on the *dietary intake of antioxidant nutrients*. The results of several of these surveys are publicly available[5] and have been summarized previously with respect to patterns of antioxidant intake.[6,7] Less commonly, large-scale studies or surveys have provided information on *blood levels of antioxidant nutrients*.[5,8] They demonstrate that regardless of mean values, examination of the distribution of such levels reveals that substantial minorities of the population have very low blood levels of antioxidant nutrients.[7]

However, essentially no data exist on the *oxidative stress status* of human populations from large-scale surveys or large epidemiologic studies. Even the large surveys that provide information on blood levels of antioxidant nutrients do not provide separate estimates of the proportion in the oxidized and reduced forms (e.g. dehydroascorbic acid and reduced ascorbic acid), which could be a useful proxy for oxidative stress status. While a number of small clinical studies have included OSS measurements, they have typically been supplementation studies, or studies in unrepresentative groups such as diabetics, a number of which have been reviewed elsewhere.[9]

This paucity of data on OSS in the normal human population represents a large gap in our knowledge about the distribution of, and causative factors of, oxidative stress status in human populations. Consequently, rather than attempting to review such data, this chapter will describe data we have obtained on OSS in a sample of 306 healthy persons, and physiological and behavioral factors associated with it.

We investigated the impact of age and other factors contributing to lipid peroxidation, one measure of OSS, in a sample of persons aged 19 to 78 years.

3. Methods

Subjects were recruited from among members of the Kaiser Permanente Health Plan of Oakland, California, as well as through fliers and advertisements to the general public. Subjects were excluded if they consumed iron supplements, or vitamin E supplements of 800 IU/day or more, or reported consumption of four or more servings of fruits and vegetables per day, intake of more than 2 alcoholic drinks per day, history of alcohol problems less than one year ago, pregnancy, use of blood thinning drugs, hemochromatosis, history of kidney stones or any other kidney problems, cancer, stroke, heart attack within the last five years, or known HIV infection, as well as any kind of hepatitis or diabetes. Persons who consumed vitamin supplements were required to stop taking them for at least five weeks preceding the collection of the data described here. Three hundred six subjects were recruited, including 138 cigarette smokers, 71 persons exposed to second-hand smoke ("passive smokers") and 97 nonsmokers.

Body weight was measured by study personnel. Extensive data were obtained, including the number of cigarettes smoked per day, and numbers of years smoked, and for those exposed to second-hand smoke, the number of cigarettes they were exposed to. All participants completed a Block98 food frequency questionnaire that produced estimates of macro- and micro-nutrients including dietary carotenoids.

Blood samples were obtained, after an overnight 12 h fast. Subjects refrained from smoking for at least 1 h prior to their clinic visit. Venous blood was drawn into EDTA-vacutainers and centrifuged at 5°C for 10 min at 1200 x g. The plasma was immediately removed from the blood cells and aliquoted into cryovials.

Plasma aliquots for ascorbic acid measurement were mixed 1:1 with 10% (w/v) meta-phosphoric acid to stabilize ascorbic acid. Meta-phosphoric acid was prepared fresh on a weekly basis and kept refrigerated. The blood and plasma samples were refrigerated and protected from light during the entire process. All cryovials were immediately stored at −70°C. Plasma samples were assayed for malondialdehyde (MDA), C-reactive protein, cotinine, ascorbic acid, tocopherols, carotenoids, cholesterol and triglycerides, and transferrin saturation.

MDA was determined using commercially available lipid peroxidation analysis kits (Oxis International, Inc., Portland, OR). Final plasma MDA concentrations were derived by applying 3rd derivative spectroscopy following measurement of each sample's UV-Vis absorption spectra. Concentrations of cotinine were determined by gas chromatography with nitrogen-phosphorus detection, in the laboratory of Dr. Neal Benowitz. Tocopherols and carotenoids were measured by reversed-phase HPLC.[10] Ascorbic acid was determined spectrophotometrically using 2,4-dinitrophenylhydrazine as chromogen.[11] C-reactive protein concentrations were measured using commercially available radial immunodiffusion assay kits (The Binding Site Ltd., San Diego, CA). Total cholesterol and triglycerides were measured using commercially available analysis kits (Sigma-Aldrich, St. Louis, MO). Transferrin saturation was analyzed at a commercial clinical laboratory (SmithKline Beecham Clinical Laboratories, Norristown, PA).

Statistical analyses were conducted using SAS Version 6.12. The dependent variable, MDA, was square-root transformed to improve normality. Univariate statistics are expressed as means ± SD. Effects were examined with Analysis of Covariance. Covariates examined included age, race, and other demographic characteristics; smoking status (active, passive, non); serum cotinine level; plasma antioxidant levels including carotenoids, alpha- and gamma-tocopherol, and total ascorbic acid; plasma lipids including serum cholesterol and triglycerides; dietary intake of vegetables and fruits; and C-reactive protein and transferrin saturation. A *p*-value of 0.05 or less was considered statistically significant.

Table 1. Characteristics of the Sample. $N = 306$

	Mean or Percent
Sex (% male)	40.8%
Current smokers (%)	45.1%
Exposed to 2nd-hand smoke (%)	23.2%
Age (years)	46.6 +/− 13.7
Weight (males) (lbs)	191.2 +/− 37.8
Weight (females) (lbs)	162.6 +/− 38.3
Dietary % of calories from fat (%)	38.3 +/− 7.5
Daily servings of fruit or juice (servings)	1.1 +/− 0.9
Daily servings of vegetables (servings)	2.6 +/− 1.9

Table 2. Univariate Relationships between MDA Level and Physical and Plasma Characteristics. $n = 306$

	n	Mean MDA	SD	p^*
Sex				
Female	177	0.97	0.57	
Male	121	0.57	0.43	< 0.0001
Age group (female)				
19–42	69	0.91	0.53	
43–53	62	1.07	0.48	
54–78	46	0.95	0.71	0.55
Age group (male)				
19–42	37	0.49	0.36	
43–53	34	0.58	0.42	
54–78	50	0.62	0.48	0.16
Current smoking				
No	164	0.65	0.52	
Yes	134	1.01	0.53	< 0.0001
Plasma cholesterol				
Q1 (low)	72	0.61	0.44	
Q2	77	0.83	0.56	
Q3	75	0.84	0.53	
Q4 (high)	74	0.95	0.62	< 0.0001
BMI quartile				
Q1 (low)	71	0.69	0.50	
Q2	70	0.86	0.56	
Q3	71	0.81	0.56	
Q4 (high)	71	0.89	0.57	0.05
C-reactive protein				
Q1 (low)	72	0.66	0.50	
Q2	78	0.72	0.52	
Q3	74	0.81	0.53	
Q4 (high)	74	1.05	0.59	< 0.0001
Ascorbic acid				
Q1 (low)	74	1.14	0.57	
Q2	75	0.86	0.53	
Q3	75	0.64	0.48	
Q4 (high)	74	0.60	0.48	< 0.0001
Alpha-tocopherol				
Q1 (low)	72	0.81	0.49	
Q2	74	0.82	0.59	
Q3	80	0.88	0.60	
Q4 (high)	72	0.72	0.53	0.54

Table 2 (*Continued*)

	n	Mean MDA	SD	*p**
Gamma-tocopherol				
Q1 (low)	77	0.63	0.51	
Q2	69	0.82	0.59	
Q3	76	0.91	0.49	
Q4 (high)	76	0.87	0.60	0.002
Alpha-carotene				
Q1 (low)	78	0.98	0.56	
Q2	69	0.83	0.54	
Q3	77	0.79	0.57	
Q4 (high)	74	0.63	0.48	0.0004
Beta-carotene				
Q1 (low)	74	0.90	0.56	
Q2	75	0.87	0.47	
Q3	74	0.85	0.58	
Q4 (high)	75	0.61	0.56	0.005
Beta-cryptoxanthin				
Q1 (low)	72	0.92	0.57	
Q2	76	0.98	0.54	
Q3	77	0.67	0.54	
Q4 (high)	73	0.66	0.50	0.0003
Lutein/Zeaxanthin				
Q1 (low)	72	0.80	0.49	
Q2	78	0.94	0.58	
Q3	74	0.79	0.61	
Q4 (high)	74	0.69	0.50	0.10
Lycopene				
Q1 (low)	74	0.93	0.49	
Q2	74	0.73	0.55	
Q3	75	0.72	0.56	
Q4 (high)	75	0.86	0.59	0.61
Transferrin saturation				
Q1 (low)	74	0.94	0.51	
Q2	71	0.80	0.54	
Q3	77	0.79	0.56	
Q4 (high)	75	0.70	0.58	0.008

*For two-level variables, *p* represents *t*-test. For 3-level or 4-level variables, *p* represents trend test.

4. Results

Subject characteristics are shown in Table 1. The mean age was 46.6 years, and approximately 40% were male. The average number of servings of fruits and vegetables, 3.7 servings, is somewhat lower than the US national average, and the percent of energy from fat is somewhat higher, due to the inclusion criterion we imposed (consuming less than four servings of fruits and vegetables).

Univariate relationships between MDA level and several factors are shown in Table 2. MDA level is strongly dependent on gender, with women having significantly higher MDA than men ($p < 0.0001$). Within each gender, there is no apparent relationship between age and MDA. Several other factors are strongly predictive of MDA level in univariate analyses, however. These include serum cholesterol level, body mass index (BMI), and C-reactive protein. Several plasma antioxidants were negatively associated with MDA, including ascorbic acid, alpha- and beta-carotene and beta-cryptoxanthin.

Because some factors are correlated with each other, such as the plasma antioxidant nutrient levels, univariate analyses of a single variable at a time can be misleading. Consequently, multivariate analysis was carried out, using Analysis of Covariance. Consistent with the univariate analyses, some of the same factors were found to be statistically significant (Table 3). The F value (which measures

Table 3. Multivariable Model of Factors Associated with MDA Level. $n = 284$ with no Missing Data*

Factor	F statistic[†]	P-value
Sex[‡]	74.78	< 0.0001
Current smoking[‡]	24.32	< 0.0001
Plasma ascorbic acid[¶]	15.13	< 0.0001
C-reactive protein[‡]	14.57	0.0002
Plasma cholesterol[‡]	9.15	0.0027
Cotinine[‡]	3.40	0.0663
Age	0.88	0.4160

*Model $R^2 = 0.43$. No other variables contributed to the model at $p < 0.05$. Variables examined included BMI, weight, alcohol consumption, dietary polyunsaturated fat intake, plasma alpha-tocopherol, gamma-tocopherol, alpha-carotene, beta-carotene, beta-cryptoxanthin, lutein/zeaxanthin, lycopene, and transferrin saturation.
[†]The F statistic measures the contribution of the variable to the prediction of the dependent variable.
[‡]Females, current smokers, and those with higher levels of C-reactive protein, plasma cholesterol and cotinine had significantly higher MDA.
[¶]Higher plasma ascorbic acid was associated with significantly lower MDA.

the contribution of that factor to the prediction of the dependent variable) was strongest for sex; current smoking was also a major predictor, as was plasma ascorbic acid in the inverse direction. Age was not associated with MDA after other predictors were controlled. A number of other potential predictive factors were examined and were not statistically significant. For example, BMI, apparently important in the univariate analysis, was not significant once serum cholesterol and other factors were included in the model. The same was true for weight. Alcohol consumption similarly was unrelated to MDA after control for other factors. When other plasma antioxidants were included in the model, only ascorbic acid remained statistically significant, while none of the other antioxidants was significant or contributed significantly to the model.

5. Discussion

We examined a number of factors for their contribution to lipid peroxidation, as measured by malondialdehyde, in a healthy adult population. Current smoking was important, not surprisingly, and serum cholesterol level also was an important factor in MDA level. Plasma ascorbic acid level was the only factor identified that was significantly inversely associated with MDA. As in the univariate analyses, age was not associated with increasing MDA in the multivariate analysis controlling for factors such as smoking and plasma lipid level.

The strong effect of gender was surprising, and we could demonstrate no explanatory factors. For example, body mass index or estimates of body fat did not reduce the gender effect. We hypothesize that the higher percentage of body fat in women may be a factor, but we were not able to measure this accurately.

Our results are consistent with those of Coudray *et al.*,[12] one of the only other studies to examine lipid peroxidation in a large sample of respondents ($n = 1389$). They too found women to have higher levels of oxidative stress than men, and found a significant positive correlation with plasma cholesterol and BMI, and no correlation with plasma carotenoids. Age was not significantly associated with lipid peroxidation in their sample of persons aged 59–71 years, although the mean in that cohort was higher than the mean in a previous younger cohort they had studied. No multivariate analyses were performed.

In a smaller study ($n = 66$, ranging in age from 25 to 93 years), Mecocci *et al.*[13] found a significantly higher level of MDA with age among subjects undergoing hip replacement or surgery for bone trauma. No multivariate analyses were performed to control for possible confounding effects of cholesterol level or other factors.

Our study has a number of features that may be relevant to our results. The large number of subjects ($n = 306$) made it possible to examine a number of potential predictive factors. The sample design selected for a certain number of active smokers (45.1% in these data), passive smokers (23.2%), and persons neither actively nor passively exposed (31.7%). All reported consuming fewer than four servings of fruits and vegetables. And except for an intentionally selected subcohort of the nonsmokers ($n = 57$), none had consumed vitamin supplements for at least the prior five weeks.

It has been asserted that MDA is too imprecise to be a useful measure of oxidative stress status.[14–16] Our data suggest, however, that MDA is quite sensitive to cigarette smoking as well as to ascorbic acid status and other factors. It would seem that if the predictive factors found in our study are controlled for in the design or analysis of other studies, MDA may be a perfectly sensitive marker of lipid peroxidation. The results in Table 3 suggest that useful information about risk factors in health research can be obtained with this assay.

Similarly, emphasis is often placed on the superior precision of HPLC or other methods of measuring ascorbic acid and MDA, to the detriment of less-expensive and simpler methods that can be more easily applied to large-scale studies. Our study demonstrates that spectrophotometric assays of ascorbic acid are useful, and several other studies have shown good correlations between HPLC and spectrophotometric methods for ascorbic acid.[17–20] Commercially available kits for measuring MDA and C-reactive protein also proved to be useful, in our hands. "The perfect is enemy to the good", and large-scale research on these factors is likely to go forward only if less difficult and expensive methods are deemed acceptable.

The "free-radical theory of aging" was proposed by Harman and others,[4] and has found extensive support in more recent biochemical research. Our results contribute to this body of literature only in suggesting that *lipid peroxidation* may not be an inevitable component in that process. That is, aging *per se* may not inevitably involve increasing lipid peroxidation; rather, such increases may be a function of the increasing cholesterol level with age, smoking when present, and inflammation (reflected in C-reactive protein) that may increase with arthritis and other conditions of aging. On the other hand, the evidence for age-dependent increases in oxidative damage to DNA, possibly resulting from mitochondrial changes with aging, appears to be considerable,[3, 21] and our study did not examine this.

There too, however, it may be worthwhile to examine such aging-related oxidative DNA damage after control for other factors that are not inevitable concomitants of age, such as cholesterol level and inflammation. In addition, it would appear to be very important to simultaneously examine ascorbic acid level. Given its importance in our data, it is possible that even DNA damage may not increase substantially in persons who have maintained a high antioxidant intake and status throughout their life. This would seem to be an important area for future research.

6. Conclusion

In order to understand the relationship between OSS and human health, it is necessary to have large-scale survey and epidemiologic study data on OSS. Such data have been extremely limited thus far, in part because of uncertainty about the "best" marker of OSS. These data have shown that one relatively simple assay can provide a useful marker of OSS. Equally important, it is clear that studies investigating causal mechanisms, or investigating the association between OSS and health, must include careful consideration of correlated variables related to OSS, such as ascorbic acid status, gender, inflammation, and lipid levels.

References

1. Halliwell, B. and Chirico, S. (1993). Lipid peroxidation: its mechanism, measurement, and significance. *Am. J. Clin. Nutri.* **57**: 715S–725S.
2. Pryor, W. A. (1987). The free-radical theory of aging revisited: a critique and a suggested disease-specific theory. *In* "Modern Biological Theories of Aging" (H. R. Warner, R. N. Butler, and R. L. Sprott, Eds.), pp. 89–112, Raven Press, New York.
3. Richter, C., Park, J.-W. and Ames, B. (1988). Normal oxidative damage to mitochondrial and nuclear DNA is extensive. *Proc. Natl. Acad. Sci. USA* **52**: 515–520.
4. Harman, D. (1956). Ageing: theory based on free radical and radiation chemistry. *J. Gerontol.* **11**: 298–300.
5. US Department of Health and Human Services (DHHS). National Center for Health Statistics. (1998). Third National Health and Nutrition Examination Survey, 1988–1994, NHANES III. (CD-ROM Series 11, No. 2A). Hyattsville, MD: Centers for Disease Control and Prevention.
6. Block, G. (1993). Antioxidant intake in the US. *Toxicol. Ind. Health* **9**: 295–301.
7. Dickinson, V. A., Block, G. and Russek-Cohen, E. (1994). Supplement use, other dietary and demographic variables, and serum vitamin C in NHANES II. *J. Am. Coll. Nutri.* **13**: 22–32.
8. Fulwood, R., Johnson, C. L. and Bryner, J. D. (1983). Hematological and nutritional biochemistry reference data for persons 6 months to 74 years of age: United States, 1976–1980. National Center for Health Statistics. Vital and Health Statistics Series 11-No. 232. DHHS Pub. No. (PHS) 83-1682.
9. McCall, M. R. and Frei, B. (1999). Can antioxidant vitamins materially reduce oxidative damage in humans? *Free Radic. Biol. Med.* **26**: 1034–1053.
10. Sowell, A. L., Huff, D. L., Yeager, P. R., Caudill, S. P. and Gunter, E. W. (1994). Retinol, α-tocopherol, lutein/zeaxanthin, β-cryptoxanthin, lycopene, α-carotene, trans β-carotene, and four retinyl esters in serum determined

simultaneously by reversed-phase HPLC with multiwavelength detection. *Clin. Chem.* **40**: 411–416.

11. Laboratory Procedures used by the Clinical Chemistry Division, Centers for Disease Control, for the 2nd Health and Nutrition Survey (NHanes II) 1976–1980, USDHHS, Public Health Services. (1979). IV. Analytical Methods, vitamin C, pp. 17–19, Atlanta, GA.

12. Coudray, C., Roussel, A. M., Mainard, F., Arnaud, J., Favier, A. and the EVA Study Group. (1997). Lipid peroxidation level and antioxidant micronutrient status in a pre-aging population; correlation with chronic disease prevalence in a French epidemiological study (Nantes, France). *J. Am. Coll. Nutri.* **16**: 584–591.

13. Mecocci, P., Fano, G., Fulle, S., MacGarvey, U., Shinobu, L., Polidori, M. C., Cherubini, A., Vecchiet, J., Senin, U. and Beal, M. F. (1999). Age-dependent increases in oxidative damage to DNA, lipids and proteins in human skeletal muscle. *Free Radic. Biol. Med.* **26**: 303–308.

14. Janero, D. R. (1990). Malondialdehyde and thiobarbituric acid-reactivity as diagnostic indices of lipid peroxidation and peroxidative tissue injury. *Free Radic. Biol. Med.* **9**: 515–540.

15. Bowen, P. E. and Mobarhan, S. (1995). Evidence from cancer intervention and biomarker studies and the development of biochemical markers. *Am. J. Clin. Nutri.* **62**: 1403S–1409S.

16. Esterbauer, H. (1996). Estimation of peroxidative damage. A critical review. *Pathol. Biol.* **44**: 25–28.

17. Schaus, E. E., Kutnink, M. A., O'Connor, D. K. and Omaye, S. T. (1986). A comparison of leukocyte ascorbate levels measured by the 2,4-dinitro-phenylhydrazine method with high-performance liquid chromatography using electrochemical detection. *Biochem. Med. Metab. Biol.* **36**: 369–376.

18. Sauberlich, H. E., Kretsch, M. J., Taylor, P. C., Johnson, H. L. and Skala, J. H. (1989). Ascorbic acid and erythorbic acid metabolism in nonpregnant women. *Am. J. Clin. Nutri.* **5**: 1039–1049.

19. Tessier, F., Birlouez-Aragon, I., Tjani, C. and Guilland, J.-C. (1996). Validation of a micromethod for determining oxidized and reduced vitamin C in plasma by HPLC-fluorescence. *Int. J. Vitam. Nutri. Res.* **66**: 166–170.

20. Otles, S. (1995). Comparative determination of ascorbic acid in bass (Morone lebrax) liver by HPLC and DNPH methods. *Int. J. Food Sci. Nutri.* **46**: 229–232.

21. Wei, Y.-H., Lu, C.-Y., Lee, H.-C., Pang, C.-Y. and Ma, Y.-S. (1998). Oxidative damage and mutation to mitochondrial DNA and age-dependent decline of mitochondrial respiratory function. *Ann. NY Acad. Sci.* **854**: 155–170.

SECTION 8

Oxidative Stress Related Diseases

Chapter 51

Oxidative Stress Related Diseases — Overview

Joe M. McCord

Joe M. McCord • Webb-Waring Institute, Box C321, University of Colorado Health Sciences Center, Denver, CO 80262
Tel: 303 315-6257, E-mail: joe.mccord@uchsc.edu

1. Introduction

The most striking observation concerning the role of oxidative stress in human disease is that it is nearly universal. As free radical pathology has unfolded during the past three decades, we have been continually surprised as evidence has been presented for disease after disease, implicating involvement of free radicals and related oxidants and the systems designed to keep them in check. There are several recurring biochemical mechanisms that underlie these relationships between oxidative stress and disease, as well as a number of novel and unique ones. The purpose of this review is not examine any particular oxidant-related disease in detail, but rather to provide a framework for categorizing them with regard to what has gone awry.

Key to the understanding of the role of oxidative stress in disease is the realization that oxidants are not all bad. Rather, we live in a redox-based world where oxidation and reduction form the basis of our ability to power the reactions upon which our lives depend. Because our cells are full of reducing substances

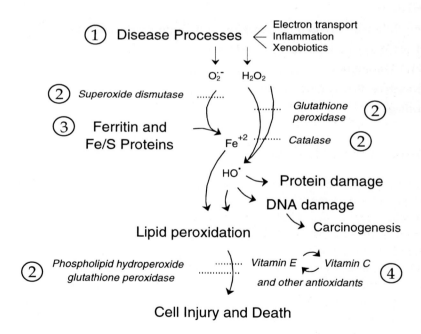

Fig. 1. A general scheme depicting ways in which oxidative balance may be upset to cause disease. (1) Production of $O_2^{\cdot-}$ and H_2O_2 may be increased in damaged mitochondria, by the NADPH oxidase of inflammatory cells, or by redox-active xenobiotics. (2) Elimination of oxidants may be impaired by decreased expression of antioxidant enzymes. (3) Damage may be increased if more stored iron is available for release and subsequent redox participation. (4) Damage may be modulated by the availability of dietary antioxidants and phytochemicals.

and oxidizing substances, unwanted reactions are minimized by kinetic barriers. For example, we have carefully designed one of our most potent reductants, NADH, such that it does not spontaneously react with molecular oxygen, one of our most potent oxidizing substances, even though the reaction is very favorable thermodynamically. Thus, we have multiple pools of compounds coexisting within our cells, both oxidants and reductants, that have great potential for reaction — they do not react until or unless we want them to do so, because enzymes maintain the kinetic control over which compounds may react with which, aided in some cases by subcellular structure and compartmentalization. When cells are sick or injured, these kinetic or structural barriers may be breached. Ion pumps may fail to keep electrolytes in their proper compartments. Osmotic imbalance may cause swelling and stretching of delicate cellular architecture, especially that of the mitochondria. Figure 1 diagrams a very general scheme whereby disease or trauma leads to increased production of cellular oxidants. There are various checks and balances in place, but these also represent sites for upsetting (or for helping to maintain) the overall balance. It is important to remember that superoxide radical appears to have important roles in normal cellular metabolism as a terminator of lipid peroxidation, as a signaling molecule, and as a molecule that can affect the concentration of nitric oxide, another free radical with important regulatory roles in the maintenance of homeostasis. The production of too much superoxide is bad; the complete elimination of superoxide may be worse. Any therapeutic attempts aimed at diseases involving oxidative stress must face this daunting paradox. It is *balance* that must be restored.

2. Mitochondrial Injury

The most fundamental mission of every cell in the human body is to generate the energy required for whatever else it may need to do. Our bodies do not have a centralized facility for the generation and distribution of power — every cell is responsible for its own needs. This puts a substantial burden on the cell, because the chemistry involved in the aerobic combustion of foodstuffs and the generation of ATP via mitochondrial electron transport is potentially nasty chemistry that requires carefully constructed containment to insure that things do not get out of hand. In a healthy tissue the situation is well in hand, and the low but unavoidable production of noxious byproducts by oxidative metabolism is offset by the cell's antioxidant enzymes (SODs, catalases, glutathione peroxidases) and its intake of antioxidant vitamins and related phytochemicals. When conditions become stressful, however, the power generation system may suddenly become a liability, just as when a major earthquake strikes a city. Gas leaks, downed power lines, and the uncontrolled oxidative processes (fires) that result may produce more damage than the initiating trauma. If the cell's structural integrity is compromised or altered by physical trauma, metabolic stress, or by energy starvation, then adequate containment of the electron transport system becomes problematic and increased production of noxious byproducts such as superoxide and hydrogen peroxide may exceed the detoxification capacity of the cell. The superoxide-mediated release of redox-active transition metal ions serves to exacerbate the situation, and uncontrolled oxidative processes such as lipid peroxidation, protein oxidation, and DNA damage may ensue. This damage may initiate the self-destruct mechanism of apoptosis, or may be sufficient to cause direct cell death via necrosis. Thus, a healthy cell is actually in a precarious steady-state balancing act, and it may be vulnerable to seemingly innocuous perturbations of that steady-state. Relatively minor insults may precipitate a catastrophic failure of the power generation system from which the cell may not recover.

Oxidative stress is now thought to make a significant contribution to all diseases that involve transient interruptions of blood supply, and hence of energy generation. These include ischemic diseases involving virtually any organ, but especially heart disease,[1] stroke,[2] intestinal ischemia,[3] and organ preservation associated with transplantation.[4]

3. Inflammation

The second major reason for the common involvement of oxidative stress in human disease stems from our immune system's reliance on the generation of superoxide by phagocyte NADPH oxidase as a means of keeping us free from microbial infection. The importance of this role of superoxide is difficult to overemphasize. It was first proposed by Bernard Babior and coworkers in 1973.[5] The cells of the immune system are ever vigilant and overly responsive by nature. If they fail to recognize an invasion by a microbe, the mistake may be fatal. If they respond inappropriately when no real threat of infection exists, however, as in the case of a minor but tissue-damaging contusion, e.g. the only price to be paid is a transiently sore muscle, perhaps. More serious "false positive" reactions by the surveillance team do occur, and may result in chronic autoimmune diseases such as rheumatoid arthritis or systemic lupus erythematosus. Even so, the immediate threat to our continued existence is less than that of an undetected infection. Thus, the system reacts with superoxide production not only to foreign antigens that may (infection) or may not (allergy) represent a real threat, but also to internal signals indicating tissue injury that may correlate with and presage a vulnerability to infection. This hyperresponsiveness of the immune system, coupled with the fact that its ability to discriminate self from non-self is not always perfect, leads to the result that immune system-generated superoxide is often involved in a very broad spectrum of human diseases, extending far beyond infectious diseases. Oxidative stress is now thought to make a significant contribution to arthritis, vasculitis, glomerulonephritis, lupus erythrematosus, adult respiratory distress syndrome,[6] scleroderma[7] and many other diseases characterized by local accumulations of activated inflammatory cells. Indeed, most diseases attributable to a disruption of the power-generating system (e.g. ischemia/reperfusion) will involve as well an inflammatory response initiated by factors signaling that tissue injury has occurred.

4. The Balance between Superoxide and Superoxide Dismutases

The importance of balance between superoxide and superoxide dismutases was suggested early on by the work of Yoram Groner and coworkers, who produced transfected cells overexpressing SOD1, the cytosolic Cu, Zn-SOD, and later, transgenic mice.[8, 9] To the surprise of many, too much SOD turned out to be problematic. Transfected cells showed higher rates of lipid peroxidation, not lower. Patients with Down syndrome, whose cells contain 50% more SOD1 due to the gene dosage effect of having a third copy of chromosome 21, similarly show increased rates of lipid peroxidation. Could it be that superoxide is actually somehow beneficial? Further demonstration of the phenomenon came with studies of isolated perfused hearts. In these studies the organ was injured by a period of no-flow ischemia; upon reperfusion the rate of superoxide production exceeded the ability of endogenous SODs to maintain an optimal level of the radical. Adding exogenous SOD to the perfusate provided substantially improved recovery of developed pressure. Adding more SOD beyond the optimal concentration, however, resulted in an ever increasing failure to recover, with the damage finally exceeding that observed when no exogenous SOD at all was added to the perfusate.[10] That is, bell-shaped dose response curves were observed. In subsequent studies we found that lipid peroxidation was inversely related to the bell-shaped dose response, seen in Fig. 2. Functional recovery was maximal at a single concentration of SOD, and lipid peroxidation was minimal at this point. Lipid peroxidation increased when SOD concentration in the perfusate was either increased or decreased from this optimal value.[11] The explanation for this rather bizarre observation, we believe, lies in the fact that lipid peroxidation is a free radical chain reaction. It must be initiated by a free radical (or by a one-electron redox event), and can only be terminated by a free radical (or by a one-electron redox event). Superoxide can act, indirectly at least, to initiate the process by reducing and liberating iron from its storage protein ferritin. This ferrous ion can then effect a reductive lysis of a lipid hydroperoxide molecule (LOOH), generating an alkoxy radical ($LO^\bullet$) capable of initiating a new chain reaction. Alternatively, superoxide can drive Fenton chemistry to produce hydroxyl radical, also capable of initiating a new chain by abstracting an allylic hydrogen from a lipid molecule. At the same time, superoxide is capable of *terminating* the free radical chain reaction via a radical–radical annihilation with the propagating

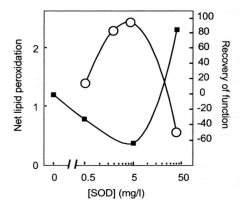

Fig. 2. Relationship between net lipid peroxidation and functional recovery of isolated rabbit hearts, in response to administered dosages of superoxide dismutase. Recovery of developed pressure correlates inversely with net lipid peroxidation. Maximal protection is seen at a dosage of 5 mg/l in the coronary perfusate. Data are replotted from Ref. 11.

lipid dioxyl radical (LOO•), reducing it to a lipid hydroperoxide. The termination of lipid peroxidation by superoxide *in vivo* appears to be a significant physiological function of the radical, and explains the puzzling observation that cells from individuals with Down syndrome show increased rates of lipid peroxidation despite their 50% higher content of SOD.[8] The single, optimal concentration of SOD in any system, then, depends on the rate at which superoxide is being produced and results from the best compromise between maintaining good termination of lipid peroxidation without excessive initiation of the process.[11] One might predict that this optimal concentration of SOD would also be a function of the concentrations of other inhibitors of lipid peroxidation in the system, such as vitamin E.

5. Other Ways of Upsetting the Balance

Apart from the two very common mechanisms discussed above which can cause abrupt increases in the rates of superoxide production within a cell, there are a number of less common mechanisms for upsetting the balance between oxidants and reductants. These mechanisms account for smaller classes of diseases, or for unique diseases that may not represent significant concerns from a global perspective, but which have nevertheless been very instrumental in furthering our understanding of oxidative imbalance. One general mechanism for increasing oxidative stress is the ingestion of redox-active xenobiotics that can change the pathways of normal electron flow with the cell, resulting in increased flux of superoxide production. Perhaps the first of these to be investigated was the herbicide paraquat (1,1'-dimethyl-4,4'-bipiridylium dichloride). Paraquat can be reduced intracellularly by an NADPH-dependent process, and the reduced compound can, in turn, reduce molecular oxygen to form superoxide and a cationic paraquat radical, both of which can liberate iron from its storage protein ferritin.[12] Cell damage appears to result from a combination of toxicities due to $O_2^{\cdot-}$, H_2O_2, and $\cdot OH$, with partial protection provided by overexpression of SOD.[13]

A similar mechanism has provided a model for Parkinson's disease. Several young heroin abusers in the California drug culture developed the clinical symptoms of severe Parkinson's disease after injecting a synthetic product, 1-methyl-4-phenyl-1,2,3,6-tetrahydropyridine (MPTP), a by-product of the synthesis of 1-methyl-4-propionoxypyridine, a "designer drug" with heroin-like effects.[14] The compound was found to be oxidized by monoamine oxidases to 1-methyl-4-phenylpyridinium (MPP$^+$), causing complex I to become a site for superoxide production.[15] Transgenic mice overproducing SOD are dramatically resistant to MPTP toxicity.[16]

Hemochromatosis is a disease of abnormally high iron absorption, and is associated with a specific Cys282Tyr mutation in the HFE gene.[17] While homozygotes become severely iron-loaded, heterozygotes (about 14% of the population) show moderate degrees of iron overload. Heterozygotes are at twice the normal risk for acute myocardial infarction[18] and for Type-2 diabetes.[19] Why do excess iron stores increase the risk of oxidative stress-related diseases? The most generally destructive action of superoxide radical may be bringing about the reductive release of iron from its storage protein, ferritin.[20] Because of its redox activity, iron is normally handled very carefully by cells and organisms. In the healthy state, there is never an appreciable concentration of "free" iron (or iron chelated by low molecular weight compounds). Any released Fe(II) is immediately chelated by compounds such as citrate or ADP, but these complexes readily participate in redox reactions, catalyzing the formation of $\cdot OH$ which can initiate lipid peroxidation or cause DNA strand breaks. Most small chelators of iron can accommodate the coordination geometry of either Fe(II) or Fe(III), such that they provide little hindrance to its cyclical reduction or oxidation. The macromolecular chelators of iron such as transferrin and ferritin, as well as drugs such as desferrioxamine, provide binding sites of such rigid specificity that Fe(III) is bound extremely tightly, but Fe(II) is not bound at all. Due to kinetic restrictions as well as the thermodynamics of binding, the iron in transferrin and ferritin is very difficult to reduce by cellular reductants, and is thus shielded from release.

Why does excess stored iron increase the risk of oxidative stress? As ferritin concentration increases within a cell, it presents, in effect, a bigger target for any superoxide available and will capture a larger fraction of that superoxide, with the deleterious result of releasing catalytically active iron. When faced with the same oxidative insult, iron-loaded tissues suffer much greater damage and loss of function than tissues with normal iron stores.[21] As long as we remain healthy, our stored iron appears to pose no great problem because superoxide concentration is low. Even so, one might predict that the iron-dependent "background" damage due to unscavenged superoxide would be proportionately higher in iron-overloaded individuals, perhaps contributing to the accumulation of random damage associated with aging. As shown in Fig. 1, ferritin is a central target for attack by superoxide radical leading to the release of its stored metal. For this reason, many forms of oxidative stress are ameliorated by iron chelating agents such as desferrioxamine that recapture released iron and discourage its participation in redox activities. Certain other iron-containing proteins within the cell share this vulnerability to superoxide. These proteins, however, contain iron in functional or catalytic settings and serve as enzymes or regulatory proteins rather than storage proteins. Quantitatively, they represent less iron than is typically present in ferritin.

6. SOD Deficiency

One of the more interesting and obvious possibilities for oxidative stress-related disease is the underproduction of SOD activity. Mammals have three known genes encoding SODs: a Cu, Zn-SOD found in the cytosol (SOD1); a Mn-SOD in the mitochondria (SOD2); and a secreted extracellular SOD (SOD3). Knockout mice have now been produced for each with surprising results. SOD1 knockouts and SOD3 knockouts get along well, especially under non-stressful conditions. SOD1 knockouts are more vulnerable to ischemia/reperfusion injury[22] and they display a motor axonopathy and vulnerability to axonal injury.[23, 24] SOD3 knockouts are more sensitive to hyperoxic lung injury,[25] to ischemia/reperfusion injury,[26] and to alloxan-induced diabetes.[27] Perhaps the most interesting feature of SOD3 deficiency, however, is impairment in spatial learning and radial maze performance.[28] These characteristics are also seen in SOD3 transgenics which overexpress the activity, underscoring once again the importance of balance in free radical metabolism. In this case, extracellular $O_2^{\cdot-}$ is thought to modulate the level of nitric oxide, a neurotransmitter. A polymorphism (R213G) in SOD3 has been found in 3–5% of the population. This arginine-to-glycine replacement is in the hydrophilic C-terminal "tail" of the enzyme, which is responsible for the binding of SOD3 to negatively charged endothelial surfaces and macromolecules such as collagen fibrils in the extracellular spaces. The mutation weakens the binding of SOD3 to polyanionic surfaces, and results in about 10-fold more SOD3 activity circulating in the plasma. The frequency of the mutation is about twice as high in patients on renal dialysis as in the population at large, suggesting that diminished "stickiness" of SOD3 may be a risk factor for renal failure.[29]

The complete elimination of SOD2 activity is lethal shortly after birth,[30] illustrating the profound importance of mitochondrial SOD. The animals die of neurodegeneration and dilated cardiomyopathy. Heterozygotes expressing about half the normal amount of MnSOD, however, appear normal and do not show increased sensitivity to hyperoxic injury[31] like the SOD3 knockouts, showing the importance of compartmentalization.

7. Oxidative Stress and Malignancy

Reining in a cell's biological imperative to proliferate and placing constraints on the natural inclination to replicate DNA and divide is no small feat. Indeed, it may require more sophisticated cellular engineering to squelch proliferation than to promote it. For a normal postmitotic cell to become malignantly transformed, several conditions may have to be met. It may be necessary to relieve certain evolutionary constraints which tell the cell not to enter into mitosis. It may be necessary to provide mild oxidative stress to serve as the driving force for proliferation. It may even be necessary to disable yet another set of evolutionary constraints designed to prevent cells from running amok by triggering apoptosis.[32] This latter set of constraints can, in fact, be triggered by oxidative stress *per se*. While wild proliferation may be a mark of success for a bacterium, it is very dangerous in higher organisms. The entire organism can be brought down by what begins as a single errant cell that has broken free of its evolutionary constraints. Thus, we have evolved a failsafe system which can detect out-of-control proliferation, and can cause programmed self destruction in any cell showing this behavior.

How might a "wannabe" cancer cell achieve and maintain the condition of mild oxidative stress necessary to drive its proliferation? It has been shown that many types of human cancer cells have reduced MnSOD.[33] In most cases, the reduced activity has been assumed to be due to defective expression of the gene (i.e. changes in the promoter region of the gene).[34] Oberley, St. Clair, *et al.* have observed in numerous studies that transfection with the gene for human MnSOD can reverse the malignant phenotype of tumor cells, suggesting that MnSOD functions as a tumor suppressor.[35] Very recently Xu *et al.*[36] reported finding a variant sequence containing a cluster of three mutations in the promoter regions of the MnSOD genes from 5 of 14 human cancer cell lines examined. All 5 cell lines were heterozygous for the variant sequence. The mutations change the binding pattern of transcription factor AP-2 and cause a marked diminution in the efficiency of the promoter using a luciferase reporter assay system. While MnSOD is inducible by interleukin-1 or by tumor necrosis factor-α, there are no recognition sequences in the promoter for the expected involvement of transcriptional factors NF-kappa-B or C/EBP-β. Such recognition sites do exist, however, in an enhancer region located in intron 2 of the gene.[37] Thus, it is interesting to speculate that serious, tumor-promoting oxidative stress may result in a cell that is unable to respond to cytokine signals, even though its basal expression of MnSOD may be perfectly normal. An additional alternative is that mutations in the coding region of MnSOD may adversely affect catalytic efficiency or the stability of the protein. In a preliminary examination of genomic DNA isolated from lung and prostate cancers, we have MnSOD genes containing mutations in all the categories described above. Impaired expression of MnSOD may be a nearly universal characteristic of transformed cells.

References

1. Omar, B. A. and McCord, J. M. (1991). Interstitial equilibration of superoxide dismutase correlates with its protective effect in the isolated rabbit heart. *J. Mol. Cell. Cardiol.* **23**: 149–159.
2. Baker, K., Marcus, C. B., Huffman, K., Kruk, H., Malfroy, B. and Doctrow, S. R. (1998). Synthetic combined superoxide dismutase/catalase mimetics are protective as a delayed treatment in a rat stroke model: a key role for reactive oxygen species in ischemic brain injury. *J. Pharmacol. Exp. Ther.* **284**: 215–221.
3. Parks, D. A., Bulkley, G. B., Granger, D. N., Hamilton, S. R. and McCord, J. M. (1982). Ischemic injury in the cat small intestine: role of superoxide radicals. *Gastroenterology* **82**: 9–15.
4. Biasi, F., Bosco, M., Chiappino, I., Chiarpotto, E., Lanfranco, G., Ottobrelli, A., Massano, G., Donadio, P. P., Vaj, M., Andorno, E. *et al.* (1995). Oxidative damage in human liver transplantation. *Free Radic. Biol. Med.* **19**: 311–317.
5. Babior, B. M., Kipnes, R. S. and Curnutte, J. T. (1973). Biological defense mechanisms. The production by leukocytes of superoxide, a potential bactericidal agent. *J. Clin. Invest.* **52**: 741–744.
6. Gonzalez, P. K., Zhuang, J., Doctrow, S. R., Malfroy, B., Benson, P. F., Menconi, M. J. and Fink, M. P. (1996). Role of oxidant stress in the adult respiratory distress syndrome: evaluation of a novel antioxidant strategy in a porcine model of endotoxin-induced acute lung injury. *Shock* **6(Suppl. 1)**: S23–S26
7. Stein, C. M., Tanner, S. B., Awad, J. A., Roberts, L. J. and Morrow, J. D. (1996). Evidence of free radical-mediated injury (isoprostane overproduction) in scleroderma. *Arthritis. Rheum.* **39**: 1146–1150.
8. Elroy-Stein, O., Bernstein, Y. and Groner, Y. (1986). Overproduction of human Cu/Zn-superoxide dismutase in transfected cells: extenuation of paraquat-mediated cytotoxicity and enhancement of lipid peroxidation. *EMBO J.* **5**: 615–622.
9. Epstein, C. J., Avraham, K. B., Lovett, M., Smith, S., Elroy-Stein, O., Rotman, G., Bry, C. and Groner, Y. (1987). Transgenic mice with increased Cu, Zn-superoxide dismutase activity: animal model of dosage effects in Down syndrome. *Proc. Natl. Acad. Sci. USA* **84**: 8044–8048.

10. Omar, B. A., Gad, N. M., Jordan, M. C., Striplin, S. P., Russell, W. J., Downey, J. M. and McCord, J. M. (1990). Cardioprotection by Cu, Zn-superoxide dismutase is lost at high doses in the reoxygenated heart. *Free Radic. Biol. Med.* **9**: 465–471.

11. Nelson, S. K., Bose, S. K. and McCord, J. M. (1994). The toxicity of high-dose superoxide dismutase suggests that superoxide can both initiate and terminate lipid peroxidation in the reperfused heart. *Free Radic. Biol. Med.* **16**: 195–200.

12. Thomas, C. E. and Aust, S. D. (1986). Reductive release of iron from ferritin by cation free radicals of paraquat and other bipyridyls. *J. Biol. Chem.* **261**: 13 064–13 070.

13. St. Clair, D. K., Oberley, T. D. and Ho, Y. S. (1991). Overproduction of human Mn-superoxide dismutase modulates paraquat-mediated toxicity in mammalian cells. *FEBS Lett.* **293**: 199–203.

14. Langston, J. W., Ballard, P., Tetrud, J. W. and Irwin, I. (1983). Chronic Parkinsonism in humans due to a product of meperidine-analog synthesis. *Science* **219**: 979–980.

15. Hasegawa, E., Takeshige, K., Oishi, T., Murai, Y. and Minakami, S. (1990). 1-methyl-4-phenylpyridinium (MPP$^+$) induces NADH-dependent superoxide formation and enhances NADH-dependent lipid peroxidation in bovine heart submitochondrial particles. *Biochem. Biophys. Res. Commun.* **170**: 1049–1055.

16. Przedborski, S., Kostic, V., Jackson-Lewis, V., Naini, A. B., Simonetti, S., Fahn, S., Carlson, E., Epstein, C. J. and Cadet, J. L. (1992). Transgenic mice with increased Cu, Zn-superoxide dismutase activity are resistant to N-methyl-4-phenyl-1,2,3,6-tetrahydropyridine-induced neurotoxicity. *J. Neurosci.* **12**: 1658–1667.

17. Feder, J. N., Gnirke, A., Thomas, W., Tsuchihashi, Z., Ruddy, D. A., Basava, A., Dormishian, F., Domingo, R., Jr., Ellis, M. C., Fullan, A., Hinton, L. M., Jones, N. L., Kimmel, B. E., Kronmal, G. S., Lauer, P., Lee, V. K., Loeb, D. B., Mapa, F. A., McClelland, E., Meyer, N. C., Mintier, G. A., Moeller, N., Moore, T., Morikang, E., Wolff, R. R. *et al.* (1996). A novel MHC class I-like gene is mutated in patients with hereditary haemochromatosis. *Nature Genetics* **13**: 399–408.

18. Tuomainen, T. P., Kontula, K., Nyyssonen, K., Lakka, T. A. and Salonen, J .T. (1998). Hemochromatosis gene HFE Cys-282-Tyr polymorphism is associated with two-fold increased risk at acute myocardial infarction in men. *Circulation* **98(Suppl: Abst 2417)**: I-459(Abstract).

19. Kwan, T., Leber, B., Ahuja, S., Carter, R. and Gerstein, H. C. (1998). Patients with Type-2 diabetes have a high frequency of the C-282-Y mutation of the hemochromatosis gene. *Clin. Invest. Med.* **21**: 251–257.

20. Biemond, P., van Eijk, H. G., Swaak, A. J. G. and Koster, J. F. (1984). Iron mobilization from ferritin by superoxide derived from stimulated polymorphonuclear leukocytes. Possible mechanism in inflammation diseases. *J. Clin. Invest.* **73**: 1576–1579.

21. Van der Kraaij, A. M. M., Mostert, L. J., van Eijk, H. G. and Koster, J. F. (1988). Iron-load increases the susceptibility of rat hearts to oxygen reperfusion damage: protection by the antioxidant (+)-cyanidanol-3 and deferoxamine. *Circulation* **78**: 442–449.

22. Yoshida, T., Maulik, N., Engelman, R. M., Ho, Y. S. and Das, D. K. (2000). Targeted disruption of the mouse Sod-I gene makes the hearts vulnerable to ischemic reperfusion injury. *Circ. Res.* **86**: 264–269.

23. Shefner, J. M., Reaume, A. G., Flood, D. G., Scott, R. W., Kowall, N. W., Ferrante, R. J., Siwek, D. F., Upton-Rice, M. and Brown, R. H. J. (1999). Mice lacking cytosolic Cu, Zn-superoxide dismutase display a distinctive motor axonopathy. *Neurology* **53**: 1239–1246.

24. Reaume, A. G., Elliott, J. L., Hoffman, E. K., Kowall, N. W., Ferrante, R. J., Siwek, D. F., Wilcox, H. M., Flood, D. G., Beal, M. F., Brown, R. H., Jr., Scott, R. W. and Snider, W. D. (1996). Motor neurons in Cu, Zn-superoxide dismutase-deficient mice develop normally but exhibit enhanced cell death after axonal injury. *Nature Genet.* **13**: 43–47.

25. Carlsson, L. M., Jonsson, J., Edlund, T. and Marklund, S. L. (1995). Mice lacking extracellular superoxide dismutase are more sensitive to hyperoxia. *Proc. Natl. Acad. Sci. USA* **92**: 6264–6268.

26. Sheng, H., Brady, T. C., Pearlstein, R. D., Crapo, J. D. and Warner, D. S. (1999). Extracellular superoxide dismutase deficiency worsens outcome from focal cerebral ischemia in the mouse. *Neurosci. Lett.* **267**: 13–16.

27. Sentman, M. L., Jonsson, L. M. and Marklund, S. L. (1999). Enhanced alloxan-induced beta-cell damage and delayed recovery from hyperglycemia in mice lacking extracellular-superoxide dismutase. *Free Radic. Biol. Med.* **27**: 790–796.

28. Levin, E. D., Brady, T. C., Hochrein, E. C., Oury, T. D., Jonsson, L. M., Marklund, S. L. and Crapo, J. D. (1998). Molecular manipulations of extra-cellular superoxide dismutase: functional importance for learning. *Behav. Genet.* **28**: 381–390.

29. Yamada, H., Yamada, Y., Adachi, T., Goto, H., Ogasawara, N., Futenma, A., Kitano, M., Hirano, K. and Kato, K. (1995). Molecular analysis of extracellular-superoxide dismutase gene associated with high level in serum. *Japan J. Human Genet.* **40**: 177–184.

30. Li, Y., Huang, T. T., Carlson, E. J., Melov, S., Ursell, P. C., Olson, J. L., Nobel, L. J., Yoshimura, M. P., Berger, C., Chan, P. H., Wallace, D. C. and Epstein, C. J. (1995). Dilated cardiomyopathy and neonatal lethality in mutant mice lacking manganese superoxide dismutase. *Nature Genet.* **11**: 376–381.

31. Tsan, M. F., White, J. E., Caska, B., Epstein, C. J. and Lee, C. Y. (1998). Susceptibility of heterozygous MnSOD gene-knockout mice to oxygen toxicity. *Am. J. Respir. Cell Mol. Biol.* **19**: 114–120.

32. McCord, J. M. and Flores, S. C. (1996). The human immunodeficiency virus and oxidative balance. *In* "Oxidative Processes and Antioxidants" (R. Paoletti, Ed.), Raven Press, New York.

33. Oberley, L. W. and Buettner, G. R. (1979). Role of superoxide dismutase in cancer: a review. *Cancer Res.* **39**: 1141–1149.

34. St. Clair, D. K. and Holland, J. C. (1991). Complementary DNA encoding human colon cancer manganese superoxide dismutase and the expression of its gene in human cells. *Cancer Res.* **51**: 939–943.

35. Zhong, W., Oberley, L. W., Oberley, T. D., Yan, T., Domann, F. E. and St. Clair, D. K. (1996). Inhibition of cell growth and sensitization to oxidative damage by overexpression of manganese superoxide dismutase in rat glioma cells. *Cell Growth. Differ.* **7**: 1175–1186.

36. Xu, Y., Krishnan, A., Wan, X. S., Majima, H., Yeh, C. C., Ludewig, G., Kasarskis, E. J. and St. Clair, D. K. (1999). Mutations in the promoter reveal a cause for the reduced expression of the human manganese superoxide dismutase gene in cancer cells. *Oncogene* **18**: 93–102.

37. Jones, P. L., Ping, D. and Boss, J. M. (1997). Tumor necrosis factor alpha and interleukin-1 beta regulate the murine manganese superoxide dismutase gene through a complex intronic enhancer involving C/EBP-beta and NF-kappa B. *Mol. Cell Biol.* **17**: 6970–6981.

Chapter 52

Role of Oxidative Stress and Other Novel Risk Factors in Cardiovascular Disease

Peter Reaven

Peter Reaven • Department of Medicine, Carl T. Hayden VAMC (111E), 650 E. Indian School Road, Phoenix, AZ 85012
Tel: (602) 277-5551, E-mail: Peter.Reaven@med.va.gov

1. Introduction

Traditional cardiovascular risk factors only partly explain the excess risk of developing atherosclerosis in most individuals. This result is not surprising because it has become increasingly evident that atherogenesis is a very complex event with multiple etiologies and processes that operate separately or in concert to elicit cardiovascular disease (CVD). Lipoprotein infiltration, retention and modification in the arterial intima, activation of cells within the artery wall, endothelial cell dysfunction, local and systemic inflammation, immunological responses, abnormal coagulation, and fibrinolysis may all be important contributors to the development of atherosclerosis. As a consequence of this broader under-standing of the multitude of mechanisms responsible for atherosclerosis, there is increasing appreciation for the many less widely recognized or "novel" cardio-vascular risk factors (CVRF) that contribute to these events or at least reflect the activity of these processes. The number of potential risk factors continues to expand rapidly and a partial list of those that are measurable in the blood is illustrated in Table 1. A common thread underlying many of these risk factors is the presence of enhanced oxidative stress. The goals of this chapter are several fold: (1) to provide a general overview of the many mechanisms of atherosclerosis (2) to review the contributing role of oxidative stress in the pathophysiology of many of these processes in high risk subjects.

Table 1. Novel Cardiovascular Risk Factors

Subfractions of LDL, triglyceride-rich and HDL lipoproteins
Dysfunctional (non-protective) HDL
Oxidized LDL
Oxidized LDL autoantibodies
Lipoprotein (a)
F2-isoprostanes
LDL immune complexes
Soluble adhesion molecules
Plasminogen activator inhibitor-1
Tissue plasminogen activator inhibitor-1
Nitrites/nitrates
Asymmetric dimethylarginine
C-reactive protein
Fibrinogen
Interleukin-6
Serum amyloid A
Homocysteine
Titers of infectious agents
Advanced glycosylation endproducts

2. Mechanisms of Atherosclerosis and Cardiovascular Events

2.1. Role of Modified Lipoproteins

Histopathology studies have consistently demonstrated that early lesions are characterized by the infiltration of monocytes/macrophages and T-cells, intra- and extra-cellular lipid and lipoprotein accumulation, and eventually by proliferation and migration of smooth muscle cells into the intimal space.[1] The initiator(s) of these events have been intensively investigated.[2, 3] An early and important step in the development of atherosclerosis appears to be accumulation of lipoproteins within the artery wall.[4, 5] Retained lipoproteins may then subsequently undergo enzymatic or oxidative modification,[5] generating particles with enhanced proatherogenic properties. Identification of one form of such modified lipoproteins, oxidized LDL (OxLDL), in the artery wall provided an important clue to the puzzle of early atherosclerosis.[6] *In vitro* studies demonstrated that these oxidatively modified lipoproteins were not internalized through the relatively tightly regulated LDL receptors but could instead be taken-up via scavenger receptors in an uncontrolled fashion, leading to foam cell formation.[7] Subsequent studies have identified several families of putative scavenger receptors for OxLDL[8] and have demonstrated a wide variety of proatherogenic effects of

Table 2. Mechanisms by which Mildly or Extensively Oxidized LDL may be Atherogenic

---

It is chemotactic for circulating mononuclear cells.

It inhibits the motility of endothelial cells and tissue macrophages.

It has enhanced uptake by macrophages leading to cholesteryl ester enrichment and foam cell formation.

It induces macrophage proliferation.

It is cytotoxic and can induce cell apoptosis.

It can alter arterial wall cells expression of genes such as MCP-1, colony stimulating factors, IL-1 and 8, and adhesion molecules.

It can induce a variety of proinflammatory genes and their products such as hemoxygenase, SAA, and ceruloplasmin.

It is immunogenic and can elicit autoantibody formation and activate T-cells.

It is more readily retained in the artery wall, increasing its likelihood of undergoing modification.

It is more susceptible to aggregation, which leads to enhanced macrophage uptake.

It can adversely alter coagulation pathways, such as by inhibition of protein C and induction of tissue factor and alteration of platelet aggregation.

It can adversely alter vasomotor properties of coronary arteries.

OxLDL, including stimulation of monocyte chemotaxis, monocyte differentiation, inhibition of macrophage migration, foam cell formation, and cell toxicity.[3, 9, 10] (Table 2). A large body of evidence indicates that most of the atherogenic effects of OxLDL are a consequence of its oxidized lipid components.[11] Much of this interest has been focused on the potential role of various products of oxidized phosholipids. Lysophosphatidylcholine is chemotactic for both monocytes and lymphocytes[9, 12] and may also induce their adhesion by enhancing expression of endothelial cell adhesion molecules.[11, 13, 14] In addition, Berliner, Fogelman and colleagues demonstrated that specific oxidized phosholipids in mildly OxLDL induced vascular endothelial cell (EC) activation and resulted in the expression of several adhesion molecules and the secretion of multiple proinflammatory cytokines/chemokines, such as MCP-1 and MCSF.[3] Aggregated LDL, LDL modified by enzymes such as 12/15-lipoxygenase, lipase or myeloperoxidase,[15, 16] and AGE-proteins also appear to induce similar proinflammatory and/or proatherogenic responses. One consequence of EC or other vascular cell activation is their enhanced capacity to further modify lipoproteins. Thus, a vicious cycle of lipoprotein oxidation, inflammation and further oxidation of lipoproteins can occur.

With more extensive oxidation of lipoproteins, other lipid oxidation products, such as oxysterols, increase in concentration and have also been shown to promote a variety of proatherogenic events. Of the oxysterols, 7-ketocholesterol (7-KC) has been found in relatively high concentrations in OxLDL produced *in vitro*[17–19] and is enriched in arterial foam cells and atherosclerotic plaque *in vivo*.[20, 21] Furthermore, 7-KC induces apoptosis in vascular cells,[22, 23] stimulates adhesion molecule expression in endothelial cells[24] and inhibits cholesterol efflux in macrophages.[25] Moreover, in recent *in vitro* studies, our laboratory has demonstrated that 7-KC stimulated monocytes to display morphologic features of differentiated macrophages; this effect was time and dose dependent and was markedly more potent than treatment with unoxidized cholesterol, in press. 7-KC-differentiated cells also expressed M-CSF and scavenger receptors CD36 and CD68, phagocytized latex beads, and formed lipid laden foam cells after exposure to acetylated LDL (AcLDL). In contrast to 7-KC, oxysterols with known cell regulatory effects such as 25-hydroxycholesterol, 7β-hydroxycholesterol and 22(R)-hydroxycholesterol did not effectively promote THP-1 differentiation. These results demonstrated that 7-KC, a prominent oxysterol in OxLDL, may play an important role in promoting monocyte differentiation and foam cell formation.

2.2. Artery Wall Inflammation

One surprising consequence of the chronic inflammatory response in the artery wall is vascular calcification.[26] The production of osteopontin, matrix gla-proteins, alkaline phosphatase as well other products characteristic of osteoblasts belies earlier dogma that the formation of calcium deposits in the artery wall was solely the result of passive processes.[26] In fact, the formation of hydroxyapatite in the artery wall is an active, regulated process that presumably reflects the surrounding proinflammatory events, and may explain in part the relationship between atherosclerotic plaque burden and coronary artery calcium.[27] Of note, specific products of OxLDL and oxidized phospholipids induce calcification of smooth muscle cells in culture,[28] suggesting that these products may contribute to artery wall calcification.

Modified lipoproteins also appear to activate the immune system. Antibodies to oxidatively modified LDL, as well as immune complexes with OxLDL, have been detected in the plasma and artery wall lesions of both animals and humans,[29, 30] and increase in both during hypercholesterolemia.[29] Activated macrophages express class II histocompatibility antigens, (e.g. HLA-DR) that allow presentation of antigens such as products of OxLDL to T-cells,[1] thus presumably explaining the presence of both activated CD4 and CD8 T-cells in the artery wall. Activated T-cells secrete a variety of cytokines, including interferon-γ and tumor necrosis factor (α and β), that can continue to propagate the inflammatory process in the artery wall.[1]

2.3. Endothelial Cell Activation/Injury

In vitro studies of vascular cell activation have also clearly demonstrated that EC are not simply passive structural conduits for circulating blood but are in fact active and integral participants in the early inflammatory response of the artery wall. EC activation facilitates mononuclear cell chemotaxis, adhesion, and diapedesis; macrophage differentiation, platelet adhesion, as well as generation of free radicals and propagation of the inflammatory milieu in the artery wall.[3] Moreover, by virtue of their location in the artery wall, EC are susceptible to activation/injury by a variety of factors such as native and modified lipoproteins, shear stress, hyperglycemia, hyperinsulinemia, plasma free radicals, AGE-proteins, elevated homocysteine concentrations, and infectious microorganisms such as herpes virus and chlamydia pneumonia.[1]

The EC changes that result from these insults lead to responses that alter normal homeostatic properties of the endothelium, a process that has been characterized as endothelial dysfunction. Although the term endothelial dysfunction is widely used, it is rather imprecise and is commonly used to describe impairments of endothelial anticoagulant and anti-inflammatory activity as well as to abnormalities of vasodilation. For example, enhanced oxidative stress may result in enhanced EC secretion of factors such as PAI-1, t-PA, urokinase, and tissue factor that promote procoagulant, versus anticoagulant, properties.[31] Although the mechanisms underlying altered endothelium dependent vaso-relaxation are undoubtedly multifactorial, nitric oxide (NO) appears to play an important role in smooth muscle cell mediated relaxation of vascular tone. Exposure to oxidants such as superoxide anion or OxLDL reduce nitric oxide formation and/or release from EC. In addition, the kinetically favored interaction of superoxide anion and NO generates peroxynitrite, a particularly potent and stable oxidant.[32] These events induce a variety of consequences including increased adhesiveness of EC to platelets and leukocytes as well as decreased vasodilation and paradoxical vasoconstriction of the artery wall in response to vasodilating agents such as acetylcholine.[31, 33] Although EC dysfunction has frequently been demonstrated in vessels with evidence of atherosclerosis, similar findings have been noted in vessels free from atherosclerosis as well as in subjects with hypertension or diabetes.[32, 34–36] These latter findings have led to the suggestion that EC dysfunction may be an early component of the atherosclerotic process and perhaps is an important risk factor in the development of clinical events. This has been borne out by several recent studies that demonstrate that EC dysfunction predicts future coronary artery disease (CAD) events in individuals with angina or angiographic evidence of mild atherosclerosis.[37, 38]

2.4. Systemic Inflammation and Plasma Markers

One consequence of artery wall inflammation is the focal generation and/or release of active components and byproducts of this process into the plasma. Soluble adhesion molecules that are shed from artery wall cells, autoantibodies to modified lipoproteins, lipoprotein: autoantibody immune complexes, are all readily detected in plasma.[1, 29] In addition, prothrombogenic factors, such as PAI-1 that are released from EC are also detectable in plasma. Inflammation may also generate systemic responses. Hepatic stimulation by inflammatory cytokines such as interleukin-6 induces release and elevation of plasma levels of acute phase proteins, such as C-reactive protein (CRP) and fibrinogen, under several inflammatory conditions, including atherosclerosis. It has therefore been suggested that plasma levels of these novel CVRF may provide additional insight into the development of atherosclerotic lesions and occurrence of macrovascular events.

3. Evidence of Enhanced Oxidative Stress and Increased Novel CVRF in Individuals at High Risk for Cardiovascular Disease

Unfortunately, our understanding of the mechanisms that underlie the excess risk of macrovascular disease in high-risk populations is incomplete. One group of individuals that have consistently demonstrated increased morbidity and mortality related to atherosclerotic diseases including CAD, ischemic strokes, and peripheral artery disease are those with diabetes mellitus (DM) Type-2.[39] It is now evident that the risk of developing the "initial" cardiovascular event for an individual with Type-2 diabetes approaches the level of risk that a nondiabetic individual with a prior history of CAD has in developing a subsequent cardiovascular event.[40] Although diabetes is frequently associated with CVD risk factors such as increased levels of total cholesterol and blood pressure, epidemiologic studies suggest that much of the increased risk for CVD is not completely explained by these or other standard cardiovascular risk factors.[39, 41] The remainder of this chapter will discuss the possibility that other less well-established risk factors may play an important role in individuals with diabetes as a result of the insulin resistance and hyperglycemia and that typically are present in individuals with Type-2 DM.[42]

These risk factors include lipid abnormalities such as increased triglyceride rich lipoproteins (TGRL) and dense (D)-LDL, increased oxidative stress, and enhanced formation of advanced glycation endproducts (AGE)-proteins. As described above, these factors may also induce proinflammatory events in the artery wall that may lead to additional proatherogenic outcomes such as increased levels of PAI-1 and endothelial dysfunction. A further consequence of these events may be systemic inflammation that stimulates increased fibrinogen and CRP production. All of these factors may act individually or in concert to enhance atherosclerosis.

3.1. Diabetic Dyslipidemia

Diabetic dyslipidemia is characterized by three main disorders: elevated levels of TGRL (such as IDL), elevated levels of D-LDL, and reduced levels of HDL.[42, 43] This "triad" of lipid abnormalities appears to be a particularly atherogenic lipoprotein profile in both diabetic and nondiabetic individuals and occurs frequently in individuals with angiographically documented CAD[44] as well as in individuals who have developed clinical CAD events.[45] Although there have been numerous animal and human studies demonstrating the atherogenic and antiatherogenic properties of LDL and HDL, respectively; until recently, most studies of TGRL and atherosclerosis have been limited to animal studies.[46, 47] Many mechanisms have been proposed to explain the proatherogenic potential of TGRL. Beta VLDL, in contrast to LDL, can be taken up in an unregulated fashion by macrophages *in vitro*, leading to foam cell formation,[48] IDL particles avidly bind proteoglycans,[49] and VLDL remnant particles, like LDL particles, are readily oxidized.[50, 51] Several angiographic studies have documented a strong and independent relationship between levels of TGRL, including IDL, with extent or progression of atherosclerosis and development of CAD events.[52–54] Moreover, intervention trials that have utilized medications that effectively reduce TGRL, (i.e. the BECAIT study,[55] the Helsinki trial,[56] and the HIT study[57]) have illustrated the potential benefits of lowering these lipoproteins on progression of athero-sclerosis and development of clinical CAD events. Treatment with gemfibrozil was equally effective in the 25% of the HDL Intervention Trial cohort that was diabetic.[57] However, it must also be noted that HDL elevation may account for at least some of the benefits demonstrated in the above trials and that these medications may have antiatherosclerotic effects that are not a result of lipid lowering. Similarly, D-LDL has been found to be associated with both extent and progression of coronary lesion stenosis and CAD events in several (but not all) cross-sectional and prospective studies.[45, 58–60] *In vitro* studies demonstrate that D-LDL may be more atherogenic because of its altered lipid and antioxidant composition[61, 62] reduced capability to bind to the LDL receptor,[63] and enhanced interaction with artery wall proteoglycans.[49] Moreover, D-LDL lipoproteins are particularly susceptible to oxidation.[64, 65] Once oxidized, they demonstrate marked proinflammatory activity and enhance monocyte chemotaxis and adherence to EC.[51]

To more fully understand the proatherogenic potential of these lipoproteins we evaluated the susceptibility of D-LDL, IDL, VLDL, and VLDL and chylomicron remnant particles to oxidation and their subsequent ability to stimulate monocyte chemotaxis and adhesion to EC.[51] To induce mild oxidation, lipoproteins were exposed overnight to either fibroblasts overexpressing 15-lipoxygenase, an enzyme implicated in cell mediated oxidation of lipoproteins *in vivo*, or a smooth muscle/ EC coculture model of the artery wall. When mildly oxidized by either cell system, D-LDL induced greater monocyte chemotaxis and adhesion than did buoyant LDL. Similarly, IDL and post-prandial remnant particles were also readily oxidized and induced monocyte chemotaxis and adhesion. These data demonstrated that several of the lipoproteins commonly elevated in Type-2 diabetes readily undergo oxidation and become proinflammatory when exposed to mild oxidative stress. The extent of lipid oxidation was not significantly different between bioactive and inactive lipoproteins, raising the possibility that oxidation of lipoproteins was generating unique types or quantities of bioactive products.

In a related series of experiments, we noted that platelet activating factor acetylhydrolase (PAF-AH), an enzyme capable of degrading select oxidized phospholipids, was present in moderate quantities not only in LDL, but in all other apoB-100 containing lipoproteins, and that this enzyme could limit development of bioactivity when lipoproteins were exposed to oxidative stress.[51]

These findings were of particular interest as other investigators have recently demonstrated that oxidation of lipoproteins led to the progressive destruction of other enzymes carried within lipoproteins.[66] To investigate the potential mechanisms involved in this process we added to LDL and HDL lipoproteins *in vitro* several reactive aldehydes commonly formed during lipid oxidation. There was a concentration dependent decrease in PAF-AH and paraoxanase (an additional enzyme with purported capacity to inactivate bioactive lipid peroxides) activity with the addition of the more reactive aldehydes that appeared dependent on the modification of free lysine groups within the enzymes. This raised the possibility that other reactive aldehydes, such as glucose, that could be present

Table 3. Parameters of Glucose Control and LDL Oxidation

	Controls	Type-1 DM Poor Control	Type-1 DM Good Control
HbA1c (%)	4.0 ± 0.7	$11.9 \pm 2.9^*$	$6.3 \pm 0.5^{*\dagger}$
Lag-time (min)	119.1 ± 15.2	$95. \pm 9.7^*$	112.1 ± 12.3
MDA (nmoles/mg protein)	22.1 ± 4.5	$39.4 \pm 7.9^*$	$22.3 \pm 5.8^\dagger$
LPO (nmoles/mg protein)	53.9 ± 27.8	$164.4 \pm 18.2^*$	$82.9 \pm 12.6^\dagger$

LEGEND: Values are mean $\pm$ SD. $^*p < 0.05$ versus healthy controls; $^\dagger p < 0.05$ versus diabetes mellitus Type-1 in poor glycemic control; MDA: malondialdehyde; Dienes: conjugated dienes; LPO: lipid peroxides. (modified from data submitted for publication to Diabetologia). This paper was eventually published in European Heart Journal.

in elevated concentrations may have similar effects. This was in fact the case, as the prolonged addition of glucose to lipoproteins led to AGE formation, lysine modification and decreased enzyme activity.[67]

These data suggested that lipoproteins from poorly controlled diabetic individuals may be more susceptible to oxidation. Several prior studies have evaluated the effects of diabetes on antioxidant status and lipoprotein susceptibility to oxidation. Enhanced lipoprotein oxidation was reported in most,[68–71] but not all studies.[72] However, most studies evaluated small numbers of subjects or had not specifically evaluated the contribution of hyperglycemia, in contrast to other metabolic manifestations of insulin resistance on these events. To address these issues we have recently analyzed plasma and LDL from 100 poorly controlled Type-1 diabetics before and after intensive insulin therapy and from age and sex matched controls. LDL isolated from poorly controlled subjects was more rapidly and more extensively oxidized when exposed to the same oxidative stress. (Table 3). The time to oxidation (lag time) and formation of lipid peroxides was strongly correlated with glycemic control ($r = 0.57$ and $r = 0.79$, respectively). Vitamin E content in LDL was also reduced in these subjects. After 3 months of intensive insulin therapy, LDL susceptibility to oxidation was dramatically reduced and vitamin E content in LDL rose. These data are consistent with the concept that hyperglycemia increases oxidative stress thus leading to reduced vitamin E levels in LDL which in turn increases the susceptibility of LDL to oxidize.

3.2. Oxidative Stress and Lipoprotein Oxidation

Plasma from diabetic animals and humans contain higher levels of lipid peroxides[73–75] and develop greater peroxide levels when exposed to oxidative stress.[76] In contrast, plasma and cellular antioxidant levels are frequently lower in diabetic compared to nondiabetic individuals.[69] Importantly, antioxidant supplementation has successfully reduced plasma measures of oxidative stress in several studies,[77] suggesting that these two events may be related. As lipoproteins from individuals with diabetes have been more susceptible to oxidation in most,[69, 70, 78] but not all studies,[71] this may be one source of enhanced oxidative stress *in vivo*. Hybridomas secreting specific antibodies to oxidized phospholipid epitopes have been isolated from spleen cells from hyper-cholesterolemic mice.[79] These antibodies have been used to establish very sensitive assays to quantitate OxLDL epitopes in plasma (E06 epitope assay). In one angiographic study, titers of OxLDL measured by this assay reflected the extent of coronary vessel lesion burden.[80] Measurements of autoantibody titers to OxLDL have also been proposed as indirect markers of lipoprotein oxidation *in vivo*. These autoantibodies are elevated in hypercholesterolemic mice and positively correlate with the extent of aortic atherosclerosis[29] in several,[30, 81, 82] but not all studies.[83–85] Studies in humans have demonstrated that levels of autoantibodies to OxLDL are related to the extent of atherosclerosis or presence of clinical cardiovascular events. To establish whether specific autoantibodies to OxLDL are associated with measures of atherosclerosis and development of cardiovascular events, our laboratory has, in collaboration with Dr. Joseph Witztum (University of California at San Diego) evaluated levels of both IgG and IgM classes of autoantibodies to malondialdehyde-modified LDL (a form of OxLDL) in several large population studies. In the Physicians Health Study we determined whether baseline levels of autoantibodies predicted subsequent myocardial infarction in 400 cases and 400 age and smoking matched controls.[85] Autoantibodies of both immunoglobulin classes were not higher in cases compared to controls. This result was not substantially altered after adjustment for multiple other standard risk factors. Surprisingly, levels of antibodies to MDA-LDL were significantly lower in diabetics rather than in nondiabetics (IgG: $33\,660 \pm 16\,054$ relative light units (RLU) versus $44\,447 \pm 26\,228$ RLU, $p < 0.001$; IgM: $54\,361 \pm 26\,630$ RLU versus $67\,195 \pm 32\,299$ RLU, $p < 0.027$).

We have also recently completed measurement of autoantibodies in a large cohort of patients in the Insulin Resistance Atherosclerosis Study (IRAS). Preliminary analysis in this study also demonstrated that autoantibodies to MDA-LDL were lower in diabetics than in individuals with either impaired or normal glucose intolerance. Correlation analysis also revealed inverse relationships between autoantibody levels and fasting and 2 h glucose values (statistically significant for IgM). These studies are consistent with the concept, reviewed below that autoantibodies to MDA-LDL may be deposited in the artery wall or cleared more rapidly in individuals with diabetes, perhaps a consequence of greater oxidized LDL/immune complex formation.

In investigations performed by other laboratories, autoantibody titers to forms of OxLDL have been elevated in individuals with diabetes in some[86–88] but not all studies.[83, 89] These discrepancies suggest the possibility that the relationship between autoantibodies to OxLDL and atherosclerosis, particularly in diabetes, is not a straightforward one. This is not surprising, as plasma levels of autoantibodies are dependent on multiple factors including exposure to antigen (OxLDL), production of antibodies, and clearance of antibodies or immune complexes. In addition, the presence of immune complexes presumably may interfere with assay detection of the complexed antibodies, leading to an underestimation of the total autoantibodies. As demonstrated by Wiklund *et al.*,[90] immunization with glycated-LDL leads to increased clearance of glycated-LDL particles, demonstrating that immune complex formation may be responsible for accelerated clearance by the liver and spleen. It is also conceivable that many antibodies to OxLDL in diabetics may be trapped in the artery wall. The presence of increased levels of modified LDL: autoantibody immune complexes in plasma of individuals with diabetes and macrovascular disease supports these possibilities.[84, 89] Moreover, diabetic individuals with low autoantibody levels to oxidized LDL frequently have high LDL immune complex levels.[84] These data illustrate the importance of measuring immune complex levels in conjunction with autoantibody levels to modified LDL.

Other markers of *in vivo* oxidative stress such as plasma and urine levels of hydrogen peroxide and F_2-isoprostanes ($iPF_{2\alpha}$), respectively, have been elevated in conditions that predispose to atherosclerosis such as insulin resistance, hypertension and hypercholesterolemia in both animals and humans.[91, 92] A particularly well studied isoprostane, 8-epi-$PGF_{2\alpha}$ (or $iPF_{2\alpha}$-III), accumulates in coronary arteries from patients with atherosclerotic heart disease, but not in arteries from patients with nonischemic cardiomyopathy.[93] This isoprostane also correlates with extent of lesion formation in mice models of atherosclerosis.[94] Although $iPF_{2\alpha}$ were initially thought to be formed only from oxidative degradation of arachidonic acid, $iPF_{2\alpha}$-III has recently been shown to also be formed by enzymatic degradation.[95] However, the contribution of this pathway to urine or plasma levels is believed small. *In vitro*, formation of $iPF_{2\alpha}$ by smooth muscle cells is increased by elevated glucose levels, demonstrating that hyperglycemia can enhance the oxidative degradation of arachidonic acid.[96] Not surprisingly, isoprostane levels in urine appear elevated in both DM Type-1 and Type-2[77, 97] and are reduced by improved glucose control.[77] However, there are no large cross-sectional or prospective studies evaluating the relationship between isoprostanes levels and atherosclerosis in either diabetic or nondiabetic individuals.

3.3. Advanced Glycation Endproducts

One consequence of prolonged hyperglycemia is the enhanced formation of both the reversible Amidori intermediate products and the irreversible AGEs on particles such as LDL (short-lived) and collagen (long-lived) within the artery wall. *In vitro* studies demonstrate that AGEs have many proatherogenic effects. AGEs have been shown to enhance endothelial permeability,[98] increase the expression of adhesion molecules by vascular cells,[99] stimulate cytokine release and enhance uptake of LDL by macrophages,[100] as well as increase platelet aggregation and fibrin stabilization. AGE proteins may also contribute to cellular oxidative stress through stimulation of free radical production by macrophages and other vascular cells.[101] Perfusion of AGE-albumin into rat lungs induces heme-oxygenase and the formation of malondialdehyde (MDA)-lysine epitopes, two markers of oxidative stress.[101] In addition, a direct consequence of the molecular rearrangements leading to AGE formation is the production of free radicals.[102] Conversely, products of lipid oxidation can induce AGE formation. It has therefore been suggested that these two processes, lipid peroxidation and AGE formation, may be mutually reinforcing and that this interaction may accelerate development of atherosclerosis. One potential consequence of this interaction of pathways is illustrated by the formation of AGEs on apolipoprotein B (apoB) particles. The AGE-apoB particle may not only increase lipoprotein retention in the artery wall through crosslinking to collagen, but may also directly enhance lipoprotein oxidation. Similar to OxLDL, proteins modified by nonenzymatic glycation, including LDL, also appear immunogenic. Antibodies to AGE-albumin, as well as to a specific epitope of AGE, 2-furoyl4(5)-(2-furanyl)-1H-imadazole (FFI) have been used to demonstrate the presence of these epitopes in the atherosclerotic lesions of euglycemic, hypercholesterolemic rabbits[103] and have been elevated in plasma of diabetic mice[104] and humans. These data demonstrate that just as there is an immune response to epitopes of OxLDL, there appears to be a humoral and cell mediated response to Amidori and AGE products that form as a result of hyperglycemia.

3.4. Endothelial Cell Activation/Injury and Dysfunction

Many of the cardiovascular risk factors prominent in individuals with diabetes can have profound effects on the endothelium, either through direct injury or via proinflammatory activity. Of particular interest is PAI-1 which is also secreted from non-EC such as liver and adipose cells. PAI-1 expression is stimulated by hyperglycemia, insulin and its precursors, TGRL, and oxidized lipoproteins,[105, 106] all factors more prevalent in Type-2 diabetes. The presence of multiple metabolic abnormalities in diabetes may be particularly relevant as several combinations of these factors have been shown to have additive or synergistic effects on PAI-1 secretion.[107] In fact, the consistent association of PAI-1 with insulin resistance and the metabolic abnormalities present in this syndrome has led to its inclusion in the insulin resistance syndrome.[108] Elevated levels of PAI-1 have been demonstrated in obese nondiabetics and Type-2 DM.[105] In cross-sectional studies of nondiabetic subjects PAI-1 has been associated with peripheral artery disease (PAD) and CVD events,[109, 110] and adjustment for PAI-1 markedly attenuated the association between proinsulin and intima-media wall thickness in one study, suggesting that PAI-1 may mediate this association.[111] In several prospective studies, PAI-1 has predicted CAD events in univariate analysis; although these associations have not always persisted after adjustment for other multiple risk factors.[112–114] To date, no large studies have specifically evaluated the relationship of PAI-1 to atherosclerotic disease in individuals with DM Type-2.

EC activation and dysfunction are also characterized by increased expression of adhesion molecules. As noted above, induction of adhesion molecule expression is increased by many factors including oxidized lipoproteins, AGE proteins and a variety of cytokines.[3, 115] Up-regulation of adhesion molecules in response to several of the above factors has been inhibited by water soluble antioxidants, suggesting that intracellular oxidative stress may play an important role in this process.[116, 117] We, and others, have shown that artery wall expression of adhesion molecules, such as vascular cell adhesion molecule (VCAM-1) corresponds to the extent of atherosclerotic plaque formation in animal models of atherosclerosis and aging.[118–120] Moreover, suppression of VCAM-1 expression by antioxidant therapy attenuates the development of atherosclerosis in rabbits.[119] Several membrane associated adhesion molecules expressed on EC can readily shed into cell culture media and plasma *in vivo*.[121, 122] This has lead to the evaluation of plasma levels of different soluble adhesion molecules as markers of endothelial activation and early atherosclerosis. Several studies have documented that ICAM-1 and VCAM-1 are independently associated with atherosclerosis and CAD events.[121, 123, 124] These adhesion molecules and P-selectin have also been elevated in patients with peripheral artery disease (PAD).[121, 125] In several studies one or more soluble adhesion molecules have been elevated in diabetic subjects compared to matched nondiabetic controls.[126–128] In at least one study, elevation of adhesion molecules in Type-2 diabetes was particularly evident when vasculopathy was present.[129]

As noted above, diabetes is also associated with diminished EC mediated vasodilation. Inactivation of NO by AGE has been previously demonstrated[130] and evidence suggests that superoxide anion may also contribute to this process.[131] Treatment of diabetic vascular tissue with superoxide dismutase or antioxidants successfully reduces endothelial dysfunction.[132] Sources of increased oxidant stress in diabetes have not been fully elucidated, but at least one study has shown increased NADPH activity in retina of diabetic rats.[133]

Additional studies[134] have suggested that insulin stimulation of NO production and glucose transport are closely linked and may be mediated through similar signaling pathways. It therefore appears that insulin resistance may be causally related to diminished NO production and EC-mediated vasomotor dysfunction in DM Type-2. This may be a very early manifestation of insulin resistance as endothelial dysfunction is detectable even in young normotensive first-degree relatives of individuals with Type-2 diabetes.[135] Thus, individuals with DM Type-2 may suffer from both decreased production and enhanced degradation of NO.

3.5. Systemic Inflammation

As described above, systemic inflammation can occur in response to local artery wall events. For example, oxidant stress and modified lipoproteins can induce release of interleukin-6, a known stimulator of acute phase proteins such as fibrinogen and CRP.[136-138] These two proteins have proven to be sensitive markers of inflammatory events. It has been suggested that fibrinogen, which influences plasma viscosity, platelet aggregation and fibrin deposition, is not only a key component in clot formation but may also have a causal role in atherosclerosis and CAD events.[139] Fibrinogen has been shown to integrate into the arteriosclerotic lesions, where it is converted to fibrin and fibrinogen degradation products. Both fibrinogen and its degradation products are able to stimulate smooth muscle cell proliferation and migration, raising the possibility that they are involved in early lesion formation.[139] Fibrinogen and its degradation product have also been associated with atherosclerosis of coronary, carotid and peripheral arteries in many studies[110, 140, 141] and are increasingly viewed as major independent risk factors.[141] A recent meta-analysis of 22 studies confirmed that fibrinogen is an independent risk factor for cardiovascular disease and that it enhances the prediction of future events in secondary prevention trials.[142] Similar to PAI-1, fibrinogen levels are frequently elevated in Type-2 diabetes.[143-146] This is not surprising as multiple determinants of plasma fibrinogen (such as hyperinsulinemia, hyperglycemia, hypertension, and low HDL) cluster in this disease.[105, 144] In fact, hyperfibrinogenemia has been suggested as a new component of the insulin resistance syndrome.[144] Although fibrinogen has not been adequately studied in diabetics as a cardiovascular risk factor, there is some data to suggest that it may be an independent predictor of vascular complications in Type-2 diabetes.[147, 148]

In contrast to fibrinogen, there is less evidence to suggest that CRP may directly induce atherosclerosis. However, CRP has been demonstrated to induce tissue factor release from human peripheral monocytes,[149] and thus contribute to thrombogenesis in areas of inflammation and monocyte infiltration. In addition, CRP has been demonstrated to activate complement, modulate monocyte chemotaxis and respiratory burst activity, and bind and aggregate apoB containing lipoproteins.[149-152] Levels of CRP are frequently elevated in Type-2 diabetes and are highest when multiple components of the insulin resistance syndrome are present.[153, 154] We have recently demonstrated that 2 different modalities of reducing insulin resistance (with glucose lowering medications[155] or with weight loss (unpublished observations)) appear to reduce CRP levels independently of weight or BMI. This supports the concept that CRP may be modulated, in part, by insulin resistance. CRP is a particularly sensitive and stable marker of inflammation that has consistently been associated with CVD in large cross-sectional studies.[156, 157] In addition, at least 6 prospective studies have confirmed that elevated levels of high-sensitivity CRP in both men and women free of known cardiovascular disease are potent, and in many cases, independent predictors of cardiovascular events.[158, 159] In the Physician's Health Study elevated CRP levels were also predictive of thromboembolic strokes.[160] This study also measured multiple standard and other novel CVRF in the same population and demonstrated that CRP remained one of the strongest independent predictors of CAD events. A critical aspect to the assessment of novel CVRF is whether they may improve upon preexisting clinical risk prediction models based on standard risk factors. Several prospective studies have demonstrated that both CRP and fibrinogen levels substantially added to the estimation of cardiovascular risk based on plasma lipid values.[161] Despite the strong relationship to insulin resistance and Type-2 diabetes, CRP levels and their association with CVD have not been adequately assessed in this population.

4. Conclusion

Atherosclerosis is a complex process that involves multiple pathogenetic mechanisms. The appreciation of this fact has led to increasing interest in identifying novel risk factors that can be readily measured in plasma and enhance the identification of individuals at high risk for CVD. This is particularly relevant in DM Type-2 because the combination of hyperglycemia, hyperinsulinemia, insulin resistance and enhanced oxidative stress commonly present in this condition may negatively modulate a wide array of CVRF. The association between novel CVRF and Type-2 DM is intriguing, and may provide new information towards understanding the excess risk of CVD in diabetes. However, as illustrated in this review, large-scale studies of these novel risk factors need to be conducted to demonstrate their contribution to CVD in individuals with and without Type-2 diabetes.

References

1. Ross, R. (1999). Atherosclerosis — an inflammatory disease. *New England J. Med.* **340**: 115–126.
2. Steinberg, D., Carew, T. E., Fielding, C., Fogelman, A. M., Mahley, R. W., Sniderman, A. D. and Zilversmit, D. B. (1989). Lipoproteins and the pathogenesis of atherosclerosis. *Circulation* **80**: 719–723.
3. Navab, M., Berliner, J. A., Watson, A. D., Hama, S. Y., Territo, M. C., Lusis, A. J., Shih, D. M., van Lenten, B. J., Frank, J. S., Demer, L. L. *et al.* (1996). The Yin and Yang of oxidation in the development of the fatty streak. A review based on the 1994 George Lyman Duff Memorial Lecture. *Arterioscler. Thromb. Vasc. Biol.* **16**: 831–842.

4. Schwenke, D. C. and Carew, T. E. (1989). Initiation of atherosclerotic lesions in cholesterol-fed rabbits. II. Selective retention of LDL versus selective increases in LDL permeability in susceptible sites of arteries. *Arteriosclerosis* **9**: 908–918.

5. Williams, K. and Tabas, I. (1998). The response-to-retention hypothesis of atherogenesis reinforced. *Curr Opin Lipidol* **9**: 471–474.

6. Yla-Herttuala, S., Palinski, W., Rosenfeld, M. E., Parthasarathy, S., Carew, T. E., Butler, S., Witztum, J. L. and Steinberg, D. (1989). Evidence for the presence of oxidatively modified low density lipoprotein in atherosclerotic lesions of rabbit and man. *J. Clin. Invest.* **84**: 1086–1095.

7. Steinberg, D., Parthasarathy, S., Carew, T. E., Khoo, J. C. and Witztum, J. L. (1989). Beyond cholesterol modifications of low-density lipoprotein that increase its atherogenicity. *New England J. Med.* **320**: 915–924.

8. Yamada, Y., Doi, T., Hamakubo, T. and Kodama, T. (1998). Scavenger receptor family proteins: roles for atherosclerosis, host defence and disorders of the central nervous system. *Cell Mol. Life Sci.* **54**: 628–640.

9. Parthasarathy, S., Quinn, M. T. and Steinberg, D. (1988). Is oxidized low density lipoprotein involved in the recruitment and retention of monocyte/macrophages in the artery wall during the initiation of atherosclerosis? *Basic Life Sci.* **49**: 375–380.

10. Hessler, J. R., Morel, D. W., Lewis, L. J. and Chisolm, G. M. (1983). Lipoprotein oxidation and lipoprotein-induced cytotoxicity. *Arteriosclerosis* **3**: 215–222.

11. Parthasarathy, S., Santanam, N., Ramachandran, S. and Meilhac, O. (1999). Oxidants and antioxidants in atherogenesis. An appraisal. *J. Lipid Res.* **40**: 2143–2157.

12. Quinn, M. T., Parthasarathy, S., Fong, L. G. and Steinberg, D. (1987). Oxidatively modified low density lipoproteins: a potential role in recruitment and retention of monocyte/macrophages during atherogenesis. *Proc. Natl. Acad. Sci. USA* **84**: 2995–2998.

13. Erl, W., Weber, P. C. and Weber, C. (1998). Monocytic cell adhesion to endothelial cells stimulated by oxidized low density lipoprotein is mediated by distinct endothelial ligands. *Atherosclerosis* **136**: 297–303.

14. Kita, T., Kume, N., Ochi, H., Nishi, E., Sakai, A., Ishii, K., Nagano, Y. and Yokode, M. (1997). Induction of endothelial platelet-derived growth factor-β-chain and intercellular adhesion molecule-1 by lysophosphatidylcholine. *Ann. NY Acad. Sci.* **811**: 70–75.

15. Sigari, F., Lee, C., Witztum, J. L. and Reaven, P. D. (1997). Fibroblasts that overexpress 15-lipoxygenase generate bioactive and minimally modified LDL. *Arterioscler. Thromb. Vasc. Biol.* **17**: 3639–3645.

16. Heinecke, J. W. (1997). Mechanisms of oxidative damage of low density lipoprotein in human atherosclerosis. *Curr. Opin. Lipidol.* **8**: 268–274.

17. Brown, A. J., Leong, S. L., Dean, R. T. and Jessup, W. (1997). 7-Hydroperoxy-cholesterol and its products in oxidized low density lipoprotein and human atherosclerotic plaque. *J. Lipid Res.* **38**: 1730–1745.

18. Chang, Y. H., Abdalla, D. S. and Sevanian, A. (1997). Characterization of cholesterol oxidation products formed by oxidative modification of low density lipoprotein. *Free Radic. Biol. Med.* **23**: 202–214.

19. Patel, R. P., Diczfalusy, U., Dzeletovic, S., Wilson, M. T. and Darley-Usmar, V. M. (1996). Formation of oxysterols during oxidation of low density lipoprotein by peroxynitrite, myoglobin, and copper. *J. Lipid Res.* **37**: 2361–2371.

20. Hulten, L. M., Lindmark, H., Diczfalusy, U., Bjorkhem, I., Ottosson, M., Liu, Y., Bondjers, G. and Wiklund, O. (1996). Oxysterols present in athero-sclerotic tissue decrease the expression of lipoprotein lipase messenger RNA in human monocyte-derived macrophages. *J. Clin. Invest.* **97**: 461–468.

21. Brown, A. J. and Jessup, W. (1999). Oxysterols and atherosclerosis. *Athero-sclerosis* **142**: 1–28.

22. Lizard, G., Moisant, M., Cordelet, C., Monier, S., Gambert, P. and Lagrost, L. (1997). Induction of similar features of apoptosis in human and bovine vascular endothelial cells treated by 7-ketocholesterol. *J. Pathol.* **183**: 330–338.

23. Miyashita, Y., Shirai, K., Ito, Y., Watanabe, J., Urano, Y., Murano, T. and Tomioka, H. (1997). Cytotoxicity of some oxysterols on human vascular smooth muscle cells was mediated by apoptosis. *J. Atheroscler. Thromb.* **4**: 73–78.

24. Lemaire, S., Lizard, G., Monier, S., Miguet, C., Gueldry, S., Volot, F., Gambert, P. and Neel, D. (1998). Different patterns of IL-1 beta secretion, adhesion molecule expression and apoptosis induction in human endothelial cells treated with 7alpha-, 7beta-hydroxycholesterol, or 7-ketocholesterol. *FEBS Lett.* **440**: 434–439.

25. Gelissen, I. C., Brown, A. J., Mander, E. L., Kritharides, L., Dean, R. T. and Jessup, W. (1996). Sterol efflux is impaired from macrophage foam cells selectively enriched with 7-ketocholesterol. *J. Biol. Chem.* **271**: 17 852–17 860.

26. Watson, K. E. and Demer, L. L. (1996). The atherosclerosis-calcification link? *Curr. Opin. Lipidol.* **7**: 101–104.

27. Wexler, L., Brundage, B., Crouse, J., Detrano, R., Fuster, V., Maddahi, J., Rumberger, J., Stanford, W., White, R. and Taubert, K. (1996). Coronary artery calcification: pathophysiology, epidemiology, imaging methods, and clinical implications. A statement for health professionals from the American Heart Association. Writing Group. *Circulation* **94**: 1175–1192.

28. Parhami, F., Morrow, A. D., Balucan, J., Leitinger, N., Watson, A. D., Tintut, Y., Berliner, J. A. and Demer, L. L. (1997). Lipid oxidation products have opposite effects on calcifying vascular cell and bone cell differentiation. A possible explanation for the paradox of arterial calcification in osteoporotic patients. *Arterioscler. Thromb. Vasc. Biol.* **17**: 680–687.

29. Palinski, W., Tangirala, R. K., Miller, E., Young, S. G. and Witztum, J. L. (1995). Increased autoantibody titers against epitopes of oxidized LDL in LDL receptor-deficient mice with increased atherosclerosis. *Arterioscler. Thromb. Vasc. Biol.* **15**: 1569–1576.

30. Yla-Herttuala, S. (1998). Is oxidized low-density lipoprotein present in vivo? *Curr. Opin. Lipidol.* **9**: 337–344.

31. Luscher, T. F. and Barton, M. (1997). Biology of the endothelium. *Clin. Cardiol.* **20**: II-3–II-10.

32. Munzel, T., Heitzer, T. and Harrison, D. G. (1997). The physiology and pathophysiology of the nitric oxide/superoxide system. *Herz.* **22**: 158–172.

33. Cosentino, F. and Luscher, T. F. (1998). Endothelial dysfunction in diabetes mellitus. *J. Cardiovasc. Pharmacol.* **32**: S54–S61.

34. Heras, M., Sanz, G., Roig, E., Perez-Villa, F., Recasens, L., Serra, A. and Betriu, A. (1996). Endothelial dysfunction of the non-infarct related, angiographically normal, coronary artery in patients with an acute myocardial infarction. *Eur. Heart J.* **17**: 715–720.

35. Gordon, J. B., Ganz, P., Nabel, E. G., Fish, R. D., Zebede, J., Mudge, G. H., Alexander, R. W. and Selwyn, A. P. (1989). Atherosclerosis influences the vasomotor response of epicardial coronary arteries to exercise. *J. Clin. Invest.* **83**: 1946–1952.

36. Wilcox, J. N., Subramanian, R. R., Sundell, C. L., Tracey, W. R., Pollock, J. S., Harrison, D. G. and Marsden, P. A. (1997). Expression of multiple isoforms of nitric oxide synthase in normal and atherosclerotic vessels. *Arterioscler. Thromb. Vasc. Biol.* **17**: 2479–2488.

37. Suwaidi, J., Hamasaki, S., Higano, S., Velianou, J., Araujo, N. and Lerman, A. (1999). Long-term follow-up of patients with mild coronary artery disease and endothelial dysfunction. *Circ. Supl.* **100**: I–48.

38. Neunteufl, T., Heher, S., Katzenschlager, R., Wslfl, G. and Maurer, G. (1999). Long-term prognostic value of flow-mediated vasodilation in the brachial artery of patients with angina pectoris: results of a 5-year follow-up study. *Circulation* **100**: I–48.

39. Pyorala, K., Laakso, M. and Uusitupa, M. (1987). Diabetes and atherosclerosis: an epidemiologic view. *Diabetes Metab. Rev.* **3**: 463–524.

40. Haffner, S. M., Lehto, S., Ronnemaa, T., Pyorala, K. and Laakso, M. (1998). Mortality from coronary heart disease in subjects with Type-2 diabetes and in nondiabetic subjects with and without prior myocardial infarction. *New England J. Med.* **339**: 229–234.

41. Uusitupa, M. I., Niskanen, L. K., Siitonen, O., Voutilainen, E. and Pyorala, K. (1990). 5-year incidence of atherosclerotic vascular disease in relation to general risk factors, insulin level, and abnormalities in lipoprotein composition in non-insulin-dependent diabetic and nondiabetic subjects. *Circulation* **82**: 27–36.

42. Reaven, G. M. (1988). Banting lecture 1988. Role of insulin resistance in human disease. *Diabetes* **37**: 1595–1607.

43. DeFronzo, R. A. and Ferrannini, E. (1991). Insulin resistance. A multifaceted syndrome responsible for NIDDM, obesity, hypertension, dyslipidemia, and atherosclerotic cardiovascular disease. *Diabetes Care* **14**: 173–194.

44. Young, M. H., Jeng, C. Y., Sheu, W. H., Shieh, S. M., Fuh, M. M., Chen, Y. D. and Reaven, G. M. (1993). Insulin resistance, glucose intolerance, hyperinsulinemia and dyslipidemia in patients with angiographically demonstrated coronary artery disease. *Am. J. Cardiol.* **72**: 458–460.

45. Stampfer, M. J., Krauss, R. M., Ma, J., Blanche, P. J., Holl, L. G., Sacks, F. M. and Hennekens, C. H. (1996). A prospective study of triglyceride level, low-density lipoprotein particle diameter, and risk of myocardial infarction. *JAMA* **276**: 882–888.

46. Mahley, R. W., Weisgraber, K. H., Innerarity, T., Brewer, H. B., Jr. and Assmann, G. (1975). Swine lipoproteins and atherosclerosis. Changes in the plasma lipoproteins and apoproteins induced by cholesterol feeding. *Biochemistry* **14**: 2817–2823.

47. Plump, A. S., Smith, J. D., Hayek, T., Aalto-Setala, K., Walsh, A., Verstuyft, J. G., Rubin, E. M. and Breslow, J. L. (1992). Severe hypercholesterolemia and atherosclerosis in apolipoprotein E-deficient mice created by homologous recombination in ES cells. *Cell* **71**: 343–353.

48. Goldstein, J. L., Ho, Y. K., Brown, M. S., Innerarity, T. L. and Mahley, R. W. (1980). Cholesteryl ester accumulation in macrophages resulting from receptor-mediated uptake and degradation of hypercholesterolemic canine beta-very low density lipoproteins. *J. Biol. Chem.* **255**: 1839–1848.

49. Anber, V., Millar, J. S., McConnell, M., Shepherd, J. and Packard, C. J. (1997). Interaction of very-low-density, intermediate-density, and low-density lipoproteins with human arterial wall proteoglycans. *Arterioscler. Thromb. Vasc. Biol.* **17**: 2507–2514.

50. Parthasarathy, S., Quinn, M. T., Schwenke, D. C., Carew, T. E. and Steinberg, D. (1989). Oxidative modification of beta-very low density lipoprotein. Potential role in monocyte recruitment and foam cell formation. *Arteriosclerosis* **9**: 398–404.

51. Lee, C., Sigari, F., Segrado, T., Horkko, S., Hama, S., Subbaiah, P. V., Miwa, M., Navab, M., Witztum, J. L. and Reaven, P. D. (1999). All ApoB-containing lipoproteins induce monocyte chemotaxis and adhesion when minimally modified. Modulation of lipoprotein bioactivity by platelet-activating factor acetylhydrolase. *Arterioscler. Thromb. Vasc. Biol.* **19**: 1437–1446.

52. Reardon, M. F., Nestel, P. J., Craig, I. H. and Harper, R. W. (1985). Lipoprotein predictors of the severity of coronary artery disease in men and women. *Circulation* **71**: 881–888.

53. Phillips, N. R., Waters, D. and Havel, R. J. (1993). Plasma lipoproteins and progression of coronary artery disease evaluated by angiography and clinical events. *Circulation* **88**: 2762–2770.
54. Hodis, H. N., Mack, W. J., Dunn, M., Liu, C., Selzer, R. H. and Krauss, R. M. (1997). Intermediate-density lipoproteins and progression of carotid arterial wall intima-media thickness. *Circulation* **95**: 2022–2026.
55. Ruotolo, G., Ericsson, C. G., Tettamanti, C., Karpe, F., Grip, L., Svane, B., Nilsson, J., de Faire, U. and Hamsten, A. (1998). Treatment effects on serum lipoprotein lipids, apolipoproteins and low density lipoprotein particle size and relationships of lipoprotein variables to progression of coronary artery disease in the Bezafibrate Coronary Atherosclerosis Intervention Trial (BECAIT). *J. Am. Coll. Cardiol.* **32**: 1648–1656.
56. Manninen, V., Elo, M. O., Frick, M. H., Haapa, K., Heinonen, O. P., Heinsalmi, P., Helo, P., Huttunen, J. K., Kaitaniemi, P., Koskinen, P. *et al.* (1988). Lipid alterations and decline in the incidence of coronary heart disease in the Helsinki Heart Study. *JAMA* **260**: 641–651.
57. Rubins, H. B., Robins, S. J., Collins, D., Fye, C. L., Anderson, J. W., Elam, M. B., Faas, F. H., Linares, E., Schaefer, E. J., Schectman, G. *et al.* (1999). Gemfibrozil for the secondary prevention of coronary heart disease in men with low levels of high-density lipoprotein cholesterol. Veterans Affairs High-Density Lipoprotein Cholesterol Intervention Trial Study Group. *New England J. Med.* **341**: 410–418.
58. Austin, M. A., Breslow, J. L., Hennekens, C. H., Buring, J. E., Willett, W. C. and Krauss, R. M. (1988). Low-density lipoprotein subclass patterns and risk of myocardial infarction. *JAMA* **260**: 1917–1921.
59. Tornvall, P., Bavenholm, P., Landou, C., de Faire, U. and Hamsten, A. (1993). Relation of plasma levels and composition of apolipoprotein B-containing lipoproteins to angiographically defined coronary artery disease in young patients with myocardial infarction. *Circulation* **88**: 2180–2189.
60. Mykkanen, L., Kuusisto, J., Haffner, S. M., Laakso, M. and Austin, M. A. (1999). LDL size and risk of coronary heart disease in elderly men and women. *Arterioscler. Thromb. Vasc. Biol.* **19**: 2742–2748.
61. Reaven, P. D., Grasse, B. J. and Tribble, D. L. (1994). Effects of linoleate-enriched and oleate-enriched diets in combination with alpha-tocopherol on the susceptibility of LDL and LDL subfractions to oxidative modification in humans. *Arterioscler. Thromb.* **14**: 557–566.
62. Reaven, P. D., Herold, D. A., Barnett, J. and Edelman, S. (1995). Effects of vitamin E on susceptibility of low-density lipoprotein and low-density lipoprotein subfractions to oxidation and on protein glycation in NIDDM. *Diabetes Care* **18**: 807–816.
63. Chapman, M. J., Guerin, M. and Bruckert, E. (1998). Atherogenic, dense low-density lipoproteins. Pathophysiology and new therapeutic approaches. *Eur. Heart J.* **19(Suppl. A)**: A24–A30.

64. Chait, A., Brazg, R. L., Tribble, D. L. and Krauss, R. M. (1993). Susceptibility of small, dense, low-density lipoproteins to oxidative modification in subjects with the atherogenic lipoprotein phenotype, Pattern B. *Am. J. Med.* **94**: 350–356.

65. Tribble, D. L., Holl, L. G., Wood, P. D. and Krauss, R. M. (1992). Variations in oxidative susceptibility among six low density lipoprotein subfractions of differing density and particle size. *Atherosclerosis* **93**: 189–199.

66. Bielicki, J. K. and Forte, T. M. (1999). Evidence that lipid hydroperoxides inhibit plasma lecithin: cholesterol acyltransferase activity. *J. Lipid Res.* **40**: 948–954.

67. Dudl, E., Barnett, J. and Reaven, P. D. (1996). Mechanism of oxidation induced inhibition of lipoprotein associated paraoxonase and platelet activity factor acetylhydrolase activity. *Circulation* **94**: I–221.

68. Picard, S., Talussot, C., Serusclat, A., Ambrosio, N. and Berthezene, F. (1996). Minimally oxidised LDL as estimated by a new method increase in plasma of Type-2 diabetic patients with atherosclerosis or nephropathy. *Diabetes Metab.* **22**: 25–30.

69. Tsai, E. C., Hirsch, I. B., Brunzell, J. D. and Chait, A. (1994). Reduced plasma peroxyl radical trapping capacity and increased susceptibility of LDL to oxidation in poorly controlled IDDM. *Diabetes* **43**: 1010–1014.

70. Moro, E., Zambon, C., Pianetti, S., Cazzolato, G., Pais, M. and Bittolo Bon, G. (1998). Electronegative low density lipoprotein subform (LDL–) is increased in Type-2 (non-insulin-dependent) microalbuminuric diabetic patients and is closely associated with LDL susceptibility to oxidation. *Acta Diabetol.* **35**: 161–164.

71. Jenkins, A. J., Klein, R. L., Chassereau, C. N., Hermayer, K. L. and Lopes-Virella, M. F. (1996). LDL from patients with well-controlled IDDM is not more susceptible to in vitro oxidation. *Diabetes* **45**: 762–767.

72. Oranje, W. A., Rondas-Colbers, G. J., Swennen, G. N., Jansen, H. and Wolffenbuttel, B. H. (1999). Lack of effect on LDL oxidation and antioxidant status after improvement of metabolic control in Type-2 diabetes. *Diabetes Care* **22**: 2083–2084.

73. Morel, D. W. and Chisolm, G. M. (1989). Antioxidant treatment of diabetic rats inhibits lipoprotein oxidation and cytotoxicity. *J. Lipid Res.* **30**: 1827–1834.

74. Berg, T. J., Nourooz-Zadeh, J., Wolff, S. P., Tritschler, H. J., Bangstad, H. J. and Hanssen, K. F. (1998). Hydroperoxides in plasma are reduced by intensified insulin treatment. A randomized controlled study of IDDM patients with microalbuminuria. *Diabetes Care* **21**: 1295–1300.

75. Freitas, J. P., Filipe, P. M. and Rodrigo, F. G. (1997). Lipid peroxidation in Type-2 normolipidemic diabetic patients. *Diabetes Res. Clin. Pract.* **36**: 71–75.

76. Haffner, S. M., Agil, A., Mykkanen, L., Stern, M. P. and Jialal, I. (1995). Plasma oxidizability in subjects with normal glucose tolerance, impaired glucose tolerance, and NIDDM. *Diabetes Care* **18**: 646–653.

77. Davi, G., Ciabattoni, G., Consoli, A., Mezzetti, A., Falco, A., Santarone, S., Pennese, E., Vitacolonna, E., Bucciarelli, T., Costantini, F. *et al.* (1999). In vivo formation of 8-iso-prostaglandin F$_2$alpha and platelet activation in diabetes mellitus: effects of improved metabolic control and vitamin E supplementation. *Circulation* **99**: 224–229.

78. Diwadkar, V. A., Anderson, J. W., Bridges, S. R., Gowri, M. S. and Oelgten, P. R. (1999). Postprandial low-density lipoproteins in Type-2 diabetes are oxidized more extensively than fasting diabetes and control samples. *Proc. Soc. Exp. Biol. Med.* **222**: 178–184.

79. Palinski, W., Horkko, S., Miller, E., Steinbrecher, U. P., Powell, H. C., Curtiss, L. K. and Witztum, J. L. (1996). Cloning of monoclonal autoantibodies to epitopes of oxidized lipoproteins from apolipoprotein E-deficient mice. Demonstration of epitopes of oxidized low density lipoprotein in human plasma. *J. Clin. Invest.* **98**: 800–814.

80. Ben-Yehuda, O., Witztum, J. L. and Keaney, J. F. J. (1996). Autoantibody titers to malondialdehyde modified LDL correlates with extent of coronary artery disease. *Circulation* **94(Suppl. 1)**: I-638.

81. Bui, M. N., Sack, M. N., Moutsatsos, G., Lu, D. Y., Katz, P., McCown, R., Breall, J. A. and Rackley, C. E. (1996). Autoantibody titers to oxidized low-density lipoprotein in patients with coronary atherosclerosis. *Am. Heart J.* **131**: 663–667.

82. Maggi, E., Finardi, G., Poli, M., Bollati, P., Filipponi, M., Stefano, P. L., Paolini, G., Grossi, A., Clot, P., Albano, E. *et al.* (1993). Specificity of auto-antibodies against oxidized LDL as an additional marker for atherosclerotic risk. *Coron. Artery Dis.* **4**: 1119–1122.

83. Uusitupa, M. I., Niskanen, L., Luoma, J., Vilja, P., Mercuri, M., Rauramaa, R. and Yla-Herttuala, S. (1996). Autoantibodies against oxidized LDL do not predict atherosclerotic vascular disease in non-insulin-dependent diabetes mellitus. *Arterioscler. Thromb. Vasc. Biol.* **16**: 1236–1242.

84. Lopes-Virella, M. F., Virella, G., Orchard, T. J., Koskinen, S., Evans, R. W., Becker, D. J. and Forrest, K. Y. (1999). Antibodies to oxidized LDL and LDL-containing immune complexes as risk factors for coronary artery disease in diabetes mellitus. *Clin. Immunol.* **90**: 165–172.

85. Gaziano, J. M., Hennekens, C. H., Ben-Yehuda, O., Stampfer, M. J., Witztum, J. L. and Reaven, P. D. (1997). Levels of antibodies to an epitope of oxidized LDL and risk of myocardial infarction (MI) in the Physicians' Health Study. *Circulation* **96**: I337.

86. Bellomo, G., Maggi, E., Poli, M., Agosta, F. G., Bollati, P. and Finardi, G. (1995). Autoantibodies against oxidatively modified low-density lipoproteins in NIDDM. *Diabetes* **44**: 60–66.

87. Makimattila, S., Luoma, J. S., Yla-Herttuala, S., Bergholm, R., Utriainen, T., Virkamaki, A., Mantysaari, M., Summanen, P. and Yki-Jarvinen, H. (1999). Autoantibodies against oxidized LDL and endothelium-dependent vasodilation in insulin-dependent diabetes mellitus. *Atherosclerosis* **147**: 115–122.

88. Leinonen, J. S., Rantalaiho, V., Laippala, P., Wirta, O., Pasternack, A., Alho, H., Jaakkola, O., Yla-Herttuala, S., Koivula, T. and Lehtimaki, T. (1998). The level of autoantibodies against oxidized LDL is not associated with the presence of coronary heart disease or diabetic kidney disease in patients with non-insulin-dependent diabetes mellitus. *Free Radic. Res.* **29**: 137–141.

89. Orchard, T. J., Virella, G., Forrest, K. Y., Evans, R. W., Becker, D. J. and Lopes-Virella, M. F. (1999). Antibodies to oxidized LDL predict coronary artery disease in Type-1 diabetes: a nested case-control study from the Pittsburgh Epidemiology of Diabetes Complications Study. *Diabetes* **48**: 1454–1458.

90. Wiklund, O., Witztum, J. L., Carew, T. E., Pittman, R. C., Elam, R. L. and Steinberg, D. (1987). Turnover and tissue sites of degradation of glucosylated low density lipoprotein in normal and immunized rabbits. *J. Lipid Res.* **28**: 1098–1109.

91. Laight, D. W., Desai, K. M., Gopaul, N. K., Anggard, E. E. and Carrier, M. J. (1999). F2-isoprostane evidence of oxidant stress in the insulin resistant, obese Zucker rat: effects of vitamin E. *Eur. J. Pharmacol.* **377**: 89–92.

92. Reilly, M. P., Pratico, D., Delanty, N., di Minno, G., Tremoli, E., Rader, D., Kapoor, S., Rokach, J., Lawson, J. and FitzGerald, G. A. (1998). Increased formation of distinct F2 isoprostanes in hypercholesterolemia. *Circulation* **98**: 2822–2828.

93. Reza Mehrabi, M., Ekmekcioglu, C., Tatzber, F., Oguogho, A., Ullrich, R., Morgan, A., Tamaddon, F., Grimm, M., Glogar, H. D. and Sinzinger, H. (1999). The isoprostane, 8-epi-PGF2 alpha, is accumulated in coronary arteries isolated from patients with coronary heart disease. *Cardiovasc. Res.* **43**: 492–499.

94. Pratico, D., Tangirala, R. K., Rader, D. J., Rokach, J. and FitzGerald, G. A. (1998). Vitamin E suppresses isoprostane generation in vivo and reduces atherosclerosis in ApoE-deficient mice. *Nat. Med.* **4**: 1189–1192.

95. Lawson, J. A., Rokach, J. and FitzGerald, G. A. (1999). Isoprostanes: formation, analysis and use as indices of lipid peroxidation in vivo. *J. Biol. Chem.* **274**: 24 441–24 444.

96. Natarajan, R., Lanting, L., Gonzales, N. and Nadler, J. (1996). Formation of an F2-isoprostane in vascular smooth muscle cells by elevated glucose and growth factors. *Am. J. Physiol.* **271**: H159–H165.

97. Gopaul, N. K., Anggard, E. E., Mallet, A. I., Betteridge, D. J., Wolff, S. P. and Nourooz-Zadeh, J. (1995). Plasma 8-epi-PGF-2alpha levels are elevated in individuals with non-insulin dependent diabetes mellitus. *FEBS Lett.* **368**: 225–229.

98. Esposito, C., Gerlach, H., Brett, J., Stern, D. and Vlassara, H. (1989). Endothelial receptor-mediated binding of glucose-modified albumin is associated with increased monolayer permeability and modulation of cell surface coagulant properties. *J. Exp. Med.* **170**: 1387–1407.

99. Vlassara, H., Fuh, H., Donnelly, T. and Cybulsky, M. (1995). Advanced glycation endproducts promote adhesion molecule (VCAM-1, ICAM-1) expression and atheroma formation in normal rabbits. *Mol. Med.* **1**: 447–456.

100. Thornalley, P. J. (1998). Cell activation by glycated proteins. AGE receptors, receptor recognition factors and functional classification of AGEs. *Cell Mol. Biol. (Noisy-le-grand)* **44**: 1013–1023.

101. Yan, S. D., Schmidt, A. M., Anderson, G. M., Zhang, J., Brett, J., Zou, Y. S., Pinsky, D. and Stern, D. (1994). Enhanced cellular oxidant stress by the interaction of advanced glycation end products with their receptors/binding proteins. *J. Biol. Chem.* **269**: 9889–9897.

102. Wolff, S. P., Bascal, Z. A. and Hunt, J. V. (1989). "Autoxidative glycosylation": free radicals and glycation theory. *Prog. Clin. Biol. Res.* **304**: 259–275.

103. Palinski, W., Koschinsky, T., Butler, S. W., Miller, E., Vlassara, H., Cerami, A. and Witztum, J. L. (1995). Immunological evidence for the presence of advanced glycosylation end products in atherosclerotic lesions of euglycemic rabbits. *Arterioscler. Thromb. Vasc. Biol.* **15**: 571–582.

104. Reaven, P., Merat, S., Casanada, F., Sutphin, M. and Palinski, W. (1997). Effect of streptozotocin-induced hyperglycemia on lipid profiles, formation of advanced glycation endproducts in lesions, and extent of atherosclerosis in LDL receptor-deficient mice. *Arterioscler. Thromb. Vasc. Biol.* **17**: 2250–2256.

105. Festa, A., D'Agostino, R., Jr., Mykkanen, L., Tracy, R. P., Zaccaro, D. J., Hales, C. N. and Haffner, S. M. (1999). Relative contribution of insulin and its precursors to fibrinogen and PAI-1 in a large population with different states of glucose tolerance. The Insulin Resistance Atherosclerosis Study (IRAS). *Arterioscler. Thromb. Vasc. Biol.* **19**: 562–568.

106. Akanji, A. O. and Al-Shayji, I. (1998). The relationships between insulin and plasminogen activator inhibitor-1 levels: assessment in groups of subjects with dyslipidaemia and hypertension. *Clin. Chim. Acta* **274**: 41–52.

107. Calles-Escandon, J., Mirza, S. A., Sobel, B. E. and Schneider, D. J. (1998). Induction of hyperinsulinemia combined with hyperglycemia and hyper-triglyceridemia increases plasminogen activator inhibitor-1 in blood in normal human subjects. *Diabetes* **47**: 290–293.

108. Juhan-Vague, I., Thompson, S. G. and Jespersen, J. (1993). Involvement of the hemostatic system in the insulin resistance syndrome. A study of 1500 patients with *angina pectoris*. The ECAT Angina Pectoris Study Group. *Arterioscler. Thromb.* **13**: 1865–1873.

109. Levy, P. J., Gonzalez, M. F., Hornung, C. A., Chang, W. W., Haynes, J. L. and Rush, D. S. (1996). A prospective evaluation of atherosclerotic

risk factors and hypercoagulability in young adults with premature lower extremity atherosclerosis. *J. Vasc. Surg.* **23**: 36–43, discussion 43–35.

110. Salomaa, V., Stinson, V., Kark, J. D., Folsom, A. R., Davis, C. E. and Wu, K. K. (1995). Association of fibrinolytic parameters with early atherosclerosis. The ARIC Study. Atherosclerosis Risk in Communities Study. *Circulation* **91**: 284–290.

111. Haffner, S. M., D'Agostino, R., Mykkanen, L., Hales, C. N., Savage, P. J., Bergman, R. N., O'Leary, D., Rewers, M., Selby, J., Tracy, R. *et al.* (1998). Proinsulin and insulin concentrations in relation to carotid wall thickness: insulin resistance atherosclerosis study. *Stroke* **29**: 1498–1503.

112. Cortellaro, M., Cofrancesco, E., Boschetti, C., Mussoni, L., Donati, M. B., Cardillo, M., Catalano, M., Gabrielli, L., Lombardi, B., Specchia, G. *et al.* (1993). Increased fibrin turnover and high PAI-1 activity as predictors of ischemic events in atherosclerotic patients. A case-control study. The PLAT Group. *Arterioscler. Thromb.* **13**: 1412–1417.

113. Thogersen, A. M., Jansson, J. H., Boman, K., Nilsson, T. K., Weinehall, L., Huhtasaari, F. and Hallmans, G. (1998). High plasminogen activator inhibitor and tissue plasminogen activator levels in plasma precede a first acute myocardial infarction in both men and women: evidence for the fibrinolytic system as an independent primary risk factor. *Circulation* **98**: 2241–2247.

114. Juhan-Vague, I., Pyke, S. D., Alessi, M. C., Jespersen, J., Haverkate, F. and Thompson, S. G. (1996). Fibrinolytic factors and the risk of myocardial infarction or sudden death in patients with *angina pectoris*. ECAT Study Group. European Concerted Action on Thrombosis and Disabilities. *Circulation* **94**: 2057–2063.

115. Khan, B. V., Parthasarathy, S. S., Alexander, R. W. and Medford, R. M. (1995). Modified low density lipoprotein and its constituents augment cytokine-activated vascular cell adhesion molecule-1 gene expression in human vascular endothelial cells. *J. Clin. Invest.* **95**: 1262–1270.

116. Marui, N., Offermann, M. K., Swerlick, R., Kunsch, C., Rosen, C. A., Ahmad, M., Alexander, R. W. and Medford, R. M. (1993). Vascular cell adhesion molecule-1 (VCAM-1) gene transcription and expression are regulated through an antioxidant-sensitive mechanism in human vascular endothelial cells. *J. Clin. Invest.* **92**: 1866–1874.

117. Conner, E. M. and Grisham, M. B. (1996). Inflammation, free radicals, and antioxidants. *Nutrition* **12**: 274–277.

118. Li, H., Cybulsky, M. I., Gimbrone, M. A., Jr. and Libby, P. (1993). An atherogenic diet rapidly induces VCAM-1, a cytokine-regulatable mononuclear leukocyte adhesion molecule, in rabbit aortic endothelium. *Arterioscler. Thromb.* **13**: 197–204.

119. Fruebis, J., Gonzalez, V., Silvestre, M. and Palinski, W. (1997). Effect of probucol treatment on gene expression of VCAM-1, MCP-1, and M-CSF in

the aortic wall of LDL receptor-deficient rabbits during early atherogenesis. *Arterioscler. Thromb. Vasc. Biol.* **17**: 1289–1302.

120. Reaven, P. D., Merat, S., Fruebis, J., Sutphin, M. and Silvestre, M. (2000). Effect of aging on aortic expression of vascular cell adhesion molecule and atherosclerosis in murine models of atherosclerosis. *J. Gerontol. Biol. Sci.* **55**(2): B85–B99.

121. De Caterina, R., Basta, G., Lazzerini, G., Dell'Omo, G., Petrucci, R., Morale, M., Carmassi, F. and Pedrinelli, R. (1997). Soluble vascular cell adhesion molecule-1 as a biohumoral correlate of atherosclerosis. *Arterioscler. Thromb. Vasc. Biol.* **17**: 2646–2654.

122. Schleiffenbaum, B., Spertini, O. and Tedder, T. F. (1992). Soluble L-selectin is present in human plasma at high levels and retains functional activity. *J. Cell Biol.* **119**: 229–238.

123. Zeitler, H., Ko, Y., Zimmermann, C., Nickenig, G., Glanzer, K., Walger, P., Sachinidis, A. and Vetter, H. (1997). Elevated serum concentrations of soluble adhesion molecules in coronary artery disease and acute myocardial infarction. *Eur. J. Med. Res.* **2**: 389–394.

124. Bitsch, A., Klene, W., Murtada, L., Prange, H. and Rieckmann, P. (1998). A longitudinal prospective study of soluble adhesion molecules in acute stroke. *Stroke* **29**: 2129–2135.

125. Blann, A. D. and McCollum, C. N. (1994). Circulating endothelial cell/leukocyte adhesion molecules in atherosclerosis. *Thromb. Haemost.* **72**: 151–154.

126. Steiner, M., Reinhardt, K. M., Krammer, B., Ernst, B. and Blann, A. D. (1994). Increased levels of soluble adhesion molecules in Type-2 (non-insulin dependent) diabetes mellitus are independent of glycaemic control. *Thromb. Haemost.* **72**: 979–984.

127. Ceriello, A., Falleti, E., Motz, E., Taboga, C., Tonutti, L., Ezsol, Z., Gonano, F. and Bartoli, E. (1998). Hyperglycemia-induced circulating ICAM-1 increase in diabetes mellitus: the possible role of oxidative stress. *Horm. Metab. Res.* **30**: 146–149.

128. Schmidt, A. M., Hori, O., Chen, J. X., Li, J. F., Crandall, J., Zhang, J., Cao, R., Yan, S. D., Brett, J. and Stern, D. (1995). Advanced glycation endproducts interacting with their endothelial receptor induce expression of vascular cell adhesion molecule-1 (VCAM-1) in cultured human endothelial cells and in mice. A potential mechanism for the accelerated vasculopathy of diabetes. *J. Clin. Invest.* **96**: 1395–1403.

129. Schmidt, A. M., Crandall, J., Hori, O., Cao, R. and Lakatta, E. (1996). Elevated plasma levels of vascular cell adhesion molecule-1 (VCAM-1) in diabetic patients with microalbuminuria: a marker of vascular dysfunction and progressive vascular disease. *Br. J. Haematol.* **92**: 747–750.

130. Bucala, R. (1996). What is the effect of hyperglycemia on atherogenesis and can it be reversed by aminoguanidine? *Diabetes Res. Clin. Pract.* **30(Suppl.)**: 123–130.

131. Somers, M. J. and Harrison, D. G. (1999). Reactive oxygen species and the control of vasomotor tone. *Curr. Hypertens. Rep.* **1**: 102–108.

132. Ting, H. H., Timimi, F. K., Boles, K. S., Creager, S. J., Ganz, P. and Creager, M. A. (1996). Vitamin C improves endothelium-dependent vasodilation in patients with non-insulin-dependent diabetes mellitus. *J. Clin. Invest.* **97**: 22–28.

133. Ellis, E. A., Grant, M. B., Murray, F. T., Wachowski, M. B., Guberski, D. L., Kubilis, P. S. and Lutty, G. A. (1998). Increased NADH oxidase activity in the retina of the BBZ/Wor diabetic rat. *Free Radic. Biol. Med.* **24**: 111–120.

134. Zeng, G. and Quon, M. J. (1996). Insulin-stimulated production of nitric oxide is inhibited by wortmannin. Direct measurement in vascular endothelial cells. *J. Clin. Invest.* **98**: 894–898.

135. Balletshofer, B. M., Rittig, K., Enderle, M. D., Volk, A., Maerker, E., Jacob, S., Matthaei, S., Rett, K. and Haring, H. U. (2000). Endothelial dysfunction is detectable in young normotensive first-degree relatives of subjects with Type-2 diabetes in association with insulin resistance [In Process Citation]. *Circulation* **101**: 1780–1784.

136. Yoshida, Y., Maruyama, M., Fujita, T., Arai, N., Hayashi, R., Araya, J., Matsui, S., Yamashita, N., Sugiyama, E. and Kobayashi, M. (1999). Reactive oxygen intermediates stimulate interleukin-6 production in human bronchial epithelial cells. *Am. J. Physiol.* **276**: L900–L908.

137. Massy, Z. A., Kim, Y., Guijarro, C., Kasiske, B. L., Keane, W. F. and O'Donnell, M. P. (2000). Low-density lipoprotein-induced expression of interleukin-6, a marker of human mesangial cell inflammation: effects of oxidation and modulation by lovastatin. *Biochem. Biophys. Res. Commun.* **267**: 536–540.

138. Klouche, M., Rose-John, S., Schmiedt, W. and Bhakdi, S. (2000). Enzymatically degraded, nonoxidized LDL induces human vascular smooth muscle cell activation, foam cell transformation, and proliferation. *Circulation* **101**: 1799–1805.

139. Smith, E. B. and Thompson, W. D. (1994). Fibrin as a factor in atherogenesis. *Thromb. Res.* **73**: 1–19.

140. Heinrich, J., Schulte, H., Schonfeld, R., Kohler, E. and Assmann, G. (1995). Association of variables of coagulation, fibrinolysis and acute-phase with atherosclerosis in coronary and peripheral arteries and those arteries supplying the brain. *Thromb. Haemost.* **73**: 374–379.

141. Kannel, W. B. (1997). Influence of fibrinogen on cardiovascular disease. *Drugs* **54**: 32–40.

142. Maresca, G., di Blasio, A., Marchioli, R. and di Minno, G. (1999). Measuring plasma fibrinogen to predict stroke and myocardial infarction: an update. *Arterioscler. Thromb. Vasc. Biol.* **19**: 1368–1377.

143. Schmitz, A. and Ingerslev, J. (1990). Haemostatic measures in Type-2 diabetic patients with microalbuminuria. *Diabet. Med.* **7**: 521–525.
144. Imperatore, G., Riccardi, G., Iovine, C., Rivellese, A. A. and Vaccaro, O. (1998). Plasma fibrinogen: a new factor of the metabolic syndrome. A population-based study. *Diabetes Care* **21**: 649–654.
145. Ganda, O. P. and Arkin, C. F. (1992). Hyperfibrinogenemia. An important risk factor for vascular complications in diabetes. *Diabetes Care* **15**: 1245–1250.
146. Juhan-Vague, I., Alessi, M. C. and Vague, P. (1996). Thrombogenic and fibrinolytic factors and cardiovascular risk in non-insulin-dependent diabetes mellitus. *Ann. Med.* **28**: 371–380.
147. Ganda, O. and Arkin, C. (1992). Hyperfibrinogenemia:a major determinant of vascular complications in diabetes. *In* "Fibrinogen, a 'New' Cardiovascular Risk Factor" (E. Ernst, Ed.), Blackwell, Oxford.
148. Keen, H., Clark, C. and Laakso, M. (1999). Reducing the burden of diabetes: managing cardiovascular disease. *Diabetes Metab. Res. Rev.* **15**: 186–196.
149. Cermak, J., Key, N. S., Bach, R. R., Balla, J., Jacob, H. S. and Vercellotti, G. M. (1993). C-reactive protein induces human peripheral blood monocytes to synthesize tissue factor. *Blood* **82**: 513–520.
150. Whisler, R. L., Proctor, V. K., Downs, E. C. and Mortensen, R. F. (1986). Modulation of human monocyte chemotaxis and procoagulant activity by human C-reactive protein (CRP). *Lymphokine Res.* **5**: 223–228.
151. Zeller, J. M., Landay, A. L., Lint, T. F. and Gewurz, H. (1986). Enhancement of human peripheral blood monocyte respiratory burst activity by aggregated C-reactive protein. *J. Leukoc. Biol.* **40**: 769–783.
152. Bhakdi, S., Torzewski, M., Klouche, M. and Hemmes, M. (1999). Complement and atherogenesis: binding of CRP to degraded, nonoxidized LDL enhances complement activation. *Arterioscler. Thromb. Vasc. Biol.* **19**: 2348–2354.
153. Hak, A. E., Stehouwer, C. D., Bots, M. L., Polderman, K. H., Schalkwijk, C. G., Westendorp, I. C., Hofman, A. and Witteman, J. C. (1999). Associations of C-reactive protein with measures of obesity, insulin resistance, and sub-clinical atherosclerosis in healthy, middle-aged women. *Arterioscler. Thromb. Vasc. Biol.* **19**: 1986–1991.
154. Pickup, J. C., Mattock, M. B., Chusney, G. D. and Burt, D. (1997). NIDDM as a disease of the innate immune system: association of acute-phase reactants and interleukin-6 with metabolic syndrome X. *Diabetologia* **40**: 1286–1292.
155. Chu, N. V., Kong, A. P. S., Kim, D. D., Debra Armstrong, R. N., Baxi, S., Deutsch, R., Caulfield, M., Mudaliar, S. R., Reitz, R., Henry, R. R. and Reaven, P. D. (2002). Differential effects of metformin and troglitazone on cardiovascular risk factors in patients with Type-2 diabetes mellitus. *Diabetes Care.*, *in press*.
156. Tracy, R. P., Lemaitre, R. N., Psaty, B. M., Ives, D. G., Evans, R. W., Cushman, M., Meilahn, E. N. and Kuller, L. H. (1997). Relationship of

C-reactive protein to risk of cardiovascular disease in the elderly. Results from the Cardiovascular Health Study and the Rural Health Promotion Project. *Arterioscler. Thromb. Vasc. Biol.* **17**: 1121–1127.

157. Thompson, S. G., Kienast, J., Pyke, S. D., Haverkate, F. and van de Loo, J. C. (1995). Hemostatic factors and the risk of myocardial infarction or sudden death in patients with *angina pectoris*. European Concerted Action on Thrombosis and Disabilities Angina Pectoris Study Group. *New England J. Med.* **332**: 635–641.

158. Kuller, L. H., Tracy, R. P., Shaten, J. and Meilahn, E. N. (1996). Relation of C-reactive protein and coronary heart disease in the MRFIT nested case-control study. Multiple risk factor intervention trial. *Am. J. Epidemiol.* **144**: 537–547.

159. Ridker, P. M. and Haughie, P. (1998). Prospective studies of C-reactive protein as a risk factor for cardiovascular disease. *J. Invest. Med.* **46**: 391–395.

160. Ridker, P. M., Cushman, M., Stampfer, M. J., Tracy, R. P. and Hennekens, C. H. (1997). Inflammation, aspirin, and the risk of cardiovascular disease in apparently healthy men. *New England J. Med.* **336**: 973–979.

161. Ridker, P. M., Glynn, R. J. and Hennekens, C. H. (1998). C-reactive protein adds to the predictive value of total and HDL cholesterol in determining risk of first myocardial infarction. *Circulation* **97**: 2007–2011.

Chapter 53

Overview of Oxidative Stress and Cancer

Steven A. Akman

Steven A. Akman • Department of Cancer Biology, Wake Forest University, Comprehensive Cancer Center, Medical Center Boulevard, Winston-Salem, North Carolina 27157
Tel: (336) 716-0230, E-mail: sakman@wfubmc.edu

1. Introduction

The potential role of oxidative stress in the development of human cancers has been recognized in the past two decades. Cancer is a multi-step process, often involving changes in a variety of cell regulatory functions, including those that control cellular architecture, growth rate, programmed death, replication fidelity, motility, signal recognition and transduction, and gene expression. Oxidative stress can affect any of these processes by modifying lipids, proteins, and nucleic acids. The effects of oxidative stress on these molecules, and the consequences of their alteration with regard to cellular homeostasis, are reviewed elsewhere in this volume. This article will review the data suggesting that oxidative stress and human cancer are causally linked.

2. Oxidative Stress and Cell Pathophysiology

2.1. Oxidative DNA Damage Induced by Endogenous and Environmental Carcinogens

Physiologic sources of ROS and RNS have been detailed elsewhere in this volume. Here the data are reviewed that indicate that exposure to endogenous and environmental carcinogens cause DNA damage indicative of ROS production. Tobacco smoke causes increased accumulation of 8-oxoguanine in leukocyte DNA;[1-5] tobacco smoke also consistently induces increased urinary excretion of the damage product 8-oxodeoxyguanosine (see below),[2,6-8] as do asbestos[9,10] and air pollutants.[11-13] The leukemogen benzene induces formation of 8-oxoguanosine in cultured leukemic cells and in bone marrow *in vivo*.[14,15] The carcinogens hydroxyaminoquinoline,[16] phenylhydrazine,[17] and nitrosobenzene[18] interact with Cu(II) ions to generate oxidative damage in target DNA. Carcinogenic transition metal ion complexes, e.g. Ni(II) complexes[19-27] and Fe(III)-nitriloacetate (NTA),[28-33] induce oxidative DNA damage *in vivo* in tissues that accumulate theses complexes. Radical intermediates formed by metal ion-catalyzed metabolism of tumor promoting organic hydroperoxides, e.g. benzoyl peroxide, induce a range of oxidative DNA base damage products in target DNA.[34] Redox cycling of carcinogenic quinones, e.g. the anti-cancer agent doxorubicin[35,36] or the pentachlorophenol metabolite tetrachloro-p-hydroquinone,[37] induces oxidative DNA damage *in vitro*[35] and *in vivo*.[36] Carcinogens, e.g. topically-applied 7,12-dimethylbenz[a]anthracene,[38,39] can also cause oxidative damage indirectly by inducing an inflammatory response.[40,41]

2.2. Mutagenesis

One potential mechanism of carcinogenesis initiated by oxidatively-modified DNA bases is via induction of mutations in critical target genes. A number of

ROS- and RNS-generating reagents are mutagenic in model human mutagenesis systems. For example, mutagenesis induced by H_2O_2,[42-44] activated white blood cells,[45] singlet oxygen generators,[46, 47] ·ROO generators;[48] ·NO generators,[49-51] and peroxynitrite[52] has been demonstrated after exposure of the pZ189 family of mutation-reporting plasmids. H_2O_2 and activated white blood cells induce primarily G:C → A:T transitions.[42-45] The lesion(s) responsible for these transitions is still controversial. A study suggesting that that 5-OHCyt is the primary promutagenic lesion induced by H_2O_2[53] has been refuted by one suggesting that the deamination product 5-OHUra is primarily responsible.[54] ·ROO induces equal proportions of G:C → T:A and G:C → C:G transversions.[48] ·NO induces transitions at both A:T and C:G base pairs,[49-51] suggesting a role for deamination of exocyclic amines; however, this mechanism has been disputed.[55, 56] Peroxynitrite induces primarily G:C → T:A transversions, with G:C → C:G transversions occurring at a lesser frequency.[52]

The mutagenicity of a number of oxidative DNA base modifications has been studied in site-specific mutagenicity assays. The major guanine oxidation product 8-oxoguanine is mutagenic in procaryotic[57-60] and eucaryotic[61-63] mutagenicity assays, causing primarily G:C → T:A transversions; however, the other principal guanine oxidation product 2,6-diamino-4-oxo-5-formamidopyrimidine (FapyGua) is primarily a DNA-blocking lesion.[64, 65] G:C → T:A transversions induced by 8-oxoguanine activate codon 12 of the c-Ha-*ras* or K-*ras* oncogenes in model mammalian systems.[24, 61, 66] The adenine oxidation product 8-oxoadenine produces A → G transitions and A → C transversions.[67] The presence of 2-oxoadenine (isoguanine) is associated with base substitutions and −1 frameshifts.[68-70]

With regard to the pyrimidines, the oxidized, deaminated cytosine products 5-hydroxyuracil and uracil glycol are highly mutagenic, producing primarily C → T transitions.[54] The major oxidized thymine product thymine glycol[71-79] and the further ring fragmented products urea and β-ureidobutyric acid[80, 81] are primarily strong replication blocks; however, they can be bypassed in a sequence context-dependent manner, producing T → C transitions and T → A transversions.

Oxidative stress can also induce mutagenic lesions in DNA through secondary mechanisms, e.g. via attack by products of lipid peroxidation. These lesions include the cyclic modifications 3,N4-etheno-2′-deoxycytidine (causes C → A transversions and C → T transitions,[82]), 1,N2-(1,3-propano)-2′-deoxyguanosine (causes G → T transversions, G → A transitions and a substantial frequency of −1, −2 frameshifts[83-86]), 1,N6-etheno-2′-deoxyadenosine (causes primarily A → G transitions[87] and the guanine-derived lesion pyrimido[1,2-a]purin-10(3H)-one (causes equivalent number of G → A transitions and G → T transversions[88]).

One of the difficulties encountered when trying to ascertain whether cancer-associated mutations originated from oxidatively-modified DNA is attributable to the overlap of oxidation-associated mutation spectra with the spectra associated with a broad range of other mutagens. For example, Hussain *et al.*[89] observed that exposure of human fibroblasts to H_2O_2/Fe(III) induced mutations in hotspot codons 248 and 249 of the p53 gene typical of those observed in hepatocellular carcinoma. However, the same mutagenic specificity is produced by exposure to aflatoxin B1, benzo[a]pyrene, or heterocyclic amines. This difficulty has led to a search for oxidative DNA damage-associated "signature" mutations. Reid and Loeb[90] proposed that tandem CC → TT transitions might be such a signature mutation in cells not exposed to UV radiation; however, other investigators have not consistently observed this tandem transition. Recently, Turker *et al.*[91] have proposed that discontinuous loss of heterozygosity in microsatellite-containing regions of chromosomes is a marker of reactive oxygen-induced genetic damage.

A second difficulty encountered is that posed by examination of the evidence that ROS-induced DNA damage is an important contributor to spontaneous mutagenesis. Bacteria[92, 93] or yeast[63] deficient in antioxidant enzymes exhibit higher spontaneous mutation rates than do wild type cells. Moreover, Rossman and colleagues[94–97] have observed that the spontaneous mutation rate of Chinese hamster ovary cells is increased under conditions of serum deprivation and that the hypermutable phenotype can be reversed by antioxidants and by over-expression of the metal-binding protein metallothionein. However, the "million-fold misunderstanding", i.e. the difference between mutation rates predicted from the measured frequency of oxidatively-modified DNA bases in human genomic DNA and the measured spontaneous mutation rates at various loci, remains unexplained. Also, two studies that directly compared the mutation spectrum induced by oxidative stress to the spectrum in the same cells under basal conditions did not observe commonality of the spectra. Gille *et al.*[98] observed primarily large deletions at the *tk* locus in human lymphoblast TK6 cells under hyperbaric oxygen, which contrasted with the small mutations observed under normoxic conditions. Oller *et al.*[99] studied hotspot mutations in TK6 cells under normoxic and hyperoxic conditions, as well as after exposure to H_2O_2. These investigators also did not observe a correlation of hotspot mutations produced under conditions of oxidative stress with those produced under normoxia.

2.3. Growth Promotion

It was demonstrated in the 1980s that not only were ROS cytotoxic and mutagenic, but, under the appropriate conditions in appropriate cells, they could be growth stimulatory.[100–107] ROS-generating compounds were shown to be tumor promoters in rodents[100, 108–116] (see below). Growth stimulation by ROS suggested epigenetic effects, e.g. on gene expression or signal transduction pathways. Initially, ROS were shown to stimulate transcription of *c-fos* and *c-jun*, components of the AP1 transcription factor, and *c-myc*[101, 103, 117–121] by a mechanism possibly involving poly ADP-ribose polymerase[118] and arachidonic acid.[119] Subsequently, over the past decade evidence has emerged that ROS function as signalling intermediates of cellular responses.[122, 123] Transient ROS production is stimulated by a variety of external stimuli, including growth factors and cytokines[124–127] and G-protein-coupled receptors.[121] "Second messenger" functions of ROS result in activation of important signal transduction components, including mitogen-activated protein kinase (MAPK),[128–130] extracellular-regulated protein kinase (ERK1/2),[131, 132] *c-jun* amino terminal kinase,[133, 134] STAT kinases[135] and the ribosomal S6 protein kinases.[136, 137] Exposure to ROS induces protein tyrosine phosphorylation in some cells,[138, 139] including phosphorylation of growth factor receptors[140, 141] and *c-src* protein.[120, 131] ROS may be the common mediator of activation of transcription factors NF-κB[142–144] and AP-1[101, 103, 117–121] for a variety of transcriptional activators.

These observations suggest that ROS are used by many cells as physiologic regulators of cell cycle progression, growth, and response to the environment. The opportunity for altered conditions of ROS exposure to foster aberrant cell cycle progression, replication, or cell division, or inappropriate responses to environmental controls is evident.

3. Animal Models-Induction of Tumors by Oxidants

Administration of certain transition metal ion complexes causes cancer in rodent models. Ni(II) complexes are renal carcinogens in rats. Kasprzak and colleagues have studied the mechanism of Ni(II)-induced renal cancer and have provided strong evidence that oxidative DNA damage is involved. Ni(II) complexes induce oxidative DNA damage *in vitro*[20–22] via a Fenton-like mechanism. Moreover, such damage in exposed rats *in vivo* accumulates to higher levels[23, 27] and is repaired more slowly[27] in the carcinogenicity target organ, the kidney, as compared to other organs such as liver. Furthermore, Ni(II)-induced renal cancer is associated with a G:C → T:A transversion in the K-*ras* oncogene;[24] the oxidative damage product 8-oxoguanine causes primarily G:C → T:A transversions.

Another well-characterized rodent renal cancer model is that induced by injection of Fe(III)-NTA. Fe-NTA causes renal proximal tubule necrosis in male rats and mice; tubular necrosis is a consequence of iron ion-mediated oxidative damage that finally leads to a high incidence of renal adenocarcinoma. Exposure of rats to Fe-NTA causes increased levels of a range of oxidatively-modified DNA bases[29–32] in a time frame concomitant with tubular necrosis.

A third well-characterized model for ROS-mediated rodent renal cancer is the estrogen-induced model, studied extensively by Liehr and colleagues. Exposure to high-dose catechol estrogens or diethylstilbestrol induces renal cancer in male Syrian hamsters. The proposed mechanism of ROS generation is enzymatic (CYP1B1) hydroxylation of estradiol to the 4-hydroxy metabolite, followed by further oxidation to the semi-quinone or quinone form.[145–148] Redox cycling of the semi-quinone/quinone generates ROS.[149] Iron supplementation enhances estradiol-mediated tumorigenesis in hamsters.[150] Acute exposure of hamsters to diethylstilbestrol induces 8-oxoguanine formation in both liver and kidney; however, chronic exposure is associated with elevated 8-oxoguanine levels in kidney only.[151] The demonstration of estradiol 4-hydroxylating activity in human breast tissue and the elevated 4:2-hydroxylating activity ratio observed in human breast cancers[152] suggests that the rodent kidney cancer model may be applicable to human female breast cancer.[153]

A number of organic hydroperoxides are tumor promoters in mouse skin keratinocytes. These hydroperoxides form alkyl and alkoxyl radicals *in vitro*;[154–156] Slaga and colleagues,[108, 114] Kensler and colleagues,[115, 156, 157] and Malkinson and colleagues[116] have presented evidence that the radical intermediates mediate tumor promotion by peroxides. Much of this evidence is based on the inhibition of tumor promotion and/or progression by superoxide dismutase-mimicking agents or natural or synthetic antioxidants.[108, 110, 158–164] Moreover, promotable clones of murine epidermal keratinocytes exhibit enhanced resistance to oxidant cytotoxicity than do non-promotable clones.[165] The tumor promoter benzoyl peroxide has been shown to generate Cu(I)-dependent DNA strand breaks in cultured keratinocytes and epithelial cells,[156, 157] oxidative DNA base damage,[34] and mutations typical of those caused by 8-oxoguanine[166] in plasmid DNA.

Wei and Frenkel[163, 164, 167–169] have observed oxidative DNA damage induced in the epidermis of SENCAR and C57BL/6J mice by topical application of tumor promoters, e.g. 12-O-tetradecanoylphorbol-13-acetate (TPA). The inflammatory response induced by these compounds causes much of the damage.[163, 164, 168, 169] TPA also induces ROS-generating xanthine oxidase activity in mouse keratinocytes.[170, 171] These investigators have linked oxidative DNA damage and tumor promoting activity by observing that: (i) the tumor promoting activity of several promoters correlates with the amount of oxidative damage induced;[163, 169] (ii) the sensitivity of different mouse strains to tumor promotion also correlates with amount of damage induction; (iii) certain chemopreventive agents in this model diminish the inflammatory response and limit oxidative DNA damage.[162, 163, 172, 173]

Not only can ROS-inducing agents act as complete carcinogens, initiators, or tumor promoters in animal models, they may also stimulate malignant progression. Exposure to benzoyl peroxide causes conversion of benign skin papillomas in mice to malignant tumors,[114] an event that rarely occurs spontaneously.

4. Evidence for Role of Oxidative Stress in Human Carcinogenesis

4.1. Intermediate Markers of Oxidative Stress

4.1.1. *Oxidative DNA Damage*

Oxidative DNA base modifications can be detected in normal human genomic DNA by a variety of measurement techniques. These measurement techniques and the controversy surrounding the relative merits of each are reviewed in other chapters of this volume and will not be discussed further here. Several studies have indicated that genomic DNA derived from cancerous or precancerous human tissue contains elevated amounts of certain modified bases as compared to normal tissues. Using gas chromatography/mass spectrometry with selected ion monitoring to assay damage products, Olinski and colleagues[174, 175] observed elevated amounts of 11 different oxidized DNA base modifications in chromatin derived from colon, stomach, ovary, brain, and lung carcinomas as compared to histologically normal tissue removed at surgery. The highest levels were observed in lung cancer tissue from smokers. Okamoto *et al.*, using HPLC-ECD, also found elevated levels of 8-oxodeoxyguanosine (8-OH-dG) in DNA derived from human renal carcinoma as compared to non-tumorous tissue from the same kidney.[176]

Malins and colleagues were the first to report elevation of modified purines including 8-oxoguanine, 8-oxoadenine, and FapyGua in DNA retrieved from human breast cancer specimens, as well as adjacent non-malignant tissue, as compared to DNA from normal breasts.[177] This group subsequently expanded their observations to suggest that the progression from normal to malignant breast tissue was accompanied by a shift from a high ratio of ring-open (Fapy) purines to 8-oxo adducts to one favoring the 8-oxo adducts.[178] Such a shift could reflect a change from a reducing nuclear environment to an oxidizing one.[179] Furthermore, DNA derived from invasive ductal breast cancer specimens of patients with metastatic disease had a higher total content of modified purines than did DNA derived from non-metastatic breast cancers,[180] suggesting a role for oxidative DNA damage in the progression to the metastatic state. Immunochemical detection of 8-oxoguanine in DNA confirmed the marked increase of this modified base in breast cancer DNA as compared to DNA retrieved from reduction mammoplasty specimens;[181] however, this difference has been disputed by HPLC-ECD measurements of 8-oxoguanine by Nagashima *et al.*[182] Also, this group did not observe elevation of 8-oxoguanine levels in DNA derived from hepatocellular carcinomas as compared to non-malignant hepatocytes.[183]

Djuric *et al.* bolstered the relationship between oxidative DNA damage and breast cancer by observing that a cohort of breast cancer patients had elevated amounts of 5-(hydroxymethyl)-2′-deoxyuridine in DNA obtained from peripheral blood lymphocytes than did matched control women without breast cancer.[184] These data suggest a global alteration in production and/or repair of oxidative DNA damage in breast cancer patients, possibly diet or body composition related. However, this study does not distinguish cause from effect.

4.1.2. *Urinary Excretion of Damage Products*

Urinary excretion of oxidized DNA bases and/or nucleosides has been studied as a biomarker of "whole-body" DNA damage. Attention has been principally focused on urinary excretion of 8OH-dG, although several other DNA damage products, including 5-(hydroxymethyl)uracil,[185, 186] thymine glycol,[187] deoxythymidine glycol, 8-oxoguanine, 8-oxoadenine, and 7-methyl-8-hydroxyguanine,[188] have been detected in human urine. Urinary 8OH-dG reflects both the rates of creation and removal of this modified base in DNA. Some uncertainty is introduced by the possibilities that 8OH-dG may also reflect dephosphorylation of oxidized nucleotides.[189] Also, concern has been raised that 8OH-dG is itself easily further oxidized.[189] Poulsen, Loft and colleagues have characterized the factors affecting urinary excretion of 8OH-dG in detail.[2, 4, 6, 7, 190–194] The principal modulating factors are resting metabolic rate, oxygen consumption, and body mass index. Age, dietary composition, antioxidant supplementation, and energy restriction have little influence. Exposure to carcinogenic tobacco smoke, ionizing radiation, and environmental pollution markedly increases excretion of 8OH-dG and 5-(hydroxymethyl)uracil,[185] whereas smoking cessation decreases it. Urinary 8OH-dG excretion also diminishes rapidly after initiation of a diet supplemented with anticarcinogenic Brussels sprouts.[195] Furthermore, elevated urinary excretion of 8OH-dG was observed in a variety of cancer patient specimens as compared to a cohort of individuals without cancer.[190]

These studies do not distinguish changes in steady state levels of 8OH-dG in DNA from reciprocal changes in the repair rate *per se*. However, carcinogen-induced elevation of oxidized DNA base modifications in humans has been supported by *in vivo* measurement of steady-state damage levels in leukocyte DNA. For example, 2–4-fold elevation of 8OH-dG levels has been observed in leukocytes harvested from individuals exposed to ionizing radiation.[196, 197] Also, leukocyte[1–5] and lung[198, 199] 8OH-dG, as assayed by HPLC-ECD, is circa 1.5-fold higher in smokers than non-smokers.

4.1.3. *Serum Antibodies*

Frenkel and colleagues[200] have measured circulating autoantibodies to the oxidative DNA damage product 5-hydroxymethyl-2'-deoxyuridine (HMdU) coupled to bovine serum albumen in human sera. Circulating antibody titers in healthy women subsequently diagnosed with colorectal or breast cancer within 0.5–6 years were significantly elevated as compared to controls. Women developing rectal cancer had the highest circulating autoantibody titers > women developing colon or breast cancer. Sera from benign breast and gastrointestinal tract disease patients also contained elevated autoantibody titers, although not to the extent of cancer patients. Interestingly, women with a strong family history of breast cancer also exhibit high anti-HMdU antibody titers. Ashok and Ali[201] have also observed enhanced serum reactivity to ROS-modified DNA in some breast cancer patients. These studies suggest the possibility that enhanced generation of oxidative DNA damage precedes and stimulates neoplasia.

4.2. Human Epidemiologic Studies Attempting to Link Oxidative Stress and Cancer

The epidemiologic evidence marshalled when considering the involvement of oxidative stress in human carcinogenesis fits into three general categories: (1) the relationship of age to cancer risk; (2) the influence of diet; (3) the role of chronic infection and inflammation. The age-specific death rates from many human cancers are a power function of age (see Fig. 3 of Ref. 202). Oxidative lesions in DNA also accumulate with age (at least in the rat).[3, 40, 203–206] Moreover, the species-specific steady-state levels of 8-oxoguanine in DNA and the amount of urinary excretion of 8OH-dG correlate inversely with mean species life span of rodents and primates.[202]

Numerous epidemiologic studies have consistently correlated inadequate intake of fruits and vegetables with elevated cancer risk.[207–210] The risk of aerodigestive tumor development is particularly elevated, with the lowest fruit and vegetable consumption quartile sustaining circa twice the risk of the highest consumption quartile. The negative correlation between fruit and vegetable consumption and cancer risk has focused attention on the chemopreventive activity of antioxidants; however, it has been difficult to separate the role of antioxidants from other potential chemopreventive compounds in these foods by epidemiologic studies. Studies correlating plasma levels have produced equivocal results. A meta-analysis of five prospective case-control studies indicated that high plasma levels of α-tocopherol were associated with a small decrease in the incidence of colorectal cancer.[211]

Several large-scale prospective trials of antioxidant supplements have also generally yielded negative results. The exception to this negative picture was the study based in Linxian, China, which has extremely high rates of esophageal and gastric cancer. This study, involving more than 29 000 participants, indicated that combined supplementation with β-carotene, α-tocopherol (60 IU), and selenium lowered mortality attributable to stomach cancer.[212] The relative risk for all cancer in this study was 0.87, 95% CI = 0.75–1.00. However, the Alpha-Tocopherol, Beta-Carotene (ATBC) Cancer Prevention trial, conducted in over 29 000 male Finnish smokers ages 50–69 failed to show a reduction in the risk of lung or gastrointestinal malignancies attributable to consumption of modest doses of β-carotene, α-tocopherol, or both.[213] There was a small reduction in the incidence of prostate cancer in men supplemented with α-tocopherol. Moreover, in 1996, Hennekens *et al.* reported the 12 year follow up of the participants of the Physician's Health Study, which showed no effect of β-carotene supplementation on the incidence of malignancies.[214] Simultaneously, the CARET study results were reported[215] that indicated an increased risk of lung cancer development in smokers attributable to combined supplementation with β-carotene plus retinol.

The Iowa Women's Health Study, which estimated vitamin E intake by means of a food frequency questionnaire, suggested a strong inverse correlation between vitamin E intake and colon cancer risk, particularly in women aged 55–59.[216] However, the Polyp Prevention Study Group trial of supplementation of vitamins E, C, and β-carotene did not reduce the incidence of new colonic adenomas in patients with a prior history of adenoma.[217] Supplements of vitamins E and C also failed to reduce the incidence of colonic polyps in the study of McKeown-Eyssen *et al.*[218]

The relationship between vitamin E and breast cancer risk is also unclear. Eleven case control and five cohort studies have examined the association between vitamin E consumption and breast cancer risk (reviewed in Ref. 219). The majority of these studies assessed consumption by food frequency questionnaire. Eight of the eleven case control studies found an inverse association between vitamin E intake and breast cancer risk; however, none of the prospective studies found a statistically significant association. These negative studies included the Iowa Women's Health Study[220] and the Netherlands Cohort Study.[221] Furthermore, of the four case control and five prospective studies looking at the association between plasma vitamin E levels and breast cancer risk to date, only three found an inverse association.[219] Two studies observed a direct association.

Plasma levels of vitamin E may be misleading, considering that 90% of the body pool is concentrated in adipose tissue.[222, 223] Adipose tissue concentration may be a better marker of long-term vitamin E exposure than serum because of the markedly slower turnover kinetics.[224] Two of three case control studies evaluating the effect of adipose tissue concentration of vitamin E on breast cancer risk found an inverse relationship. The study conducted by Chajes *et al.*[225] observed a lower breast adipose vitamin E concentration in breast cancer patient specimens as compared to patients with benign breast diseases, whereas Zhu *et al.*[226] in Finland observed such a relationship only in postmenopausal women. The European Community Multicentre Study on Antioxidants, Myocardial Infarction, and Cancer of the Breast found no association between buttock fat vitamin E content and breast cancer risk.[227]

Antioxidant supplementation trials have been criticized for the wrong choice of antioxidants, the lack of dose-response information, the selection of study populations where no prevention is possible, and the difficulty of assessing synergy between antioxidants. Supplementation trials employing intermediate marker measurements, usually either some measure of oxidative DNA damage or urinary excretion of 8OH-dG, have been conducted in an attempt to overcome these difficulties. Studies by Poulsen, Loft *et al.*[6, 7, 193] have not shown a correlation between either plasma levels of antioxidant vitamins or supplementation with vitamins C or E, β-carotene, or coenzyme Q10 and urinary excretion of 8OH-dG. In contrast, Duthie *et al.*[228] observed diminished oxidation of pyrimidines in lymphocyte DNA, as measured by the Comet assay, of males aged 50–59 after supplementation with a combination of vitamins C, E, and β-carotene. Also, Fraga et al.[229] observed an inverse correlation between dietary vitamin C content and seminal fluid ascorbic acid content and the 8-oxoguanine content of sperm cell DNA in men on a high vitamin C diet. Some of the discrepancies between these intermediate marker studies may relate to individual tissue exposure and uptake of antioxidants.

Redox transitions of transition metal ions may generate ROS. Epidemiologic evidence from the National Health and Nutrition Examination Survey suggested that higher total body burden of iron, the most abundant cellular transition metal ion, is associated with increased cancer risk in men.[230] Lung, bladder, esophagus, and colon were the organs most at risk. In a case-control study of diet and rectal cancer, dietary iron intake was associated with cancer risk in men only.[231] In contrast, two case-control studies of colonic adenomatous polyps showed an inverse correlation between iron ingestion and risk of polyps.[232, 233] However, these studies measured only dietary iron and not supplemental vitamin intake.

Chronic inflammation, a source of persistent ROS and RNS flux, is a cancer risk factor. Chronic inflammatory bowel disease is a potent risk factor for the development of colon cancer;[234–238] cancer risk rises with increasing duration of disease. A strong epidemiologic association exists between the risk of stomach cancer development and the presence of infection with *Helicobacter pylori* (reviewed in Ref. 239). In 1994, the International Agency for Research on Cancer of the World Health Organization classified *H. pylori* as a class 1 human carcinogen.[240] *Helicobacter pylori* infection induces a robust chronic inflammatory response in the stomach. Gastric mucosa is chronically infiltrated by neutrophils, macrophages, as well as T and B lymphocytes, which, paradoxically, do not clear the infection, but leave the host prone to complications, including cancer.

5. Conclusions

Considerable theoretical and model system data suggest an important role for oxidative stress in human carcinogenesis. However, a number of challenges remain to be overcome in order to obtain a detailed understanding of the contributions of oxidative stress. Among these challenges are: (1) establishment of cause and effect relationships between oxidative stress and carcinogenesis; (2) clarification of the relationship of cellular redox stress with specific steps in the multi-step process of carcinogenesis; (3) resolution of the ambiguous epidemiologic evidence linking antioxidant defense capacity and cancer risk. This effort will require identification of appropriate intermediate markers, earlier interventions, and longer follow-up studies.

References

1. Kiyosawa, H., Suko, M., Okudaira, H., Murata, T., Miyamoto, T., Chung, M.-H., Kasai, H. and Nishimura, S. (1990). Cigarette smoking induces formation of 8-hydroxygeoxyguanosine, one of the oxidative DNA damages in human peripheral leukocytes. *Free Radic. Res. Commun.* **11**: 23–27.

2. Loft, S., Vistisen, K., Ewertz, M., Tjonneland, A., Overvad, K. and Poulsen, H. E. (1992). Oxidative DNA damage estimated by 8-hydroxydeoxyguanosine excretion in humans: influence of smoking, gender and body mass index. *Carcinogenesis* **13**: 2241–2247.

3. Lagorio, S., Tagesson, C., Forastiere, F., Axelson, O. and Carere, A. (1994). Exposure to benzene and urinary concentrations of 8-hydroxydeoxyguanosine, a biological marker of oxidative damage to DNA. *Occup. Env. Med.* **51**: 739–743.

4. Loft, S., Astrup, A., Buemann, B. and Poulsen, H. E. (1994). Oxidative DNA damage correlates with oxygen consumption in humans. *FASEB J.* **8**: 534–537.

5. Asami, S., Hirano, T., Yamaguchi, R., Tomioka, Y., Itoh, H. and Kasai, H. (1996). Increase of a type of oxidative DNA damage, 8-hydroxyguanine, and its repair activity in human leukocytes by cigarette smoking. *Cancer Res.* **56**: 2546–2549.

6. Van Poppel, G., Poulsen, H.E., Loft, S. and Verhagen, H. (1995). No influence of beta carotene on oxidative DNA damage in male smokers. *J. Natl. Cancer Inst.* **87**: 310–311.

7. Prieme, H., Loft, S., Nyyssonen, K., Salonen, J. T. and Poulsen, H. E. (1997). No effect of supplementaion with vitamin E, ascorbic acid, or coenzyme Q10 on oxidative DNA damage estimated by 8-oxo-7,8-dihydor-2′-deoxyguanosine excretion in smokers. *Am. J. Clin. Nutri.* **65**: 503–507.

8. Prieme, H., Loft, S., Klarlund, M., Gronback, K., Tonnesen, P. and Poulsen, H. E. (1998). Effect of smoking cessation on oxidative DNA modification estimated by 8-oxo-7,8-dihydro-2′-deoxyguanosine excretion. *Carcinogenesis* **19**: 347–351.

9. Tagesson, C., Chabiuk, D., Axelson, O., Baranski, B., Palus, J. and Wyszynska, K. (1993). Increased urinary excretion of the oxidative DNA adduct, 8-hydroxydeoxyguanosine, as a possible early indicator of occupational cancer hazards in the asbestos, rubber, and azo-dye industries. *Polish J. Occup. Med. Env. Health* **6**: 357–368.

10. Adachi, S., Yoshida, S., Kawamura, K., Takahashi, T., Kasai, H., Natori, Y. and Watanabe, S. (1994). Inductions of oxidative DNA damage and mesothelioma by crocidolite, with special reference to the presence of iron inside and outside of asbestos fiber. *Carcinogenesis* **15**: 753–758.

11. Suzuki, J., Inoue, Y. and Suzuki, S. (1995). Changes in the urinary excretion level of 8-hydroxyguanine by exposure to reactive oxygen-generating substances. *Free Radic. Biol. Med.* **18**: 431–436.

12. Suzuki, J., Inoue, Y. and Suzuki, S. (1995). Changes in urinary level of 8-hydroxyguanine by exposure to reactive oxygen-generating substances. *Free Radic. Biol. Med.* **18**: 431–436.

13. Loft, S., Poulsen, H. E., Vistisen, K. and Knudsen, L. E. (1999). Increased urinary excretion of 8oxo-2'-deoxyguanosine, a biomarker of oxidative DNA damage, in urban bus drivers. *Mutat. Res.* **441**: 11–19.

14. Kolachana, P., Subrahmanyam, V. V., Meyer, K. B., Zhang, L. and Smith, M. T. (1993). Benzene and its phenolic metabolites produce oxidative DNA damage in HL60 cells in vitro and in the bone marrow in vivo. *Cancer Res.* **53**: 1023–1026.

15. Zhang, L., Robertson, M. L., Kolachana, P., Davison, A. J. and Smith, M. T. (1993). Benzene metabolite, 1,2,4-benzenetriol, induces micronuclei and oxidative DNA damage in human lymphocytes and HL60 cells. *Env. Mol. Mutagen.* **21**: 339–348.

16. Yamamoto, K., Inoue, S. and Kawanishi, S. (1993). Site-specific DNA damage and 8-hydroxydeoxyguanosine formation by hydroxylamine and 4-hydroxyaminoquinoline 1-oxide in the presence of Cu(II): role of active oxygen species. *Carcinogenesis* **14**: 1397–1401.

17. Yamamoto, K. and Kawanishi, S. (1992). Site-specific DNA damage by phenylhydrazine and phenelzine in the presence of Cu(II) ion or Fe(III) complexes: roles of active oxygen species and carbon radicals. *Chem. Res. Toxicol.* **5**: 440–446.

18. Ohkuma, Y. and Kawanishi, S. (2000). Oxidative DNA damage by a metabolite of carcinogenic and reproductive toxic nitrobenzene in the presence of NADH and Cu(II). *Biochem. Biophys. Res. Commun.* **257**: 555–560.

19. Kasprzak, K. S. and Hernandez, L. (1989). Enhancement of hydroxylation and deglycosylation of 2'-deoxyguanosine by carcinogenic nickel compounds. *Cancer Res.* **49**: 5964–5968.

20. Nackerdien, Z., Kasprzak, K. S., Rao, G., Halliwell, B. and Dizdaroglu, M. (1991). Nickel(II)- and cobalt(II)-dependent damage by hydrogen peroxide to the DNA bases in isolated chromatin. *Cancer Res.* **51**: 5837–5842.

21. Datta, A. K., Riggs, C. W., Fivash, M. J. and Kasprzak, K. (1991). Mechanisms of nickel carcinogenesis. Interaction of Ni(II) with 2'-deoxynucleosides and 2'-deoxynucleotides. *Chem. Biol. Interact.* **79**: 323–334.

22. Dutta, A. K., Misra, M., North, S. L. and Kasprzak, K. (1992). Enhancement by nickel(II) and L-histidien of 2'-deoxyguanosine oxidation with hydrogen peroxide. *Carcinogenesis* **19**: 283–287.

23. Kasprzak, K., Diwan, B. A., Rice, J. M., Misra, M., Olinski, R. and Dizdaroglu, M. (1992). Nickel(II)-mediated oxidative DNA base damage

in renal and hepatic chromatin of pregnant rats and their fetuses. Possible relevance to carcinogenesis. *Chem. Res. Toxicol.* **5**: 809–815.

24. Higinbotham, K. G., Rice, J. M., Diwan, B. A., Kasprzak, K. S., Reed, C. D. and Perantoni, A. O. (1992). GGT to GTT transversions in codon 12 of the K-ras oncogene in rat renal sarcomas induced with nickel subsulfide or nickel subsulfide/iron are consistent with oxidative damage to DNA. *Cancer Res.* **52**: 4747–4751.

25. Misra, M., Olinski, R., Dizdaroglu, M. and Kasprzak, K. S. (1993). Enhancement by L-histidine of nickel(II)-induced DNA-protein cross-linking and oxidative DNA base damage in the rat kidney. *Chem. Res. Toxicol.* **6**: 33–37.

26. Datta, A. K., North, S. L. and Kasprzak, K. (1994). Effect of nickel(II) and tetraglycine on hydroxylation of the guanine moiety in 2'-deoxyguanosine, DNA, and nucleohistone by hydrogen peroxide. *Sci. Total Env.* **148**: 207–216.

27. Kasprzak, K., Jaruga, P., Zastawny, T. H., North, S. L., Riggs, C. W., Olinski, R. and Dizdaroglu, M. (1997). Oxidative DNA base damage and its repair in kidneys and livers of nickel(II)-treated male F344 rats. *Carcinogenesis* **18**: 271–277.

28. Inoue, S. and Kawanishi, S. (1987). Hydroxyl radical production and human DNA damage induced by ferric nitrilotriacetate and hydrogen peroxide. *Cancer Res.* **47**: 6522–6527.

29. Umemura, T., Sai, K., Takagi, A., Hasegawa, R. and Kurokawa, Y. (1990). Formation of 8-hydroxydeoxyguanosine (8OH-dG) in rat kidney DNA after intraperitoneal administration of ferric nitrilotriacetate (Fe-NTA). *Carcinogenesis* **11**: 345–347.

30. Umemura, T., Sai, K., Tagaki, A., Hasegawa, R. and Kurokawa, Y. (1990). Oxidative DNA damage, lipid peroxidation and nephrotoxicity induced in the rat kidney after ferric nitrilotriacetate administration. *Cancer Lett.* **54**: 95–100.

31. Toyokuni, S., Mori, T. and Dizdaroglu, M. (1994). DNA base modifications in renal chromatin of Wistar rats treated with a renal carcinogen, ferric nitrilotriacetate. *Int. J. Cancer* **57**: 123–128.

32. Okada, S. (1995). Iron-induced tissue damage and cancer: the role of reactive oxygen species-free radicals. *Pathol. Int.* **19**: 339–347.

33. Bahnemann, R., Leibold, E., Kittel, B., Mellert, W. and Jackh, R. (1998). Different patterns of kidney toxicity after subacute administration of Na-nitrilotriacetic acid and Fe-nitrilotriacetic acid to Wistar rats. *Toxicol. Sci.* **46**: 166–175.

34. Akman, S. A., Kensler, T. W., Doroshow, J. H. and Dizdaroglu, M. (1993). Copper ion-mediated modification of bases in DNA in vitro by benzoyl peroxide. *Carcinogenesis* **14**: 1971–1974.

35. Akman, S. A., Doroshow, J. H., Burke, T. G. and Dizdaroglu, M. (1992). DNA base modifications induced in isolated human chromatin by NADH

dehydrogenase-catalyzed reduction of doxorubicin. *Biochemistry* **31**: 3500–3506.

36. Faure, H., Mousseau, J., Cadet, J., Guimmier, C., Tripier, M., Hida, H. and Favier, A. (1998). Urine 8-oxo-7,8-diihydro-2'-deoxyguanosie versus 5-(hydroxymethyl)uracil as DNA oxidation marker in adriamycin-treated patients. *Free Radic. Res.* **28**: 377–381.

37. Dahlhaus, M., Almstadt, E. and Appel, K. E. (1994). The pentachlorophenol metabolite tetrachloro-p-hydroquinone induces the formation of 8-hydroxy-2'-deoxyguanosine in liver DNA of male B6C3F1 mice. *Toxicol. Lett.* **74**: 265–274.

38. Frenkel, K., Wei, H. and Wei, L. (1995). 7,12-dimethylbenz[a]anthracene induces oxidative DNA modification in vivo. *Free Radic. Biol. Med.* **19**: 373–380.

39. Fischer, W. H. and Lutz, W. K. (1995). Correlation of induvidual papilloma latency time with DNA adducts, 8-hydroxy-2'-deoxyguanosine, and the rate of DNA synthesis in the epidermis of mice treated with 7,12-dimethylbenz[a]-anthracene. *Proc. Natl. Acad. Sci. USA* **92**: 5900–5904.

40. Wang, Y. J., Ho, Y. S., Lo, M. J. and Lin, J. K. (1995). Oxidative modification of DNA bases in rat liver and lung during chemical carcinogenesis and aging. *Chem. Biol. Interact.* **94**: 135–145.

41. Liu, Z., Lu, Y., Rosenstein, B., Lebwohl, M. and Wei, H. (1998). Benzo[a]pyrene enhances the formation of 8-hydroxy-2'-deoxyguanosine by ultraviolet A radiation in calf thymus DNA and human epidermoid carcinoma cells. *Biochemistry* **37**: 10 307–10 312.

42. Moraes, E. C., Keyse, S. M., Pidoux, M. and Tyrrell, R. M. (1989). The spectrum of mutations generated by passage of a hydrogen peroxide damaged shuttle vector plasmid through a mammalian host. *Nucleic Acids Res.* **17**: 8301–8312.

43. Moraes, E. C., Keyse, S. M. and Tyrrell, R. M. (1990). Mutagenesis by hydrogen peroxide treatment of mammalian cells: a molecular analysis. *Carcinogenesis* **11**: 283–293.

44. Akman, S., Forrest, G., Doroshow, J. and Dizdaroglu, M. (1990). Mutations of plasmid pZ189(pZ) replicating in CV-1 cells caused by $H2O2/Fe^{2+}$(HF) treatment. *Proc. Am. Assoc. Cancer Res.* **31**: 146–146. (**Abstract.**)

45. Akman, S. A., Sander, F. and Garbutt, K. (1996). In vivo mutagenesis of the reporter plasmid pSP189 induced by exposure of host Ad293 cells to activated polymorphonuclear leukocytes. *Carcinogenesis* **17**: 2137–2142.

46. Sies, H. and Menck, C. F. (1992). Singlet oxygen induced DNA damage. *Mutat. Res.* **275**: 367–375.

47. Bessho, T., Tano, K., Nishimura, S. and Kasai, H. (1993). Induction of mutations in mouse FM3A cells by treatment with riboflavin plus visible light and its possible relation with formation of 8-hydroxyguanine (7,8-dihydor-8-oxoguanine) in DNA. *Carcinogenesis* **14**: 1069–1071.

48. Valentine, M. R., Rodriguez, H., and Termini, J. (1998). Mutagenesis by peroxy radical is dominated by transversions at deoxyguanosine: evidence for the lack of involvement of 8-oxo-dG and/or abasic site formation. *Biochemistry* **37**: 7030–7038.

49. Nguyen, T., Brunson, D., Crespi, C. L., Penman, B. W., Wishnok, J. S. and Tannenbaum, S. R. (1992). DNA damage and mutation in human cells exposed to nitric oxide in vitro. *Proc. Natl. Acad. Sci. USA* **89**: 3030–3034.

50. Routledge, M. N., Wink, D. A., Keefer, L. K. and Dipple, A. (1994). DNA sequence changes induced by two nitric oxide donor drugs in the supF assay. *Chem. Res. Toxicol.* **7**: 628–632.

51. Kelman, D. J., Christodoulou, D., Wink, D. A., Keefer, L. K., Srinivasan, A. and Dipple, A. (1997). Relative mutagenicites of gaseous nitrogen oxides in the supF gene of pSP189. *Carcinogenesis* **18**: 1045–1048.

52. Juedes, M. J. and Wogan, G. N. (1996). Peroxynitrite-induced mutation spectra of pSP189 following replication in bacteria and in human cells. *Mutat. Res.* **349**: 51–61.

53. Feig, D. I., Sowers, L. C. and Loeb, L. A. (1994). Reverse chemical mutagenesis: identification of the mutagenic lesions resulting from reactive oxygen species damage to DNA. *Proc. Natl. Acad. Sci. USA* **91**: 6609–6613.

54. Kreutzer, D. A. and Essigmann, J. M. (1998). Oxidized, deaminated cytosines are a source of C–T transitions in vivo. *Proc. Natl. Acad. Sci. USA* **95**: 3578–3582.

55. Schmutte, C., Rideout III, W. M., Shen, J.-C. and Jones, P. A. (1994). Mutagenicity of nitric oxide is not caused by deamination of cytosine or 5-methylcytosine in double-stranded DNA. *Carcinogenesis* **15**: 2899–2903.

56. Hartman, Z., Henrikson, E. N., Hartman, P. E. and Cebula, T. A. (1994). Molecular models that may account for nitrous acid mutagenesis in organisms containing double-stranded DNA. *Env. Mol. Mutagen.* **24**: 168–175.

57. Wood, M. L., Dizdaroglu, M., Gajewski, E. and Essigmann, J. M. (1990). Mechanistic studies of ionizing radiation and oxidative mutagenesis: genetic effects of a single 8-hydroxyguanine (7-hydro-8-oxoguanine) residue inserted at a unique site in a viral genome. *Biochemistry* **29**: 7024–7032.

58. Moriya, M., Ou, C., Bodepudi, V., Johnson, F., Takeshita, M. and Grollman, A. P. (1991). Site-specific mutagenesis using a gapped duplex vector: a study of translesion synthesis past 8-oxodeoxyguanosine in *Escherichia coli*. *Mutat. Res.* **254**: 281–288.

59. Cheng, K. C., Cahill, D. S., Kasai, H., Nishimura, S. and Loeb, L. A. (1992). 8-hydroxyguanine, an abundant form of oxidative DNA damage, causes G–T and A–C substitutions. *J. Biol. Chem.* **267**: 166–172.

60. Wood, M. L., Esteve, A., Morningstar, M. L., Kuziemko, G. M. and Essigmann, J. M. (1992). Genetic effects of oxidative DNA damage:

comparative mutagenesis of 7,8-dihydro-8-oxoguanine and 7,8-dihydro-8-oxoadenine in *Escherichia coli*. *Nucleic Acids Res.* **20**: 6023–6032.

61. Kamiya, H., Miura, H., Ishikawa, H., Inoue, H., Nishimura, S. and Ohtsuka, E. (1992) c-Ha-ras containing 8-hydroxyguanine at codon 12 induces point mutations at the modified and adjacent positions. *Cancer Res.*, **52**: 3483-3485.

62. Moriya, M. (1993) Single-stranded shuttle phademid for mutagenesis studies in mammalian cells: 8-oxoguanine in DNA induces targeted G:C → T:A transversions in simian kidney cells. *Proc. Natl. Acad. Sci. USA* **90**: 1122–1126.

63. Thomas, D., Scot, A. D., Barbey, R., Padula, M. and Boiteux, S. (1997). Inactivation of OGG1 increases the incidence of G:C to T:A transversions in *Saccharomyces cerevisiae*: evidence for endogenous oxidative damage to DNA in eukaryotic cells. *Mol. Gen. Genet.* **254**: 171–188.

64. Boiteux, S. and Laval, J. (1983). Imidazole open ring 7-methylguanine: an inhibitor of DNA synthesis. *Biochem. Biophys. Res. Commun.* **110**: 552–558.

65. O'Connor, T. R., Boiteux, S. and Laval, J. (1988) R.ing-opened 7-methylguanine residues in DNA are a block to in vitro DNA synthesis. *Nucleic Acids Res.* **16**: 5879–5894.

66. Kasprzak, K. S., Higinbotham, K., Diwan, B. A., Perantoni, A. O. and Rice, J. M. (1990). Correlation of DNA base oxidation with the activation of K-ras oncogene in nickel-induced renal tumors. *Free Radic. Biol. Med.* **9(Suppl. 1)**: 172–172. **(Abstract.)**

67. Kamiya, H., Miura, H., Murata-Kamiya, N., Ishikawa, H., Sakaguchi, T., Inoue, H., Sasaki, T., Masutani, C., Hanaoka, F. and Nishimura, S. (1995). 8-hydroxyadenine (7,8-dihydro-8-oxoadenine) induces misincorporation in in vitro DNA synthesis and mutations in NIH 3T3 cells. *Nucleic Acids Res.* **23**: 2893–2899.

68. Kamiya, H. and Kasai, H. (1996). Effect of sequence contexts on mis-incorporation of nucleotides opposite 2-hydroxyadenine. *FEBS Lett.* **391**: 113–116.

69. Kamiya, H. and Kasai, H. (1997). Substitution and deletion mutations induced by 2-hydroxyadenine in *Escherichia coli*: effects of sequence contexts in leading and lagging strands. *Nucleic Acids Res.* **25**: 304–311.

70. Kamiya, H. and Kasai, H. (1997). Mutations induced by 2-hydroxyadenine on a shuttle vector during leading and lagging strand syntheses in mammalian cells. *Biochemistry* **36**: 11 125–11 130.

71. Rouet, P. and Essigmann, J. M. (1985). Possible role for thymine glycol in the selective inhibition of DNA synthesis on oxidized DNA templates. *Cancer Res.* **45**: 6113–6118.

72. Ide, H., Kow, Y. W. and Wallace, S. S. (1985). Thymine glycols and urea residues in M13 DNA constitute replicative blocks in vitro. *Nucleic Acids Res.* **13**: 8035–8052.

73. Clark, J. M. and Beardsley, G. P. (1986). Thymine glycol lesions terminate chain elongation by DNA polymerase I in vitro. *Nucleic Acids Res.* **14**: 737–749.

74. Hayes, R. C. and LeClerc, J. E. (1986). Sequence dependence for bypass of thymine glycols in DNA by DNA polymerase I. *Nucleic Acids Res.* **14**: 1045–1061.

75. Clark, J. M. and Beardsley, G. P. (1987). Functional effects of *cis*-thymine glycol lesions on DNA synthesis in vitro. *Biochemistry* **26**: 5398–5403.

76. Hayes, R. C., Petrullo, L. A., Huang, H., Wallace, S. S. and LeClerc, J. E. (1988). Oxidative damage in DNA: lack of mutagenicity by thymine glycol lesions. *J. Mol. Biol.* **201**: 239–246.

77. Clark, J. M. and Beardsley, G. P. (1989). Template length, sequence context, and 3′-5′ exonuclease activity modulate replicative bypass of thymine glycol lesions in vitro. *Biochemistry* **28**: 775–779.

78. Basu, A. K., Loechler, E. L., Leadon, S. A. and Essigmann, J. M. (1989). Genetic effects of thymine glycol: site-specific mutagenesis and molecular modeling studies. *Proc. Natl. Acad. Sci. USA* **86**: 7677–7681.

79. Essigmann, J. M., Basu, A. K. and Loechler, E. L. (1989). Mutagenic specificity of alkylated and oxidized DNA bases as determined by site-specific mutagenesis. *Ann. Ist. Super. Sanita.* **25**: 155–161.

80. Evans, J., Maccabee, M., Hatahet, Z., Courcelle, J., Bockrath, R., Ide, H. and Wallace, S. (1993). Thymine ring saturation and fragmentation products: lesion bypass, misinsertion and implications for mutagenesis. *Mutat. Res.* **299**: 147–156.

81. Maccabee, M., Evans, J. S., Glackin, M. P., Hahahet, Z. and Wallace, S. S. (1994). Pyrimidine ring fragmentation products: effect of lesion structure and sequence context on mutagenesis. *J. Mol. Biol.* **236**: 514–530.

82. Moriya, M., Zhang, W., Johnson, F. and Grollman, A. P. (1994). Mutagenic potency of exocyclic DNA adducts: marked differences between *Escherichia coli* and simian kidney cells. *Proc. Natl. Acad. Sci. USA* **91**: 11 899–11 903.

83. Benamira, M., Singh, U. and Marnett, L. J. (1992). Site-specific frameshift mutagenesis by a propanodeoxyguanosine adduct postiioned in the (CpG)4 hot-spot of *Salmonella typhimurium* hisD3052 carried on an M13 vector. *J. Biol. Chem.* **267**: 22 392–22 400.

84. Burcham, P. and Marnett, L. J. (2000). Site-specific mutagenesis by a propanodeoxyguanosine adduct carried on an M13 genome. *J. Biol. Chem.* **269**: 28 844–28 850.

85. Hashim, M. F. and Marnett, L. J. (1996). Sequence-dependent induction of base pair substitutions and frameshifts by propanodeoxyguanosine during in vitro DNA replication. *J. Biol. Chem.* **271**: 9160–9165.

86. Hashim, M. F., Schnetz-Boutard, N. C. and Marnett, L. J. (1997). Replication of template-primers containing propanodeoxyguanosine by DNA polymerase

beta. Induction of base pair substitution and frameshift mutations by template slippage and deoxynucleoside triphosphate stabilization. *J. Biol. Chem.* **272**: 20 205–20 212.

87. Pandya, G. A. and Moriya, M. (1996). 1,N6-ethenodeoxyadenosine, a DNA adduct highly mutagenic in mammalian cells. *Biochemistry* **35**: 11 487–11 492.

88. Fink, S. P., Reddy, G. R. and Marnett, L. J. (1997). Mutagenicity in *Escherichia coli* of the major DNA adduct derived from the endogenous mutagen malondialdehyde. *Proc. Natl. Acad. Sci. USA* **94**: 8652–8657.

89. Hussain, S. P., Aguilar, F., Amstad, P. and Cerutti, P. (1994). Oxy-radical induced mutagenesis of hotspot codons 248 and 249 of the human p53 gene. *Oncogene* **9**: 2277–2281.

90. Reid, T. M. and Loeb, L. A. (1993). Tandem double CC-TT mutations are produced by reactive oxygen species. *Proc. Natl. Acad. Sci. USA* **90**.

91. Turker, M. S., Gage, B. M., Rose, J. A., Elroy, D., Ponomareva, O. N., Stambrook, P. J. and Tischfield, J. A. (1999). A novel signature mutation for oxidative damage resembles a mutational pattern found commonly in human cancers. *Cancer Res.* **59**: 1837–1839.

92. Farr, S. B., D'ari, R. and Touati, D. (1986). Oxygen-dependent mutagenesis in *Escherichia coli* lacking superoxide dismutase. *Proc. Natl. Acad. Sci. USA* **83**: 8268–8272.

93. Storz, G., Christman, M. F., Sies, H., and Ames, B. N. (1987). Spontaneous mutagenesis and oxidative damage to DNA in *Salmonella typhimurium*. *Proc. Natl. Acad. Sci. USA* **84**: 8917–8921.

94. Goncharova, E. I. and Rossman, T. G. (1994). A role for metallothionein and zinc in spontaneous mutagenesis. *Cancer Res.* **54**: 5318–5323.

95. Goncharova, E. I., Nadas, A. and Rossman, T. G. (1996). Serum deprivation, but not inhibition of growth *per se*, induces a hypermutable state in Chinese hamster G12 cells. *Cancer Res.* **56**: 752–756.

96. Rossman, T. G., Goncharova, E. I., Nadas, A. and Dolzhanskaya, N. (1997). Chinese hamster cells expressing antisense to metallothionein become spontaneous mutators. *Mutat. Res.* **373**: 75–85.

97. Rossman, T. G. and Goncharova, E. I. (1998). Spontaneous mutagenesis in mammalian cells is caused mainly by oxidative events and can be blocked by antioxidants and metallothionein. *Mutat. Res.* **402**: 103–110.

98. Gille, J. J., van Berkel, C. G. and Joenje, H. (1994). Muagenicity of metabolic oxygen radicals in mammalian cell cultures. *Carcinogenesis* **15**: 2695–2699.

99. Oller, A. R. and Thilly, W. G. (1992). Mutational spectra in human B-cells. Spontaneous, oxygen and hydrogen peroxide-induced mutations at the hprt gene. *J. Mol. Biol.* **228**: 813–826.

100. Zimmerman, R. and Cerutti, P. (1984). Active oxygen acts as a promoter of transformation in mouse embryo C3H/10T1/2/C18 fibroblasts. *Proc. Natl. Acad. Sci. USA* **81**: 2085–2087.

101. Shibanuma, M., Kuroki, T. and Nose, K. (1988). Induction of DNA replication and expression of protooncogenes *c-myc* and *c-fos* in quiescent Balb/3T3 cells by xanthine/xanthine oxidase. *Oncogene* **3**: 17–21.

102. Murrell, G., Francis, M. and Bromley, L. (1990). Modulation of fibroblast proliferation by oxygen free radicals. *Biochem. J.* **265**: 659–665.

103. Amstad, P., Crawford, D., Muehlematter, D., Zbinden, I., Larsson, R. and Cerutti, P. (1990). Oxidants stress induces the proto-oncogenes, *c-fos* and *c-myc* in mouse epidermal cells. *Bull. Cancer* **77**: 501–502.

104. Anonymous. (1992). Active oxygen species stimulate vascular smooth muscle cell growth and proto-oncogene expression. *Circ. Res.* **70**: 593–599.

105. Duque, I., Rodriguez-Puyol, M., Ruiz, P., Gonzalez-Rubio, M., Diez-Marques, M. L. and Rodriguez-Puyol, D. (1993). Calcium channel blockers inhibit hydrogen peroxide-induced proliferation of cultured rat mesangial cells. *J. Pharmacol. Exp. Ther.* **267**: 612–616.

106. Burdon, R. H. and Gill, V. (1993). Cellularly generated active oxygne species and HeLa cell proliferation. *Free Radic. Res. Commun.* **19**: 203–213.

107. Li, N., Oberley, T. D., Oberley, L. W. and Zhong, W. (1998). Inhibition of cell growth in NIH/3T3 fibroblasts by overexpression of manganese superoxide dismutase: mechanistic studies. *J. Cell Physiol.* **175**: 359–369.

108. Slaga, T. J., Klein-Szanto, A. P., Triplett, L. and Yotti, L. P. (1981). Skin tumor-promoting activity of benzoyl peroxide, a widely used free radical generating compound. *Science* **213**: 1023–1025.

109. Solanki, V., Rana, R. S. and Slaga, T. J. (1981). Diminution of mouse epidermal superoxide dismutase and catalase activities by tumor promoters. *Carcinogenesis* **2**: 1141–1146.

110. Kensler, T. W., Bush, D. M. and Kozumbo, W. J. (1983). Inhibition of tumor promotion by a biomimetic superoxide dismutase. *Science* **221**: 75–77.

111. Reiners, J. J., Nesnow, S. and Slaga, T. J. (1984). Murine susceptibility to two-stage skin carcinogenesis is influenced by the agent used for promotion. *Carcinogenesis* **5**: 301–307.

112. Kensler, T. W. and Trush, M. A. (1984). Role of oxygen radicals in tumor promotion. *Env. Mutagen.* **6**: 593–612.

113. Kozumbo, W. J., Trush, M. A. and Kensler, T. W. (1985). Are free radicals involved in tumor promotion? *Chem. Biol. Interactions* **54**: 199–207.

114. O'Connell, J. F., Klein-Szanto, A. P., di Giovanni, D. M., Fries, J.W. and Slaga, T. J. (1986). Enhanced malignant progression of mouse skin tumors by the free-radical generator benzoyl peroxide. *Cancer Res.* **46**: 2863–2865.

115. Taffe, B. G. and Kensler, T. W. (1988). Tumor promotion by a hydroperoxide metabolite of butylated hydroxytoluene, 2,6-di-tert-4-hydroperoxyl-4-methyl-2,5-cyclohexadienone, in mouse skin. *Res. Commun. Chem. Pathol. Pharmacol.* **61**: 291–303.

116. Thompson, J. A., Schullek, K. M., Fernandez, C. A. and Malkinson, A. M. (1989). A metabolite of butylated hydroxytoluene with potent tumor promoting activity in mouse lung. *Carcinogenesis* **10**: 773–775.
117. Hollander, M. and Fornace, A. (1989). Induction of *fos* RNA by DNA damaging agents. *Cancer Res.* **49**: 1687–1692.
118. Amstad, P., Krupitza, G. and Cerutti, P. (1992). Mechanism of *c-fos* induction by active oxygen. *Cancer Res.* **52**: 3952–3960.
119. Rao, G. N., Lassegue, B., Griendling, K. K. and Alexander, R. W. (1993). Hydrogen peroxide stimulates transcription of *c-jun* in vascular smooth muscle cells: role of arachidonic acid. *Oncogene* **8**: 2759–2764.
120. Lee, S. F., Huang, Y. T., Wu, W. S. and Lin, J. K. (1996). Induction of *c-jun* protooncogene expression by hydrogen peroxide through hydroxyl radical generation and p60SRC tyrosine kinase activation. *Free Radic. Biol. Med.* **21**: 437–448.
121. Kim, J. H., Kwack, H. J., Choi, S.E., Kim, B. C., Kim, Y. S., Kang, I. J. and Kumar, C. C. (1997). Essential role of Rac GTP ase in hydrogen peroxide-induced activation of *c-fos* serum response element. *FEBS Lett.* **406**: 93–96.
122. Schreck, R. and Baeuerle, P. A. (1991). A role for oxygen radicals as second messengers. *Trends Cell Biol.* **1**: 39–42.
123. Rhee, S. G. (1999). Redox signaling: hydrogen peroxide as intracellular messenger. *Exp. Mol. Med.* **31**: 53–59.
124. Ohba, M., Shibanuma, M., Kuroki, T. and Nose, K. (1994). Production of hydrogen peroxide by transforming growth factor-beta 1 and its involvement in induction of egr-1 in mouse osteoblastic cells. *J. Cell Biol.* **126**: 1079–1088.
125. Sundaresan, M., Yu, Z. X., Ferrans, V. J., Irani, K. and Finkel, T. (1995). Requirement for generation of H2O2 for platelet-derived growth factor signal transduction. *Science* **270**: 296–299.
126. Lee, K. S., Buck, M., Houglum, K. and Chojkier, M. (1995). Activation of hepatic stellate cells by TGF alpha and collagen Type-1 is mediated by oxidative stress through *c-myb* expression. *J. Clin. Invest.* **96**: 2461–2468.
127. Sattler, M., Winkler, T., Verma, S., Byrne, C. H., Shrikhande, G., Salgia, R. and Griffin, J. D. (1999). Hematopoetic growth factors signal through the formation of reactive oxygen species. *Blood* **93**: 2928–2935.
128. Chen, Q., Olashaw, N. and Wu, J. (1995). Participation of reactive oxygen species in the lysophosphatidic acid-stimulated mitogen-activated protein kinase kinase activation pathway. *J. Biol. Chem.* **270**: 28 499–28 502.
129. Whisler, R. L., Goyette, M. A., Grants, I. S. and Newhouse, Y. G. (1995). Sublethal levels of oxidant stress stimulate multiple serine/threonine kinases and suppress protein phosphatases in Jurkat T cells. *Arch. Biochem. Biophys.* **319**: 23–35.
130. Mendelson, K. G., Contois, L. R., Tevosian, S. G., Davis, R. J. and Paulson, K. E. (1996). Independent regulation of JNK/p38 mitogen-activated protein

kinases by metabolic oxidative stress in the liver. *Proc. Natl. Acad. Sci. USA* **93**: 12 908–12 913.

131. Aikawa, R., Komura, I., Yamazaki, I., Zou, Y., Kudoh, S., Tanaka, M., Shiojima, I., Hiroi, Y. and Yazaki, Y. (1997). Oxidative stress activates extracellular signal-regulated kinases through Src and Ras in cultured cardiac myocytes of neonatal rats. *J. Clin. Invest.* **100**: 1813–1821.

132. Peus, D., Vasa, R. A., Beyerle, A., Meves, A., Krautmacher, C. and Pittelkow, M. R. (1999). UVB activates ERK1/2 and p38 signaling pathways via reactive oxygen species in cultured keratinocytes. *J. Invest. Derm.* **112**: 751–756.

133. Lee, L., Irani, K. and Finkel, T. (1998). Bcl-2 regulates nonapoptotic signal transduction: inhibition of *c-Jun* N-termianl kinase (JNK) activation by IL-1 beta and hydrogen peroxide. *Mol. Genet. Metabol.* **64**: 19–24.

134. Roberts, M. L. and Cowsert, L. M. (1998). Interleukin-1 beta and reactive oxygen species mediate activation of *c-Jun* NH2-terminal kinases, in human epithelial cells, by tow independent pathways. *Biochem. Biophys. Res. Commun.* **251**: 166–172.

135. Simon, A. R., Rai, U., Fanburg, B. L. and Cochran, B. H. (1998). Activation of the JAK-STAT pathway by reactive oxygen species. *Am. J. Physiol.* **275**: C1640–C1652.

136. Bae, G.-U., Seo, D.-W., Kwon, H.-K., Lee, H. Y., Hong, S., Lee, Z.-W., Ha, K.-S., Lee, H.-W. and Han, J.-W. (1999). Hydrogen peroxide activates p70S6k signaling pathway. *J. Biol. Chem.* **274**: 32 596–32 602.

137. Abe, J., Okuda, M., Huang, Q., Yoshizumi, M. and Berk, B. C. (2000). Reactive oxygen species activate p90 ribosomal S6 kinase via Fyn and Ras. *J. Biol. Chem.* **275**: 1739–1748.

138. Bhat, N. R. and Zhang, P. (1999). Hydrogen peroxide activation of mutiple mitogen-activated protein kinases in an oligodendrocyte cell line: role of extracellular signal-regulated kinase in hydrogen peroxide-induced cell death. *J. Neurochem.* **72**: 112–119.

139. Zent, R., Ailenberg, M., Downey, G. P. and Silverman, M. (1999). ROS stimulate reorganization of mesangial cell-collagen gels by tyrosine kinase signaling. *Am. J. Physiol.* **276**: F278–F287.

140. Gonzalez-Rubio, M., Voit, S., Rodriguez-Puyol, D., Weber, M. and Marx, M. (1996). Oxidative stress induces tyrosine phosphorylation of PDGF alpha and beta receptors and pp60c-Src in mesangial cells. *Kidney Int.* **50**: 134–173.

141. Peus, D., Meves, A., Vasa, R. A., Beyerle, A., O'Brien, R, and Pittwlkow, M. R. (1999). H2O2 is required for UVB-induced EGF receptor and down-stream signaling pathway activation. *Free Radic. Biol. Med.* **27**: 1197–1202.

142. Schreck, R., Rieber, P. and Baeuerle, P. A. (1991). Reactive oxygen inter-mediates as apparently widely used messengers in the activation of the NF-kappa B transcription factor and HIV-1. *EMBO J.* **10**: 2247–2258.

143. Schmidt, K. N., Amstad, P., Cerutti, P. and Bauerle, P. A. (1995). The roles of hydrogen peroxide and superoxide as messengers in the activation of transcription factor NF-kappa B. *Chem. Biol.* **2**: 12

144. Schmidt, K. N., Amstad, P., Cerutti, P. and Bauerle, P. A. (1996). Identification of hydrogen peroxide as the relevant messenger in the activation pathway of transcription factor NF-κB. *In* "Biological Reactive intermediates V" (R. Snyder, Ed.), pp. 63–68, Plenum Press, New York.

145. Liehr, J. G., Ulubelen, A. A. and Strobel, H. W. (1986). Cytochrome P450 mediated redox cycling of estrogens. *J. Biol. Chem.* **261**: 16 865–16 870.

146. Roy, D., Bernhardt, A., Strobel, H. W. and Liehr, J. G. (1992). Catalysis of the oxidation of steroid and stilbene estrogens to estrogen quinone metabolites by the beta-naphthoflavone-inducible cytochrome P450 IA family. *Arch. Biochem. Biophys.* **296**: 450–456.

147. Hammond, D. K., Zhu, B. T., Wang, M. Y., Ricci, M. J. and Liehr, J. G. (1997). Cytochrome P450 metabolism of estradiol in hamster liver and kidney. *Toxicol. Appl. Pharmacol.* **145**: 54–60.

148. Sarabia, S. F., Zhu, B. T., Kurosawa, T., Tohma, M. and Liehr, J. G. (1997). Mechanism of cytochrome P450-catalyzed aromatic hydroxylation of estrogens. *Chem. Res. Toxicol.* **10**: 767–771.

149. Liehr, J. G. and Roy, D. (1990). Free radical generation by redox cycling of estrogens. *Free Radic. Biol. Med.* **8**: 415–423.

150. Wyllie, S. and Liehr, J. G. (1998). Enhancement of estrogen-induced renal tumorigenesis in hamsters by dietary iron. *Carcinogenesis* **19**: 1285–1290.

151. Roy, D., Floyd, R. A. and Liehr, J. G. (1991). Elevated 8-hydroxydeoxyguanosine levels in DNA of diethylstilbestrol-treated Syrian hamsters: covalent DNA damage by free radicals generated by redox cycling of diethylstilbestrol. *Cancer Res.* **51**: 3882–3885.

152. Liehr, J. G. and Ricci, M. J. (1996). 4-hydroxylation of estrogens as marker of human mammary tumors. *Proc. Natl. Acad. Sci. USA* **93**: 3294–3296.

153. Liehr, J. G. (1997). Hormone-associated cancer: mechanistic similarities between human breast cancer and estrogen-induced kidney carcinogenesis in hamsters. *Env. Health Perspect.* **105(Suppl. 3)**: 565–569.

154. Taffe, B. G., Takahashi, N., Kensler, T. W. and Mason, R. P. (1987). Generation of free radicals from organic hydroperoxide tumor promoters in isolated mouse keratinocytes. Formation of alkyl and alkoxyl radicals from tert-butyl hydroperoxide and cumene hydroperoxide. *J. Biol. Chem.* **262**: 12 143–12 149.

155. Taffe, B. G., Zweier, J. L., Pannell, L. K. and Kensler, T. W. (1989). Generation of reactive intermediates from the tumor promoter butylated hydroxytoluene hydroperoxide in isolated murine keratinocytes or by hematin. *Carcinogenesis* **10**: 1261–1268.

156. Swauger, J. E., Dolan, P. M., Zweier, J. L., Kuppusamy, P. and Kensler, T. W. (1991). Role of the benzoyloxyl radical in DNA damage mediated by benzoyl peroxide. *Chem. Res. Toxicol.* **4**: 223–228.

157. Kensler, T. W., Guyton, K. Z., Egner, P. A., McCarthy, T., Lesko, S. A. and Akman, S. (1995). Role of reactive intermediates in tumor promotion and progression. *Prog. Clin. Biol. Res.* **391**: 103–116.

158. Troll, W., Frenkel, K. and Teebor, G. (1983). Free oxygen radicals: necessary contributors to tumor promotion and cocarcinogenesis. *Int Symp. Princess. Takamatsu. Cancer Res. Fund.* **14**: 207–218.

159. Marnett, L. (1987). Peroxyl free radicals: potential mediators of tumour initiation and promotion. *Carcinogenesis* **8**: 1365–1373.

160. O'Brien, P. J. (1988). Radical formation during the peroxidase catalyzed metabolism of carcinogens and xenotiotics: the reactivity of these radicals with GSH, DNA, and unsaturated lipid. *Free Radic. Biol. Med.* **4**: 169–183.

161. Perchellet, J.-P. and Perchellet, E. M. (1989). Antioxidants and multistage carcinogenesis in mouse skin. *Free Radic. Biol. Med.* **7**: 377–408.

162. Wei, H. and Frenkel, K. (1992). Suppression of tumor promoter-induced oxidative events and DNA damage in vivo by Sarcophytol A: a possible mechanism of antipromotion. *Cancer Res.* **52**: 2298–2303.

163. Wei, H. and Frenkel, K. (1993). Relationship of oxidative events and DNA oxidation in SENCAR mice to in vivo promoting activity of phorbol ester-type tumor promoters. *Carcinogenesis* **14**: 1195–1201.

164. Wei, H., Wei, L., Frenkel, K., Bowen, R. and Barnes, S. (1993). Inhibition of tumor promoter-induced hydrogen peroxide formation in vitro and in vivo by genistein. *Nutri. Cancer* **20**: 1–12.

165. Muehlematter, D., Larsson, R. and Cerutti, P. (1988). Active oxygen induced DNA strand breakage and poly ADP-ribosylation in promotable and non-promotable JB6 mouse epidermal cells. *Carcinogenesis* **9**: 239–245.

166. Akman, S. A., Doroshow, J. H. and Kensler, T. W. (1992). Copper-dependent site-specific mutagenesis by benzoyl peroxide in the supF gene of the mutation reporter plasmid pS189. *Carcinogenesis* **13**: 1783–1787.

167. Wei, H. and Frenkel, K. (1991). In vivo formation of oxidized DNA bases in tumor promoter-treated mouse skin. *Cancer Res.* **51**: 4443–4449.

168. Wei, H. (1992). Relationship of oxidative events and DNA damage to in vivo promoting activity. *Diss. Abstr. Int.* **B53**: 1272–1272.

169. Wei, L., Wei, H. and Frenkel, K. (1993). Sensitivity to tumor promotion of SENCAR and C57BL/6J mice correlates with oxidative events and DNA damage. *Carcinogenesis* **14**: 841–847.

170. Reiners, J. J., Pence, B. C., Barcus, M. C. and Cantu, A. R. (1987). 12-O-tetradecanoylphorbol-13-acetate-dependent induction of xanthine dehydrogenase and conversion to xanthine oxidase in murine epidermis. *Cancer Res.* **47**: 1775–1779.

171. Reiners, J. J., Thai, G., Rupp, T. and Cantu, A. R. (1991). Assessment of the antioxidant/prooxidant status of murine skin following topical treatment with 12-O-tetradecanoylphorbol-13-acetate and throughout the ontogeny of

skin cancer. Part I: quantitation of superoxide sidmutase, catalase, glutathione peroxidase and xanthine oxidase. *Carcinogenesis* **12**: 2337–2343.

172. Huang, M. T., Ma, W., Yen, P., Xie, J. G., Han, J., Frenkel, K., Grunberger, D. and Conney, A. H. (1997). Inhibitory effects of topical application of low doses of curcumin on 12-O-tetradecanoylphorbol-13-acetate-induced tumor promotion and oxidized DNA bases in mouse epidermis. *Carcinogenesis* **18**: 83–88.

173. Wei, H., Bowen, R., Zhang, X. and Lebwohl, M. (1998). Isoflavone genistein inhibits the initiation and promotion of two-stage skin carcinogenesis in mice. *Carcinogenesis* **19**: 1509–1514.

174. Olinski, R., Zastawny, T. H., Budzon, J., Skokowski, J., Zegarski, W. and Dizdaroglu, M. (1992). DNA base modifications in chromatin of human cancerous tissues. *FEBS Lett.* **309**: 193–198.

175. Jaruga, P., Zastawny, T. H., Skokowski, J., Dizdaroglu, M. and Olinski, R. (1994). Oxidative DNA base damage and antioxidant enzyme activities in human lung cancer. *FEBS Lett.* **341**: 59–64.

176. Okamoto, K., Tokoyuni, S., Uchida, K., Ogawa, O., Takenewa, J., Kakehi, Y., Kinoshita, H., Hattori, Y., Hiai, H. and Yoshida, O. (1994). Formation of 8-hydroxy-2'-deoxyguanosine and 4-hydroxy-2-nonenal-modified proteins in human renal-cell carcinoma. *Int. J. Cancer* **58**: 825–829.

177. Malins, D. C. and Haimanot, R. (1991). Major alterations in the nucleotide structure of DNA in cancer of the female breast. *Cancer Res.* **51**: 5430–5432.

178. Malins, D. C., Holmes, E. H., Polissar, N. L. and Gunselman, S. J. (1993). The etiology of breast cancer: characteristic alterations in hydroxyl radical-induced DNA base lesions during oncogenesis with potential for evaluating incidence risk. *Cancer* **71**: 3036–3043.

179. Steenken, S. (1989). Purine bases, nucleosides, and nucleotides: aqueous solution redox chemistry and transformation reactions of their radical cations and e- and OH adducts. *Chem. Rev.* **89**: 503–520.

180. Malins, D. C., Polissar, N. L. and Gunselman, S. J. (1996). Progression of human breast cancers to the metastatic state is linked to hydroxyl radical-induced DNA damage. *Proc. Natl. Acad. Sci. USA* **93**: 2557–2563.

181. Musarrat, J., Arezina-Wilson, J. and Wani, A. A. (1996). Prognostic and aetiological relevance of 8-hydroxyguanosine in human breast carcinogenesis. *Eur. J. Cancer* **32A**: 1209–1214.

182. Nagashima, M., Tsuda, H., Takenoshita, S., Nagamachi, Y., Hirohashi, S., Yokota, J. and Kasai, H. (1995). 8-hydroxydeoxyguanosine levels in DNA of human breast cancers are not significantly different from those of non-cancerous breast tissues by the HPCL-ECD method. *Cancer Lett.* **90**: 157–162.

183. Shimoda, R., Nagashima, M., Sakamoto, M., Yamaguchi, N., Hirohashi, S., Yokota, J. and Kasai, H. (1994). Increased formation of oxidative DNA damage, 8-hydroxydeoxyguanosine, in human livers with chronic hepatitis. *Cancer Res.* **54**: 3171–3172.

184. Djuric, Z., Simon, M. S., Luongo, D. A., LoRusso, P. M. and Martino, S. (1994). Levels of 5-hydroxymethyl-2′-deoxyuridine in blood DNA as a marker of breast cancer risk. *Proc. Am. Assoc. Cancer Res.* **35**: 286–286.

185. Bianchini, F., Donato, F., Faure, H., Ravanat, J.-L., Hall, J. and Cadet, J. (1998). Urinary excretion of 5-(hydroxymethyl) uracil in healthy volunteers: effect of active and passive tobacco smoke. *Int. J. Cancer* **77**: 40–46.

186. Ravanat, J.-L., Guicherrd, P., Tuce, Z. and Cadet, J. (1999). Simultaneous determination of five oxidative DNA lesions in human urine. *Chem. Res. Toxicol.* **12**: 802–808.

187. Cathcart, R., Schwiers, E., Saul, R. L. and Ames, B. N. (1984). Thymine glycol and thymidine glycol in human and rat urine: a possible assay for oxidative DNA damage. *Proc. Natl. Acad. Sci. USA* **81**: 5633–5637.

188. Stillwell, W. G., Xu, H.-X., Adkins, J. A., Wishnok, J. S. and Tannenbaum, S. R. (1989). Analysis of methylated and oxidized purines in urine by capillary gas chromatography-mass spectrometry. *Chem. Res. Toxicol.* **2**: 94–99.

189. Halliwell, B. (1998). Can oxidative DNA damage be used as a biomarker of cancer risk in humans? Problems, resolutions and preliminary results from nutritional supplementation studies. *Free Radic. Res.* **29**: 469–486.

190. Tagesson, C., Kallberg, M., Klintenberg, C. and Starkhammar, H. (1995). Determination of urinary 8-hydroxydeoxyguanosine by automated coupled-column high performance liquid chromatography: a powerful technique for assaying in vivo oxidative DNA damage in cancer patients. *Eur. J. Cancer* **31A**: 934–940.

191. Loft, S., Velthuis-te Wierik, E. J., van den Berg, H. and Poulsen, H. E. (1995). Energy restriction and oxidative DNA damage in humans. *Cancer Epi. Bio. Prev.* **4**: 515–519.

192. Velthuis-te Wierik, E. J., van Leeuwen, R. E., Hendriks, H. F., Verhagen, H., Loft, S., Poulsen, H. E. and van den Berg, H. (1995). Short-term moderate energy restriction does not affect indicators of oxidative stress and genotoxicity in humans. *J. Nutri.* **125**: 2631–2639.

193. Loft, S. and Poulsen, H. E. (1998). Estimation of oxidative DNA damage in man from urinary excretion of repair products. *Acta Biochimica Polonica* **45**: 133–144.

194. Poulsen, H. E., Loft, S., Prieme, H., Vistisen, K., Lykkesfeldt, J., Nyyssonen, K. and Salonen, J. T. (1998). Oxidative DNA damage in vivo: relationship to age, plasma antioxidants, drug metabolism, glutathione-S-transferase activity and urinary creatinine excretion. *Free Radic. Res.* **29**: 565–571.

195. Verhagen, H., Poulsen, H. E., Loft, S., van Poppel, G., Willems, M. I. and van Bladeren, P. J. (1995). Reduction of oxidative DNA-damage in humans by brussels sprouts. *Carcinogenesis* **16**: 970.

196. Wilson, V. L., Taffe, B. G., Shields, P. G., Powey, A. C. and Harris, C. C. (1993). Detection and quantification of 8-hydroxydeoxyguanosine adducts

in peripheral blood of people exposed to ionizing radiation. *Env. Health Perspect.* **99**: 261–263.

197. Povey, A. C., Wilson, V. L., Weston, A., Doan, V. T., Wood, M. L., Essigmann, J. M. and Shields, P. G. (1993). Detection of oxidative damage by 32P-postlabelling: 8-hydroxydeoxyguanosine as a marker of exposure. *IARC. Sci. Publ.*, 105–114.

198. Kiyosawa, H., Suko, M., Okudaira, H., Murata, T., Miyamoto, T., Chung, M.-H., Kasai, H. and Nishimura, S. (1990). The effect of smoking on 8-hydroxydeoxyguanosine, one of the oxidative DNA damages in human peripheral leukocytes. *Free Radic. Res. Commun.* **1**: 50.

199. Asami, S., Manabe, H., Miyake, J., Tsurodome, Y., Hirano, T., Yamaguchi, H., Itoh, H. and Kasai, H. (1997). Cigarette smoking induces an increase in oxidative DNA damage, 8-hydroxydeoxyguanosine, in a central site of the human lung. *Carcinogenesis* **18**: 1763–1766.

200. Frenkel, K., Karkoszka, J., Glassman, T., Dubin, N., Toniolo, P., Taioli, E., Mooney, L. A. and Kato, I. (1998). Serum autoantibodies recognizing 5-hydroxymethyl-2'-deoxyuridine, an oxidized DNA base, as biomarkers of cancer risk in women. *Cancer Epi. Bio. Prev.* **7**: 49-57.

201. Ashok, B. T. and Ali, R. (1998). Binding of human anti-DNA autoantibodies to reactive oxygne species modified-DNA and probing oxidative DNA damage in cancer using monoclonal antibody. *Int. J. Cancer* **78**: 404–409.

202. Cutler, R. G. (1991). Human longevity and aging: possible role of reactive oxygen species. *Ann. NY Acad. Sci.* **621**: 1–28.

203. Fraga, C., Shigenaga, M. K., Park, J.-W., Degan, P. and Ames, B. N. (1990). Oxidative damage to DNA during aging: 8-hydroxy-2'-deoxyguanosine in rat organ DNA and urine. *Proc. Natl. Acad. Sci. USA* **87**: 4533–4537.

204. Hayakawa, M., Hattori, K., Sugiyama, S. and Ozawa, T. (1992). Age-associated oxygen damage and mutations in mitochondrial DNA in human hearts. *Biochem. Biophys. Res. Commun.* **189**: 979–985.

205. Wallace, D. C., MacGarvey, U., Kauffman, A. E., Koontx, D., Shoffner, J. M. and Beal, M. F. (1993). Oxidative damage to mitochondrial DNA shows marked age-dependent increases in human brain. *Ann. Neurol.* **34**: 609–616.

206. Bohr, U. A., Anson, R. M., Mazur, S. and Dianov, G. L. (1998). Oxidative DNA damage processing and changes with aging. *Toxicol. Lett.* **102**: 47–52.

207. Steinmetz, K. A. and Potter, J. D. (1991). Vegetables, fruit, and cancer. *Cancer Causes Contr.* **2**: 325–357.

208. Block, G., Patterson, B. and Subar, A. (1992). Fruit, vegetables, and cancer prevention: a review of the epidemiological evidence. *Nutri. Cancer* **18**: 1–29.

209. Lu, M., Guo, Q. and Kallenbach, N. R. (1992). Structure and stability of sodium and potassium complexes of dT4G4 and DT4G4T. *Biochemistry* **31**: 2455–2459.

210. Potter, J. D. (1997). Cancer prevention: epidemiology and experiment. *Cancer Lett.* **114**: 7–9.

211. Longnecker, M. P., Martin-Moreno, J.-M., Knekt, P., Nomura, A. M. Y., Schober, S. E., Stahelin, H. B., Wald, N., Gey, K. F. and Willet, W. C. (1992). Serum alpha-tocopherol concentration in relation to subsequent colorectal cancer: pooled data from five cohorts. *J. Natl. Cancer Inst.* **84**: 430–435.

212. Blot, W. J., Li, J. Y., Taylor, P. R., Guo, W., Dawsey, S. and Wang, G. Q. (1993). Nutrition intervention trials in Linxian, China: supplementation with specific vitamin/mineral combinations, cancer incidence, and disease-specific mortality in the general population. *J. Natl. Cancer Inst.* **85**: 1483–1492.

213. (1994). The alpha tocopherol, b.c.c.p.s.g. The effect of vitamin E and beta carotene on the incidence of lung cancer and other cancers in male smokers. *New England J. Med.* **330**: 1029–1035.

214. Hennekens, C. H., Buring, J. E., Mansson, J. E., Stampfer, M. R., Ross, A. B., Cook, N. R., Belanger, C., LaMotte, F., Gaziano, J. M., Ridker, P. M., Willet, W. C. and Peto, R. (1996). Lack of effect of long-term supplementaion with beta carotene on the incidence of malignant neoplasms and cardiovascular disease. *New England J. Med.* **334**: 1145–1149.

215. Omenn, G. S., Goodman, G. E., Thornquist, M. D., Balmes, J., Cullen, M. R., Glass, A., Keogh, J. P., Meyskens, F. L., Valanis, B., Williams, J. H., Barnhart, S. and Hammar, S. (1996). Effects of a combination of beta carotene and vitamin A on lung cancer and cardiovascular disease. *New England J. Med.* **334**: 1150–1155.

216. Bostick, R. M., Potter, J. D., McKenzie, D. R., Sellers, T. A., Kushi, L. H., Steinmetz, K. A. and Folsom, A. R. (1993). Reduced risk of colon cancer with high intake of vitamin E. The Iowa Women's Health Study. *Cancer Res.* **53**: 4230–4237.

217. Greenberg, E. R., Baron, J. A., Tosteson, T. D., Freeman, D. H., Jr., Beck, G. J., Bond, J. H., Colacchio, T. A., Coller, J. A., Frankl, H. D. and Haile, R. W. (1994). A clinical trial of antioxidant vitamins to prevent colorectal adenoma. Polyp Prevention Study Group. *New England J. Med.* **331**: 141–147.

218. McKeown-Eyssen, G., Holloway, C., Jazmaji, V., Bright-See, E., Dion, P. and Bruce, W. R. (1988). A randomized trial of vitamins C and E in the prevention of recurrence of colorectal polyps. *Cancer Res.* **48**: 4701–4705.

219. Kimmick, G. G., Bell, R. A. and Bostick, R. M. (1997). Vitamin E and breast cancer: a review. *Nutri. Cancer* **27**: 109–117.

220. Kushi, L. H., Fee, R. M., Sellers, T. A., Zheng, W. and Folsom, A. R. (1996). Intake of vitamins A, C, and E and postmenopausal breast cancer. The Iowa Women's Health Study. *Am. J. Epidemiol.* **144**: 165–174.

221. Verhoeven, D. T., Assen, N. and Goldhohm, R. A. (1997). Vitamins C and E, retinol, beta-carotene and dietary fibre in relation to breast cancer risk: a prospective cohort study. *Br. J. Cancer* **75**: 149–155.

222. Kayden, H. J. (1983). Tocopherol content of adipose tissue from vitamin E deficient humang. *In* "Biology of vitamin E" (R. Porter, and J. Whelan, Eds.), pp. 70–91, The Pittman Press, London.

223. Traber, M. G. and Kayden, H. J. (1987). Tocopherol distribution and intracellular localization in human adipose tissue. *Am. J. Clin. Nutri.* **46**: 488–495.

224. Handelman, G. J., Epstein, W. L., Peerson, J., Spiegelman, D., Machlin, L. J. and Dratz, E. A. (1994). Human adipose alpha tocopherol and gamma tocopherol kinetics during and after 1 year of alpha tocopherol supplementation. *Am. J. Clin. Nutri.* **59**: 1025–1032.

225. Chajes, V., Lhuillery, C., Sattler, W., Kostner, G. M. and Bougnoux, P. (1996). Alpha tocopherol and hydroperoxide content in breast adipose tissue from patients with breast tumors. *Int. J. Cancer* **67**: 170–175.

226. Zhu, Z., Parviainen, M., Mannisto, S., Pietinen, P., Eskelinen, M., Syrjanen, K. and Uusitupa, M. (1996). Vitamin E concentration in breast adipose tissue of breast cancer patients. *Cancer Causes Contr.* **7**: 591–595.

227. Van't Veer, P., Strain, J. J., Fernandez-Crehuet, J., Martin, B. C., Thamm, M., Kardinaal, A. F., Kohlmeier, L., Huttinen, J. K., Martin-Moreno, J.-M. and Kok, F. J. (1996). Tissue antioxidants and postmenopausal breast cancer. The European Community Mutlicentre Study on Antioxidants, Myocardial Infarction, and Cancer of the Breast (EURAMIC). *Cancer Epi. Bio. Prev.* **5**: 441–447.

228. Duthie, S. J., Ma, A., Ross, M. A. and Collins, A. R. (1996). Antiixodant supplementation decreases oxidative DNA damage in human lymphocytes. *Cancer Res.* **56**: 1291–1295.

229. Fraga, C., Motchnik, P. A., Shigenaga, M. K., Helbock, H. J., Jacob, R. A. and Ames, B. N. (1991). Ascorbic acid protects against endogenous oxidative DNA damage in human sperm. *Proc. Natl. Acad. Sci. USA* **88**: 11 003–11 006.

230. Stevens, R. G., Jones, D. Y., Micozzi, M. S. and Taylor, P. R. (1988). Body iron stores and the risk of cancer. *New England J. Med.* **319**: 1047–1052.

231. Freudenheim, J. L., Graham, S., Marshall, J. R., Haughey, B. P. and Wilkinson, G. (1990). A case-control study of diet and rectal cancer in Western New York. *Am. J. Epidemiol.* **131**: 612–624.

232. Hoff, G., Moen, I. E, Trygg, K., Froelich, W., Sauar, J., Vatn, M., Gjone, E. and Larsen, S. (1986). Epidemiology of polyps in the rectum and sigmoid colon. *Scand. J. Gastroent.* **21**: 199–204.

233. Macquart-Moulin, G., Riboli, E., Cornee, J., Kaaks, R. and Berthezene, P. (1987). Colorectal polyps and diet: a case control study in Marseilles. *Int. J. Cancer* **40**: 179–188.

234. Rosenqvist, H., Ohrling, H., Lagercrantz, R. and Edling, N. (1959). Ulcerative colitis and carcinoma coli. *Lancet* i: 906–908.

235. MacDougall, I. P. M. (1964). The cancer risk in ulcerative colitis. *Lancet* ii: 655–658.

236. Devroede, G. J., Taylor, W. F., Sauer, W. G., Jackman, R. J. and Stickler, G. B. (1971). Cancer risk and life expectancy of children with ulcerative colitis. *New England J. Med.* **285**: 17–17.

237. Weedon, D. D., Shorter, R. G., Illstrup, D. M., Huizenga, K. A. and Taylor, W. F. (1973). Crohn's disease and cancer. *New England J. Med.* **289**: 1099–1103.
238. Ekbom, A., Helmich, O., Zack, M. and Adami, H.-O. (1990). Ulcerative colitis and colorectal cancer: a population-based study. *New England J. Med.* **323**: 1228–1233.
239. Alexander, G. A. and Brawley, O. W. (2000). Association of *Helicobacter pylori* infection with gastric cancer. *Military Med.* **165**: 21–28.
240. Anonymous. (1994). *Schistosomes, Liver Flukes and Helicobacter pylori.* International Agency for Research on Cancer, Lyon.

Chapter 54

Oxidative Stress and Skin Cancer

Arthur K. Balin* and R.G. Allen

Arthur K. Balin and **R.G. Allen** • Sally Balin Medical Center for Dermatology, Cosmetic Surgery and Longevity Medicine, 110 Chesley Drive, Media, PA 19063

*Corresponding Author.
Tel: (610) 565-3300, E-mail: akbalin@aol.com

1. Introduction

The skin has the largest surface area of any organ and is the most vulnerable to environmental insults. The incessant bombardment of this organ by environmental challenges contributes either directly or indirectly to a variety of pathological conditions. Of particular importance are the effects of exposure to sunlight, which have been linked to increases in the rate of skin aging and in the incidence of certain epidermal malignancies. Although still a subject of some controversy, the role of oxidation in skin aging is probably better understood than its role in the formation of tumors. In at least some cases, oxidation may be directly involved in the formation of sequelae that subsequently produce skin tumors, but other effects of cellular oxidant/antioxidant balance also appear to play a significant role in tumor formation and progression. In the following discussion, we examine the roles of oxidants and antioxidants in the formation of several types of skin tumors including basal cell carcinoma (BCC), squamous cell carcinoma (SCC) and melanoma. The contribution of skin malignancies to the overall cancer rate is significant; in the United States alone more than a million new cases of skin malignancies are reported annually which is comparable to the incidence of all other cancers combined.

2. Light, Oxidation and DNA Damage

Skin is exposed to a broad range of wavelengths in the ultraviolet (UV) and visible light spectrum; however, it is those wavelengths in the UV portion of the spectrum that appear to be of principle importance to carcinogenesis. UV light is designated UVA (320–400 nm), UVB (290–320 nm) and UVC (200–290 nm). UVC can form oxidants by photolysis of H_2O; UVA and UVB generate oxidants through photodynamic action such as dissociation of H_2O_2. All three forms of UV can photoactivate light sensitive molecules such as riboflavin.[1] A variety of oxidative reactions are stimulated by exposure to UV light; however, it is unclear that these or subsequent reactions are principally responsible for the correlation between skin cancer and exposure to UV radiation. Instead, direct absorption of energy by DNA and the formation of thymidine dimers appear to be more significant to skin carcinogenesis. Spectral analysis of UV absorption by DNA reveals that relatively little energy is actually absorbed in the UVA range. The amount of energy absorbed increases progressively throughout the UVB-UVC ranges. Due to the earth's ozone layer, relatively low levels of UVC reach the surface and it is thus UVB exposure that has been implicated in most types of skin cancers.[2] Of course, oxidative injury can also produce mutation and increase the probability of damage to critical genes; however, the role of oxidation in skin cancer formation is probably more strongly correlated with survival of transformed cells than with causing the initial mutations that lead to these diseases (see discussion below).

3. Basal Cell Carcinoma (BCC)

BCC is the most common human cancer. Nearly all basal cell epitheliomas arise on sun-exposed bodily areas. This type of lesion is found most commonly on the head and neck of men. The frequency of BCC incidence correlates with both geographical latitude and the relative level of sun exposure. Other predisposing factors include: exposure to X-rays, burn scars and xeroderma pigmentosum. Basal cell nevus syndrome (BCNS) is an inherited disorder that results in a high frequency of basal cell carcinomas.

4. Genetics

It has been shown that mutations in human homolog of the *Drosophila* patched (*ptc*) gene is inherited by BCNS patients,[3] furthermore somatic mutations in ptc are also found in many sporadic BCC.[4-7] Patched inhibits the sonic hedgehog signal transduction pathway, which is chronically active in the absence of *ptc*. Whether or not mutations are observed, the *ptc* gene is nearly always overexpressed in BCC.[8] The *ptc* gene encodes a transmembrane protein that represses the smoothened (Smo) signal transduction protein. In mammals Smo is released (activated) by interaction of sonic hedgehog (*Shh*) with *Ptc*;[9,10] it activates transcription of hedgehog targets through the action of the transcription factor Gli.[11] Presently, no direct relationship between cellular oxidation levels and this pathway is known to exist.

5. Oxidative Stress and Apoptosis

Aberrant activation of signal transduction pathways frequently leads to apoptosis. Apoptotic cell death is a natural defense against cancer, which simply kills transformed cells before they can multiply. It is stimulated through release of mitochondrial cytochrome *c*, which results in activation of a death protease (caspase-3) and increased free radical generation due to uncoupled respiration.[12,13] The increased oxidant generation stimulates a mitochondrial permeability transition that causes further release of cytochrome *c* from mitochondria and activates a second apoptosis inducing factor (AIF[13]). The proapoptotic protein bax is elevated in BCC as well as other skin cancers;[14] in absence of other changes this should limit growth. The fact that growth is not successfully limited may result partly from an elevation of *bcl-2*, which can block apoptosis. The anti-apoptotic activity of *bcl-2* stems partly from its ability to upregulate an antioxidant pathway in cells.[15] Bcl-2 also stabilizes mitochondrial membranes and blocks cytochrome *c* release[13] as well as preventing or delaying activation of *c-jun*, a member of the AP-1 pathway involved in apoptosis[16] that is also redox sensitive.[17]

The *bcl-2* family of proteins exhibits other redox-independent functions that may also be important to prevention of apoptosis; however, the magnitude of the effect of this protein on cellular antioxidant status should not be underestimated. In fact, overexpression of *bcl-2* is sufficient to block lipid peroxidation in some types of cells following experimental treatments with oxidants.[15] *Bcl-2* tends to be greatly elevated in BCC.[14, 18–20] Interestingly, *bcl-2* expression also rises in skin cells overexpressing *shh*.[21] Although the increased expression of *bcl-2* does not directly lead to BCC it does promote the survival of tumor cells by preventing apoptosis. Mutations in the anti-tumor *p53* gene have also been reported to occur frequently in BCC and may also permit survival of tumor cells by preventing the normal stimulation of apoptosis by *p53*.

6. Actinic Keratoses (AK) and Squamous Cell Carcinoma (SCC)

Actinic keratoses are clones of anaplastic keratinocytes; they are confined to the epidermis and occur very commonly on sun-damaged areas of skin in elderly people. When left untreated they can invade through the basement membrane of the epidermal-dermal junction to become squamous cells carcinoma (SCC[22]). The prevalence of AK in the white population is increased in both sexes with age irrespective of sun exposure.[23] Nevertheless, male sex as well as sun exposure predisposed individuals to a greater number of actinic keratoses. Progression of AK to invasive squamous cell carcinoma occurs when the buds of atypical keratinocytes extend into the dermis and develop into detached nests of abnormal cells capable of autonomous growth. Aside from sun exposure, several other factors can also predispose individuals to SCC. These include exposure to X-rays, previous injuries such as burns or chronic leg ulcers and xeroderma pigmentosum. Bowen's disease is a form of SCC that occurs with higher frequency on covered (non-sun-exposed) areas of the skin. A strong correlation is known to exist between Bowen's SCC and ingestion of inorganic arsenic. The sources of the arsenic include contaminated water, insecticides and Fowler's solution (a drug that was commonly used to treat asthma[24]).

7. Treatment

A variety of effective treatments can be used for BCC and SCC including: excisional surgery, curettage and electrodessication, radiotherapy, intra-lesional interferon injections cryosurgery and Mohs micrographic surgery. Treatment must be tailored for each patient particularly in the elderly. In most cases surgical removal of cutaneous neoplasms provides the most direct and definitive treatment. Mohs micrographic surgery provides the highest cure rate and preserves the most normal tissue. In this technique, multiple thin horizontal layers of the cancer are removed. Each layer is examined microscopically for cancer cells. The surgeon continues to remove and examine layers until no cancer cells are found. This method ensures complete removal of the tumor.

8. p53

The precise cause of AK and SCC is unknown. A very high frequency of mutations in the anti-oncogene *p53* (> 90%) are frequently associated with both of these lesions suggesting that damage to this gene may result in very early stages of tumor development.[25-28] UV damaged skin that exhibits *p53* mutations also tends to exhibit telomerase activity,[29] an enzyme that can confer immortality in cells cultured *in vitro*.[30] Nevertheless, the primary effects of *p53* mutations are probably linked to failure to stimulate oxidative changes that lead to apoptosis.

When functioning normally p53 protein can induce apoptosis. Further analysis using a rapid screening method revealed that only 14 of 7202 transcripts were induced by p53 prior to the onset of apoptosis.[31] Most of the affected genes encoded proteins that could either generate or respond to oxidants. On the basis of these results it was suggested that p53 protein stimulates apoptosis through (i) transcriptional induction of redox-related genes, (ii) increased formation of oxidants, and (iii) oxidative degeneration of mitochondrial components. Defects in p53 might be expected to diminish apoptosis in tumor cells. In contrast to the observations made in BCC, a decrease or loss of bcl-2 has been reported both in AK[20] and SCC.[14] Aside from decreased p53 activity, other bcl-2 family proteins are elevated in some cases and may also impart resistance to apoptosis in SCC by limiting cellular oxidation.[32]

9. Melanoma

The incidence of skin cancer has been rising since the 1950s. Currently about 75% of skin cancer-associated deaths are caused by malignant melanoma.[33] Unlike the relatively strong correlation between sun exposure and the incidences of BCC and squamous cell carcinoma, the relationship between UV-exposure and melanoma is less clear. Only lentigo melanoma exhibits a relationship to sun exposure that is similar to non-melanoma skin cancers.[2] The occurrences of other types of melanoma are not as prominent in sun-exposed areas of the skin. Furthermore, the occurrence of melanomas is independent of occupation and cumulative lifetime sun exposure.[34] Nevertheless, UV light in combination with other environmental factors can greatly increase the probability of melanoma formation.[2] For example, in laboratory studies with mice, UV exposure does not appear to be sufficient (has not been reported) to induce melanoma; however, UV irradiation combined with tumor promoters or chemical carcinogens strongly stimulate melanoma formation.[35, 36]

10. Treatment

In early stages, surgical removal is the most effective treatment of melanoma. In later stages more aggressive, though much less effective, radiation and chemo-therapeutic treatments can be used.

11. Melanoma and Oxidative Metabolism

The precise cause of melanoma is presently unknown; however, there is a very strong link between this type of cancer and deletions from the long arm of chromosome 6.[37] Introduction of a normal copy of chromosome 6 into melanoma cells using microcell hybridization has been used to suppress the transformed phenotype of the cells.[38] The manganese-containing form of superoxide dismutase (SOD-2) is localized to 6q25 which is near a region believed to be important to controlling the transition from normal to a trans-formed phenotype in some cases.[39, 40] Loss of SOD-2 activity has frequently been associated with increased metastatic activity in melanoma.[41, 42] In fact, introduction of SOD-2 into cultures of human melanoma cells stimulates differentiation.[39] As in the case of BCC and SCC, the level of p53 mutations is elevated in melanoma cells. Interestingly, the suppression of the transformed phenotype by introduction of chromosome 6 into melanoma cells or by over-expression of antisense cyclin D is associated with an elevation of both SOD-2 and p53.[43, 44] Although it has been suggested that SOD-2 may act as a type of anti-oncogene and may lead to possible therapeutic applications, it has proven difficult to provide consistent support for this premise.[45] Not all melanoma cells appear to be affected by SOD-2.[46] Furthermore, in some cases, overexpression of SOD can actually promote tumor survival and increase resistance to other treatments such as IL-1, TNF and ionizing radiations.[47] In spite of this, under-standing the relationship between tumor formation and SOD-2 expression/activity could yield much greater insight into the mechanisms of melanoma formation and survival.

Finally it should be noted that the basal levels of oxidants are frequently altered in melanoma cells. In view of the lower SOD activity that is frequently associated with melanoma, it is not surprising that the levels of superoxide detected in these cells tends to be elevated.[48] Surprisingly, this increase may aid in melanoma survival. Hydrogen peroxide tends to be low in melanoma cells and can induce apoptosis if added to cells in culture; conversely, superoxide appears to have a protective effect and limits apoptosis in these cells.[49] Consistent with this observation was the report that the free radical generating herbicide paraquat conferred radiation resistance on melanoma cells.[50] At present it appears that a relationship does exist between SOD-2 activity, superoxide levels and the malignant phenotype of some melanomas; however, much more investigation will be required to define the relationship and the various signaling pathways it affects.

12. Conclusion

We have discussed the potential role of oxidants in several types of common skin tumors. Although light exposure stimulates production of oxidants; the effects of light on mutagenesis appear to frequently result from direct energy absorption by DNA rather than oxidant production. Hence, while oxidants can directly participate in causing mutations that lead to malignant transformations, changes in redox balance that promote antioxidation, prevent apoptosis and permit cellular survival may be a far more significant role for redox changes in the generation of cancerous cells. The principal pathways that are important for oxidant/antioxidant effects on tumor survival are the p53 and bcl-2 pathways. Both of these genes as well as other members of the pathways they regulate tend to be mutated or exhibit altered expression in cancer cells. Nevertheless, the relatively large changes in oxidative metabolism associated with some cutaneous neoplasms may affect many pathways that ultimately influence tumor growth and survival; however, the importance of redox changes to cellular regulatory pathways is only beginning to be understood and it is still too early to know all of the effects of redox changes in the formation and promotion of skin neoplasms.

References

1. Fuchs, J. (1992). *Oxidative Injury in Dermatopathology*, Springer-Verlag, New York, p. 360.
2. Kripke, M. L. (1999). Carcinogenesis: ultraviolet radiation. *In* "Dermatology in General Medicine" (I. M. Freedberg, A. Z. Eisen, K. Wolff, K. F. Austen, L. A. Goldsmith, S. I. Katz, and T. B. Fitzpatric, Eds.), pp. 465–472, McGraw Hill, New York,
3. Johnson, R. L., Rothman, A. L., Xie, J., Goodrich, L. V., Bare, J. W., Bonifas, J. M., Quinn, A. G., Myers, R. M., Cox, D. R., Epstein, E. H., Jr. and Scott, M. P. (1996). Human homolog of *patched*, a candidate gene for the basal cell nevus syndrome. *Science* **272**: 1668–1671.
4. Bodak, N., Queille, S., Avril, M. F., Bouadjar, B., Drougard, C., Sarasin, A. and Daya-Grosjean, L. (1999). High levels of patched gene mutations in basal-cell carcinomas from patients with xeroderma pigmentosum. *Proc. Natl. Acad. Sci. USA* **96**: 5117–5122.

5. Aszterbaum, M., Rothman, A., Johnson, R. L., Fisher, M., Xie, J., Bonifas, J. M., Zhang, X., Scott, M. P. and Epstein, E. H., Jr. (1998). Identification of mutations in the human PATCHED gene in sporadic basal cell carcinomas and in patients with the basal cell nevus syndrome. *J. Invest. Dermatol.* **110**: 885–888.

6. Aszterbaum, M., Epstein, J., Oro, A., Douglas, V., LeBoit, P. E., Scott, M. P. and Epstein, E. H., Jr. (1999). Ultraviolet and ionizing radiation enhance the growth of BCCs and trichoblastomas in patched heterozygous knockout mice. *Nature Med.* **5**: 1285–1291.

7. Gailani, M. R., Stahle-Backdahl, M., Leffell, D. J., Glynn, M., Zaphiropoulos, P. G., Pressman, C., Unden, A. B., Dean, M., Brash, D. E., Bale, A. E. and Toftgard, R. (1996). The role of the human homologue of *Drosophila* patched in sporadic basal cell carcinomas. *Nature Genet.* **14**: 78–81.

8. Unden, A. B., Zaphiropoulos, P. G., Bruce, K., Toftgard, R. and Stahle-Backdahl, M. (1997). Human patched (PTCH) mRNA is overexpressed consistently in tumor cells of both familial and sporadic basal cell carcinoma. *Cancer Res.* **57**: 2336–2340.

9. Stone, D. M., Hynes, M., Armanini, M., Swanson, T. A., Gu, Q., Johnson, R. L., Scott, M. P., Pennica, D., Goddard, A., Phillips, H., Noll, M., Hooper, J. E., de Sauvage, F. and Rosenthal, A. (1996). The tumour-suppressor gene patched encodes a candidate receptor for Sonic hedgehog. *Nature* **384**: 129–134.

10. Kallassy, M., Toftgard, R., Ueda, M., Nakazawa, K., Vorechovsky, I., Yamasaki, H. and Nakazawa, H. (1997). Patched (ptch)-associated preferential expression of smoothened (smoh) in human basal cell carcinoma of the skin. *Cancer Res.* **57**: 4731–4735.

11. Ghali, L., Wong, S. T., Green, J., Tidman, N. and Quinn, A. G. (1999). Gli-1 protein is expressed in basal cell carcinomas, outer root sheath keratinocytes and a subpopulation of mesenchymal cells in normal human skin. *J. Invest. Dermatol.* **113**: 595–599.

12. Cai, J. and Jones, D. P. (1998). Superoxide in apoptosis. Mitochondrial generation triggered by cytochrome *c* loss. *J. Biol. Chem.* **273**: 11 401–11 404.

13. Cai, J., Yang, J. and Jones, D. P. (1998). Mitochondrial control of apoptosis: the role of cytochrome *c*. *Biochem. Biophys. Acta* **1366**: 139–149.

14. Delehedde, M., Cho, S. H., Sarkiss, M., Brisbay, S., Davies, M., El-Naggar, A. K. and McDonnell, T. J. (1999). Altered expression of bcl-2 family member proteins in nonmelanoma skin cancer. *Cancer* **85**: 1514–1522.

15. Hockenbery, D. M., Oltvai, Z. N., Yin, X.-M., Milliman, C. L. and Korsmeyer, S. J. (1993). Bcl-2 functions in an antioxidant pathway to prevent apoptosis. *Cell* **75**: 241–251.

16. Bossy-Wetzel, E., Bakiri, L. and Yaniv, M. (1997). Induction of apoptosis by the transcription factor *c-Jun*. *EMBO J.* **16**: 1695–1708.

17. Allen, R. G. and Tresini, M. (2000). Oxidative stress and gene regulation. *Free Radic. Biol. Med.* **28**: 463–499.

18. Chang, C. H., Tsai, R. K., Chen, G. S., Yu, H. S. and Chai, C. Y. (1998). Expression of bcl-2, p53 and Ki-67 in arsenical skin cancers. *J. Cutan. Pathol.* **25**: 457–462.

19. Morales-Ducret, C. R., van de Rijn, M., LeBrun, D. P. and Smoller, B. R. (1995). Bcl-2 expression in primary malignancies of the skin. *Arch. Dermatol.* **131**: 909–912.

20. Nakagawa, K., Yamamura, K., Maeda, S. and Ichihashi, M. (1994). Bcl-2 expression in epidermal keratinocytic diseases. *Cancer* **74**: 1720–1724.

21. Fan, H., Oro, A. E., Scott, M. P. and Khavari, P. A. (1997). Induction of basal cell carcinoma features in transgenic human skin expressing Sonic hedgehog. *Nature Med.* **3**: 788–792.

22. Balin, A. K., Lin, A. N. and Pratt, L. (1988). Actinic keratoses. *J. Cutan. Aging Cosmetic Dermatol.* **1**: 77–86.

23. Engel, A., Johnson, M. L. and Haynes, S. G. (1988). Health effects of sunlight exposure in the United States. Results from the first National Health and Nutrition Examination Survey, 1971–1974. *Arch. Dermatol.* **124**: 72–79.

24. Braverman, I. (1981). *Signs of Systemic Disease*, pp. 67–89, W.B. Saunders, Philadelphia

25. Nomura, T., Nakajima, H., Hongyo, T., Taniguchi, E., Fukuda, K., Li, L. Y., Kurooka, M., Sutoh, K., Hande, P. M., Kawaguchi, T., Ueda, M. and Takatera, H. (1997). Induction of cancer, actinic keratosis, and specific p53 mutations by UVB light in human skin maintained in severe combined immunodeficient mice. *Cancer Res.* **57**: 2081–2084.

26. Brash, D. E., Ziegler, A., Jonason, A. S., Simon, J. A., Kunala, S. and Leffell, D. J. (1996). Sunlight and sunburn in human skin cancer: p53, apoptosis, and tumor promotion. *J. Invest. Dermatol. Symp. Proc.* **1**: 136–142.

27. Brash, D. E., Rudolph, J. A., Simon, J. A., Lin, A., McKenna, G. J., Baden, H. P., Halperin, A. J. and Ponten, J. (1991). A role for sunlight in skin cancer: UV-induced p53 mutations in squamous cell carcinoma. *Proc. Natl. Acad. Sci. USA* **88**: 10 124–10 128.

28. Reiss, M., Brash, D. E., Munoz-Antonia, T., Simon, J. A., Ziegler, A., Vellucci, V. F. and Zhou, Z. L. (1992). Status of the p53 tumor suppressor gene in human squamous carcinoma cell lines. *Oncol. Res.* **4**: 349–357.

29. Ueda, M., Ouhtit, A., Bito, T., Nakazawa, K., Lubbe, J., Ichihashi, M., Yamasaki, H. and Nakazawa, H. (1997). Evidence for UV-associated activation of telomerase in human skin. *Cancer Res.* **57**: 370–374.

30. Bodnar, A. G., Ouellette, M., Frolkis, M., Holt, S. E., Chiu, C.-P., Morin, G. B., Harley, C. B., Shay, J. W., Lichtsteiner, S. and Wright, W. E. (1998). Extension of life-span by introduction of telomerase into normal human cells. *Science* **279**: 349–352.

31. Polyak, K., Xia, Y., Zweier, J. L., Kinzler, K. W. and Vogelstein, B. (1997). A model for p53-induced apoptosis. *Nature* **389**: 300–305.
32. Kojima, H., Endo, K., Moriyama, H., Tanaka, Y., Alnemri, E. S., Slapak, C. A., Teicher, B., Kufe, D. and Datta, R. (1998). Abrogation of mitochondrial cytochrome *c* release and caspase-3 activation in acquired multidrug resistance. *J. Biol. Chem.* **273**: 16 647–16 650.
33. Greulich, K. M., Utikal, J., Peter, R. U. and Krahn, G. (2000). *c-myc* and nodular malignant melanoma. A case report. *Cancer* **89**: 97–103.
34. Fears, T. R., Scotto, J. and Schneiderman, M. A. (1977). Mathematical models of age and ultraviolet effects on the incidence of skin cancer among whites in the United States. *Am. J. Epidemiol.* **105**: 420–427.
35. Epstein, J. H., Epstein, W. L. and Nakai, T. (1967). Production of melanomas from DMBA-induced "blue nevi" in hairless mice with ultraviolet light. *J. Natl. Cancer Inst.* **38**: 19–30.
36. Romerdahl, C. A., Stephens, L. C., Bucana, C. and Kripke, M. L. (1989). The role of ultraviolet radiation in the induction of melanocytic skin tumors in inbred mice. *Cancer Commun.* **1**: 209–216.
37. Millikin, D., Meese, E., Vogelstein, B., Witkowski, C. and Trent, J. (1991). Loss of heterozygosity for loci on the long arm of chromosome 6 in human malignant melanoma. *Cancer Res.* **51**: 5449–5453.
38. Trent, J. M., Stanbridge, E. J., McBride, H. L., Meese, E. U., Casey, G., Araujo, D. E., Witkowski, C. M. and Nagle, R. B. (1990). Tumorigenicity in human melanoma cell lines controlled by introduction of human chromosome 6. *Science* **247**: 568–571.
39. Church, S. L., Grant, J. W., Ridnour, L. A., Oberley, L. W., Swanson, P. E., Meltzer, P. S. and Trent, J. M. (1993). Increased manganese superoxide dismutase expression supresses the malignant phenotype of human melanoma cells. *Proc. Natl. Acad. Sci. USA* **90**: 3113–3117.
40. Ozer, H. L., Banga, S. S., Dasgupta, T., Houghton, J., Hubbard, K., Jha, K. K., Kim, S. H., Lenahan, M., Pang, Z., Pardinas, J. R. and Patsalis, P. C. (1996). SV40-mediated immortalization of human fibroblasts. *Exp. Gerontol.* **31**: 303–310.
41. Kwee, J. K., Mitidieri, E. and Affonso, O. R. (1991). Lowered superoxide dismutase in highly metastatic B16 melanoma cells. *Cancer Lett.* **57**: 199–202.
42. Borrello, S., de Leo, M. E. and Galeotti, T. (1993). Defective gene expression of MnSOD in cancer cells. *Mol. Aspects Med.* **14**: 253–258.
43. Alvarez, M., Strasberg Rieber, M. and Rieber, M. (1998). Chromosome 6-mediated suppression of metastatic ability increases basal expression of UV-inducible superoxide dismutase and induction of p53. *Int. J. Cancer* **77**: 586–591.
44. Rieber, M. and Rieber, M. S. (1999). Tumor suppression without differentiation or apoptosis by antisense cyclin D1 gene transfer in K1735 melanoma involves

induction of p53, p21^{WAF1} and superoxide dismutases. *Cell Death Differ.* **6**: 1209–1215.

45. Varachaud, A., Berthier-Vergnes, O., Rigaud, M., Schmitt, D. and Bernard, P. (1998). Variable expression of MnSOD in three different human melanoma cell lines. *Eur. J. Dermatol.* **8**: 90–94.

46. Miele, M. E., McGary, C. T. and Welch, D. R. (1995). SOD-2 (MnSOD) does not suppress tumorigenicity or metastasis of human melanoma C8161 cells. *Anticancer Res.* **15**: 2065–2070.

47. Hirose, K., Longo, D. L., Oppenheim, J. J. and Matsushima, K. (1993). Overexpression of mitochondrial manganese superoxide dismutase promotes the survival of tumor cells exposed to interleukin-1, tumor necrosis factor, selected anticancer drugs, and ionizing radiation. *FASEB J.* **7**: 361–368.

48. Bittinger, F., Gonzalez-Garcia, J. L., Klein, C. L., Brochhausen, C., Offner, F. and Kirkpatrick, C. J. (1998). Production of superoxide by human malignant melanoma cells. *Melanoma Res.* **8**: 381–387.

49. Clement, M. V., Ponton, A. and Pervaiz, S. (1998). Apoptosis induced by hydrogen peroxide is mediated by decreased superoxide anion concentration and reduction of intracellular milieu. *FEBS Lett.* **440**: 13–18.

50. Jaworska, A., Stojcevic-Lemic, N. and Nias, A. H. (1993). The effect of paraquat on the radiosensitivity of melanoma cells: the role of superoxide dismutase and catalase. *Free Radic. Res. Commun.* **18**: 139–145.

Chapter 55

Modification of Endogenous Antioxidant Enzymes in the Brain as Well as Extra-Brain Tissues by Propargylamines: Is it Related to the Life Prolonging Effect of (–)Deprenyl?

Kenichi Kitani*, Chiyoko Minami, Wakako Maruyama,
Setsuko Kanai, Gwen O. Ivy and Maria-Cristina Carrillo

Kenichi Kitani, Chiyoko Minami and **Wakako Maruyama** • National Institute for Longevity Sciences, 36-3, Gengo, Moriokacho, Obu, Aichi, 4748522, Japan
Setsuko Kanai • Tokyo Metropolitan Institute of Gerontology, Tokyo, Japan
Gwen O. Ivy • Life Science Division, University of Toronto at Scarborough, Scarborough, Ontario, Canada, M1C 1A4
Maria-Cristina Carrillo • National University of Rosario, Rosario, Argentina
*Corresponding Author.
Tel: 81562450183, E-mail: kitani@nils.go.jp

1. Summary

(–)Deprenyl, a monoamine oxydase B (MAO B) inhibitor is known to upregulate antioxidant enzymes such as superoxide dismutase (SOD) and catalase (CAT) activities in brain dopaminergic regions. The drug is also the sole chemical which was repeatedly shown to increase life spans of several animal species including rats, mice, hamsters and dogs.

Two other propargylamines, one, rasagiline and the other, R-N-(2-heptyl)-N-methyl-propargylamine (R-2HMP), one of the aliphatic propargylamines, were investigated for their effects on antioxidant enzymes. It turned out that all three of the propargylamines share properties of enhancing antioxidant enzyme activities not only in brain dopaminergic regions but in extra-brain tissues such as the heart, kidneys, adrenal glands and the spleen. An apparent extension of life spans of experimental animals reported in the past may be better explained by these new observations that these drugs, (–)deprenyl in particular, upregulate SOD and CAT activities not only in the brain but in extra-brain vital organs including the heart and kidneys. These observations may also help explain an anti-tumorigenic effect (as well as immunomodulatory effect) reported in the past for (–)deprenyl which may contribute to the extension of apparent life spans of animals prone to spontaneous tumor development during aging.

2. Introduction

Despite the growing support for the "Free radical theory of aging",[1] direct evidence for this theory is as yet lacking. Although some of the earlier attempts to prolong life spans of animals by feeding so-called antioxidant chemicals [summarized by Harman[2]] claim that some animals treated with antioxidants lived significantly longer than control diet fed counterparts, the past difficulty in maintaining animals in so-called specific pathogen-free conditions to the present standards has made these results inconclusive.

A recent work which failed to demonstrate a significantly longer survival of mice given diets containing five different types of antioxidants[3] is an example showing that this type of attempt has still not achieved unanimous success. At this time of crossing the border of the two centuries (2000–2001), it is still the general consensus of experimental gerontologists that the only reproducible means to significantly prolong life spans of animals is the dietary restriction paradigm (for review, see Ref. 4).

On the other hand, studies in the last two decades have provided a number of successes in preventing (or retarding the development of) many experimentally or naturally induced so-called age-associated disorders including atherosclerosis,[5] hypercholesterolemia,[6] hypertension,[7] cancer in particular,[8] and specific oxidative tissue damages[9, 10] etc. Most of these chemicals are from Chinese herb medicines or micronutrients contained in foods (for review, see Ref. 11). A recent study by Joseph *et al.*[12] further provided a promise in the future that this type of approach may be rewarding not only for prevention or retardation of onsets of many disorders but for deterioration of age-induced brain functions. These recent successful studies are also increasingly supported by epidemiological studies (*e.g.* Ref. 13), suggesting that antioxidant strategies by means of intake of various kinds of antioxidant chemicals, pharmaceuticals or nutrients are a practical and promising approach for the prevention of so-called age-associated disorders in which the role of oxidative-tissue damage is increasingly demonstrated.

As discussed earlier, however, a direct intervention in aging *per se* by means of this type of approach (*i.e.* administration of antioxidants) still remains unsuccessful.

The failure in obtaining a successful intervention in aging does not disprove the validity of the free radical theory but certainly does not prove it. Reasons for this failure have been discussed previously by one of the authors.[11] Another type of anti-oxidant strategy is at least theoretically a modification of endogenous antioxidant defense system(s). Among them, upregulation of antioxidant enzymes has been much less studied in the past.[14, 15] One of the major reasons is that the means for upregulating these enzymes are quite limited. At least experimentally, genetic manipulations have revealed that this is possible.[16] However, upregulation of Cu, Zn-SOD in rodents has produced pathologic animals mimicking human Down's syndrome.[17, 18] The only success so far is the study of Orr and Sohal[16] demonstrating the production of fruit fly linages which lived for significantly longer times when their genes were upregulated for Cu, Zn-SOD together with catalase (CAT).[16] Although this study still needs back-up studies to be validated, the study has suggested that this type of approach is at least experimentally possible.* Obviously, this type of genetic modulation does not allow a straight clinical application, however. Another more practical (and realistic) approach is the modification of these enzyme activities by pharmacological means. In this chapter, we summarize results of one of these types of approach, namely the use of "propargylamines". A series of these chemicals have been recently examined by our group and were found to share properties in common of elevating activities of both Cu, Zn- and Mn-SODs as well as CAT primarily in brain regions of dopaminergic nature.[14, 19, 20] Our recent works further have

---

*The study of Orr and Sohal[16] has been recently criticized for its statistical analysis. However, at least two recent studies have shown that a significant life span extension can be achieved for *Drosophila* by means of upregulating Cu, Zn-SOD only.[74, 75] These new observations may have important implications for the pharmacological manipulation of antioxidant defense mechanisms for the purpose of prolonging life spans of animals as has been discussed in this chapter.[4–6]

shown that other propargylamines so far tested increase SOD and CAT activities not only in brain regions as we observed with (–)deprenyl but also in extra-brain tissues such as the heart and kidneys[19, 20] but not in the liver. At the same time (–)deprenyl has been shown to prolong life spans of at least four different animal species such as rats,[21–23] mice,[24] hamsters[25] and dogs.[26] The causal relationship of these two seemingly different effects of (–)deprenyl remains unelucidated, however, we will discuss this possibility mainly based on results of our recent studies.

3. Changes in Endogenous Antioxidant Enzyme Activities with Age

Many studies in the past have reported changes with age in antioxidant enzyme activities with age. For example, there are several studies reporting an age-dependent decline in SOD and CAT activities in the liver which were interpreted as a cause for age-induced declines in cellular and organ functions.[27–30] In contrast, most studies in skeletal muscles have shown an increase with age in enzyme activities[31] which have been interpreted as the results of adaptation to perpetual oxidative stress during aging. Could this difference be explained on the basis of difference in organs studied? Figure 1 shows our own studies in the liver on CAT activities with age which shows an age-dependent decline in male rat livers, while in females the activities were significantly higher in aged than young animals.[32] A similar sex difference in CAT activities with age was reported previously by Rikans *et al.*[33] Changes in liver SOD activities with age are not so clear in the past literature showing increased, decreased and unchanged values with age depending on different studies reported (for review, see Ref. 34).

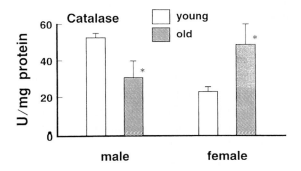

Fig. 1. Catalase activities in livers of young (7 months old) and old (27–30 months old) Fischer 344 (F344) rats of both sexes. Reproduced from Carrillo *et al.*[32] with permission. *Significantly different from corresponding values in young rats.

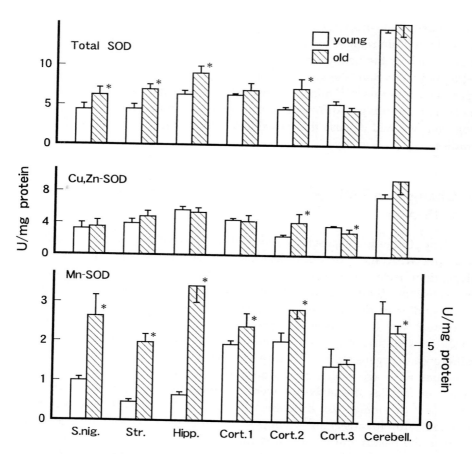

Fig. 2. Superoxide dismutase enzyme activities in several brain regions in young (7-month-old) and old (27- to 30-month-old) male *F344* rats. *Significantly different from corresponding values in young rats. S.nigra; substantia nigra: Str.; striatum: Hipp.; hippocampus: Cort. 1; frontal cortex: Cort. 2; parietotemporal cortex: Cort. 3; occipital cortex: Cerebell; cerebellum. Reproduced from Carrillo *et al.*[32] with permission.

Figure 2 compares SOD and CAT activities in different brain regions in young and old male Fischer 344 (F344) rats.[32] A marked increase in Mn-SOD activities in some brain regions in old animals can be seen. However, in female rat brains enzyme values were almost identical for young and old animals. An increase with age in SOD activities in male rat brains agrees with several past studies with age showing an increase with age in old rat brains[35-37] as well as one study by Williams *et al.*[38] showing an increase in messenger RNA levels for Mn-SOD with age in different brain regions of old *F344* rats. However, some studies reported no change (*e.g.* Refs. 39 and 40) or even a decline with age in male rat brains.[29, 41]

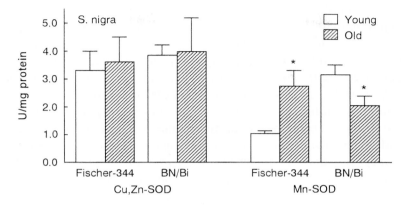

Fig. 3. Comparison of Cu, Zn- and Mn-SOD activities in striata of young and old male *F344* and BN/Bi rats. Activities in BN/Bi rats which were measured by the method of McCord and Friedovich[42] were converted to corresponding values by the method by Elstner and Heupel[43] for comparison by the correction factor shown in Fig. 4. *Significantly different from corresponding values in young animals ($P < 0.05$).

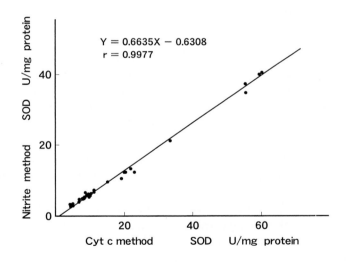

Fig. 4. Superoxide dismutase activities determined by two different methods on the same tissue samples. X-axis, enzyme activities measured by the method of McCord and Friedovich;[42] Y-axis, enzyme activities measured by the method by Elstner and Heupel.[43]

We have recently examined SOD and CAT activities in brain regions in BN/Bi rats and were startled to see practically no increase with age in enzyme activities even in males as well as in females (Fig. 3). One caution should be taken for a comparison between enzyme values obtained by different studies. Figure 4 shows a relationship between enzyme values obtained in the same samples by different methods.[42, 43] Although a highly significant linear relationship could be observed, absolute values were different between different methods. Unless we have a correction factor such as Fig. 4, it is not possible to compare enzyme values obtained in different studies. In sum, antioxidant enzyme activities vary depending on methods used, sexes, age and strains even in the same species. Interpretation of studies comparing enzyme values found in different studies has to be done with great caution. What is more important is that age-related changes in antioxidant enzyme activities are variable again depending on sexes, organs, strains and species, and that reported results are quite variable even in the same organ of the same sex of animals. Again age-related changes in enzyme activities should be interpreted very cautiously. Although many authors in the past have discussed and interpreted their own observations in this regard (*e.g.* Refs. 27 and 29), it is the opinion of the authors that no rational interpretation in general term can be made in terms of general mechanism(s) of aging.

4. Antioxidant Enzyme Upregulation: Common Properties of Propargylamines

Very limited information is available with regard to the possibility of pharmacological modifications of endogenous antioxidant enzyme activities. Several Chinese herb medicines including Gingseng,[44] ursolic acid,[45] etc. have been shown to increase activities of SOD or CAT in the liver. Except for this limited information propargylamines are unique in their properties of increasing SOD and CAT activities selectively in tissues of primarily dopaminergic nature.

Figure 5 summarizes results of one of our studies which demonstrates clearly that CAT as well as both types of SOD activities in selective brain regions are significantly increased when a proper dose of (−)deprenyl is administered.[46] Since in the liver, enzyme activities were never increased with (−)deprenyl in our repeated studies,[46–49] we (and others) have believed until recently that the increase in antioxidant enzyme activities is selective only in certain regions in the brain (for review, see Refs. 14 and 15).

Figure 6 shows our recent study using rasagiline.[19] Rasagiline[51, 52] also increased SOD and CAT activities as (−)deprenyl does in certain brain regions of primarily dopaminergic nature. In this study, we had a look at extra-brain tissues of dopaminergic nature (the heart and kidneys) for the first time and found upregulation of SOD and CAT activities by the drug is not exclusive for brain tissues but occurs in these extra-brain tissues. Subsequent studies have shown similar results with R-2(heptyl)-methyl propargylaine (R-2HMP) revealing upregulation in the brain as well as outside of the brain.[20] R-2HMP is one of the aliphatic propargylamines all of which possess an antiapoptotic effect as well as

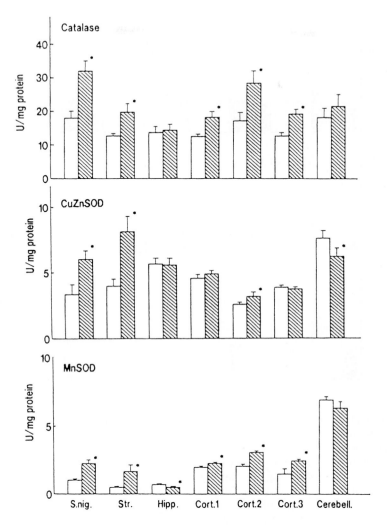

Fig. 5. Catalase (top panel) and superoxide dismutase (lower two panels) enzyme activities in young control (white columns) and deprenyl-treated (shaded columns) male rats. The dose of deprenyl is 2.0 mg/kg/day, s.c. continuous infusion for 3 weeks. Reproduced from Carrillo *et al.*[46] with permission.

a MAO B inhibitory effect[53, 54] as do (–)deprenyl and rasagiline.[51, 52] Furthermore, we have confirmed that (–)deprenyl upregulates SOD and CAT activities in the heart and kidneys as we observed with rasagiline and R-2HMP as well as in adrenal glands, and the spleen.[20] These increases were accompanied by concomitant increases in mRNA levels for Cu, Zn-SOD as well as MnSOD.[20]

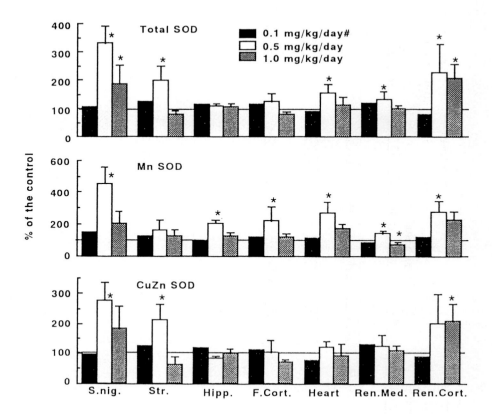

Fig. 6. Effect of rasagiline pretreatment on superoxide dismutase activities on different brain regions and tissues in 8-month-old male F344 rats. All values are expressed as percentages relative to respective control values in rats given a saline solution infusion. *Significantly different from respective control values ($p < 0.05$). Black bars indicate values in a rat given a dose of 0.1 mg/kg/day for 3.5 weeks, white bars indicate values in rats given a dose of 0.5 mg/kg/day and hatched bars values in rats given a dose of 1.0 mg/kg/day for 3.5 weeks. S.nig., substantia nigra; Str., striatum; Hipp., hippocampus; F. Cort, frontal cortex; Ren. Med., kidney medula; Ren. Cort., kidney cortex. Reproduced from Carrillo et al.[19] with permission.

Figure 7 compares dose efficacy relationships with (–)deprenyl[46] and rasagiline.[19] Qualitatively, both propargylamines show a similar effect on the brain, however, quantitatively the magnitude of increase as well as the effective dose range is greater with (–)deprenyl than those with rasagiline.[19] Interestingly, R-2HMP was even smaller in its effect than rasagiline.[20] Thus, although these properties are shared by all three propargylamines tested, the magnitudes of increase in activities appear to be different depending on the rest of the chemical moiety of each propargylamine.

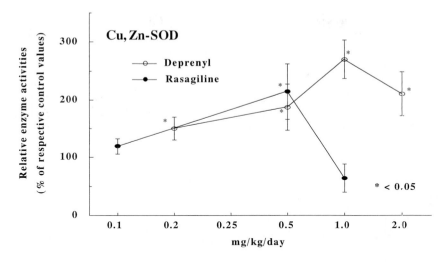

Fig. 7. Relative Cu, Zn-SOD activities in striata of male F344 rats treated with (−)deprenyl and rasagiline at various doses (continuous s.c. infusion for 3–3.5 weeks). Values are expressed as percentage of respective control values. *Significantly different from respective control values ($P < 0.05$). The figure was drawn from results previously reported by the authors.[19, 46]

5. Variability of an Optimal Dose of (−)Deprenyl on Antioxidant Enzyme Upregulation

As is clear from Fig. 6 (and Fig. 8), there exists an optimal dose range for increasing antioxidant enzyme activities. An important point is that this optimal dose varies widely depending on animal species, strains, sexes and age and most importantly the duration of drug treatment (for review, see Refs. 14, 15 and 50). Some studies which reported an absence of upregulation of SODs may be explained on the basis of the variability of the optimal dose. For example, Gallagher *et al.*[55] recently reported an absence of upregulation of SOD activities in male Wistar rats treated for 9 to 15 months at a dose of 0.5 mg/kg/injection s.c. 3 times a week. This is most likely due to the long period of treatment with (−)deprenyl which reduces the optimal dose and narrows the optimal dose range. In fact, we ourselves have recently reported that the dose of 1.0 mg/kg/injection 3 times a week, which is in the middle of the optimal dose range (Fig. 8), when animals were treated for one month, lost the effect, when the treatment was continued for 13 months in aging F344 rats (Fig. 9).[56] Sporadic studies which also failed in increasing SOD activities for a short period of 3 weeks[57, 58] may also be explained as due to the possible higher P450 enzyme activities which eventually inactivate (−)deprenyl by metabolizing it. Indeed,

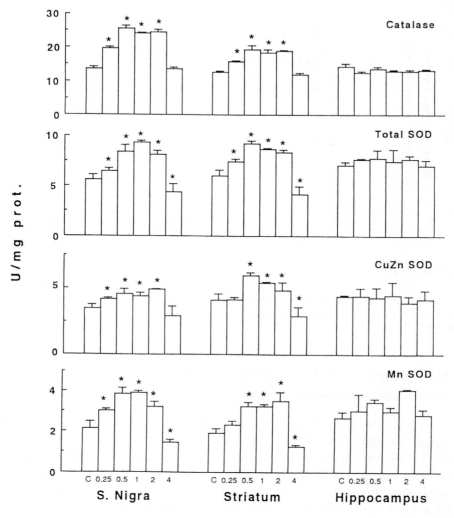

Fig. 8. Catalase (CAT) and superoxide dismutase (SOD) activities in three different brain regions of 27-month-old male F344 rats treated with various doses of deprenyl for 1 month before sacrifice. C: control rats given saline injections, 0.25–4 indicate doses of (−)deprenyl (mg/kg/injection, 3 times per week) for 1 month. *Significantly different from respective control values. ($P < 0.05$, Scheffe's F test). Reproduced from Carrillo et al.[60] with permission.

several fold difference in the rate of formation of (−)amphetamine, a major metabolite of (−)deprenyl between different rat strains was previously reported by Yoshida et al.[59]

A ten times greater optimal dose found in young male F344 rats (2.0 mg/ kg/day, s.c. 3-week continuous infusion) than female rats (0.2 mg/kg/day)[46]

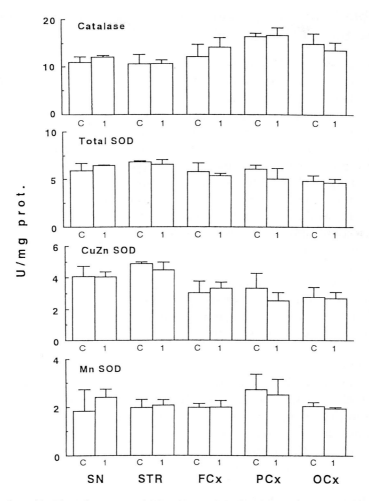

Fig. 9. Catalase (CAT) and superoxide dismutase (SOD) activities in substantia nigra (SN), striatum (STR) and hippocampus in 31-month-old male F344 rats treated with saline (c) or deprenyl solution at a dose of 1.0 mg/kg/injection,[1] three times a week for 13 months beginning at the age of 18 months. Reproduced from Carrillo *et al.*[60] with permission.

can easily be explained mostly by the difference in P450 activities which are usually several fold to more than 10 times greater in male than in female animals of this strain.[60]

As far as the pharmacology of propargylamines is concerned, these variables which practically modify and sometimes critically affect the results must be carefully taken into account, when results are discussed in comparison with results reported in the past.

6. Possible Roles of Upregulation of Antioxidant Enzymes in Life Span Extension of Animals

The mechanism(s) for the extension of apparent life spans reported for different animal species (for review, see Refs. 14 and 15) remains unknown. However we have suggested that the upregulation of SOD and CAT activities may be causally related to its effect on life spans of animals. A major basis for this suggestion, although circumstantial, is the apparent parallelism between dose-efficacy relationships for effects on antioxidant enzymes and life spans.[14, 15] Our study which examined the survival of male F344 rats receiving (−)deprenyl injections at a dose of 1.0 mg/kg/day (3 times a week) beginning at the age of 18 months has shown that the survival of treated animals became smaller (3/12) than control animals (7/12) at the age of 31 months,[56] with deprenyl apparently losing its positive effect on survivals that was observed when a smaller dose (0.5 mg/kg/day injection) was applied.[21]

At the same time in surviving animals SOD and CAT activities were almost identical for treated and control animals (Fig. 8).[60] The results of Gallagher *et al.*[61] which also showed a shorter life span in (−)deprenyl treated rats may also be explained in the same way: a long term 0.5 mg/kg/injection became excessive for the life span as well as for SOD upregulation as indeed the same authors later reported.[55]

On the other hand, some other studies[62, 63] which also failed in modifying life spans of animals may be explained as due to too small doses in these studies, if SOD and CAT upregulation are causally related to life span prolongation. The dose used in these studies (1.0 mg/kg/day) in mice and rats both by means of oral administration may have been too small to upregulate enzyme activities, since the first pass effect by the liver reduces the systemic availability to less than 10% of that of the same dose systemically introduced (*e.g.* subcutaneous administration).

In our hands, 0.1 mg/kg/day (s.c. infusion) for 3 weeks caused a very minor change in SOD activities in mouse brain.[64] The failure in obtaining a positive effect on life span prolongation in these studies[61–63] may be explained as due to an inappropriate dose used, if the life prolonging effect is causally related to its effect on antioxidant enzyzmes. Accordingly, pharmacological effects of (-)deprenyl such as MAO B inhibition and antiapoptosis that can be achieved with a dose at least one order of magnitude lower than the optimal dose for antioxidant enzymes do not appear to affect the life span of animals. The only exception for this rule is the study in female hamsters in which a relatively low oral dose increased the life span.[25] However, no information is available on the optimal dose of (−)deprenyl on antioxidants enzyme activities in this species.

Ruehl *et al.* reported a surprizingly marked effect in prolonging life spans of aging (10–15 year old) female Beagle dogs with treatment of (−)deprenyl.[26] We have previously confirmed that SOD activities are also increased in *S. nigra* and *striatum* (but not in hippocampus) in this species[65] suggesting that both effects are achieved over a wide variaty of animal species, if an appropriate dose is selected.

7. Neurohumoral Events Affected by (–)Deprenyl

Assuming that our contention that effects on SOD and CAT and on life spans of (–) deprenyl treated rats are causally related, the mechanisms underlying these effects remain unknown. (–)Deprenyl has been suggested to be antitumorigenic. The authors originally sugested in our first suvival study that the drug may have an effect in reducing or at least delaying the onset of subcutaneous tumors in F344 rats.[21] These tumors are benign in nature and are not a direct cause for deaths of animals, however, tumors eventually develop skin erosion and ulcers leading to death due to the loss of body fluid through these skin lesions. More direct evidence that (–) deprenyl can reduce the incidence of breast cancer spontaneously occurring during aging or prevent the development of experimentally induced cancer in rats have been recently reported.[66–69] Ruehl *et al.*[26] also reported a marked difference in the incidence of breast cancer in control and (–)deprenyl treated aging female Beagle dogs than in controls.

Another pharmacological effect of (–)depreyl possibly related to the life prolonging effect of the drug is an immunomodulation. Freisleben *et al.*[70] reported a striking prolongation of survivals of immunodeficient mice treated with (–)deprenyl. ThyagaRajan *et al.*[68] also reported the immunological involvement of (–)deprenyl. These past results coupled with other studies reporting the involvement of humoral factors such as NGF,[71] interleukins[72] or neurotrophic factors[73] suggest that (–)deprenyl treatment probably affects the neuro-immuno-endocrine system presumably by means of involving humoral factors and thereby modifying the life span of animals.**

Causal relationships between these humoral events and antioxidant enzyme upregulation should clearly be pursued in the future.

8. Conclusions

Although many past studies have reported age-related alterations in antioxidant enzyme activities such as SOD, CAT and GSH peroxidase, a careful examination

---

**In relation to the above arguments, we have recently observed that (–)deprenyl treatment increases concentrations of tumor necrosis factor (TNF)-α and interleukin (IL)-1β in some extra-brain organs where SOD and CAT activities were increased. These observations will be discussed in our more recent chapter in press (Kitani K. *et al.*, *Ann. N.Y. Acad. Sci.*).

of the past studies reveals that results are very contradictory, reporting increased, decreased and unchanged enzyme activities with age. These variabilities are usually explained as due to sex, strain and species differences. Our recent two studies in F344 and BN/Bi rats confirm these variabilities and strongly suggest that no rational interpretation of these age-related changes in enzyme activities in terms of mechanism(s) of aging is possible at present.

Propargylamines including (–)deprenyl were found to have common properties of enhancing SOD and CAT activities in selective dopaminergic brain regions as well as extra-brain tissues such as the heart, kidneys, adrenals and the spleen. These new observations may help to explain reported effects of (–)deprenyl such as anti-tumorigenic as well as immunomodulatory effect of the drug. The mechanisms of the effect of (–)deprenyl to prolong survivals of animals, as reported in the past, must be further pursued with these systemic and pharmacologic effects of the drug in mind.

Acknowledgments

A part of studies discussed in this chapter was supported by grants in aid from the Ministry of Health and Welfare (Comprehensive Research on Aging and Health, H1108) and Science and Technology Agency in Japan. The skillful secretarial work by Mrs. T. Ohara is gratefully appreciated.

References

1. Harman, D. (1956). Aging: a theory based on free radical and radiation chemistry. *J. Gerontol.* **12**: 257–263.
2. Harman, D. (1994). Free-radical theory of aging. Increasing the functional life span. *In* "Pharmacology of Aging Processes. Methods of Assessment and Potential Interventions" (I. Zs-Nagy, D. Harman, and K. Kitani, Eds.), Vol. 717, pp. 1–15, *Ann. NY Acad. Sci.*, New York.
3. Lipman, R. D., Bronson, R. T., Wu, D., Smith, D. E., Prior, R., Cao, G., Han, S. N., Martin, K. R., Meydani, S. N. and Meydani, M. (1998). Disease incidence and longevity are unaltered by dietary antioxidant supplementation initiated during middle age in C57BL/6 mice. *Mech. Ageing Dev.* **103**: 269–284.
4. Bertrand, H. A., Herliky, J. T., Ikeno, Y. and Yu, B. P. (1999). Dietary restriction. *In* "Method in Aging Research" (B. P. Yu, Ed.), pp. 271–300, CRC Press, Bocaraton, Boston, London, New York, Washington D.C.
5. Yokozawa, T., Dong, E., Nakagawa, T., Kim, D. W., Hattori, M. and Nakagawa, H. (1998). Effects of Japanese black tea on atherosclerotic disorders. *Biosci. Biotechnol. Biochem.* **62**: 44–48.

6. Yokozawa, T. and Dong, E. (1997). Influence of green tea and its three major components upon low-density lipoprotein oxidation. *Exp. Toxicol. Pathol.* **49**: 329–335.

7. Inoue, M., Watanabe, N., Matsuno, K., Morino, Y., Tanaka, Y., Amachi, T. and Sasaki, T. (1990). Inhibition of oxygentoxicity by targeting superoxide dismutase to endothelial cell surface. *FEBS Lett.* **269**: 89–92.

8. Kim, J. M., Araki, S., Kim D. J., Park, C. B., Takasuka, N., Baba-Toriyama, H., Ota, T., Nir, Z., Khachik, F., Shimidzu, N., Tanaka, Y., Osawa, T., Uraji, T., Murakoshi, M., Nishino, H. and Tsuda, H. (1998). Chemopreventive effect of carotenoids and curcumins on mouse colon carcinogenesis after 1,2-dimethylhydrazine initiation. *Carcinogenesis* **18**: 81–85.

9. Ando, Y., Inoue, M., Hirota, M., Morino, Y. and Araki, S. (1989). Effect of a SOD deprivative on cold induced brain edema. *Brain Res.* **477**: 286–291.

10. Dong, E., Yokozawa, T., Liu, Z. W., Oda, S., Muto, Y., Hattori, M. and Watanabe, H. (1997). Protective effects of Daio-botampi-to and its three major components on rat kidney and renal proximal tubule cells subjected to ischemia (hypoxia)-reperfusion (reoxygenation). *J. Trad. Med.* **14**: 41–48.

11. Kitani, K. (1998). Antioxidant strategies against aging and age-associated disorders. Part I: supplementation with chemicals, pharmaceuticals and nutrients (nutriceuticals). *Korean J. Gerontol.* **8**: 66–70.

12. Joseph, J. A., Shuhitt-Hale, B., Denisova, N., Prior, R. L., Cao, G., Martin, A., Taglialatela, G. and Bickford, P. C. (1998). Long-term dietary strawberry, spinach or vitamin E, supplementation retards the onset of age-related neuronal signal-transduction and cognitive behavioral deficits. *J. Neurosci.* **18**: 8047–8055.

13. Hertog, M. G., Feskens, E. J., Hollman, P. C., Katan, M. B. and Kromhout, D. (1993). Dietary antioxidant flavonoids and risk of coronary heart disease, The Zutphen Elderly Study. *Lancet* **342**: 1007–1011.

14. Kitani K., Kanai, S., Ivy, G. O. and Carrillo, M. C. (1998). Assessing the effects of deprenyl on longevity and antioxidant defences in different animal models. *Ann. NY Acad. Sci.* **854**: 291–306.

15. Kitani, K., Kanai, S., Ivy, G. O. and Carrillo, M. C. (1999). Pharmacological modifications of endogenous antioxidant enzymes with special reference to the effect of deprenyl: a possible antioxidant strategy. *Mech. Ageing Dev.* **111**: 211–221.

16. Orr, W. C. and Sohal, R. S. (1994). Extension of life span by overexpression of superoxide dismutase and catalase in *Drosophila melanogaster*. *Science* **263**: 1128–1130.

17. Avraham, K. B., Schickler, D., Sapaznikov, D., Yaron, R. and Groner, Y. (1988). Down's syndrome: abnormal neuromuscular junction in tongue of transgenic mice with elevated levels of human Cu, Zn-superoxide dismutase. *Cell* **51**: 823–829.

18. Epstein, C. J., Avraham, K. B., Covett, M., Smith, S., Elory-Stein, O., Rotman, G., Bry, C. and Groner, Y. (1987). Transgenic mice with increased Cu, Zn-superoxide dismutase activity: animal model of dosage effects in Down syndrome. *Proc. Natl. Acad. Sci. USA* **84**: 8044–8048.

19. Carrillo, M. C., Minami, C., Kitani, K., Maruyama, W., Ohashi, K., Yamamoto, T., Naoi, M., Kanai, S. and Youdim, M. B. H. (2000). Enhancing effect of rasagiline on superoxide dismutase and catalase activities in the dopaminergic system in the rat. *Life Sci.* **67**: 577–585.

20. Minami, C., Kitani, K., Maruyama, W., Yamamoto, T., Carrillo, M. C. and Ivy, G. O. (2000a). Three different propargylamines share properties of increasing antioxidant enzyme activities in dopaminergic brain regions as well as extra-brain tissues in rats. Abstract 29th Annual Scientific Meeting, American Aging Association, p. 39.

21. Kitani, K., Kanai, S., Sato, Y., Ohta, M., Ivy, G. O. and Carrillo, M. C. (1993). Chronic treatment of (–)deprenyl prolongs the life span of male Fischer 344 rats: further evidence. *Life Sci.* **52**: 281–288.

22. Knoll, J. (1988). The striatal dopamine dependency of life span in male rats: longevity study with (–)deprenyl. *Mech. Ageing Dev.* **46**: 237–262.

23. Milgram, N. W., Racine, R. J., Nellis, P., Mendonca, A. and Ivy, G. O. (1990). Maintenance of L-deprenyl prolongs life in aged male rats. *Life Sci.* **47**: 415–420.

24. Archer, J. R. and Harrison, D. E. (1996). L-deprenyl treatment in aged mice slightly increases life spans, and greatly reduces fecundity by aged males. *J. Gerontol.* **31A**: B448–B453.

25. Stoll, S., Hafner, U., Kraenzlin, B. and Mueller, W. E. (1997). Chronic treatment of Syrian hamsters with low-dose selegiline increases life span in females but not males. *Neurobiol. Aging* **18**: 205–211.

26. Ruehl, W. W., Entriken, T. L., Muggenburg, B. A., Bruyette, D. S., Griffith, W. G. and Hahn, F. F. (1997). Treatment with L-deprenyl prolongs life in elderly dogs. *Life Sci.* **61**: 1037–1044.

27. Reiss, U. and Gershon, D. (1978). Methionine sulfoxide reductase: a novel protective enzyme in liver and its potentially significant role in aging. *In* "Liver and Aging, 1978" (K. Kitani, Ed.), pp. 55–61, Elsevier North-Holland, Amsterdam.

28. Rao, G., Xia, E. and Richardson, A. (1990). Effect of age on the expression of antioxidant enzymes in male Fischer 344 rats. *Mech. Ageing Dev.* **53**: 49–60.

29. Semsei, I., Rao, G. and Richardson, A. (1991). Expression of superoxide dismutase and catalase in rat brain as a function of age. *Mech. Ageing Dev.* **58**: 13–19.

30. Xia, E., Rao, G., van Remmen, H., Heydari, A. E. and Richardson, A. (1995). Activities of antioxidant enzymes in various tissues of male Fischer 344 rats are altered by food restricion. *J. Nutri.* **125**: 195–201.

31. Leeuwenburgh, C., Fiebig, R., Chandwaney, J. and Ji, L. L. (1994). Aging and exercise training in skeletal muscle: responses of glutathione and antioxidant enzyme systems. *Am. J. Physiol.* **267**: R439–R445.

32. Carrillo, M. C., Kanai, S., Sato, Y. and Kitani, K. (1992). Age-related changes in antioxidant enzyme activities are region and organ selective as well as sex in the rat. *Mech. Ageing Dev.* **65**: 187–198.

33. Rikans, L. E., Snowden, C. D. and Moore, D. R. (1991). Sex-dependent differences in the effect of aging on antioxidant defense mechanisms of rat liver. *Biochim. Biophys. Acta* **1074**: 195–200.

34. Matsuo, M., Gomi, F. and Dooley, M. M. (1992). Age-related alterations in antioxidant capacity and lipid peroxidation in brain, liver, and lung homogenates of normal and vitamin E deficient rats. *Mech. Ageing Dev.* **64**: 273–292.

35. Cao-Danh, H., Benedetti, M. S. and Dostert, P. (1983). Differential changes in superoxide dismutase activity in brain and liver of old rats and mice. *J. Neurochem.* **40**: 1003–1007.

36. Santiago, L. A., Osato, J. A., Liu, J. and Mori, A. (1993). Age-related increases in superoxide dismutase activity and thiobarbituric acid-reactive substances: Effect of bio-catalyzer in aged rat brain. *Neurochem. Res.* **18**(6): 711–717.

37. Vannella, A., Geremia, E., D'Urso, G., Tiriolo, P., di Silvestro, I., Grimaldi, R. and Pinturo, R. (1982). Superoxide dismutase activities in aging rat brain. *Gerontology* **28**: 108–113.

38. Williams, L. R., Carter, D. B., Dunn, E. and Conner, J. R. (1995). Indicators of oxidative stress in aged Fischer 344 rats: potential for neurotrophic treatment. *In* "Alzheimer's and Parkinson's Disease: Recent Advances" (I. Hannin, A. Fischer, and M. Yoshida, Eds.), pp. 641–646, Plenum Pub. Corp., New York.

39. Ceballos-Picot, I., Nicole, A., Clément, M., Bourre, J.-M. and Sinet, P.-M. (1992). Age-related changes in antioxidant enzymes and lipid peroxidation in brains of control and transgenic mice overexpressing copper-zinc superoxide dismutase. *Mutat. Res.* **275**: 281–293.

40. Kellog, E. W. and Friedovich, I. (1976). Superoxide dismutase in the rat and mouse as a function of age and longevity. *J. Gerontol.* **31**: 405–408.

41. Gupta, A., Hasan, M., Chander, R. and Kapoor, N. K. (1991). Age-related elevation of lipid peroxidation products: diminution of superoxide dismutase activity in the central nervous system of rats. *Gerontology* **37**: 303–309.

42. McCord, E. F. and Friedovich, I. (1969). Superoxide dismutase. An enzymic function for erythrocuprein (hemocuprein). *J. Biol. Chem.* **244**: 6049–6055.

43. Elstner, E. F. and Heupel, A. (1976). Inhibition of nitrite formation from hydroxylammoniumchloride: a simple assay for superoxide dismutase. *Anal. Biochem.* **70**: 616–620.

44. Chung, H. T., Shim, K. H. and Song, S. H. (1997). Action mechanisms of ginseng against oxidative stress and aging. *Prog. Med.* **17**: 903–914.
45. Kim, J. S., Huh, J. I., Song, S. H., Shim, K. H., Back, B. S., Kim, K. W., Lee, K. Y. and Chung, H. Y. (1996). The antioxidative mechanism of ursolic acid. *Korean J. Gerontol.* **6**: 52–56.
46. Carrillo, M. C., Kanai, S., Nokubom M., Ivy, G. O., Sato, Y. and Kitani, K. (1992). Deprenyl increases activities of superoxide dismutase and catalase in striatum but not in hippocampus: the sex and age-related differences in the optimal dose in the rat. *Exp. Neurol.* **116**: 286–294.
47. Carrillo, M. C., Kanai, S., Nokubo, M. and Kitani, K. (1991). Deprenyl induces activities of both superoxide dismutase and catalase but not of glutathione peroxidase in the striatum of young male rats. *Life Sci.* **48**: 517–521.
48. Carrillo, M. C., Kanai, S., Sato, Y., Ivy, G. O. and Kitani, K. (1992). Sequential changes in activities of superoxide dismutase and catalase in brain regions and liver during (–)deprenyl infusion in female rats. *Biochem. Pharmacol.* **44**: 2185–2189.
49. Carrillo, M. C., Kitani, K., Kanai, S., Sato, Y. and Ivy, G. O. (1992). The ability of (–)deprenyl to increase superoxide dismutase activities in the rat is tissue and brain region selective. *Life Sci.* **50**: 1985–1992.
50. Kitani, K., Miyasaka, K., Kanai, S., Carrillo, M. C. and Ivy, G. O. (1996). Upregulation of antioxidant enzyme activities by deprenyl: implications for life span extension. *Ann. NY Acad. Sci.* **786**: 391–409.
51. Finberg, J. P. M., Takashima ,T., Johnston, J. M. and Commissiong, J. W. (1998). Increased survival of dopaminergic neurons by rasagiline, a monoamine oxidase B-inhibitor. *Neuro. Rep.* **9**: 703–707.
52. Sterling, J., Veinberg, A., Lerner, D., Goldenberg, W., Levy, R., Youdim, M. and Finberg, J. (1998). (R)(+)-N-Propargyl-1-aminoindan (rasagiline) and derivatives; highly selective and potent inhibitorsof monoamine oxidase B. *J. Neural. Transm.* **52(Suppl.)**: 301–305.
53. Boulton, A. A. (1999). Symptomatic and neuroprotective properties of the aliphatic proargylamines. *Mech. Ageing Dev.* **111**: 201–209.
54. Boulton, A. A., Davis, B. A., Durden, D. A., Dyck, L. E., Juorio, A. V., Li, X. M., Paterson, I. A. and Yu, P. H. (1997). Aliphatic propargylamines: new antiapoptotic drugs. *Drug Dev. Res.* **42**: 150–156.
55. Gallagher, I. M., Clow, A., Jenner, P. and Glover, V. (1999). Effect of long-term administration of pergolide and (–)deprenyl on age related decline in hole board activity and antioxidant enzymes in rats. *Biogenic Amines* **15**: 379–393.
56. Carrillo, M. C., Kanai, S., Kitani, K. and Ivy, G. O. (2000). A high dose of long term treatment with deprenyl loses its effect on antioxidant enzyme activities as well as on survivals of Fischer-344 rats. *Life Sci.* **in press**.

57. Knoll, J. (1989). The pharmacology of selegiline/(–)deprenyl. New aspects. *Acta Neurol. Scand.* **126**: 83–91.

58. Lai, C. T., Zuo, D. M. and Yu, P. H. (1994). Is brain superoxide dismutase activity increased following chronic treatment with l-deprenyl? *J. Neural. Transm.* **41(Suppl.)**: 221–229.

59. Yoshida, T., Oguro, T. and Kuroiwa, Y. (1987). Hepatic and extrahepatic metabolism of deprenyl, a selective monoamine oxidase (MAO)B inhibitor of amphetamines in rats: sex and strain differences. *Xenobiotica* **17**: 957–963.

60. Kamataki, T., Maeda, K., Shimada, M., Kitani, K., Nagai, T. and Kato, R. (1985). Age-related alteration in the activities of drug-metabolizing enzymes and contents of sex-specific forms of cytochrome P450 in liver microsomes from male and female rats. *J. Pharmacol. Exp. Ther.* **233**: 222–228.

61. Gallagher, I. M., Clow, A. and Glover, V. (1998). Long term administration of (–)deprenyl increases mortality in male Wistar rats. *J. Neural. Transm.* **52(Suppl.)**: 315–320.

62. Bickford, P. C., Adams, C. E., Boyson, S. J., Curella, P., Gerhardt, G. A., Heron, C., Ivy, G. O., Lin, A. M. L. Y., Murphy, M. P., Poth, K., Wallace, D. R., Young, D. A., Zahniser, N. R. and Rose, G. M. (1997). Long-term treatment of male F344 rats with deprenyl: assessment of effects on longevity, behavior, and brain function. *Neurobiol. Aging* **18**: 309–318.

63. Ingram, D. K., Wiener, H. L., Chachich, M. E., Long, J. M., Hengemihle, J. and Gupta, M. (1993). Chronic treatment of aged mice with L-deprenyl produces marked MAOB inhibition but no beneficial effects on survival, motor performance, or nigral lipofuscin accumulation. *Neurobiol. Aging* **14**: 431–440.

64. Carrillo, M. C., Kitani, K., Kanai, S., Sato, Y., Ivy, G. O. and Miyasaka, K. (1996). Long term treatment with (–)deprenyl reduces the optimal dose as well as the effective dose range for increasing antioxidant enzyme activities in old mouse brain. *Life Sci.* **59**: 1047–1057.

65. Carrillo, M. C., Milgram, N. W., Wu, P., Ivy, G. O. and Kitani, K. (1994). (–)Deprenyl increases activities of superoxide dismutase (SOD) in striatum of dog brain. *Life Sci.* **54**: 1483–1489.

66. ThyagaRajan, S., Meites, J. and Auadri, S. K. (1995). Deprenyl reinitiates estrous cycles, reduces serum prolactin, and decreases the incidence of mammary and pituitary tumors in old acyclic rats. *Endocrinology* **136**: 1103–1110.

67. ThyagaRajan, S., Felten, S. Y. and Felten, D. L. (1998). Antitumor effect of L-deprenyl in rats with carcinogen-induced mammary tumors. *Cancer Lett.* **123**: 177–183.

68. ThyagaRajan, S., Madden, K. S., Stevens, S. Y. and Felten, D. L. (1999). Effects of L-deprenyl treatment on noradrenergic innervation and immune reactivity in lymphoid organs of young F344 rats. *J. Neuroimmunol.* **96**: 57–65.

69. ThyagaRajan, S. and Quadri, S. K. (1999). L-deprenyl inhibits tumor growth, reduces serum prolactin, and suppresses brain monoamine metabolism in rats with carcinogen-induced mammary tumors. *Endocrine* **10**: 225–232.

70. Freisleben, H. J., Neeb, A., Lehr, F. and Ackermann, H. (1997). Influence of selegiline and lipoic acid in the life expectancy of immunosuppressed mice. *Arzheim-Forsch* **47**: 776–780.

71. Li, X.-M., Juorio, A. V., Qi, J. and Boulton, A. A. (1998). L-deprenyl potentiates NGF-induced changes in superoxide dismutase mRNA in PC12 cells. *J. Neurosci. Res.* **53**: 235–238.

72. Ruehl, W. W., Bice, D., Muggenburg, B., Bruyette, D. D. and Stevens, D. R. (1994). L-Deprenyl and canine longevity: evidence for an immune mechanism, and implications for human aging. 2nd Conference on Anti-Aging Medicine, Las Vegas. (Abstract). pp. 25–26.

73. Tang, Y. P., Ma, Y. L., Chao, C. C., Chen, K. Y. and Lee, E. H. Y. (1998). Enhanced glial cell line-derived neurotrophic factor mRNA expression upon (–)deprenyl and melatonin treatments. *J. Neurosci. Res.* **53**: 593–604.

74. Parks, T. L., Elia, A. J., Dickson, D., Hilliber, A. T., Phillips, A. J. and Bonlianne, G. L. (1998). Extension of *Drosophila* life spans by over expression of human SOD-1 in motorneurons. *Nature Genet.* **19**: 171–174.

75. Sun, J. and Tower, J. (1999). FLP recombinase mediated induction of Cu, Zn-superoxide dismutase transgene expression can extend the life span of adult *Drosophila melanogaster* flies. *Mol. Cell. Biol.* **19**: 216–228.

Chapter 56

Oxidative Stress and
Age-Related Macular Degeneration

Jiyang Cai, Paul Sternberg*, Jr. and Dean P. Jones

Jiyang Cai and **Dean P. Jones** • Department of Biochemistry, Emory University School of Medicine, Atlanta, GA 30322, USA
Paul Sternberg, Jr. • Department of Ophthalmology, Emory Eye Center, Emory University, 1365-B Clifton Rd, N. E., Atlanta, GA 30322, USA
*Corresponding Author.
Tel: (404) 778-4120, E-mail: ophtps@emory.edu

1. Introduction

Age-related macular degeneration (AMD) is a chronic and progressive degeneration of photoreceptors and their underlying retinal pigment epithelium (RPE) in the macular area of the retina.[1, 2] In primate eyes, the macula is a slightly pigmented area located about 3 mm temporal to the optic disk where the optic nerve joins the outer retina [Fig. 1(a)]. It is the thinnest part of the retina and provides high resolution of central and color vision. The center of the macula, termed the fovea, is lined with only cone photoreceptors and has the highest visual discrimination. Progressive loss of photoreceptors, especially in the fovea, results in severe central vision impairment and even permanent blindness. The loss of the ability to read, drive and live independently is particularly traumatic for otherwise healthy and active individuals.

The early pathological changes of AMD often start with drusen formation in the macular area.[3] Ninety percent of elderly people have small, hard drusen with well-defined edges that are usually not considered as pathologic. However, large, soft drusen with diameters > 63 mm are diagnosed as either early AMD or age-related maculopathy (ARM).[4] [Fig. 1(b)] The plaques can merge, become confluent and lead to geographic atrophy of the central retina. The nature of drusen remains poorly defined. Histologically, drusen appear as an aggregation of hyaline materials deposited between the Bruch's membrane and the retina and are considered to be accumulated debris of degenerated RPE cells.[5]

Clinically, AMD can be classified into two types, a non-exudative form (dry form) [Fig. 1(c)] and an exudative form (wet form) [Fig. 1(d)]. The former is due to a slow and progressive degeneration of photoreceptors and the RPE, while the

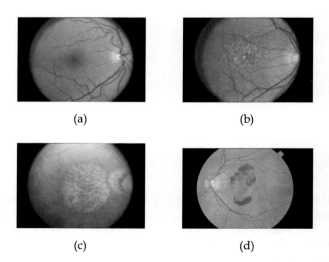

(a)   (b)

(c)   (d)

Fig. 1

latter involves choroidal neovascularization and often is associated with hemorrhage, scar formation and rapid loss of central vision. These two types of AMD are likely to have different etiology and different risk factors.

The incidence of AMD has been investigated by several large epidemiological studies and the data have been summarized by several comprehensive reviews.[6, 7] According to the Framingham Eye Study, almost 11% of people aged 65 to 74 have varying degrees of AMD/ARM, and the incidence increases to 28% between ages of 75 and 85 years.[8] Results from the Beaver Dam Eye Study reported the incidence of AMD is 18 and 30% in age groups 65 to 74 and beyond 75, respectively.[9] Two percent of people aged from 75 to 84 are legally blind, and AMD is the leading cause of vision loss leading to blindness in this population.[9]

Despite the high incidence and the severe vision impairment that the patients may suffer, only a limited number of AMD patients are amenable to treatment. At present, there is no effective treatment for the non-exudative form of AMD. Laser photocoagulation treatment can be used in some choroidal neovascularization cases if the neovascular complex has well-defined boundaries.[10] Photocoagulation therapy, however, has a high recurrence rate and leaves permanent damage to the retina.[11] Recently, photodynamic therapy has been proven beneficial for some cases of subfoveal choroidal neovascularization. Other alternative treatments, such as surgical macular translocation, submacular surgery, radiation and pharmacological interventions are still at investigational stages. As the percentage of elderly people increases in developed countries, AMD will continue to increase in social and economic impact. Much more effort will be required in both mechanistic studies to determine the etiology and therapeutic approaches for effective treatments for AMD.

2. Multiple Factors Contribute to AMD

Although vision loss in the central retina results from photoreceptor damage, the initial pathogenesis of AMD involves the degeneration of RPE.[3, 6, 12] The biochemical mechanisms responsible for RPE degeneration and the underlying genetic determinants of risk have not been defined. Most likely, multiple factors are involved.[13] These can include environmental factors and diet as well as different susceptibilities to these external factors based upon different genetic backgrounds.

Accumulating evidence shows that there is a tendency of familial aggregation of AMD cases. A higher prevalence of AMD is found among monozygotic twins[14-16] and first-degree relatives with AMD[17] than among unaffected individuals. Relatives of exudative AMD patients have a much higher likelihood to develop AMD. Compared to the relatives of control probands, the odds ratio is 3.1.[17] However, environmental factors, such as dietary intake and smoking habits, could also be shared among family members in addition to the inherited genetic information.

Several genes have been studied for their association with AMD, based upon their involvement in inherited retinal dystrophies that have similar clinical manifestations to AMD.[18-21] To date, genetic epidemiological studies cannot securely attribute the etiology of AMD to mutations of any single gene, mainly due to the lack of their functional analysis. Like other age-related degenerative diseases, the incidence of AMD rises exponentially with age. In principle, a mutated gene could start to show its abnormal phenotype soon after birth. If so, the functional abnormality should correlate with age linearly, rather than exponentially. Therefore, it is difficult to explain why inherited genetic mutations affect the function of RPE cells only during the last 1/3 of an individual's life, unless accumulation of damage caused by environmental factors is superimposed on the genetic background.

3. Oxidative Stress and AMD

Similar to other age-related diseases that are reviewed in this volume, the development of AMD appears to involve oxidative stress. The prevailing hypothesis is that under normal physiological conditions, RPE cells are subject to a relatively high oxidative stress, and this can be further exaggerated by an age-related decrease of the antioxidant function. The lesions induced by oxidative injury may accumulate with time, and once a certain threshold is reached, the affected RPE cells may be triggered to undergo apoptosis and be eliminated from the epithelium layer. Because of the close interaction between RPE and photoreceptors in both nutritional and metabolic aspects, the progressive RPE loss and dysfunction could cause a secondary degeneration of rods and cones.

The unique physiology of RPE creates a high burden of oxidative stress. The oxygen tension in the macular area is > 70 mm Hg and, during daylight time, the macular area is exposed to focused light illumination. As suggested in a study of Chesapeake Bay watermen,[22] extreme light exposure in the presence of high oxygen tension can accelerate the development of AMD. Long-term exposure to visible light, especially blue light, has been found to induce damage to RPE and the choriocapillaris in mouse models[23] as well as in cultured cells.[24]

RPE cells are responsible for the clearance of photoreceptor debris. With constant shedding of the outer segments and synthesis of new membranes, a photoreceptor can achieve total renewal within 10 days. The shed membranes have the highest concentration of polyunsaturated fatty acids (PUFAs) of any human tissue and are promptly phagocytosed by RPE. Phagocytosis and lipid peroxidation generate large amount of reactive oxygen species,[25] and the non-degradable particles accumulate within RPE cells and form lipofuscin. A major fluorescent component of lipofuscin has been identified as A2-E (N-retinylidene-N-tetinylethanol-amine), a product formed from all-trans-retinaldehyde and ethanolamine. A2-E appears to be detrimental to RPE cells by a variety of mechanisms, including inhibition of lysosomal degradative capacity, loss of membrane integrity and phototoxicity.[26] With age, RPE cells accumulate more lipofuscin than any other tissue of the body and may export the excessive amount into the extracellular space, contributing to drusen formation. Thus, components of lipofuscin could cause changes in cell biology that accelerate dysfunction with time and also serve as a photosensitizer that transduces the energy from light to RPE cells, or generate free radicals when illuminated by light, causing cell death.[26]

Various environmental factors, especially cigarette smoking, exaggerate the oxidative stress to RPE. Smokers have been found to have lower plasma vitamin C and carotenoids.[27] Their red blood cells have increased tendency to peroxidize *in vitro*.[28] Toxic compounds in the tobacco smoke, including carbon monoxide and nicotine, can induce both hypoxia and generation of reactive oxygen species.[29]

The antioxidant capacity of RPE decreases with age. When assayed in human donor eyes, catalase activity in both macular and peripheral RPE was found to be negatively correlated with donor age.[30] Illumination of isolated RPE cells with blue light induced an age-dependent increase of oxygen consumption and hydrogen peroxide production.[31] Because of the high level of oxidative stress in RPE, once the counterbalance from the antioxidant system is compromised with age, lesions can accumulate at an exponential rate.

4. Age-Related Changes of Plasma Antioxidants

Age-related eye diseases such as cataract and AMD are often associated with a more oxidized plasma redox state. Different antioxidants, including carotenoids, vitamin C, vitamin E and glutathione, have been measured and compared between subjects in different age groups. Although the results are not always consistent, which could be either due to the study design or the sample analysis, there is clearly a trend showing a systemic oxidation associated with age.

4.1. Glutathione

Glutathione is a very abundant water-soluble antioxidant that exists as either a reduced form (GSH) or an oxidized form (GSSG). GSH functions as an antioxidant both enzymatically and non-enzymatically. Principal enzymatic reactions are catalyzed by glutathione peroxidases and glutathione S-transferases. Non-enzymatic reactions include reduction of free radicals and thiol-disulfide exchanges that preserve protein thiols. In these reactions, GSSG is produced. GSSG is reduced back to GSH by the NADPH-dependent GSSG reductase, which is known to be present in the retina. Because the conversion of GSH to GSSG is a function of the oxidant load and the conversion of GSSG back is a function of the antioxidant capacity, the balance of GSH to GSSG provides a dynamic indicator of oxidative stress.

Concentration of GSH in cells is usually in the millimolar range while GSSG is 10 to 50 µM. Both GSH and GSSG are continuously released into extracellular fluids, and, therefore, the balance of GSH and GSSG in the extracellular fluids provides a measure of tissue oxidant/antioxidant balance. This balance can be expressed in terms of the redox potential (E_h) for the GSH/GSSG redox couple, calculated with the Nernst equation, because the electron transfer is a two-electron reaction (GSSG + 2e⁻ + 2H⁺ → 2GSH).[32] For this calculation, $E_h = E_0 + RT/F$ ln([GSSG]/[GSH]²), where R is the gas constant, T is the absolute temperature and F is Faradays constant. E_0 was taken as −240 mV.[33] The E_h becomes oxidized about 1 mV with each year of age, indicating that a generalized oxidation occurs with age.[34] Although the specific contribution of this oxidation to age-related disease has not been established, a pilot study of 40 AMD patients and 27 age-matched non-AMD individuals showed that there was a trend toward oxidation in the AMD subjects.[34] This question is currently undergoing more detailed investigation as a study that is ancillary to the multi-center Age-Related Eye Disease Study that is being sponsored by the National Eye Institute. A generalized oxidation in association with age could predispose RPE cells to induction of apoptosis by environmental or infectious agents. Thus, therapeutic or preventive measures to maintain the GSH antioxidant system could provide an important strategy to protect against AMD.

4.2. Carotenoids

The carotenoids lutein and zeaxanthin are the major pigmented components of the macula.[35, 36] They filter the incoming visible light and absorb blue light with wavelength shorter than 460 nm. Retinal carotenoids also scavenge singlet oxygen.[37] High dietary intake of carotenoids was reported to be associated with a lower risk of AMD.[38] A large clinical observational study showed a negative correlation between total plasma carotenoids and the risk of neovacular AMD.[39] Smokers are at greater risk from AMD and have less macular pigment.[40] Among the individual carotenoids, high plasma β-carotene and lycopene concentrations are associated with lower risk of AMD.[41]

5. Mitochondria of RPE as a Target of Oxidative Injury

Although most mitochondrial proteins are encoded in the nuclear DNA, the mRNA's for 13 proteins that are essential for oxidative phosphorylation, as well as the mitochondrial transfer and ribosomal RNA's are encoded in a small circular mitochondrial DNA. Compared to nuclear DNA, mitochondrial DNA (mtDNA) is more susceptible to oxidative injury. The mitochondrial DNA is not protected by histones, does not have such an extensive network of DNA repair systems and is exposed to more oxidative conditions. mtDNA damage has been found to accumulate with age, especially in post-mitotic tissues, and this accumulation of mtDNA lesions appears to associate with various age-related human diseases.[42]

In SV-40 transformed RPE cells treated with 50 to 200 μM hydrogen peroxide, the total yield of full length mtDNA after PCR amplification was decreased while the amplification of nuclear DNA fragments of similar length was not affected.[43] This suggests that mtDNA in human RPE is sensitive to oxidant-induced deletions and rearrangements. Similar results have been found in fibroblasts.[44] Using quantitative PCR to measure the "common" 4977 bp deletion of mtDNA, Barreau reported an accumulation of deletions in people older than 60.[45] Although the full spectrum of mitochondrial DNA deletions and mutations remains to be determined, it appears that RPE cells accumulate age-related oxidative damage to their mtDNA.

Recent *in vitro* studies suggest that oxidative stress may induce apoptosis in RPE cells via a mitochondria-dependent mechanism.[46] Treatment with a chemical oxidant, *t*-butylhydroperoxide, induces apoptosis in cultured RPE cells. An early decrease of mitochondrial membrane potential was observed before caspase activation and DNA fragmentation. This sequence of events suggests that mitochondria are the primary target of oxidative injury.

Mitochondria play critical roles in regulating apoptosis in higher organisms. In a disease like AMD that is associated with chronic oxidative injury, the age-related accumulation of lesions in only a small population of mitochondria could potentiate other stress-induced toxicity. Mitochondrial antioxidants, including coenzyme Q10 and lipoic acid, may be useful to protect mitochondria in RPE and provide an alternate strategy for protection against AMD.

6. Protection of RPE by Detoxification Enzyme Inducers

If oxidative stress is involved in the etiology of AMD, the development of AMD may be prevented or delayed by strengthening the antioxidant capacity of RPE. Since no effective treatment strategy is available today, many patients have already been prescribed with vitamins and zinc. The protective effects of long-term supplementation with antioxidants and zinc against AMD was recently demonstrated in the Age Related Eye Disease Study (AREDS).[46a] However, several other studies have not supported the AREDS findings.[47, 48]

An alternative way to provide protection against oxidative injury is by induction of glutathione-dependant antioxidant systems within RPE cells. A number of chemopreventive agents used in cancer research have been shown to induce GSH synthesis and induce activities of detoxification enzymes both in animal studies and in cultured cells. Among them, a naturally occurring compound, dimethylfumarate (DMF), has been tested on its protective effect against peroxide-induced damage in cultured human RPE cells.[49] DMF is rich in certain fruits, such as apples. It has relatively low toxicity even at large doses and is used as a food additive. When cultured human RPE cells were treated with 200 µM DMF for 24 h, the GSH concentration increased 2.5 fold and was accompanied by a moderate increase in glutathione S-transferase activity. DMF pretreatment renders RPE cells resistant to subsequent peroxide-induced toxicity.

A variety of other agents may be more effective than DMF in inducing detoxification enzymes and may be more useful for protection against AMD. A variety of these inducers are present in the diet, with allyl sulfide (onion, garlic) and sulforaphane (broccoli) being ones that have been extensively studied. Oltipraz, an anti-schistosomal agent, has been used as a chemopreventive agent in colon cancer and also protects against oxidant-induced apoptosis in human RPE.[50] While these agents are promising, it must be noted that detoxification enzyme inducers also induce apoptosis in some epithelial cells.[51] This implies that there is a risk of exacerbating AMD if amounts consumed are sufficient to induce apoptosis in RPE. Thus, carefully controlled trials are needed to ascertain whether treatment with detoxification enzyme inducers can provide a safe way to protect against oxidant-induced apoptosis in RPE and thereby prevent or delay development of AMD.[52]

7. Conclusion

Age-related macular degeneration (AMD) is the leading cause of vision loss in elderly Americans. Substantial circumstantial evidence indicates that oxidative stress is an important contributing factor in the development of AMD. At present, there is no effective treatment or preventive strategy for this disease. Antioxidant supplementation appears likely to provide some protection but available data are not yet definitive. Other strategies include use of specific mitochondrial antioxidants and agents that induce the activity of detoxification enzymes in the retinal pigment epithelium. However, additional studies to assess safety and efficacy are needed to develop recommendations for reducing risk from this debilitating disease of aging.

Acknowledgments

Financial support was provided by National Institutes of Health (USA) grants EY07892, EY06360, ES09047, Foundation Fighting Blindness, and Research to Prevent Blindness, Inc.

References

1. Klein, B. E. and Klein, R. (1982). Cataracts and macular degeneration in older Americans. *Arch. Ophthalmol.* **100**: 571–573.
2. National Advisory Eye Council. (1982). Vision research: a national plan. Executive Summary, 1983–1987. Bethesda, MD. US Department of Health and Human Services, Public Health Services; NIH Publication No. 82-2469.
3. Green, W. R. and Key, S. N. (1977). Senile macular degeneration: a histopathologic study. *Trans. Am. Ophthalmol. Soc.* **75**: 180–254.
4. The International ARM Epidemiological Study Group. (1995). An international classification and grading system for age-related maculopathy and age-related macular degeneration. *Surv. Ophthalmol.* **39**: 367–374.
5. Spraul, C. W., Lang, G. E. and Grossniklaus, H. E. (1996). Morphometric analysis of the choroid, Bruch's membrane and retinal pigment epithelium in eyes with age-related macular degeneration. *Invest. Ophthalmol. Vis. Sci.* **37**: 2724–2735.
6. Beatty, S., Boulton, M., Henson, D., Koh, H. H. and Murray, I. J. (1999). Macular pigment and age-related macular degenration. *Br. J. Ophthalmol.* **83**: 867–877.
7. Christen, W. G. (1999). Antioxidant vitamins and age-related eye disease. *Proc. Assoc. Am. Physicians.* **111**: 16–21.
8. Leibowitz, H., Krueger, D., Maunder, L. *et al.* (1980). The Framingham Eye Study Monograph. An ophthalmological and epidemiological study of cataract, glaucoma, diabetic retinopathy, macular degeneration, and visual acuity in a general population of 2631 adults, 1973–1975. *Surv. Ophthalmol.* **24**: 335–610.
9. Klein, R., Klein, B. E. F. and Linton, K. L. P. (1992). Prevalence on age-related maculopathy. The Beaver Dam Eye Study. *Ophthalmology* **99**: 933–943.
10. Ciulla, T. A., Danis, R. P. and Harris, A. (1998). Age-related macular degeneration: a review of experimental treatments. *Surv. Ophthalmol.* **43**: 134–146.
11. Macular Photocoagulation Study Group. (1991). Subfoveal neovascular lesions in age-related macular degeneration: guidelines for evaluation and treatment in the macular photocoagulation study. *Arch. Ophthalmol.* **109**: 1242–1257.
12. Green, W. R., McDonnel, P. J. and Yeo, J. H. (1985). Pathologic features of senile macular degeneration. *Ophthalmology* **92**: 615–627.
13. Snodderly, D. M. (1995). Evidence for protection against age-related macular degeneration by carotenoids and antioxidant vitamins. *Am. J. Clin. Nutri.* **62(Suppl.)**: 1148S–1161S.
14. Melrose, M., Sagargal, L. E. and Lucier, A. C. (1985). Monozygotic twins with age-related macular degeneration. *Ophthal. Surg.* **16**: 648–651.
15. Meyers, S. M. and Zachary, A. A. (1988). Monozygotic twins with age-related macular degeneration. *Arch. Ophthalmol.* **106**: 651–653.

16. Klein, M. L., Mauldin, W. M. and Stoumbos, V. D. (1994). Heredity and age-related macular degeneration: observations in monozygotic twins. *Arch. Ophthalmol.* **112**: 932–937.

17. Seddon, J. M., Ajani, U. A. and Mitchell, B. D. (1997). Familial aggregation of age-related maculopathy. *Am. J. Ophthalmol.* **123**: 199–206.

18. Pertukhin, K., Koisti, M. J., Bakall, B., Li, W., Xie, G., Marknell, T., Sandgren, O., Forsman, K., Holmgren, G., Andreasson, S., Vujic, M., Bergen, A. A. B., McGarty-Dugan, V., Figueroa, D., Austin, C. P., Metzker, M. L., Caskey, C. T. and Wadelius, C. (1998). Identification of the gene responsible for Best macular dystrophy. *Nature Genet.* **19**: 241–247.

19. Allikmets, R., Singh, N., Sun, H., Shroyer, N. F., Hutchinson, A., Chidambaram, A., Gerrard, B., Baird, L., Stauffer, D., Peiffer, A., Rattner, A., Smallwood, P., Li, Y., Anderson, K. L., Lewis, R. A., Nathans, J., Leppert, M., Dean, M. and Lupski, J. R. (1997). A photoreceptor cell-specific ATP-binding transporter gene (ABCR) is mutated in recessive Stargardt macular dystrophy. *Nature Genet.* **15**: 236–245.

20. Martinez-Mir, A., Paloma, E., Allikmets, R., Ayuso, C., Rio, T. D., Dean, M., Vilageliu, L., Gonzalez-Duarte, R. and Balcells, S. (1998) Retinitis pigmentosa caused by a homozygous mutation in the Stargardt disease gene ABCR. *Nature Genet.* **18**: 11–12.

21. Taylor, H. R., Munoz, B., West, S., Bressler, N. M., Bressler, S. B. and Rosenthal, F. S. (1990). Visible light and risk of age-related macular degeneration. *Trans. Am. Ophthalmol. Soc.* **88**: 163–178

22. West, S. K., Rosenthal, F. S., Bressler, N. M. *et al.* Exposure to sunlight and other risk factors for age-related macular degeneration. *Arch. Ophthalmol.* **107**: 875–879.

23. Dorey, C. K., Delori, F. C. and Akeo, K. (1990). Growth of cultured RPE and endothelial cells is inhibited by blue light but not green or red light. *Curr. Eye Res.* **9**: 549–559.

24. Tate, D. J., Jr., Miceli, M. V. and Newsome, D. A. (1995). Phagocytosis and H_2O_2 induce catalase and metallothionein gene expression in human retinal pigment epithelial cells. *Invest. Ophthalmol. Vis. Sci.* **36**: 1271–1279.

25. Hjelmeland, L. M., Cristofolo, V. J., Funk, W., Rakoczy, E. and Katz, M. L. (1999). Senescence of the retinal pigment epithelium. *Mol. Vis.* **5**: 33–36.

26. Schutt, F., Davies, S., Kopitz, J., Holz, F. G. and Boulton, M. E. (2000). Photodamage to human RPE cells by A2-E, a retinoid component of lipofuscin. *Invest. Ophthal. Vis. Sci.* **41**: 2303–2308.

27. Chow, C. K., Thacker, R. R., Changchit, C., Bridges, R. B., Rehm, S. R., Humble, J. and Turber, J. (1986). Lower levels of vitamin C and carotenes in plasma of cigarette smokers. *J. Am. Coll. Nutri.* **5**: 305–312.

28. Duthie, G. G., Arthur, J. R. and James, W. P. (1991). Effects of smoking and vitamin E on blood antioxidant status. *Am. J. Clin. Nutri.* **53**: 1061S–1063S.

29. Church, D. F. and Pryor, W. A. (1985). Free-radical chemistry of cigarette smoke and its toxicological implications. *Env. Health Perspect.* **64**: 111–126.
30. Liles, M. R., Newsome, D. A. and Oliver, P. D. (1991). Antioxidant enzymes in the aging human retinal pigment epithelim. *Arch. Ophthalmol.* **109**: 1285–1288.
31. Rozanowska, M., Jarvis-Evans, J., Korytowski, W., Boulton, M. E., Burke, J. M. and Sarna, T. (1995). Blue light-induced reactivity of retinal pigment, in vitro generation of oxygen-reactive species. *J. Biol. Chem.* **270**: 18 825–18 830.
32. Jones, D. P., Carlson, J. L., Mody, V.C., Jr., Cai, J., Lynn, M. J., Sternberg, P., Jr. (2000). Redox state of glutathione in human plasma. *Free Radic. Biol. Med.* **28**: 625–635.
33. Rost, J. and Rapoport, S. (1964). Reduction potential of glutathione. *Nature* **201**: 185–187.
34. Samiec, P. S., Drews-Botsch, C., Flagg, E. W., Kurtz, J. C., Sternberg, P., Jr., Reed, R. L. and Jones, D. P. (1998). Glutathione in human plasma: decline in association with aging, age-related macular degeneration, and diabetes. *Free Radic. Biol. Med.* **24**: 699–704.
35. Handelman, G. J., Dratz, E. A., Reay, C. C., van Kuijk, F. J. G. M. (1988). Carotenoids in the human macula and whole retina. *Invest. Ophthalmol. Vis. Sci.* **29**: 850–855.
36. Hammond, B. R., Jr., Wooten, B. R. and Snodderly, D. M. (1998). Individual variations in the spatial profile of human macular pigment. *J. Am. Opt. Soc. Am.* **14**: 1187–1196
37. Khachik, F., Beecher, G. R. and Smith, J. C., Jr. (1997). Lutein, lycopene and their oxidative metabolites in chemoprevention of cancer. *J. Cell. Biochem.* **22**: 236–246.
38. Goldberg, J., Flowerdew, G., Smith, E., Brody, J. A. and Tso, M. O. (1988). Factors associated with age-related macular degeneration. An analysis of data from the First National Health and Nutrition Examination Survey. *Am. J. Epidemiol.* **128**: 700–710.
39. Eye Disease Case-Control Study Group. (1992). Risk factors for neovascular age-related macular degeneration. *Arch. Ophthalmol.* **110**: 1701–1708.
40. Hammond, B. R., Jr., Wooten, B. R. and Snodderly, D. M. (1996). Cigarette smoking and retinal carotenoids: implications for age-related macular degeneration. *Vis. Res.* **36**: 3003–3009.
41. Mares-Perlman, J. A., Brady, W. E., Klein, R., Klein, B. E., Bowen, P., Stacewicz-Sapuntzakis, M. and Palta, M. (1995). Serum antioxidants and age-related macular degeneration in a population-based case-control study. *Arch. Ophthalmol.* **113**: 1518–1523.
42. Wallace, D. C. (1999). Mitochondrial diseases in man and mouse. *Science* **283**: 1482–1488.

43. Ballinger, S. W., van Houten, B., Jin, G.-F., Conklin, C. A. and Godley, B. F. (1999). Hydrogen peroxide causes significant mitochondrial DNA damage in human RPE cells. *Exp. Eye Res.* **68**: 765–772.

44. Yakes, F. M. and van Houten, B. (1997). Mitochondrial DNA damage is more extensive and persists longer than nuclear DNA damage in human cells following oxidative stress. *Proc. Natl. Acad. Sci. USA* **94**: 514–519.

45. Barreau, E., Brossa, J. Y., Courtois, Y. and Treton, J. A. (1996). Accumulation of mitochondrial DNA deletions in human retina during aging. *Invest. Ophthalmol. Vis. Sci.* **37**: 384–391.

46. Cai, J., Wu, M., Nelson, K. C., Sternberg, P., Jr. and Jones, D. P. (1999). Oxidant-induced apoptosis in cultured human retinal pigment epithelial cells. *Invest. Ophthalmol. Vis. Sci.* **40**: 959–966.

46a. Age-Related Eye Disease Study Research Group. (2001). A randomized, placebo-controlled, clinical trial of high-dose supplementation with vitamins C and E, beta carotene, and zinc for age-related macular degeneration and vision loss. *Arch. Ophthalmol.* **119**: 1417–1436.

47. Christen, W. G., Ajani, U. A., Glynn, R. J., Manson, J. E., Schaumberg, D. A., Chew, E. C., Buring, J. E. and Hennekens, C. H. (1999). Prospective cohort study of antioxidant vitamin supplement use and the risk of age-related maculopathy. *Am. J. Epidemiol.* **149**: 476–484.

48. Stur, M., Tittl, M., Reitner, A. and Meisinger, V. (1996). Oral zinc and the second eye in age-related macular degeneration. *Invest. Ophthalmol. Vis. Sci.* **37**: 1225–1235.

49. Nelson, K. C., Carlson, J. L., Newman, M. L., Sternberg, P., Jr., Jones, D. P., Kavanagh, T. J., Diaz, D., Cai, J. and Wu, M. (1999). Effect of dietary inducer dimethylfumarate on glutathione in cultured human retinal pigment epithelial cells. *Invest. Ophthalmol. Vis. Sci.* **40**: 1927–1935.

50. Sternberg, P., Jr., Nelson, K. C. and Jones, D. P. (1999). Oltipraz increases glutathione in human retinal pigment epithelial cells. *Invest. Ophthalmol. Vis. Sci.* **40**: S224. **(Abstract.)**

51. Kirlin, W. G., Cai, J., DeLong, M. J., Patten, E. J. and Jones, D. P. (1999). Dietary compounds that induce cancer preventive phase 2 enzymes activate apoptosis at comparable doses in HT29 colon carcinoma cells. *J. Nutri.* **129**: 1827–1835.

52. Watson, B., Cai, J. and Jones, D. P. (2000). Diet and apoptosis. *Ann. Rev. Nutri.* **20**: 485–505.

Chapter 57

Preeclampsia and Antioxidants

Rhoda Wilson

Rhoda Wilson • Department of Medicine, Glasgow Royal Infirmary, 10 Alexandra Parade, Glasgow G31 2ER
Tel: 0141 2215418, E-mail gcl025@clinmed.gla.ac.uk

1. Introduction

Preeclampsia, which affects 5–10% of all pregnancies in the western world, is a leading cause of maternal morbidity and mortality and contributes significantly to preterm delivery and fetal growth retardation. It occurs in the third trimester of pregnancy and is characterized by hypertension, reduced utero-placental blood flow, protein urea, platelet aggregation, increased vasoconstriction and oedema. It is also associated with an imbalance of increased thomboxane and reduced prostacyclin.[1] The rate of progression of the disease and the organ systems affected can vary in different women.[2] The causes of preeclampsia are still unknown. Research has shown that oxidative stress may have a role to play in the development of preeclampsia. Oxidative stress is defined as the imbalance between pro-oxidant and antioxidant forces in favour of pro-oxidants. This results in increased formation of reactive oxygen species (ROS) that can cause lipid peroxidation and protein modification.[3] This, in turn, causes change in membrane properties and cell dysfunction. Such damage is usually prevented by an extensive multilayered antioxidant system. The degree to which the mother can respond to the changes in antioxidant production may have an influence on the severity of the disease.[2]

This review considers the roles played by antioxidants, in the development of preeclampsia and whether there is any role for antioxidant supplements in the treatment of the disease.

2. Antioxidant Levels in Peripheral Blood

2.1. Lipid Peroxidation

The body has a multilayered antioxidant system including enzymic and non-enzymic components to cope with increased levels of ROS. Antioxidants are a diverse family of compounds designed to prevent overproduction of, and damage by, free radicals. There is widespread evidence of inflammatory cell and antioxidant activity in preeclampsia. Lipid peroxidation occurs in all cells and tissues and involves the oxidative conversion of unsaturated fatty acids to lipid peroxides and secondary metabolites. Lipid peroxides are toxic compounds, which can increase peripheral vasoconstriction and thromboxane synthesis and decrease prostacyclin synthesis. Levels are increased in normal pregnancy implying that this is associated with some degree of oxidative stress. Many studies have now found levels of lipid peroxides to be further increased in preeclampsia[4, 5] with levels correlating with increased blood pressure, but not perinatal outcome.[4] The increased levels of lipid peroxides and glutathione peroxidase activity combined with the reduced levels of antioxidants are thought to contribute to the development of preeclampsia. Antioxidants are known to inhibit lipid peroxidation and therefore protect cells and enzymes from destruction by peroxides. A study by Wang *et al.*[6] investigated whether there was an imbalance between lipid peroxides and antioxidant activity in preeclampsia. The results showed lipid peroxides were significantly increased in mild preeclampsia and further increased in severe preeclampsia. Levels of vitamin E were also reduced in women with severe preeclampsia. The ratio of lipid peroxide to vitamin E was increased in mild preeclampsia and the increase was more significant in women with severe preeclampsia. It is possible that such an imbalance could be the cause of the endothelial and platelet damage seen in preeclampsia. Lipid peroxidation can be initiated by superoxide anions from activated neutrophils.

2.2. Neutrophils

Neutrophils have also been implicated in the pathogenesis of preeclampsia as they can produce ROS, which can cause changes in membrane phospholipids. This can cause changes in membrane fluidity, integrity and permeability of a variety of cell types. Increased vascular permeability and reactivity also occurs. Neutrophils are increased in the peripheral blood of women with preeclampsia.[7] Not all studies have found such an increase. Crocker *et al.*[8] measured neutrophil function more directly than in some earlier studies and found that while superoxide anion generation was significantly reduced in normal pregnancy this was not observed in preeclamptic patients. They found circulating neutrophils in healthy pregnant and preeclamptic women were neither activated nor primed *in vivo*. The increased ROS generation seen in the preeclamptic group may highlight a role for neutrophils in the oxidative stress and pathophysiology of the disease.

2.3. Glutathione

Glutathione is the central component of the antioxidant defence system. Alexa *et al.*[9] found reduced levels of glutathione and glutathione peroxidase in women with preeclampsia and whilst lipid peroxide levels were not measured the authors speculated that due to the glutathione changes levels must be raised. They concluded that whilst lipid peroxides are not a cause of preeclampsia they do contribute to the pathogenesis of the disease. Further evidence that preeclamptic women lack protective antioxidant mechanisms comes from a study of Vitoratos *et al.*[10] who found the plasma of women with preeclampsia showed a total loss of caeruloplasmin ferroxidase activity.

Thiols are effective antioxidants, which can be synthesized by red blood cells. As they have only a limited capacity to repair, this makes them useful for investigating oxidative stress. Using this system Chen *et al.*[11] found plasma thiol and plasma glutathione to be significantly reduced in women with preeclampsia compared to healthy pregnant women. Intracellular glutathione and superoxide dismutase (SOD) were also significantly reduced. The intracellular antioxidant buffering level is thought to be how the system removes adventitious ROS from the system where as the extracellular system serves to control the release of ROS for essential functions.[10] This reduced intracellular buffering found in red cells from preeclamptic women probably reflects increased ROS activity. However no correlation was found between the antioxidant buffering level and blood pressure in the preeclamptic patients. The authors concluded that reduced antioxidant buffering could account for several of the pathological features of preeclampsia such as increased intracellular calcium, red cell deformability and endothelial damage.

Whilst decreased superoxide activity has been found in women with preeclampsia increased superoxide formation has also been reported.[12] This imbalance in the pro and antioxidant systems has been implicated in the pathogenesis of the disease. As genetic factors have also been associated with preeclampsia,[2, 13] Chen *et al.*[14] considered whether there might be a mutation in a critical region of the SOD gene or whether its abnormal expression may result in alteration of SOD activity. No significant difference was found in the size of the CU, Zn SOD gene or its expression in preeclamptic patients and controls, suggesting that the reduced levels of SOD seen in preeclampsia were not due to gene abnormalities but occurred during the development of the disease.

Spickett *et al.*[15] used non invasive nuclear magnetic resonance (NMR) spectroscopy, to measure the oxidative balance of glutathione in intact erythrocytes from preeclamptic and normotensive patients. The glutathione redox balance was monitored using the ratio of the B-cysteinyl (g_2) to the γ-glutamyl (g_4) resonance. An increased g_2/g_4 ratio shows a shift towards a more reduced glutathione pool. The results showed that while there was no significant difference in the g_2/g_4 ratio between healthy pregnant and preeclamptic patients, levels were significantly lower than in non-pregnant women. However the moderate — severe preeclamptic group showed a clear split with about half the patients having g_2/g_4 ratios comparable with normal pregnancy, whilst the remainder had very low g_2/g_4 ratios indicating severe oxidative stress. The division of the subgroups may be related to the intrinsic susceptibility of erythrocytes to oxidative stress. It is possible that in some patients there are differences in erythrocyte morphology, metabolism or membrane structure that can cause increased sensitivity to oxidative stress. These results do not support previous work using conventional assays[11, 16] and this may be because these assays involved cell destruction that may upset the system's delicate oxidative balance. Erythrocytes from preeclamptic women have been found to be more fragile and to show increased susceptibility to lysis under hypo-osmotic stress.[15] This is comparable with the hypothesis that increased oxidative stress in preeclampsia results in perioxidative damage to cell membranes and a reduction in membrane fluidity. These changes are consistent with increased lipid peroxidation in the membrane as peroxidation of polyunsaturated fatty acids leads to membrane disruption and decreased fluidity.

3. Placental Antioxidant Levels

3.1. Lipid Peroxides

Preeclampsia only occurs in the presence of the placenta and the symptoms disappear after delivery. Endothelial cell dysfunction has been hypothesized to be a major contributor in the pathogenesis of preeclampsia.[1] This hypertensive disease of pregnancy is characterized by changes in the placenta and utero-placental vasculataure.[17] It has been hypothesized that the placenta could be the source of the increased circulating levels of lipid peroxides seen in preeclamptic women. Production of lipid peroxides is increased in the placenta of women with preeclampsia.[18] Lipid peroxides will vasoconstrict human placenta by stimulating thromboxane production.[18] Walsh and Wang[19] carried out a series of experiments to try and establish whether the placenta was the source of the elevated levels of lipid peroxides found in the maternal blood. They found that the human placenta does secrete lipid peroxides, with more being secreted on the maternal side of the placenta than the foetal side. As the lipid peroxides secreted by the placenta had a half-life of about 3 h they would therefore be stable enough to function as circulating compounds. Following delivery of the placenta levels of maternal lipid peroxides fell significantly. The placenta is made up of trophobast cells and core tissue. The latter contains vascular and stromal tissue consisting of macrophages and fibroblasts. Either could be the source of the increased lipid peroxide and thromboxane production seen in preeclampsia. Walsh and Wang[21] found that both tissue types produced abnormally high lipid peroxide and thromboxane levels. However, lipid peroxides and thromboxane were produced primarily by the trophoblast cells and stromal tissue, where as PGI2 was produced by vascular tissue.[21]

3.2. The Role of Superoxide in Lipid Peroxide Generation

Why lipid peroxide levels should rise so significantly in preeclampsia is not fully understood but antioxidants and glutathione peroxidase in particular, are known to limit their generation and inactivate them once formed. Levels of glutathione peroxidase and SOD levels are known to be reduced in preeclampsia[18, 19] and as glutathione peroxidase inactivates lipid peroxide and superoxide dismutase inactivates the superoxide anion, deficiencies in the levels of these compounds could raise levels of oxygen free radicals and lipid peroxides. Further evidence for superoxide having a role to play in raising placental lipid peroxide levels comes from a study by Wang and Walsh.[20] They used placental mitochondria that are enriched with polyunsaturated fatty acids and are known to be susceptible to peroxidation. They are also a source of oxygen radicals. They found that the superoxide generated by the mitochondria could contribute to the oxidative stress seen in preeclampsia and that the mitochondria could cause the rise in placental lipid peroxide levels.

4. Antioxidants and the Treatment of Preeclampsia

4.1. Vitamin E

Uncontrolled lipid peroxidation has been proposed as a cause of endothelial dysfunction damage seen in atherosclerosis and preeclampsia.[22] Giving anti-oxidants that act at a variety of sites could interrupt this process. A high intake of vitamin E, a potent lipid soluble antioxidant, has been associated with a lower risk of coronary artery disease.[23] It has been proposed that uncontrolled lipid peroxidation may be the cause of the endothelial damage seen in preeclampsia.[24] Wang *et al.*[6] found that the imbalance between thromboxane and prostacyclin in preeclampsia correlated with the imbalance in lipid peroxides and antioxidants such as vitamin E. In rats lipid peroxides have been shown to cause changes in uteroplacental perfusion similar to the pathophysiological changes seen in preeclampsia.[25] These findings have led to studies being carried out to evaluate the use of antioxidants in the treatment of preeclampsia.

Schiff *et al.*[26] undertook a study to evaluate the hypothesis that a lack of vitamin E, a lipid soluble antioxidant with potent scavenging ability against a variety of free radicals, may be associated with the development of preeclampsia. They found that dietary consumption of vitamin E was similar in preeclamptic and control groups. However if the analysis was extended to include dietary vitamin E supplements then consumption was significantly higher in the PE patients. These findings clearly refute the cause effect relationship between vitamin E deprivation and PE. It is possible that the vitamin E supplements may have forestalled the appearance of PE. Plasma levels of vitamin E were also significantly higher in the preeclamptic group and whilst this is in agreement with some previous studies[27] it disagrees with others.[6] As the highest vitamin E levels were found in the patients with the most severe disease, it is possible that the increase is a compensatory response to the primary oxidative stress seen in preeclamptic patients.

4.2. Vitamin C

A randomized trial was initiated[28] where patients between 24 and 32 weeks gestation, with early onset preeclampsia received combined antioxidant treatment with allopurinol, vitamin E and vitamin C or a placebo. Lipid peroxide levels were not significantly different between the two groups. However, the authors did find a trend towards later delivery in those patients who received treatment with 52% of the women (14/27) treated with antioxidants delivering within 14 days compared to 72% (22/29) in the untreated group. The authors concluded that these results did not encourage routine use of antioxidants in the treatment of preeclampsia, but it is possible that the late intervention may have precluded the patients deriving maximum benefit.

A later study[29] assessed the benefit of antioxidant supplements on markers of endothelial and placental function. Unlike the study of Gulmezoglu *et al.*[28] the supplements were given between 18 and 22 weeks gestation. The results showed the supplements (vitamins E and C) lowered the biochemical indicators of the disease in women at risk.

5. Antioxidants and the Immune System

Cytokines and ROS are frequent companions at the site of inflammation. Cytokines are produced by vascular endothelium cells, circulating neutrophils and monocytes, and by trophoblast and hofbauer cells in the placenta. Endothelial cell dysfunction has been proposed as a major contributor to the pathogenesis of preeclampsia[1] and free radicals and inflammatory cytokines are known to be potent activators of vascular endothelium. Recent results[30] suggest that the redox status of the cell is a crucial determinant in the regulation of the chemokine system.

5.1. TNF-α

IL-1 and TNF-α have been shown to induce ROS production[22] and recent studies have shown that ROS can up-regulate certain chemokine receptors.[30] This effect is antagonized by antioxidants. In a recent study Rinehart *et al.*[31] examined cytokine expression in placental tissue from normotensive and PE women and found enhanced expression of TNF-α and IL-1β in preeclamptic third trimester placental tissue suggesting that these cytokines may be involved in the underlying aetiology of preeclampsia. Increased levels of TNF-α have been found in the serum of preeclamptic women.[32] This increase was only found after the disease had been clinically established suggesting that the rise is a consequence and not an initiating factor. Increased levels of TNF-α and IL-1 will cause functional alterations to endothelial cells and will up-regulate the expression of platelet derived growth factor, endothelium 1 and plasminogen activator all of which are known to be increased in preeclampsia.[33] The increased levels of TNF-α are also known to activate neutrophils, previously reported in preeclamptic women.[7]

5.2. Glutathione

Levels of glutathione are reduced in preeclampsia.[7, 8] Glutathione has been shown to be an endogenous regulator of TNF production *in vivo* with increased levels of glutathione inhibiting TNF production and reduced levels of glutathione increased TNF production.[30] Whilst this comparison was not made in preeclamptic women, it shows glutathione may be involved in regulating cytokine production. Perhaps studies are required to see whether glutathione has such a role to play in preeclampsia. Within mitochondria TNF is involved in the release of oxidized free radicals, which are then involved in the reduction of oxidized glutathione and the subsequent formation of lipid peroxides. This process results in further toxicity due to increased levels of lipid peroxides, the oxidized glutathione and further TNF production.[31]

5.3 Nitric Oxide

Free radicals also impair endothelial cell nitric oxide (NO) production and inhibit macrophage inducible synthetase.[34] The role of NO in pregnancy and preeclampsia is conflicting. Maternal results have reported plasma nitrite and nitrate levels to be decreased, unchanged and increased.[35, 36] In an attempt to establish whether systemic NO synthase was altered in normal pregnancy and preeclampsia Bioccardo et al.[37] found that while systemic levels of CGMP, the NO second messenger, was increased in normal pregnancy, preeclampsia was not associated with any changes in fetal placenta NO synthesis. Activated leucocytes are a major source of free radicals in preeclampsia for they produce cytokines that irreversibly convert endothelial cell xanthine dehydrogenase to xanthine oxidase, forming oxygen free radicals.[38]

5.4. IL-6

Interleukin-6 (IL-6) is known to induce an acute-phase response and pree-clampsia shows many of the features of an acute phase response. Oxygen free radicals have also been show to induce IL-6 synthesis in endothelium cells. Using ^{1}H spin echo NMR it was found that monocytes and erythrocytes, incubated in the presence of IL-6, could significantly reduce the erythrocyte glutathione pool within a 3 h time course. In the absence of monocytes the erythrocytes showed little sign of oxidation. It would appear that oxidants were released by the monocytes in response to the IL-6.[39] The oxidation of the erythrocyte glutathione was inhibited by indomethicin, suggesting this compound may be of value in the oxidative pathology of preeclampsia. A wide variety of circulating levels of IL-6 have been reported in PE[40, 41] but some studies have reported elevated levels which could be the cause of the increased oxidative damage seen in the monocytes,[42] resulting in turn in depleted antioxidant levels. Vince et al.[43] found significantly higher circulating levels of IL-6, TNF-α and TNF receptors in preeclamptic compared to normotensive controls. The highest levels of IL-6 were found in women with the lowest platelet counts, that is those with the most severe disease. Whilst these, and other authors have found high circulating levels of IL-6,[42] levels of IL-6 within the placenta are reduced.[44]

6. Conclusion

Abnormal production of antioxidants, free radicals and cytokines has been reported in the circulation and placenta of preeclamptic women. All three clearly have a role to play in the development of the condition, but it is not clear which acts as the trigger in a complex process. Oxidative stress may begin in the placenta with reduced antioxidant production that, in turn, causes the uncontrolled production of lipid peroxides and raised levels of thromboxane that activate leucocytes as they circulate through the intervillous space. It is these leucocytes that are responsible for linking the oxidative stress in the placenta with the maternal circulation. Following delivery of the placenta levels of maternal lipid peroxides fall significantly suggesting that the placenta is the source of the elevated levels of lipid peroxides found in the maternal circulation. There is no doubt that oxidative stress and antioxidants have a role to play in the development of preeclampsia. Whilst it seems unlikely that they are involved in the primary events of the disease, they may be involved in the secondary events responsible for the pathophysiology of the condition.

References

1. Roberts, J. M., Taylor, R. N., Musci, T. J., Rodgers, G. M., Hubel, C. A. and Mclaughlin, M. K. (1989). Preeclampsia: an endothelial cell disorder. *Am. J. Obstet. Gynecol.* **161**: 1200–1204.
2. Walker, J. J. (1998). Antioxidants and inflammatory cell response in preeclampsia. *Sem. Reprod. Endocrinol.* **16**: 47–55.
3. Sinclair, J. M., Barnett, A. H. and Lunec, J. (1999). Free radicals and antioxidant systems in health and disease. *Br. J. Hosp. Med.* **43**: 334–344.
4. Walsh, S. W. (1994). Lipid peroxidation in pregnancy. *Hyper. Preg.* **13**: 1–32.
5. Poranen, A. K., Ekblad, U., Uotila, P. and Ahotupa, M. (1996). Lipid peroxidation and antioxidants in normotensive and preeclamptic pregnancy. *Placenta* **17**: 401–405.
6. Wang, Y., Walsh, S. W., Gaou, J. and Zhang, J. (1991). The imbalance between thromboxane and prostacyclin in preeclampsia is associated with an imbalance between lipid peroxides and vitamin E in maternal blood. *Am. J. Obstet. Gynecol.* **165**: 1695–1700.
7. Greer, I. A., Haddad, N. G., Dawes, J., Johnstone, F. D. and Calder, A. A. (1989). Neutrophil activation in pregnancy induced hypertension. *BROG* **96**: 978–982.

8. Crocker, I. P., Wellings, R. P., Fletcher, J. and Baker, P. N. (1999). Neutrophil function in women with preeclampsia. *BROG* **106**: 822–888.

9. Alexa, I. D. and Jerca, L. (1996). The role of oxidative stress in the aetiology of preeclampsia at the GSH and GSH-PX levels in normal pregnancy and preeclampsia. *Rev. Med. Chir. Soc. Nat. IASI* **100**: 131–135.

10. Vitoratos, N., Salamalekis, E., Dalamaga, N. and Kassanos, D. (1999). Defective antioxidant mechanisms via changes in serum caeruloplasmin total iron binding capacity of serum in women with preeclampsia *Eur. J. Obstet. Gynecol. Rep. Biol.* **84**: 63–67.

11. Chen, G., Wilson, R., Cumming, G., Walker, J. J., Smith, W. E. and McKillop, J. H. (1994). Antioxidant and immunological markers in PIH and essential hypertension in pregnancy. *J. Matern. Fetal Med.* **3**: 132–138.

12. Rice-Evan, C. and Bruckdorfer, K. R. (1992). Free radicals, lipoproteins and cardiovascular dysfunction. *Mol. Aspects Med.* **13**: 1–111.

13. Arngrimsson, R., Bjornsson, S., Geirsson, R. T., Bjornsson, H., Walker, J. J. and Snaedal, G. (1990). Genetic and familial predisposition to eclampsia and preeclampsia in a defined population. *BROG* **97**: 762–769.

14. Chen, G., Wilson, R., Boyd, P., McKillop, J. H., Leitch, C., Walker, J. J. and Burdon, R. H. (1994). Normal superoxide dismutase (SOD) gene in pregnancy induced hypertension: is the reduced SOD activity a secondary phenomenon? *Free Radic. Res.* **21**: 59–66.

15. Spickett, C. M, Reglinski, J., Smith, W. E., Wilson, R., Walker, J. J. and McKillop, J. H. (1998). Erythrocyte glutathione balance and membrane stability during preeclampsia. *Free Radic. Biol. Med.* **24**: 1049–1055.

16. Wisdom, S., Wilson, R. W., McKillop, J. H. and Walker, J. J. (1991). Antioxidant activity in normal pregnancy and pregnancy-induced hyperension. *Am. J. Obstet. Gynecol.* **165**: 1701–1704.

17. Redman, C. W. (1991). Current topic: preeclampsia and the placenta. *Placenta* **12**: 309–325.

18. Wang, Y., Walsh, S. W. and Kay, H. H. (1992). Placental lipid peroxides and thromboxane are increased and prostaglandin is decreased in women with preeclampsia. *Am. J. Obst. Gynecol.* **167**: 946–949.

19. Walsh, S. W. and Wang, Y. (1993). Deficient glutathione peroxidase activity in preeclampsia is associated with increased placental production of thromboxane and lipid peroxides. *Am. J. Obstet. Gynecol.* **169**: 1456–1481.

20. Wang, Y. and Walsh, S. W. (1998). Placental mitochondria as a source of oxidative stress in preeclampsia. *Placenta* **19**: 581–586.

21. Walsh, S. W. and Wang Y. (1995). Trophoblast and placenta villous core production of lipid peroxides, thromboxane and prostacyclin in preeclampsia. *J. Clin. Endo. Metab.* **80**: 1888–1893.

22. Dekker, G. and Sibai, B. M. (1998). Etiology and pathology of preeclampsia: current concepts. *Am. J. Obstet. Gynecol.* **179**: 1359–1375.

23. Rimm, E. B., Stampfer, M. J., Ascherio, A., Giovannuci, E., Colditz, G. A. and Willett, W. C. (1993). Vitamin E consumption and the risk of coronary heart disease in men. *New England J. Med.* **328**: 1450–1456.

24. Hubel, C. A., Roberts, J. M., Taylor, R. N., Musci, T. J., Rogers, G. M. and McLaughlin, M. K. (1989). Lipid peroxidation in pregnancy. New perspectives on preeclampsia. *Am. J. Obstet. Gynecol.* **161**: 1025–1034.

25. Hubel, C. A., Griggs, K. C. and McLaughlin, M. K. (1989). Lipid peroxidation can alter vascular function in vitamin E deficient rats. *Am. J. Physiol.* **256**: 25–26.

26. Schiff, E., Friedman, S. A, Stampfer, M., Kao, L., Barrett, P. H. and Sibai, B. M. (1996). Dietary consumption and plasma concentration of vitamin E in pregnancies complicated by preeclampsia. *Am. J. Obstet. Gynecol.* **175**: 1024–1028.

27. Uotila, J. T., Tuimala, R. J., Aarino, T. M., Pyykko, K. A. and Ahotupa, M. O. (1993). Findings on lipid peroxidation and antioxidant function in hypertensive complications of pregnancy. *BROG* **100**: 270–276.

28. Metin Gulmezoglu, A., Justus Hofmayr, G. and Oosthuisen, M. M. J. (1997). Antioxidants in the treatment of severe preeclampsia: an exploratory control trial. *BROG* **104**: 689–696.

29. Chappell, L. C., Seed, P. T., Briley, A. L., Kelly, F. J., Lee, R., Hunt, B. J., Parma, K., Bewley, S. J., Shennan, A. H., Steer, P. J. and Poston, L. (1999). The effect of antioxidants on occurrence of preeclampsia in women at increased risk; a randomized trial. *Lancet* **354**: 810–816.

30. Saccani, A., Saccani, S., Orlando, S., Sironi, M., Bernasconi, S., Chezzi, P., Mantovani, A. and Sica, A. (2000). Redox regulation of chemokine receptor expression. *Proc. Natl. Acad. Sci.* **97**: 2761–2766.

31. Rinehart, B. K., Terrone, D. A., Lagoo–Deenadayalan, S., Barber, W., Hale, E. A., Marin, J. N. and Bennet, W. A. *Am. J. Obstet. Gynecol.* **181**: 915–920.

32. Mekins, J. W., McLaughlin, P. J., West, P. C., McFadyen, I. J. and Johnson, P. M. (1994). Endothelial cell activation of TNF-α and the development of preeclampsia. *Clin. Exp. Immunol.* **98**: 110–114.

33. Fiers, W. (1991). TNF characterization of the molecular and cellular in vivo levels. *FEBS Lett.* **285**: 199–212.

34. Yang, X., Cai, B., Sciacca, R. R. and Cannon, P. J. (1994). Inhibition of inducible NO synthase in macrophages by oxidized low-density lipoproteins. *Circ. Res.* **74**: 318–328.

35. Smarason, A. K., Allman, K. G., Young, D. and Redman, C. W. G. (1997). Elevated levels of serum nitrate, a stable end product of nitric oxide in women with preeclampsia. *BROG* **104**: 538–543.

36. Curtis, N. E., Gude, N. M., King, R. G., Marriot, P. J., Rook, T. T. and Brennecke, S. P. (1995). Nitric oxide metabolites in normal human pregnancy and preeclampsia. *Hyper. Preg.* **14**: 339–349.

37. Boccardo, P., Soregarol, M., Aiello, S., Norris, M., Donadelli, R., Lojacono, A. and Benigini A. (1996). Systemic and fetal-maternal nitric oxide synthesis in normal pregnancy and preeclampsia. *BROG* **103**: 879–886.
38. Friedl, H. P., Till, G. O., Ryan, U. S. and Ward, P. A. (1989). Mediator-induced activation of xanthine oxidase in endothelial cells. *FASEB J.* **3**: 2512–2518.
39. Spickett, C. M., Smith, W. E., Reglinski J., Wilson, R. and Walker, J. J. (1998). Oxidation of erythrocyte glutathione by monocytes stimulated with IL-6. Analysis by ^{1}H spin echo NMR. *Clin. Chem. Acta* **270**: 115–124.
40. Kauma, S. W., Wang, Y. and Walsh, Y. (1995). Preeclampsia is associated with decreased placental IL-6 production. *J. Soc. Gynaecol. Invest.* **2**: 614–617.
41. Opsjon, S. L., Augustgulan, R. and Waage, A. (1995). Interleukin-1, Interleukin-6 and tumour necrosis factor at delivery in pre-eclamptic disorders. *Acta Obstet. Gynecol. Scand.* **74**: 19–26.
42. Kopferminc, M. J. Pearceman, A. M., Aderka, D., Wallace, D. and Socol, M. (1996). Soluble TNF receptors and IL-6 levels in patients with severs preeclampsia. *Obstet. Gynecol.* **88**: 420–427.
43. Vince, G. S., Starkey, P. M., Austgulen, R., Kwiatkowski D. and Redman, C. W. G. (1995). Il-6, TNF and soluble TNF receptors in women with pre-eclampsia. *BROG* **102**: 20–25.
44. Conrad, K. P. and Benyo, D. F. (1997). Placental cytokines and the pathogenesis of preeclampsia. *Am. J. Reprod. Immunol.* **37**: 240–249.

Chapter 58

Iron Regulation, Hemochromatosis, and Cancer

Shinya Toyokuni*, Munetaka Ozeki,
Tomoyuki Tanaka and Hiroshi Hiai

Shinya Toyokuni, Munetaka Ozeki, Tomoyuki Tanaka and **Hiroshi Hiai** • Department
of Pathology and Biology of Diseases, Graduate School of Medicine, Kyoto University,
Sakyo-ku, Kyoto 606-8501, Japan

*Corresponding Author.
Tel: +81-75-753-4423, E-mail: toyokuni@path1.med.kyoto-u.ac.jp.

1. Introduction

Redox cycling is a characteristic of transition metals including iron. Iron is an essential metal involved in oxygen transport by hemoglobin in mammals and activity of many enzymes including catalase and cytochromes.[1] Both deficiency and overload induce pathologic conditions in humans as microcytic anemia and hemochromatosis, respectively. Therefore, iron metabolism should be and is finely regulated. Recently, the understanding of iron metabolism has been enormously expanded by new findings regarding new iron transporters, mRNA-based regulation, and discovery of the hemochromatosis gene HFE. There is a growing body of evidence that suggests a role of iron in carcinogenesis.[2] This has been clearly demonstrated in animal models, and now the precise molecular mechanisms is to be elucidated.

2. Fenton Chemistry and "Free" Iron

Iron present in heme, iron–sulfur clusters or closely associated with proteins plays an important role in a variety of fundamental cellular functions such as oxygen transport, energy metabolism, electron transport and modulation of H_2O_2 levels. On the other hand, non-protein bound "free" or "catalytic" iron works differently.

A Nobel laureate Christian deDuve hypothesizes that iron was essential for the origin of life on earth in his book entitled "Blueprint for a cell"[3] as follows: "Thanks to the UV-supported photooxidation of Fe(II), CO_2 and other inorganic precursors were reduced to prebiotic building blocks with the consumption of protons. Oxidation of the synthesized materials takes place with Fe(III) ions as electron acceptors and is coupled to thioester-dependent substrate-level phosphorylations, capable, in turn, of supporting work. Thanks to the iron cycle, UV-light energy is made to support vital work". This hypothesis clearly describes the important character of iron, redox cycling.

Iron is the most abundant transition metal in human body (approximately 2–6 g).[1] Redox cycling of iron has closely associated with the production of reactive oxygen species (ROS). Fenton reported as early as in 1894 that ferrous sulfate and H_2O_2 causes oxidation of tartaric acid that gives a beautiful violet color on the addition of caustic alkali.[4] This has been the discovery of Fenton reaction that produces hydroxyl radicals ($^{\bullet}OH$) [Eq. (1)].

$$Fe(II) + H_2O_2 \rightarrow Fe(III) + {}^{\bullet}OH + OH^-. \tag{1}$$

In order to connect this chemical reaction with biological systems, the concept of "catalytic" or "free" iron presented by Gutteridge[5] is important. The original detection of "free" iron is by thiobarbituric acid (TBA) method in an *in vitro* system that contains a sample, calf thymus DNA, bleomycin, ascorbic acid and H_2O_2. Bleomycin in the presence of Fe(II) degrades DNA to form TBA-reactive substances. The chemistry of this system was analyzed by mass spectrometry.[6] Characteristics of "free" iron consist of the two components: (1) redox-activity and (2) diffusibility. In biological environments at neutral pH, Fe(III) is more stable than Fe(II) because of its low energy potential. However, Fe(III) dissolves in water at a very low concentration (10^{-17} M) at neutral pH. Most of Fe(III) precipitates as hydroxides at neutral pH.[7] On the other hand, iron chelated with citrate, ADP, ATP or GTP can stay as "free" iron at neutral pH.[8] In these iron chelates, at least one of the six ligands of iron are left free to maintain catalyzing activity.[9] It was shown that the fewer the number of ligands involved in chelation, the higher the catalyzing activity for ROS production is preserved.[10]

3. "Free" Iron in the Biological Environment

Limited amount of data is currently available concerning the localization of "free" iron in the cytoplasm or nucleus of cells because of the deficiency in appropriate methods. It has been believed that there exists a minute cellular pool of iron that is solubilized via chelation to low molecular weight biomolecules such as citrate and the adenine nucleotides.[11, 12] It is this pool of iron that is considered available for pathological free radical reactions in most instances. Recently, it was shown that nuclei of rat liver take up iron from ferric citrate by a process that is dependent on ATP.[13]

On the other hand, more data is available regarding extracellular "free" iron. Clinical significance of "non-transferrin plasma iron" ("free" iron) has been discussed.[14] Plasma transferrin is a considerable reserve for coping with increasing amounts of incoming iron. However, in acute iron poisoning, "free" iron concentrations ranging from 128 to over 800 μmol/l have been documented, exceeding many times the total biding capacity of transferrin.[15] Similarly, in severe idiopathic hemochromatosis and Bantu siderosis, acute episodes of abdominal pain and shock have been observed in extremely high serum iron measurements exceeding 2000 μmol/l.[16]

Another important concept for iron-dependent oxidative damage is site-specific mechanism. Fe(III) ions that are loosely bound to biological molecules such as DNA and proteins can undergo cyclic reduction and oxidation. This concept is different from "free" iron in that iron is not diffusible, but explains the funneling of free radical damage to specific sites and the possible "multi-hit" effect on the molecule.[17]

4. Significance of Superoxide in Iron-Dependent Oxidative Damage

Whereas reactivity of superoxide is relatively low, superoxide can reduce Fe(III) as shown in Eq. (2). The sum of Eqs. (1) and (2) is Eq. (3).

$$Fe(III) + O_2^- \rightarrow Fe(II) + O_2 \tag{2}$$

$$H_2O_2 + O_2^- \rightarrow {}^{\bullet}OH + OH^- + O_2. \tag{3}$$

Equation (3) is called Haber–Weiss reaction which was first postulated by Haber and Weiss in 1934.[18] Furthermore, superoxide has a potential to release iron from lactoferrin,[19] saturated transferrin,[20] ferritin[21, 22] or hemosiderin[22] in a catalytically-active form. These reactions are thought to be important in situations that increase superoxide generation such as in inflammation[23] or ischemia-reperfusion.[24]

5. Iron Transporter

How iron crosses cellular membrane is critical for its absorption and redistribution. There has been until recently little molecular information available on the mechanisms how metal ions are taken up by mammalian cells. In the presence of oxygen, Fe(III) is the favored species, but in the organism Fe(II) is required. The uptake and transport of iron under physiological conditions require special mechanisms since Fe(III) has a very low solubility at neutral pH as described above.[7] While the process of transferrin receptor (TfR)-mediated endocytosis has been well studied,[25] this was not the pathway by which iron is taken up into the body.

In 1997, mouse Nramp2/DCT1 was identified as an iron transporter by studying microcytic anemia mice (*mk*) with a genetic approach. The anemia of *mk* mouse is unresponsive to increased dietary iron, and iron injections did not reverse the anemia, suggesting a block of iron entry into red blood cell precursors as well.[26] Independently, this gene was identified with an expression cloning from a duodenal cDNA library prepared from mRNA from rats fed a low-iron diet. DCT1 was isolated by screening this library using a radiotracer assay of Fe(II) uptake in Xenopus oocytes. These experiments further revealed that DCT1 transports not only Fe(II) but also Zn(II), Mn(II), Cu(II), Co(II), Cd(II) and Pb(II).[27]

Nramp1 was cloned from functionally impaired macrophages.[28] Localization studies showed that Nramp1 protein is present in the lysosomal compartment of macrophages, and in phagosomal membranes during phagocytosis. Nramp1 may play a role in resistance to infections by depleting the phagosome of Fe(II) and other essential divalent metal cations.[29] Recently, it was reported in West Africans that four Nramp1 polymorphisms were each significantly associated with tuberculosis infection. Subjects who were heterozygous for two Nramp1 polymorphisms in intron 4 and the 3' untranslated region (UTR) of the gene were particularly overrepresented among those with tuberculosis, as compared with those with the most common Nramp1 genotype.[30]

Intestinal epithelial cells have two different iron transporters, one in the apical and one in the basolateral membrane. This is supported by the findings in *sla* mice that show normal uptake of iron into the villus cells, which is probably mediated by DCT1/Nramp2, but the release into the blood stream is impaired. Recent attempts to clone the basolateral membrane transporter suggest that it consists of at least two subunits, one for Fe(II) transport and the other for oxidation of Fe(II) to Fe(III).[31]

In addition to these iron transporters, frataxin, which is defective in the mitochondria of patients with Friedreich's ataxia has been isolated and shown to be an iron transport exit mechanism for mitochondria. Its defects lead to iron accumulation in the myocardium of patients.[32, 33]

6. Post-transcriptional Regulation of Iron Metabolism

The expression of modulator proteins in the iron metabolism of mammalian cells is controlled by intracellular iron levels. Recently, it was shown that this regulation is mediated at a post-transcriptional level, namely by specific mRNA-protein interactions in the cytoplasm. Particular hairpin structures, called as iron-responsesive elements (IREs) in the respective mRNAs, are recognized by transacting proteins, called iron-regulatory proteins (IRPs) that can control the rate of mRNA translation or stability. IREs are recognized not only in the 5' UTR of ferritin H- and L-chain[34, 35] but also in the 3' UTR of transferrin receptor (TfR) mRNA.[36]

The predicted interaction of IRP-1 with the TfR IRE was readily demonstrated in cells treated with iron chelator to remove iron and a clear correlation was obtained with the induction of TfR mRNA and protein after iron deprivation.[37] Now it has been established that IRP-1 plays a dual role as IRE-binding form without [4Fe-4S] cluster in iron deficiency and as cytoplasmic aconitase with [4Fe-4S] cluster in iron sufficiency. In iron deficiency, while translation of ferritin is blocked by the interaction of IRE and IRP-1 in the 5' UTR region, mRNA of TfR is stabilized by the same interaction in the 3' UTR region. Furthermore, it was discovered that signals other than iron levels can regulate IRP-1 and IRP-2 and modulate iron metabolism. Nitric oxide or oxidative stress transforms an inactive form of IRP-1 with [4Fe-4S] cluster to an active form without [4Fe-4S] cluster although the required time is different (~ 15 h versus < 1 h). The molecular mechanism of this activation has been reviewed in detail.[38]

7. Hemochromatosis

Hereditary hemochromatosis is an iron overload disorder that in years could not be diagnosed until the progressive accumulation of iron, mainly in the form of ferritin and hemosiderin, caused organ injury, particularly to liver, heart and endocrine pancreas. The disease has been diagnosed on the basis of classic triad: (1) a micronodular pigment liver cirrhosis in all the cases, (2) diabetes mellitus in about 75–80% of cases, and (3) skin pigmentation in about 75–85% of the cases. However, the disease can now be discovered much earlier by biochemical studies of blood before cirrhosis and other organ injuries have developed.[39] Hemochromatosis, a condition following hemosiderosis, is also used for a secondary excessive iron deposition with functional deterioration of any affected organ that is induced by genetic or non-genetic etiology such as thalassemia major and repeated transfusion.

Recently, HFE, the responsible gene for hereditary hemochromatosis was identified by a positional cloning approach. HFE is related to major histocompatibility complex class I proteins and is mutated in hereditary hemochromatosis.[40] The structure of its product was analyzed by X-ray crystallography.[41] It was shown that HFE protein binds to the transferrin receptor (TfR), a receptor by which cells acquire iron-loaded transferrin, and in the case of hemochromatosis that this interaction is disrupted.[42] The 2.8 Å crystal structure of a complex between the extracellular portions of HFE and TfR shows two HFE molecules which grasp each side of a twofold symmetric TfR dimer.[43] Intestinal crypt cells express HFE and TfR, whereas mature villus cells express DCT1 (see Sec. 5), but not HFE. These results lead to a speculation that HFE and TfR together sense serum iron in crypt cells. It was proposed that HFE and TfR in crypt cells regulate the expression of the proteins involved in iron absorption in villus cells, including DCT1, via the IRE/IRP system.[44]

It is of note that the major cause of death today in hereditary hemochromatosis is hepatocellular carcinoma.[39] Indeed, there is 219, 240 or 92.9 times greater risk, respectively, for primary hepatocellular carcinoma than the age-matched control population in three independent studies.[45–47] In general, hepatocelllular carcinoma is preceded by cirrhosis. A high incidence of cancers originating from the other organs (esophageal cancer, skin melanoma, etc.) has also been reported.[47–49] Further, cases of hepatocellular carcinoma in the absence of cirrhosis[50, 51] and after reversal of cirrhosis with therapy[52] have recently been reported. These suggest that irreversible genetic alteration may have occurred early in the course of the disease.

It has been suggested that lipid peroxidation is one of the factors important for the etiology of high incidence of hepatocarcinogenesis in hereditary hemochromatosis. The mechanisms include increased lysosomal membrane fragility and peroxidative damage of organelle such as microsomes and mitochondria.[53] It was shown that serum "free" form of iron in advanced hemochromatosis patients exists largely as complexes with citrate.[54] It was shown that ferric citrate efficiently induces oxidative single- and double-strand breaks in plasmid DNA *in vitro*.[55]

8. Iron and Carcinogenesis in Humans

Iron deficiency is an important nutritional problem in the developing countries because it reduces work performance by inducing anemia and reducing iron-containing respiratory enzymes required for muscle performance. In meat-eating countries, however, iron excess may be more of a problem than iron deficiency.[56, 57] There is a hypothesis that increased body iron stores are associated with an increased risk of cancer and with overall death rates. Two lines of evidence provide a biological rationale for the hypothesis. Iron can catalyze the production of ROS, and these species may be proximate carcinogens. Secondly, iron may increase the chances that cancer cells will survive and flourish.[58] There are several human epidemiological studies to support this hypothesis.[59, 60]

Nelson proposed a hypothesis that intestinal exposure to ingested iron may be a principal determinant of human colorectal cancer risk,[61] and they recently demonstrated a dose-response relationship for serum ferritin level and colon adenoma risk.[62] Indeed, large amounts of unabsorbed iron reach the colonic lumen especially in individuals who consume processed foods adulterated with inorganic iron and vitamin C supplements. This issue has been reviewed.[63]

Occupational exposure to asbestos is thought to pose an increased risk for the development of lung cancer (i.e. diffuse malignant mesothelioma, bronchogenic carcinoma). Approximately 30% of asbestos fiber is made of iron by weight. There is a close correlation between the incidence of tumors and iron content of asbestos fiber inhaled. Chrysotile and crocidolite are the two major responsible asbestos fibers that contain high amount of iron.[64]

Furthermore, increased body iron stores (as judged from serum ferritin level, transferrin concentration or transferrin saturation) were associated with poor prognosis of several human malignant neoplasms (neuroblastoma, childhood Hodgkin's disease, acute lymphocytic leukemia).[65–67]

9. Iron and Carcinogenesis in Animals

In 1959, Richmond first reported that an iron compound induced malignant tumor in animals.[68] In 1982, Okada first showed that an iron compound induces malignant tumor at sites different from those of injections.[69] Repeated intraperitoneal injections of an iron chelate, ferric nitrilotriacertate (Fe-NTA) induce a high incidence (60–92%) of renal cell carcinoma (RCC) in rats[70] and mice.[71] Nitrilotriacetic acid (NTA) is a synthetic aminotricarboxylic acid that efficiently forms water-soluble chelate complexes with several metal cations at neutral pH and has been used as a substitute for polyphosphates in detergents for household and hospital use in the US, Canada and Europe.[72] Fe-NTA works as an efficient "free" iron *in vitro* at neutral pH.[73] This model is characterized by (1) induction of only carcinoma, not sarcoma, (2) highly malignant potential as shown by pulmonary metastasis or peritoneal invasion, (3) marked increase in oxidatively modified biological molecules such as 8-oxoguanine, thymine-tyrosine cross-link, 4-hydroxy-2-nonenal and its modified proteins especially at an early stage, and (4) significant reduction of mortality and cancer incidence by vitamin E pretreatment.[2, 74]

10. Target Genes of Oxidative Stress-Induced Carcinogenesis

In contrast to rigorous antigen-antibody relationship of immunological reactions, free radical reactions reveals no such specificity. Therefore, it was hypothesized that genes are randomly injured and that there will be no specific target genes in oxidative stress-induced cancer. However, we have doubted this hypothesis, and used four independent strategies to answer the question whether there are target genes in Fe-NTA-induced cancer: (1) candidate gene approach (*ras*, *p53*, *vhl* and *tsc2*), (2) genes associated with metabolism of ROS (glutathione S-transferase, etc.), (3) genetic analysis using F1 hybrid rats, and finally (4) differential display technique.

Approaches (1) and (2) have not been successful. Thus, we have undertaken a strategy to scan the whole genome of Fe-NTA-induced RCCs in F1 hybrid rats in search of loss of heterozygosity (LOH) with microsatellite polymorphic markers by PCR. A preliminary experiment revealed a significantly high frequency of LOH (> 30%) on chromosomes 5 and 8. We then focused on chromosome 5 and collected data on all the RCCs registered. All the microsatellite markers appropriate for analysis showed LOH of > 40%.[75] Since LOH suggests the presence of target tumor suppressor genes according to the Knudson's "two hit theory",[76] we searched for candidate genes from the map position. p15 [INK4B] (p15) and p16 [INK4A] (p16) were the only candidate genes reported thus far at the map position indicated. We then evaluated whether these two genes are one of the targets for carcinogenesis by Southern blot analysis, PCR/SSCP analysis, Northern blot analysis as well as methylation-specific PCR analysis. In conclusion, 30.7% of p15 and 53.8% of p16 were affected either by point mutation, deletion or methylation of the promoter region. This is the first report that showed the presence of any target gene in oxidative stress-induced cancer model.[77] The biological significance of this finding is immense since p16 is associated not only with RB pathway as a cyclin-dependent kinase-4 inhibitor but also with p53 pathway via p19 ARF and MDM2.[77, 78]

In another study, we have applied a modified fluorescent differential display technique to the tumors. We have screened more than 80 000 PCR products. Reverse Northern blotting confirmed differential expression of 20 transcripts, which showed either significant increase, decrease or lack of expression in the RCCs. Fifteen cDNA clones were identified by homology search, which included annexin II, Y-box binding protein, ribosomal proteins, heat shock proteins, DNA polymerase, nonmuscle caldesmon (increased); protein tyrosine phosphatase (decreased); selenoprotein P, stromal cell-derived factor 1, intestinal trefoil protein, NADH dehydrogenase and insulin-like growth factor binding protein 7 (deleted). Most of the identified genes were associated with stress response or cellular proliferation. These results suggest that multiple, interactive genetic and epigenetic pathways are involved in carcinogenesis induced by iron-mediated oxidative stress.[79]

11. Conclusion

Iron plays an important role in free radical-induced tissue damage and carcinogenesis. In the past few years, our understanding of iron metabolism and iron-induced carcinogenesis has been enormously expanded. Modulation of iron intake and iron metabolism might be helpful for the prevention of aging and carcinogenesis.

Acknowledgments

This work was supported in part by a Grant-in-Aid from the Japanese Ministry of Education, Science, Sports and Culture, and a grant from the program for Promotion of Basic Research Activities for Innovative Bioscience (PROBRAIN).

References

1. Wriggleworth, J. M. and Baum, H. (1980). The biochemical function of iron. *In* "Iron in Biochemistry and Medicine, II" (A. Jacobs, and M. Worwood, Eds.), pp. 29–86, Academic Press, London.
2. Toyokuni, S. (1996). Iron-induced carcinogenesis: the role of redox regulation. *Free Radic. Biol. Med.* **20**: 553–566.
3. De Duve, C. (1991). *Blueprint for a Cell: The Nature and Origin of Life*, Neil Patterson Publishers, Burlington, NC.
4. Fenton, H. J. H. (1894). Oxidation of tartaric acid in presence of iron. *J. Chem. Soc.* **65**: 899–910.
5. Gutteridge, J. M. C., Rowley, D. A. and Halliwell, B. (1982). Superoxide-dependent formation of hydroxyl radicals and lipid peroxidation in the presence of iron salts: detection of "catalytic" iron and anti-oxidant activity in extracellular fluids. *Biochem. J.* **206**: 605–609.
6. Gajewski, E., Aruoma, O. I., Dizdaroglu, M. and Halliwell, B. (1991). Bleomycin-dependent damage to the bases in DNA is a minor side reaction. *Biochemistry* **30**: 2444–2448.

7. Lippard, S. J. and Berg, J. M. (1994). *Principles of Bioorganic Chemistry*, University Science Books, Mill Valley, CA.

8. Gutteridge, J. M. C. (1990). Superoxide-dependent formation of hydroxyl radicals from ferric-complexes and hydrogen peroxide: an evaluation of fourteen iron chelators. *Free Radic. Res. Commun.* **9**: 119–125.

9. Graf, E., Mahoney, J. R., Bryant, R. G. and Eaton, J. W. (1984). Iron-catalyzed hydroxyl radical formation: stringent requirement for free iron coordination site. *J. Biol. Chem.* **259**: 3620–3624.

10. Toyokuni, S. and Sagripanti, J. L. (1992). Iron-mediated DNA damage: sensitive detection of DNA strand breakage catalyzed by iron. *J Inorg. Biochem.* **47**: 241–248.

11. Mulligan, M., Althaus, B. and Linder, M. C. (1986). Non-ferritin, non-heme iron pools in rat tissues. *Int. J. Biochem.* **18**: 791–798.

12. Weaver, J. and Pollack, S. (1989). Low-Mr iron isolated from guinea pig reticulocytes as AMP-Fe and ADP-Fe complexes. *Biochem. J.* **261**: 787–792.

13. Gurgueira, S. and Meneghini, R. (1996). An ATP-dependent iron transport system in isolated rat liver nuclei. *J. Biol. Chem.* **271**: 13 616–13 620.

14. Hershko, C. and Peto, T. E. A. (1987). Anotation: non-transferrin plasma iron. *Br. J. Haematol.* **66**: 149–151.

15. Reynolds, L. G. and Klein, M. (1985). Iron-poisoning: a preventable hazard of childhood. *South African Med. J.* **67**: 680–683.

16. Buchannan, W. M. (1971). Shock in Bantu siderosis. *Am. J. Clin. Pathol.* **55**: 401–406.

17. Chevion, M. (1988). A site-specific mechanism for free radical induced biological damage: the essential role of redox-active transtion metals. *Free Radic. Biol. Med.* **5**: 27–37.

18. Haber, F. and Weiss, J. (1934). The catalytic decomposition of hydrogen peroxide by iron salts. *Proc. Roy. Soc. London* **A147**: 332–351.

19. Bannister, J. V., Bannister, W. H., Hill, H. A. O. and Thornalley, P. J. (1982). Enhanced production of hydroxyl radicals by the xanthine-xanthine oxidase reaction in the presence of lactoferrin. *Biochem. Biophys. Acta* **715**: 116–120.

20. Motohashi, N. and Mori, I. (1983). Superoxide-dependent formation of hydroxyl radical catalyzed by transferrin. *FEBS Lett.* **157**: 197–199.

21. Biemond, P., van Eijk, H. G., Swaak, J. G. and Koster, J. F. (1984). Iron mobilization from ferritin by superoxide derived from stimulated polymorphonuclear leukocytes. *J. Clin. Invest.* **73**: 1576–1579.

22. O'Connell, M. J., Halliwell, B., Moorehouse, C. R., Aruoma, O. I., Baum, O. I. and Peters, T. J. (1986). Formation of hydroxyl radicals in the presence of ferritin and haemosiderin: is haemosiderin formation a biological protective mechanism? *Biochem. J.* **234**: 727–731.

23. Babior, B. M., Kipnes, R. S. and Curnutte, J. T. (1973). Biological defence mechamism: the production by leukocytes of superoxide, a potent bactericidal agent. *J. Clin. Invest.* **52**: 741–744.

24. McCord, J. M. (1985). Oxygen-derived free radicals in postischemic tissue injury. *New England J. Med.* **312**: 159–163.

25. Richardson, D. and Ponka, P. (1997). The molecular mechanisms of the metabolism and transport of iron in normal and neoplastic cells. *Biochim. Biophys. Acta* **1331**: 1-40.

26. Fleming, M., Trenor, C. R., Su, M., Foernzler, D., Beier, D., Dietrich, W. and Andrews, N. (1997). Microcytic anaemia mice have a mutation in Nramp2, a candidate iron transporter gene. *Nature Genet.* **16**: 383–386.

27. Gunshin, H., Mackenzie, B., Berger, U., Gunshin, Y., Romero, M., Boron, W., Nussberger, S., Gollan, J. and Hediger, M. (1997). Cloning and characterization of a mammalian proton-coupled metal-ion transporter. *Nature* **388**: 482–488.

28. Atkinson, P. and Barton, C. (1998). Ectopic expression of Nramp1 in COS-1 cells modulates iron accumulation. *FEBS Lett.* **425**: 239–242.

29. Gruenheid, S., Pinner, E., Desjardins, M. and Gros, P. (1997). Natural resistance to infection with intracellular pathogens: the Nramp1 protein is recruited to the membrane of the phagosome. *J. Exp. Med.* **185**: 717–730.

30. Bellamy, R., Ruwende, C., Corrah, T., McAdam, K., Whittle, H. and Hill, A. (1998). Variations in the NRAMP1 gene and susceptibility to tuberculosis in West Africans. *New England J. Med.* **338**: 640–644.

31. Vulpe, C., Kuo, Y., Murphy, T., Cowley, L., Askwith, C., Libina, N., Gitschier, J. and Anderson, G. (1999). Hephaestin, a ceruloplasmin homologue implicated in intestinal iron transport, is defective in the sla mouse. *Nature Genet.* **21**: 195–199.

32. Babcock, M., de Silva, D., Oaks, R., Davis-Kaplan, S., Jiralerspong, S., Montermini, L., Pandolfo, M. and Kaplan, J. (1997). Regulation of mitochondrial iron accumulation by Yfh1p, a putative homolog of frataxin. *Science* **276**: 1709–1712.

33. Foury, F. and Cazzalini, O. (1997). Deletion of the yeast homologue of the human gene associated with Friedreich's ataxia elicits iron accumulation in mitochondria. *FEBS Lett.* **411**: 373–377.

34. Hentze, M., Caughman, S., Rouault, T., Barriocanal, J., Dancis, A., Harford, J. and Klausner, R. (1987) Identification of the iron-responsive element for the translational regulation of human ferritin mRNA. *Science* **238**: 1570–1573.

35. Hentze, M., Rouault, T., Caughman, S., Dancis, A., Harford, J. and Klausner, R. (1987). A *cis*-acting element is necessary and sufficient for translational regulation of human ferritin expression in response to iron. *Proc. Natl. Acad. Sci. USA* **84**: 6730–6734.

36. Casey, J., Hentze, M., Koeller, D., Caughman, S., Rouault, T., Klausner, R. and Harford, J. (1988). Iron-responsive elements: regulatory RNA sequences that control mRNA levels and translation. *Science* **240**: 924–928.

37. Koeller, D., Casey, J., Hentze, M., Gerhardt, E., Chan, L., Klausner, R. and Harford, J. (1989). A cytosolic protein binds to structural elements within the

iron regulatory region of the transferrin receptor mRNA. *Proc. Natl. Acad. Sci. USA* **86**: 3574–3578.

38. Hentze, M. and Kuhn, L. (1996). Molecular control of vertebrate iron metabolism: mRNA-based regulatory circuits operated by iron, nitric oxide, and oxidative stress. *Proc. Natl. Acad. Sci. USA* **93**: 8175–8182.

39. Cotran, R. S., Kumar, V. and Collins, T. (1999). *Robbins Pathologic Basis of Disease*, Sixth Edition, W.B. Saunders, Philadelphia.

40. Feder, J., Gnirke, A., Thomas, W., Tsuchihashi, Z., Ruddy, D., Basava, A., Dormishian, F., Domingo, R. J., Ellis, M., Fullan, A., Hinton, L., Jones, N., Kimmel, B., Kronmal, G., Lauer, P., Lee, V., Loeb, D., Mapa, F., McClelland, E., Meyer, N., Mintier, G., Moeller, N., Moore, T., Morikang, E., Wolff, R. *et al.* (1996). A novel MHC class I like gene is mutated in patients with hereditary haemochromatosis. *Nature Genet.* **13**: 399–408.

41. Lebron, J., Bennett, M., Vaughn, D., Chirino, A., Snow, P., Mintier, G., Feder, J. and Bjorkman, P. (1998). Crystal structure of the hemochromatosis protein HFE and characterization of its interaction with transferrin receptor. *Cell* **93**: 111–123.

42. Parkkila, S., Waheed, A., Britton, R., Bacon, B., Zhou, X., Tomatsu, S., Fleming, R. and Sly, W. (1997). Association of the transferrin receptor in human placenta with HFE, the protein defective in hereditary hemochromatosis. *Proc. Natl. Acad. Sci. USA* **94**: 13 198–13 202.

43. Bennett, M., Lebron, J. and Bjorkman, P. (2000). Crystal structure of the hereditary haemochromatosis protein HFE complexed with transferrin receptor. *Nature* **403**: 46–53.

44. Waheed, A., Parkkila, S., Saarnio, J., Fleming, R., Zhou, X., Tomatsu, S., Britton, R., Bacon, B. and, Sly W. (1999). Association of HFE protein with transferrin receptor in crypt enterocytes of human duodenum. *Proc. Natl. Acad. Sci. USA* **96**: 1579–1584.

45. Niederau, C., Fischer, R., Sonnenberg, A., Stremmel, W., Trampisch, H. J. and Strohmyer, G. (1985). Survival and causes of death in cirrhotic and in noncirrhotic patients with primary hemochromatosis. *New England J. Med.* **313**: 1256–1262.

46. Bradbear, R. A., Bain, C., Siskind, V., Schofield, F. D., Webb, S., Azelsen, E. M., Halliday, J. W., Bassett, M. L. and Powell, L. W. (1985). Cohort study of internal malignancy in genetic hemochromatosis and other chronic non-alcoholic liver diseases. *J. Natl. Cancer Inst.* **75**: 81–84.

47. Hsing, A. W., McLaughlin, J. K., Olsen, J. H., Mellemkjar, L., Wacholder, S. and Fraumeni, J. F. J. (1995). Cancer risk following primary hemochromatosis: a population-based cohort study in Denmark. *Int. J. Cancer* **60**: 160–162.

48. Ammann, R. W., Muller, E., Bansky, J., Schuler, G. and Hacki, W. H. (1980). High incidence of extrahepatic carcinomas in idiopathic hemochromatosis. *Scand. J. Gastroenterol.* **15**: 733–736.

49. Tiniakos, G. and Williams, R. (1988). Cirrhotic process, liver cell carcinoma and extrahepatic malignant tumors in idiopathic hemochromatosis. *Appl. Pathol.* **6**: 128–138.

50. Fellows, I. W., Stewart, M., Jeffcoate, W. J. and Smith, P. G. (1988). Hepatocellular carcinoma in primary hemochromatosis in the absence of cirrhosis. *Gut* **29**: 1603–1609.

51. Kew, M. D. (1990). Pathogenesis of hepatocellular carcinoma in hereditary hemochromatosis: ocurrence in noncirrhotic patients. *Hepatology* **6**: 1086–1087.

52. Blumberg, R. S., Chopra, S., Ibrahim, R., Crawford, J., Farraye, F. A., Zeldis, J. R. and Berman, M. D. (1988). Primary hepatocellular carcinoma in idiopathic hemochromatosis after reversal of cirrhosis. *Gastroenterology* **95**: 1399–1402.

53. Niemela, O., Parkkila, S., Britton, R., Brunt, E., Janney, C. and Bacon, B. (1999). Hepatic lipid peroxidation in hereditary hemochromatosis and alcoholic liver injury. *J. Lab. Clin. Med.* **133**: 451–460.

54. Grootveld, M., Bell, J. D., Halliwell, B., Aruoma, O. I., Bomford, A. and Sadler, P. J. (1989). Non-transferrin-bound iron in plasma or serum from patients with idiopathic hemochromatosis: characterization by high performance liquid chromatography and nuclear magnetic resonance spectroscopy. *J. Biol. Chem.* **264**: 4417–4422.

55. Toyokuni, S. and Sagripanti, J. L. (1993). Induction of oxidative single- and double-strand breaks in DNA by ferric citrate. *Free Radic. Biol. Med.* **15**: 117–123.

56. Conrad, M. E., Uzel, C., Berry, M. and Latour, L. (1994). Ironic catastrophes: One's food-anather's poison. *Am. J. Med. Sci.* **307**: 434–437.

57. Herbert, V., Shaw, S., Jayatilleke, E. and Stopler-Kasdan, T. (1994). Most free-radical injury is iron-related: it is promoted by iron, hemin, holoferritin and vitamin C, and inhibited by deferoxamine and apoferritin. *Stem Cells* **12**: 289–303.

58. Stevens, R. G., Jones, D. Y., Micozzi, M. S. and Taylor, P. R. (1988). Body iron stores and the risk of cancer. *New England J. Med.* **319**: 1047–1052.

59. Stevens, R. G., Graubard, B. I., Micozzi, M. S., Neriishi, K. and Blumberg, B. S. (1994). Moderate elevation of body iron level and increased risk of cancer occurrence and death. *Int. J. Cancer* **56**: 364–369.

60. Knekt, P., Reunanen, H., Takkunen, H., Aromaa, A., Heliovaara, M. and Hakulinen, T. (1994). Body iron stores and risk of cancer. *Int. J. Cancer* **56**: 379–382.

61. Nelson, R. L. (1992). Dietary iron and colorectal cancer risk. *Free Radic. Biol. Med.* **12**: 161–168.

62. Nelson, R. G., Davis, F. G., Sutter, E., Sobin, L. H., Kikendall, J. W. and Bowen, P. (1994). Body iron stores and risk of colonic neoplasia. *J. Natl. Cancer Inst.* **86**: 455–460.

63. Weinberg, E. D. (1994). Association of iron with colorectal cancer. *Biometals* **7**: 211–216.

64. Mossman, B. T., Bignon, J., Corn, M., Seaton, A. and Gee, J. B. L. (1990). Asbestos: scientific developments and implications for public policy. *Science* **247**: 294–301.

65. Evans, J. E., D'angio, G. J., Propert, K., Anderson, J. and Hann, H.-W. L. (1987). Prognostic factors in neuroblastoma. *Cancer* **59**: 1853–1859.

66. Hann, H.-W. L., Lange, B., Stahlhut, M. W. and McGlynn, K. A. (1990). Prognostic importance of serum transferrin and ferritin in childhood Hodgkin's disease. *Cancer* **66**: 313–316.

67. Potaznic, D., Groshen, S., Miller, D., Bagin, R., Bhalla, R., Schwartz, M. and de Sousa, M. (1987). Association of serum iron, serum transferrin saturation, and serum ferritin with survival in acute lymphocytic leukemia. *Am. J. Pediatr. Hematol/Oncol.* **9**: 350–355.

68. Richmond, H. G. (1959). Induction of sarcoma in the rat by iron-dextran complex. *Br. Med. J.* **1**: 947–949.

69. Okada, S. and Midorikawa, O. (1982). Induction of rat renal adenocarcinoma by Fe-nitrilotriacetate (Fe-NTA). *Japan Arch. Int. Med.* **29**: 485–491.

70. Ebina, Y., Okada, S., Hamazaki, S., Ogino, F., Li, J. L. and Midorikawa, O. (1986). Nephrotoxicity and renal cell carcinoma after use of iron- and aluminum-nitrilotriacetate complexes in rats. *J. Natl. Cancer Inst.* **76**: 107–113.

71. Li, J. L., Okada, S., Hamazaki, S., Ebina, Y. and Midorikawa, O. (1987). Subacute nephrotoxicity and induction of renal cell carcinoma in mice treated with ferric nitrilotriacetate. *Cancer Res.* **47**: 1867–1869.

72. Anderson, R. L., Bishop, W. E. and Campbell, R. L. (1985). A review of the environmental and mammalian toxicology of nitrilotriacetic acid. *Crit. Rev. Toxicol.* **15**: 1–102.

73. Hamazaki, S., Okada, S., Li, J.-L., Toyokuni, S. and Midorikawa, O. (1989). Oxygen reduction and lipid peroxidation by iron chelates with special reference to ferric nitrilotriacetate. *Arch. Biochem. Biophys.* **272**: 10–17.

74. Zhang, D., Okada, S., Yu, Y., Zheng, P., Yamaguchi, R. and Kasai, H. (1997). Vitamin E inhibits apoptosis, DNA modification, and cancer incidence induced by iron-mediated peroxidation in Wistar rat kidney. *Cancer Res.* **57**: 2410–2414.

75. Tanaka, T., Iwasa, Y., Kondo, S., Hiai, H. and Toyokuni, S. (1999). High incidence of allelic loss on chromosome 5 and inactivation of p15[INK4B] and p16[INK4A] tumor suppressor genes in oxystress-induced renal cell carcinoma of rats. *Oncogene* **18**: 3793–3797.

76. Knudson, A. G., Jr., Hethcote, H. W. and Brown, B. W. (1975). Mutation and childhood cancer: a probabilistic model for the incidence of retinoblastoma. *Proc. Natl. Acad. Sci. USA* **72**: 5116–5120.

77. Kamijo, T., Weber, J. D., Zambetti, G., Zindy, F., Roussel, M. F. and Sherr, C. J. (1998). Functional and physical interactions of the ARF tumor suppressor with p53 and Mdm2. *Proc. Natl. Acad. Sci. USA* **95**: 8292–8297.

78. Tao, W. and Levine, A. J. (1999). P19(ARF) stabilizes p53 by blocking nucleo-cytoplasmic shuttling of Mdm2. *Proc. Natl. Acad. Sci. USA* **96**: 6937–6941.
79. Tanaka, T., Kondo, S., Iwasa, Y., Hiai, H. and Toyokuni, S. (2000). Expression of stress-response and cell proliferation genes in renal cell carcinoma induced by oxidative stress. *Am. J. Pathol.* **156**: 2149–2157.

SECTION 9

Aging and Oxidative Stress

Chapter 59

Aging and Oxidative Stress in Transgenic Mice

Holly Van Remmen*, James Mele,
XinLian Chen and Arlan Richardson

Holly Van Remmen and **Arlan Richardson** • Department of Physiology, and GRECC,
Audie Murphy VA Hospital, 7400 Merton Minter Blvd., San Antonio, TX 78229
James Mele and **XinLian Chen** • Department of Cellular and Structural Biology, The
University of Texas Health Science Center at San Antonio, South Texas Veterans Health
Care System, San Antonio, Texas 78284-7756

*Corresponding Author.
Tel: 210/617-5300, Ext: 5673, E-mail: vanremmen@uthscsa.edu

1. Summary

The role of oxidative stress/damage in aging can now be studied directly in mammals using transgenic or knockout mice. Over the past two decades, a large number of transgenic and knockout mouse models with alterations in genes involved in antioxidant protection have been developed. Currently, there are only a few studies that have used transgenic/knockout mice to study the role of oxidative stress in aging. Transgenic mice that have increased activity of Cu, Zn superoxide dismutase show no alteration in life span compared to non-transgenic mice. However, disruption of the p66^{shc} gene, which plays a role in response of cells to oxidative stress, increases life span. In contrast, pilot data with heterozygous Mn superoxide dismutase knockout mice suggest that reduced expression of Mn superoxide dismutase results in a shorter life span.

2. Introduction

Oxidative damage to cellular macromolecules has long been implicated in the aging process. One of the most popular theories in aging research, the free radical hypothesis of aging, was first introduced by Harman in 1956. He proposed that free radicals produced during the course of normal aerobic metabolism caused random deleterious damage that accumulated over time and contributed to the process of aging and various age-associated diseases.[1, 2] Over the past two decades, it has become apparent that many reactive oxygen species (ROS), such as peroxides, which are not free radicals, also play a role in oxidative damage in cells; therefore, the free radical hypothesis of aging has been modified to the oxidative stress hypothesis of aging. In its current form, the oxidative stress hypothesis of aging can be stated as follows:[3-5] *A chronic state of oxidative stress exists in cells of aerobic organisms even under normal physiological conditions because of an imbalance of prooxidants and antioxidants. This imbalance results in a steady-state accumulation of oxidative damage in a variety of macromolecules. Oxidative damage increases during aging, which results in a progressive loss in the functional efficiency of various cellular processes.* Although this theory is currently one of the most popular explanations for how aging occurs at the biochemical level, most of the evidence in support of this theory is correlative.

One consistent line of evidence to support the oxidative stress hypothesis of aging has been the large number of studies that have shown an age-related increase in oxidative damage to a variety of molecules (lipid, protein, and DNA) in organisms ranging from invertebrates to humans.[3, 4, 6] Another strong line of evidence in support of this theory is the fact that dietary restriction, which is the only experimental manipulation that has been shown to retard aging and extend life span in rodents, alters the age-related accumulation of oxidative damage in rodent tissues.[7-9] Dietary restriction has been shown to reduce the level of oxidative damage in liver and other tissues as measured by a decrease in lipofuscin,[10-12] lipid peroxidation,[9, 12-18] protein oxidation,[19, 20] and DNA oxidation.[21-23]

Currently, the strongest evidence for the oxidative stress theory of aging has come from studies with invertebrates. The age-1 mutants of *C. elegans*, which were selected for increased longevity, have been shown to be more resistant to hydrogen peroxide and paraquat and to have increased superoxide dismutase and catalase activities.[24, 25] Recently, Honda and Honda[26] showed that the daf-2 mutants of *C. elegans*, which have extended lifespan, are resistance to oxidative stress and have increased levels of Mn superoxide dismutase. Taub *et al.*[27] also showed that mutations in the cytosolic catalase gene (ctl-1) eliminated the daf-c and clk-1 mediated extension of adult lifespan in *C. elegans* suggesting that the extension of lifespan in the daf-c and clk-1 mutants was due to reduced oxidative damage. In addition, Arking's laboratory found that the extended longevity of long lived *Drosophila* was correlated to an increased expression of superoxide dismutase and catalase.[28, 29] Rose's laboratory found that the longer life span in their strains of *Drosophila* covaries with increased resistance to various kinds of stress and with the increased frequency of expression of a more active form of superoxide dismutase.[30]

Investigators are now using *Drosophila* that have been genetically engineered to overexpress various antioxidant enzymes to study the role of oxidative stress in aging. The observation by Orr and Sohal in 1994 that overexpression of both CuZn superoxide dismutase (CuZnSOD) and catalase resulted in a significant increase (14 to 34%) in the life span of Drosophila and resulted in lower levels of protein and DNA oxidation, is the most direct evidence that the age-related accumulation of oxidative damage is responsible for aging.[31, 32] However, transgenic studies in *Drosophila* using P-element mediated transformation are limited by the fact that the control and experimental lines have different genetic backgrounds, a factor that has been shown to alter lifespan independently apart from any transgenetic manipulation.[33-35] Therefore, it is not clear if this initial exciting observation will be confirmed when larger numbers of lines are studied. To overcome the problem of positional effects, Tower's laboratory used an inducible yeast FLP/FRT recombination system to induce the expression of various antioxidant enzymes.[36] Overexpression of CuZnSOD resulted in an increase in lifespan of up to 48%. However, they found no added benefit of overexpressing both catalase and CuZnSOD. Parkes[37] recently targeted the overexpression of CuZnSOD in *Drosophila* to motorneurons using a yeast UAS element that was regulated by a GAL-4 activator and found that overexpression of CuZnSOD in motorneourons resulted in an increase in lifespan (40%) as well as an increase in resistance to paraquat and γ-irradiation.

3. Transgenic/Knockout Mouse Models with Alterations in the Antioxidant Defense System

3.1. Transgenic Mice Overexpressing Antioxidant Genes

Over the past two decades various transgenic and knockout mouse models have been generated that have alterations in a wide variety of genes involved in antioxidant protection. Tables 1 lists transgenic mouse models that overexpress genes coding for proteins that play a role in antioxidant protection. A major problem encountered by investigators in generating transgenic models to study aging is the selection of the enhancer/promoter to drive the expression of the transgene. Because of the global nature of the aging process (i.e. aging affects essentially all cells and no one tissue is responsible for aging), transgenic experiments designed to test mechanisms of aging require that the gene or process of interest be altered in most, if not all, tissues of the transgenic mice. Therefore, the best promoters/enhancers for these types of aging studies would be those that direct the ubiquitous expression of a gene. Unfortunately, most of the promoters and enhancers that have been characterized in transgenic mice direct the expression of genes to one or a few tissues.[38] While these models are useful for studying the effect of oxidative stress on a particular tissue, these models are of limited usefulness in studying the role of oxidative stress/damage in aging.

An alternative approach to using a ubiquitous promoter is to use a large genomic DNA fragment containing the gene (introns and exons) of interest and the 5'- and 3'-flanking regions with the regulatory sequences necessary to drive the expression of the transgene. Cloning genomic DNA fragments that are large enough to span an entire gene became routine in the early 1990s with the introduction of the P1 bacteria phage vector, which permits cloning of genomic DNA fragments of 80 to 100 kb in length.[39, 40] Yan et al.[41] produced transgenic mice using a 45-kb human renin genomic fragment that contained approximately 25-kb 5'-flanking DNA and 6-kb 3'-flanking DNA. The expression of the renin transgene in these mice was comparable to the endogenous renin gene. In contrast, the expression of the renin transgene was quite different from the endogenous gene when transgenic mice were made with a renin gene construct that contained only a few kb of 5-flanking DNA.

The first transgenic mouse that overexpressed an antioxidant enzyme was produced using a 15-kb fragment containing the human 10-kb Sod-1 gene.[42] The initial study reported that the expression of CuZnSOD was dramatically increased in brain of this transgenic line; however, the expression of CuZnSOD was only marginally higher than non-transgenic mice in many of the other tissues studied. Later studies with other lines of transgenic mice reported enhanced expression of CuZnSOD in all tissues of the transgenic mice.[42, 43] Recently, our laboratory used very large fragments (60- to 80-kb) of human genomic DNA containing the intact genes for CuZnSOD and catalase to generate transgenic mice. These genes

Table 1. Transgenic Mice Overexpressing Proteins Involved in Antioxidant Protection

Catalase

Genetic Background	Promoter	Transgene Expression	Phenotype
FVB	Mouse α cardiac myosin heavy chain	Heart: 2- to 630-fold	Suppression of doxorubicin induced cardiac lipid peroxidation, elevation of serum creatine phosphokinase, and functional changes in the isolated atrium.[58]
			Resistant to ischemia-reperfusion injury in the heart.[59]
			Repression of hypoxia-reoxygenation injury in the heart.[60]
			Overexpression of catalase in heart is limited to cardiomyocytes.[61]
	Rat insulin I	Pancreas islet: 10- to 50-fold	Protect beta cell against hydrogen peroxide toxicity and reduce the diabetogenic effect of streptozocin *in vivo.*[64]
C57BL/6 x C3H	Mouse α-AFP enhancer 1/human β globin	Liver: 3- to 4-fold, Gut: 1.9-fold	Catalase overexpression increased the concentrations of malondialdehyde (in untreated mice only) and conjugated dienes (in both untreated and ciprofibrate-fed mice).[62]
			Inhibition of NF-κB activation and DNA synthesis induced by the peroxisome proliferator ciprofibrate.[63]
			Catalase overexpression did not affect the concentration of 8 OH-dG induced by ciprofibrate.[63]
C57BL/6	Endogenous human catalase	2- to 4-fold in all tissues	Hepatocytes are resistant to hydrogen peroxide.[65]

Table 1 (*Continued*)

Glutathione Peroxidase 1

Genetic Background	Promoter	Transgene Expression	Phenotype
C57BL/6 x CBA/J	Mouse HMGCoA Reductase	Brain: 4-fold Muscle: 1.6-fold Liver: 1.3-fold Kidney: 1.5-fold Skin: 2.1-fold	Transgenic mice are thermosensitive.[66] Decreased sensitivity of dopamine neurons to damage due to energy impairment.[67] Reduced polyploidy and increased production of hydrogen peroxide in hepatocytes after partial hepatectomy.[68] Protection from ischemia/reperfusion injury in kidney.[69] Increased sensitivity to acetaminophen toxicity.[66] Protection against 6-hydroxydopamine-induced toxicity in dopaminergic neurons.[70] Protection against focal cerebral ischemia/reperfusion damage.[71] Increased tumorigenesis in a DMBA/TPA two-stage skin carcinogenesis model.[72]
B6C3F1	2.0 kb mouse GPx1 endogenous	Heart: 5-fold	Transgenic mice are less susceptible to ischemia/reperfusion-induced apoptosis in the heart.[73] Resistance to myocardial ischemia reperfusion injury.[74]

Table 1 (*Continued*)

Genetic Background	Promoter	Transgene Expression	Phenotype
B6C3F1	Human metallothionein-II A	GPx1 protein level: Brain: Substantia nigra: 3.6-fold Stratium: 1.8-fold Cortex: 2.3-fold Cerebellum: 4.8-fold	Characterization of the model.[75] Increased resistance to tumor promoter-induced loss of glutathione peroxidase activity in skin.[76] Overexpression of glutathione peroxidase prevents impairment of synaptic transmission by transient hypoxia in hippocampal slices.[77]
B6C3F1	Mouse endogenous GPX1 (2.2 kb)	Lens: 5-fold GPX1 mRNA level: 24% in kidney Heart: 6-fold Lung: 5.8-fold Muscle: 2.6-fold	Protection against the lethal oxidative stress caused by high levels of paraquat.[78] Overexpression of GPx1 does not affect the expression of plasma glutathione peroxidase or phospholipid hydroperoxide glutathione peroxidase in mice offered diets adequate or deficient in selenium.[79] The lenses of the transgenic mice were able to resist the cytotoxic effect (damage to cell morphology, DNA strand breaks) of hydrogen peroxide.[80] Overexpression of GPx1 in lens epithelial cells does not significantly change the response to hydrogen peroxide stress.[81]

Table 1 (*Continued*)

CuZn Superoxide Dismutase

Genetic Background	Promoter	Transgene Expression	Phenotype
CBYB/6XB6D/2	Endogenous	Brain: 2 to 10-fold	Not sufficient to extend the life spans of transgenic mice.[43]
		Heart: 3 to 4.7-fold	
		Lung: 1.4 top 3.3-fold	Airways resistant to allergen-induced changes.[82]
		Liver: 1.2 to 2.3-fold	Increased SOD is effective on the amelioration of vasospasm after subarachnoid hemorrage.[83]
		RBCs: 2 to 4.8-fold	
		Muscle: 3.5-fold	
		Tongue: 1 to 4-fold	Deficient in spatial memory.[84]
		Pancreas: 1.7-fold	Gross atrophy of the quadriceps muscle; other hindlimb muscles variably affected.[85]
		Spleen: 3-fold	
		Thymus: 2 to 5-fold	
		Bone marrow: 1 to 5-fold	Decreased damage following ischemia/reperfusion in the gut.[86]
			Transgenic mice overexpressing the human gene for Cu, Zn SOD are not protected against experimentally induced endotoxemia.[87]
			Decreased neurotoxicity after treatment with 3-NP.[88]
			Reduction in infarct volume and neurological deficits after a focal stroke.[89]
			Homozygous SOD-Tg mice showed resistance to the lethal effects of MDA and MDMA.[90]
			Decreased secretion of prostaglandin by kidney and cerebellum.[91]
			Morphological changes in neuromuscular junction similar to changes in aging muscle.[92]
			More resistant to cold induced endothelial cell injury.[93]

Table 1 (*Continued*)

Genetic Background	Promoter	Transgene Expression	Phenotype
			Less sensitive to kainic acid-induced behavioral seizures than control mice.[94]
			Cerebroprotective effects during cerebral ischemia-reperfusion mediated by astrocyte glutamate transport.[95]
			Reduced cytosolic release of cytochrome c after transient focal cerebral ischemia.[96]
			Reduced neuronal death in response to peroxynitrite formation.[97]
			Increased production of H_2O_2 following hypoxemia ischemia in brain.[98]
			Decreased pulmonary oxygen toxicity and associated histologic damage and mortality.[99]
			Higher concentrations of the biogenic amines in specific brain regions.[100]
			Higher sensitivity to infection by *Plasmodium berghei*.[101]
			Thymus and bone marrow abnormalities.[102]
			Impairment of muscle function.[103]
			Increased susceptibility to kainic acid-mediated excitotoxicity.[104]
			Similar changes to those observed in tongues of patients with Down's syndrome.[105]
			Morphological remodeling and increased complexity in the neuromuscular.[106]
			Less susceptible to acute edematous pancreatitis.[107]
			Protects against glutamate neurotoxicity *in vitro*.[108]
			No significant difference in either brain distribution or in concentrations of [3H]-MPTP binding.[109]
			Reduces the level of ischemic damage in brain.[110]
			Resistance against MPTP-induced neurotoxicity.[111]
			Decreased serotonin uptake by platelets.[112]
			Morphological changes in the neuromuscular junction.[113]

Table 1 (*Continued*)

Genetic Background	Promoter	Transgene Expression	Phenotype
		Embryos: 4-12x	Protective effect against diabetes-associated embryopathy.[114]
B6D2F1	Endogenous	Brain: 1.93x	Increased TBA-reactive material was significantly higher in transgenic brains.[115]
		Heart: 1.69x	
		Thymus: 1.49x	
		Muscle: 1.25 fold	Decreased concentrations of 4-hydroxy-2-nonenal in brain.[116]
		Liver: 1.19 fold	
		Kidney: 1.18 fold	Premature involution of the thymus.[117]
		Spleen: 1.35 fold	
		Lung: 1.26x	Modification of the cellular architecture and morphology associated with a lipidic invasion and signs of a premature involution of the thymus.[118]
		Erythrocytes: 1.09x	
			Decreased mossy fiber innervation in hippocampus.[119]
C57BL/6 x CBA/J	Mouse HMGCoA Reductase	Intestine: 2-fold	Protects tissues from neutrophil infiltration and lipid peroxidation during intestinal ischemia-reperfusion.[120]
		Brain: 2.5-fold	
		Muscle: 3.9-fold	
		Liver: 3.2-fold	
CBYB/6XB6D/2XCD-1	Endogenous	3x in intraperitoneal macrophages	Intracellular production and release of H_2O_2 in macrophages from transgenic mice activated by PMA was found to be significantly increased.[121]
		2-fold pancreas	Increased tolerance of pancreatic beta cells to oxidative stress-induced diabetogenesis.[122]
		1–5-fold in all other tissues	
C57L/6J x C3H/HeJ	Endogenous	Heart: 10-fold	Protects the heart from this injury.[123]
CBYB/6XB6D/2; CD-1 for Sod2	Endogenous	Brain: 3-fold	Overexpression of CuZnSOD does not prevent neonatal lethality in mice that lack MnSOD.[124]
		Liver: 2-fold	
		Heart: 2-4-fold	
		Blood 5-6-fold	

Table 1 (*Continued*)

Mn Superoxide Dismutase

Genetic Background	Promoter	Transgene Expression	Phenotype
B6C3	β-actin	Heart: 1.7-3-fold, Brain: 1.3-fold Lung: 1.6-to 2-fold Muscle: 1.9-fold	Protection against peroxidative damage to membrane lipids.[125]
			Increased resistance to ischemia/reperfusion injury.[126]
			Increased survival following exposure to 90% O_2.[127]
			Protected against adriamycin cardiotoxicity.[128]
FVB/N	human surfactant protein C	Lung: 4-fold	Increased survival following exposure to 95% O_2.[129]

Extracellular Superoxide Dismutase

C57BL/ 6 x C3H	human surfacent protein C	Lung: 3-fold	Protective response to hyperoxia.[130]
C57BL/ 6 x C3H	β-actin	Brain: 5-fold Heart, muscle: 2-3-fold	Reduced injury from global cerebral ischemia.[131]
			Resistance to cognitive effects of an NO synthase inhibitor and impaired learning.[132]
			Improves preservation of myocardial function after ischemia and reperfusion injury.[133]
			Increased sensitivity to O_2 toxicity.[134]

Thioredoxin

C57BL/6	β-actin	Heart, kidney, skin, liver, lung, brain	Attenuation of focal ischemic brain damage.[135]

contained 22- to 25-kb of the 5'-flanking region to these two genes. Transgenic mice generated with these large genomic DNA fragments overexpressed CuZnSOD and catalase 2- to 3-fold in all tissues of the transgenic mice. In other words, a similar magnitude of overexpression of these genes was achieved in all tissues of the transgenic mice. Therefore, it appears that global overexpression of a transgenic can be obtained using large genomic DNA fragments containing the transgene. However, there are problems associated with generating transgenic mice using large fragments of DNA. For example, large DNA fragments are more difficult to prepare and are more susceptible to shearing, and it is not uncommon to have rearrangements when the large DNA fragments integrate into the genome. In addition, one must be concerned that genes other that the gene of interest might be present in the large genomic DNA fragments.

3.2. Knockout Mice Showing Reduced Expression of Antioxidant Genes

Using germline transmission of targeted mutations, various types of mutations can be generated in specific genes, e.g. point mutations, in which a slightly altered protein product is produced, or, more commonly, null (knockout) mutations where a portion of the gene is deleted and the expression of the gene reduced. Mice heterozygous for the mutation can then be bred to obtain mice homozygous for the mutated allele. However, in some cases, mice homozygous for a null mutation are not viable (e.g. Mn Superoxide Dismutase (MnSOD) knockout mice). The fetuses die during embryonic development or the mice survive for only a short time after birth. Although most of the research using knockout mice has been conducted with homozygous null knockout mice, mice heterozygous for a mutated gene are a potentially powerful resource for aging research that has been underutilized. Heterozygous knockout mice generally exhibit one half of the expression of the mutated gene found in the tissues of wild type mice. However, this must be established for each knockout mouse model because it is possible that up-regulation of the expression of the non-mutated allele occurs to compensate for the reduced copy number of the gene. A particular strength of heterozygous knockout mice is that a 50% decrease in the activity of an enzyme or a biochemical process is physiologically relevant to the aging process. Most enzymes that decrease with age show less than a 50% decrease.[44-47] In addition, the changes in gene expression that occur with dietary restriction, which has been shown to modulate the aging process in rodents,[7, 48] are generally in the range of 30 to 50%.[49, 50] Therefore, the changes in gene expression that occur with age or are altered by dietary restriction are in the same range as the changes in expression found in transgenic mice heterozygous for a null mutation.

Table 2 lists all the studies in which mice with targeted mutations in a gene coding for a protein involved in antioxidant protection have been generated. For example, knockout mice have been generated in which the genes coding for the

Table 2. Knockout Mice with Reduced Expression of Proteins Involved in Antioxidant Protection

Glutathione Peroxidase 1

Genetic Background	Phenotype
129SVJ x C57BL/6	GPx1 knockout mice are fertile, appear normal, and show no increase in sensitivity to hyperoxia.[136]
	Lens epithelial cells of the GPx1 knockout mice do not show significant difference in response to H_2O_2 compared to wild type mice.[81, 136]
	Lenses of the GPx1 knockout mice are sensitive to cytotoxicity of hydrogen peroxide.[137]
	Increased sensitivity to photochemical stress.[138]
	Increased sensitivity to paraquat.[78]
	Increased sensitivity to paraquat-induced oxidative destruction of lipids and protein.[139]
	Susceptible to viral-induced myocarditis.[140]
	GPx1 knockout mice express normal level of plasma and phospholipid hydroperoxide glutathione peroxidases in various tissues.[79]
	Increased vulnerability to malonate, 3-nitropropionic acid, and 1-methyl-4-phenyl-1,2,5,6-tetrahydropyridine.[141]
	Increased sensitivity to diquat-induced oxidative stress.[142]
	Increased sensitivity to N-methyl-4-phenyl-1,2,5,6-tetrahydropyridine toxicity.[143]
	Protection from γ-irradiation damage in jejunum crypts.[144]
	Elevated basal brain oxidative levels and resistance to kainic acid-induced seizure activity and neurodegeneration.[145]
129SVJ	Highly sensitive to paraquat and the neurons are more sensitive to H_2O_2.[146]
	Increased susceptibility to myocardial ischemia reperfusion injury.[147]
	Hearts show increased ischemia/reperfusion injury.[73]
129Svs3 x C57BL/6	GPx1 knockout mice show 20% decrease in body weight at the age of 8 months; increased lipid peroxides and liver mitochondria release more H_2O_2.[148]

CuZn Superoxide Dismutase

C57BL/6	Hearts from heterozygous and homozygous knockout mice vulnerable to ischemia reperfusion injury.[149]

Table 2 (*Continued*)

Genetic Background	Phenotype
CD-1	Axonal sprouting and reinnervation of denervated muscle fibers are functionally impaired in the absence of SOD-1.[150]
	Knockout mice are more susceptible to noise-induced permanent threshold shifts than wild-type and heterozygous control mice.[151]
	Increased infarct volume, brain swelling, increased apoptotic neuronal cell death and neurological deficits 24 h following transient focal cerebral ischemia.[152]
	Cu, Zn SOD deficient mice exhibit marked vulnerability to motor neuron loss after axonal injury.[153]
	Changes in ultrastructural calcium distribution during maturation in spinal- and oculo-motor neurons.[154]
	Homozygous mutant mice developed significantly more lymphocyte apoptosis than did heterozygous knockout mice or wildtype mice.[155]
	Increased superoxide production and hippocampal injury in the knockouts after global ischemia.[156]
129/CD-1	Mild muscle denervation and behavioral and physiological motor deficits.[157]
	Compared with wild-type mice, homozygous and heterozygous knockout mice exhibited significant threshold elevations and greater hair cell loss.[158, 159]
129SvJxC57BL/6	Female homozygous knock-out mice showed a markedly reduced fertility compared with that of wild-type and heterozygous knock-out mice.[160]

Mn Superoxide Dismutase

Not stated	Several novel pathologic phenotypes including severe anemia, degeneration of neurons in the basal ganglia and brainstem, and progressive motor disturbances characterized by weakness, rapid fatigue, and circling behavior; 10% of the knockout mice exhibit markedly enlarged and dilated hearts.[52]
CD-1	Tissue-specific inhibition of the respiratory chain enzymes NADH-dehydrogenase (complex I) and succinate dehydrogenase (complex II), inactivation of the tricarboxylic acid cycle enzyme aconitase, partial defect in 3-hydroxy-3-methylglutaryl-CoA lyase, and accumulation of oxidative DNA damage.[51]
	Treatment with the superoxide dismutase mimetic MnTBAP rescues the mutant mice from systemic pathology and dramatically prolongs their survival.[161]
	More susceptible to hyperoxia injury.[162]
	Normal resistance to 100% O_2 toxicity in heterozygous knockout mice.[163]

Table 2 (*Continued*)

Mn Superoxide Dismutase

Genetic Background	Phenotype
	Increase in infarct size and hemisphere enlargement following cerebral ischemia resulting in advanced neurological deficits but without inducing DNA fragmentation.[164]
C57BL/6	Characterization of the model.[55]
	Decrease in the respiratory control ratio and increased rate of induction of the permeability transition was observed in liver mitochondria isolated from heterozygous knockout mice.[57]

Extracellular Superoxide Dismutase

C57BL/6	Low EC-SOD activity may contribute to high alloxan susceptibility of beta-cells.[165]
	Reduction in survival after exposure to greater than 99% oxygen compared to wild-type mice and an earlier onset of severe lung edema.[166]

Thioredoxin

129sv x C57BL/6	$TRX^{-/-}$ mice are embryonic lethal; $TRX^{+/-}$ mice are viable, fertile and appear normal.[167]

p66[shc]

Not stated	30% increase in lifespan and increased resistance to paraquat.[54]

major antioxidant enzymes has been disrupted, e.g. Sod-1, Sod-2, Sod-3, and GPx1. Interestingly, only the homozygous, null mutation in the Sod-2 gene (i.e. mice lacking MnSOD) is lethal. Mice lacking MnSOD die within 1 to 18 days from dilated cardiomyopathy or neurodegeneration depending on the genetic background.[51–53] In contrast, homozygous mutations in Sod-1, Sod-3, and GPx1 are not lethal, and mice with these null mutations appear normal except for an increased sensitivity to certain types of oxidative stress (Table 2).

4. Transgenic/Knockout Mouse Models Used to Study the Role of Oxidative Stress in Aging

4.1. Transgenic Mice Overexpressing CuZnSOD

The transgenic mice generated by Epstein's and Groner's laboratories using the 15-kb genomic fragment containing the Sod-1 gene have been extensively studied respect to resistance to oxidative stress and a variety of other biological parameters (see Table 1). In general, these mice, which overexpress CuZnSOD, show increased resistance to oxidative stress, e.g. focal cerebral ischemia, pulmonary oxygen toxicity and reactive oxygen species such as peroxynitrite. However, several negative effects have also been reported in these transgenic mice, e.g. reduced macrophage function and thymus evolution. Recently, Huang et al.[43] reported the results of a detailed study of the life span of CuZnSOD hemizygous and homozygous transgenic mice, which showed 1.5- to 3-fold and 2- to 5-fold, respectively, higher activities of CuZnSOD in various tissues compared to nontransgenic, control mice. In a pilot study, in which a total of 45 mice (15 nontransgenic, 13 hemizygous, and 17 homozygous) were studied in conventional animal housing, the life span (mean survival) of the homozygous transgenic mice was observed to be significantly shorter (24%) compared to the control, nontransgenic mice. In a large life span study, in which a total of 417 mice were studied (198 nontransgenic mice, 200 hemizygous mice, and 98 homozygous mice) under barrier conditions, no significanct difference in life span was observed between the control, nontransgenic mice and the hemizygous and homozygous transgenic mice. The mean (± SEM) and the maximum (when 95% had diet) survival for the nontransgenic, hemizygous, and homozygous transgenic mice were: 20.1 ± 0.6 and 31 months; 19.8 ± 0.5 and 31 months; 18.5 ± 0.7 and 30 months, respectively. Thus, in contrast to what has been reported in *Drosophila*, overexpression of CuZnSOD does not appear to enhance the life span of mice. In fact, under conventional housing conditions, when the mice are exposed to more pathogenic insults, the life span of mice overexpressing CuZnSOD is significantly reduced.

4.2. Transgenic Mice Deficient in p66[shc]

Migliaccio et al.[54] reported that homozygous knockout mice null for p66[shc] showed increased resistance to oxidative stress and an increase in life span. The Shc locus encodes three proteins distinguished by differential splicing: the 46, 52, and 66 kD forms. The 46 kD and 52 kD forms are involved in signal transduction through the Ras and MAP kinase pathways while the 66 kD form, p66[shc], appears to play a role in response of cells to oxidative stress, e.g. the protein becomes phosphorylated when cells are exposed to UV light or hydrogen peroxide. Cells from the mice homozygous for the null mutation in p66[shc] were observed to be more resistant to apoptosis induced by either UV irradiation or hydrogen peroxide. In addition, mice with the disrupted p66[shc] gene were found to be resistant to paraquat toxicitiy. The mice lacking p66[shc] lived approximately 30% longer than the control, wild type littermates i.e. the survival increased from approximately 850 days in the wildtype mice to almost 1100 days in the homozygous knockout mice null for p66[shc]. Although this exciting observation is consistent with free radical/oxidative stress theory of aging, it should be noted that the survival data were generated with only a limited number of mice (14 wildtype and 15 homozygous p66[shc] knockout mice) and the mean survival of the mice was relatively short, i.e. the quality of the animal husbandry conditions might be a poor. It is critical that additional survival studies be conducted with these mice under barrier conditions to establish that the mice are living longer because they are aging more slowly.

4.3. Transgenic Mice Deficient in MnSOD

Knockout mice heterozygous for the Sod-2 gene (Sod-2$^{-/+}$) show reduced (30–80%) MnSOD activity in all tissues studied with out any compensatory up expression in other antioxidant enzymes.[55] Thus, these mice have a compromised

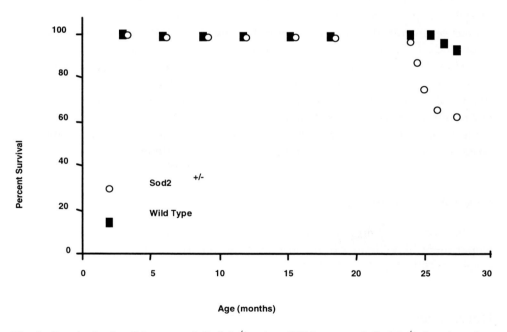

Fig. 1. Survival of wildtype and Sod-2$^{-/+}$ mice. Wildtype and Sod-2$^{-/+}$ female mice, backcrossed on a C57Bl/6 genetic background (B6-Sod-2 < tml > Cje) for 13 generations were raised and maintained for 27 months under barrier conditions. The graph represents the results of a preliminary study comparing the survival of the wildtype and Sod-2$^{-/+}$ mice. This cohort of mice was originally established with 28 wildtype and 18 Sod-2$^{-/+}$ mice. At approximately 27 months of age, 3 wildtype mice (~11%) had died compared to 11 (~39%) of the Sod-2$^{-/+}$ mice. All remaining mice were sacrificed for cross sectional studies.

antioxidant defense system. Because MnSOD is located in the mitochondrial matrix, it plays a critical role in protecting the mitochondria from oxidant stress by enzymatically scavenging superoxide anions that are produced as a by-product of the respiratory chain. Superoxide anions are charged molecules and do not readily cross membranes; therefore, if not destroyed, they can directly (or indirectly by formation of other reactive oxygen species) result in oxidative damage to molecules in the mitochondria. Because the mitochondria are a major site of production of reactive oxygen species, it has been suggested that oxidative damage to the mitochondria can exert a compounding effect with age whereby damaged mitochondria in turn release more reactive oxygen species causing increased oxidative damage to mitochondrial, as well as cytosolic and nuclear proteins, lipids and DNA, leading eventually to dysfunctional or defective mitochondria and the age-related decline in physiologic function.[56] Studies from our laboratory show that the Sod-2$^{-/+}$ mice have altered mitochondrial function (e.g. reduced respiration and an increased sensitivity of the mitochondrial permeability transition to oxidative stress) and increased oxidative damage.[57] Thus the Sod-2$^{-/+}$ mouse model is a potentially valuable model for studying how reduced antioxidant protection in mitochondria could affect aging. Figure 1 shows the results of a pilot study in which the survival of wild type, control mice and Sod-2$^{-/+}$ mice was studied over 27 months. The first cross sectional cohort of our aging study consisted of 18 Sod-2$^{-/+}$ mice and 28 wildtype mice. By 27 months of age, 11 of the Sod-2$^{-/+}$ mice had died compared to 3 mice from the wildtype group. Thus, our preliminary data suggest that Sod-2$^{-/+}$ mice, which show reduced antioxidant protection, have a shorter life span. This experiment needs to be confirmed in a larger study in which mice are followed throughout their life span.

Acknowledgments

This work was supported by a Merit Review and a Center grant from the Department of Veteran Affairs, a grant from the Texas Heart Association, NIH grants AG16998 and AG15908.

References

1. Harman, D. (1956). Aging: a theory based on free radical and radiation chemistry. *J. Gerontol.* **11**: 298–300.
2. Harman, D. (1991). The aging process: major risk factor for disease and death. *Proc. Natl. Acad. Sci. USA* **88**: 5360–5363.
3. Warner, H. R. (1994). Superoxide dismutase, aging, and degenerative disease. *Free Radic. Biol. Med.* **17**: 249–258.

4. Sohal, R. S. and Orr, W. C. (1992). Relationship between antioxidants, prooxidants, and the aging process. *Ann. NY Acad. Sci.* **663**: 74–84.
5. Sohal, R. and Weindruch, R. (1996). Oxidative stress, caloric restriction, and aging. *Science* **273**: 59–67.
6. Bohr, V. A. and Anson, R. M. (1995). DNA damage, mutation and fine structure DNA repair in aging. *Mutat. Res.* **338**: 25–34.
7. Masoro, E. J. (1984). Nutrition as a modulator of the aging process. *Physiologist* **27**: 98–101.
8. Walford, R. L., Harris, S. B. and Weindruch, R. (1987). Dietary restriction and aging: historical phases, mechanisms and current directions. *J. Nutri.* **117**: 1650–1654.
9. Yu, B. P., Lee, D.-W. and Choi, J.-H. (1991). Prevention of free radical damage by food restriction. *In* "Biological Effects of Dietary Restriction" (L. Fishbein, Ed.), pp. 191–197, Springer-Verlag, New York.
10. De, A. K., Chipalkatti, S. and Aiyar, A. S. (1983). Some biochemical parameters of ageing in relation to dietary protein. *Mech. Ageing Dev.* **21**: 37–48.
11. Enesco, H. E. and Kruk, P. (1981). Dietary restriction reduces fluorescent age pigment accumulation in mice. *Exp. Gerontol.* **16**: 357–361.
12. Rao, G., Xia, E., Nadakavukaren, M. J. and Richardson, A. (1990). Effect of dietary restriction on the age-dependent changes in the expression of antioxidant enzymes in rat liver. *J. Nutri.* **120**: 602–609.
13. Xia, E., Rao, G., Van Remmen, H., Heydari, A. R. and Richardson, A. (1995). Activities of antioxidant enzymes in various tissues of male Fischer 344 rats are altered by food restriction. *J. Nutri.* **125**: 195–201.
14. Chipalkatti, S., De, A. K. and Aiyar, A. S. (1983). Effect of diet restriction on some biochemical parameters related to aging in mice. *J. Nutri.* **113**: 944–950.
15. Davis, L. J., Tadolini, B., Biagi, P. L., Walford, R. L. and Licastro, F. (1993). Effect of age and extent of dietary restriction on hepatic microsomal lipid peroxidation potential in mice. *Mech. Ageing Dev.* **72**: 155–163.
16. Laganiere, S. and Yu, B. P. (1987). Anti-lipoperoxidation action of food restriction. *Biochem. Biophys. Res. Commun.* **145**: 1185–1191.
17. Levin, G., Cogan, U. and Mokady, S. (1992). Food restriction and membrane fluidity. *Mech. Ageing Dev.* **62**: 137–141.
18. Pieri, C., Falasca, M., Marcheselli, F., Moroni, F., Recchioni, R., Marmocchi, F. and Lupidi, G. (1992). Food restriction in female Wistar rats: Part V. Lipid peroxidation and antioxidant enzymes in liver. *Arch. Gerontol. Geriatr.* **14**: 93–99.
19. Youngman, L. D., Park, J. K. and Ames, B. N. (1992). Protein oxidation associated with aging is reduced by dietary restriction of protein or calories. *Proc. Natl. Acad. Sci. USA* **89**: 9112–9116.
20. Sohal, R. S., Ku, H.-H., Agarwal, S., Forster, M. J. and Lal, H. (1994). Oxidative damage, mitochondrial oxidant generation and antioxidant defenses during

aging and in response to food restriction in the mouse. *Mech. Ageing Dev.* **74**: 121–133.

21. Chen, L. H. and Snyder, D. L. (1992). Effects of age, dietary restriction and germ-free environment on glutathione-related enzymes in Loubund-Wistar rats. *Arch. Gerontol. Geriatr.* **14**: 17–26.

22. Youngman, L. D. (1993). Protein restriction (PR) and caloric restriction (CR) compared: effects on DNA damage, carcinogenesis, and oxidative damage. *Mutat. Res. DNAging Genet. Instability Aging* **295**: 165–179.

23. Sohal, R. S., Agarwal, S., Candas, M., Forster, M. J. and Lal, H. (1994). Effect of age and caloric restriction on DNA oxidative damage in different tissues of C57BL/6 mice. *Mech. Ageing Dev.* **76**: 215–224.

24. Vanfleteren, J. R. (1993). Oxidative stress and ageing in *Caenorhabditis elegans*. *Biochem. J.* **292**: 605–608.

25. Larsen, P. L. (1993). Aging and resistance to oxidative damage in *Caenorhabditis elegans*. *Proc. Natl. Acad. Sci. USA* **90**: 8905–8909.

26. Honda, Y. and Honda, S. (1999). The daf-2 gene network for longevity regulates oxidative stress resistance and Mn superoxide dismutase gene expression in *Caenorhabditis elegans*. *FASEB J.* **13**: 1385–1393.

27. Taub, J., Lau, J. F., Ma, C., Hahn, J. L., Hoque, R., Rothblatt, J. and Chalfie, M. (1999). A cytosolic catalase is needed to extend adult lifespan in *Caenorhabditis elegans* daf-c and clk-1 mutants. *Nature* **399**: 162–166.

28. Arking, R., Dudas, S. P. and Baker, G. T., III (1993). Genetic and environmental factors regulating the expression of an extended longevity phenotype in a long lived strain of *Drosophila*. *Genetica* **91**: 127–142.

29. Dudas, S. P. and Arking, R. (1994). A coordinate up-regulation of antioxidant gene activities is associated with the delayed onset of senescence in a long-lived strain of *Drosophila*. *J. Gerontol.* **in press**.

30. Tyler, R. H., Brar, H., Singh, M., Latorre, A., Graves, J. L., Mueller, L. D., Rose, M. R. and Ayala, F. J. (1993). The effect of superoxide dismutase alleles on aging in *Drosophila*. *Genetica* **91**: 143–149.

31. Sohal, R. S., Agarwal, A., Agarwal, S. and Orr, W. C. (1995). Simultaneous overexpression of copper- and zinc-containing superoxide dismutase and catalase retards age-related oxidative damage and increases metabolic potential in *Drosophila melanogaster*. *J. Biol. Chem.* **270**: 15 671–15 674.

32. Orr, W. C. and Sohal, R. S. (1994). Extension of life span by overexpression of superoxide dismutase and catalase in *Drosophila melanogaster*. *Science* **263**: 1128–1130.

33. Kaiser, M., Gasser, M., Ackermann, R. and Stearns, S. C. (1996). P-element inserts in transgenic flies: a cautionary tale. *Heredity* **78**: 1–11.

34. Stearns, S. C., Kaiser, M. and Hillesheim, E. (1993). Effects on fitness components of enhanced expression of elongation factor EF1-alpha in *Drosophila melanogaster*. Part I. The contrasting approaches of molecular and population biologists. *Am. Nature* **142**: 961–993.

35. Stearns, S. C. and Kaiser, M. (1993). The effects of enhanced expression of elongation factor EF1-alpha on lifespan in *Drosophila melanogaster. Genetica* **91**: 167–182.

36. Sun, J. and Tower, J. (1999). FLP recombinase-mediated induction of CuZn superoxide dismutase transgene expression can extend the life span of adult *Drosophila melanogaster* flies. *Mol. Cell Biol.* **19**: 216–228.

37. Parkes, T. L., Elia, A. J., Dickinson, D., Hilliker, A. J., Phillips, J. P. and Boulianne, G. (1998). Extension of *Drosophila* lifespan by overexpression of human SOD-1 in motorneurons. *Nature Genet.* **19**: 171–174.

38. Richardson, A., Morgan, W. W., Nelson, J. F., Sharp, Z. D., Heydari, A. R. and Walter, C. A. (1997). The use of transgenic mice in aging research. *ILAR J.* 124–136.

39. Pierce, J. C. and Sternberg, N. L. (1992). Using bacteriophage P1 system to clone high molecular weight genomic DNA. *Meth. Enzymol.* **216**: 549–574.

40. Pierce, J. C., Sauer, B. and Sternberg, N. (1992). A positive selection vector for cloning high molecular weight DNA by the bacteriophage P1 system: improved cloning efficacy. *Proc. Natl. Acad. Sci. USA* **89**: 2056–2060.

41. Yan, Y., Lufei, H., Chen, R., Sealey, J. E., Laragh, J. H. and Catanzaro, D. F. (1998). Appropriate regulation of human renin gene expression and secretion in 45-kb human renin transgenic mice. *Hypertension* **32**: 205–214.

42. Epstein, C. J., Avraham, K. B., Lovett, M., Smith, S., Elroy-Stein, O., Rotman, G., Bry, C. and Groner, Y. (1987). Transgenic mice with increased CuZn superoxide dismutase activity: animal model of dosage effects in down syndrome. *Proc. Natl. Acad. Sci. USA* **84**: 8044–8048.

43. Huang, T. T., Carlson, E. J., Gillespie, A. M., Shi, Y. and Epstein, C. J. (2000). Ubiquitous overexpression of CuZn superoxide dismutase does not extend life span in mice. *J. Gerontol. A Biol. Sci. Med. Sci.* **55**: B5–B9.

44. Finch, C. E. (1972). Enzyme activities, gene function, and aging in mammals. *Exp. Gerontol.* **7**: 53–67.

45. Richardson, A., Roberts, M. S. and Rutherford, M. S. (1985). Aging and gene expression. *In* "Review Biological Research in Aging" (M. Rothstein, Ed.), pp. 395–419, Alan R. Liss, Inc., New York.

46. Richardson, A., Birchenall-Sparks, M. C. and Staecker, J. L. (1983). Aging and transcription. *In* "Biological Research in Aging" (M. Rothstein, Ed.), pp. 275–294, Alan R. Liss, Inc., New York.

47. Richardson, A. and Semsei, I. (1987). Effect of aging on translation and transcription. *In* "Review of Biological Research in Aging" (M. Rothstein, Ed.), pp. 467–483, Alan R. Liss, Inc., New York.

48. Masoro, E. J. (1985). Nutrition and aging: acurrent assessment. *J. Nutri.* **115**: 842–848.

49. Pahlavani, M. A., Haley-Zitlin, V. and Richardson, A. (1994). Influence of dietary restriction on gene expression: changes in the transcription of specific

genes. *In* "Modulation of the Aging Process by Dietary Restriction" (B. P. Yu, Ed.), pp. 143–156, CRC Press, Boca Raton, FL.

50. Van Remmen, H., Ward, W., Sabia, R. V. and Richardson, A. (1995). Effect of age on gene expression and protein degradation. *In* "Handbook of Physiology Volume on Aging" (E. J. Masoro, Ed.), pp. 171–234, Oxford University Press, New York.

51. Li, Y., Huang, T.-T., Carlson, E. J., Melov, S., Ursell, P. C., Olson, J. L., Noble, L. J., Yoshimura, M. P., Berger, C., Chan, P. H., Wallace, D. C. and Epstein, C. J. (1995). Dilated cardiomyopathy and neonatal lethality in mutant mice lacking manganese superoxide dismutase. *Nature Genet.* **11**: 376–381.

52. Lebovitz, R. M., Zhang, H., Vogel, H., Cartwright, J., Dionne, L., Lu, N., Huang, S. and Matzuk, M. M. (1996). Neurodegeneration, myocardial injury, and perinatal death in mitochondrial superoxide dismutase deficient mice. *Proc. Natl. Acad. Sci. USA* **93**: 9782–9787.

53. Huang, T.-T., Carlson, E. J., Gillespie, A. M. and Epstein, C. J. (1998). Genetic modification of the dilated cardiomyopathy and neonatal lethality phenotype of mice lacking mangnese superoxide dismutase. *Age* **21**: 83–84.

54. Migliaccio, E., Giorgio, M., Mele, S., Pelicci, G., Reboldi, P., Pandolfi, P. P., Lanfrancone, L. and Pelicci, P. G. (1999). The p66[shc] adaptor protein controls oxidative stress response and life span in mammals. *Nature* **402**: 309–313.

55. Van Remmen, H., Salvador, C., Epstein, C. J. and Richardson, A. (1998). Characterization of the antioxidant status of the heterozygous manganese superoxide dismutase knockout mouse. *Arch. Biochem. Biophys.* **363**: 91–97.

56. Harman, D. (1972). The biologic clock: the mitochondria? *J. Am. Geriatr. Soc.* **20**: 145–147.

57. Williams, M. D., Van Remmen, H., Conrad, C. C., Huang, T.-T., Epstein, C. J. and Richardson, A. (1998). Increased oxidative damage is correlated to altered mitochondrial function in heterozygous manganese superoxide dismutase knockout mice. *J. Biol. Chem.* **273**: 28 510–28 515.

58. Kang, Y. J., Chen, Y. and Epstein, P. N. (1996). Suppression of doxorubicin cardiotoxicity by overexpression of catalase in the heart of transgenic mice. *J. Biol. Chem.* **271**: 12 610–12 616.

59. Li, G., Chen, Y., Saari, J. T. and Kang, Y. J. (1997). Catalase-overexpressing transgenic mouse heart is resistant to ischemia-reperfusion injury. *Am. J. Physiol.* **273**: H1090–H1095.

60. Chen, Y., Yu, A., Saari, J. T. and Kang, Y. J. (1997). Repression of hypoxia-reoxygenation injury in the catalase-overexpressing heart of transgenic mice. *Proc. Soc. Exp. Biol. Med.* **216**: 112–116.

61. Zhou, Z. and Kang, Y. J. (2000). Cellular and subcellular localization of catalase in the heart of transgenic mice. *J. Histochem. Cytochem.* **48**: 585–594.

62. Nilakantan, V., Li, Y., Glauert, H. P. and Spear, B. T. (1996). Increased liver-specific catalase activity in transgenic mice. *DNA Cell Biol.* **15**: 625–630.

63. Nilakantan, V., Spear, B. T. and Glauert, A. M. (1998). Effect of the peroxisome proliferator ciprofibrate on lipid peroxidation and 8-hydroxydeoxyguanosine formation in transgenic mice with elevated hepatic catalase activity. *Free Radic. Biol. Med.* **24**: 1430–1436.

64. Xu, B., Moritz, J. T. and Epstein, P. N. (1999). Overexpression of catalase provides partial protection to transgenic mouse beta cells. *Free Radic. Biol. Med.* **27**: 830–837.

65. Chen, X., Mele, J., Giese, H., Dollé, M. E. T., van Remmen, H., Richardson, A. and Vijg, J. (2000). Ubiquitous overexpression of human Cu, Zn superoxide dismutase and catalase gene in mice, **in preparation**.

66. Mirochnitchenko, O., Palnitkar, U., Philbert, M. and Inouye, M. (1995). Thermosensitive phenotype of transgenic mice overproducing human glutathione peroxidases. *Proc. Natl. Acad. Sci. USA* **92**: 8120–8124.

67. Zeevalk, G. D., Bernard, L. P., Albers, D. S., Mirochnitchenko, O., Nicklas, W. J. and Sonsalla, P. K. (1997). Energy stress-induced dopamine loss in glutathione peroxidase-overexpressing transgenic mice and in glutathione-depleted mesencephalic cultures. *J. Neurochem.* **68**: 426–429.

68. Nakatani, T., Inouye, M. and Mirochnitchenko, O. (1997). Overexpression of antioxidant enzymes in transgenic mice decreases cellular ploidy during liver regeneration. *Exp. Cell Res.* **236**: 137–146.

69. Ishibashi, N., Weisbrot-Lefkowitz, M., Reuhl, K., Inouye, M. and Mirochnitchenko, O. (1999). Modulation of chemokine expression during ischemia/reperfusion in transgenic mice overproducing human glutathione peroxidases. *J. Immunol.* **163**: 5666–5677.

70. Bensadoun, J. C., Mirochnitchenko, O., Inouye, M., Aebischer, P. and Zurn, A. D. (1998). Attenuation of 6-OHDA-induced neurotoxicity in glutathione peroxidase transgenic mice. *Eur. J. Neurosci.* **10**: 3231–3236.

71. Weisbrot-Lefkowitz, M., Reuhl, K., Perry, B., Chan, P. H., Inouye, M. and Mirochnitchenko, O. (1998). Overexpression of human glutathione peroxidase protects transgenic mice against focal cerebral ischemia/reperfusion damage. *Brain Res. Mol. Brain Res.* **53**: 333–338.

72. Lu, Y. P., Lou, Y. R., Yen, P., Newmark, H. L., Mirochnitchenko, O. I., Inouye, M. and Huang, M. T. (1997). Enhanced skin carcinogenesis in transgenic mice with high expression of glutathione peroxidase or both glutathione peroxidase and superoxide dismutase. *Cancer Res.* **57**: 1468–1474.

73. Maulik, N., Yoshida, T. and Das, D. K. (1999). Regulation of cardiomyocyte apoptosis in ischemic reperfused mouse heart by glutathione peroxidase. *Mol. Cell Biochem.* **196**: 13–21.

74. Yoshida, T., Watanabe, M., Engelman, D. T., Engelman, R. M., Schley, J. A., Maulik, N., Ho, Y. S., Oberley, T. D. and Das, D. K. (1996). Transgenic mice overexpressing glutathione peroxidase are resistant to myocardial ischemia reperfusion injury. *J. Mol. Cell Cardiol.* **28**: 1759–1767.

75. Mirault, M., Tremblay, A., Furling, D., Trepanier, G., Dugre, F., Puymirat, J. and Pothier, F. (1994). Transgenic glutathione peroxidase mouse models for neuroprotection studies. *Ann. NY Acad. Sci.* **738**: 104–115.

76. Bilodeau, J. F. and Mirault, M. E. (1999). Increased resistance of GPx-1 transgenic mice to tumor promoter-induced loss of glutathione peroxidase activity in skin. *Int. J. Cancer* **80**: 863–867.

77. Furling, D., Ghribi, O., Lahsaini, A., Mirault, M. E. and Massicotte, G. (2000). Impairment of synaptic transmission by transient hypoxia in hippocampal slices: improved recovery in glutathione peroxidase transgenic mice. *Proc. Natl. Acad. Sci. USA* **97**: 4351–4356.

78. Cheng, W. H., Ho, Y. S., Valentine, B. A., Ross, D. A., Combs, G. F., Jr. and Lei, X. G. (1998). Cellular glutathione peroxidase is the mediator of body selenium to protect against paraquat lethality in transgenic mice. *J. Nutri.* **128**: 1070–1076.

79. Cheng, W. H., Ho, Y. S., Ross, D. A., Valentine, B. A., Combs, G. F. and Lei, X. G. (1997). Cellular glutathione peroxidase knockout mice express normal levels of selenium-dependent plasma and phospholipid hydroperoxide glutathione peroxidases in various tissues. *J. Nutri.* **127**: 1445–1450.

80. Reddy, V. N., Lin, L. R., Ho, Y. S., Magnenat, J. L., Ibaraki, N., Giblin, F. J. and Dang, L. (1997). Peroxide-induced damage in lenses of transgenic mice with deficient and elevated levels of glutathione peroxidase. *Ophthalmologica* **211**: 192–200.

81. Spector, A., Yang, Y., Ho, Y.-S., Magnenat, J.-L., Wang, R.-R., Ma, W. and Li, W.-C. (1996). Variation in cellular glutathione peroxidase activity in lens epithelial cells, transgenics and knockouts does not significantly change the response to H_2O_2 stress. *Exp. Eye Res.* **62**: 521–540.

82. Larsen, G. L., White, C. W., Takeda, K., Loader, J. E., Nguyen, D. D., Joetham, A., Groner, Y. and Gelfand, E. W. (2000). Mice that overexpress CuZn superoxide dismutase are resistant to allergen-induced changes in airway control. *Am. J. Physiol. Lung Cell Mol. Physiol.* **279**: L350–L359.

83. Kamii, H., Kato, I., Kinouchi, H., Chan, P. H., Epstein, C. J., Akabane, A., Okamoto, H. and Yoshimoto, T. (1999). Amelioration of vasospasm after subarachnoid hemorrhage in transgenic mice overexpressing CuZn superoxide dismutase. *Stroke* **30**: 867–871.

84. Gahtan, E., Auerbach, J. M., Groner, Y. and Segal, M. (1998). Reversible impairment of long-term potentiation in transgenic CuZnSOD mice. *Eur. J. Neurosci.* **10**: 538–544.

85. Rando, T. A., Crowley, R. S., Carlson, E. J., Epstein, C. J. and Mohapatra, P. K. (1998). Overexpression of CuZn superoxide dismutase: a novel cause of murine muscular dystrophy. *Ann. Neurol.* **44**: 381–386.

86. Horie, Y., Wolf, R., Flores, S. C., McCord, J. M., Epstein, C. J. and Granger, D. N. (1998). Transgenic mice with increased CuZn superoxide dismutase

activity are resistant to hepatic leukostasis and capillary no-reflow after gut ischemia/reperfusion. *Circ. Res.* **83**: 691–696.

87. De Vos, S., Epstein, C. J., Carlson, E., Cho, S. K. and Koeffler, H. P. (1995). Transgenic mice overexpressing human CuZn superoxide dismutase (CuZnSOD) are not resistant to endotoxic shock. *Biochem. Biophys. Res. Commun.* **208**: 523–531.

88. Beal, M. F., Ferrante, R. J., Henshaw, R., Matthews, R. T., Chan, P. H., Kowall, N. W., Epstein, C. J. and Schulz, J. B. (1995). 3-nitropropionic acid neurotoxicity is attenuated in CuZn superoxide dismutase transgenic mice. *J. Neurochem.* **65**: 919–922.

89. Yang, G., Chang, P. H., Chen, J., Carlson, E., Chen, S. F., Weinstein, P., Epstein, C. J. and Kamii, H. (1994). Human CuZn superoxide dismutase transgenic mice are highly resistant to reperfusion injury after focal cerebral ischemia. *Stroke* **25**: 165–170.

90. Cadet, J. L., Ladenheim, B., Baum, I., Carlson, E. and Epstein, C. J. (1994). CuZn superoxide dismutase (CuZnSOD) transgenic mice show resistance to the lethal effects of methylenedioxyamphetamine (MDA) and of methylenedioxymethamphetamine (MDMA). *Brain Res.* **655**: 259–262.

91. Minc-Golomb, D., Knobler, H. and Groner, Y. (1991). Gene dosage of CuZnSOD and Down's syndrome: diminished prostaglandin synthesis in human trisomy 21, transfected cells and transgenic mice. *EMBO J.* **10**: 2119–2124.

92. Groner, Y., Elroy-Stein, O., Avraham, K. B., Yarom, R., Schickler, M., Knobler, H. and Rotman, G. (1990). Down syndrome clinical symptoms are manifested in transfected cells and transgenic mice overexpressing the human CuZn superoxide dismutase gene. *J. Physiol.* **84**: 53–77.

93. Morita-Fujimura, Y., Fujimura, M., Gasche, Y., Copin, J. C. and Chan, P. H. (2000). Overexpression of CuZn superoxide dismutase in transgenic mice prevents the induction and activation of matrix metalloproteinases after cold injury-induced brain trauma. *J. Cereb. Blood Flow Metab.* **20**: 130–138.

94. Levkovitz, Y., Avignone, E., Groner, Y. and Segal, M. (1999). Upregulation of GABA neurotransmission suppresses hippocampal excitability and prevents long-term potentiation in transgenic superoxide dismutase-overexpressing mice. *J. Neurosci.* **19**: 10 977–10 984.

95. Chen, Y., Ying, W., Simma, V., Copin, J. C., Chan, P. H. and Swanson, R. A. (2000). Overexpression of CuZn superoxide dismutase attenuates oxidative inhibition of astrocyte glutamate uptake. *J. Neurochem.* **75**: 939–945.

96. Fujimura, M., Morita-Fujimura, Y., Noshita, N., Sugawara, T., Kawase, M. and Chan, P. H. (2000). The cytosolic antioxidant CuZn superoxide dismutase prevents the early release of mitochondrial cytochrome c in ischemic brain after transient focal cerebral ischemia in mice. *J. Neurosci.* **20**: 2817–2824.

97. Ying, W., Anderson, C. M., Chen, Y., Stein, B. A., Fahlman, C. S., Copin, J. C., Chan, P. H. and Swanson, R. A. (2000). Differing effects of CuZn superoxide dismutase overexpression on neurotoxicity elicited by nitric oxide, reactive oxygen species, and excitotoxins. *J. Cereb. Blood Flow Metab.* **20**: 359–368.

98. Fullerton, H. J., Ditelberg, J. S., Chen, S. F., Sarco, D. P., Chan, P. H., Epstein, C. J. and Ferriero, D. M. (1998). CuZn superoxide dismutase transgenic brain accumalates hydrogen peroxide after perinatal hypoxia ischemia. *Ann. Neurol.* **44**: 357–364.

99. White, C. W., Avraham, K. B., Shanley, P. F. and Groner, Y. (1991). Transgenic mice with expression of elevated levels of CuZn superoxide dismutase in the lungs are resistant to pulmonary oxygen toxicity. *J. Clin. Invest.* **87**: 2162–2168.

100. Schwartz, P. J., Berger, U. V. and Coyle, J. T. (1995). Mice transgenic for CuZn superoxide dismutase exhibit increased markers of biogenic amine function. *J. Neurochem.* **65**: 660–669.

101. Golenser, J., Peled-Kamar, M., Schwartz, E., Friedman, I., Groner, Y. and Pollack, Y. (1998). Transgenic mice with elevated level of CuZnSOD are highly susceptible to malaria infection. *Free Radic. Biol. Med.* **24**: 1504–1510.

102. Peled-Kamar, M., Lotem, J., Okon, E., Sachs, L. and Groner, Y. (1995). Thymic abnormalities and enhanced apoptosis of thymocytes and bone marrow cells in transgenic mice overexpressing CuZn superoxide dismutase: implications for Down syndrome. *EMBO J.* **14**: 4985–4993.

103. Peled-Kamar, M., Lotem, J., Wirguin, I., Weiner, L., Hermalin, A. and Groner, Y. (1997). Oxidative stress mediates impairment of muscle function in transgenic mice with elevated level of wild-type CuZn superoxide dismutase. *Proc. Natl. Acad. Sci. USA* **94**: 3883–3887.

104. Bar-Peled, O., Korkotian, E., Segal, M. and Groner, Y. (1996). Constitutive overexpression of CuZn superoxide dismutase exacerbates kainic acid-induced apoptosis of transgenic CuZn superoxide dismutase neurons. *Proc. Natl. Acad. Sci. USA* **93**: 8530–8535.

105. Yarom, R., Sapoznikov, D., Havivi, Y., Avraham, K. B., Schickler, M. and Groner, Y. (1988). Premature aging changes in neuromuscular junctions of transgenic mice with an extra human CuZnSOD gene: a model for tongue pathology in Down's syndrome. *J. Neurol. Sci.* **88**: 41–53.

106. Avraham, K. B., Sugarman, H., Rotshenker, S. and Groner, Y. (1991). Down's syndrome: morphological remodelling and increased complexity in the neuromuscular junction of transgenic CuZn superoxide dismutase mice. *J. Neurocytol.* **20**: 208–215.

107. Kikuchi, Y., Shimosegawa, T., Moriizumi, S., Kimura, K., Satoh, A., Koizumi, M., Kato, I., Epstein, C. J. and Toyota, T. (1997). Transgenic CuZn superoxide dismutase ameliorates caerulein-induced pancreatitis in mice. *Biochem. Biophys. Res. Commun.* **233**: 177–181.

108. Chan, P. H., Chu, L., Chen, S. F., Carlson, E. J. and Epstein, C. J. (1990). Reduced neurotoxicity in transgenic mice overexpressing human CuZn superoxide dismutase. *Stroke* **21**: 80–82.

109. Przedborski, S., Kostic, V., Jackson-Lewis, V., Carlson, E., Epstein, C. J. and Cadet, J. L. (1991). Quantitative autoradiographic distribution of [^{3}H]-MPTP binding in the brains of superoxide dismutase transgenic mice. *Brain Res. Bull.* **26**: 987–991.

110. Kinouchi, H., Epstein, C. J., Mizui, T., Carlson, E., Chen, S. F. and Chan, P. H. (1991). Attenuation of focal cerebral ischemic injury in transgenic mice overexpressing CuZn superoxide dismutase. *Proc. Natl. Acad. Sci. USA* **88**: 11 158–11 162.

111. Przedborski, S., Kostic, V., Jackson-Lewis, V., Naini, A. B., Simonetti, S., Fahn, S., Carlson, E., Epstein, C. J. and Cadet, J. L. (1992). Transgenic mice with increased CuZn superoxide dismutase activity are resistant to *N*-methyl-4-phenyl-1,2,3,6-tetrahydropyridine-induced neurotoxicity. *J. Neurosci.* **12**: 1658–1667.

112. Schickler, M., Knobler, H., Avraham, K. B., Elroy-Stein, O. and Groner, Y. (1989). Diminished serotonin uptake in platelets of transgenic mice with increased CuZn superoxide dismutase activity. *EMBO J.* **8**: 1385–1392.

113. Avraham, K. B., Schickler, M., Sapoznikov, D., Yarom, R. and Groner, Y. (1988). Down's syndrome: abnormal neuromuscular junction in tongue of transgenic mice with elevated levels of human CuZn superoxide dismutase. *Cell* **54**: 823–829.

114. Hagay, Z. J., Weiss, Y., Zusman, I., Peled-Kamar, M., Reece, E. A., Eriksson, U. J. and Groner, Y. (1995). Prevention of diabetes-associated embryopathy by overexpression of the free radical scavenger CuZn superoxide dismutase in transgenic mouse embryos. *Am. J. Obstet. Gynecol.* **173**: 1036–1041.

115. Ceballos-Picot, I., Nicole, A., Briand, P., Grimber, G., Delacourte, A., Flament, S., Blouin, J. L., Thevenin, M., Kamoun, P. and Sinet, P. M. (1991). Expression of human CuZn superoxide dismutase gene in transgenic mice: model for gene dosage effect in Down syndrome. *Free Radic. Res. Commun.* **12–13(Part 2)**: 581–589.

116. Bonnes-Taourel, D., Guérin, M.-C., Torreilles, J., Ceballos-Picot, I. and de Paulet, A. C. (1993). 4-hydroxynonenal content lower in brains of 25 month old transgenic mice carrying the human CuZn superoxide dismutase gene than in brains of their non-transgenic littermates. *J. Lipid Mediat.* **8**: 111–120.

117. Nabarra, B., Casanova, M., Paris, D., Nicole, A., Toyama, K., Sinet, P., Ceballos, I. and London, J. (1996). Transgenic mice overexpressing the human CuZnSOD gene: ultrastructural studies of a premature thymic involution model of Down's Syndrome (Trisomy 21). *Lab. Invest.* **74**: 617–626.

118. Nabarra, B., Casanova, M., Paris, D., Paly, E., Toyoma, K., Ceballos, I. and London, J. (1997). Premature thymic involution, observed at the ultrastructural

level, in two lineages of human SOD-1 transgenic mice. *Mech. Ageing Dev.* **96**: 59–73.

119. Barkats, M., Bertholet, J.-Y., Venault, P., Ceballos-Picot, I., Nicole, A., Phillips, J., Moutier, R., Roubertoux, P., Sinet, P.-M. and Cohen-Salmon, C. (1993). Hippocampal mossy fiber changes in mice transgenic for the human CuZn superoxide dismutase gene. *Neurosci. Lett.* **160**: 24–28.

120. Deshmukh, D. R., Mirochnitchenko, O., Ghole, V. S., Agnese, D., Shah, P. C., Reddell, M., Brolin, R. E. and Inouye, M. (1997). Intestinal ischemia and reperfusion injury in transgenic mice overexpressing CuZn superoxide dismutase. *Am. J. Physiol.* **273**: C1130–C1135.

121. Mirochnitchenko, O. and Inouye, M. (1996). Effect of overexpression of human CuZn superoxide dismutase in transgenic mice on macrophage functions. *J. Immunol.* **156**: 1578–1586.

122. Kubisch, H. M., Wang, J., Luche, R., Carlson, E., Bray, T. M., Epstein, C. J. and Phillips, J. P. (1994). Transgenic CuZn superoxide dismutase modulates susceptibility to Type-I diabetes. *Proc. Natl. Acad. Sci. USA* **91**: 9956–9959.

123. Wang, P., Chen, H., Qin, H., Sankarapandi, S., Becher, M. W., Wong, P. C. and Zweier, J. L. (1998). Overexpression of human CuZn superoxide dismutase (SOD-1) prevents postischemic injury. *Proc. Natl. Acad. Sci. USA* **95**: 4556–4560.

124. Copin, J., Gasche, Y. and Chan, P. H. (2000). Overexpression of CuZn superoxide dismutase does not prevent neonatal lethality in mutant mice that lack manganese superoxide dismutase [In Process Citation]. *Free Radic. Biol. Med.* **28**: 1571–1576.

125. Ibrahim, W., Lee, U. S., Yen, H. C., St. Clair, D. K. and Chow, C. K. (2000). Antioxidant and oxidative status in tissues of manganese superoxide dismutase transgenic mice. *Free Radic. Biol. Med.* **28**: 397–402.

126. Chen, Z., Siu, B., Ho, Y. S., Vincent, R., Chua, C. C., Hamdy, R. C. and Chua, B. H. (1998). Overexpression of MnSOD protects against myocardial ischemia/reperfusion injury in transgenic mice. *J. Mol. Cell Cardiol.* **30**: 2281–2289.

127. Ho, Y. S., Vincent, R., Dey, M. S., Slot, J. W. and Crapo, J. D. (1998). Transgenic models for the study of lung antioxidant defense: enhanced manganese-containing superoxide dismutase activity gives partial protection to B6C3 hybrid mice exposed to hyperoxia. *Am. J. Respir. Cell Mol. Biol.* **18**: 538–547.

128. Yen, H.-C., Oberley, T. D., Vichitbandha, S., Ho, Y.-S. and St. Clair, D. K. (1996). The protective role of manganese superoxide dismutase against adriamycin-induced acute cardiac toxicity in transgenic mice. *J. Clin. Invest.* **98**: 1253–1260.

129. Wispe, J. R., Warner, B. B., Clark, J. C., Dey, C. R., Neuman, J., Glasser, S. W., Crapo, J. D., Chang, L. Y. and Whitsett, J. A. (1992). Human Mn superoxide dismutase in pulmonary epithelial cells of transgenic mice confers protection from oxygen toxicity. *J. Biol. Chem.* **267**: 23 937–23 941.

130. Folz, R. J., Abushamaa, A. M. and Suliman, H. B. (1999). Extracellular superoxide dismutase in the airways of transgenic mice reduces inflammation and attenuates lung toxicity following hyperoxia. *J. Clin. Invest.* **103**: 1055–1066.

131. Sheng, H., Kudo, M., Mackensen, G. B., Pearlstein, R. D., Crapo, J. D. and Warner, D. S. (2000). Mice overexpressing extracellular superoxide dismutase have increased resistance to global cerebral ischemia. *Exp. Neurol.* **163**: 392–398.

132. Levin, E. D., Brady, T. C., Hochrein, E. C., Oury, T. D., Jonsson, L. M., Marklund, S. L. and Crapo, J. D. (1998). Molecular manipulations of extracellular superoxide dismutase: functional importance for learning. *Behav. Genet.* **28**: 381–390.

133. Chen, E. P., Bittner, H. B., Davis, R. D., Van Trigt, P. and Folz, R. J. (1998). Physiologic effects of extracellular superoxide dismutase transgene overexpression on myocardial function after ischemia and reperfusion injury. *J. Thorac. Cardiovasc. Surg.* **115**: 450–458.

134. Oury, T. D., Ho, Y.-S., Piantadosi, C. A. and Crapo, J. D. (1992). Extracellular superoxide dismutase, nitric oxide, and central nervous system O_2 toxicity. *Proc. Natl. Acad. Sci. USA* **89**: 9715–9719.

135. Takagi, Y., Mitsui, A., Nishiyama, A., Nozaki, K., Sono, H., Gon, Y., Hashimoto, N. and Yodoi, J. (1999). Overexpression of thioredoxin in transgenic mice attenuates focal ischemic brain damage. *Proc. Natl. Acad. Sci. USA* **96**: 4131–4136.

136. Ho, Y.-S., Magnenat, J. L., Bronson, R. T., Cao, J., Gargano, M., Sugawara, M. and Funk, C. D. (1997). Mice deficient in cellular glutathione peroxidase develop normally and show no increased sensitivity to hyperoxia. *J. Biol. Chem.* **272**: 16 644–16 651.

137. Reddy, V. N., Lin, L.-R., Ho, Y.-S., Magnenat, J.-L., Ibaraki, N., Giblin, F. J. and Dang, L. (1997). Peroxide-induced damage in lenses of transgenic mice with deficient and elevated levels of glutathione peroxidase. *Ophthalmologica* **211**: 200.

138. Spector, A., Kuszak, J. R., Ma, W., Wang, R. R., Ho, Y. and Yang, Y. (1998). The effect of photochemical stress upon the lenses of normal and glutathione peroxidase-1 knockout mice. *Exp. Eye Res.* **67**: 457–471.

139. Cheng, W., Fu, Y. X., Porres, J. M., Ross, D. A. and Lei, X. G. (1999). Selenium-dependent cellular glutathione peroxidase protects mice against a pro-oxidant-induced oxidation of NADPH, NADH, lipids, and protein. *FASEB J.* **13**: 1467–1475.

140. Beck, M. A., Esworthy, R. S., Ho, Y. S. and Chu, F. F. (1998). Glutathione peroxidase protects mice from viral-induced myocarditis. *FASEB J.* **12**: 1143–1149.

141. Klivenyi, P., Andreassen, O. A., Ferrante, R. J., Dedeoglu, A., Mueller, G., Lancelot, E., Bogdanov, M., Andersen, J. K., Jiang, D. and Beal, M. F.

(2000). Mice deficient in cellular glutathione peroxidase show increased vulnerability to malonate, 3-nitropropionic acid, and 1-methyl-4-phenyl-1,2,5,6-tetrahydropyridine. *J. Neurosci.* **20**: 1–7.

142. Fu, Y., Cheng, W. H., Porres, J. M., Ross, D. A. and Lei, X. G. (1999). Knockout of cellular glutathione peroxidase gene renders mice susceptible to diquat-induced oxidative stress. *Free Radic. Biol. Med.* **27**: 605–611.

143. Zhang, J., Graham, D. G., Montine, T. J. and Ho, Y. S. (2000). Enhanced N-methyl-4-phenyl-1,2,3,6-tetrahydropyridine toxicity in mice deficient in CuZn superoxide dismutase or glutathione peroxidase. *J. Neuropathol. Exp. Neurol.* **59**: 53–61.

144. Esworthy, R. S., Mann, J. R., Sam, M. and Chu, F. F. (2000). Low glutathione peroxidase activity in Gpx-1 knockout mice protects jejunum crypts from gamma-irradiation damage. *Am. J. Physiol. Gastrointest. Liver Physiol.* **279**: G426–G436.

145. Jiang, D., Akopian, G., Ho, Y. S., Walsh, J. P. and Andersen, J. K. (2000). Chronic brain oxidation in a glutathione peroxidase knockout mouse model results in increased resistance to induced epileptic seizures. *Exp. Neurol.* **164**: 257–268.

146. De Haan, J. B., Bladier, C., Griffiths, P., Kelner, M., O'Shea, R. D., Cheung, N. S., Bronson, R. T., Silvestro, M. J., Wild, S., Zheng, S. S., Beart, P. M., Hertzog, P. J. and Kola, I. (1998). Mice with a homozygous null mutation for the most abundant glutathione peroxidase, Gpx-1, show increased susceptibility to the oxidative stress-inducting agents paraquat and hydrogen peroxide. *J. Biol. Chem.* **273**: 22 528–22 536.

147. Yoshida, T., Maulik, N., Engelman, R. M., Ho, Y.-S., Magnenat, J. L., Rousou, J. A., Flack III, J. E., Deaton, D. and Das, D. K. (1997). Glutathione peroxidase knockout mice are susceptible to myocardial ischemia reperfusion injury. *Circulation* **96**: II216–II220.

148. Esposito, L. A., Kokoszka, J. E., Waymire, K. G., Cottrell, B., MacGregor, G. R. and Wallace, D. C. (2000). Mitochondrial oxidative stress in mice lacking the glutathione peroxidase-1 gene [In Process Citation]. *Free Radic. Biol. Med.* **28**: 754–766.

149. Yoshida, T., Maulik, N., Engelman, R. M., Ho, Y. S. and Das, D. K. (2000). Targeted disruption of the mouse SOD-1 gene makes the hearts vulnerable to ischemic reperfusion injury. *Circ. Res.* **86**: 264–269.

150. Shefner, J. M., Reaume, A. G., Flood, D. G., Scott, R. W., Kowall, N. W., Ferrante, R. J., Siwek, D. F., Upton-Rice, M. and Brown, R. H., Jr. (1999). Mice lacking cytosolic CuZn superoxide dismutase display a distinctive motor axonopathy. *Neurology* **53**: 1239–1246.

151. Ohlemiller, K. K., McFadden, S. L., Ding, D. L., Flood, D. G., Reaume, A. G., Hoffman, E. K., Scott, R. W., Wright, J. S., Putcha, G. V. and Salvi, R. J. (1999). Targeted deletion of the cytosolic CuZn superoxide dismutase gene

(SOD-1) increases susceptibility to noise-induced hearing loss. *Audiol. Neurootol.* **4**: 237–246.

152. Kondo, T., Reaume, A. G., Huang, T. T., Murakami, K., Carlson, E. J., Chen, S., Scott, R. W., Epstein, C. J. and Chan, P. H. (1997). Edema formation exacerbates neurological and histological outcomes after focal cerebral ischemia in CuZn superoxide dismutase gene knockout mutant mice. *Acta Neurochir. Suppl. (Wien)* **70**: 62–64.

153. Reaume, A. G., Elliott, J. L., Hoffman, E. K., Kowall, N. W., Ferrante, R. J., Siwek, D. F., Wilcox, H. M., Flood, D. G., Beal, M. F., Brown, R. H., Jr., Scott, R. W. and Snider, W. D. (1996). Motor neurons in CuZn superoxide dismutase deficient mice develop normally but exhibit enhanced cell death after axonal injury. *Nature Genet.* **13**: 43–47.

154. Siklos, L., Engelhardt, J. I., Reaume, A. G., Scott, R. W., Adalbert, R., Obal, I. and Appel, S. H. (2000). Altered calcium homeostasis in spinal motoneurons but not in oculomotor neurons of SOD-1 knockout mice. *Acta Neuropathol. (Berl)* **99**: 517–524.

155. Freeman, B. D., Reaume, A. G., Swanson, P. E., Epstein, C. J., Carlson, E. J., Buchman, T. G., Karl, I. E. and Hotchkiss, R. S. (2000). Role of CuZn superoxide dismutase in regulating lymphocyte apoptosis during sepsis. *Crit. Care Med.* **28**: 1701–1708.

156. Kawase, M., Murakami, K., Fujimura, M., Morita-Fujimura, Y., Gasche, Y., Kondo, T., Scott, R. W. and Chan, P. H. (1999). Exacerbation of delayed cell injury after transient global ischemia in mutant mice with CuZn superoxide dismutase deficiency. *Stroke* **30**: 1962–1968.

157. Flood, D. G., Reaume, A. G., Gruner, J. A., Hoffman, E. K., Hirsch, J. D., Lin, Y. G., Dorfman, K. S. and Scott, R. W. (1999). Hindlimb motor neurons require CuZn superoxide dismutase for maintenance of neuromuscular junctions. *Am. J. Pathol.* **155**: 663–672.

158. McFadden, S. L., Ding, D., Burkard, R. F., Jiang, H., Reaume, A. G., Flood, D. G. and Salvi, R. J. (1999). CuZnSOD deficiency potentiates hearing loss and cochlear pathology in aged 129, CD-1 mice. *J. Comp. Neurol.* **413**: 101–112.

159. McFadden, S. L., Ding, D., Reaume, A. G., Flood, D. G. and Salvi, R. J. (1999). Age-related cochlear hair cell loss is enhanced in mice lacking CuZn superoxide dismutase. *Neurobiol. Aging* **20**: 1–8.

160. Ho, Y.-S., Gargano, M., Cao, J., Bronson, R. T., Heimler, I. and Hutz, R. J. (1998). Reduced fertility in female mice lacking CuZn superoxide dismutase. *J. Biol. Chem.* **273**: 7765–7769.

161. Melov, S., Schneider, J. A., Day, B. J., Hinerfeld, D., Coskun, P., Mirra, S. S., Crapo, J. D. and Wallace, D. C. (1998). A novel neurological phenotype in mice lacking mitochondrial manganese superoxide dismutase [see comments]. *Nature Genet.* **18**: 159–163.

162. Jackson, R. M., Helton, E. S., Viera, L. and Ohman, T. (1999). Survival, lung injury, and lung protein nitration in heterozygous MnSOD knockout mice in hyperoxia. *Exp. Lung Res.* **25**: 631–646.

163. Tsan, M.-F., White, J. E., Caska, B., Epstein, C. J. and Lee, C. Y. (1998). Susceptibility of heterozygous MnSOD gene-knockout mice to oxygen toxicity. *Am. J. Respir. Cell Mol. Biol.* **19**: 114–120.

164. Murakami, K., Kondo, T., Kawase, M., Li, Y., Sato, S., Chen, S. F. and Chen, P. S. (1998). Mitochondrial susceptibility to oxidative stress exacerbates cerebral infarction that follows permanent focal cerebral ischemia in mutant mice with manganese superoxide dismuatse deficiency. *J. Neurosci.* **18**: 205–213.

165. Sentman, M. L., Jonsson, L. M. and Marklund, S. L. (1999). Enhanced alloxan-induced beta-cell damage and delayed recovery from hyperglycemia in mice lacking extracellular-superoxide dismutase. *Free Radic. Biol. Med.* **27**: 790–796.

166. Carlsson, L. M., Jonsson, J., Edlund, T. and Marklund, S. L. (1995). Mice lacking extracellular superoxide dismutase are more sensitive to hyperoxia. *Proc. Natl. Acad. Sci. USA* **92**: 6264–6268.

167. Matsui, M., Oshima, M., Oshima, H., Takaku, K., Maruyama, T., Yodoi, J. and Taketo, M. M. (1996). Early embryonic lethality caused by targeted disruption of the mouse thioredoxin gene. *Dev. Biol.* **178**: 179–185.

Chapter 60

Caloric Restriction, Aging and Oxidative Stress

Kevin C. Kregel

Kevin C. Kregel • Integrative Physiology Laboratory, Department of Exercise Science, 532 FH, The University of Iowa, Iowa City, IA. 52242
Tel: (319) 335-7596, E-mail: kevin-kregel@uiowa.edu

1. Introduction

Aging produces a progressive and irreversible decline in physiological function in most multicellular animals. However, molecular mechanisms for this senescence-associated loss of functional capacity have yet to be fully elucidated. Several viable mechanisms have been proposed, including telomere shortening in replicating cells, an accumulation of DNA damage that results in genomic instability, and cumulative damage to macromolecules caused by reactive oxygen species.[3, 10–12] A variety of experimental approaches, including genetic manipulation, are being pursued in an attempt to offset the effects of aging.[26] However, caloric restriction (CR), which involves reduced caloric intake without essential nutrient deficiency, is the most widely studied and efficacious intervention for extending maximal life span in short-lived mammals.[33, 35] When properly executed, CR can also delay a variety of age-related pathological conditions and reduces the incidence of many age-related diseases. Researchers have concluded that CR increases survival in species such as mice and rats by retarding the aging process and, as a result, CR has become an effective tool for studying aging.[19, 27, 30, 33, 35]

While numerous investigators have documented the beneficial effects of CR on increasing longevity and reducing age-related pathologies, the biological mechanisms underlying the retardation of aging and disease processes is still unknown. A wide variety of cutting-edge molecular and cellular biology techniques are currently being utilized in an effort to delineate these mechanisms. In addition, research efforts are focused on the possibility that CR can have beneficial effects on longer-lived species such as primates and humans.

2. Extension of Maximum Life-Span

The initial observations that laboratory rats not only live longer but also have fewer age-associated diseases were made in the 1930s. McCay and colleagues demonstrated that rats on calorically-restricted, nutritionally sound diets lived longer than rats allowed to eat *ad libitum*.[25] Interest in the CR phenomenon increased substantially in the 1970s[35] and has been an active area of research over the past two decades.[19, 33, 35] CR is the only consistently proven method of extending both mean life-span and maximum life-span (MLS; defined as the mean survival of the longest-lived decile) in rodents and nonrodent species (e.g. protozoa, fish, spiders), strongly suggesting that there is a relationship between aging and energy intake.[20, 25, 31, 35] To date, no definitive information is available regarding the possibility that CR can also extend life span and blunt the deleterious changes associated with aging in humans. However, ongoing studies on nonhuman primates suggest that CR can be successfully administered in primates and that these animals exhibit a broad array of physiological alterations.[5, 13, 16, 30]

The majority of CR experiments have been performed in mice and rats and in most studies food intake was reduced by 25 to 50 percent. With this regimen, an increase of ~40% in MLS has been documented.[21, 33, 35, 37] Importantly, the life-extending benefits of CR were not due to any particular nutrient. Instead, these benefits were dependent on the prevention of malnutrition and a reduction in overall caloric intake.

There are several postulated mechanisms by which CR asserts its actions on MLS and physiological function. These mechanisms include depression of metabolic rate, blunted growth, reductions in body temperature, a decrease in body fat, alterations in neuroendocrine and immunologic function, alterations in gene expression, increased DNA repair, enhanced apoptosis, and decreased oxygen radical damage.[19, 33, 35, 37] It should be pointed out that several of these potential mechanisms have fallen out of favor over the past decade. Regardless, despite the increased interest and research efforts in this area over the past decade, the mechanisms underlying the retardation of aging and disease processes have not been fully elucidated.

3. Attenuation of Oxidative Damage and the Associated Decline in Function

Because CR can extend MLS, it is being increasingly used as a model paradigm for understanding basic mechanisms of aging. For example, there is accumulating evidence to suggest that cumulative oxidative damage to macromolecules such as lipids, proteins, and DNA significantly contributes to the lowered functional capacity observed in aged organisms.[1, 33] One proposed mechanism for these observations is that the increase in oxidative stress and subsequent biomolecular damage associated with aging are the result of an increased rate of reactive oxygen species generation and a greater susceptibility of tissues to oxidative injury.

In support of this postulate, CR has been shown to attenuate the age-associated cellular accumulation of oxidatively damaged molecules.[9, 33, 35] For example, CR can blunt the increase in lipid peroxidation, accumulation of oxidized proteins, and oxidative damage to DNA that occurs with aging.[7, 23, 32] Based in part on observations such as these, it has been hypothesized by Sohal and Weindruch that the mechanism by which CR retards aging processes is by decreasing cellular oxidative stress.[33] This hypothesis is based on the premise that a chronic state of oxidative stress exists in cells because of an imbalance between prooxidants (e.g., reactive oxygen species such as superoxide radical and hydrogen peroxide) and antioxidant enzymes (e.g. MnSOD, CuZnSOD, catalase, glutathione peroxidase). As an organism ages, the amount of oxidative damage increases, and it is postulated that the cumulative effect of this damage is a primary factor contributing to the aging process.

Studies in support of this hypothesis have documented that CR can lower steady-state levels of oxidative stress and retard physiological alterations generally associated with aging.[33] Sohal and Weindruch also suggest that the beneficial effects of CR in rodents is associated with a decrease in metabolic rate. However, this postulate remains controversial, due in part to observations that CR can blunt age-associated changes in rodents without decreasing the rate of metabolism.[8, 19, 24]

In addition to a decline in the rate of generation of reactive oxygen species, CR could also reduce the age-associated accumulation of oxidatively damaged molecules by other postulated mechanisms. One possibility is that CR enhances intracellular antioxidant defenses in order to offset the elevated levels of prooxidants generated within senescence. Studies focused on the activities of individual antioxidant enzymes in a variety of tissues during aging or in response to CR have generated conflicting results.[22, 29, 32] However, more recent findings have demonstrated that CR can enhance some aspects of antioxidant enzyme function.[2, 6, 9]

Another possible mechanism for the reduced oxidative damage associated with CR involves the repair of oxidatively damaged molecules. Investigators have noted that CR attenuates the age-associated decline in DNA repair and promotes both the proteolytic removal of damaged proteins and their replacement by newly synthesized proteins.[34, 36]

In summary, current research efforts are aimed at delineating the contribution of oxidative stress and damage to the process of aging. There is compelling evidence that CR reduces the accrual of oxidative damage in rodents, although the responsible mechanisms are not well-defined. However, these finding suggest that the anti-aging effects of CR are associated with an attenuation of oxidative damage. Recent research supports the hypothesis that CR may act by reducing the total amount of oxidative stress within an animal, thereby reducing the rate of accrual of age-associated damage to cellular constituents such as proteins, lipids, and DNA.

4. New Directions in Caloric Restriction Research

4.1. Caloric Restriction and Environmental Stress

Although extensive data support the tenet that CR offsets the age-associated accrual of oxidative injury,[33, 35] the majority of these experiments have been conducted in isolated tissues from nonstressed animals. As a result, little information exists regarding the ability of CR animals to respond to a physical challenge or adapt to repeated stresses. This is an important consideration since there is a well-documented decline in the ability of an aged organism to cope with a variety of environmental stresses.[4, 14, 15, 28]

Hall *et al.*[9] recently investigated the influence of long-term CR and aging on stress tolerance in senecent CR rats and their control-fed counterparts exposed to an environmental heating protocol on two consecutive days. It was hypothesized that CR would enhance heat tolerance by reducing the magnitude of cellular oxidative stress and subsequent accrual of tissue injury. The CR animals had reductions in heat-induced radical generation and stress protein accumulation, along with significant declines in cellular injury in the liver compared to the control group. In addition, heat stress stimulated a marked induction of antioxidant enzymes in the CR rats, while the stress-related induction of antioxidant enzymes was blunted in the controls. These observations suggest that a CR regimen protects rats from the pathogenesis of normal age-related injury, as well as providing significant protection from the cytotoxic effects of environmental stress. The improvement in stress tolerance in CR animals was also associated with blunted production of radical species and a preservation of cellular ability to adapt to stress through a distinctive pattern of antioxidant enzyme induction.

4.2. Gene Expression Profiles

One of the most provocative areas of research in the CR field involves studies investigating whether alterations in gene expression occur with aging and whether the expression profile is modified with a long-term CR regimen. For instance, experimental evidence has suggested that CR increases MLS and reduces physiological dysfunction by attenuating the age-related decrease in immunologic function. The observation that CR attenuates immunosenescence has led to several studies examining whether CR exerts its actions through modulation of gene expression. Findings in the area of CR and immunologic function, with an emphasis on potential molecular mechanisms, are reviewed in a recent paper by Pahlavani.[27]

Investigators have also taken advantage of the recently developed oligonucleotide-based array ("gene chip") technology to investigate changes in gene expression with CR and aging. The use of oligonucleotide arrays is an attractive strategy for elaborating an unbiased molecular profile of a large number of genes during an intervention such as CR. This experimental approach offers the potential to identify molecules or cellular pathways not previously associated with CR.

In a study by Lee *et al.*,[17] the expression profile of over 6000 genes was analyzed CR for skeletal muscle samples obtained from young, old, and age-matched old mice using oligonucleotide arrays. Aging resulted in a gene expression pattern suggestive of a marked stress response and lower expression of a variety of metabolic and biosynthetic genes. Most of these alterations were either partially or completely prevented in CR mice. Interestly, less than 2% of the genes analyzed had greater than a two-fold difference in gene expression as a function of age, while CR modified the expression of approximately 1% of the genes surveyed. These results suggest that both aging *per se* and CR can alter the gene expression profile in mammals. Further, the relatively small percentage of genes with differential expression suggests that neither the aging process nor the changes manifested with a CR regimen is due to large, widespread alterations in gene expression.

In a subsequent study, the Weindruch and Prolla team again utilized gene chip technology to evaluate approximately 6300 genes in the neocortex and cerebellum of mice, providing the first global view of gene expression patterns in the aging mammalian brain and its modulation by CR.[18] The gene expression profile in both regions of the aging brain was indicative of an increased inflammatory response, a greater level of cellular stress (e.g., oxidative stress-inducible genes and heat shock factors), and an attenuation in neural growth and trophic factors. These investigators noted that the transcriptional alterations observed in the brain of aging mice parallels the patterns previously noted with human neurodegenerative disorders. The gene expression profile of senescent CR mice was also compared to age-matched control-fed mice in the neocortex. Interestly, long-term CR produced an attenuation in the age-associated induction of genes encoding inflammatory and cellular stress responses.

4.3. Nonhuman Primate Studies

Investigations of the effects of CR on nonhuman primate models have been ongoing in several laboratories since the late 1980s. Because of the longitudinal nature of these studies, and the relatively long life spans of most primate species, the outcome of this research will not be fully elucidated for several years. However, the available results suggest that the physiological changes in monkeys in response to CR are similar to those observed in rodents. For example, circulating levels of insulin and glucose are decreased, insulin sensitivity improves, and both body temperature and energy expenditure are reduced.[5, 13, 16, 30] In addition, evaluations based on several disease biomarkers suggest that monkeys undergoing CR will be less likely to develop diabetes, cardiovascular disease, obesity, and immune dysfunction.[30] As these studies go forward, and before findings in primates can be applied to humans, it will also be important to assess the overall health and fitness status of these animals in order to evaluate their ability to cope with physiologically-relevant stressors such as infection, exercise, hyperthermia and hypothermia.

5. Conclusions

It has been known for many years that CR, if properly executed without malnutrition, will greatly extend the MLS of rodents and decrease the incidence of late-life diseases such as cancer. However, the mechanisms by which CR works to increase life span and decrease the incidence of chronic disease remain unclear. Researchers in this field are now faced with two primary questions. First, what are the biological mechanisms underlying the retardation of aging and disease in lower mammalian species such as rodents? Second, will CR exert similar beneficial effects in longer-lived mammals, including nonhuman and human primates? Although much has been learned about the effects of CR on physiological function, our understanding of how CR alters cellular and molecular processes and its potential applicability to humans is far from complete.

Acknowledgments

This work is supported in part by National Institutes of Health grants R01-AG12350 and R01-AG14687.

References

1. Ames, B., Shinenaga, M. K. and Hagen, T. M. (1993). Oxidants, antioxidants, and the degenerative diseases of aging. *Proc. Natl. Acad. Sci. USA* **90**: 7915–7922.
2. Armeni, T., Pieri, C., Marra, M., Saccucci, F. and Principato, G. (1998). Studies on the life prolonging effect of food restriction: glutathione levels and glyoxylase enzymes in the liver. *Mech. Ageing Dev.* **101**: 101–110.
3. Beckman, K. B. and Ames, B. N. (1998). The free radical theory of aging matures. *Physiol. Rev.* **78**: 547–581.
4. Blake, M. J., Gershon, D., Fargnoli, J. and Holbrook, N. J. (1990). Discordant expression of heat shock protein mRNAs in tissues of heat-stressed rats. *J. Biol. Chem.* **25**: 15 275–15 279.
5. Bodkin, N. L., Ortmeyer, H. K. and Hansen, B. C. (1995). Long-term dietary restriction in older-aged rhesus monkeys: effects on insulin resistance. *J. Gerontol. Biol. Sci.* **50A**: B142–B147.
6. Cook, C. J. and Yu, B. P. (1998). Iron accumulation in aging: modulation by dietary restriction. *Mech. Ageing Dev.* **102**: 1–13.
7. Dubey, A., Forster, M. J., Lal, H. and Sohal, R. S. (1996). Effect of age and caloric intake on protein oxidation in different brain regions and on behavioral functions of the mouse. *Arch. Biochem. Biophys.* **333**: 189–197.
8. Duffy, P. H., Feuers, R. F., Leakey, J. E. A. and Hart, R. W. (1991). Chronic caloric restriction to old female mice: changes in circadian rhythms of physiological and behavioral variables. *In* "Biological Effects of Dietary Restriction" (L. Fishbein, Ed.), pp. 245–263, Springer-Verlag, Berlin.
9. Hall, D. M., Oberley, T. D., Moseley, P. L., Buettner, G. R., Oberley, L. W., Weindruch, R. and Kregel, K. C. (2000). Caloric restriction improves

thermotolerance and reduces hyperthermia-induced cellular damage in old rats. *FASEB J.* **14**: 78–86.

10. Jazwinski, S. (1996). Longevity, genes, and aging. *Science* **273**: 54–59.

11. Johnson, B. F., Sinclair, D. A. and Guarente, L. L. (1999). Molecular biology of aging. *Cell* **96**: 291–302.

12. Johnson, T. E. (1997). Genetic influences on aging. *Exp. Gerontol.* **32**: 11–22.

13. Kemnitz, J. W., Roecker, E. B., Weindruch, R., Elson, D. F., Baum, S. T. and Bergman, R. N. (1994). Dietary restriction increases insulin sensitivity and lowers blood glucose in rhesus monkeys. *Am. J. Physiol.* **266**: E540–E547.

14. Kregel, K. C., Moseley, P. L., Skidmore, R., Gutierrez, J. A. and Guerriero, V. (1995). HSP70 accumulation in tissues of heat-stressed rats is blunted with advancing age. *J. Appl. Physiol.* **79**: 1673–1678.

15. Kregel, K. C., Tipton, C. M. and Seals, D. R. (1990). Thermal adjustments to nonexertional heat stress in mature and senescent Fischer 344 rats. *J. Appl. Physiol.* **68**: 1337–1342.

16. Lane, M. A., Baer, D. J., Rumpler, W. V., Weindruch, R., Ingram, D. K., Tilmont, E. M., Cutler, R. G. and Roth, G. S. (1996). Calorie restriction lowers body temperature in rhesus monkeys, consistent with a postulated anti-aging mechanism in rodents. *Proc. Natl. Acad. Sci. USA* **93**: 4159–4164.

17. Lee, C.-K., Klopp, R. G., Weindruch, R. and Prolla, T. A. (1999). Gene expression profile of aging and its retardardation by caloric restriction. *Science* **285**: 1390–1393.

18. Lee, C.-K., Weindruch, R. and Prolla, T. A. (2000). Gene-expression profile of the ageing brain in mice. *Nature Genet.* **25**: 294–296.

19. Masoro, E. J. (2000). Caloric restriction and aging: an update. *Exp. Gerontol.* **35**: 299–305.

20. Masoro, E. J. (1988). Food restriction in rodents: an evaluation of its role in the study of aging. *J. Gerontol.* **43**: B59–B64.

21. Masoro, E. J., McCarter, R. J. M., Katz, M. S. and McMahan, C. A. (1992). Dietary restriction alters characteristics of glucose fuel use. *J. Gerontol. Biol. Sci.* **47**: B202–B208.

22. Matsuo, M., Gomi, F. and Dooley, M. M. (1992). Age-related alterations in antioxidant capacity and lipid peroxidation in brain, liver, and lung homogenates of normal and vitamin E deficient rats. *Mech. Ageing Dev.* **64**: 273–292.

23. Matsuo, M., Gomi, F., Furamoto, K. and Sagai, M. (1993). Food restriction suppresses an age-dependent increase in exhalation rate of pentane from rats: a longitudinal study. *J. Gerontol. Biol. Sci.* **48**: B133–B138.

24. McCarter, R. J. M. and Palmer, J. (1992). Energy metabolism and aging: a lifelong study of Fischer 344 rats. *Am. J. Physiol.* **263**: E448–E452.

25. McCay, C. M., Crowell, M. F. and Maynard, L. A. (1935). The effect of retarded growth upon the length of life span and upon the ultimate body size. *J. Nutri.* **10**: 63–79.

26. Orr, W. C. and Sohol, R. S. (1994). Extension of life-span by overexpression of superoxide dismutase and catalase in *Drosophila melanogaster*. *Science* **263**: 1128–1130.

27. Pahlavani, M. A. (2000). Caloric restriction and immunosenescence: a current perspective. *Front. Biosci.* **5**: 580–587.

28. Papaconstantinou, J. (1994). Unifying model of the programmed (intrinsic) and stochastic (extrinsic) theories of aging. *Ann. NY Acad. Sci.* **719**: 195–211.

29. Rikans, L. E., Moore, D. R. and Snowden, C. D. (1991). Sex-dependent differences in the effects of aging on antioxidant defense mechanisms of rat liver. *Biochim. Biophys. Acta* **1074**: 195–200.

30. Roth, G. S., Ingram, D. K. and Lane, M. A. (1999). Calorie restriction in primates: will it work and how will we know? *J. Am. Geriat. Soc.* **47**: 896–903.

31. Rudzinski, M. A. (1952). Overfeeding and lifespan in *Tokophyra infusionum*. *J. Gerontol.* **7**: 544–552.

32. Sohal, R. S., Agarwal, S., Candas, M., Forster, M. J. and Lal, H. (1994). Effect of age and caloric restirction on DNA oxidative damage in different tissues of C56BL/6-mice. *Mech. Ageing Dev.* **76**: 215–224.

33. Sohal, R. S. and Weindruch, R. (1996). Oxidative stress, caloric restriction, and aging. *Science* **273**: 59–63.

34. Van Remmen, H., Ward, W. F., Sabia, R. V. and Richardson, A. (1995). Gene expression and protein degradation. *In* "Handbook of Physiology-Aging" (E. J. Masoro, Ed.), pp. 171–234, Oxford University Press, New York.

35. Weindruch, R. and Walford, R. L. (1988). *In* "The Retardation of Aging and Disease by Dietary Restriction" (C. C. Thomas), Springfield, IL

36. Weraarchakul, N., Strong, R., Wood, W. G. and Richardson, A. (1989). Effect of aging and dietary restriction on DNA repair. *Exp. Cell Res.* **181**: 197–204.

37. Yu, B. P. (1994). How diet influences the aging process of the rat. *Proc. Soc. Exp. Biol. Med.* **205**: 97–105.

Chapter 61

Hormones and Oxidative Stress

Holly M. Brown-Borg and S. Mitchell Harman*

Holly M. Brown-Borg • Assistant Professor, Department of Physiology, University of North Dakota School of Medicine, P.O. Box 9037, 501 North Columbia Road, Grand Forks, ND 58202-9037
Tel: (701) 777-3949, E-mail: brownbrg@medicine.nodak.edu
S. Mitchell Harman • Director, Kronos Longevity Research Institute, 4455 East Camelback Road, Phoenix, AZ 85018
*Corresponding Author.
Tel: 602-778-7484, E-mail: harman@kronosinstitute.org

1. Introduction

Existing data suggest that, in general, aging organisms tend to be in a pro-oxidant state. Therefore, maintenance of an efficient antioxidative defense system may be critical for extending lifespan. Evidence for a role of the endocrine system in oxidative stress and aging is slowly emerging. The relationships of the physiology of the endocrine system to oxidative damage processes are complex, incompletely understood, and variable, depending on the hormone in question. Moreover, recent reports suggest that the hormonal contribution to the state of oxidant/antioxidant balance changes as hormone profiles change with age.

There are a number of mechanisms by which hormones may affect the process of oxidative stress and oxidative damage. First, hormones may alter the rate at which oxygen free radicals are produced by the mitochondrial cytochrome system. For example, thyroid hormones up-regulate a large number of genes involved in oxidative metabolism,[1] with a net effect of increasing mitochondrial oxidative processes and thus, indirectly, oxygen free radical production. Second, hormones may modulate the production or activation state of enzymes and other compounds involved in detoxifying oxygen free radicals. For example, Growth Hormone (GH) appears to down-regulate the production of copper and zinc superoxide dismutase (Cu, Zn SOD) and catalase[2] in some tissues of mice. Third, hormones and hormone-related compounds may themselves act as antioxidants or oxidants, directly reducing or producing oxygen free radicals, as has been shown for estrogens[3,4] and DHEA.[5] Finally, hormones may influence rates of repair of oxidized tissue components by altering production or availability of molecules involved in repair of protein damage or modulating amounts or activity of repair enzymes. Examples of this type of interaction are the reduction in activity of the 70 kD heat shock protein (HSP70) by glucocorticoids[6] and the finding that activation of the insulin receptor increases transcription of XPD, a gene for a DNA helicase involved in transcription and nucleotide excision repair.[7] In the following chapter we will review what is known about the relationship of several different hormone axes to oxygen free radical production, protection, and oxidative damage and repair.

2. Growth Hormone and IGF-I

Growth hormone (GH) has received much attention as a possible anti-aging therapy since levels of this hormone decline with age and are known to be important in the maintenance of normal body composition in adults. However, experimental data suggest that GH and its downstream effector, Insulin-like Growth Factor I (IGF-I), may alter cellular oxidative processes, thereby contributing to oxidative stress and actually promoting aging.

In humans and animals, plasma GH levels decline with age reflecting reduced GH release from the pituitary as a consequence of decreasing stimulatory and increasing inhibitory inputs from the hypothalamus.[8] This age-related decrease in GH leads to a decline in plasma levels of IGF-I, which is the main mediator of GH action.[9, 10] Changes in body composition that occur with aging are similar to those found in GH deficiency.[11-13] A well-known study by Rudman and coworkers[14] showed that GH replacement in elderly men with low IGF-I levels reduced adiposity and increased lean body mass, with equivocal improvement in bone mineral density. Clinical use of GH has been suggested to benefit adults with GH deficiency by partially reversing the observed loss of muscle mass and increased adiposity and reducing known cardiovascular consequences of GH deficiency.[15] Although recent reports suggest no real benefit of GH therapy in adults (beyond that of exercise),[16] at present there are no definitive answers with regard to the efficacy or safety of long-term GH therapy in humans. In support of GH as an anti-aging hormone, however, Caloric Restriction (CR), which extends life span in different species including rodents,[17] has been shown to delay the age-related decrease in GH,[18] even while circulating IGF-I levels remain suppressed throughout CR.[19]

However, both pathological and experimentally-induced GH excess can lead to reductions in lifespan. Significantly increased incidence of cardiovascular disease, diabetes and tumors[20, 21] are found in *acromegalic* patients and the death rate among these individuals is significantly higher.[22] Furthermore, growth hormone therapy significantly increased mortality in a study in severely ill patients.[23] Finally, a circumstantial positive association between higher levels of IGF-I and prostate[24, 25] and breast[26] cancer have been reported. In agreement with the evidence in humans, the life expectancy of GH transgenic mice is also compromised with transgenics living less than 50% as long as wild-type mice.[27-30] These mice do not become diabetic, but are insulin resistant and hyperinsulinemic.[31, 32] They appear normal and healthy until approximately six to eight months of age when they begin to show signs of premature aging including scoliosis, weight loss, and general loss of body condition. While these mice have been shown to die of renal failure related to GH excess,[33] chronic exposure to high levels of GH also appears to accelerate aging. Physiological indicators of premature aging observed in these mice include reduced replicative potential of cells *in vitro*[28] and reduced reproductive lifespan.[34] Additionally, indications of premature central nervous system aging include reduced catecholamine turnover,[35] increased astrogliosis,[36] and impaired learning and memory.[37]

Importantly, GH transgenic mice also exhibit signs of oxidative stress that may be related to the noted accelerated aging. Elevated levels of superoxide radical and lipid peroxidation are found in tissues of mice exposed to high levels of GH.[38] SOD gene expression is depressed (Brown–Borg, unpublished data), and catalase, the enzyme responsible for eliminating H_2O_2, is decreased

40–80% in tissues from GH transgenic mice[39] when compared to wild-type siblings. Catalase protein is reduced significantly in aged transgenic animals and activity of the enzyme is decreased 25–50%. Elevation of levels of liver ascorbate by more than 50%[2] are also observed in GH transgenic mice. Both superoxide radical and ascorbate have been shown to decrease catalase activity[40, 41] and thus may have direct effects on catalase in these mice. In addition, both ascorbate and IGF-I can act as pro-oxidants under certain conditions.[42, 43] Furthermore, there is evidence suggesting that several proteins involved in oxidative phosphorylation (mitochondrial; Complexes I, II, V) are significantly suppressed in these mice,[44] again suggesting that major alterations in oxidative metabolism are present. Others have reported that Complex I activity declines with aging and leads to increased production of superoxide radicals.[45, 46]

In vitro evidence also suggests that GH and IGF-I modulate the expression of enzymes involved in the elimination of free radicals. Catalase activity is significantly suppressed in the presence of GH and IGF-I in primary mouse hepatocyte cultures.[47] In addition, manganese SOD, the enzyme that converts the superoxide radical into hydrogen peroxide (H_2O_2) in mitochondria, is also decreased with the addition of GH in this *in vitro* cell system. Further studies are underway to substantiate the role of GH on the antioxidative defense capacity of cells. Although the neuroprotective effects of IGF-I are well documented,[48–50] this growth factor may also aggravate neuronal injury, as has been found for other neuronal growth factors (nerve growth factor, brain-derived growth factor).[51] IGF-I has been shown to potentiate free radical-induced neurotoxicity and neuronal cell death.[43, 52] The existence of conflicting data compromises the therapeutic potential of the IGF's.

In addition to an increase in oxidative membrane damage (lipid peroxidation) noted above,[38] oxidative DNA and protein damage are significantly increased in liver and brain tissues of animals exposed to high levels of GH.[38a] Both the pathological GH excess and hyperinsulinemia may account for the increased oxidative damage and accelerated aging observed in these animals. High serum insulin concentrations are strongly correlated with decreased liver catalase activity via a suppression in catalase synthesis.[53] Insulin not only stimulates H_2O_2 production in human fat cells,[54] but also has been found to inhibit *proteasome*, an enzyme responsible for degrading oxidized proteins.[55, 56] Decreased proteasome activity would result in an increase in the net rate of accumulation of free radical-mediated damage. Therefore, several mechanisms are present that depress antioxidant defenses and may contribute to the premature aging observed under conditions of GH excess.

In contrast to the reductions in life span observed with GH overexpression, GH deficiency is associated with significant extensions of life span in mice.[57] Growth hormone-deficient *Ames* dwarf mice live more than a year longer than wild-type mice, with 49% and 64% extensions of life span for males and females, respectively.

Phenotypically identical *Snell* dwarf mice also live significantly longer than their wild-type counterparts[58] (Flurkey and Harrison, personal communication). Both *Ames* and *Snell* dwarf mice have complete primary deficiency of GH, prolactin, and thyroid stimulating hormone due to mutations in the *Prophet* of Pit-1 (Prop-1) gene or the Pit-1 gene (respectively), a transcription factor that is responsible for the differentiation of pituitary somatotrophs, lactotrophs, and thyrotrophs.[59, 60] Plasma insulin tends to be lower,[61] and IGF-I levels in peripheral blood are also undetectable[62] in these mice.

In studies conducted in the *Ames* dwarf mouse, differences have been found in antioxidant defense mechanisms when compared with wild-type mice. At several ages, the Ames dwarf exhibits higher levels of catalase (mRNA, protein, and activity) in various tissues.[2, 39] In conjunction with higher liver catalase activity, the level of liver inorganic peroxide (a measure of H_2O_2) and mitochondrial-generated H_2O_2, are reduced in the dwarf mouse.[38a, 63, 64] The activity and level of Cu, Zn SOD is elevated in liver and brain tissue when compared to wild-type animals. Oxidative DNA and protein damage are significantly lower in dwarf tissues[38a] suggesting that these animals are exposed to fewer reactive oxygen species (ROS) compared with their GH-sufficient counterparts. In addition, the total pool of glutathione (GSH), a non-enzymatic antioxidant, is significantly elevated in dwarf liver tissues compared with wild-type livers.[64] Low ascorbate levels in liver tissue of dwarf mice are possibly related to catalase levels.[41] Mitochondrial proteins involved in oxidative phosphorylation (Complexes I, II, V) are significantly elevated in dwarf mice.[44] Additional observations suggest that Ames dwarf mice have a biological advantage over normal wild-type mice with regard to enzymatic scavenging of toxic metabolic byproducts, which may contribute to enhanced longevity.

Several physical and biochemical characteristics exhibited by the dwarf mice suggest associations between aging, antioxidants, and hormone status. First, body size has been shown to play a role in aging. For example, within species (e.g. dogs and horses), smaller breeds live longer than larger breeds.[65, 66] Short individuals tend to live longer than tall individuals.[67] Some of these differences have been linked to lower IGF-I levels in smaller breeds.[68–70] *Ames* dwarf mice are smaller than normal mice,[71, 72] and plasma levels of IGF-concentrations are below radioimmunoassay detectability.[62] The interactions between GH and the IGF's in determining both body size and metabolism may contribute to possible alterations in antioxidative defenses. Early evidence showed that neonatal hypophysectomy in rats moderately extended life span if glucocorticoids and L-thyroxine (T_4) were replaced.[73, 74] In humans, there is a lack of data on life expectancy in patients with GH deficiency/resistance. Importantly, however, there is evidence[75] that shows increased longevity in hypopituitary patients with mutations at the Prop-1 locus, the same gene that is mutated in the *Ames* dwarf mouse. Furthermore, GH receptor knockout mice[76] have been shown to outlive their normal counterparts[77] and maintain learning and memory skills that decline in wild-type mice.[78] When IGF-I or IGF-I receptor genes are knocked out, the mice are not viable or exhibit severe abnormalities.[79] The available data suggest that defects in GH synthesis or signaling have profound significant effects on aging. Prop-1, Pit-1, and GH receptor/GH binding protein genes are involved in delaying aging processes and maybe considered "gerontogenes" in mammals.

There is evidence in other species implicating the GH or IGF signaling pathway in longevity and oxidative stress. Mutations in the daf-2 gene, a homologue of the mammalian insulin-like growth factor receptor, have been shown to increase life span in the nematode.[80] Consistent with findings in the dwarf mouse, reduced signaling in these pathways increases SOD activity, decreases accumulation of oxidized protein and decreases oxidative stress in this species.[81–85] The function of these mutated genes is unclear; however, they may alter antioxidant activities that affect longevity.

The evidence clearly implicates GH modulation of free radical processes and longevity. The increased growth in transgenic animals may generate more free radicals and saturate the normal antioxidative defense mechanisms. While in dwarf animals, the slow maturation process may allow the defense mechanisms to operate more efficiently in the absence of GH, producing less oxidative damage and permitting longer life. GH augments ROS production, decreases ROS elimination processes, and may impair repair processes, all of which lead to increased oxidative stress. As individuals age, the decline in endocrine function may contribute to alterations in antioxidative enzyme activities and thus modify the rate of accumulation of oxidative damage. A growing body of evidence suggests that GH plays an important role in these processes.

3. Insulin, Glucose, and Diabetes

Insulin, a polypeptide produced by the beta cells of the pancreatic islets, is a key metabolic hormone. Insulin promotes storage of glucose as glycogen in liver and of triglycerides in adipose tissue and mediates uptake and utilization of glucose by skeletal muscle. In the absence of adequate insulin action, whether due to failure of insulin secretion or to peripheral insulin resistance, excessive amounts of glucose accumulate in the circulation (hyperglycemia). The classical action of insulin occurs via binding to the insulin receptor, a transmembrane tyrosine kinase. There is evidence that hyperglycemia and hyperinsulinemia contribute both interactively and independently to oxidative tissue damage.

With regard to a direct action of insulin on ROS generation, it has been shown that insulin can cause a rapid transient increase in H_2O_2 accumulation in human fat cell suspensions. Facchini, *et al.*[86] have speculated that hyperinsulinemia could, by its effects on antioxidative enzymes and on free-radical generators, enhance oxidative stress. If hyperinsulinemia accelerates oxidative damage, this effect could mediate, in part, the life-prolonging effect of calorie restriction and help explain why mutations decreasing the overall activity of insulin-like receptors increase lifespan in *Caenorhabditis elegans*.[83]

Most work in the area of oxidative stress, insulin, and glucose suggests that it is not insulin itself, but the high levels of glucose, which result from a deficiency of insulin action, that are responsible for ROS generation. Thus, both Type-1 (insulin deficient) and Type-2 (insulin resistant) forms of diabetes are associated with increased levels of oxidative stress and oxidative damage.[87, 88] Moreover, it has been shown that brief elevations of glucose can increase oxidative stress, as during a Glucose Tolerance Test (GTT).[88] After a test meal plasma malondialdehyde and vitamin C levels increased, while protein SH groups, uric acid, vitamin E, total plasma radical-trapping activity decreased, both in control subjects and, to a significantly greater extent, in Type-2 diabetic subjects.[89] The authors concluded that because glucose, but not insulin, rose higher in diabetic than in control subjects, glucose itself plays an important role in the generation of oxygen free radicals. [89]

More evidence that glucose plays an independent role in ROS generation in human diabetes comes from a study of short-term intensive insulin therapy in chronically hyperglycemic Type-2 patients. At baseline, in patients compared with control subjects, erythrocyte free and total malondialdehyde (MDA) and Percent Polyunsaturated Fatty Acids (PUFAs) were increased, serum vitamin E and GSH were reduced, and no difference was observed in activities of GSH peroxidase (GPX), SOD, or catalase. After euglycemic insulin therapy no effects were observed on PUFAs, GSH, or total MDA, but free MDA decreased and vitamin E increased suggesting, that normoglycemia reduced ROS production.[90]

Several mechanisms by which glucose can lead to the generation of ROS have been explored. A continuous flux of H_2O_2 produced in SV40-transformed human fibroblasts *in vitro* by a glucose/glucose oxidase system provokes both mitochondrial and nuclear DNA damage.[91] Exposure of cultured Smooth Muscle Cells or Endothelial Cells to high glucose increased ROS production. This increase was inhibited by an NADPH oxidase inhibitor and inhibition reversed with Protein Kinase C (PKC) inhibitors.[92] An *in vivo* glucose challenge increased ROS generation and levels of p47phox, an NADPH component, and decreased alpha-tocopherol levels in human Polymorphonuclear Neutrophil Cells (PMN) and Mononuclear Cells (MNC).[93] The above findings suggest that glucose causes ROS production via PKC-dependent activation of NADPH oxidase, a conclusion consistent with the finding that the exogenous addition of NADPH to Red Blood Cell (RBC) lysates induced membrane lipid peroxidation.[94] Glucose-induced membrane lipid peroxidation and osmotic fragility were blocked in RBC pretreated with fluoride, an inhibitor of glucose metabolism or with inhibitors of the Cytochrome P450 system, suggesting that oxidative metabolism of glucose leads directly to ROS generation, which then damages lipid membranes.[94]

Another mechanism by which glucose may cause oxidative damage involves protein glycation and the formation of Advanced Glycation End products (AGE's).[95] The accelerated non-enzymatic glycation of proteins during hyperglycemia and its potential role in the pathogenesis of diabetic complications was first brought to light by the work of Cerami and co-workers.[96, 97] Antioxidants can dissociate structural damage to proteins from the incorporation of monosaccharide into protein, indicating a potential therapeutic role for antioxidants in the prevention of protein damage by glycation.[98]

EPR spectroscopy indicates that glycation of protein generates active centers for catalyzing one-electron oxidation-reduction reactions and mimics the characteristics of metal-catalyzed oxidation systems.[99] In addition EPR analysis of glycation has implicated the glucose fragment, glycolaldehyde, in the generation of a pyrazinium free-radical cation in reactions both with free amino acids and the protein, histone H1, *in vitro* and *in vivo*.[100] These results together indicate that glycated proteins accumulating *in vivo* provide stable active sites for catalyzing the formation of free radicals. Lipid glycation may also contribute to membrane lipid peroxidation involved in the pathogenesis of diabetic complications and aging.[101]

There is, in fact, considerable evidence that high glucose levels result in oxygen free radical damage to various cell components. For example, when RBC's are incubated with high concentrations of glucose, lipid peroxidation and osmotic fragility are increased, which can be inhibited by oxygen radical scavengers.[94] This *in vitro* finding is consistent with a report that, compared with normal volunteers, recently-diagnosed Type-1 diabetics had elevated plasma levels of conjugated lipid dienes, and, in longstanding diabetics, there was also a marked increase in Thiobarbituric Acid-Reactive Substances (TBARS), conjugated dienes, and lipid hydroperoxide levels.[87] *In vivo* lipid peroxidative damage, as determined by measurement of exhaled ethane, is also increased in rats made diabetic with Streptozocin or acutely hyperglycemic by intraperitoneal dextrose administration, while insulin treatment of diabetic rats attenuated ethane production.[102]

Mitochondrial DNA (mtDNA) is another important cell component damaged by oxidative stress. Studies employing long-extension Polymerase Chain Reaction (PCR) combined with a quantitative PCR have shown a 5.7-fold increase in heterogeneous mutations in mtDNA in the muscle tissue of humans, ages 55–75 years, with either Impaired Glucose Tolerance (IGT) or diabetes compared with age- and sex-matched control subjects.[103] A similar study in an obese rat strain (cp) with insulin resistance, hyperinsulinemia, and hyperlipidemia showed that the cp genotype confers susceptibility to mtDNA deletions *in vivo* and that high glucose concentrations induce mtDNA mutations *in vitro*.[104] Urinary 8-hydroxy-2'-deoxyguanosine (8OH-dG) has been reported to serve as a sensitive biomarker of oxidative DNA damage and is linked to an increase in mtDNA deletions. In patients with Type-2 diabetes mellitus, the total 24 h urinary excretion of 8OH-dG was markedly higher than in non-diabetic age- and sex-matched control subjects. In addition, high glycosylated hemoglobin levels were associated with greater urinary 8OH-dG.[105] In another study,[106] mtDNA deletions and 8OH-dG content in muscle from Type-2 diabetic patients were significantly higher than those of the control subjects. There was a significant correlation between mtDNA deletions and the 8OH-dG content ($P < 0.0001$). Both mtDNA deletions and the 8OH-dG content were correlated with duration of diabetes and were greater in muscle from patients with diabetic nephropathy or retinopathy. Thus, oxidative mtDNA damage may contribute to the pathogenesis of diabetic complications.

There may also be an association between Type-2 diabetes and oxidative stress in the pancreatic beta cells. In GK rats, a model of non-obese Type-2 diabetes, levels of 8OH-dG and 4-Hydroxy-2-Nonenal (HNE)-modified proteins by quantitative immunohistochemical analyses were higher in pancreatic beta-cells than in those of control *Wistar* rats.[107] These results indicate that chronic hyperglycemia might be responsible for oxidative stress in the pancreatic beta cells leading to the secondary beta-cell failure characteristic of Type-2 diabetes.

An early study reported increased ROS production by MNC from the blood of patients with diabetes mellitus.[108] HPLC assays of 8OH-dG on DNA extracted from MNC of patients with Type-1 diabetes mellitus, Type-2 diabetes, and age-matched healthy volunteers showed diabetic patients to have significantly more 8OH-dG than controls. ROS generation in these cells measured by chemilumine-scence was also significantly greater in diabetic patients.[109] Consistent with these findings is a report showing significantly elevated levels of DNA damage, using the alkaline comet assay which measures single-stranded DNA breaks and alkali-labile sites, in the PMN fraction, and nonsignificant increases in the lymphocyte, MNC, and whole blood fractions from Type-1 diabetic subjects compared to controls.[110] Because MNC play an important role in atherogenesis, the above data suggest that oxidative stress could play a role in the acceleration of atherogenesis in diabetes and in the defective PMN function, which contributes to the susceptibility of diabetic patients to infections.[111, 112]

The role of oxidative stress in producing vascular complications of diabetes has also been investigated in various studies. High glucose increased H_2O_2 production and apoptosis in cultured human umbilical vein endothelial cells. High glucose also up-regulated Nitric Oxide Synthase (NOS) protein expression early but gradually reduced it after longer exposure.[113] The increased apoptosis was reversed by vitamin C or the Nitric Oxide (NO) donor (Sodium Nitroprusside), but enhanced by an NOS inhibitor, suggesting that NO plays a protective role and that long-term high glucose exposure leads to an imbalance of NO and ROS. The above-cited finding by Inoguchi et al.[92] that glucose also accelerates ROS production in isolated vascular smooth muscle cells and endothelial cells also suggests a role for local oxidative stress in the acceleration of atherosclerosis in patients with diabetes and insulin-resistance syndrome. In addition, morphological studies in rats revealed severe alterations of myocardial structure after a diabetes duration of three months that were clearly improved if animals were treated with tocopherol-acetate.[114]

The role of oxidative damage, as it affects risk of atherosclerotic macro-vascular disease via oxidation of circulating lipids, is dealt with in more detail in another chapter in this volume (see chapter by Reaven, "Oxidative Stress and Cardiovascular Risk"). However there are also acute effects of glucose on vascular function which appear to be mediated via glucose-generated ROS actions on the NO pathway, and which may contribute to the vascular pathology of diabetes.

As reviewed by Tesfamariam *et al.*,[115] several studies have shown that increased oxidative stress may contribute to the impairment of endothelium-dependent relaxation and increase the release of vasoconstrictor prostanoids in arteries of diabetic animals and humans. Endothelium-dependent coronary flow response is impaired in diabetic rat hearts, a defect prevented by perfusion of the hearts with SOD or pretreatment of the rats with tocopherol-acetate.[114] Transient H_2O_2 induced contraction was more pronounced in endothelium-containing rings of aorta from diabetic versus control rats. Removal of the endothelium or pretreatment of rings with N(G)-Nitro-L-Arginine Methyl Ester (L-NAME, 100 μM) reversed this difference, indicating that increased H_2O_2 could be an important factor in the development of vascular hyperreactivity associated with diabetes.[116]

In patients with both Type-1[117] and Type-2[118] diabetes mellitus, but not control subjects, arterial infusion of vitamin C improved endothelium-dependent vasodilation. In diabetic dogs, impaired coronary microvascular responses to Acetyl Choline were restored to normal by topical application of SOD and catalase.[119] In healthy human volunteers acute elevations of plasma glucose caused significant increments in systolic and diastolic blood pressure. Infusion of the somatostatin analogue, *octreotide*, to avoid the possible confounding vascular actions of insulin, did not influence the hemodynamic effects of hyperglycemia, whereas intravenous GSH prevented these effects.[120] Thus, even acute hyperglycemia can produce relevant systemic hemodynamic changes and alter baroreflex activity via a free-radical-mediated, pathway. The above findings suggest that NO degradation by oxygen-derived free radicals produced by metabolism of excess glucose contributes to impaired vascular reactivity in diabetes.

Another serious problem encountered in pregnant diabetic patients is hyperglycemia-induced embryonic malformations leading to birth defects. This complication may be due to a glucose-induced increase in ROS formation.[121] In rat embryos cultured under diabetic conditions, there was a high rate of congenital anomalies, decreased growth and protein content, and a decrease in the activity of both SOD and catalase.[122] In addition there were decreases in vitamins C and E, and an increase in uric acid. Supplementation with vitamins C and E abolished the deleterious effects of diabetic conditions on the embryos.[122]

In a study of pregnant diabetic, compared with normal rats, the formation of intracellular ROS, estimated by flow cytometry, was increased in isolated embryonic cells and the concentration of intracellular GSH was lower. Administration of a specific inhibitor of y-glutamyl cysteine synthetase, the rate-limiting GSH synthesizing enzyme, to diabetic rats further reduced GSH levels and increased the frequency of embryonic lesions. Administration of GSH or insulin restored GSH concentration in the embryos, reduced the formation of ROS, and led to improvement of lesions. These results indicate that glucose-induced GSH depletion by ROS may have an important role in the development of embryonic malformations in diabetes.[121]

In rat embryos, both hyperglycemia *in vitro* and maternal diabetes *in vivo* caused embryonic dysmorphogenesis and increased lipid peroxidation with the addition of N-Acetylcysteine (NAC) normalizing both. This antioxidant treatment also restored low PGE2 concentrations, suggesting that diabetes-induced oxidative stress aggravates loss of COX-2 activity.[123] These same investigators showed that PGE2 supplementation or the antioxidants SOD and NAC normalized the development of embryos cultured with COX inhibitors at high glucose concentrations. They suggested that high glucose concentrations disturb embryonic development via increases in ROS and altered metabolism of arachidonic acid.[124] Alpha-Cyano-4-Hydroxycinnamic Acid (CHC), a mitochondrial pyruvate transport inhibitor, *in vitro,* or maternal ingestion of vitamin E or 2,6-di-tert-Butyl-4-Methylphenol (BHT) *in vivo* prevented glucose-induced mitochondrial swelling in rat embryos.[125]

The above findings suggest that antioxidant supplementation during pregnancy with diabetes could help prevent fetal malformations and birth defects. In pregnant Streptozotocin diabetic rats, administration of oral vitamin E[126, 127] increased concentrations of the vitamin in maternal, embryonic, and fetal tissues, completely normalized the fivefold increase of TBARS found in fetal liver, and decreased the rate of malformations and increased size and maturation of embryos. In the same animal model, vitamin C treatment led to higher levels of ascorbic acid in the placenta, maternal and fetal liver, greater alpha-tocopherol concentration in the placenta, reduced concentrations of TBARS in maternal serum, and lowered the rates of embryonic resorptions and malformations in proportion to the dose administered.[128]

In a study assessing nutritional antioxidant status and lipid peroxidation in diabetic compared with healthy pregnant women eating *ad lib,*[129] no significant differences in the dietary intake of the major antioxidant vitamins (retinol, vitamin C, or vitamin E) or beta-carotene, but greater intakes of a number of other micronutrients (including Se, Zn, Mg, Mn, riboflavin, thiamin, niacin, and folate) were seen in diabetic patients. Serum levels of alpha-tocopherol, beta-carotene, and lycopene were also greater in diabetic patients. There was no evidence of greater lipid peroxidation in diabetic patients, and total antioxidant capacity was similar in the two groups. Thus, in this single uncontrolled study, nutritional antioxidant status was as good or better in diabetic expectant mothers than in control non-diabetic subjects. Despite the numerous animal studies cited above suggesting a role of glucose-generated ROS in diabetic fetal malformations, we could find no published study examining the effects of antioxidant supplementation on birth defects in babies of women with diabetes.

Causal relationships between chronic hyperglycemia and both diabetic microvascular and macrovascular disease have been convincingly demonstrated.[130–132] Nishikawa et al.[133] have suggested that increased production of ROS is a unifying mechanism linking elevated glucose and the major pathways responsible for diabetic damage to end organs. Thus, the question whether antioxidant treatment could help prevent diabetic complications should be considered.

In elderly nonobese human subjects with normal glucose tolerance[134] vitamin E potentiated insulin-mediated whole-body glucose disposal and changes in plasma vitamin E concentrations were correlated with changes in insulin-stimulated whole-body glucose disposal, suggesting that antioxidant therapy may improve glucose tolerance. In patients with Type-2 diabetes both total GSH and vitamin E levels were increased and malonaldehyde levels were reduced after vitamin E supplementation as well as after improved glycemic control.[135] In another study of, Type-2 diabetic patients, randomly assigned to vitamin E supplementation or placebo for eight weeks,[136] there was a significant improvement in the percent change in brachial artery responsiveness, oxidative stress indices, and intracellular cation content in the vitamin E treated group. These results suggest that antioxidant supplementation has the potential for preventing the long-term complications of diabetes.

Given the above-cited data suggesting that ROS production is implicated in causation of chronic complications of diabetes, it might be useful to perform clinical measurements of oxidative stress for prognosis, as well as to evaluate preventive actions of antioxidative therapy. Villa Caballero et al.[137] have suggested quantifying ratios of oxidized and reduced GSH (GSSG/GSH) and TBARS in diabetics. However, no ideal method is currently available for this purpose, and the usefulness of current methods needs to be confirmed.

In summary, the evidence from a wide variety of experiments conducted *in vitro* and in animal models as well as in diabetic patients shows that high levels of glucose can and do generate ROS. Clearly more needs to be done to investigate the role of this phenomenon in causing complications of diabetes mellitus and whether measures to reduce or counteract ROS would be protective.

4. Estrogens

The central structure of steroid hormones is a cholanthrene ring that consists of three adjacent 6-carbon rings (A, B, and C rings) and a 5-carbon ring (D), which shares two carbons with the C ring. Variation in steroid hormone structure and action is due to the presence of hydroxyl, keto, and radical side-chain groups on various of the ring carbons or the desaturation of one or more of the ring bonds. The estrogenic female steroid hormones and related metabolites have a desaturated (phenolic) A ring with a hydroxyl group on carbon 3. This configuration makes these molecules relatively polar at the A ring side, since the hydrogen of the 3-hydroxyl group can dissociate, making estrogens weakly acidic. That is, they can function as proton donors or receptors depending on their state. There is evidence that estrogens can function as antioxidants and that some estrogen metabolites (catechol estrogens) may also be involved in free radical generation. Estrogens may also regulate oxidative processes in some target cells via their hormonal actions.

One clue that estrogens may be involved in oxidative processes comes from the finding that male rats have more malondialdehyde in their livers than do females.[138] In liver tissue *in vitro*, estradiol-17-beta (E_2) suppressed free-radical generation by decreasing xanthine oxidase activity and completely inhibiting aldehyde oxidase activity leading to reduced formation of lipid peroxides.[138] Pretreatment with E_2 or its metabolite 2-hydroxy-E_2 inhibits NADPH and ADP-Fe^{3+}-dependent lipid peroxidations and inhibited oxygen uptake in rat liver microsomes.[139] Results obtained with other free-radical generating systems have suggested that 2-hydroxy-E_2 may interact with alkoxyl rather than with peroxyl radicals.[139] In the rat liver microsome system, E_2 was a more effective inhibitor of peroxidation of LDL and DNA damage than all other ROS scavengers tested, with no additional effect when selective ROS scavengers were added. Lipid protection appeared to be due to E_2 inhibition of superoxide generation, and DNA damage protection appeared to be due to inhibition of chain propagation.[140] When estrogens were compared with a variety of other steroid hormones in an *in vitro* system, 17 beta-E_2, 17 alpha-E_2, and Estriol reduced ROS accumulation by about 65%, whereas Estrone (E_1) and other steroids had no significant antioxidant properties.[141]

The physiologic and potential pharmacologic role of estrogen's protective effects against oxidative damage is the subject of intense investigation. For example, the effect of estrogen to prevent coronary heart disease is only partially due to decreased LDL and increased HDL cholesterol.[142, 143] Estrogens also have direct effects on the vascular wall that may help prevent atherosclerosis.[144, 145] Oxygen-derived free radicals in the arterial wall are known to modify LDL, increasing its atherogenicity.[140] When arterial LDL flux was measured in carotid arteries from ovariectomized three-month-old rats, both E_2 at physiological concentrations and alpha-tocopherol attenuated LDL accumulation caused by Tumor Necrosis Factor (TNF stimulates ROS generation), supporting an antioxidant role for E_2 in protection against atherosclerosis.[146] In postmenopausal women taking estrogen replacement, activity and amount of granulocyte Myeloperoxidase (MPO) increased significantly.[147] MPO activity consumes H_2O_2 and could decrease ROS accumulation. If a similar effect occurs in macrophages this might reduce HDL-oxidation in arterial foam cells and improve the outflow of cholesterol, presenting another possible mechanism by which estrogen could protect against atherosclerosis. It would also appear that plant estrogens (phytoestrogens) have antioxidant effects similar to those observed with mammalian estrogens. In a recent study, the phytoestrogens, daidzein, genistein, and resveratrol reduced oxidative DNA damage (measured as 8OH-dG) induced by treatment with Advanced Glycation End-products (AGE's) and also increased total GSH levels in rat vascular SMC *in vitro*.[148]

Probably the most active area of investigation of the protective effects of estrogen against oxidative damage are related to the observations that men have higher rates of Alzheimer's Disease (AD) than women, and that postmenopausal women on estrogen replacement therapy have a lower risk and/or later onset of AD.[149, 150] Various explanations for this effect have been suggested,[151] including estrogen interaction with neurotrophins and neurotransmitter systems relevant to AD, estrogenic modulation of synaptic plasticity, estrogen effects on beta-amyloid and apolipoprotein E, estrogen blunting of neurotoxic adrenal stress, augmentation of cerebral glucose utilization, and enhanced cerebral blood flow. In a recent review Green *et al.*[152] also cited activation of the nuclear estrogen receptor, altered expression of bcl-2 and related proteins, activation of the mitogen activated kinase pathway, activation of cAMP signal transduction pathways, modulation of intracellular calcium homeostasis, and direct antioxidant activity.

It has been shown that concentrations as low as 0.2 nM of E_2 reduce death of neuroblastoma cells *in vitro* in response to the neurotoxic fragment of beta-amyloid (A beta 25–35) cells *in vitro*.[153] In a study comparing various steroids for protection of estrogen-responsive human neuroblastoma cells *in vitro* against serum deprivation, only E_2 and its nonestrogenic stereoisomer 17 alpha-E_2 provided protection, which was antagonized just about one-third by tamoxifen, indicating that most neuroprotective effect of estrogen is not attributable to the general steroid structure and may not be mediated via a tamoxifen-antagonized receptor mechanism.[154] In a confirming experiment employing a variety of oxidative stressors including beta-amyloid, E_2, 17 alpha-E_2, and some other E_2 derivatives prevented intracellular peroxide accumulation and neuronal cell death. Analysis suggested that the neuroprotective antioxidant activity was dependent on the presence of a hydroxyl group in the C3 position on the A ring and was independent of estrogen receptors.[155] Further studies demonstrated that all estrogens, including the non-steroidal estrogen Diethylstilbestrol (DES), having an intact phenolic ring protected neuroblastoma cells from the toxic effects of serum-deprivation, whereas 3-O-methyl ether congeners were inactive.[156] In similar studies, brief preincubation with E_2 or 17alpha-E_2, but not other steroids (corticosterone, testosterone, and cholesterol), protected rat ventral mesencephalic neurons in primary culture against superoxide anion, H_2O_2, and glutamate-induced neurotoxicity, which effect was not significantly blocked either by an estrogen-receptor antagonist or a protein-synthesis inhibitor.[157] There also is a strong synergy between E_2 and GSH to increase neuroprotective potency, which was not attenuated by the addition of an estrogen-receptor antagonist.[158] Thus, findings suggest that neuroprotection is at least partially mediated by mechanisms which do not require estrogen receptors or activation of genome transcription. The presence of reduced GSH in culture media increases the neuroprotective potency of estrogens against neurotoxic beta-amyloid peptide damage to a murine neuronal cell line by an average of 400-fold.[4] The above data suggest the possibility that combined estrogen-antioxidant therapy may help prevent or ameliorate neurodegenerative diseases such as Alzheimer's Disease.

In an entirely different model, E_2 at high concentrations protected against oxidative stress-induced apoptosis and cell death in pig luteal and follicular tissues exposed *in vitro* to H_2O_2, an effect not attenuated by actinomycin D, suggesting a predominantly receptor-independent, nongenomic mode of action. Ovarian cells were not protected by nonaromatizable steroids.[159] The latter experiment suggests that high intraovarian levels of E_2 may exert part of their known protective action against follicular atresia[160] by a free-radical-scavenging mechanism.

Much of the ischemic damage to tissues is caused by oxygen free radicals that are released during reperfusion. In a study assessing whether physiologic levels of E_2 can prevent ischemic brain injury *in vivo*, rats were implanted either immediately after Ovariectomy (Ovx) or at the onset of ischemia, with capsules that produced physiologically low or physiologically high serum 17 beta-E_2 levels. When the middle cerebral artery was occluded two weeks post-Ovx, E_2 pretreatment, but not acute treatment, significantly reduced cortical infarct volume compared with controls.[161] In dogs subjected to occlusion of the left anterior descending coronary artery, 2 weeks of 17 beta-E_2 pretreatment decreased arrhythmias, maintained systolic shortening, preserved coronary flow response to acetylcholine, and prevented lipid peroxidation (exhaled *n*-Pentane) during the reperfusion stage.[162] Estrogen also appears to protect rodent cardiac and skeletal muscle from basal and iron-stimulated lipid peroxidation, measured as TBARS, and from cell damage, measured as creatine kinase efflux.[3] The antioxidant effects of estrogen to prevent toxic damage from chemotherapeutic agents may be another fruitful area for research. For example, in an *in vivo* study in rats, Adriamycin oxidative-stress-induced nephropathy was prevented by simultaneous administration of E_2.[163]

As reviewed by Liehr and Roy,[164] in addition to their established antioxidant effects *in vitro* and *in vivo*, prooxidant effects of estrogens have also been reported. For example, 8-hydroxylation of guanine bases of DNA have been induced by stilbene and the steroid estrogen metabolites (catechol estrogens) 4-hydroxyestradiol (4-OHE_2) or 4-hydroxyestrone (4-OHE_1) in a hamster kidney microsomal activating system *in vitro*.[165] In another *in vitro* system, lipid peroxidation (conjugated diene formation) of human LDL was accelerated by low concentrations of the catecholestrogens, 2- or 4-hydroxyE$_2$ and 2- or 4-methoxyE$_1$ and slowed by the estrogens, E$_2$ or estriol.[166] In an *in vivo* study, treatment of hamsters with E$_2$ led to formation of 2- and 4-hOHE$_2$ in approximately equal amounts and decreased the 2-hydroxylase and 4-hydroxylase activity in liver and kidney. The activity of Cytochrome P450, which oxidizes estrogen hydroquinones to quinones, normalized for specific content of Cytochrome P450, was 2.5-fold higher in kidney compared with liver and increased in kidney with chronic estrogen treatment. E$_2$ also increased concentrations of superoxide in kidney by 40%.[167] The latter study shows that catechol estrogens are generated *in vivo* and that E$_2$ or its metabolites may regulate the activity of enzymes leading to catechol estrogen production.

E$_2$ can undergo a one-electron oxidation to its reactive phenoxyl radical. Electron spin resonance spectroscopy has been used to demonstrate the production of semiquinone free radicals from the oxidation of the catechol estrogens 2- and 4-OH-E$_2$ and 2,6- and 4,6-OH$_{(2)}$-E$_2$. These radicals abstract hydrogen from reduced GSH, generating the GSH thiel radical and from reduced beta-Nicotinamide-Adenine Dinucleotide (NADH) to generate the NAD. The superoxide generated may react with another NADH to form NAD, thus propagating a chain reaction leading to oxygen consumption and H$_2$O$_2$ accumulation.[168] The resulting accumulation of intracellular H$_2$O$_2$ could explain the hydroxyl radical-induced DNA base lesions reported in female breast cancer tissue.[169] It has also been found that the catechol estrogens 2-OH-E$_2$ and 4-OH-E$_2$ can be oxidized by copper in a process that generates singlet oxygen ROS that causes DNA damage, as can be detected by the appearance of strand breaks in plasmid DNA. Since copper is present in the nucleus in association with guanines, these results suggest an alternate site-specific mechanism for the formation of oxidative DNA damage associated with estrogen treatment.[170]

It is clear that estrogens play a role in the genesis of breast cancer in both human and animal model systems,[171] but whether catechol estrogen-induced generation of ROS is involved in carcinogenesis is a question under active investigation.[172, 173] The production of H$_2$O$_2$, the hydroxyl radical, and the E$_1$ 3,4-semiquinone has been demonstrated in E$_1$ 3,4-quinone treated human breast cancer subcellular fractions.[174] In another study, microdialysis determination of *in vivo* concentrations of GSH in breast tissue and fat during the menstrual cycle in healthy women (23–32 years old) showed increased GSH concentrations in the mid-luteal phase compared to the follicular phase of the menstrual cycle in both tissues, suggesting that progesterone may antagonize the antioxidant effects of estrogen, but not that high estrogen levels are themselves prooxidant.[175]

The synthetic steroids Ethynylestradiol (EE) and Norethindrone (NET) commonly used in oral contraceptives and known to induce hepatic neoplasia, appear to cause DNA damage by generating free radicals in the liver. The treatment of rats with EE and NET caused elevated 8OH-dG levels, decreased GPX, and led to the formation of hepatic microcarcinomas in liver. Vitamin C and beta-carotene coadministration reduced the elevation of 8OH-dG levels and suppressed the formation of hepatic cancers.[176]

The artificial non-steroid stilbene estrogen, Diethylstilbestrol (DES) has also been demonstrated to be a carcinogen. Metabolic redox cycling between DES and DES',4''-quinone and reduction of DES quinones to their hydroquinones by xanthine oxidase occurs by both one electron transfer to the quinone and by formation of superoxide which then reduces the quinone.[177] DES forms superoxide radicals and induces DNA strand breaks in the presence of horseradish peroxidase/H_2O_2 metabolism in a cell-free system and in Syrian hamster embryo cells *in vitro*.[178] In another experiment, SOD or catalase suppressed DES, but not EE — induced proliferation of Syrian hamster renal proximal tubular cells *in vitro*, suggesting a role for ROS in the DES proliferative response.[179] Treatment of hamsters with DES resulted in an increase in oxidative damage to renal proteins, measured as protein carbonyl formation.[180] All of the above suggests that DES generation of ROS may be one mechanism by which DES induces cancers.

5. Thyroid Hormones

Thyroid hormones are secreted by the thyroid gland under the control of a pituitary glycoprotein hormone, Thyroid Stimulating Hormone (TSH). TSH secretion, in turn, is stimulated by hypothalamic neurosecretion of Thyroid Regulating Hormone (TRH) into the pituitary portal venous system and inhibited by thyroid hormone in the blood (a classic negative feedback servo-control loop). Thyroid hormones are synthesized by iodination, at the 3- and 5-carbon positions, of tyrosine residues on the protein thyroglobulin, which fills the thyroid follicles. Two molecules of iodinated tyrosine are linked by an ether bond to form Iodothyronine. The two active thyroid hormone molecules are L-thyroxine (3, 5, 3', 5' tetraidothyronine), also known as T_4, and 3, 5, 3' triiodothyronine, also known as T_3. T_4 is the major circulating hormone. T_3 is found in the serum in lower quantities. Most (99.8%) circulating T_4 is tightly bound to a binding protein, Thyroid Binding Globulin (TBG) and loosely bound to plasma albumin. T_3 has a lower affinity for TBG by about one order-of-magnitude. Only the non-protein bound or "free" T_4 and T_3 are available to cross cell membranes and interact with thyroid hormone receptors within the cell. Once inside the cell T_4 serves as a precursor and is monodeoiodinated to produce the active hormone, which is T_3. It is T_3 that binds to nuclear receptor proteins and then, after interacting with a variety of other nuclear regulatory proteins, to the DNA strand itself at gene-specific regulatory sites (thyroid regulatory elements or TRE's). Thus, the major actions of thyroid hormone are mediated by gene activation with transcription, and *de novo* synthesis of active protein molecules.

Thyroid hormone is a prime regulator of intermediary metabolism in nearly every cell type in the body. It accelerates the rate at which cells oxidize fuel, leading to increased thermogenesis. It is now known that this primary thyroid hormone action and other secondary and complementary effects are mediated by the activation of a wide variety of genes involved in multiple metabolic pathways.[1] In particular, thyroid hormone up-regulates enzymes and cytochromes involved in mitochondrial function such as Cytochrome c oxidase[181] and mitochondrial β FI ATP'ase,[182] proteins involved in glucose transport and metabolism, such as glucose transporter,[183] glucokinase,[184] and 6-phosphofructo-2-kinase[185] and in fatty acid metabolism, such as acetyl CoA carboxylase[186] and fatty acid synthetase.[186, 187] Arguably most important, is the effect of thyroid hormone to stimulate production of thermogenins, the uncoupling proteins,[188] leading to increased heat production.

Thermogenesis is coupled to Adenosine-Diphosphate (ADP) Phosphorylation via a proton gradient across the inner mitochondrial membrane. Proton leaks through this membrane uncouple respiration from Adenosine-Triphosphate (ATP) synthesis, dissipating energy as heat. This mechanism was first identified in thermogenic brown adipose tissue mitochondria, which contain a unique proton carrier referred to as Uncoupling Protein 1 (UCP1).[189] In other types of cells, mitochondria also release heat showing that coupling of substrate oxidation to ADP phosphorylation is less than 100%. Proton leak accounts for approximately 26% of the total oxygen consumption rate in isolated rat hepatocytes and 52% of the oxygen consumption rate of resting perfused muscle, so that up to 38% of the basal metabolic rate of a rat is due to proton-leak thermogenesis.[190] These findings suggest that proton leak heat production is an important function in homeotherms. Studies of mitochondria from tissues other than brown fat led to the identification of UCP2 and UCP3, homologues of the brown fat UCP (renamed UCP1). The UCP2 gene is widely expressed, whereas the UCP3 gene is mainly expressed in skeletal muscle tissue (and brown fat in mice).[191] It is now known that both UCP2[192] and UCP3[193, 194] are up-regulated by thyroid hormone.

The effect of *in vivo* thyroid status on membrane potential [delta psi(m)] of mitochondria isolated from rat hepatocytes has been studied by means of a cytofluorometric technique and the delta psi(m)-specific probe JC-1.[195] The delta psi(m) level was higher in hypothyroid and lower in hyperthyroid than in euthyroid animals. Hepatocyte respiratory rates showed an opposite trend with the highest respiratory rate in mitochondria of hyperthyroid and the lowest in those from hypothyroid animals. The authors interpreted these findings to mean that mitochondrial energy coupling is most efficient in hypothyroid and lowest in hyperthyroid hepatocytes. In another study of rat hepatocytes,[196] employing top-down elasticity analysis, data revealed that 43% of the increase in the resting respiration rate in hyperthyroid hepatocytes compared with euthyroid hepatocytes, was due to differences in the proton leak, and 59% was due to differences in the activity of the phosphorylating subsystem. There were no significant effects on the substrate oxidation subsystem. The kinetics of the proton leak can also be determined indirectly, by measuring the oxygen consumption of mitochondria in the presence of *oligomycin* which blocks phosphorylation of ATP. Using this system, Porter *et al.*[197] showed that the proton leak in isolated liver mitochondria is not significantly different in a comparison of young and old rats but that *in situ* hepatocyte mitochondria of old rats showed an apparently greater proton leak when compared with young rats. Administration of T_3 increased the proton leak in rat muscle mitochondria. In another study, proton leak, as determined by the relationship between respiration rate and membrane potential, was of lesser magnitude in mitochondria from hypothyroid rats than in those from euthyroid controls.[198] Finally, it appears that the kinetics of the protonmotive-force generators (e.g. dicarboxylate carrier, succinate dehydrogenase and the respiratory chain) are unchanged in mitochondria from hypothyroid animals, but that the kinetics of the protonmotive-force consumers (e.g. adenine nucleotide and phosphate carriers) are altered, supporting the hypothesis that there are important effects of thyroid hormone on mitochondrial ATP synthase or Adenine Nucleotide Translocator.[199]

It has been an open question whether the thyroid hormone-induced increase in mitochondrial membrane proton leak might be associated with accelerated generation of ROS and an increase in the rate of oxidative damage to mitochondria and other cell components. In an investigation of the possible oxidative phenomena in patients with Graves' Disease (autoimmune hyperthyroidism), 30 hyperthyroid patients, both before initiation of therapy and after attainment of the euthyroid state, were compared with 30 age-matched healthy control subjects.[200] Patients with untreated Graves' Disease showed significant increases in serum lipid peroxidation activity indices. These changes were accompanied by a decrease in plasma thiol and erythrocyte lysate thiol concentrations. Hyperthyroidism also resulted in marked increases in intracellular SOD, catalase, and GPX compared with controls. However, extracellular anti-free-radical scavenging system potential, as measured by GSH reductase activity and total antioxidant status level, were significantly decreased in untreated Graves' patients. Successful treatment normalized the free radical and antioxidant activity indices. These results indicate greater generation of ROS and impairment of *extracellular* antioxidant defense systems along with the induction of *intracellular* antioxidant defense systems in hyperthyroidism.

One possible site of oxidative damage with thyroid abnormalities is the myocardium. Both hyperthyroid and hypothyroid patients often complain of cardiovascular symptoms, which have been suggested to relate both to changes in metabolism and also to specific hyper- and hypo-thyroid cardiomyopathies.[201, 202] In a study designed to evaluate myocardial oxidative metabolism in hyper-thyroidism,[203] dynamic Positron Emission Tomography (PET) with 11C-acetate was performed in 19 patients who had not yet undergone treatment and in nine normal subjects. The results suggested excessive myocardial oxygen consumption. In another study also employing PET with 11C-acetate, patients with a history of thyroidectomy for thyroid cancer were investigated in the hypothyroid state and again 4–6 weeks later after oral thyroid hormone replacement to euthyroidism.[204] In addition, magnetic resonance imaging was applied to determine left ventricular geometry and Stroke Work Index (SWI), and an estimate of myocardial efficiency, the Work Metabolic Index (WMI), was obtained. Systemic vascular resistance and left ventricular mass were higher in hypothyroidism, whereas ejection fraction and SWI were lower. Despite the reduction of oxygen consumption, the WMI was significantly lower. Estimates of cardiac work are more severely suppressed than those of oxidative metabolism, suggesting decreased efficiency of oxidative metabolism in the hypothyroid state. Thus, both hyperthyroid and hyperthyroid states could potentially be associated with increased oxidative stress in the heart.

The relationship of thyroid status to cardiac oxygen consumption and oxygen radical production by functional mitochondria and oxidative DNA damage have also been studied experimentally in rats rendered hyper- and hypo-thyroid by chronic T_3 and 6-*n*-propyl-2-thiouracil treatments, respectively.[205] Hypothyroidism decreased heart mitochondrial H_2O_2 production by 50% in State 4 and 80% in State 3. This decrease in oxygen radical generation occurred mainly at Complexes III and I and was due to a reduction in the free-radical leak in the respiratory chain. Heart genomic DNA damage, measured as 8OH-dG decreased in hypothyroid animals to 40% of control. In this study hyperthyroidism did not significantly change heart mitochondrial H_2O_2 production nor was heart 8OH-dG increased in hyperthyroid animals. The lack of increase in H_2O_2 production per unit of mitochondrial protein appeared to protect mitochondria against self-inflicted damage during hyperthyroidism.

In contrast, in another study in which rats were made hyper- or hypo-thyroid for four weeks,[206] an increase in cardiac muscle lipid peroxide was observed in the hyperthyroid rats. This finding was accompanied by an increase in mitochondrial SOD and mitochondrial oxidative marker enzymes (Cytochrome c oxidase and fumarase). In this same study, hypothyroidism led to a reduction in oxidative markers and mitochondrial SOD, but only a marginal decrease in lipid peroxidation. Cytosolic SOD did not change in relation to either oxidative metabolism or lipid peroxidation.

In a study of thyroid effects on free radical defenses in the heart,[207] myocardial tissue levels of CoQ10, CoQ9, vitamin E, catalase, GPX, and SOD, as well as serum levels of CoQ9 and SOD were measured in rats rendered hyper- or hypo-thyroid by four weeks of T_4 or methimazole treatment. Compared with untreated controls, reductions in CoQ9 levels were observed in hearts, but not serum, of both hyper- and hypo-thyroid rats. Cardiac levels of vitamin E and SOD increased in hyperthyroid, but were unchanged in hypothyroid rats. Heart muscle GPX levels were lower in hyperthyroid and greater in hypothyroid rats. No differences in catalase levels were observed.

Heart tissue from rats with chronic thyrotoxicosis shows a high degree of nonspecific subcellular structural damage, consistent with either hypoxia, reperfusion injury, or toxic effects. In a group of hyperthyroid rats given ascorbic acid as a free radical scavenger, damage was less intensive and occurred more rarely, indicating a possible role of free radicals in the generation of the observed alterations.[208]

The results of the above studies, taken together, appear to indicate that thyroid excess can lead to myocardial damage via a mechanism involving free radical generation, probably due to increased proton leak at the mitochondrial membrane. The observed compensatory increase in antioxidant defenses in the hyperthyroid heart is also consistent with such a mechanism. With hypothyroidism, there is either no increase or possibly even a decrease in evidence of oxidative stress in the heart.

Hyperthyroidism in human patients is also associated with a symptomatic myopathy of skeletal muscle.[209] However, little is known regarding the mechanism of muscle damage in this condition. Rats treated with high doses of T_4 showed an increase in the oxygen consumption in State 3 respiration of mitochondria isolated from Soleus (slow twitch) and Extensor Digitorum Longus (EDL, fast twitch) muscles, without modification of either their phosphorylation efficiency (ADP/O ratio) or their coupling. Muscle fiber alterations observed included atrophy and changes to the structure and shape of mitochondria. Also, abundant autophagic vacuoles with mitochondrial debris, myelin-like figures, glycogenosomes, and primary lysosomes were found.[210] In contrast, thyroidectomy provoked a reduction in the oxygen consumption rate in State 3 respiration in both Extensor Digitorum Longus (EDL, fast twitch) and Soleus (slow twitch) muscles of the rat, the last one being the most affected. Morphological alterations after eight weeks of thyroidectomy were also found in mitochondria. These organelles exhibited cristae swelling and formation of autophagic vacuoles in subsarcolemmal and intermyofibrillar spaces. Altered mitochondria were also seen in the axon terminal and the postjunctional region of the motor endplate.[211]

In a study in which were rats were rendered hyper- or hypo-thyroid for four weeks and then killed, Asayama et al.[206] measured free radical scavengers (Cu, Zn SOD, MnSOD, GPX, and catalase), mitochondrial oxidative marker enzymes (Cytochrome c oxidase and fumarase), and lipid peroxide in soleus (slow oxidative), extensor digitorum longus (fast glycolytic) muscles, and the liver. An increase in lipid peroxide was observed in the soleus muscles of hyperthyroid rats accompanied by an increase in mitochondrial SOD and oxidative markers. No such change was observed in either fast glycolytic muscle or liver. GPX decreased in all tissues of hyperthyroid rats, and there was a parallel decrease in catalase in most tissues. Hypothyroidism induced a reduction in oxidative markers and mitochondrial SOD in heart and skeletal muscles, but only a marginal change in lipid peroxidation. The cytosolic SOD was unaltered. These results suggest that the enhanced oxidative metabolism and decreased GPX in hyperthyroidism result in an increase in lipid peroxidation in slow oxidative and heart muscle, with possible organ damage. No adverse reaction mediated by active oxygen species was found in hypothyroid rat tissues.[206]

In their article reviewing the relevance of metabolic derangements and oxidative stress to thyrotoxic myopathy, Asayama and Kato[212] summarized a large volume of circumstantial evidence indicating that hyperthyroid muscle tissues undergo multiple biochemical changes that predispose to free-radical-mediated injury. These include a profound effect on mitochondrial oxidative activity, synthesis and degradation of proteins and vitamin E, the sensitivity of the tissues to catecholamines, the differentiation of muscle fibers, and the levels of antioxidant enzymes. The authors concluded that hyperthyroidism accelerates lipid peroxidation in both the heart and slow-oxidative muscles, suggesting a contribution of ROS to the muscular injury caused by thyroid hormones.

Another potential site of thyroid hormone-mediated ROS generation is the liver. The effect of thyroid hormones on hepatic Xanthine Oxidase activity and lipid peroxides determined by measurement of TBARS were studied in rats after intraperitoneal injections of T_3 and T_4 in a 1:4 ratio for three days.[213] The concentration of lipid peroxides and Xanthine Oxidase activity were increased in treated, compared with euthyroid, rats. In another study of rat liver protein oxidation and lipid peroxidative effects,[214] T_3 administration for 1–3 days elicited a progressive increase in body temperature and in the rate of hepatic O_2 uptake. Hepatic TBARS showed a maximal 3.1-fold increase at two days of treatment, whereas protein oxidation, measured by content of protein hydrazone derivatives, exhibited a maximal 88% increase at three days. Finally, in a third study[215] a single dose of T_3 led to an early and transient calorigenic response and a fall in liver GSH levels, with a maximal effect at two days. Addition of the gamma-Glutamyltransferase (gamma-GT) inhibitor DL-serineborate (4 mM) to the perfusate abolished the hepatic loss of GSH elicited by T_3, and enhanced the concentration of GSH, studied at two days after hormone administration. The above findings suggest that enhanced Xanthine Oxidase activity depletes GSH content in hyperthyroid rats, resulting in greater oxidative stress. The differential time course of oxidative changes in lipids versus proteins may represent differences in the susceptibility of target molecules to attack by ROS and/or in the efficiency of repair mechanisms.

In cells of the immune system, oxidative processes are involved in bacterial killing, cytotoxic activity, and Programmed Cell Death (*apoptosis*).[216] Immune system cells have thyroid hormone receptors, and the immune system responds to changes in levels of thyroid hormones. It has been shown that PMN from hyperthyroid patients have an increase in mitochondrial oxygen consumption and superoxide anion generation.[217] This phenomenon could have clinical implications. For example, bronchial asthma worsens after the development of hyperthyroidism.[218] In a recent study, T_4 enhanced antigen-stimulated production of superoxide anion by alveolar neutrophils and macrophages obtained from asthmatic patients, suggesting that T_4 enhanced production of ROS by alveolar immune cells might play an important role in the hyperthyroid exacerbation of asthma.[218]

In rats, administration of single doses of T_3 for three days produced a 3.8-fold increment in respiratory burst activity, detected by chemiluminescence, of zymosan-stimulated PMN's and enhanced NADPH oxidase and myeloperoxidase activity. The observed increase in rate of O_2-generation occurred with no change in SOD activity.[219] Treatment of T lymphocytes with T_3 and T_4 *in vitro* induced reduction of mitochondrial transmembrane potential (delta psi), increased production of ROS, and enhanced apoptosis.[220] In these cells expression of anti-apoptotic Bcl-2 protein was also reduced. Consistent with the above, lymphocytes from patients with Graves' Disease showed enhanced apoptosis compared with those from normal individuals.[220] These results suggest that thyroid hormones increase generation of ROS in both major lineages of immune cells and may induce apoptotic cell death in human lymphocytes.

The variety of findings consistent with oxidative stress as a mediator of pathophysiological effects of hyperthyroidism suggest that patients might benefit from exogenous antioxidant protection in the interval before euthyroidism can be established by definitive therapy. However, antioxidant intervention has not been extensively investigated in this condition. In a single experiment,[221] vitamin E was found to reduce bone resorption, a known effect of hyperthyroidism[222] in thyrotoxic rats. Survival rates were also significantly increased in thyrotoxic rats given vitamin E, suggesting the role of free radicals in the overall morbidity and mortality in thyrotoxicosis.

In summary, thyroid hormone appears to increase mitochondrial production of ROS in a variety of cells and tissues, an effect variably associated with compensatory increases in antioxidant defenses. This phenomenon should be further investigated as a possible pathophysiologic mechanism in thyrotoxicosis, and the potential role of antioxidant interventions to improve outcomes for hyperthyroid patients should be studied.

6. Adrenal Hormones

6.1. DHEA

Dehydroepiandrosterone (DHEA) is a steroid molecule produced in large quantities, (15–30 mg/day) mainly by the adrenal cortex and found in relatively (compared with other plasma steroids) high concentrations in human blood. Most

DHEA circulates as the sulfate conjugate (DHEA-S; range of normal concentration, 0.5–2.5 µg/ml). Free DHEA concentrations are much lower (range of normal, 0.02–0.09 µg/ml). Lower primates have relatively little circulating DHEA, and other mammals secrete almost none. Moreover, DHEA levels decline monotonously with age from puberty to senescence[223] and do so more dramatically than any other circulating steroid measured (except for E_2 during menopause). These two facts have given rise to speculation that DHEA secretion in large amounts may be a unique adaptation contributing to the relative longevity of the human species, and that the age-related decline in DHEA in some way presages or contributes to senescence. A number of studies attempting to correlate endogenous human DHEA levels with age-related changes in body composition or function have generally shown either weak or non-significant relationships. Such studies have examined immune function and cytokine secretion,[224] body composition and strength,[225–229] insulin sensitivity,[230, 231] and cognitive function,[232, 233] among others. Despite numerous experiments showing beneficial effects of DHEA administration in various animal models, DHEA treatment of older men and women has not been convincingly demonstrated to have significant clinical benefits, with the exception of a single controlled study by Morales *et al.*[234] showing improvement in self-reported physical and psychological well-being in post-menopausal women. At present, the only clinical demonstration of therapeutic value for DHEA comes from a study in which women with pathologic adrenal failure, treated with DHEA, showed improved sex hormone-binding globulin and total and HDL cholesterol, as well as well-being and sexuality, probably as a result of normalization of testosterone levels, which are deficient in hypoadrenal females.[235] Probably the best discussion of the relationship of DHEA to aging is the exposition by Hornsby[236] in which the author makes the case that the most likely function for DHEA is as a precursor for conversion to potent androgens, which mediate adrenarchy, a unique sexual signaling mechanism occurring just before puberty (gonadarchy) in higher primates. The subsequent shrinkage of the DHEA-secreting tissue zone of the adrenal cortex and the decrease in DHEA levels might then be viewed as the post-pubertal involution of an organ whose function has been fulfilled, rather than as a phenomenon of aging/senescence.

The concept that DHEA plays an important role as an "anti-aging hormone" is confounded by the fact that no specific receptor protein for the DHEA molecule has been definitively identified in any tissue. Classical steroid hormone actions are receptor-mediated. This has led to the hypothesis that, like estrogens (see above), DHEA may act via a non-receptor mediated pathway, perhaps as an antioxidant, to protect cells and tissues from oxidative stress. Not all studies have found effects of DHEA on oxidative processes. In one system, in which fluorescence of phycoerythrin was monitored after addition of a peroxy radical generator, cortisone and corticosterone appeared to have mild pro-oxidant properties, whereas, DHEA, progesterone, androstenedione, cortisol, and aldosterone had no significant antioxidant properties.[141] However, in other studies, DHEA has been demonstrated to behave both as an antioxidant and as a pro-oxidant, depending on the conditions of the experiment. In a human liver cell line, subjected to a pro-oxidant stimulus (CuOOH), DHEA protected against lipid peroxidation (TBARS) and cell death at low concentrations (0.1 μM, 1 μM), which effects disappeared at 10 μM, and were intensified at 50 μM. Neither of these effects of DHEA was evident except after a "lag-phase".[237]

Decreased free radical damage in response to DHEA has been seen in a number of *in vitro* models. Pretreatment of rat primary hippocampal cell suspensions and human hippocampal tissue with DHEA protected against the toxicity induced by H_2O_2 and sodium nitroprusside and prevented H_2O_2/$FeSO_4$-stimulated lipid oxidation.[238] DHEA added *in vitro* to alveolar macrophages, recovered from lungs of asbestos workers, significantly reduced the exaggerated superoxide anion release characteristic of these cells.[239] In a similar experiment DHEA appeared to inhibit glucose-6-phosphate dehydrogenase, leading to reduced production of superoxide anion (O_{2-}) by human neutrophils treated with 12-O-tetradecanoylphorbol-13-acetate.[240] As noted above, high glucose concentrations induce oxidative stress and may contribute to the tissue damage associated with diabetes. In primary cultures of rat kidney mesangial cells, DHEA reversed the impairment of cell growth induced by high glucose levels, attenuated lipid peroxidation (TBARS generation and 4-hydroxynonenal concentration), and preserved the cellular content of reduced GSH as well as the membrane Na(+)/K(+) ATPase activity.[241] Similarly, cell death of bovine retinal pericytes cultured at high glucose concentration is completely and specifically reversed by addition of concentrations of DHEA similar to those found in human plasma (100 nM) to the culture medium.[242]

Various *in vivo* studies have also suggested an antioxidant action of DHEA. For example, a single dose of DHEA given to rats *in vivo* makes both liver and brain microsomes and plasma constituents (LDL) more resistant to lipid peroxidation triggered by copper.[243] DHEA administered to rats by a single injection is concentrated in liver microsomes, which become resistant to lipid peroxidation induced by incubation with Carbon Tetrachloride (CCl₄). DHEA also reduces both the increase in malondialdehyde production and the formation of fluorescent lipid peroxidation products in liver.[244] Protection against CCl₄ damage occurs despite protein covalent binding of a CCl₃ radical similar to that seen in microsomes from unsupplemented rats.[245] In rats subjected to repeated immobilization stress, DHEA administration partly reversed stress-induced inhibition of body weight gain, increased adrenal weight and glucocorticoid receptor levels, and decreased lipid peroxidation, suggesting that DHEA may act as an anti-stress hormone by reducing free radical generation.[246] In rats subjected to hyperglycemic stress, DHEA pretreatment decreased lipid peroxidation in liver, brain, and kidney, but did not modify cytosolic levels of alpha-tocopherol or GSH, nor the activities of GPX, GSH-reductase, or GSH-transferase.[247] DHEA also prevented oxidative injury, evaluated as levels of ROS (H_2O_2, hydroxyl radical) and membrane integrity and function (synaptic function, membrane Na/K-ATPase activity, lactate dehydrogenase release) in synaptosomes from brains of diabetic rats subjected to transient ischemia/reperfusion from bilateral carotid artery occlusion.[248] DHEA pretreatment also protected rat skeletal muscle flaps from reperfusion injury, as assessed by microscopic measures of vascular hemo-dynamics.[249] In another experiment, DHEA injections significantly increased liver, kidney and adrenal weights and Hepatic Total Sulfhydryl (SH) groups and non-protein SH contents, while decreasing *ex vivo* and iron-induced lipid peroxidation, in vitamin E-deficient, but not vitamin E-replete, rats.[250]

Findings such as the above have given rise to various hypotheses as to the mechanisms of action of DHEA as an oxidant antagonist. For example, DHEA treatment both *in vitro*[251, 252] and *in vivo*[253] has been reported to lower respiratory rates of mitochondria isolated from rat adrenals, heart, kidneys, brain and brown adipose tissue.

The pro-oxidant effects of DHEA have been less thoroughly explored. Like catechol estrogens, DHEA can cause lipid peroxidation in rat liver microsomes and mitochondria and may induce liver cancers. In *in vivo* experiments, the rate of hepatic peroxisomal fatty acid oxidation was approximately 240%, and the activities of GSH reductase, GSH transferase, and catalase were significantly higher in the livers of DHEA-treated rats compared with controls. In these studies a high vitamin E diet decreased DHEA-induced microsomal and mitochondrial lipid peroxidation.[254, 255] In rat skeletal muscle[256] and heart[257] aerobic exercise increased TBARS, appearing to be a mild oxidative stressor. DHEA exacerbated and vitamin E diminished this effect.

6.2. Glucocorticoids

Arguably the most important products of the adrenal cortex are the glucocorticoids, of which cortisol is the primary human hormone. Glucocorticoid receptors are found in nearly every cell type and mediate a wide variety of physiologic effects. Notably, glucocorticoids are stress hormones with actions causing increased blood glucose, maintenance of vascular tone and blood pressure, and protection of cells against various types of stress. Glucocorticoids also modulate immune function, tending to decrease inflammation and thus serve as a "governor" on otherwise potentially overexuberant immune responses that can damage tissue. Despite the results of a study by Kodama[258] suggesting that cortisol, among other steroids, has a weak potential for physiological formation of free radicals, there is little evidence that this molecule acts directly as either an oxidant or antioxidant. Rather cortisol and other glucocorticoids appear to modulate free-radical production, scavenging, and repair via their cellular effects mediated through the glucocorticoid nuclear receptor that interacts with chromosomal Glucocorticoids Regulatory Elements (GRE's) to activate or suppress gene transcription, and hence specific protein synthesis. Actions of glucocorticoids on oxidation-related processes have been particularly well studied in CNS neurons and immune system cells.

Oxygen free radical generation is an important component of the response of a variety of immune system cells, including monocytes and neutrophils, after contact with foreign antigens. This mediates both destruction of bacteria and local tissue inflammation and damage. Superoxide radicals have been implicated in the pathogenesis of tissue injury in several forms of inflammation and arthritis *in vivo*.[259] In human subjects, ROS generation by PMN and MNC was inhibited by the intravenous injection of hydrocortisone[260, 261] or dexamethasone, this latter steroid also increased the plasma concentration of Interleukin-10 (IL-10), an immunomodulatory cytokine that inhibits T(H)1 cells.[262] Antigen-stimulated release of superoxide anion in human MNC *in vitro* was augmented after either IFN-gamma or lipopolysaccharide priming but inhibited by various active glucocorticoids.[263] In a similar study, superoxide generation in response to a synthetic chemotactic factor by human peripheral neutrophils *in vitro* was inhibited by hydrocortisone.[259]

A number of experiments have investigated whether glucocorticoids exert tissue protective effects against immune-cell-generated oxidative stress *in vivo*. Oxidative damage is important in the pathogenesis of Chronic Obstructive Pulmonary Disease (COPD). However, pretreatment with fluticasone propionate, an inhalable glucocorticoid, failed to inhibit ROS production by pulmonary cells recovered by bronchoalveolar lavage from smokers with COPD. In rat distal ileum-mucosal damage due to ischemia and reperfusion is associated with increased malondialdehyde and myeloperoxidase activity which is reduced by SOD or the xanthine oxidase inhibitor, allopurinol. In contrast to the pulmonary study cited above, pretreatment with hydrocortisone prevented malondialdehyde accumulation, mucosal damage, and neutrophil infiltration, suggesting that neutrophil generation of ROS is an important factor in postperfusion injury and can be attenuated by means of a glucocorticoid.[264] Similar results were obtained in studies of arteriovenous ischemia- and cold ischemia-reperfusion of rat skin flaps, in which dexamethasone treatment prior to reperfusion increased flap survival and reduced edema and neutrophil infiltration.[265, 266] Pretreatment with methylprednisolone, a potent glucocorticoid, also protected left ventricular function and coronary flow of excised rat hearts against reperfusion and H_2O_2-induced oxidative stress.[267]

Glucocorticoids are also catabolic hormones. When present in excess, they induce bone loss and muscle wasting. Rats fed a diet supplemented with vitamin E showed less muscle weight loss in response to injections of corticosterone than those fed a basal diet. Muscle protein carbonyl content and lipid oxidation (TBARS) were increased by corticosterone and reduced by vitamin E, results suggesting that increased ROS may play a role in glucocorticoid-induced muscle wasting.[268] Cortisol treatment has been shown to increase superoxide anion production by liver microsomes from rats, with the amount formed being proportional to the amount of cortisol given.[269] A possible mechanism by which glucocorticoids may increase oxidative damage is by inhibiting the production of protective factors. For example, dexamethasone inhibits the induction of MnSOD messenger RNA and MnSOD protein production in response to lipopolysaccharide or TNF-α in a glucocorticoid-sensitive cell line, an effect counteracted by actinomycin-D, and so requiring protein synthesis (i.e. GR activation).[270] Glucocorticoids may also protect or foster damage to cell proteins by their interactions with heat shock proteins. In the above-cited cardioprotection experiments,[271] glucocorticoid treatment induced the cardioprotective heat shock protein, HSP72. In contrast, in the prostate of castrated rats, castration-induced increases of the transcripts for HSP70 were substantially reduced by cortisol,[272] and long-term dexamethasone treatment reduced the increase in expression of HSP70 in the adrenal following restraint stress by fourfold.[6]

Many neurons have endogenous Glucocorticoid Receptors (GR) and considerable data suggest that GR activation and may render neurons more vulnerable to oxidative damage. Oxidative injury to primary cultured hippocampal neurons induced by glutamate, $FeSO_4$, and Alzheimer's Disease-associated Amyloid beta Protein (A-beta) is exacerbated by corticosterone.[273] Glucocorticoid treatment also increases oxidative cell death induced by A-beta or glutamate in clonal mouse hippocampal cells and rat primary embryonal neurons.[274] These increases in cell death were blocked by a specific GR antagonist.[275] In rat primary hippocampal neuron culture, neuronal susceptibility to adriamycin toxicity and adriamycin-induced ROS generation were increased by physiological levels of glucocorticoids in culture media.[276] However, in cortical neuron cultures, which contain lesser amounts of GR, glucocorticoids had no effect on the adriamycin dose-response.[277]

7. Conclusion

Hormones are important modulators and mediators of oxidative stress. Hormonally active molecules play important roles both in protecting cells against oxidative stress and in generating oxygen free radical damage or sensitizing cells to such damage. These actions may occur via hormone receptor interactions, but are, in some cases, due to the redox characteristics of the molecules themselves, which are then mediated by non-receptor-dependent mechanisms.

References

1. Jameson, J. L. and DeGroot, L. J. (1995). Mechanisms of thryoid hormone action. *In* "Endocrinology" (L. J. DeGroot, Ed.), vol. 1, pp. 587, Harcourt Brace Co., Philadelphia.

2. Brown-Borg, H. M., Bode, A. M. and Bartke, A. (1999). Antioxidative mechanisms and plasma growth hormone levels: potential relationship in the aging process. *Endocrine* **11**: 41–48.

3. Persky, A. M., Green, P. S., Stubley, L., Howell, C. O., Zaulyanov, L., Brazeau, G. A. and Simpkins, J. W. (2000). Protective effect of estrogens against oxidative damage to heart and skeletal muscle in vivo and in vitro. *Proc. Soc. Exp. Biol. Med.* **223**: 59–66.

4. Green, P. S., Gridley, K. E. and Simpkins, J. W. (1998). Nuclear estrogen receptor-independent neuroprotection by estratrienes: a novel interaction with glutathione. *Neuroscience* **84**: 7–10.

5. Tamagno, E., Aragno, M., Boccuzzi, G., Gallo, M., Parola, S., Fubini, B., Poli, G. and Danni, O. (1998). Oxygen free radical scavenger properties of dehydroepiandrosterone. *Cell Biochem. Funct.* **16**: 57–63.

6. Udelsman, R., Blake, M. J., Stagg, C. A. and Holbrook, N. J. (1994). Endocrine control of stress-induced heat shock protein 70 expression in vivo. *Surgery* **115**: 611–666.

7. Perfetti, R. and Aggarwal, S. (1999). Signalling via receptor tyrosine kinase modulates the expression of the DNA repair enzyme XPD in cultured cells. *Mol. Cell Endocrinol.* **157**: 171–180.

8. Giustina A. and Veldhuis J. D. (1998). Pathophysiology of the neuroregulation of growth hormone secretion in experimental animals and the human. *Endocrin. Rev.* **19**: 717–797.

9. D'Costa, A. P., Ingram, R. L., Lenham, J. E. and Sonntag, W. E. (1993). The regulation and mechanisms of action of growth hormone and insulin-like growth factor-1 during normal ageing. *J. Reprod. Fertil. Suppl.* **46**: 87–98.

10. Kelijman, M. (1991). Age-related alterations of the growth hormone/insulin-like-growth factor-I axis. *J. Am. Geriatr. Soc.* **39**: 295–307.

11. Rudman, D. (1985). Growth hormone, body composition, and aging. *J. Am. Geriatr. Soc.* **33**: 800–807.

12. Corpas, E., Harman, S. M. and Blackman, M. R. (1993). Human growth hormone and human aging. *Endocrin. Rev.* **14**: 20–39.

13. Vance, M. L. and Mauras, N. (1999). Growth hormone therapy in adults and children. *New England J. Med.* **341**: 1206–1216.

14. Rudman, D., Feller, A. G., Nagraj, H. S., Gergans, G. A., Lalitha, P. Y., Goldberg, A. F., Schlenker, R. A., Cohn, L., Rudman, I. W. and Mattson, D. E. (1990). Effects of human growth hormone in men over 60 years old. *New England J. Med.* **323**: 1–6.

15. Sacca, L., Cittadini, A. and Fazio, S. (1994). Growth hormone and the heart. *Endocrin. Rev.* **15**: 555–573.

16. Lamberts, S. W. (2000). The somatopause: to treat or not to treat? *Horm. Res.* **53**: 42–43.

17. Weindruch, R. and Sohal, R. S. (1997). Seminars in medicine of the Beth Israel Deaconess Medical Center. Caloric intake and aging. *New England J. Med.* **337**: 986–994.

18. Sonntag, W. E., Xu, X., Ingram, R. L. and D'Costa, A. (1995). Moderate caloric restriction alters the subcellular distribution of somatostatin mRNA and increases growth hormone pulse amplitude in aged animals. *Neuroendocrinology* **61**: 601–608.

19. Sonntag, W. E., Lynch, C. D., Cefalu, W. T., Ingram, R. L., Bennett, S. A., Thornton, P. L. and Khan, A. S. (1999). Pleiotropic effects of growth hormone and insulin-like growth factor (IGF)-1 on biological aging: inferences from moderate caloric-restricted animals. *J. Gerontol. A: Biol. Sci. Med. Sci.* **54**: B521–B538.

20. Orme, S. M., McNally, R. J., Cartwright, R. A. and Belchetz, P. E. (1998). Mortality and cancer incidence in acromegaly: a retrospective cohort study. United Kingdom Acromegaly Study Group. *J. Clin. Endocrinol. Metab.* **83**: 2730–2734.

21. Bengtsson, B. A., Eden, S., Ernest, I., Oden, A. and Sjogren, B. (1988). Epidemiology and long-term survival in acromegaly. A study of 166 cases diagnosed between 1955 and 1984. *Acta Med. Scand.* **223**: 327–335.

22. Alexander, L., Appleton, D., Hall, R., Ross, W. M. and Wilkinson, R. (1980). Epidemiology of acromegaly in the Newcastle region. *Clin. Endocrinol. (Oxford)* **12**: 71–79.

23. Takala, J., Ruokonen, E., Webster, N. R., Nielsen, M. S., Zandstra, D. F., Vundelinckx, G. and Hinds, C. J. (1999). Increased mortality associated with growth hormone treatment in critically ill adults [see comments]. *New England J. Med.* **341**: 785–792.

24. Chan, J. M., Stampfer, M. J., Giovannucci, E., Gann, P. H., Ma, J., Wilkinson, P., Hennekens, C. H. and Pollak M. (1998). Plasma insulin-like growth factor-I and prostate cancer risk: a prospective study. *Science* **279**: 563–566.

25. Harman, S. M., Metter, E. J., Blackman, M. R., Landis, P. K. and Carter, H. B. (2000). Serum levels of insulin-like growth factor-I (IGF-I), IGF-II, IGF- binding protein 3, and prostate-specific antigen as predictors of clinical prostate cancer. *J. Clin. Endocrinol. Metab.* **85**: 4258–4265.

26. Hankinson, S. E., Willett, W. C., Colditz, G. A., Hunter, D. J., Michaud, D. S., Deroo, B., Rosner, B., Speizer, F. E. and Pollak, M. (1998). Circulating concentrations of insulin-like growth factor-I and risk of breast cancer [see comments]. *Lancet* **351**: 1393–1396.

27. Wolf, E., Kahnt, E., Ehrlein, J., Hermanns, W., Brem, G. and Wanke, R. (1993). Effects of long-term elevated serum levels of growth hormone on life expectancy of mice: lessons from transgenic animal models. *Mech. Ageing Dev.* **68**: 71–87.

28. Pendergrass, W. R., Li, Y., Jiang, D. and Wolf, N. S. (1993). Decrease in cellular replicative potential in "giant" mice transfected with the bovine growth hormone gene correlates to shortened life span. *J .Cell Physiol.* **156**: 96–103.

29. Cecim, M., Bartke, A., Yun, J. S. and Wagner, T. E. (1994). Expression of human but not bovine growth hormone genes promotes development of mammary tumors in transgenic mice. *Transgenics* **1**: 431–437.

30. Kajiura, L. J. and Rollo, C. D. (1994). A mass budget for transgenic "supermice" engineered with extra rat growth hormone genes: evidence for energetic limitation. *Can. J. Zool.* **72**: 1010–1017.

31. Balbis, A., Bartke, A. and Turyn, D. (1996). Overexpression of bovine growth hormone in transgenic mice is associated with changes in hepatic insulin receptors and in their kinase activity. *Life Sci.* **59**: 1363–1371.

32. Balbis, A., Dellacha, J. M., Calandra, R. S., Bartke, A. and Turyn, D. (1992). Down regulation of masked and unmasked insulin receptors in the liver of transgenic mice expressing bovine growth hormone gene. *Life Sci.* **51**: 771–778.

33. Yang, C. W., Striker, L. J., Kopchick, J. J., Chen, W. Y., Pesce, C. M., Peten, E. P. and Striker, G. E. (1993). Glomerulosclerosis in mice transgenic for native or mutated bovine growth hormone gene. *Kidney Int. Suppl.* **39**: S90–S94.

34. Bartke, A., Chandrashekar, V., Turyn, D., Steger, R. W., Debeljuk, L., Winters, T. A., Mattison, J. A., Danilovich, N. A., Croson, W., Wernsing, D. R. and Kopchick, J. J. (1999). Effects of growth hormone overexpression and growth hormone resistance on neuroendocrine and reproductive functions in transgenic and knock-out mice. *Proc. Soc. Exp. Biol. Med.* **222**: 113–123.

35. Steger, R. W., Bartke, A. and Cecim, M. (1993). Premature ageing in transgenic mice expressing different growth hormone genes. *J. Reprod. Fertil. Suppl.* **46**: 61–75.

36. Miller, D. B., Bartke, A. and O'Callaghan, J. P. (1995). Increased glial fibrillary acidic protein (GFAP) levels in the brains of transgenic mice expressing the bovine growth hormone (BGH) gene. *Exp. Gerontol.* **30**: 383–400.

37. Meliska, C. J., Burke, P. A., Bartke, A. and Jensen, R. A. (1997). Inhibitory avoidance and appetitive learning in aged normal mice: comparison with transgenic mice having elevated plasma growth hormone levels. *Neurobiol. Learn Mem.* **68**: 1–12.

38. Rollo, C. D., Carlson, J. and Sawada, M. (1996). Accelerated aging of giant transgenic mice is associated with elevated free radical processes. *Can. J. Zool.* **74**: 606–620.

38a. Brown-Borg, H. M., Johnson, W. T., Rakoczy, S. G. and Romanick, M. A. (2002). Mitochondrial oxidant production and oxidative damage in Ames dwarf mice. *J. Am. Aging Assoc.,* **in press**.

39. Brown-Borg, H. M. and Rakoczy, S. G. (2000). Catalase expression in delayed and premature aging mouse models. *Exp. Gerontol.* **35**: 199–212.

40. Kono, Y. and Fridovich, I. (1982). Superoxide radical inhibits catalase. *J. Biol. Chem.* **257**: 5751–5754.

41. Nemoto, S., Otsuka, M. and Arakawa, N. (1997). Effect of high concentration of ascorbate on catalase activity in cultured cells and tissues of guinea pigs. *J. Nutri. Sci. Vitaminol. (Tokyo)* **43**: 297–309.

42. Herbert, V., Shaw, S. and Jayatilleke, E. (1996). Vitamin C driven free radical generation from iron [published errata appear in (June 1996), *J. Nutri.* **126**(6): 1746 and (July 1996), **126**(7): 1902]. *J. Nutri.* **126**: 1213S–1220S.

43. Ryu, B. R., Ko, H. W., Jou, I., Noh, J. S. and Gwag, B. J. (1999). Phosphatidylinositol 3-kinase-mediated regulation of neuronal apoptosis and necrosis by insulin and IGF-I. *J. Neurobiol.* **39**: 536–546.

44. Brown-Borg, H. M., Rakoczy, S. G. and Romanick, M. (2000). Mitochondrial proteins in brains and livers of premature and delayed aging mouse models. *Fifth Int. Symp. Neurobiol. Neuroendocrin. Aging.*

45. Pitkanen, S. and Robinson, B. H. (1996). Mitochondrial complex I deficiency leads to increased production of superoxide radicals and induction of superoxide dismutase. *J. Clin. Invest.* **98**: 345–351.

46. Lenaz, G., Bovina, C., Castelluccio, C., Fato, R., Formiggini, G., Genova, M. L., Marchetti, M., Pich, M. M., Pallotti, F., Parenti Castelli, G. and Biagini, G. (1997). Mitochondrial complex I defects in aging. *Mol. Cell Biochem.* **174**: 329–333.

47. Brown-Borg, H. M., Rakoczy, S. G., Romanick, M. A. and Kennedy, M. A. (2002). Effects of growth hormone and insulin like growth factor-1 on hepatocyte antioxidative enzymes. *Exp. Biol. Med.* **in press**.

48. Torres-Aleman, I., Naftolin, F. and Robbins, R. J. (1990). Trophic effects of insulin-like growth factor-I on fetal rat hypothalamic cells in culture. *Neuroscience* **35**: 601–608.

49. Bozyczko-Coyne, D., Glicksman, M. A., Prantner, J. E., McKenna, B., Connors, T., Friedman, C., Dasgupta, M. and Neff, N. T. (1993). IGF-1 supports the survival and/or differentiation of multiple types of central nervous system neurons. *Ann. NY Acad. Sci.* **692**: 311–313.

50. Zackenfels, K., Oppenheim, R. W. and Rohrer, H. (1995). Evidence for an important role of IGF-I and IGF-II for the early development of chick sympathetic neurons. *Neuron* **14**: 731–741.

51. Koh, J. Y., Gwag, B. J., Lobner, D. and Choi, D. W. (1995). Potentiated necrosis of cultured cortical neurons by neurotrophins [see comments]. *Science* **268**: 573–575.

52. Gwag, B. J., Koh, J. Y., Chen, M. M., Dugan, L. L., Behrens, M. M., Lobner, D. and Choi, D. W. (1995). BDNF or IGF-I potentiates free radical-mediated injury in cortical cell cultures. *Neuroreport* **7**: 93–96.

53. Xu, L. and Badr, M. Z. (1999). Enhanced potential for oxidative stress in hyperinsulinemic rats: imbalance between hepatic peroxisomal hydrogen peroxide production and decomposition due to hyperinsulinemia. *Horm. Metab. Res.* **31**: 278–282.

54. Krieger-Brauer, H. I. and Kather, H. (1992). Human fat cells possess a plasma membrane-bound H_2O_2-generating system that is activated by insulin via a mechanism bypassing the receptor kinase. *J. Clin. Invest.* **89**: 1006–1013.

55. Hamel, F. G., Bennett, R. G., Harmon, K. S. and Duckworth, W. C. (1997). Insulin inhibition of proteasome activity in intact cells. *Biochem. Biophys. Res. Commun.* **234**: 671–674.

56. Hamel, F. G., Bennett, R. G. and Duckworth, W. C. (1998). Regulation of multicatalytic enzyme activity by insulin and the insulin-degrading enzyme. *Endocrinology* **139**: 4061–4066.

57. Brown-Borg, H. M., Borg, K. E., Meliska, C. J. and Bartke, A. (1996). Dwarf mice and the ageing process [letter]. *Nature* **384**: 33.

58. Miller, R. A. (1999). Kleemeier award lecture: are there genes for aging? *J. Gerontol. A: Biol. Sci. Med. Sci.* **54**: B297–B307.

59. Sornson, M. W., Wu, W., Dasen, J. S., Flynn, S. E., Norman, D. J., O'Connell, S. M., Gukovsky, I., Carriere, C., Ryan, A. K., Miller, A. P., Zuo, L., Gleiberman, A. S., Andersen, B., Beamer, W. G. and Rosenfeld, M. G. (1996). Pituitary lineage determination by the Prophet of Pit-1 homeodomain factor defective in Ames dwarfism. *Nature* **384**: 327–333.

60. Li, S., Crenshaw, E. B. D., Rawson, E. J., Simmons, D. M., Swanson, L. W. and Rosenfeld, M. G. (1990). Dwarf locus mutants lacking three pituitary cell types result from mutations in the POU-domain gene Pit-1. *Nature* **347**: 528–533.

61. Borg, K. E., Brown-Borg, H. M. and Bartke, A. (1995). Assessment of the primary adrenal cortical and pancreatic hormone basal levels in relation to plasma glucose and age in the unstressed Ames dwarf mouse. *Proc. Soc. Exp. Biol. Med.* **210**: 126–133.

62. Chandrashekar, V. and Bartke, A. (1993). Induction of endogenous insulin-like growth factor-I secretion alters the hypothalamic-pituitary-testicular function in growth hormone-deficient adult dwarf mice. *Biol. Reprod.* **48**: 544–551.

63. Bartke, A., Brown-Borg, H. M., Bode, A. M., Carlson, J., Hunter, W. S. and Bronson, R. T. (1998). Does growth hormone prevent or accelerate aging? *Exp. Gerontol.* **33**: 675–687.

64. Brown-Borg, H. M., Rakoczy, S. G., Romanick, M. A. and Kennedy, M. A. (2001). Relationship between plasma growth hormone, antioxidants and oxidative damage in premature and delayed aging mice. 83rd Annual Endocrine Society Meeting, Denver. (Abstract # P1-418, p. 237.)

65. Promislow, D. E. (1993). On size and survival: progress and pitfalls in the allometry of life span. *J. Gerontol.* **48**: B115–B123.

66. Patronek, G. J., Waters, D. J. and Glickman, L. T. (1997). Comparative longevity of pet dogs and humans: implications for gerontology research. *J. Gerontol. A: Biol. Sci. Med. Sci.* **52**: B171–B178.

67. Samaras, T. T. and Elrick, H. (1999). Height, body size and longevity. *Acta Med. Okayama* **53**: 149–169.

68. Comfort, A. (1961). Life span of animals. *Sci. Am.* **205**: 114.

69. Eigenmann, J. E., Patterson, D. F. and Froesch, E. R. (1984). Body size parallels insulin-like growth factor-I levels but not growth hormone secretory capacity. *Acta Endocrin. (Copenh)* **106**: 448–453.

70. Eigenmann, J. E., Patterson, D. F., Zapf, J. and Froesch, E. R. (1984). Insulin-like growth factor-I in the dog: a study in different dog breeds and in dogs with growth hormone elevation. *Acta Endocrin. (Copenh)* **105**: 294–301.

71. Bartke, A. (1964). Histology of the anterior hypophysis, thyroid and gonads of two types of dwarf mice. *Anat. Rec.* **149**: 225–236.

72. Slabaugh, M. B., Lieberman, M. E., Rutledge, J. J. and Gorski, J. (1981). Growth hormone and prolactin synthesis in normal and homozygous Snell and Ames dwarf mice. *Endocrinology* **109**: 1040–1046.

73. Everitt, A. V., Seedsman, N. J. and Jones, F. (1980). The effects of hypophysectomy and continuous food restriction, begun at ages 70 and 400 days, on collagen aging, proteinuria, incidence of pathology and longevity in the male rat. *Mech. Ageing Dev.* **12**: 161–172.

74. Denckla, W. D. (1974). Role of pituitary and thyroid glands in the decline of minimal O_2 consumption with age. *J. Clin. Invest* **53**: 582.

75. Krzisnik, C., Kolacio, Z., Battelino, T., Brown, M., Parks, J. S. and Laron, Z. (1999). The "little people" of the Island of Krk — revisited. Etiology of hypopituitarism revealed. *J. Endo. Genet.* **1**: 9–19.

76. Zhou, Y., Xu, B. C., Maheshwari, H. G., He, L., Reed, M., Lozykowski, M., Okada, S., Cataldo, L., Coschigamo, K., Wagner, T. E., Baumann, G. and Kopchick, J. J. (1997). A mammalian model for Laron syndrome produced by targeted disruption of the mouse growth hormone receptor/binding protein gene (the Laron mouse). *Proc. Natl. Acad. Sci. USA* **94**: 13 215–13 220.

77. Kopchick, J. J. and Laron, Z. (1999). Is the Laron mouse an accurate model of Laron syndrome? *Mol. Genet. Metab.* **68**: 232–236.

78. Kinney, B. A., Coshigano, K. T., Kopchick, J. J., Steger, R. W. and Bartke, A. (2000). Evidence that middle-aged growth hormone resistant GH-R-KO mice have increased memory retention ompared to their normal siblings: implications for delayed aging. 82nd Annual Endocrine Society, Toronto. (Abstract # 252, p. 70.)

79. Baker, J., Liu, J. P., Robertson, E. J. and Efstratiadis, A. (1993). Role of insulin-like growth factors in embryonic and postnatal growth. *Cell* **75**: 73–82.

80. Kenyon, C., Chang, J., Gensch, E., Rudner, A. and Tabtiang R. (1993). A *Caenorhabditis elegans* mutant that lives twice as long as wild type [see comments]. *Nature* **366**: 461–464.

81. Morris, J. Z., Tissenbaum, H. A. and Ruvkun, G. (1996). A phosphatidy-linositol-3-OH kinase family member regulating longevity and diapause in *Caenorhabditis elegans*. *Nature* **382**: 536–539.

82. Kenyon, C. (1997). Environmental factors and gene activities that influence life span. In *"Caenorhabiditis Elegans* II" (D. L. Riddle, T. Blumenthal, B. J. Meyer, and J. R. Press, Eds.), p. 791, Cold Spring Harbor, New York.

83. Kimura, K. D., Tissenbaum, H. A., Liu, Y. and Ruvkun, G. (1997). Daf-2, an insulin receptor-like gene that regulates longevity and diapause in *Caenorhabditis elegans* [see comments]. *Science* **277**: 942–946.

84. Lin, K., Dorman, J. B., Rodan, A. and Kenyon, C. (1997). Daf-16: an HNF-3/forkhead family member that can function to double the life-span of *Caenorhabditis elegans* [see comments]. *Science* **278**: 1319–1322.

85. Mihaylova, V. T., Borland, C. Z., Manjarrez, L., Stern, M. J. and Sun, H. (1999). The PTEN tumor suppressor homolog in *Caenorhabditis elegans* regulates longevity and dauer formation in an insulin receptor-like signaling pathway. *Proc. Natl. Acad. Sci. USA* **96**: 7427–7432.

86. Facchini, F. S., Hua, N. W., Reaven, G. M. and Stoohs, R. A. (2000). Hyper-insulinemia: the missing link among oxidative stress and age-related diseases? [In Process Citation]. *Free Radic. Biol. Med.* **29**: 1302–1306.

87. Guzel, S., Seven, A., Satman, I. and Burcak, G. (2000). Comparison of oxidative stress indicators in plasma of recent-onset and long-term Type-1 diabetic patients. *J. Toxicol. Env. Health* **59**: 7–14.

88. Ceriello, A., Bortolotti, N., Crescentini, A., Motz, E., Lizzio, S., Russo, A., Ezsol, Z., Tonutti, L. and Taboga, C. (1998). Antioxidant defences are reduced during the oral glucose tolerance test in normal and non-insulin-dependent diabetic subjects. *Eur. J. Clin. Invest.* **28**: 329–333.

89. Ceriello, A., Bortolotti, N., Motz, E., Crescentini, A., Lizzio, S., Russo, A., Tonutti, L. and Taboga, C. (1998). Meal-generated oxidative stress in Type-2 diabetic patients. *Diabetes Care* **21**: 1529–1533.

90. Peuchant, E., Delmas-Beauvieux, M. C., Couchouron, A., Dubourg, L., Thomas, M. J., Perromat, A., Clerc, M. and Gin, H. (1997). Short-term insulin therapy and normoglycemia. Effects on erythrocyte lipid peroxidation in NIDDM patients. *Diabetes Care* **20**: 202–207.

91. Salazar, J. J. and Van Houten, B. (1997). Preferential mitochondrial DNA injury caused by glucose oxidase as a steady generator of hydrogen peroxide in human fibroblasts. *Mutat. Res.* **385**: 139–149.

92. Inoguchi, T., Li, P., Umeda, F., Yu, H. Y., Kakimoto, M., Imamura, M., Aoki, T., Etoh, T., Hashimoto, T., Naruse, M., Sano, H., Utsumi, H. and Nawata, H. (2000). High glucose level and free fatty acid stimulate reactive oxygen species production through protein kinase C — dependent activation

of NAD(P)H oxidase in cultured vascular cells [In Process Citation]. *Diabetes* **49**: 1939–1945.

93. Mohanty, P., Hamouda, W., Garg, R., Aljada, A., Ghanim, H. and Dandona, P. (2000). Glucose challenge stimulates reactive oxygen species (ROS) generation by leucocytes. *J. Clin. Endocrin. Metab.* **85**: 2970–2973.

94. Jain, S. K. (1989). Hyperglycemia can cause membrane lipid peroxidation and osmotic fragility in human red blood cells. *J. Biol. Chem.* **264**: 21 340–21 345.

95. Halliwell, B. and Gutteridge, J. M. C. (1999). *Free Radicals in Biology and Medicine*, Third Edition, New York.

96. Vlassara, H., Brownlee, M. and Cerami, A. (1986). Nonenzymatic glycosylation: role in the pathogenesis of diabetic complications. *Clin. Chem.* **32**: B37–B41.

97. Cerami, A., Vlassara, H. and Brownlee, M. (1986). Role of nonenzymatic glycosylation in atherogenesis. *J. Cell Biochem.* **30**: 111–120.

98. Hunt, J. V., Dean, R. T. and Wolff, S. P. (1988). Hydroxyl radical production and autoxidative glycosylation. Glucose autoxidation as the cause of protein damage in the experimental glycation model of diabetes mellitus and ageing. *Biochem. J.* **256**: 205–212.

99. Yim, M. B., Kang, S. O. and Chock, P. B. (2000). Enzyme-like activity of glycated cross-linked proteins in free radical generation. *Ann NY Acad. Sci.* **899**: 168–181.

100. Wondrak, G. T., Varadarajan, S., Butterfield, D. A. and Jacobson, M. K. (2000). Formation of a protein-bound pyrazinium free radical cation during glycation of histone H1 [In Process Citation]. *Free Radic. Biol. Med.* **29**: 557–567.

101. Oak, J., Nakagawa, K. and Miyazawa, T. (2000). Synthetically prepared aamadori-glycated phosphatidylethanolaminecan trigger lipid peroxidation via free radical reactions. *FEBS Lett.* **481**: 26–30.

102. Habib, M. P., Dickerson, F. D. and Mooradian, A. D. (1994). Effect of diabetes, insulin, and glucose load on lipid peroxidation in the rat. *Metabolism* **43**: 1442–1445.

103. Liang, P., Hughes, V. and Fukagawa, N. K. (1997). Increased prevalence of mitochondrial DNA deletions in skeletal muscle of older individuals with impaired glucose tolerance: possible marker of glycemic stress [published erratum appears in (September 1997), *Diabetes* **46**(9): 1532]. *Diabetes* **46**: 920–923.

104. Fukagawa, N. K., Li, M., Liang, P., Russell, J. C., Sobel, B. E. and Absher, P. M. (1999). Aging and high concentrations of glucose potentiate injury to mitochondrial DNA. *Free Radic. Biol. Med.* **27**: 1437–1443.

105. Leinonen, J., Lehtimaki, T., Toyokuni, S., Okada, K., Tanaka, T., Hiai, H., Ochi, H., Laippala, P., Rantalaiho, V., Wirta, O., Pasternack, A. and Alho, H.

(1997). New biomarker evidence of oxidative DNA damage in patients with non-insulin-dependent diabetes mellitus. *FEBS Lett.* **417**: 150–152.

106. Suzuki, S., Hinokio, Y., Komatu, K., Ohtomo, M., Onoda, M., Hirai, S., Hirai, M., Hirai, A., Chiba, M., Kasuga, S., Akai, H. and Toyota, T. (1999). Oxidative damage to mitochondrial DNA and its relationship to diabetic complications. *Diabetes Res. Clin. Pract.* **45**: 161–168.

107. Ihara, Y., Toyokuni, S., Uchida, K., Odaka, H., Tanaka, T., Ikeda, H., Hiai, H., Seino Y. and Yamada Y. (1999). Hyperglycemia causes oxidative stress in pancreatic beta-cells of GK rats, a model of Type-2 diabetes. *Diabetes* **48**: 927–932.

108. Hiramatsu, K. and Arimori, S. (1988). Increased superoxide production by mononuclear cells of patients with hypertriglyceridemia and diabetes. *Diabetes* **37**: 832–837.

109. Dandona, P., Thusu, K., Cook, S., Snyder, B., Makowski, J., Armstrong, D. and Nicotera, T. (1996). Oxidative damage to DNA in diabetes mellitus. *Lancet* **347**: 444–445.

110. Hannon-Fletcher, M. P., O'Kane, M. J., Moles, K. W., Weatherup, C., Barnett, C. R. and Barnett, Y. A. (2000). Levels of peripheral blood cell DNA damage in insulin dependent diabetes mellitus human subjects. *Mutat. Res.* **460**: 53–60.

111. Rayfield, E. J., Ault, M. J., Keusch, G. T., Brothers, M. J., Nechemias, C. and Smith, H. (1982). Infection and diabetes: the case for glucose control. *Am. J. Med.* **72**: 439–450.

112. Naghibi, M., Smith, R. P., Baltch, A. L., Gates, S. A., Wu, D. H., Hammer, M. C. and Michelsen, P. B. (1987). The effect of diabetes mellitus on chemotactic and bactericidal activity of human polymorphonuclear leukocytes. *Diabetes Res. Clin. Pract.* **4**: 27–35.

113. Ho, F. M., Liu, S. H., Liau, C. S., Huang, P. J., Shiah, S. G. and Lin-Shiau, S. Y. (1999). Nitric oxide prevents apoptosis of human endothelial cells from high glucose exposure during early stage. *J. Cell Biochem.* **75**: 258–263.

114. Rosen, P., Ballhausen, T., Bloch, W. and Addicks, K. (1995). Endothelial relaxation is disturbed by oxidative stress in the diabetic rat heart: influence of tocopherol as antioxidant. *Diabetologia* **38**: 1157–1168.

115. Tesfamariam, B. (1994). Free radicals in diabetic endothelial cell dysfunction. *Free Radic. Biol. Med.* **16**: 383–391.

116. Karasu, C. (1999). Increased activity of H_2O_2 in aorta isolated from chronically streptozotocin-diabetic rats: effects of antioxidant enzymes and enzymes inhibitors. *Free Radic. Biol. Med.* **27**: 16–27.

117. Timimi, F. K., Ting, H. H., Haley, E. A., Roddy, M. A., Ganz, P. and Creager, M. A. (1998). Vitamin C improves endothelium-dependent vasodilation in patients with insulin-dependent diabetes mellitus. *J. Am. Coll. Cardiol.* **31**: 552–557.

118. Ting, H. H., Timimi, F. K., Boles, K. S., Creager, S. J., Ganz, P. and Creager, M. A. (1996). Vitamin C improves endothelium-dependent vasodilation in patients with non-insulin-dependent diabetes mellitus. *J. Clin. Invest.* **97**: 22–28.

119. Ammar, R. F., Jr., Gutterman, D. D., Brooks, L. A. and Dellsperger, K. C. (2000). Free radicals mediate endothelial dysfunction of coronary arterioles in diabetes. *Cardiovasc. Res.* **47**: 595-601.

120. Marfella, R., Verrazzo, G., Acampora, R., La Marca, C., Giunta, R., Lucarelli, C., Paolisso, G., Ceriello, A. and Giugliano, D. (1995). Glutathione reverses systemic hemodynamic changes induced by acute hyperglycemia in healthy subjects. *Am. J. Physiol.* **268**: E1167–E1173.

121. Sakamaki, H., Akazawa, S., Ishibashi, M., Izumino, K., Takino, H., Yamasaki, H., Yamaguchi, Y., Goto, S., Urata, Y., Kondo, T. and Nagataki, S. (1999). Significance of glutathione-dependent antioxidant system in diabetes-induced embryonic malformations. *Diabetes* **48**: 1138–1144.

122. Ornoy, A., Zaken, V. and Kohen, R. (1999). Role of reactive oxygen species (ROS) in the diabetes-induced anomalies in rat embryos in vitro: reduction in antioxidant enzymes and low-molecular-weight antioxidants (LMWA) may be the causative factor for increased anomalies. *Teratology* **60**: 376–386.

123. Wentzel, P., Welsh, N. and Eriksson, U. J. (1999). Developmental damage, increased lipid peroxidation, diminished cyclooxygenase-2 gene expression, and lowered prostaglandin E_2 levels in rat embryos exposed to a diabetic environment. *Diabetes* **48**: 813–820.

124. Wentzel, P. and Eriksson, U. J. (1998). Antioxidants diminish developmental damage induced by high glucose and cyclooxygenase inhibitors in rat embryos in vitro. *Diabetes* **47**: 677–684.

125. Yang, X., Borg, L. A., Siman, C. M. and Eriksson, U. J. (1998). Maternal antioxidant treatments prevent diabetes-induced alterations of mitochondrial morphology in rat embryos. *Anat. Rec.* **251**: 303–315.

126. Viana, M., Herrera, E. and Bonet, B. (1996). Teratogenic effects of diabetes mellitus in the rat. Prevention by vitamin E. *Diabetologia* **39**: 1041–1046.

127. Siman, C. M. and Eriksson, U. J. (1997). Vitamin E decreases the occurrence of malformations in the offspring of diabetic rats. *Diabetes* **46**: 1054–1061.

128. Siman, C. M. and Eriksson, U. J. (1997). Vitamin C supplementation of the maternal diet reduces the rate of malformation in the offspring of diabetic rats. *Diabetologia* **40**: 1416–1424.

129. Bates, J. H., Young, I. S., Galway, L., Traub, A. I. and Hadden, D. R. (1997). Antioxidant status and lipid peroxidation in diabetic pregnancy. *Br. J. Nutri.* **78**: 523–532.

130. No authors. (1993). Implications of the diabetes control and complications trial. American Diabetes Association. *Diabetes* **42**: 1555–1558.

131. No authors. (1995). Effect of intensive diabetes management on macrovascular events and risk factors in the diabetes control and complications trial. *Am. J. Cardiol.* **75**: 894–903.

132. Kuusisto, J., Mykkanen, L., Pyorala, K. and Laakso, M. (1994). Non-insulin-dependent diabetes and its metabolic control are important predictors of stroke in elderly subjects. *Stroke* **25**: 1157–1164.

133. Nishikawa, T., Edelstein, D. and Brownlee, M. (2000). The missing link: a single unifying mechanism for diabetic complications. *Kidney Int.* **58**: 26–30.

134. Paolisso, G., di Maro, G., Galzerano, D., Cacciapuoti, F., Varricchio, G., Varricchio, M. and D'Onofrio, F. (1994). Pharmacological doses of vitamin E and insulin action in elderly subjects. *Am. J. Clin. Nutri.* **59**: 1291–1296.

135. Sharma, A., Kharb, S., Chugh, S. N., Kakkar, R. and Singh, G. P. (2000). Effect of glycemic control and vitamin E supplementation on total glutathione content in non-insulin-dependent diabetes mellitus. *Ann. Nutri. Metab.* **44**: 11–13.

136. Paolisso, G., Tagliamonte, M. R., Barbieri, M., Zito, G. A., Gambardella, A., Varricchio, G., Ragno, E. and Varricchio, M. (2000). Chronic vitamin E administration improves brachial reactivity and increases intracellular magnesium concentration in Type-II diabetic patients. *J. Clin. Endocrin. Metab.* **85**: 109–115.

137. Villa-Caballero, L., Nava-Ocampo, A. A., Frati-Munari, A. C. and Ponce-Monter, H. (2000). Oxidative stress. Should it be measured in the diabetic patient? *Gac. Med. Mex.* **136**: 249–256.

138. Huh, K., Shin, U. S., Choi, J. W. and Lee, S. I. (1994). Effect of sex hormones on lipid peroxidation in rat liver. *Arch. Pharm. Res.* **17**: 109–114.

139. Miura, T., Muraoka, S. and Ogiso, T. (1996). Inhibition of lipid peroxidation by estradiol and 2-hydroxyestradiol. *Steroids* **61**: 379–383.

140. Ayres, S., Abplanalp, W., Liu, J. H. and Subbiah, M. T. (1998). Mechanisms involved in the protective effect of estradiol-17-beta on lipid peroxidation and DNA damage. *Am. J. Physiol.* **274**: E1002–E1008.

141. Mooradian, A. D. (1993). Antioxidant properties of steroids. *J. Steroid. Biochem. Mol. Biol.* **45**: 509–511.

142. Bush, T. L. (1996). Evidence for primary and secondary prevention of coronary artery disease in women taking oestrogen replacement therapy. *Eur. Heart J.* **17(Suppl. D)**: 9–14.

143. Herrington, D. M., Werbel, B. L., Riley, W. A., Pusser, B. E. and Morgan, T. M. (1999). Individual and combined effects of estrogen/progestin therapy and lovastatin on lipids and flow-mediated vasodilation in postmenopausal women with coronary artery disease. *J. Am. Coll. Cardiol.* **33**: 2030–2037.

144. Waddell, T. K., Rajkumar, C., Cameron, J. D., Jennings, G. L., Dart, A. M. and Kingwell, B. A. (1999). Withdrawal of hormonal therapy for 4 weeks decreases arterial compliance in postmenopausal women. *J. Hypertens.* **17**: 413–418.

145. Ylikorkala, O., Cacciatore, B., Paakkari, I., Tikkanen, M. J., Viinikka, L. and Toivonen, J. (1998). The long-term effects of oral and transdermal postmenopausal hormone replacement therapy on nitric oxide, endothelin-1, prostacyclin, and thromboxane. *Fertil. Steril.* **69**: 883–888.

146. Walsh, B. A., Mullick, A. E., Walzem, R. L. and Rutledge, J. C. (1999). 17-beta-estradiol reduces tumor necrosis factor-alpha-mediated LDL accumulation in the artery wall. *J. Lipid Res.* **40**: 387–396.

147. Bekesi, G., Magyar, Z., Kakucs, R., Sprintz, D., Kocsis, I., Szekacs, B. and Feher, J. (1999). Changes in the myeloperoxidase activity of human neutrophilic granulocytes and the amount of enzyme deriving from them under the effect of estrogen. *Orv. Hetil.* **140**: 1625–1630.

148. Mizutani, K., Ikeda, K., Nishikata, T. and Yamori, Y. (2000). Phytoestrogens attenuate oxidative DNA damage in vascular smooth muscle cells from stroke-prone spontaneously hypertensive rats. *J. Hypertens.* **18**: 1833–1840.

149. Tang, M. X., Jacobs, D., Stern, Y., Marder, K., Schofield, P., Gurland, B., Andrews, H. and Mayeux, R. (1996). Effect of oestrogen during menopause on risk and age at onset of Alzheimer's. *Lancet* **348**: 429–432.

150. Kawas, C., Resnick, S., Morrison, A., Brookmeyer, R., Corrada, M., Zonderman, A., Bacal, C., Lingle, D. D. and Metter, E. (1997). A prospective study of estrogen replacement therapy and the risk of developing Alzheimer's disease. The Baltimore Longitudinal Study of Aging [published erratum appears in (August 1998), *Neurology* **51**(2): 654]. *Neurology* **48**: 1517–1521.

151. Henderson, V. W. (1997). The epidemiology of estrogen replacement therapy and Alzheimer's disease. *Neurology* **48**: S27–S35.

152. Green, P. S. and Simpkins, J. W. (2000). Neuroprotective effects of estrogens: potential mechanisms of action. *Int. J. Dev. Neurosci.* **18**: 347–358.

153. Green, P. S., Gridley, K. E. and Simpkins, J. W. (1996). Estradiol protects against beta-amyloid (25–35)-induced toxicity in SK-N-SH human neuroblastoma cells. *Neurosci. Lett.* **218**: 165–168.

154. Green, P. S., Bishop, J. and Simpkins, J. W. (1997). 17-alpha-estradiol exerts neuroprotective effects on SK-N-SH cells. *J. Neurosci.* **17**: 511–515.

155. Behl, C., Skutella, T., Lezoualc'h, F., Post, A., Widmann, M., Newton, C. J. and Holsboer, F. (1997). Neuroprotection against oxidative stress by estrogens: structure-activity relationship. *Mol. Pharmacol.* **51**: 535–541.

156. Green, P. S., Gordon, K. and Simpkins, J. W. (1997). Phenolic A ring requirement for the neuroprotective effects of steroids. *J. Steroid Biochem. Mol. Biol.* **63**: 229–235.

157. Sawada, H., Ibi, M., Kihara, T., Urushitani, M., Akaike, A. and Shimohama, S. (1998). Estradiol protects mesencephalic dopaminergic neurons from oxidative stress-induced neuronal death. *J. Neurosci. Res.* **54**: 707–719.

158. Gridley, K. E., Green, P. S. and Simpkins, J. W. (1998). A novel, synergistic interaction between 17-beta-estradiol and glutathione in the protection of

neurons against beta-amyloid 25–35-induced toxicity in vitro. *Mol. Pharmacol.* **54**: 874–880.

159. Murdoch, W. J. (1998). Inhibition by oestradiol of oxidative stress-induced apoptosis in pig ovarian tissues. *J. Reprod. Fertil.* **114**: 127–130.

160. Harman, S. M., Louvet, J. P. and Ross, G. T. (1975). Interaction of estrogen and gonadotrophins on follicular atresia. *Endocrinology* **96**: 1145–1152.

161. Dubal, D. B., Kashon, M. L., Pettigrew, L. C., Ren, J. M., Finklestein, S. P., Rau, S. W. and Wise, P. M. (1998). Estradiol protects against ischemic injury. *J. Cereb. Blood Flow Metab.* **18**: 1253–1258.

162. Kim, Y. D., Chen, B., Beauregard, J., Kouretas, P., Thomas, G., Farhat, M. Y., Myers, A. K. and Lees, D. E. (1996). 17-beta-estradiol prevents dysfunction of canine coronary endothelium and myocardium and re-perfusion arrhythmias after brief ischemia/reperfusion. *Circulation* **94**: 2901–2908.

163. Montilla, P., Tunez, I., Munoz, M. C., Delgado, M. J. and Salcedo, M. (2000). Hyperlipidemic nephropathy induced by adriamycin in ovariectomized rats: role of free radicals and effect of 17-beta-estradiol administration. *Nephron* **85**: 65–70.

164. Liehr, J. G. and Roy, D. (1998). Pro-oxidant and antioxidant effects of estrogens. *Meth. Mol. Biol.* **108**: 425–435.

165. Wyllie, S. and Liehr, J. G. (1997). Release of iron from ferritin storage by redox cycling of stilbene and steroid estrogen metabolites: a mechanism of induction of free radical damage by estrogen. *Arch. Biochem. Biophys.* **346**: 180–186.

166. Markides, C. S. A., Roy, D. and Liehr, J. G. (1998). Concentration dependence of prooxidant and antioxidant properties of catecholestrogens. *Arch. Biochem. Biophys.* **360**: 105–112.

167. Liehr, J. G., Roy, D., Ari-Ulubelen, A., Bui, Q. D., Weisz, J. and Strobel, H. W. (1990). Effect of chronic estrogen treatment of Syrian hamsters on microsomal enzymes mediating formation of catecholestrogens and their redox cycling: implications for carcinogenesis [published erratum appears in (March 1991), *J. Steroid Biochem.* **38**(3): 3]. *J. Steroid Biochem.* **35**: 555–560.

168. Kalyanaraman, B., Hintz, P. and Sealy, R. C. (1986). An electron spin resonance study of free radicals from catechol estrogens. *Fed. Proc.* **45**: 2477–2484.

169. Sipe, H. J., Jr., Jordan, S. J., Hanna, P. M. and Mason, R. P. (1994). The metabolism of 17-beta-estradiol by lactoperoxidase: a possible source of oxidative stress in breast cancer. *Carcinogenesis* **15**: 2637–2643.

170. Li, Y., Trush, M. A. and Yager, J. D. (1994). DNA damage caused by reactive oxygen species originating from a copper-dependent oxidation of the 2-hydroxy catechol of estradiol. *Carcinogenesis* **15**: 1421–1427.

171. Clemons, M. and Goss, P. (2001). Mechanisms of disease: estrogen and the risk of breast cancer. *New England J. Med.* **344**: 276–285.

172. Liehr, J. G. and Roy, D. (1990). Free radical generation by redox cycling of estrogens. *Free Radic. Biol. Med.* **8**: 415–423.

173. Liehr, J. G. (1997). Hormone-associated cancer: mechanistic similarities between human breast cancer and estrogen-induced kidney carcinogenesis in hamsters. *Env. Health Perspect.* **105(Suppl. 3)**: 565–569.

174. Nutter, L. M., Wu, Y. Y., Ngo, E. O., Sierra, E. E., Gutierrez, P. L. and Abul-Hajj, Y. J. (1994). An O-quinone form of estrogen produces free radicals in human breast cancer cells: correlation with DNA damage. *Chem. Res. Toxicol.* **7**: 23–28.

175. Dabrosin, C., Ollinger, K., Ungerstedt, U. and Hammar, M. (1997). Variability of glutathione levels in normal breast tissue and subcutaneous fat during the menstrual cycle: an in vivo study with microdialysis technique. *J. Clin. Endocrin. Metab.* **82**: 1382–1384.

176. Ogawa, T., Higashi, S., Kawarada, Y. and Mizumoto, R. (1995). Role of reactive oxygen in synthetic estrogen induction of hepatocellular carcinomas in rats and preventive effect of vitamins. *Carcinogenesis* **16**: 831–836.

177. Roy, D., Kalyanaraman, B. and Liehr, J. G. (1991). Xanthine oxidase-catalyzed reduction of estrogen quinones to semiquinones and hydroquinones. *Biochem. Pharmacol.* **42**: 1627–1631.

178. Epe, B., Schiffmann, D. and Metzler, M. (1986). Possible role of oxygen radicals in cell transformation by diethylstilbestrol and related compounds. *Carcinogenesis* **7**: 1329–1334.

179. Oberley, T. D., Allen, R. G., Schultz, J. L. and Lauchner, L. J. (1991). Antioxidant enzymes and steroid-induced proliferation of kidney tubular cells. *Free Radic. Biol. Med.* **10**: 79–83.

180. Winter, M. L. and Liehr, J. G. (1991). Free radical-induced carbonyl content in protein of estrogen-treated hamsters assayed by sodium boro[3H]hydride reduction. *J. Biol. Chem.* **266**: 14 446–14 450.

181. Wiesner, R. J., Kurowski, T. T. and Zak, R. (1992). Regulation by thyroid hormone of nuclear and mitochondrial genes encoding subunits of cytochrome c oxidase in rat liver and skeletal muscle. *Mol. Endocrin.* **6**: 1458–1467.

182. Izquierdo, J. M., Luis, A. M. and Cuezva, J. M. (1990). Postnatal mitochondrial differentiation in rat liver. Regulation by thyroid hormones of the beta-subunit of the mitochondrial F1-ATPase complex. *J. Biol. Chem.* **265**: 9090–9097.

183. Weinstein, S. P., Watts, J., Graves, P. N. and Haber, R. S. (1990). Stimulation of glucose transport by thyroid hormone in ARL 15 cells: increased abundance of glucose transporter protein and messenger ribonucleic acid. *Endocrinology* **126**: 1421–1429.

184. Hoppner, W. and Seitz, H. J. (1989). Effect of thyroid hormones on glucokinase gene transcription in rat liver. *J. Biol. Chem.* **264**: 20 643–20 647.

185. Wall, S. R., van den Hove, M. F., Crepin, K. M., Hue, L. and Rousseau, G. G. (1989). Thyroid hormone stimulates expression of 6-phosphofructo-2-kinase in rat liver. *FEBS Lett.* **257**: 211–214.

186. Swierczynski, J., Mitchell, D. A., Reinhold, D. S., Salati, L. M., Stapleton, S. R., Klautky, S. A., Struve, A. E. and Goodridge, A. G. (1991). Triiodothyronine-induced accumulations of malic enzyme, fatty acid synthase, acetyl-coenzyme A carboxylase, and their mRNAs are blocked by protein kinase inhibitors. Transcription is the affected step. *J. Biol. Chem.* **266**: 17 459–17 466.

187. Stapleton, S. R., Mitchell, D. A., Salati, L. M. and Goodridge, A. G. (1990). Triiodothyronine stimulates transcription of the fatty acid synthase gene in chick embryo hepatocytes in culture. Insulin and insulin-like growth factor amplify that effect. *J. Biol. Chem.* **265**: 18 442–18 446.

188. Bianco, A. C., Sheng, X. Y. and Silva, J. E. (1988). Triiodothyronine amplifies norepinephrine stimulation of uncoupling protein gene transcription by a mechanism not requiring protein synthesis. *J. Biol. Chem.* **263**: 18 168–18 175.

189. Lin, C. S. and Klingenberg, M. (1980). Isolation of the uncoupling protein from brown adipose tissue mitochondria. *FEBS Lett.* **113**: 299–303.

190. Brand, M. D., Chien, L. F., Ainscow, E. K., Rolfe, D. F. and Porter, R. K. (1994). The causes and functions of mitochondrial proton leak. *Biochim. Biophys. Acta* **1187**: 132–139.

191. Ricquier, D. (1999). Uncoupling protein 2 (UCP2): molecular and genetic studies. *Int. J. Obes. Relat. Metab. Disord.* **23(Suppl. 6)**: S38–S42.

192. Lanni, A., de Felice, M., Lombardi, A., Moreno, M., Fleury, C., Ricquier, D. and Goglia, F. (1997). Induction of UCP2 mRNA by thyroid hormones in rat heart. *FEBS Lett.* **418**: 171–174.

193. Larkin, S., Mull, E., Miao, W., Pittner, R., Albrandt, K., Moore, C., Young, A., Denaro, M. and Beaumont, K. (1997). Regulation of the third member of the uncoupling protein family, UCP3, by cold and thyroid hormone. *Biochem. Biophys. Res. Commun.* **240**: 222–227.

194. Gong, D. W., He, Y., Karas, M. and Reitman, M. (1997). Uncoupling protein 3 is a mediator of thermogenesis regulated by thyroid hormone, beta-3-adrenergic agonists, and leptin. *J. Biol. Chem.* **272**: 24 129–24 132.

195. Bobyleva, V., Pazienza, T. L., Maseroli, R., Tomasi, A., Salvioli, S., Cossarizza, A., Franceschi, C. and Skulachev, V. P. (1998). Decrease in mitochondrial energy coupling by thyroid hormones: a physiological effect rather than a pathological hyperthyroidism consequence. *FEBS Lett.* **430**: 409–413.

196. Harper, M. E. and Brand, M. D. (1994). Hyperthyroidism stimulates mitochondrial proton leak and ATP turnover in rat hepatocytes but does not change the overall kinetics of substrate oxidation reactions. *Can. J. Physiol. Pharmacol.* **72**: 899–908.

197. Porter, R. K., Joyce, O. J., Farmer, M. K., Heneghan, R., Tipton, K. F., Andrews, J. F., McBennett, S. M., Lund, M. D., Jensen, C. H. and Melia, H. P. (1999). Indirect measurement of mitochondrial proton leak and its application. *Int. J. Obes. Relat. Metab. Disord.* **23(Suppl. 6)**: S12–S18.

198. Pehowich, D. J. (1999). Thyroid hormone status and membrane *n*-3 fatty acid content influence mitochondrial proton leak. *Biochim. Biophys. Acta* **1411**: 192–200.

199. Hafner, R. P., Brown, G. C. and Brand, M. D. (1990). Thyroid-hormone control of state-3 respiration in isolated rat liver mitochondria. *Biochem. J.* **265**: 731–734.

200. Komosinska-Vassev, K., Olczyk, K., Kucharz, E. J., Marcisz, C., Winsz-Szczotka, K. and Kotulska, A. (2000). Free radical activity and antioxidant defense mechanisms in patients with hyperthyroidism due to Graves' disease during therapy. *Clin. Chim. Acta* **300**: 107–117.

201. Forfar, J. C., Muir, A. L., Sawers, S. A. and Toft, A. D. (1982). Abnormal left ventricular function in hyperthyroidism: evidence for a possible reversible cardiomyopathy. *New England J. Med.* **307**: 1165–1170.

202. Santos, A. D., Mathew, P. K. and Miller, R. P. (1980). The cardiomyopathy of hypothyroidism revisited. *Am. J. Dis. Child.* **134**: 547–559.

203. Torizuka, T., Tamaki, N., Kasagi, K., Misaki, T., Kawamoto, M., Tadamura, E., Magata, Y., Yonekura, Y., Mori, T. and Konishi, J. (1995). Myocardial oxidative metabolism in hyperthyroid patients assessed by PET with carbon-11-acetate. *J. Nucl. Med.* **36**: 1981–1986.

204. Bengel, F. M., Nekolla, S. G., Ibrahim, T., Weniger, C., Ziegler, S. I. and Schwaiger, M. (2000). Effect of thyroid hormones on cardiac function, geometry, and oxidative metabolism assessed noninvasively by positron emission tomography and magnetic resonance imaging. *J. Clin. Endocrin. Metab.* **85**: 1822–1827.

205. Lopez-Torres, M., Romero, M. and Barja, G. (2000). Effect of thyroid hormones on mitochondrial oxygen free radical production and DNA oxidative damage in the rat heart [In Process Citation]. *Mol. Cell. Endocrin.* **168**: 127–134.

206. Asayama, K., Dobashi, K., Hayashibe, H., Megata, Y. and Kato, K. (1987). Lipid peroxidation and free radical scavengers in thyroid dysfunction in the rat: a possible mechanism of injury to heart and skeletal muscle in hyperthyroidism. *Endocrinology* **121**: 2112–2118.

207. Mano, T., Sinohara, R., Sawai, Y., Oda, N., Nishida, Y., Mokuno, T., Kotake, M., Hamada, M., Masunaga, R., Nakai, A. *et al.* (1995). Effects of thyroid hormone on coenzyme Q and other free radical scavengers in rat heart muscle. *J. Endocrin.* **145**: 131–136.

208. Wajdowicz, A., Dabros, W. and Zaczek, M. (1996). Myocardial damage in thyrotoxicosis — ultrastructural studies. *Pol. J. Pathol.* **47**: 127–133.

209. Olson, B. R., Klein, I., Benner, R., Burdett, R., Trzepacz, P. and Levey, G. S. (1991). Hyperthyroid myopathy and the response to treatment. *Thyroid* **1**: 137–141.
210. Crespo Armas, A., Finol, H. J., Anchustegui, B. and Cordero, Z. (1993). Skeletal muscle ultrastructural and biochemical alterations induced by experimental hyperthyroidism. *Acta Cient. Venez.* **44**: 234–239.
211. Crespo-Armas, A., Finol, H. J., Anchustegui, B. and Cordero, Z. (1992). Effects of thyroidectomy on biochemical and ultrastructural aspects of rat slow and fast muscles. *Acta Cient. Venez.* **43**: 148–153.
212. Asayama, K. and Kato, K. (1990). Oxidative muscular injury and its relevance to hyperthyroidism. *Free Radic. Biol. Med.* **8**: 293–303.
213. Huh, K., Kwon, T. H., Kim, J. S. and Park, J. M. (1998). Role of the hepatic xanthine oxidase in thyroid dysfunction: effect of thyroid hormones in oxidative stress in rat liver. *Arch. Pharm. Res.* **21**: 236–240.
214. Tapia, G., Cornejo, P., Fernandez, V. and Videla, L. A. (1999). Protein oxidation in thyroid hormone-induced liver oxidative stress: relation to lipid peroxidation. *Toxicol. Lett.* **106**: 209–214.
215. Videla, L. A. and Fernandez, V. (1995). Effect of thyroid hormone administration on the depletion of circulating glutathione in the isolated perfused rat liver and its relationship to basolateral gamma-glutamyltransferase activity. *J. Biochem. Toxicol.* **10**: 69–77.
216. Rollet-Labelle, E., Grange, M. J., Elbim, C., Marquetty, C., Gougerot-Pocidalo, M. A. and Pasquier, C. (1998). Hydroxyl radical as a potential intracellular mediator of polymorphonuclear neutrophil apoptosis. *Free Radic. Biol. Med.* **24**: 563–572.
217. Szabo, J., Foris, G., Mezosi, E., Nagy, E. V., Paragh, G., Sztojka, I. and Leovey, A. (1996). Parameters of respiratory burst and arachidonic acid metabolism in polymorphonuclear granulocytes from patients with various thyroid diseases. *Exp. Clin. Endocrin. Diabetes* **104**: 172–176.
218. Nishizawa, Y., Fushiki, S. and Amakata, Y. (1998). Thyroxine-induced production of superoxide anion by human alveolar neutrophils and macrophages: a possible mechanism for the exacerbation of bronchial asthma with the development of hyperthyroidism. *In Vivo* **12**: 253–257.
219. Fernandez, V. and Videla, L. A. (1995). On the mechanism of thyroid hormone-induced respiratory burst activity in rat polymorphonuclear leukocytes. *Free Radic. Biol. Med.* **19**: 359–363.
220. Mihara, S., Suzuki, N., Wakisaka, S., Suzuki, S., Sekita, N., Yamamoto, S., Saito, N., Hoshino, T. and Sakane, T. (1999). Effects of thyroid hormones on apoptotic cell death of human lymphocytes. *J. Clin. Endocrin. Metab.* **84**: 1378–1385.
221. Ima-Nirwana, S., Kiftiah, A., Sariza, T., Gapor, M. T. and Khalid, B. A. (1999). Palm vitamin E improves bone metabolism and survival rate in thyrotoxic rats. *Gen. Pharmacol.* **32**: 621–626.

222. Odell, W. D. and Heath, H. D. (1993). Osteoporosis: pathophysiology, prevention, diagnosis, and treatment. *Dis. Mon.* **39**: 789–867.

223. Orentreich, N., Brind, J. L., Vogelman, J. H., Andres, R. and Baldwin, H. (1992). Long-term longitudinal measurements of plasma dehydroepiandrosterone sulfate in normal men. *J. Clin. Endocrin. Metab.* **75**: 1002–1004.

224. Straub, R. H., Konecna, L., Hrach, S., Rothe, G., Kreutz, M., Scholmerich, J., Falk, W. and Lang, B. (1998). Serum dehydroepiandrosterone (DHEA) and DHEA sulfate are negatively correlated with serum interleukin-6 (IL-6), and DHEA inhibits IL-6 secretion from mononuclear cells in man in vitro: possible link between endocrinosenescence and immunosenescence. *J. Clin. Endocrin. Metab.* **83**: 2012–2017.

225. Berr, C., Lafont, S., Debuire, B., Dartigues, J. F. and Baulieu, E. E. (1996). Relationships of dehydroepiandrosterone sulfate in the elderly with functional, psychological, and mental status, and short-term mortality: a French community-based study. *Proc. Natl. Acad. Sci. USA* **93**: 13 410–13 415.

226. Maccario, M., Ramunni, J., Oleandri, S. E., Procopio, M., Grottoli, S., Rossetto, R., Savio, P., Aimaretti, G., Camanni, F. and Ghigo, E. (1999). Relationships between IGF-I and age, gender, body mass, fat distribution, metabolic and hormonal variables in obese patients. *Int. J. Obes. Relat. Metab. Disord.* **23**: 612–618.

227. Abbasi, A. A., Mattson, D. E., Duthie, E. H., Jr., Wilson, C., Sheldahl, L., Sasse, E. and Rudman, I. W. (1998). Predictors of lean body mass and total adipose mass in community-dwelling elderly men and women. *Am. J. Med. Sci.* **315**: 188–193.

228. Abbasi, A., Duthie, E. H., Jr., Sheldahl, L., Wilson, C., Sasse, E., Rudman, I. and Mattson, D. E. (1998). Association of dehydroepiandrosterone sulfate, body composition, and physical fitness in independent community-dwelling older men and women [see comments]. *J. Am. Geriatr. Soc.* **46**: 263–273.

229. Kostka, T., Arsac, L. M., Patricot, M. C., Berthouze, S. E., Lacour, J. R. and Bonnefoy, M. (2000). Leg extensor power and dehydroepiandrosterone sulfate, insulin-like growth factor-I and testosterone in healthy active elderly people. *Eur. J. Appl. Physiol.* **82**: 83–90.

230. Haffner, S. M., Valdez, R. A., Mykkanen, L., Stern, M. P. and Katz, M. S. (1994). Decreased testosterone and dehydroepiandrosterone sulfate concentrations are associated with increased insulin and glucose concentrations in nondiabetic men. *Metabolism* **43**: 599–603.

231. Phillips, G. B. (1996). Relationship between serum dehydroepiandrosterone sulfate, androstenedione, and sex hormones in men and women. *Eur. J. Endocrin.* **134**: 201–206.

232. Kalmijn, S., Launer, L. J., Stolk, R. P., de Jong, F. H., Pols, H. A., Hofman, A., Breteler, M. M. and Lamberts, S. W. (1998). A prospective study on cortisol, dehydroepiandrosterone sulfate, and cognitive function in the elderly. *J. Clin. Endocrin. Metab.* **83**: 3487–3492.

233. Moffat, S. D., Zonderman, A. B., Harman, S. M., Blackman, M. R., Kawas C. and Resnick, S. M. (2000). The relationship between longitudinal declines in dehydroepiandrosterone sulfate concentrations and cognitive performance in older men. *Arch. Int. Med.* **160**: 2193–2198.

234. Morales, A. J., Nolan, J. J., Nelson, J. C. and Yen, S. S. (1994). Effects of replacement dose of dehydroepiandrosterone in men and women of advancing age [published erratum appears in (September 1995), *J. Clin. Endocrin. Metab.* **80**(9): 2799]. *J. Clin. Endocrin. Metab.* **78**: 1360–1367.

235. Arlt, W., Callies, F., van Vlijmen, J. C., Koehler, I., Reincke, M., Bidlingmaier, M., Huebler, D., Oettel, M., Ernst, M., Schulte, H. M. and Allolio, B. (1999). Dehydroepiandrosterone replacement in women with adrenal insufficiency. *New England J. Med.* **341**: 1013–1020.

236. Hornsby, P. J. (1997). DHEA: a biologist's perspective. *J. Am. Geriatr. Soc.* **45**: 1395–1401.

237. Gallo, M., Aragno, M., Gatto, V., Tamagno, E., Brignardello, E., Manti, R., Danni, O. and Boccuzzi, G. (1999). Protective effect of dehydroepiandrosterone against lipid peroxidation in a human liver cell line. *Eur. J. Endocrin.* **141**: 35–39.

238. Bastianetto, S., Ramassamy, C., Poirier, J. and Quirion, R. (1999). Dehydroepiandrosterone (DHEA) protects hippocampal cells from oxidative stress-induced damage. *Brain Res. Mol. Brain Res.* **66**: 35–41.

239. Rom, W. N. and Harkin, T. (1991). Dehydroepiandrosterone inhibits the spontaneous release of superoxide radical by alveolar macrophages in vitro in asbestosis. *Env. Res.* **55**: 145–156.

240. Whitcomb, J. M. and Schwartz, A. G. (1985). Dehydroepiandrosterone and 16 alpha-Br-epiandrosterone inhibit 12-O-tetradecanoylphorbol-13-acetate stimulation of superoxide radical production by human polymorphonuclear leukocytes. *Carcinogenesis* **6**: 333–335.

241. Brignardello, E., Gallo, M., Aragno, M., Manti, R., Tamagno, E., Danni, O. and Boccuzzi, G. (2000). Dehydroepiandrosterone prevents lipid peroxidation and cell growth inhibition induced by high glucose concentration in cultured rat mesangial cells. *J. Endocrin.* **166**: 401–406.

242. Brignardello, E., Beltramo, E., Molinatti, P. A., Aragno, M., Gatto, V., Tamagno, E., Danni, O., Porta, M. and Boccuzzi, G. (1998). Dehydroepiandrosterone protects bovine retinal capillary pericytes against glucose toxicity. *J. Endocrin.* **158**: 21–26.

243. Boccuzzi, G., Aragno, M., Seccia, M., Brignardello, E., Tamagno, E., Albano, E., Danni, O. and Bellomo, G. (1997). Protective effect of dehydroepiandrosterone against copper-induced lipid peroxidation in the rat. *Free Radic. Biol. Med.* **22**: 1289–1294.

244. Aragno, M., Tamagno, E., Boccuzzi, G., Brignardello, E., Chiarpotto, E., Pizzini, A. and Danni, O. (1993). Dehydroepiandrosterone pretreatment

protects rats against the pro-oxidant and necrogenic effects of carbon tetrachloride. *Biochem. Pharmacol.* **46**: 1689–1694.

245. Aragno, M., Tamagno, E., Poli, G., Boccuzzi, G., Brignardello, E. and Danni, O. (1994). Prevention of carbon tetrachloride-induced lipid peroxidation in liver microsomes from dehydroepiandrosterone-pretreated rats. *Free Radic. Res.* **21**: 427–435.

246. Hu, Y., Cardounel, A., Gursoy, E., Anderson, P. and Kalimi, M. (2000). Anti-stress effects of dehydroepiandrosterone: protection of rats against repeated immobilization stress-induced weight loss, glucocorticoid receptor production, and lipid peroxidation. *Biochem. Pharmacol.* **59**: 753–762.

247. Aragno, M., Brignardello, E., Tamagno, E., Gatto, V., Danni, O. and Boccuzzi, G. (1997). Dehydroepiandrosterone administration prevents the oxidative damage induced by acute hyperglycemia in rats. *J. Endocrin.* **155**: 233–240.

248. Aragno, M., Parola, S., Brignardello, E., Mauro, A., Tamagno, E., Manti, R., Danni, O. and Boccuzzi, G. (2000). Dehydroepiandrosterone prevents oxidative injury induced by transient ischemia/reperfusion in the brain of diabetic rats. *Diabetes* **49**: 1924–1931.

249. Lohman, R., Yowell, R., Barton, S., Araneo, B. and Siemionow, M. (1997). Dehydroepiandrosterone protects muscle flap microcirculatory hemo-dynamics from ischemia/reperfusion injury: an experimental in vivo study. *J. Trauma* **42**: 74–80.

250. Ng, H. P., Wang, Y. F., Lee, C. Y. and Hu, M. L. (1999). Toxicological and antioxidant effects of short-term dehydroepiandrosterone injection in young rats fed diets deficient or adequate in vitamin E. *Food Chem. Toxicol.* **37**: 503–508.

251. Mohan, P. F. and Cleary, M. P. (1989). Dehydroepiandrosterone and related steroids inhibit mitochondrial respiration in vitro. *Int. J. Biochem.* **21**: 1103–1107.

252. McIntosh, M. K., Pan, J. S. and Berdanier, C. D. (1993). In vitro studies on the effects of dehydroepiandrosterone and corticosterone on hepatic steroid receptor binding and mitochondrial respiration. *Comp. Biochem. Physiol. Comp. Physiol.* **104**: 147–153.

253. Mohan, P. F. and Cleary, M. P. (1991). Short-term effects of dehydroepian-drosterone treatment in rats on mitochondrial respiration. *J. Nutri.* **121**: 240–250.

254. Swierczynski, J., Kochan, Z. and Mayer, D. (1997). Dietary alpha-tocopherol prevents dehydroepiandrosterone-induced lipid peroxidation in rat liver microsomes and mitochondria. *Toxicol. Lett.* **91**: 129–136.

255. McIntosh, M. K., Goldfarb, A. H., Curtis, L. N. and Cote, P. S. (1993). Vitamin E alters hepatic antioxidant enzymes in rats treated with dehydro-epiandrosterone (DHEA). *J. Nutri.* **123**: 216–224.

256. Goldfarb, A. H., McIntosh, M. K., Boyer, B. T. and Fatouros, J. (1994). Vitamin E effects on indexes of lipid peroxidation in muscle from DHEA-treated and exercised rats. *J. Appl. Physiol.* **76**: 1630–1635.

257. Goldfarb, A. H., McIntosh, M. K. and Boyer, B. T. (1996). Vitamin E attenuates myocardial oxidative stress induced by DHEA in rested and exercised rats. *J. Appl. Physiol.* **80**: 486–490.

258. Kodama, M., Inoue, F., Saito, H., Oda, T. and Sato, Y. (1997). Formation of free radicals from steroid hormones: possible significance in environmental carcinogenesis. *Anticancer Res.* **17**: 439–444.

259. Simchowitz, L., Mehta, J. and Spilberg, I. (1979). Chemotactic factor-induced generation of superoxide radicals by human neutrophils: effect of metabolic inhibitors and antiinflammatory drugs. *Arthritis. Rheum.* **22**: 755–763.

260. Dandona, P., Thusu, K., Hafeez, R., Abdel-Rahman, E. and Chaudhuri, A. (1998). Effect of hydrocortisone on oxygen free radical generation by mononuclear cells. *Metabolism* **47**: 788–791.

261. Dandona, P., Suri, M., Hamouda, W., Aljada, A., Kumbkarni, Y. and Thusu, K. (1999). Hydrocortisone-induced inhibition of reactive oxygen species by polymorphonuclear neutrophils [see comments]. *Crit. Care Med.* **27**: 2442–2444.

262. Dandona, P., Mohanty, P., Hamouda, W., Aljada, A., Kumbkarni, Y. and Garg, R. (1999). Effect of dexamethasone on reactive oxygen species generation by leukocytes and plasma interleukin-10 concentrations: a pharmacodynamic study. *Clin. Pharmacol. Ther.* **66**: 58–65.

263. Szefler, S. J., Norton, C. E., Ball, B., Gross, J. M., Aida, Y. and Pabst, M. J. (1989). IFN-gamma and LPS overcome glucocorticoid inhibition of priming for superoxide release in human monocytes. Evidence that secretion of IL-1 and tumor necrosis factor-alpha is not essential for monocyte priming. *J. Immunol.* **142**: 3985–3992.

264. Otamiri, T. (1989). Oxygen radicals, lipid peroxidation, and neutrophil infiltration after small-intestinal ischemia and reperfusion. *Surgery* **105**: 593–597.

265. Willemart, G., Knight, K. R. and Morrison, W. A. (1998). Dexamethasone treatment prior to reperfusion improves the survival of skin flaps subjected to secondary venous ischaemia. *Br. J. Plast. Surg.* **51**: 624–628.

266. Willemart, G., Knight, K. R., Ayad, M., Wagh, M. and Morrison, W. A. (1999). The beneficial antiinflammatory effect of dexamethasone administration prior to reperfusion on the viability of cold-stored skin flaps. *Int. J. Tissue React.* **21**: 71–78.

267. Valen, G., Kawakami, T., Tahepold, P., Starkopf, J., Kairane, C., Dumitrescu, A., Lowbeer, C., Zilmer, M. and Vaage, J. (2000). Pretreatment with methylprednisolone protects the isolated rat heart against ischaemic and oxidative damage. *Free Radic. Res.* **33**: 31–43.

268. Ohtsuka, A., Kojima, H., Ohtani, T. and Hayashi, K. (1998). Vitamin E reduces glucocorticoid-induced oxidative stress in rat skeletal muscle. *J. Nutri. Sci. Vitaminol. (Tokyo)* **44**: 779–786.

269. Nelson, D. H. and Ruhmann-Wennhold, A. (1975). Corticosteroids increase superoxide anion production by rat liver microsomes. *J. Clin. Invest.* **56**: 1062–1065.

270. Valentine, J. F. and Nick, H. S. (1994). Glucocorticoids repress basal and stimulated manganese superoxide dismutase levels in rat intestinal epithelial cells. *Gastroenterology* **107**: 1662–1670.

271. Valen, G., Kawakami, T., Tahepold, P., Dumitrescu, A., Lowbeer, C. and Vaage, J. (2000). Glucocorticoid pretreatment protects cardiac function and induces cardiac heat shock protein 72. *Am. J. Physiol. Heart. Circ. Physiol.* **279**: H836–H843.

272. Rennie, P. S., Bowden, J. F., Freeman, S. N., Bruchovsky, N., Cheng, H., Lubahn, D. B., Wilson, E. M., French, F. S. and Main, L. (1989). Cortisol alters gene expression during involution of the rat ventral prostate. *Mol. Endocrin.* **3**: 703–708.

273. Goodman, Y., Bruce, A. J., Cheng, B. and Mattson, M. P. (1996). Estrogens attenuate and corticosterone exacerbates excitotoxicity, oxidative injury, and amyloid beta-peptide toxicity in hippocampal neurons. *J. Neurochem.* **66**: 1836–1844.

274. Behl, C. (1998). Effects of glucocorticoids on oxidative stress-induced hippocampal cell death: implications for the pathogenesis of Alzheimer's disease. *Exp. Gerontol.* **33**: 689–696.

275. Behl, C., Lezoualc'h, F., Trapp, T., Widmann, M., Skutella, T. and Holsboer, F. (1997). Glucocorticoids enhance oxidative stress-induced cell death in hippocampal neurons in vitro. *Endocrinology* **138**: 101–106.

276. McIntosh, L. J. and Sapolsky, R. M. (1996). Glucocorticoids may enhance oxygen radical-mediated neurotoxicity. *Neurotoxicology* **17**: 873–882.

277. McIntosh, L. J. and Sapolsky, R. M. (1996). Glucocorticoids increase the accumulation of reactive oxygen species and enhance adriamycin-induced toxicity in neuronal culture. *Exp. Neurol.* **141**: 201–206.

Chapter 62

Exercise

Helaine M. Alessio

Helaine M. Alessio • Miami University, Oxford, Ohio 45056
Tel: 513-529-2707, E-mail: alessih@muohio.edu

1. Introduction

In the 21st century, the importance of exercise to life span has been affirmed by large scale research studies that concur that regular exercise reduces the risk of disease and death.[1-3] But an underlying mechanism of exercise is that it induces oxidative stress. With aging, oxidative stress increases in skeletal muscle and vital organs. It is therefore, important to learn about the separate and combined effects of exercise and aging-especially as they might influence oxidative stress. A better understanding of exercise, aging, and oxidative stress will assist in exercise recommendations for persons of all ages.

2. Exercise Principles

A single bout of exercise increases metabolism and oxidative stress during and immediately following exercise.[4-8] This could translate into an increased production of superoxide radicals due to a mass action effect by an approximate 10-fold increase in oxygen consumption.[9] If antioxidant levels increase proportionally to oxygen consumption, then this exercise-induced change is inconsequential. Furthermore, regular or repeated bouts of exercise are associated with lower resting metabolic rate, higher antioxidant activity, and lower oxidation of LDL's and more protection against oxidation of proteins and DNA.[10,11]

Principles of exercise include: frequency, intensity, time, and type. These principles are used to precisely prescribe exercise programs for individuals having different fitness levels and health status. In this way, the stress of acute exercise would be proportional to the ability of the individual. With too little stress, little to no homeostatic changes or health benefits are likely to occur. With too much stress, homeostasis could be altered severely, resulting in an "open window" for muscle damage, disturbance in the oxidative stress balance, and possible illness.[12] When performed regularly, an appropriate exercise prescription is likely to result in improved health and life expectancy. The Surgeon General's Report on Physical Activity and Health[13] recommends participating in regular physical activity 4–6 days a week (frequency), at 50–85% of maximum ability (intensity), for 15–60 min per day (time).

Appropriate adaptive responses to exercise can include: (1) acute cellular adaptation to free radical attack, and (2) cell signaling which affects cell functioning for a longer time period. The latter category has attracted recent attention with an increase in our understanding of how free radicals trigger signal transduction pathways in the cell. Activation of the signal transduction pathway can generate active transcription factors that can bind to the promoter region of select genes and activate transcription. This would, in effect, jumpstart the process for *de nova* synthesis of antioxidants and other proteins that defend against free radicals. Three transcription factors that are currently under investigation for the role in exercise-induced oxidative stress include nuclear factor kappa B (NF-κB), activator protein 1 (AP-1), and heat shock factor.[14]

3. Aging Principles

Aging usually refers to physiological changes that occur following maturation. During the aging process, organisms undergo change, albeit at different rates. Classic studies have investigated the formation of oxidants, scavenger activities, cell damage and death across species in order to learn more about free radicals and aging. Altered free radical production may be a key determinant of maximum life span. Cutler reported that lifetime energy potential was directly related to antioxidants responsible for removing free radicals, including superoxide dismutase (SOD), serum carotenoids, and alpha tocopherol.[15] He showed that longer lived species had higher levels of these three antioxidants. But, not all antioxidants directly correlate with life spans. Some, like glutathione, glutathione peroxidase, and glutathione transferase are negatively related to life span. Sohal and colleagues have shown that antioxidant enzymes do not demonstrate consistent correlations with either life span across species or aging within species.[16-18] Aging probably does not result in a uniform decline in cellular antioxidant activity. Instead, specific adaptation patterns are revealed in aged tissues. Aging may cause a reduction of protein turnover and cell regenerating capacity, which in turn decreases some cellular antioxidant enzymes.

The free radical theory of aging holds to several basic tenets: (1) mitochondria could play a pacemaker type of role, (2) mitochondrial DNA, a major site of oxidative damage, is not protected by exogenous antioxidants, and (3) endogenous antioxidant defense is regulated to provide a net protection that is difficult to change significantly by the addition or deletion of select antioxidants. If oxidative stress does cause aging, energy expenditure is of prime importance.

4. Pleiotropic and Hormetic Effects of Exercise and Aging

Many theories agree that aging occurs in large part as the result of side effects of normal biological processes that are necessary for survival. This is referred to as pleiotropy. Put another way, pleiotropy refers to by-products of beneficial developmental and metabolic processes that, over time, can destroy an organism. Classic studies by Sacher[19,20] and Cutler[21] have described aging as the result of a gradual loss of an organism to maintain homeostatic regulation of processes that produce and remove specific metabolic by-products. They further suggest that species having higher life span potentials also have a qualitatively superior homeostatic control system.

Hormesis refers to a phenomenon whereby organisms subjected to low levels of stress appear to function better and live longer than organisms subjected to high levels of stress.[22] Luckey[23] introduced the term "hormoligosis" to describe the class of phenomena in which small quantities of a toxic agent acted in a stimulatory, not destructive way. Ordy *et al.*[24] and Reinke, Stutz and Hunstein[25] reported that moderate physiological stress can lead to a decreased mortality rate in mice. Larger doses of stress, (e.g. radiation) can shorten life. Hormetic effects of aging and longevity have been investigated by Sacher.[26] Hormetic effects can increase longevity by reducing vulnerability without altering aging processes. Hormesis may be an important key to understanding the enduring stability of organisms and how stability depends on a dynamic relation between the organism and its environment; prooxidants and antioxidants.

The pleiotrpoic properties of aging are similar in many respects to the pleiotropic nature of exercise. Exercise, when performed regularly, confers many health-related benefits. Yet, exercise, by its very nature, demands a large energy transfer that in many cases, requires large volumes of oxygen. Greater energy transfer and oxygen consumption, while necessary for exercise, can be detrimental to the organism. But how can energy metabolism and oxygen consumption be harmful? One possible explanation comes from an evolutionary perspective that essentially all metabolic processes result from a trade-off between beneficial versus harmful side-effects. Metabolic pathways in humans and other organisms are not 100% efficient. When energy is transferred from food to a working muscle, over 70% of the potential energy is released as heat. Exercise is associated with increased core temperature, and heat is known to disrupt electron transfer in the mitochondria. Whether the electron flux is 1% or 0.1%,[27] increased oxygen consumption during aerobic exercise may contribute to a mass action effect whereby more reactive oxygen species are produced. In the absence of an appropriate antioxidant response, exercise may induce oxidative stress. On the other hand, there is ample evidence that exercise can activate select antioxidant activities during or immediately following acute exercise.[28–30]

Exercise appears to have hormetic properties, but does not change the rate of aging.[31] Exhaustive and extreme exercise are known to be harmful. Moderate to moderate-high intensity exercise has been associated with reduced morbidity and mortality in large-scale studies. On a molecular level, moderate to moderate-high intensity exercise may result in health benefits by upregulating heat shock proteins, increasing prostaglindin synthesis, and increasing protein turnover.[32] The release of heat shock proteins and certain prostaglindins may have protective effects locally, and protein turnover may facilitate removal of cell fragments and synthesis of new proteins.

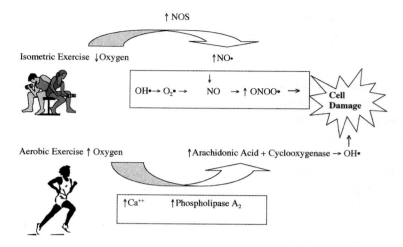

Fig. 1. Different types of exercise, aerobic and isometric, may induce cell damage via different mechanisms. See text for explanation. NOS = nitric oxide synthase, NO• = nitric oxide radical, OH• = hydroxyl radical, ONOO• = peroxynitrate radical, Ca^{++} = calcium.

5. Exercise-Induced Oxidative Stress

There are many ways that exercise can induce oxidative stress (Fig. 1). Mitochondrial respiration is an important source of oxygen radicals and hence, a potential contributor to reperfusion injury. Alessio *et al.*[33] measured biomarkers of oxidative stress in subjects following isometric exercise. The dramatic blood pressure changes observed during isometric exercise may cause ischemia-reperfusion. In ischemic-reperfused models, oxygen radicals can be generated upon reflow following vasoconstriction. Under hypoxic conditions, reducing equivalents may accumulate within the mitochondrial electron transport chain, resulting in a phenomenon known as reductive stress.[34] Upon reoxygenation, a burst of one-electron reductions may convert molecular oxygen to superoxide radicals. Mechanical stress contributes to exercise-induced damage of muscle fibers, even in the absence of increased oxygen consumption.

Another potential source of oxidative stress following exercise is muscle neutrophil and macrophage invasion and superoxide generation during the acute phase inflammatory response in high-intensity exercise. Neutrophils flow easily through cell walls and into tissue spaces and are attracted to infected areas, where they proceed to ingest bacteria and other foreign or abnormal material. This process, called phagocytosis, occurs in the presence of oxygen, and neutrophils may reduce molecular oxygen to the superoxide radical via NADPH oxidase, which is usually dormant in resting cells.[35] Lipid peroxidation by-products, including cyclooxygenase- and lipoxygenase-derived products, are potent vasoactive and chemoattractant factors. Oxidized lipids and proteins may also play a causal role in inflammation. In a study by Saxton, Donnelly, and Roper,[36] both eccentric (muscle lengthening) and concentric (muscle shortening) arm and leg exercises were performed. Immediately following concentric leg exercise in Saxton's study, protein carbonyl derivatives and serum creatine kinase activity increased, although the lipid peroxidation by-product, thiobarbituric acid derived substances, did not.

Reduced glutathione (GSH) is an important antioxidant that has recently been implicated in inflammation.[37] When GSH becomes oxidized and forms oxidized glutathione (GSSG), redox sensitive transcription factors such as NF-κB and activator protein 1 (*c-Fos/c-Jun*, AP-1) become activated. AP-1 and *c-Fos/c-Jun* have proinflammatory actions as they promote the transcription of cytokine genes (e.g. IL-1B, TNF-α, and iNOS) and upregulate adhesion molecules (e.g. I-CAM). Cells stick together more readily, causing changes in fluid retention as well as oxygen tension in damaged tissues. Tissues damaged as a result of high-intensity exercise become inflamed. Inflammation is usually not a serious concern, since it is directly linked to wound healing. Nevertheless, if the critical balance between inflammation and wound healing is left unchecked, then wound healing will decline and tissues can be destroyed. Tissues may very well be destroyed by an uncontrolled series of respiratory bursts, shown in the following reaction, and catalyzed by NADPH oxidase:

$$2 O_2 + NADPH \rightarrow 2 O_2^- + NADP^+ + H^+.$$

Hellsten[38] described two ways that xanthine oxidase may be involved in exercise-induced oxidative stress. First, xanthine oxidase is associated with lipid peroxidation via the conversion of xanthine dehydrogenase to xanthine oxidase in the presence of hypoxanthine. Second, free radical production via xanthine oxidase, may mediate the process of inflammation. Under normal skeletal conditions the enzyme xanthine dehydrogenase aids in the conversion of hypoxanthine to uric acid. Ischemia-reperfusion and temporary disruptions of ATP-dependent Ca^{++} pumps, both of which are associated with exercise, can lead to increased intracellular Ca^{++} concentrations. Increased intramuscular Ca^{++} concentrations during periods of high intensity exercise may activate calcium dependent protease, which convert xanthine dehydrogenase to xanthine oxidase. Xanthine oxidase uses molecular oxygen instead of NAD^+ as an electron acceptor and thus generates the superoxide radical.[39]

Several exercise studies have reported an increase in xanthine dehydrogenase:xanthine oxidase in muscle.[40, 41] High intensity exercise is associated with depleted energy stores, including adenine nucleotide levels. Adenine nucleotides in exhausted muscle can be further degraded to inosine monophosphates which can be degraded to hypoxanthine. Hypoxanthine is a reliable indicator of adenine nucleotide break down.[38] The decline of adenine nucleotides, which otherwise would have been used for energy, could impair the ability of the energy-requiring calcium pump, resulting in increased calcium in the cell. Muscle damage caused by some types of high-intensity exercise, in particular, eccentric exercise, can lead to an increase in hypoxanthine levels as well as activate the immune system. Muscle damage may activate neutrophils, which facilitate the conversion of xanthine dehydrogenase to xanthine oxidase. Superoxide radicals could be generated this way. Muscle damage may also attract macrophages to the injury site, causing an increase in chemotactic factors, tumor necrosis factor, and a further increase in the influx of neuthrophils. All of these events would increase superoxide radical production in the exercised muscle and would have a catabolic effect.

Two other potential mechanisms for free radical generation during high-intensity exercise include increases in the activities of cyclooxygenase and nitric oxide synthase (NOS). Increased Ca^{++} concentrations may activate the enzyme phospholipase A2, which releases arachidonic acid from phospholipids. Cyclooxygenase reacts with arachidonic acid to generate the hydroxyl radical.[42] Accelerated phospholipid degradation and associated membrane dysfunction is irreversible in ischemic liver cell injury. Hypoxic conditions have also been shown to increase NOS activity leading to the formation of nitric oxide radicals. These radicals may exert a weak prooxidant effect by themselves or combine with superoxide to form a more potent oxidant, peroxynitrate.

6. Exercise as a Prevention Against Oxidative Stress

There has not always been broad base support for health related benefits associated with exercise. In the early 1900s, animal studies by Slonaker[43] and Benedict and Sherman,[44] reported lower life expectancy in rats that were forced to exercise in drums. At the time, the Rate of Living Theory of Aging provided a mechanism for these results. Exercise increases metabolic rate and high metabolic rate is proportional to a shortened life span. What these studies did not distinguish was high versus moderate intensity exercise. High intensity exercise that is overly strenuous can be harmful to health and mortality.[45] Large scale research studies concur that when adults become moderately active they experience a 44% reduction in mortality.[3] Although adults can benefit further with vigorous exercise,[1, 2] the gains are not linear.

A paradox exists in which the same regular exercise that exposes skeletal muscle to repeated exposure to oxidative stress is also responsible for upregulating antioxidant and protective systems.[46] If aging is associated with the accumulation of free radicals and by-products of free radical reactions, then the addition of regular exercise over a life span may act as a "wild card" benefiting some but not others. Increased capillary density, increased HDL/LDL ratio, less oxidized LDL, enhanced glucose tolerance, and decreased risk of obesity are associated with regular exercise. Enhanced glucose tolerance would mean that less insulin is needed to transport glucose from the blood into the cell for energy transfer. This could spare glucose from becoming oxidized. Decreased risk of obesity leads to decreased levels of fat, which would otherwise provide primary substrate for free radical attack. Related to the lower rate of obesity often associated with those who exercise is lower body weight. Lower body weight appears to be the nexus for the mechanism by which caloric restriction imparts its health and longevity benefits as well as regular exercise. Caloric restriction can exert significant health benefits in the absence of exercise.[47] Nevertheless the same study suggests that both caloric restriction and regular exercise may act to mobilize certain cell protective agents and reactions. Previously, Kim, McCarter, and Yu[48] had described increased catalase, glutathione peroxidases, and ascorbic acid levels compared to *ad libitum* fed animals. Mitochondrial membrane fluidity deteriorated less with age in the caloric restricted and/or exercised rats compared to *ad libitum* fed animals, providing indirect evidence of lower microsomal reactive oxygen species production. A previous study by Goodrick,[49] reported health and life expectancy increased when *ad libitum* fed rats voluntarily exercised, but not when rats that were fed every other day, were exercised. A different caloric restriction protocol designed by Holloszy and Schectman,[50] in which rats were fed 40% less food, showed health and life expectancy benefits in both caloric restricted and exercised animals, with the combined effects of caloric restriction and exercise having the best impact. Even milder caloric restriction of 10% has been shown to benefit health and life expectancy of animals.[51] No changes in maximum life span were reported by any of these studies, so it has been acknowledged that while caloric restriction and exercise may impart health benefits and increase life expectancy (thus making the survival curve more rectangular), no aging or anti-aging effect has been identified.

7. Different Exercise Protocols

There is a general agreement that a single bout of exercise can change the status of some antioxidants in skeletal and heart muscle.[4, 52–57] There is a good rationale for why high-intensity exercise may induce oxidative stress. Singh[58] concisely described several mechanisms of free radical generation during high-intensity exercise: mitochondrial electron transport chain, anoxia-reoxygenation, mechanical damage to muscles, increased inhalation of environmental pollutants containing free radicals, and oxidation of catecholamines. High-intensity exercise increases prooxidant activity as well as cause damage to muscle sarcoplasmic and endoplasimic reticulum.[59] In response to increased prooxidants, high intensity exercise appears to be superior to low intensity exercise in the upregulation of skeletal muscle SOD activity.[60] Exercise-training induced changes in SOD were influenced by both time and exercise intensity, with 60–90 min per day of medium and high-intensity exercise showing a greater increase in SOD activity compared to 30 min of either low, medium, or high. A similar effect was reported for glutathione peroxidase activity, with increasing activity proportional to increased exercise training duration from 30 min per day to 90 min per day.

Most exercise protocols used in studies on exercise-induced oxidative stress included aerobic exercise. Few have studied non-aerobic exercise such as sprinting or isometric contraction. Alessio and Cutler[61] reported that 1 min of high intensity sprinting resulted in a 167% increase in thiobarbituric acid reactive substances in rat fast twitch-oxidative and 157% increase in fast twitch glycolytic muscle fibers. This was more than double the levels found in the same muscle fibers following 20 min of moderate intensity running. Sprint training resulted in increased rat muscle glutathione and upregulation of glutathione redux cycle enzyme activities in fast twitch-glycolytic and slow-twitch glycolytic muscle fibers and heart.[62]

8. Conclusion

Both exercise and aging share oxidative stress as a common underlying mechanism. The balance of prooxidants and antioxidants in exercise is disrupted when the intensity becomes too high, and this could result in oxidative stress. The oxidative stress balance in healthy aging processes is probably too subtle to measure. Regardless of age, in most healthy organisms, the antioxidant response to exercise-induced oxidative stress is proportional, resulting in a successful defense with beneficial health side effects of exercise that have positive effects on morbidity and mortality.

Acknowledgments

Thanks to Angela McClure for final preparation of manuscript.

References

1. Paffenbarger, R. S., Jr., Hyde, R. T., Wing, A. L. and Hsieh, C.-C. (1986). Physical activity, all-cause mortality, and longevity of college alumni. *New England J. Med.* **314**: 605–613.
2. Paffenbarger, R. S., Jr., Hyde, R. T., Wing, A. L., Lee, I.-M., Dung, D. L. and Kampert, J. B. (1993). The association of changes in physical activity level and other lifestyle characteristics with mortality among men. *New England J. Med.* **328**: 538–545.
3. Blair, S. N., Kohl, H. W. III, Barlow, C. E., Paffenbarger, R. S., Jr., Gibbons, L. W. and Macera, C. A. (1995). Changes in physical fitness and all-cause mortality: a prospective study of healthy and unhealthy men. *J. Am. Med. Assoc.* **273**: 1093–1098.

4. Alessio, H. M. and Goldfarb, A. H. (1988). Lipid peroxidation and scavenger enzymes during exercise. Adaptive response to training. *J. Appl. Physiol.* **64**: 1333–1336.

5. Alessio, H. M., Goldfarb, A. H. and Cutler, R. G. (1988). MDA content increases in fast- and slow-twitch skeletal muscle with intensity of exercise in a rat. *Am. J. Physiol.* **255**: C874–C877.

6. Ji, L. L. (1993). Antioxidant enzyme response to exercise and aging. *Med. Sci. Sport. Exerc.* **25**: 225–231.

7. Sen, C. K., Marin, E., Kretzschmar, M. and Hanninen, O. (1992). Skeletal muscle and liver glutathione homeostasis in response to training, exercise, and immobilization. *J. Appl. Physiol.* **73**: 1265–1272.

8. Radak, Z., Sasvari, M., Nyakas, C., Pucsok, J., Nakamoto, H. and Goto, S. (2000). Exercise preconditioning against hydrogen peroxie-induced oxidative damage in proteins of rat myocardium. *Arch. Biochem. Biophys.* **376**: 248–251.

9. Halliwell, B. (1994). Free radicals, antioxidants, and human disease: curiosity, cause, or consequence? *Lancet* **344**: 721–724.

10. Radak, Z., Sasvari, M., Nyakas, C., Pucsok, J., Nakamoto, H. and Goto, S. (2000). Changes in urine 8-hydroxydeoxyguanosine levels of super marathon runners during a four day race period. *Life Sci.* **66**: 1763–1767.

11. Vasankari, T. J., Kujala, U. M., Vasankari, T. M., Vuorimaa, T. and Ahotupa, M. (1996). Effects of acute prolonged exercise on serum and LDL oxidation and antioxidant defences. *Free Radic. Biol. Med.* **22**: 509–513.

12. Nieman, D. C. (1998). Exercise and resistance to infection *Can. J. Physiol. Pharmacol.* **76**: 573–580.

13. US Department of Health and Human Services. (1996). Physical Activity and Health: A Report of the Surgeon General. Atlanta, GA: US Department of Health and Human Services, Centers for Disease Control and Prevention, National Center for Chronic Disease Prevention and Health Promotion.

14. Sen, C. K. and Packer, L. (1996). Antioxidant and redox regulation of gene transcription. *FASEB J.* **10**: 1–12.

15. Cutler, R. G. (1984). Evolutionary biology of aging and longevity in mammalian species. *In* "Aging and Cell Function" (J. E. Johnson, Ed.), pp. 1–148, Plenum Press, New York.

16. Sohal, R. S., Svensson, I. and Brunk, U. T. (1990). Hydrogen peroxide production in liver mitochondria in different species. *Mech. Ageing Dev.* **53**: 209–215.

17. Sohal, R. S., Arnold, L. A. and Sohal, B. H. (1990). Age-related changes in antioxidant enzymes and prooxidant generation in tissues of the rat with special reference to parameters in two insect species. *Free Radic. Biol. Med.* **9**: 495–499.

18. Sohal, R. S. and Brunk, U. T. (1992). Mitochondrial production of prooxidants and cellular senesence. *Mutat. Res.* **275**: 295–300.

19. Sacher, G. A. (1968). Molecular versus systemic theories on the genesis of ageing. *Exp. Gerontol.* **3**: 265–270.
20. Sacher, G. A. (1970). Allometric and factorial analysis of brain structure in insectivores and primates. *In* "The Primate Brain" (C. R. Noback, and W. Montagna, Eds.), pp. 245–287, Appleton-Century Crofts, New York.
21. Cutler, R. G. (1983). Superoxide dismutase, longevity, and specific metabolic rate. *Gerontology* **29**: 113–119.
22. Furst, A. (1987). Hormetic effects in pharmacology. *Health Phys.* **52**: 527–530. Luckey, T.D. (1968). Insecticide hormoligosis. *J. Econ. Entomol.* **62**: 7–12.
23. Ordy, J. M., Samorajski, T., Zeman, W. and Curtis, H. J. (1967). Interaction effects of environmental stress and deuteron irradiation of the brain on mortality and longevity in C57BL/10 mice. *Proc. Soc. Exp. Biol. Med.* **126**: 184–190.
24. Reinke, U., Stutz, E. and Hunstein, W. (1970). Life span and tumor incidence in rats subjected to severe stress before whole body X-irradiation. *Proc. First Eur. Symp. Late Effects Rad.*, pp. 151–172, Äasaccia Nuclear Center.
25. Sacher, G. A. (1977). Life table modification and life prolongation. *In* "Handbook of the Biology of Aging" (C. Finch, and L. Hayflick, Eds.), pp. 582–638, Van Nostrand Reinhold, New York.
26. Beckman, K. B. and Ames, B. N. (1998). The free radical theory of aging matures. *Physiol. Rev.* **78**: 547–581.
27. Alessio, H. M. and Blasi, E. R. (1997). Physical activity as a natural antioxidant booster and its effect on a healthy life span. *Res. Quart. Exerc. Sport.* **68**: 292–302.
28. Sen, C. K. (1995). Oxidants and antioxidants in exercise. *J. Appl. Physiol.* **79**: 675–686.
29. Alessio, H. M. (1993). Exercise-induced oxidative stress. *Med. Sci. Sport. Exerc.* **25**: 218–224.
30. McCarter, R. J. M. (2000). Caloric restriction, exercise, and aging. *In* "Handbook of Oxidants and Antioxidants in Exercise" (C. K. Sen, L. Packer, and O. Hanninen, Eds.), pp. 797–829, Elsevier Science, B.V., Amsterdam.
31. Neafsey, P. J. (1990). Longevity hormesis: a review. *Mech. Ageing Dev.* **51**: 1–31.
32. Alessio, H. M., Hagerman, A. E., Fulkerson, B. K. Ambrose, J., Rice, R. E. and Wiley, R. L. (2000). Generation of reactive oxygen species after exhaustive aerobic and isometric exercise. *Med. Sci. Sports Exerc.* **32**: 1576–1581.
33. Kehrer, J. P. and Lund, L. G. (1994). Cellular reducing equivalents and oxidative stress. *Free Radic. Biol. Med.* **17**: 65–75.
34. Niess, A. M., Dickhuth, H. H., Northoff, H. and Fehrenback, E. (2000). Free radicals and oxidative stress in exercise-immuniological aspects. *Exerc. Immuniol. Rev.* **5**: 22–56.

35. Saxton, J. M., Donnelly, A. E. and Roper, H. P. (1994). Indices of free radical-mediated damage following maximum voluntary eccentric and concentric muscular work. *Eur. J. Appl. Physiol.* **68**: 189–193.

36. Rahman, I. and MacNee, W. (2000). Regulation of redox glutathione levels and gene transcription in lung inflammation: therapeutic approaches. *Free Radic. Biol. Med.* **28**: 1405–1420.

37. Hellsten, Y. (1994). The role of xanthine oxidase in exercise. *In* "Exercise and Oxygen Toxicity" (C. K. Sen, L. Packer, and O. Hanninen, Eds.), pp. 211–229, Elsevier Science B.V., Amsterdam.

38. Kuppusamy, P. and Zweier, J. L. (1989). Characterization of free radical generation by xanthine oxidase. Evidence for hydroxyl radical generation. *J. Biol. Chem.* **264**: 9880–9884.

39. Hellsten, Y., Frandsen, U., Orthenblad, N., Sjodin, B. and Richter, E.A. (1997). Xanthine oxidase in human skeletal muscle following eccentric exercise: a role in inflammation. *J. Physiol.* **498**(1): 239–245.

40. Radak, Z., Asano, K. Inoue, M., Kizaki, T., Oh-Ishi, S., Suzuki, K., Taniguchi, N. and Ohno, H. (1995). Superoxide dismutase derivative reduces oxidative damage in skeletal muscle of rats during exhaustive exercise. *J. Appl. Physiol.* **79**: 129–135.

41. Chien, K. R., Abrams, J., Serroni, A., Martin, J. T. and Farber, J. L. (1978). Accelerated phospholipid degradation and associated membrane dysfunction in irreversible, ischemic liver cell injury. *J. Biol. Chem.* **253**: 4809–4817.

42. Slonaker, J. R. (1912). *J. Animal Behav.* **2**: 20–42.

43. Benedict, F. G. and Sherman, H. C. (1937). *J. Nutri.* **14**: 179–198.

44. Holloszy, J. O. and Khort, W. (1995). *In* "Handbook of the Physiology of Aging" (E. J. Masaro, Ed.), pp. 633–666, Oxford University Press, New York.

45. Leeuwenburgh, C., Fiebig, R., Chandwaney, R. and Ji, L. L. (1994). Aging and exercise training in skeletal muscle: responses of glutathione and antioxidant enzyme systems. *J. Appl. Physiol.* **267**: R439–R445.

46. McCarter, R., Shimokawa, I., Ikeno, Y., Higami, Y., Hubbard, G. and Yu, B. P. (1997). Physical activity as a factor in the action of dietary restriction on aging: effects in Fischer 344 rats. *Ageing* **9**: 73–79.

47. Kim, J. D., McCarter, R. J. M. and Yu, B. P. (1996). Influence of age, exercise, and dietary restriction on oxidative stress in rats. *Ageing* **8**: 123–129.

48. Goodrick, C. L. (1980). Effects of long-term voluntary wheel exercise on male and female Wistar rats. I. Longevity, body weight, and metabolic rate. *Gerontology* **26**: 22–23.

49. Holloszy, J. O. and Schectman, K. B. (1991). Interaction between exercise and food restriction: effects on longevity of male rats. *J. Appl. Physiol.* **70**: 1529–1535.

50. Holloszy, J. O., Smith, E. K., Vining, M. and Adams, S. (1985). Effect of voluntary exercise on longevity in rats. *J. Appl. Physiol.* **59**: 826–831.

51. Viguie, C. A., Balz, F., Shigenaga, M. K., Ames, B. N., Packer, L. and Brooks, G. A. (1993). Antioxidant status and indexes of oxidative stress during consecutive days of exercise. *J. Appl. Physiol.* **75**: 566–572.

52. Criswell, D., Pwers, S. Dodd, S., Lawler, J., Edwards, W., Renshler, K. and Grinton, S. (1993). High intensity training induced changes in skeletal muscle antioixdant enzyme activity. *Med. Sci. Sport. Exerc.* **25**: 1135–1140.

53. Robertson, J. D., Maughan, R. J., Duthie, G. G. and Morrice, P. C. (1991). Increased blood antioxidant systems of runners in response to training load. *Clin. Sci.* **80**: 611–618.

54. Jenkins, R. R., Friedland, R. and Howald, H. (1984). The relationship of oxygen uptake to superoxide dismutase and catalase activity in human skeletal muscle. *Int. J. Sport. Med.* **5**: 11–14.

55. Ji, L. L., Fu, R. and Mitchell, E. W. (1992). Glutathione and antioxidant enzymes in skeletal muscle: effects of fiber type and exercise intensity. *J. Appl. Physiol.* **73**: 1854–1859.

56. Kanter, M. M., Nolte, L. A. and Holloszy, J. O. (1993). Effects of an antioxidant vitamin mixture on lipid peroxidation at rest and post exercise. *J. Appl. Physiol.* **74**: 965–969.

57. Sen, C. K., Rankinen, T., Vaisanen, S. and Rauramaa, R. (1994). Oxidative stress after human exercise: effect of N-acetylcysteine sypplementation. *J. Appl. Physiol.* **76**: 2570–2577.

58. Singh, V. N. (1992). A current perspective on nutrition and exercise. *J. Nutri.* **122**: 760–765.

59. Venditti, P. and di Meo, S. (1996). Antioxidants, tissue damage, and endurance in trained and untrained young male rats. *Arch. Biochem. Biophys.* **331**: 63–68.

60. Powers, S., Criswell, D., Lawler, J., Ji, L., Martin, D., Herb, R. and Dudley, G. (1994). *Am. J. Physiol.* **266**: R375–R380.

61. Alessio, H. M. and Cutler, R. G. (1990). Production and removal of lipid peroxidation by-products after exercise. *In* "Exercise Physiology: Current Selected Research, 4" (C. O. Dotson, and J. H. Humphrey, Eds.), pp. 61–70, AMC Press, New York.

62. Atalay, M., Seene, T., Hanninen, O. and Sen C. K. (1996). Skeletal muscle and heart antioxidant defenses in response to sprint training. *Acta Physiol. Scand.* **158**: 129–134.

SECTION 10

Evolutionary Comparative Biology of Aging and Longevity

Chapter 63

Evolutionary Basis of Human Aging

Thomas B.L. Kirkwood

Thomas B. L. Kirkwood • Department of Gerontology, University of Newcastle upon Tyne, Institute for Ageing and Health, Newcastle General Hospital, Newcastle upon Tyne, NE4 6BE, UK
Tel: +44 191 256 3319, E-mail: tom.kirkwood@ncl.ac.uk

1. The Evolutionary Theory of Senescence

Human aging shares many features with aging in other species. The starting point for considering the evolutionary basis of human aging is therefore the general evolutionary theory of senescence. This theory requires a definition of aging that can be generalized across species in which the physiology of aging may show marked differences. The relevant definition of aging is as a "progressive loss of function accompanied by decreasing fertility and increasing mortality with advancing age". Such a trait, which impairs survival and fertility, is obviously bad for the individual, raising deep questions about why and how it has evolved.[1]

An early explanation for evolution of aging was the idea that senescence is programmed. Such programming could, it was suggested, be advantageous if it limited population size or accelerated the turnover of generations and thereby aided adaptation to changing environments. However, the flaw in this argument is that for most species there is little evidence that senescence makes an important contribution to mortality in the wild. Mortality in natural populations is mostly due to extrinsic hazards, such as infection, predation, starvation, or cold. As a consequence, wild animals rarely live long enough to grow old and it is implausible to suggest that natural selection exerts a direct influence to bring about an active process of senescence.

In fact, it is the relative absence of senescence in wild populations that points to the important principle which underlies the current evolutionary understanding of aging. Due to extrinsic mortality, there is a progressive weakening in the force of selection with increasing age.[2] By an age when wild survivorship has declined to very low levels, the force of selection is too weak to oppose the accumulation of germ-line mutations with late-acting deleterious effects,[3] resulting in a "selection shadow" which allows late-acting deleterious mutations to accumulate unchecked. To this essentially neutral process we can add the contribution from pleiotropic genes[4] which have good effects early in life but bad effects at later ages. Since early effects influence a much larger fraction of the population than are harmed by late effects, the genes in question can be favoured by selection even if the bad effects result in senescence and death.

A different approach to explaining the evolution of aging comes from considering the optimum investment in cellular maintenance and repair, particularly in the light of the "division of labour" between germ-line and soma first recognized by Weismann.[5] While the germ-line lineage requires a high level of ongoing maintenance to secure faithful transfer of genetic information from generation to generation, maintenance and repair of somatic cells is only required to keep the organism in sound physiological condition through the period of time it is likely to survive in the wild. This is the "disposable soma" theory,[6, 7] which suggests that aging results from an evolutionary compromise between investments in survival and reproduction. Nearly all of the mechanisms required to combat intrinsic deterioration (DNA repair, antioxidant systems, etc.) require metabolic resources. These resources are often scarce and must be used with great efficiency. Even when resources are plentiful, there is no advantage to be gaining from investing in better maintenance than is necessary and fitness will generally be enhanced by channelling any surplus into extra reproduction, rather than repair.

The evolutionary theory of senescence leads to a number of testable predictions. Firstly, aging is not programmed in the sense that there exists any active mechanism to cause death. Instead, life span is influenced by "longevity assurance" genes that control the levels of somatic maintenance and repair. Secondly, longevity of different species is determined by optimizing the level of somatic maintenance according to the degree of extrinsic hazard in the species' ecological niche. An adaptation that reduces hazard should result in selection pressure to increase maintenance, which would explain the greater longevity of animals that have evolved flight (birds, bats) compared with those that must remain on the ground. Thirdly, there are many maintenance mechanisms, so aging and longevity are likely to be under highly polygenic control. The levels at which individual genes are set will influence the average rate at which damage accumulates or is repaired. Fourthly, at a fine scale the mechanisms of aging are intrinsically stochastic. Therefore chance, as well as genetic and environmental factors, may play a significant role.[8] The role of chance in aging is clearly indicated by the variance in life spans observed for genetically identical populations (e.g. inbred mice, nematodes) maintained in homogeneous environments.

2. Comparative Studies of Aging and Longevity

The prediction that safe environments (those with low extrinsic mortality) favor extended longevity whereas hazardous environments favor shorter lives has been tested in various ways. Humans have the longest life span of any mammalian species and therefore these kinds of comparative analysis are particularly valuable in understanding the reasons for the evolution of human longevity.

Adaptations that reduce extrinsic mortality (wings, protective shells, large brain) are generally linked with increased longevity (bats, birds, turtles, humans). For example, comparison of a mainland population of opossums subject to significant predation by mammals with an island population not subject to such predation, found the predicted slower aging in the island population.[9] Among social insects, those with the most protected nests exhibit the longest life spans in their queens.[10]

At the molecular and cellular levels, numerous studies support the idea that the effort devoted to cellular maintenance and repair processes will vary with longevity. The long-lived rodent species *Peromyscus leucopus* exhibits lower generation of reactive oxygen species (ROS) than the shorter-lived species *Mus musculus*[11] and this general pattern has been confirmed across a variety of other species.[12–14] DNA repair capacity has been shown to correlate with mammalian life span in numerous comparative studies (see Ref. 15), as has the level of poly(ADP-ribose) polymerase,[16] an enzyme that plays a key role in the maintenance of genomic integrity. The quality of maintenance and repair mechanisms may be revealed by the capacity to cope with external stress. Comparisons of the capacity of cultured cells to withstand a variety of imposed stressors have shown that cells taken from long-lived species have superior stress resistance to that of cells from shorter lived-species.[17, 18] In all such comparative studies, care should be taken to minimize, as far as possible, the effects of potential confounding variables. An example is the study by Kapahi *et al.*[18] where cells were derived using a carefully standardized protocol from similar biopsy sites in animals of similar ages, and where the data were analyzed using a statistical procedure that took account of the phylogenetic relationships between the species in question.[19]

A consistent finding in mammalian comparative studies has been the superior levels of cell maintenance and repair that have been detected within human cells.

3. Evolutionary Basis of Life Extension Through Calorie Restriction

It has long been known that life span in laboratory rodents is extended, typically by 20–30%, by reducing food intake,[20, 21] and there is considerable interest in the possibility that a similar mechanism could act in primates or even humans. In order to assess this possibility, it is helpful to consider what might be the evolutionary basis of the calorie-restriction effect.

Long-term calorie restriction in rodents results in a small, lean animal, with impaired fertility, but which is otherwise healthy and active. Almost without exception, somatic maintenance functions are up-regulated.[22–24] Activity levels are affected differently in rodents and primates by short-term fasting,[25] although recent data indicate that primates, like rodents, may up-regulate activity levels when subjected to long-term calorie restriction.[26]

At first sight, it seems paradoxical that an organism with less energy available can up-regulate physiological processes, which must have an associated metabolic cost. This requires that metabolic savings be made elsewhere. The most obvious saving is in reproductive effort. Not only is reproduction costly in direct physiological terms but behaviours associated with reproduction can also be expensive. Rodents typically invest a large fraction of their energy budget in reproduction. Calorie restriction results in animals that are mostly infertile, although there is some variation depending on the species, the sex, the degree of restriction and the age at which restriction is first applied.[20]

Harrison and Archer[27] suggested that the primary role of calorie restriction is to postpone reproductive senescence. If a famine lasts longer than the normal reproductive life span of the animal, any female that delays reproductive senescence and is able to breed after the famine has passed will experience a selective advantage. Harrison and Archer[27] predicted that calorie restriction should have greater effect on species with shorter reproductive life spans, and suggested a comparison between the house mouse, *Mus musculus*, and the longer-lived white-footed mouse, *Peromyscus leocupus*. They also predicted that calorie restriction should have little effect on long-lived species such as humans. A more explicit hypothesis was proposed by Holliday,[28] based on the disposable soma theory of aging. Under conditions of plentiful food supply, a constant amount of energy is allocated to maintenance. As food supply diminishes, reproduction is reduced and eventually curtailed. Holliday suggested that at low food levels where, in principle, a small amount of energy could be allocated to reproduction, it may in fact be better temporarily to increase the investment in somatic maintenance. The potential benefit is that the animal gains an increased chance of survival with a reduced intrinsic rate of senescence, thereby permitting reproductive value to be preserved for when the famine is over.

This idea was recently tested with an evolutionary model to examine whether the physiological responses to calorie restriction observed in the laboratory might reflect an adaptive resource allocation strategy that has evolved to maximize fitness under conditions of intermittent food stress in the wild.[29] The model was developed for *M. musculus*, for which extensive physiological and life history data have been recorded both in the wild and in the laboratory,[30] and was based on a resource allocation rule governing the partition of energy between maintenance functions and reproduction. In the model, resources allocated to reproduction were used to produce progeny, whereas resources allocated to maintenance were used to conserve state, i.e. to slow aging. It was found that as long as two assumptions were made, the model clearly predicted an increase in the allocation of metabolic resources to somatic maintenance during periods of famine.[29] These assumptions were (i) the presence of a "reproductive overhead", i.e. a minimum investment in reproduction that must be made in order to initiate the fertile state, and (ii) the presence of an adverse effect of famine directly on juvenile survival. In the absence of either a reproductive overhead or an effect on juvenile survival, calorie restriction had little effect on the optimal allocation of energy to maintenance. Reproduction declined as food availability diminished, but it continued as long as there was sufficient energy to support any reproduction at all.

The evolutionary analysis suggests that there may be adaptive value in having a mechanism to effect coordinate control of multiple stress response systems so as to be able to adapt metabolically to a varying environment. A somewhat analogous example of life history plasticity to that seen in rodent calorie restriction may be found in the nematode, *Caenorhabditis elegans*. Under conditions of high larval density and low food availability, *C. elegans* larvae switch into the alternative, non-feeding, non-reproducing larval form called the dauer that can survive for at least twice as long as the normal form. Dauers are resistant to a variety of environmental stresses including ROS, elevated temperature, and ionizing radiation.[31] Significantly, one of the genes involved in the switch to dauer development, daf-2, shows homology to the human insulin receptor gene,[32] suggesting that metabolic switching may have a broad role to play in regulating aging processes.

4. Evolution of Human Reproductive Senescence

A puzzling feature of the human life history is the female menopause, where fertility comes to a relatively abrupt halt at around the age of 45–50, when the impact of senescence on most other functions is still quite small. The post-reproductive phase in women is profoundly different from the post-reproductive phase in most other species, and should not be confused with the programmed cessation of fertility that occurs in semelparous species, in which individuals reproduce only once. In a semelparous species, the force of natural selection approximates a step function, being uniformly high until reproduction begins, and declining very abruptly as reproduction is completed, because the chance of surviving to breed again is effectively zero. This explains the sudden collapse of any pressure to invest in somatic maintenance and repair. In contrast, the evolution of menopause probably reflects the action of positive selection for an extended post-reproductive survival period.

Assisted fertilization shows that post-menopausal women can successfully bear children without serious complications and, although the proximate cause of menopause appears to be oocyte depletion (linked also with neuroendocrine changes), this begs the question why natural selection has not produced a store of oocytes which lasts for longer. One possibility is that during most of humanity's evolutionary history, women rarely survived beyond 45–50 years, so selection simply produced about as many oocytes as would be required. Evidence from hunter-gatherer communities suggests, however, that even though *average* life expectancy is short, women who avoid the hazards of early life and reach childbearing age have a reasonable chance of surviving to the age of menopause and beyond.[33] Early female reproductive senescence has been reported in other species (e.g. chimpanzees, macaques, toothed whales) but is generally less clear-cut, suggesting that if the menopause has an evolutionary basis, this may be found in the special circumstances of the human life history. In particular, menopause could be linked with the evolution of human longevity, notably, through the effects of increased brain size and sociality.[33–40] Increased neonatal brain size coupled with the constraint on the birth canal imposed by the mechanics of a bipedal gait has made giving birth unusually difficult for human females, and the risks of child-bearing would increase even more steeply with age if fertility were to persist during the later period of the life span. Human infants remain highly dependent for extended periods and, in the ancestral environment, their survival will have been unlikely if their mother died in childbirth. This suggests that there may have been a fitness advantage in limiting reproduction to ages when it was comparatively safe, and thus increasing the likelihood of the mother surviving to raise her existing offspring to a state of successful independence. Additionally, post-menopausal women may contribute to the successful rearing of their grandchildren, by providing assistance to their own adult offspring and thereby increasing their inclusive fitness, i.e. their overall genetic contribution to future generations.

5. Evolution of Germ-Line Immortality

Concomitant with the evolution of longevity assurance mechanisms in somatic cells, it is highly pertinent to ask about the evolution of mechanisms to preserve germ-line immortality, particularly when germ cells must preserve their viability across decades as in the case of human oocytes. The germ-line must in a fundamental sense be immortal, since damage cannot be permitted to accumulate across generations without causing extinction. The remarkable observation is that babies are born young, even though there is clear evidence that the germ cell population does undergo significant aging. In the case of the human ovary the rate of follicular loss accelerates from around age 35, and male fertility begins to decline from around age 30. There is also an increase in the frequency of chromosomal abnormalities in newborn children as a function of maternal and, to a lesser extent, paternal age.[41, 42]

The explanation for the presevation of germ-line immortality, in spite of the fact that germ cells are exposed to the same general kinds of oxidative stress as somatic cells must lie in one or a combination of the following factors. Firstly, germ cells may be endowed with special maintenance and repair systems,[6] the enzyme telomerase being a good example. Secondly, immature germ cells may be maintained in a state of relative metabolic quiescence. Thirdly, there may be intense selection at the cell or embryo level during gametogenesis, conception, and pregnancy which serves to screen out the great majority of faults. It may be relevant that during each human menstrual cycle about 20 ovarian follicles are triggered to start the process of maturation, although usually only one completes its development and is ovulated. As yet, we know rather little about the reasons for this apparently stringent selection, which may provide opportunity to screen out those follicles that have accumulated oxidative (and perhaps other) damage.

A mechanism of probable significance in the evolution of the female germ-line is the stringent bottleneck in the size of the cellular mitochondrial population in early embryogenesis. A healthy complement of mitochondria is essential for subsequent viability of the offspring, and mitochondrial DNA (mtDNA) mutations tend to accumulate with age.[43] If the mitochondrial population in the oocyte contains a fraction of organelles bearing mtDNA mutations, a bottleneck coupled with an effective quality screen might select embryos that carry only intact mitochondria[44] and thereby play an important role in the preservation of germline immortality.

6. Human Aging and Oxidative Stress

Based on our understanding of the evolution of human aging it is possible to identify the important links between oxidative stress, antioxidant protection and human longevity. It would appear that human longevity has been driven by the continual reduction of the level of extrinsic mortality that has been made possible by the evolution of the human brain. This in turn has produced selection for ever higher levels of cellular protection against oxidative stress, as revealed strikingly in comparative studies. This has been an ongoing process and there seems no reason to suppose that the current levels of our antioxidant defenses represent any kind of maximum. Thus, the clear implication from the comparative studies is that the human aging process may prove malleable.

The present human life history, as represented by our current levels of intrinsic vulnerability to age-related morbidity and mortality and by our current schedules of fertility (including menopause), reflect the environmental conditions experienced by previous generations. It is only in the last few generations that significant change has occurred in these conditions, as represented by the dramatic demographic transitions that has occurred in less than 200 years. Thus, we can anticipate that the evolutionary forces that shaped the human aging process may have altered profoundly. It is of interest to understand these evolutionary forces not only for what they may tell us about the possible future evolution of our aging processes, but also for the insights they can provide into the present-day genetic architecture of how we have evolved to cope with the ever-present threat of oxidative stress.

References

1. Kirkwood, T. B. L. and Austad, S. N. (2000). Why do we age? *Nature* **408**: 233–238.
2. Charlesworth, B. (1994). *Evolution in Age-Structured Populations*, Cambridge University Press, Cambridge.
3. Medawar, P. B. (1952). *An Unsolved Problem of Biology*, H. K. Lewis, London.
4. Williams, G. C. (1957). Pleiotropy, natural selection and the evolution of senescence. *Evolution* **11**: 398–411.

5. Kirkwood, T. B. L. and Cremer, T. (1982). Cytogerontology since 1881: a reappraisal of August Weismann and a review of modern progress. *Human Genet.* **60**: 101–121.

6. Kirkwood, T. B. L. (1977). Evolution of ageing. *Nature* **270**: 301–304.

7. Kirkwood, T. B. L. (1986). Human senescence. *Bioessays* **18**: 1009–1016.

8. Finch, C. E. and Kirkwood, T. B. L. (2000). *Chance, Development and Aging,* Oxford University Press, New York.

9. Austad, S. N. (1993). Retarded senescence in an insular population of opossums. *J. Zool.* **229**: 695–708.

10. Keller, L. and Genoud, M. (1997). Extraordinary lifespans in ants: a test of evolutionary theories of aging. *Nature* **389**: 958–960.

11. Sohal, R. S., Ku, H.-H. and Agarwal, S. (1993). Biochemical correlates of longevity in two closely-related rodent species. *Biochem. Biophys. Res. Commun.* **196**: 7–11.

12. Ku, H.-H., Brunk, U. T. and Sohal, R. S. (1993). Relationship between mitochondrial superoxide and hydrogen peroxide production and longevity of mammalian-species. *Free Radic. Biol. Med.* **15**: 621–627.

13. Herrero, A. and Barja, G. (1999). 8-oxo-deoxyguanosine levels in heart and brain mitochondrial and nuclear DNA of two mammals and three birds in relation to their different rates of aging. *Aging Clin. Exp. Res.* **11**: 294–300.

14. Barja, G. and Herrero, A. (2000). Oxidative damage to mitochondrial DNA is inversely related to maximum life span in the heart and brain of mammals. *FASEB J.* **14**: 312–318.

15. Kirkwood, T. B. L. (1989). DNA, mutations and aging. *Mutat. Res.* **219**: 1–7.

16. Grube, K. and Bürkle, A. (1992). Poly(ADP-ribose) polymerase activity in mononuclear leukocytes of 13 mammalian species correlates with species-specific life span. *Proc. Natl. Acad. Sci. USA* **89**: 11 759–11 763.

17. Ogburn, C. E., Austad, S. N., Holmes, D. J., Kiklevich, J. V., Gollahon, K., Rabinovitch, P. S. and Martin, G. M. (1998). Cultured renal epithelial cells from birds and mice: enhanced resistance of avian cells to oxidative stress and DNA damage. *J. Gerontol.* **53**: B287–B292.

18. Kapahi, P., Boulton, M. E. and Kirkwood, T. B. L. (1999). Positive correlation between mammalian life span and cellular resistance to stress. *Free Radic. Biol. Med.* **26**: 495–500.

19. Pagel, M. (1998). Inferring evolutionary processes from phylogenies. *Zoologica Scripta.* **26**: 331–348.

20. Weindruch, R. H. and Walford, R. L. (1988). *The Retardation of Aging and Disease by Calorie-Restriction,* Charles C. Thomas, Springfield, Il.

21. Sprott, R. L. (1997). Diet and calorie restriction. *Exp. Gerontol.* **32**: 205–214.

22. Masoro, E. J. (1993). Dietary restriction and aging. *J. Am. Geriatr. Soc.* **41**: 994–999.

23. Yu, B. P. (1994). How diet influences the aging process of the rat. *Proc. Soc. Exp. Biol. Med.* **205**: 97–!105.

24. Merry, B. J. (1995). Effect of calorie-restriction on aging — an update. *Rev. Clin. Gerontol.* **5**: 247–258.

25. Masoro, E. J. and Austad, S. N. (1996). The evolution of the anti-aging action of calorie-restriction: a hypothesis. *J. Gerontol. A: Biol. Sci. Med. Sci.* **51**: B387–B391.

26. Weed, J. L., Lane, M. A., Roth, G. S., Speer, D. L. and Ingram, D. K. (1997). Activity measures in rhesus monkeys on long-term calorie restriction. *Physiol. Behav.* **62**: 97–103.

27. Harrison, D. E. and Archer, J. R. (1988). Natural selection for extended longevity from food restriction. *Growth Dev. Ageing* **52**: 65.

28. Holliday, R. (1989). Food, reproduction and longevity — is the extended lifespan of calorie-restricted animals an evolutionary adaptation? *Bioessays* **10**: 125–127.

29. Shanley, D. P. and Kirkwood, T. B. L. (2000). Calorie restriction and aging: a life history analysis. *Evolution* **54**: 740–750.

30. Berry, R. J. and Bronson, F. H. (1992). Life history and bioeconomy of the house mouse. *Bio. Rev. Camb. Phil. Soc.* **67**: 519–550.

31. Lithgow, G. J. (1996). Molecular genetics of *Caenorhabditis elegans* aging. *In* "Handbook of the Biology of Aging" (E. L. Schneider, and J. W. Rowe, Eds.), pp. 55–73, Academic Press, San Diego, California.

32. Guarente, L. and Kenyon, C. (2000). Genetic pathways that regulate ageing in model organisms. *Nature* **408**: 255–262.

33. Hill, K. and Hurtado, A. M. (1996). *Ache Life History: The Ecology and Demography of a Foraging People*, Aldine de Gruyter, New York.

34. Williams, G. C. (1957). Pleiotropy, natural selection and the evolution of senescence. *Evolution* **11**: 398–411.

35. Hamilton, W. D. (1966). The moulding of senescence by natural selection. *J. Theor. Biol.* **12**: 12–14.

36. Kirkwood, T. B. L. and Holliday, R. Ageing as a consequence of natural selection. *In* "The Biology of Human Ageing" (A. J. Collins, and A. H. Bittles, Eds.), pp. 1–16, Cambridge University Press, Cambridge.

37. Austad, S. N. (1994). Menopause: an evolutionary perspective. *Exp. Gerontol.* **29**: 255–263.

38. Peccei, J. S. (1995). The origin and evolution of menopause: the altriciality-lifespan hypothesis. *Ethol. Sociobiol.* **16**: 425–449.

39. Hawkes, K., O'Connell, J. F., Jones, N. G. B., Alvarez, H. and Charnov, E. L. (1998). Grandmothering, menopause, and the evolution of human life histories. *Proc. Natl. Acad. Sci. USA* **95**: 1336–1339.

40. Shanley, D. P. and Kirkwood, T. B. L. (2001). Evolution of the human menopause. *Bioessays*, **in press**.

41. Risch, N., Reich, E. W., Wishnick, M. M. and McCarthy, J. G. (1987). Spontaneous mutation and parental age in humans. *Am. J. Human Genet.* **41**: 218–248.

42. Gavrilov, L. A. and Gavrilova, N. S. (1997). Parental age at conception and offspring longevity. *Rev. Clin. Gerontol.* **7**: 5–12.
43. Wallace, D. C. (1999). Mitochondrial diseases in man and mouse. *Science* **283**: 1482–1488.
44. Cummins, J. M. (2000). Mitochondrial dysfunction and ovarian aging. *In* "Female Reproductive Aging" (E. R. Te Velde, P. L. Pearson, and F. J. Broekmans, Eds.), pp. 207–224, Parthenon Publishing, Carnforth, UK.

Chapter 64

Genetic Stability, Dysdifferentiation, and Longevity Determinant Genes

Richard G. Cutler

Richard G. Cutler • Kronos Longevity Research Institute, 4455 East Camelback Road, Ste B-135, Phoenix, Arizona 85259, USA
Tel: 602 977 0106, Email: richard.cutler@cox.net

1. Biology of Human Aging

1.1. The Human Survival Curve

Considerable information on human aging can be obtained from an evaluation of human survival data.[1-7] This survival data is used to calculate the percentage survival of a population from age zero (birth) to age at death. When percent survival versus chronological age is plotted as a graph it is called a "Percentage Survival Curve". The age when 50% of the original population has died is called "Life Expectancy" and the maximum age an individual reaches that was in this population is called "Maximum Lifespan Potential" or simply "Lifespan".

A number of interesting observations about human aging and longevity can be derived when percent survival curves from different periods of human history are compared. Let us take for example the *Homo sapiens* species or human populations that existed about 10 000 years ago to the present time period.[8-12] Comparison of the survival curves from these time periods leads to the following observations:

(1) For most of human history, life expectancy was only about 20 to 30 years. Some individuals may have lived to 80 to 100 or so years of age, but this would be very rare. Since we are dealing here with the same *Homo sapiens* as is living today, the innate aging rate is believed to be the same 10 000 years ago as it is today. That is, as will be discussed later, the age dependent decline of maximum reserve capacity of physiological functions in humans was the same 10 000 years ago as it is today. So this short life expectancy that existed 10 000 years ago is explained as being the result of the very intense environmental hazards that existed then and not, as is often thought, a result of a high rate of aging.

(2) Over the past 500 years (1500 AD to present) life expectancy has dramatically increased from about 20 to 30 years to about 70 to 80 years today in the developed countries of the world.[3, 4, 7, 9, 11] This is the result of successfully decreasing the environmental hazard intensity including reduction of death related to predators, physical environmental hazards (cold, floods, etc.), infectious disease and improved year round nutrition. The largest contributing factors were largely a result of better sanitation, hygiene and available non-infectious or pure drinking water. Simply put we owe a lot to the plumbers. The discovery of antibiotics, vaccines and other remarkable recent advances in modern medicine has played much less of a role in advancing human life expectancy and the general health of the nation as is commonly thought. This is because the increase in life expectancy occurred in a steady manner before such medicines were even available. Such modern advances are however likely to become increasingly more important in the future.[3, 4, 7]

(3) In spite the remarkable recent increase in life expectancy, maximum lifespan potential or lifespan has remained essentially unchanged and has remained at about 100 to 125 years throughout human history. These results are understood when it is realized that death in human populations has essentially two major causes. The first is related to the exogenous environmental hazards ranging from natural disasters accidents, presence of Sabra Toothed tigers to infectious disease. The second is related to endogenous internal environmental hazards (that is inside the body), which is the major causal component of aging. Aging accelerating processes do however exist in the external environment and consist of factors as background ionization radiation (minerals and cosmic sources of gamma and X-ray) and non-ionization irradiation (UV radiation). However their contribution to accelerated aging is relatively small. *As we will see later it is the "internal environmental hazards" created by the* **normal** *functions of the body itself that is the origin of aging processes and which will eventually result in the death of all individuals even if that individual successfully escapes all external environmental hazards of life.* It is this aging process that places a limit to maximum lifespan possible, which is presently estimated to be about 100 to 125 years for humans.[3, 4, 7]

(4) Human lifespan has remained unchanged for about 100 000 years, which is the current estimate of how long the *Homo sapiens* species has existed. The survival curve was essentially a steadily declining function where life expectancy was about 25 years.[3, 4, 7] The result of life expectancy increasing over the past several thousand years of human history while lifespan has remained fixed at about 100–125 years produced a rectangularization effect of the human survival curve. Over the past 100 years, the rate of human survival curve rectangularization has slowed down to almost no change over the past 25 to 50 years.

Thus, today there has been little increase in life expectancy in the developed countries throughout the world although there has been a small and steady increase in the percentage of the very old in the population (80 years and above) and a remarkable slow but steady increase in their lifespan.[9–11, 13–16] This interesting trend has been observed in a number of different human population studies. The basis for most individuals living to 80 to 100 is because they have aged uniformly having few innate weaknesses. However those individuals living from 100 to over 120 years are likely to represent a very special population of individuals that age at a slower rate. These later groups of individuals living from 100 to 122 years have been called "super centenarians" and are likely to represent a very special population of individuals where their innate aging rate may be less than the average person. If this is true, then, perhaps they represent the new generation of super *Homo sapiens*. It would be this population of super centenarians that should be very interesting to study in an effort to identify what special genes or genetic differences they might have that may account for their remarkable long, healthy and productive lifespan.

(5) From survival data the rate of death or the rate of mortality can be calculated as a function of age of the population. When the rate of mortality versus age is graphed it is seen that the rate of mortality is very high at birth and then steadily decreases reaching a minimum at about the age of sexual maturity or from 10 to 15 years of age. What is most interesting is that the age where minimum "rate of mortality" is reached is also the same age and independent of the absolute rate of death found in different countries. For example, where life expectancy in Sweden is about 75 years and in India or Mexico it is about 40 to 50 years, the age where minimum "rate of mortality" occurs is still the same or 10 to 15 years of age. This suggest that the age where the rate of mortality reaches a minimum is independent of external environmental hazards and is instead related to innate aging processes. It is also at this age where human vigor and resistance to disease is maximum and where it could be said the human reaches its peak of health and youth. Or to put it another way, this is the age all humans throughout the world are the youngest.[3, 4, 7]

Unfortunately, this period of maximum youth is very short and does not last longer than few years. Close analysis of these survival data indicates that essentially after reaching the age of sexual maturity there is no plateau where rate of mortality is constant. Some people even think aging does not begin until the age of thirty or forty when they begin to really feel its effects. But actually it is down hill from age 10 to 15 years in terms of decrease in optimum physiological performance. In a sense you are either getting younger or older. There is no in between period where there is no age-related change in an individual's physiological status. No wonder Olympic athletic records are being broken by younger individuals. The challenge is to reach peak training at the age of peak youth.

(6) Another interesting observation on examining mortality data is that after the age of sexual maturity risk to death of all causes increases at an exponential rate not at a linear rate.[1, 2, 17] Thus, rate of mortality doubles on a regular bases as the population grows older. Some individuals have used the time it takes to double mortality rate as a measurement of biological aging rate of populations of different species. But this idea is very controversial since mortality data, no matter what age it is measured, is never completely independent of environmental hazard influences. There is also the theoretical argument that information about the biological processes of aging in individuals is not contained in acturial data consisting only of age at death from any causes. I therefore think that physiological data is the only reliable means to obtain rate of biological aging.[3, 4, 7]

(7) Today in the developed countries throughout the world, the increase in life expectancy has greatly diminished and has been essentially constant at a

value of about 75 to 80 years over the past 10 to 20 years. Thus past success in reducing external environmental hazards resulting in an increase in life expectancy has essentially been optimized. Such information leads to the conclusion that problems associated with the internal hazards of life rather than the external hazards of life now need to be addressed more seriously if we want to further increase the healthy and productive years of human life. Indeed the increase in life expectancy through reduction of external hazards of life actually was most effective in increasing the time individuals were physiologically older not younger.

In this regard, it is of interest to ask what gain in lifespan might be achieved if it was possible to decrease the intensity of the internal environmental hazards to life? This is the same as decreasing aging rate that would in turn result in an increase in lifespan. For most of human history the ratio of external to internal environmental hazards were high and the population structure consisted of mostly young people. So everyone was essentially physiologically young. But now by being able to reduce external environmental hazards and not the internal environmental hazards, this ratio has changed to where populations in developed countries today have a higher percentage of physiologically old individuals. That is, reducing external environmental hazard has allowed the human population to live deeper and deeper into old age. So we are really living older longer not younger longer. So much of this increase longevity enjoyed today consists of decreasing functional capacities and increasing incidence of disease. Of course this is still better than the death alternative. But no wonder people are quick to ask, "who wants to live 100 years". The idea of living younger longer and not older longer is a novel concept that has not happened yet.

So what would the human survival curve look like if we returned the ratio of external to internal environmental hazards back to where it was 10 000 years ago?[3] This could be achieved in theory by reducing internal environmental hazards or decreasing the rate of human aging while keeping the intensity of external environmental hazards constant as it exist today. On carrying out this exercise I discover the interesting result that if human aging rate is decreased by about six to seven fold, then few individuals would ever live long enough to suffer and die from the effects of aging. They would instead be killed by the external environmental hazards that exist today as for example transportation related accidents. Life expectancy in this case would be about 400 years and lifespan about 800 years. This is what it takes in terms of increase in lifespan to return back to a population of youth. The point here is you do not need to completely stop aging to effectively remove it from existence in the population. The job is done by increasing lifespan by a factor about eight fold, which I say is not impossible.[3, 4, 7]

This type of data can also be used to answer a common question. If we are successful in reducing aging rate, then will not this also result in an increase in the percentage of old people in the population as well as accelerated the world's population explosion? The answer to the first question is yes, the percentage of old people would be increased but only in chronological age not physiological age. Most of the people would be physiology young. As to the population explosion problem as long as people have two to three children per lifespan, then increasing lifespan in itself will have no effect on the rate of world population growth. In fact, there is already a clear indication this is true in countries having the greatest life expectancies. These countries have the least rate in population growth even though lifespan is the greatest. Japan is the prime example.[3, 4, 7]

1.2. Physiology of Human Aging

The effect of human aging process is to decrease the functional capacity of essentially all aspects of human biology.[7] *That is aging is ubiquitous in nature and effects the entire organism.* Nothing escapes. There is no known biological function or structure of the human body that remains unaffected or unchanged by the aging processes.[2-4, 7, 18, 19]

When measuring the physiology of human aging as a function of an individual's chronological age it is necessary to measure maximum reserve capacity of function.[20] This maximum reserve capacity of function is high in young individuals but is steadily lost as they grow older. When they reach the age of about 35 to 40 years then most reserve capacity is depleted and from there on out decline of maximum function and performance begins. It is at this age that individuals begin to feel old physiologically although they have actually been aging at the same rate for most of their life.

For example, we are born with two kidneys, but when we are young one kidney can still provide sufficient functional capacity for normal life and its usual stresses. However, in individuals 40 years old or older they now need both kidneys to achieve normal function even though some compensation will occur to reduce this disadvantage. Further age-related decline in kidney function would then contribute to general decline in function of the entire body.

Thus, in measuring the aging of human physiological systems maximum reserve capacity for function is always determined. Examples are maximum capacity of lung, heart or kidney function and the five senses: seeing, hearing, feeling, smelling and tasting. But many more tests are also done. When this was done for many individuals of different ages and the data normalized as percent of maximum reserve capacity for each function measured, then it is discovered that the normalized rate of decline in maximum reserve capacity is about the same regardless of what parameter or physiological function was measured. That is no physiological system appears to age any more rapidly than another. Everything was going down hill at about the same rate. Although this was quite obvious, I

have therefore defined the average negative slope of these curves to represent the physiological aging rate, which I believe is much more meaningful biologically than the time it take to double rate of mortality.[3, 4, 7]

This synchrony of the different physiological aging processes is also predicated from evolutional theory. There is no selective evolutionary advantage for any one physiological function to outlast another or to be in excess of what was needed compared to other functions. Beginning at birth, all physiological functions appear to increase in maximum reserve capacity to reach a peak at around the age of sexual maturity and then they steadily decrease in a linear manner over the remaining period of life. At the age of about 100 years functional reserve capacity is around 20% of maximum reached at the age of about 10 to 15 years of age.

It is important to realize that normal aging processes produce a linear decline in physiological functions that in turn results in an exponential increase in the occurrence of disease as cancer and probability of death.[3, 4, 7] This fact further complicates the problem to gain information about the biological nature of aging through actuarial means. Aging affects biological functions in a linear manner resulting in the whole organism becoming more sensitive to death from all causes in an exponential manner.

I have suggested that the negative slope of the declining maximum reserve capacity physiological curves for human represents the rate of human aging. A similar interpretation can be made of the maximum reserve capacity data for other non-human primate species and mammalian species. When this is done the lifespan of a mammalian species is found to be proportional to the negative slope or rate of lost of their physiological functional reserve capacities. This has been important in my own studies to estimate biological aging rate in a wide variety of mammalian species. This is because very few data are available on physiological functional capacities of different species as compared to an abundance of data on lifespan. Thus, I used lifespan, which I called maximum lifespan potential (MLSP) in my comparative animal studies to reflect actual physiological aging rates of the species.[3, 4, 6, 7, 21]

One of the great deficiencies we still have today however is the continued lack of reliable data on the physiological measurements of aging and age related diseases in different mammalian species. In particular, this data is surprisingly deficient in the great apes and old world monkeys from which we could learn so much about the physiological and molecular mechanisms involved during the evolution of increased primate longevity.

The age-dependent curve of maximum reserve capacity of physiological functions can be used to define two key major areas of interest in the biogerontology research community. The original and somewhat classic problem in the field of biogerontology has been to understand how physiological functions decline with increasing age. This research area then focuses on understanding the causes and mechanisms of the age dependent lost of physiological functions. For many years this was essentially the only goal in this field of biogerontology. The rational was if we knew more about what causes aging and the mechanisms involved, then perhaps we will be better able to fix or repair it. Here, the thinking was largely that aging is a "wear and tear" process somewhat like an automobile wearing out. The job of gerontology was to find out what parts wear out first and find the means to repair or extend their function.

This reminds me also of a comment made by a famous Nobel Prize winner about 20 years ago when asked by a US Senator if more funds should be spent on basic aging research. His quick answer was absolutely not! He then went on to explain that it would be a huge waste of funds critically needed by other fields of biology research. This is because we need to understand first how an organism functions "right" before we can possibly understand how it goes "wrong". How can you possibly fix a broken or worn out automobile if you do not understand first its basic principle and design of operation? We must therefore wait until we achieve a full understand of normal human function before we begin studies of how human age. And this may take 100 years!

Now, this makes good sense does it? It certainly did to the senator and so no extra funds were committed to basic aging research and I might say very little since then. But I believe this famous scientist was seriously mistaken.

The problem is that he did not understand that in the study of aging "what goes wrong" is not a separate problem from understanding "what goes right". For example, if much of the biology of an animal is related to the maintenance of general health and physiological functions and if longevity determinant genes as DNA repair and antioxidants exist, then how are we ever going to understand the functions of these genes if scientist were not also funded to think along the lines of aging and the mechanisms that control aging rate and general health maintenance? The answer of course is that aging is a normal and essential part of life science from conception, to birth to death. You cannot separate it out as something special that occurs only in chronologically old individuals. Aging must be studied simultaneously along with the other biological aspects of the organism.

The second area of research is just getting underway after about a 30-year incubation period when I and a few other scientists first suggested this goal.[3, 4, 7, 22] I had proposed that aging is more than a simple wear and tear processes and that protective, defense and repair process are likely to exist that are involved in governing the aging rate of the organism.[3, 4, 7, 22] In this light on looking at the slope representing the lost of physiological functions in different mammalian species we should ask not only "how these functions decline" but also "what control their rate of decline". In fact, with the suggested possibility of a relatively few key longevity determinant genes, this understanding of what controls aging rate may come much sooner that understanding what causes the aging processes itself.[4, 22] Moreover, an answer to this question of "what controls aging rate rather than what causes aging" could lead more directly to practical methods to extend the healthy and productive years of human life.

A few more points need to be made before leaving this subject. As a population of individuals becomes older, they become increasingly more heterogeneous in terms of their markers of physiological aging. This is thought to be a result of both random and different sensitivities of individuals to external environmental hazards that over a lifetime accumulation and thus become amplified with time. Thus, ten-year-old children are remarkably similar as to their physiological status of age but take the same individuals 60 years later and you will likely find much heterogeneity with some individuals being much older physiologically than others. Indeed, we know that although most individuals have an average lifespan of about 75 years some far exceed and others do not even reach this value.[1, 2]

It is also of interest that all individuals do not start off at birth with the same degree of physiology functional capacity or what could be called "youth status". For example, in measuring vital lung capacity of a group of healthy individuals at the age of 20 years a wide range of functional vital capacities are found. Most remarkably when these individuals are followed longitudinally the age dependent decline of all the individuals progresses at a similar rate thus creating a family of declining parallel curves.

These data indicate that the rate of aging (as least for lung function in the example given here) between individuals is similar, but the physiological reserve functional capacity can be substantially different. Individuals having started with the lowest functional capacity reach critical low functional levels at an earlier age and thus would be expected to have a shorter lifespan. So, rate of lost of a given physiological function is not the only factor that plays a role in determining life long general health status and lifespan. It also depends on what functional reserve capacity you are initially born with.

Also of interest here, using the case of vital lung capacity for illustration purposes, is that it appears possible to jump up from one curve to another one by exercise and/or eating a more nutritious diet. The reverse can also happen. But if an individual does succeed in improving his lung physiological functional capacity they do not also succeed in changing its rate of decline. These data emphasize the universality of aging rate and that dietary and exercise is not likely to affect aging rate itself but can have beneficial effects by increasing physiological reserve capacity.

Over many different countries, races, diet and lifestyles what is amazing is the remarkable homogeneity of aging rate and lifespan and not the differences that exist in human populations. Thus, it is important to determine if this result for lung vital functional capacity also holds true for other physiological functions. My guess is that it does. Finally, as will be noted later, the longer life expectancy for human females is likely to be a result of their greater physiological reserve capacity and not because they age at a slower rate than males.

1.3. Diseases of Human Aging

Few individuals die directly as a result of declining functional capacity of their physiological functions.[3-5] The major causes of death have changed over the years where initially infectious disease, predation, warfare and accidents were dominant. Today, major causes of death are related largely to age dependent diseases such as cardiovascular disease, cancer or diabetes. The onset frequency of cancer as well as other age-related diseases increases exponentially with age. For example the onset frequency of cancer incidence increases to about the fifth power of age for most mammalian species including human. What is most interesting here is that up to the age of about 40 years, death by cancer is very rare. So for most of human history over the past 100 000 years where life expectancy was about 25 to 30 years few individuals ever died of cancer. However, in individuals living past the age of 50 years as they do today then the chance of death by cancer rapidly becomes very significant.

The effect of aging on decreasing overall health status accumulates steadily throughout life. Individuals in their forties and fifties have less physical vigor and strength no matter how hard they work out. It is only a matter of time, for example, when the 4 minute mile becomes impossible to achieve. Also, the number of pathological diagnosis by regular examination by a physician increases steadily with patient age after 20 years of age. Thus, even though most diagnosis are not serious, their steady increase with chronological age throughout lifespan suggest that aging is taking place in relative young as well as old individuals.

There is also the interesting fact that human females live longer than males. This female advantage however did not always exist.[8-11, 14, 23, 24] Several thousand years ago life expectancy of the human female was usually shorter than males. This was a result simply because the female was essentially in a state of pregnancy for most of her life after reaching sexual maturity. Thus, the reason I believe why human females now live longer than males is that they have evolved greater physiological reserve capacity throughout their body over thousands of years of evolution to withstand the extra burden of constant pregnancy. However, this pregnancy burden had been decreasing rapidly particularly in developed countries. So, women live longer now by taking advantage of extra or unused reserved capacity that evolve for maintaining life in pregnancy.

This explanation is based on data reporting the shorter lifespan of human females even several hundred years ago as compared to males and the inverse correlation between number of children born per lifespan and female life expectancy. In addition, the stress per pregnancy has decreased as a result of better care and nutrition.[8]

Thus, female extra life expectancy is not likely as result of aging less rapidly, but simply because the female is left with extra reserve capacity such as kidney and heart function that can now be used for non-pregnancy related requirements as aging. Unfortunately for the male, tension and hazards of life does not seem to have diminish as greatly as for the female and in some cases may have actually increased. The result is the some five to eight years of additional lifespan human females have over males, at the present time.

Another point is that many individuals appear to have difficulty in accepting the fact that the brain ages just as rapidly as other body functions. In fact, the concept that aging is even important to the general health status of the nation has not been always generally accepted for one reason or another. I remember that about 20 years ago we heard from a well known director of the largest gerontology research institute in the USA that human aging processes are really "benign" for all practical purpose. He stated that aging only becomes important to the health status of the every old in our population and that the decline in function and lost of health is a result of the diseases of chronological age and not aging itself. As a consequence he pointed out that although we have no hope to control the aging process itself these diseases of aging could be controlled or even cured by traditional medical research and practice. So as a result of this decision basic research in aging was once again by-passed for the time-honored medical orientated approach to treat the problems of aging.

This represents a very serious mistake in my opinion, particularly in light of compelling evidence indicating that even the complete elimination of the major diseases causing deaths in the USA will result in negligible increase in the general health status of the nation. Moreover, we still have the same problems with diseases today 20 years later largely because it is the aging process that causes the diseases in the first place.

As to brain function we were told about the same time 20 years ago that in general brain function actually improves with old age. Examples and pictures are given in support of this statement of many remarkable successful individuals still actively working in their 70s and 80s. This was wonderful data and good politics against the national problem of "ageism" and the economic problems an increasing population of disabled people may bring to the USA. But it is not reality. The brain does have remarkable reserve and adaptability capacity, but biologically it appears to age just as rapidly as any other physiological system of the body. There are of course many means of adaptation that the human body develops to reduce the impact of primary age-related lost in brain and other organ functions. So, the progression of the aging process in the brain is not quite as apparent as, for example, facial features unless you look for it.

The implication of brain aging however goes far beyond simply losing memory or cognitive function. It is this other problem not often mentioned that may explain in part why brain aging has been so difficult to accept as compared to not being able to run the 4 minute mile any longer. This is because brain aging also results in the steady and progressive lost of self or self-realization. This lost of self is a direct result function of a decrease in functional capacity of the brain itself. This condition is even made worst because of the decrease in amount and quality of input into the brain from the sensory organs. The decline of input of sight, sound, feeling and smell with age is well established. So loss of both quality and quantity sensory input contributes to the problem of brain function. It is also hard to detect this decline in self-realization and conscientious when it occurs to oneself, for it is a slow step-by-step process taking place throughout life. Indeed, there are many similarities in normal brain aging to Alzheimer's disease. The real difference is that Alzheimer's disease represents an accelerated brain aging process. So in light of the brain aging implications, human aging in my view is completely unacceptable in a civilized world. There is no such thing as the golden years of old age or even "successful aging". I do not want to be successful in aging but instead successful in staying young! So, let us face up to the reality of the situation and do something about it.

There is another important characteristic of physiological aging process that we have already touched upon and that is compensation and adaptation. We now understand that some of the major physiological, biochemical and molecular changes that occur with age are not primary dysfunctional aspects of the aging process itself, but rather secondary beneficial adaptations required to maintain function. Thus, these compensation and adaptations work to decrease the disabilitating effects of aging. This phenomenon is observed not only at the physiological level but also at the gene level as well. Unfortunately, it is difficult to separate the secondary beneficial adaptations from the primary detrimental effects of aging. Too often these beneficial adaptations have been lumped together with the detrimental aging processes. There is also the problem of over adaptation or compensation that can then contribute to further decline in function. Chronic inflammation is a good example.

One of the common assumptions of the aging process is that it is a continuous chronological age-dependent process. However, we must remember that the continuous appearing graphs showing steady decline of physiological functions come from averaging large numbers of individuals that are frequently from a cross sectional rather than a longitudinal study. So does an individual age in a continuous manner like these graphs indicate for averaged populations suggest? I do not think so and would suggest that aging of a single individual may actually be more of a discontinuous function than a continuous function. This idea is based on the concept of the strong adaptation and compensational forces that exist to maintain homeostasis of essentially all body functions. This "quantum model of individual aging" proposes that even though the aging process *per se* may be continuous, the level of physiological function of an individual over their lifespan consists of a series of decreasing and sometimes increasing plateaus or steps. The model is that function is maintained for as long as possible and then perhaps after experiencing an unusual stress, the individuals physiological age status changes rather quickly to readjust to another lower set point level that will then be maintained until the next stress stimulation comes along. So, the true kinetics of aging is related to both the innate aging processes and the individuals' innate ability of their reserve capacity to adapt or compensate. Thus periods of large stresses as in serious illness or divorce might then result in a new lower set point explaining why in some circumstances individuals sometimes appear to age overnight and appear never come back to that previous more youthful state they were at before the stress occurred. This model predicts then that physiological biomarkers of aging may be substantially inaccurate if they do not measure adaptation and compensational process of aging.

1.4. Economic Impact of Aging

Medical and general health care cost increase dramatically after the age of about 60 years.[3–5, 7, 21, 25] In addition, the return value such expenditures provide in terms of increasing the healthy and productive years of life steadily diminishes with age. So medical and general health care costs increase but value decreases for the older individual. The principle strategy in coping with aging in the past has been to focus on specific diseases of aging. The rational is that by understanding the cause of the disease, then better means of prevention and cure of the diseases will be developed. This idea appears reasonable at first glance and even today this is still the major strategy taken where focus is on treating the many different diseases of aging rather than treating the aging process that really caused the problem in the first place. Examples of such diseases where aging has not been seriously studied as a primary means of control are the cardiovascular system, cancer, diabetes and now neurological diseases as Alzheimer disease. To date, there has been some success in reducing these problems but not much in view of the relatively huge amount of funds spent.

Of course, the major reason for lack of more significant progress is the problem of aging itself. For example, estimates have been made to determine what might be the impact in terms of an increase in life expectancy in the USA if the major causes of death were completely eliminated.[3] The result is startling to most people. For example, the complete elimination of heart and related cardiovascular disease would result in an increase in lifespan for the USA population of about 10 to 15 years for a person that was 65 years of age. The number of years gained for a new born are a couple of more years. Now 10 to 15 years of increased lifespan appears great, but then we have to remember that these additional years of lifespan will probably result in living 10 to 15 more years deeper into old physiological age, because nothing was done about the progression of diabetes, cataracts, muscle deterioration, memory lost and much more.

But even more surprising to many is the predicted increase in lifespan if all forms of cancer could be completely prevented. That is you would never have to worry about death from cancer again, no matter how old you were. In this case the gain in lifespan is only about 2 years for the USA population for either a newborn or 65 years old!

Now, how can this be? Why isn't the impact much greater when complete elimination of the major diseases occurs? To find the answer we have to remember that aging is ubiquitous. That is essentially everything in our body deteriorates or goes down-hill simultaneously. So, the result is that when one disease is eliminated such as cancer then another disease quickly takes its place. An analogy is the "wear and tear" taking place in a well-designed automobile where all parts were designed to wear out at about the same rate. Simply replacing the tires, spark plugs, pistons or the carburetor will have little impact in extending the designed lifespan of the entire car. In fact, you will soon find by using this replacement strategy that essentially the whole car will need to be replaced to get say 300 000 miles out of a car that was designed to last 150 000 miles. Unfortunately, you cannot replace all the parts in human... at least not just yet.

So clearly piece meal elimination of specific diseases of aging will not have a great impact on the general health status of the nation even if we are completely successful in eliminating all the principle diseases existing today! The greatest health problem of our nation is aging that leads to the general decline in functions of the whole body.

So why then has not more effort been focused on reducing the impact of the aging processes if that's where the problem really is? The answer is found in a complex mix of issues involving religion, politics and science. One problem is the popular believe that aging is too complex of a problem to solve at this time. If we cannot even cure the common cold or the many different human genetic diseases, then how could we possibility do anything about aging. Another problem is that there is still not a national commitment to increase both the quality and quantity of human life in the USA. The standard line of government research institutes is "we want to increase the life in your years but not the years of life". So what's wrong with more years of life? Speaking for myself I would like both frankly.

Up to the last few years it was generally believed that nothing could really be done about the aging process because of its apparent extraordinary high complexity. But is this conclusion correct? Is aging as complex as many scientists have implied over the past fifty years or so? It has been assumed for years that because the aging processes affects essentially all biological systems of the body it must be impossible to really do anything about it. And besides aging is natural so there must be something good about it. If so, I wonder what it is.

This more recent conclusion of hopelessness did not come without a long human history of trying to do something about aging. Much of human scientific and medical efforts over the past several thousand years have been dedicated on how to increase the health and productive years of human life by reducing the rate of aging. But unfortunately all efforts failed. Indeed the general failure of these efforts has lead to the origin of many religions and cultures throughout the world.[4, 7] The approach appears to be if you cannot do anything about aging and death then turn it into a good thing or "if you cannot beat it join it". So it was proposed you will need to die to live forever or be saved. Then we have the failure of medicine to cure what appear to be much less complex problems. Examples are sickle cell anemia or the common cold. So some pessimism that we cannot do anything about aging is clearly justified.

In this light the present focus of the scientific medical community on diseases of aging instead of the aging process itself does make some sense. Treatment of the diseases of aging seems to be a more achievable task with less religious, political and economic risks than treatment of aging itself.

There is another related issue here. Lifespan for human is about 120 years but very few people reach this age. Average age at death in the USA is instead only about 75 to 80 years. Why is this the case? Why don't more people live to a 100? It was often thought that lifespan of an individual was largely a matter of their aging rate. That is the slower your rate of aging the longer you would live. But then it appears that most individuals' age at the same rate.

The answer to this question is fundamental to any general health promotion research program. When autopsies are done on individuals dying in their 50s or 60s say from heart disease or cancer they do not appear to have suffered from an accelerated aging processes.[19, 26] Their general physiological status in terms of aging is found to be in good shape and similar to an individual that will live to 100 years. Instead, it is just that a specific localized problem killed them not the general effects of accelerated global aging processes.

Thus, when individuals that have died at an age of 90 to 100 years are analyzed for causes of death it is found that they have essentially **everything** wrong with them but **nothing** in particular. That is they did not age less rapidly than individuals that died much earlier but more uniformly. Their longevity was a result of not having any serious weak links in their complex physiological network and this effect allowed them to live out their full biological potential of longevity. On the other hand there is very little that can now be done for a dying centenarian because essentially everything has gone bad simultaneously and the whole body needs regeneration.

This result has indicated that centenarians represent the true health span potential as well as lifespan potential of human. Centenarians have a long period of excellent general health maintenance and productivity and usually a very short period of serious decline in health leading to death. Thus, *I believe our goal should be to achieve both the lifespan and health span of the centenarian for all people of the world*. To do this, a research goal should be to identify what these weak links are that result in early death of most individuals and to understand what is unique about centenarians that enable them to live out their full biological potential so successfully. Some data now indicate that these weak links are related to innate deficiencies in defense and repair against oxidative stress, and glycation/glyoxidation stress. We will learn more about this later and the great promise of longevity determinant genes.

1.5. Complexity of Aging

Aging at the phenotypic level is clearly very complex.[3, 4, 27–29] But these complexities at the phenotypic level do not necessary imply that its causes are equally complex. Indeed, there are many examples where single point mutations results in complex consequences in terms of biological and pathological effects. To better illustrate this concept I have used two extreme models of aging that represents the main two schools of thought on this issue. The first one represents the conventional thinking that aging is multifactoral and polygenetic and therefore has many separate and independent causes as well as control mechanisms. Here, the high complexity of aging at the phenotypic level is thought to have equal complexity at the genotypic level. There are no primary aging processes and thus no primary anti-aging processes represented in this model. This model is supported by a very popular evolutionary hypothesis of aging called the Pleiotropic Antagonistic Hypothesis of Aging.[30–37] Here, aging is proposed to be the result of essentially all genes in the organism. Individuals supporting this hypothesis had in the past strongly criticized other scientist attempts to find specific causes of aging or means to decrease aging rate. The reason being there are simply too many independent causes of aging (highly polygenetic) as predicted according to their mathematical population based model of aging to implement any significant postponement of aging.

It is also important to note here that in the pleiotropic antagonistic hypothesis of aging the focus is largely on what cases aging or how aging evolved. It does not deal with what causes longevity or how longevity evolved. Indeed, proponents of this model assumed that longevity did not evolve at all but instead it was aging that evolved.

So what regulates aging rate in the pleiotropic antagonistic model? Their hypothesis is that aging is a result in the expression of genes that are beneficial at early ages but decremental at later ages. Thus, aging rate is determined by the timing of these genes as a function of chronological age and would accordingly be coupled to developmental rate. Another contributing factor is the relative balance between the benefit and decremental effects of these aging genes. So we have aging rate being determined by development rate. This is actually true for many different species. But it is clear that the existence of longevity determinant genes is not part of the hypothesis unless they also govern development rate. And this is what I have proposed. The fundamental difficulty then with this model is that there is a lot more to aging than just pleiotropic effects.

The other model is where primary aging processes are proposed to exist. *The basic rational for the existence of primary aging processes in organisms living today is the proposal that a few primary aging processes have always existed throughout the entire evolutionary history of life.*[3, 4, 7, 21, 22, 38] Some of these internal and external environmental hazards or potential aging processes needed to be countered or reduced for the origin of life to occur and proceed. But the main point here is that they have always existed and remained essentially unchanged since the origin of the prokaryote and eukaryote cells millions of years ago. These common hazards are called primary aging processes and the means that evolved to reduce them are called antiaging or longevity determinant processes. *Thus, I have proposed that the evolution of life in all its complexity was shaped largely by the interplay between these primary aging and antiaging processes.* Primary aging processes are therefore those processes that give rise to the complex phenotypic aging processes we observe in all life forms living today. The proposed existence of primary aging processes is fundamental because in principle its makes possible that only a few primary anti-aging processes may exist that act to control most of the important aging processes of the organism.

So, we have two fundamental models to test. The non-primary and the primary models of aging. I am very pleased to be reporting here now that the primary aging model that I had proposed in 1972 is proving to be correct particularly for the potential importance of oxidative stress.[18, 39–44] Yet I believe both models are right and that the only means that evolved to deal with the non-primary aspects of aging related to pleiotropic effects is slowing down the rate of development. This may have occurred through genetic alterations related to increased neoteny, which will be discussed later.

2. Evolutionary and Comparative Biology Studies

2.1. The Pleiotropic Antagonistic Hypothesis of Aging

The heart of the pleiotropic antagonistic hypothesis of aging is the observation that selective pressure to maintain general health maintenance status of an individual in a population decreases steadily as it grows older chronologically.[30–35, 37, 45, 46] This is because natural external environmental hazards that can cause death are ever-present independent of any aging process. Thus, with increasing chronological age of the individuals there are simple less individuals living and thus fewer individuals to benefit from any genetic change that may increase healthy lifespan.

So even if aging did not originally exist it would have evolved simply because of lack of sufficient selective pressure to prevent it from occurring. What we have here is the optimization of the relative benefits of a given gene population of young individuals where there are many to benefit as opposed to older individuals where there are less to benefit. From this simple and eloquent concept experts in the field of population and evolutionary genetics have put forth an "Evolution of Aging Hypothesis" that explains why aging now exist in all living organisms.[45, 46]

I have proposed another hypothesis.[3, 4, 22, 27, 28, 47, 48] I believe the pleiotropic based hypothesis of aging is correct in principle, but it fails to predict a major class of aging and antiaging processes, which are primary in nature and have existed throughout most of the evolutionary history of life. The major flaw in the pleiotropic antagonistic hypothesis is that their argument is based on the evolution of aging, which begins with an immortal organism. If it is true, then, that aging evolved from a decrease in selection pressure in the chronologically old, then it must also be true that if such selective pressure did not decrease aging would not exist. I think this is nonsense. Although lifespan in such a hypothetical case would be infinite, I propose that the aging processes would still remain in existence within the organism. The pleiotropic antagonistic hypothesis of aging is more appropriate in addressing the issue involved in the evolution of finite lifespans in different organisms rather than the evolution of the aging process itself. But this hypothesis has much more serious problems.

I agree that selective pressure will decrease with chronological age and give rise to developmental related pleiotropic aging processes. But I argue that the original or very first forms of life already had primary aging processes in existence that are not easily subject to natural selection pressure elimination. That is there are basic physiochemical/biophysical properties of the living process that requires extraordinary high energy cost for immortality to occur thus seriously decreasing the probability of any immortal organism ever existing regardless of the amount of selective pressure to eliminate them.

Examples on the aging side are the positive driving entropy effects on any information carrying macromolecule leading to the eventual lost of that information and the natural environmental internal and external hazards leading a cell to eventual destruction. Specific examples include the toxic nature of oxygen and nitrogen free radicals and the reducing sugars, which react, with all components of a cell. Examples on the antiaging side are the repair, replacement and turnover processes and various other defense and preventive mechanisms.

In the early 1970s, I took a more global viewpoint beginning with the evolution of life and then progressing to the more advanced living forms in an effort to understand more about the basic nature of aging.[5, 22, 27, 28, 49, 50] The working hypothesis I arrived at integrates the basic pleiotropic antagonistic concepts but most importantly presents an explanation for both aging and longevity of different species at the biochemical and genetic level.

Since I have already published extensively on this work I will only briefly review the major concepts and discuss recent experimental data that support the hypothesis in preference to the pleiotropic antagonistic hypothesis of aging.[4, 22, 28, 38, 48–50] I do this by discussing briefly the origin of life beginning with the first replication forms, to free living cells, to multicellular simple organisms, and then finally to multicellular complex organisms where the body or soma is separate from the germ cells. I view the origin of life and the continued evolution of life in its many different forms as representing the general success of life maintenance process overcoming external environmental hazards and innate thermodynamic/entropy and pleiotropic factors related to the life process that contribute towards its destruction. To understand this argument it is easier to speak in terms of information theory. Life itself is not a physical entity but rather a process. This process is able to maintain the propagation and maintenance of information contained within a physical entity that enables this life process to continue to be maintained and to propagate this information without limitation. Yet this physical entity itself has a finite lifespan.

For illustration, let us assume that one of the earliest life forms consisted of a replicating RNA/protein complex.[3, 4, 7, 22] Information contained in the RNA molecule codes for proteins that provide protection and catalyzes its replication. However, the RNA molecule is in a hazard environment that will eventually lead to its alteration or destruction of the proteins or the RNA molecule itself. These hazards are both physical (as radiation) and biochemical in nature. The RNA and protein molecules also have natural instabilities related to temperature and the many molecules that can react with it forming adduct, cross-links and other types of modifications. So even under the best of circumstances the RNA/protein complex has a finite lifespan.

I have proposed that the trick that makes life possible in the beginning and ever since was the process of replication.[3, 4, 7, 22] Replication provided the means to preserve information indefinitely in a physical entity (the RNA molecule) in spite of the RNA molecule itself having a finite lifespan. It is only this information that is preserved that can in principle be immortal.

Once replication is possible it must occur before a critical amount of information is lost. When this event is achieved then the replication system is set up for its evolution. This initial origin of life processes probably started and failed many times, each time lasting a little longer or shorter until finally a system evolved that had the potential for indefinite preservation of the information that makes the replication in a given time frame possible.

I therefore define "aging processes" here as the destructive processes that act to destroy the replicating system and "life maintenance processes" as those processes as replication that makes the indefinite information preservation processes possible. Death of the living system occurs when in a replication system information is lost before a successful round of replication is completed and the physical entity no longer is able to replicate. Since this initial beginning many different protective and defense strategies evolved. For example, creating high numbers of replicating units to counter the random nature of the imperfect replication system assured on average sufficient number of error free or age free replication units to survive. This process of evolving more efficient life maintenance systems through the natural process of selection has continued for the past several billion years resulting in the diversity of living systems we have present today.

The main point here is that it is highly likely that aging and antiaging processes were present from the origin of life to the present period and that these two processes are normal characteristics then and now of all living organisms. Thus, it is nonsense to speak of the evolution of aging even when referring to complex multicellular organisms without also speaking of the evolution of longevity. Moreover, I propose that in the recent evolution of the primates and particularly during the hominid ancestral descendent sequence leading to human over the past 65 million years, that no new pleiotropic antagonistic aging process or primary aging or antiaging process has evolved. The aging and antiaging mechanism were already largely in place having evolved over the past several billions years of evolution of life.[3, 4, 7, 22, 49, 50] Moreover, I proposed that before the separation of germ cells from the soma, that the basic cellular aging and anti-aging process were already largely in place. Thus, this separation of soma and germ cells in the evolution of multicellular organism did not lead to the evolution of aging or antiaging processes but in the alteration in expression of aging and longevity processes already in place.

The other point is the prediction of primary aging processes. From what we know about the possible complex biochemical conditions necessary for the spontaneous origin of life it appears reasonable to predict the existence of primary aging processes just on this basis or rather than assuming that all aging processes are of equal importance. However, as will be discussed now, there are other more compelling reasons to believe that primary aging processes exist.

In my early evolutionary studies, I proposed oxidative stress as one of a number of potential primary aging processes. This is because oxidative stress has existed from the very origin and beginning of life and acts to destroy life. I then proposed experiments to determine if oxidative stress remained a key primary aging process in primate species.[22, 48] These experiments focused on testing for the existence of corresponding antiaging or life maintenance process to reduce oxidative stress such as DNA repair, antioxidants and information redundancy in tissues of different mammalian species as a function of their lifespan. In terms of defense mechanisms against oxidative stress, I proposed that the amount of antioxidant protection in a tissue, such as the amount of the enzyme superoxide dismutase (SOD) per free radical generation intensity or Specific Metabolic Rate (SMR) in that tissue should correlate positively with the lifespan of mammalian species. This prediction has been demonstrated to be correct in our and other laboratories and represent the first evidence supporting the existence of primary aging and antiaging processes.[18, 39–41, 43, 51, 52]

2.2. The Disposal Soma Hypothesis of Aging

I would now like to address another hypothesis called the Disposal Soma Hypothesis of Aging.[46, 53–57] It is of course well known that in the evolution of multicellular organisms that an important stage in the evolution lineage was the separation of soma from germ cells. However, I view this occurrence as representing simply another one of many different strategies taken in the evolution of protecting information involved the propagation of life and not the beginning or origin in the evolution of aging processes. Now, a key-determining factor in selecting what processes or strategies succeed best in protecting this information is energy expenditure considerations. This has always been the case beginning with the origin of life. Thus, I agree that the separation of soma from germ cells represents an energy efficient means to protect germ cell contained information as compared to protecting the many more somatic cells making up an entire organism. However, I think such energy conservation strategies are related to what was already dedicated to life maintenance processes and existed in non-metazoan organisms before germ and soma cells separated. Indeed, one might propose that the entire makeup of a multicellular organism represents one whole life maintenance processes. So, I believe it cannot be argued that the separation of germ cells from the soma represents the origins of aging in multicellular organisms and that before this time that cellular aging did not exist. This is a serious error and neglects the fact that aging already existed in the single free-living cells and in multicellular organisms before soma and germ cells were separated.

Perhaps this somewhat novel concept I am suggesting would be clearer if I digress a little here to point out that free-living cells age just as do multicellular organisms. Some individuals view free-living cells such as bacteria as being immortal since the population of such cells can divide and grow indefinitely. Also it is commonly thought that because free-living cells are always dividing and producing apparently young daughter cells that no accumulated aging effects occur.

But I believe that damage is accumulating and is pasted onto daughter cells. Cellular aging in free living dividing populations of cells can be observed by studying fixed populations.[22] This procedure permits following the lineage of a large number of daughter cells. Experiments using fix populations of *E. coli* suggested the necessity of replication as a primary antiaging process and the death of cells when replication is decreased to a critical level or inhibited.[22]

For example, in the *E. Coli* experiments it was found that the cells cannot survive without dividing at some minimum rate. Division at a minimum rate is required to maintain a minimum mutational load.[22] Similar problems are also likely to exist for eukaryote cells. Thus, as far as I am aware, all free living cells need replication and division as well as internal renewal process to survive indefinitely. And as already pointed out these single cell prokaryotic and eukaryotic organisms appears to have similar primary aging processes that have been predicted to exist in higher multicellular organisms such as glycation and oxidation reactions. Indeed, this is why the single cell organism yeast represents a valuable model to study longevity determinant genes in humans.

The increasingly popular concept that there might be organisms that do not age or that negligible senescence exists is also inconsistent to this global primary aging hypothesis.[58–60] Although organisms like the hydra or sea anemone appear at first sight not age they really do age at the cellular level.[3, 7, 22, 47] Their trick of indefinite survival is high cellular turnover in the population of cells that make up their soma. The same is true in long-lived trees with the turnover of the cells in the cambium layer.[3, 4, 7, 22] The so-called long-lived trees have no postmitoic cells that are long lived. Most of the tree is dead wood and the only living cells are in a state of constant turnover in the cambium layer. Saying a tree can live for thousands of years and we should find out how it is able to live that long is misleading. It is like saying the human race is immortal since although the individuals have finite lifespan the population at large continues to exist through reproduction.

Also, I think organisms that do appear to age very slowly, as negligible senescence should not be considered to be some type of special exception requiring special classification. This is because their lifespan should be related inversely to the intensity of their environmental hazards as is true with all other organisms. If a fish, for example, is found to live over 100 years, I would simply predict that its must be living under low environmental hazards that make the evolution of such longevity possible through natural selection. And the longevity mechanisms that did evolve are not likely to be substantially different from why human lives longer than the chimp or why the whale lives longer than human.

The disposal soma hypothesis proposed that lifespan of a species is related to how much energy expenditure will be placed in protecting the soma from aging. I fully agree with this concept but I seriously disagree that this is a new and novel idea. Certainly energy conservation is important for governing intensity of defense and repair processes for this was the very mechanism I had originally proposed explaining why different species evolved different lifespans well before the disposal soma hypothesis was published.[3–5, 7, 22, 48, 61] My idea here was based on the observation that species living in their natural ecological nitch rarely live long enough to lose performance essential to survival as a result of aging. It was also observed that an excellent inverse correlation exists between the intensity of the environment hazards of a species natural ecological nitch and its lifespan in captivity. *Since I knew there was a correlation between aging rate and lifespan this meant that longevity evolved in a species, but only to the point where further increase could not be justified in terms of increased energy expenditure related to survival gained.*

A question I have often asked to further explain this concept is "why evolve life maintenance systems that has a cost in energy consumption to enable a mouse to live ten years where in reality the mouse on the average can only live about one year or less because of the high intensity of its natural environmental hazards?" What I have learned and published is that the mouse is committed to an energy expenditure rate just enough to the keep the mouse healthy and age free to the point that senescence plays little or no role in its survival when it is living in its natural conditions. *This is the basic hypothesis I have been using for years explaining both how and why different species evolved different lifespans.*

It was from this reasoning that I predicted a positive correlation of DNA repair intensity and antioxidant status per reactive oxygen species generation intensity should exist with the lifespan of the animal.[22] *Extra energy would not be expected to be spent in producing life maintenance process (as SOD) higher than needed to ensure survival.* This is the very basis of all of my comparative studies seeking correlations of potential longevity gene products or toxic gene product with a species lifespan. Thus, I have proposed that it is the intensity of the external environment hazards that determine the extent energy will be devoted to life maintenance process not the separation of soma from germ cells. So with this concept already published and many published experiments to support it, I find it difficult to determine what new information or insight is gained from the Disposable Soma/Energy Conservation Hypothesis.[3, 4, 7, 22, 47, 49, 50]

3. The Genetic Complexity of Aging and Longevity Processes

The most influential event affecting my research in the area of biogerontology was learning that human had the greatest lifespan of all primate species and most other mammalian species other than possibly the whale. It was also of keen interest to me to learn that human's closest living relative, the chimpanzee, had a lifespan and health span of only about half that of human.[3, 5, 22, 27, 28, 48–50] Although this was already well known to primatologist it was important to me because it was well known that chimpanzee had an extraordinary similar biology compared to human and shared a common ancestor with human only about 5 million years ago.

In the past, the popular concept was that human lived longer than the chimpanzee because they had better doctors available or that chimpanzee would live equally long as human if they were not kept in captivity.[1, 62] This of course is not true. If anything chimps have better doctors than most humans. To many gerontologists the popular explanation was that the human body was built like the Mercedes Benz of mammalian species. That is, it lasted longer because of superior engineering, design and longer lasting parts. But this was also not true. The basic biology from physiological to molecular of all the primates is remarkably the same. There is no evidence that humans are built of superior materials or have superior design.

When I undertook a comparative study of lifespan, genetics, and biology of primate's species I made the following interesting observations:[22, 28, 49, 50]

(1) Lifespan of non-human primate species as determined from of over 100 zoos world wide indicated a range from 45 to 60 years for the great apes, 30 to 40 years for the old world monkeys and 15 to 25 years for the new world monkeys.

(2) Biology of primate aging processes appears remarkably similar in all primate species. The major difference was quantitative and not qualitative in nature. Thus, primate aging process appeared to be simply a difference in the rate of expression of the same aging process found in all the different species. This

similarity as to the qualitative nature of the aging processes is consistent considering the remarkable genetic and biological similarity all primates share with one another.

(3) Developmental biology is remarkable similar in the different primate species but the rate of development and the timing of the different stages of development leading to sexual maturity are proportional to their lifespan. The longer-lived primate species have common stages of developmental biology, but each stage takes a proportional longer time to develop.

(4) The longer-lived primate species depended more on learned behavior for their evolutionary success than the shorter live species where humans represent the maximum development of this characteristic. *It was therefore proposed that the primary driving force in the evolution of longevity in the Old World Monkeys, Great Apes and the Hominid species was to take advantage of learned behavior over instinctive behavior. This hypotheses has far reaching implications in our understanding of the unique nature of man who clearly represents the apex of this evolutionary strategy.*

(5) Lifespan calculated using the Sacher Brain/Body weight formula for primates, non-primates and so called *living fossil species* are in remarkable agreement with captivity lifespan data. Thus, estimate of lifespan of primates in captivity plus calculated lifespan give additional assurance that values used in our lifespan studies are reasonable.

Based on the above observations and conclusions a study of the evolution of longevity of primates was then undertaken. The key to this study was the calculation of lifespans for extinct species using the Sacher Brain/Body weight formula and scientific literature estimates of brain and body weights of extinct and living primate species.[28, 48–50] Two strategies were used to estimate the genetic and biochemical complexity of the processes controlling aging rate. I already had speculated that the genetic complexity governing aging rate in primate species would be much simpler than previously believed to be the case by other gerontologist. This was because different primate species had up to five fold differences in lifespan yet aged qualitatively in the same way. This suggested to me the possible existence of primary aging processes because it appeared too unlikely to find such homogeneity in the qualitative nature of aging processes if thousands of genes need to be altered in the same way at the same time.

However, a stronger augment for the existence of primary aging processes and primary longevity determinant genes is to estimate how fast longevity evolved. The idea here is that if aging was a result of a very large number of pleiotropic acting genes, as has been proposed by proponents of the pleiotropic antagonistic hypothesis, then longevity would evolve slower than other biological characteristics since thousands of genes would need to be changed all in the right direction at about the same time. This prediction is based on my observation that primate species living today all appeared to age physiologically the same way but at different aging rates. Thus, the antiaging mechanisms whatever they were appeared to affect the aging process the same way in all species. This observation is novel and critical and forms an important foundation of all my evolutionary studies in primates. The opposing concept is that longevity evolved in a piecemeal fashion decreasing, for example, aging rate of the brain then the heart and so forth.

This study was done first in primate species[28, 49] then the ungulates and carnivore species.[50] The more direct estimates of the rate of evolution of lifespan was made by comparison of the lifespan differences of two closely related primate species with the time when they shared a common ancestor. For example, between human and chimpanzee the difference in lifespan is about 50 years and the time when then shared a common ancestor was about 5 million years ago. So on average 50 years of lifespan evolved in 5 million years or 10 years of lifespan per million years. Similar comparison was done for many other primate species pairs. The general result was that lifespan increased during the evolution of the primates over their 65 million history of evolution and at an ever increasing rate as one progressed from new world monkeys, to old world monkeys to the great apes and human. Note here that the conclusion was that lifespan evolved not aging, and that the qualitative nature of aging processes remained the same. A conclusion in opposition to the antagonistic pleiotropic hypothesis of aging.

However, much more detailed information on the evolution of lifespan during primate evolution was possible by applying the Sacher brain/body weight formula to estimate lifespan of extinct primate species. There was not a mammalian species known where this formula did not work equally well. This was shown to be true for the called progressive as well as living fossil species. This approach proved very useful along the hominid ancestral descendent sequence where advantage was taken of the existence of considerable amount of data available on brain and body size estimates of the many different hominid species.

Results using both of these approaches were as follows:[28, 49, 50]

(1) Lifespan increased in general during primate evolution but was exceptionally rapid along the hominid ancestral descendent sequence leading to the *Homo sapiens*. Maximum rate of increase in lifespan was about 14 years per 100 000 years occurring about 100 000 years ago. This represents a rate of about 580 milli-Darwin units, which is an extraordinary high rate in comparison to evolutionary rate of other mammalian species characteristics.

(2) *This rapid rate of increase lifespan represents a uniform extension of healthy lifespan and thus, the uniform postponement of all the normal causes of aging processes.* Thus, the data strongly supports the existence of primary aging processes and in turn primary antiaging processes. The data does not support the prediction of the pleiotropic antagonistic hypothesis of aging of high genetic complexity of genes controlling aging rate.

(3) A more exact quantitative estimate of genetic complexity involved in determining human lifespan is difficult since there are many different genetic

mechanisms of how primate longevity could evolve. In terms of point mutations estimates are about 4×10^{-3} amino acid/gene/generation or about one alteration in each of 150 to 250 genes. This represent about 0.5% of the genes out of about a total of 40 000 involve in all aspects of human evolution over the past 10 000 generations. This 0.5% includes not only Longevity Determinant Genes but also Intelligence Determinant Genes and every other gene involved in recent hominid evolution.[28, 49, 50]

(4) However, we now understand more about the nature of primate evolution where not only are point mutations involved but also considerable amount of genome/chromosome rearrangement. Genome rearrangement is like the phenomena originally called jumping genes, which involve DNA transposon elements, Retro transposons elements, Alu elements, Retro-elements and the short and long interspersed repetitive elements. Since it is still generally believed that most primate evolution consisted of changes in gene regulation rather that changes in structural genes, then both mechanisms of point mutation and gene rearrangement involving largely regulatory gene elements probably played an important role.

(5) What the complexly might be in terms of the amount of gene rearrangement occurring during recent hominid evolution has not been estimated. So overall genetic complexity in this aspect could still be very high. Nevertheless, whatever the mechanism(s), it is clear that increase of lifespan did occur at a rather high rate as a result of relative few point mutations and/or unknown degree of gene rearrangements. This data supports the concept of relative few primary aging processes and the existence relative few Longevity Determinant Genes. Detailed genomic comparisons at the sequence and gene level as well as detailed microarray expression analysis and proteomic analysis of human with chimpanzee and human with the super-centenarians is likely to answer such questions in the near future as well as possibility identifying new longevity determinant genes.

Finally, it is very important to point out that Dr. Steven N. Austad has described evidence that the when opossums are exposed to less intense environmental hazard that in a period of just 5000 years their lifespan has significantly increased.[115] This was apparently a special case where the opossums migrated to an island where opossum predators were considerable less intense as compared to the mainland. Although Dr. Austad did not point this out, his data beautifully confirms the possibility of a rapid increase in lifespan I had previously predicted for the primate species when they are exposed to a lower environmental hazard.[3, 4, 7, 22, 49, 50] For the great apes and hominid species the lower environmental hazard was the indirect result of greater ability in learned versus instinctive behavior where as in the opossum it was the direct result of moving to an island having less environmental hazards. But the effect and results were the same.

Thus, the rapid evolution of lifespan observed in the opossum are completely consistent with the hypothesis that few genetic alterations are necessary to increase lifespan and also provides independent support for the existence of primary aging processes and relative few longevity determinant genes.

4. What are the Longevity Determinant Genes?

It was clear to me when I wrote my first paper on aging in 1972 that aging was not genetically programmed as were the population doubling cycles of cells in tissue culture and that aging did not evolve as was argued by some evolutionary biologist.[22] What convinced me of this is the fact that few animals living in their natural ecological nitch ever live long enough to suffer from any effects of aging, so clearly there would be no driving force present to ever evolve a genetic program to age a species for its own good. In addition, there is the remarkable inverse correlation between the lifespan or aging rate of a species and the intensity of its natural environmental hazards. Thus, these two observations suggest to me that aging was not genetically programmed as was strongly suggested by the aging *in vitro* tissue culture group. But if this was not true then the only other alternative was that aging is passive in nature and is the effect of byproducts of essential developmental and metabolic reactions that were selected for evolutionarily. Of course, if we start with the origin of life then both aging and longevity process co-evolved, but even then aging was always a by-product of essential metabolisms. So we have two separate arguments that lead to the proposal that aging is passive in nature and not active or genetically programmed.

In mammalian species, I predicted two major classes of aging processes. These are (1) the developmental-linked biosenescence processes and (2) the continuously-acting biosenescence processes. Longevity evolved during primate evolution by decreasing the intensity of these two major aging processes. The genetic mechanism of how this occurred is by changes taking place in a common set of genes I have collectively called Longevity Determinant Genes.

With this as a background, my proposal was that Longevity Determinant Genes control the intensity of general health maintenance processes and accordingly lifespan of a mammalian specie.[3, 6, 7, 28, 29, 38, 48, 49, 63] Thus, the extent general health is maintained and therefore the lifespan of an animal is dependent by the degree this common set of longevity determinant genes are expressed or repressed. Most importantly is the prediction that during the recent evolution of increased lifespan along the hominid ancestral descendent sequence the developmental-linked biosenescence processes (such as the antagonistic pleiotropic effects) or the continuously acting biosenescence processes (such as the oxidative stress effects) were uniformly decreases in intensity. This occurred by simply lowering the intensity of expression of the developmental-linked biosenescence processes by decreasing rate of development. In addition, the up regulation of protective and repair processes occurred to reduce the intensity of the continuously acting biosenescence processes. The genetic mechanisms of how this occurred are predicted to be by relative few point mutations and gene rearrangements that resulted in changes in gene regulation effecting the temporal and degree of expression of longevity determinant regulatory and structural genes already in place.

I believe that there is a good possibility that the expressions of the pleiotropic derived aging process are largely coupled to differentiation and developed processes. In view of the great genetic and biochemical complexity of the pleiotropic aging processes and their deep involvement in many different normal processes of differentiation and development that have evolved over millions of years, I have proposed that the primary means that evolved to decrease their impact on limiting lifespan has been to postpone their expression by decreasing rate of differentiation and development at all stages leading to sexual maturation.

Major primary aging processes involve the production of toxic components in the normal pathways of energy metabolism that are byproducts of normal and essential metabolism. The category called continuously acting biosenescence processes includes the reacting oxygen and nitrogen species associated with energy related metabolism. Another toxic component that is not by products of metabolism in the energy metabolism pathway is the presence of the reactive reducing sugars such as glucose. Glucose in turn is subject to glycation and glyoxylation reactions leading to the Advanced Glycated Endproducts (AGE).

It is important to emphasize here that my interest in free radicals and oxidative stress as a primary cause of aging processes as first published in 1972[22] was independently derived from the comparative biology and evolutionary studies described here and was not taken from the free radical theory previous proposed by Dr. Denham Harman in 1956.[64–67] Moreover, the proposed predictions that aging rate in different species are controlled in part by the timing and degree of endogenously expressed protective and defense processes held in common in the different species was also completely original. These new concepts were only later incorporated in Harman's Free Radical Theory of Aging after the publication of my papers in this area.[22–24, 27–29]

I have favored the principle means of dealing with reactive oxygen and nitrogen species to be preventative in nature and not repair. One strategy is to first minimize innate production of these reactive species by maximizing energy production efficiency. For example, mitochondria represent the largest single source of production of reacting oxygen and nitrogen species. Thus, preventive strategies would involve reducing the amount of reactive species produced per amount of oxygen-consumed necessary in producing a given quantity of ATP energy. Another more direct means would be by simply lowering metabolic rate or by decreasing body temperature or an increase in body size. There is some evidence that all above process have indeed occurred during the evolution of longevity in mammalian species.

A second means is a step removed from primary preventive strategies and this is to scavenge the reactive species that are produced by antioxidants before they can cause any damage. This is particularly important in the case of lipid peroxidation where one free radical can initiate a chain reaction of lipid peroxidation producing even more free radicals.

A third means would be to evolve greater resistance of cellular components, tissue components and organ components to oxidative stress and glycation stress. For example, longer-lived species may have lowering percent composition of amino acids in proteins or unsaturated fatty acids in membranes that have greater susceptibility to oxidative damage.

The fourth means would be to remove that damage that has occurred. This could be both by non-selective or selective turnover of cellular components. In non-dividing cells both general and selective renewal of essentially all cellular constituents are known to exist. There are many different repair processes that are now known that detect and remove oxidative mediated damage from nucleic acids, proteins and lipid structures. A potentially important protective mechanism for maintaining mitochondria integrity is whole mitochondrial division and turnover processes including of course the replication of the mitochondria genome itself.

Now, this was all in theory before many experiments were undertaken in my laboratory. *The guiding principle in these experiments was to determine what was unique about human biology that could account at least in part for the unusual long, healthy and productive human lifespan.* I was not focused at that time nor am I now to test the hypothesis as to its potential application to cover all forms of life although I had always planned to do so in the future.

4.1. Rate of Development as a Function of Lifespan

One of the first surveys done was a comparison of general rate of development with lifespan of mammalian species in general and primate species in particular. In general, an excellent positive correlation was found to exist with lifespan and the age of sexual maturity and many other specific stages of early development.[3, 4, 47, 49] This lead to the prediction that any non-toxic means used to decrease rate of development of an animal would always decrease its aging rate and increase its lifespan. A key example here is caloric restriction when administrated early in life decreases rate of development, postpones age of sexual maturation and increases lifespan in mice and rats. However, the primary mechanism of caloric restriction increase may not be postponement of development but a decrease in oxidative stress.

It is important to note here that extension of development period is not an obvious means of increasing reproduction fitness. In evolution theory many think that a major strategy is to reproduce as soon as possible and that after sexual maturity is completed, the individual is evolutionally dead. So in this light what possible advantage would postponement of sexual maturation have for increased survival given that a human infant requires considerable care, protection and instruction before reaching reproductive age? My answer was the advantage was an increase the degree of learned behavior traits.[7, 49, 50] This extended development period reduced the aging effects of pleiotropic acting genes by postponing the time of their expression. The resulting increase in lifespan particularly in the early growing period of life was an ideal setting for learning to take place when the individual was small, not sexually mature and had high neotenous characteristics. Thus, postponement of the age of sexual development provided more time to learn, to create and more time for the parents to teach new learned information to their progeny. Increased learned behavior then resulted indirectly to lower environmental hazards and in increased survival. But here survival was the key to success not reproduction as it is for species dependent on instinct behavior. Clearly there is also a role here for the value of parents living much beyond the age of sexual reproduction. Other means to deal with the pleiotropic aspects of aging rather than postponement in time of their expression may have involved as well, but I know of no other ideas or evidence at this time.

4.2. Evolution of Human Neotenous Traits

There is an additional reason why I had proposed the evolution of longevity in primates involved an extension of development.[3, 7, 49, 50] This is related to the unusually neotenous nature of the human.[68, 69] There is much evidence that the evolution of the many unique human characteristics such as their long period of differentiation and development, large brain size relative to body size as well as intellectual and behavioral characteristics is the direct result of alterations in gene expression governing neotenous characteristics. For example, it has already been suggested that human represents the maintenance into the adult stage of life of morphological and behavioral characteristics present in the juvenile chimpanzee. Indeed, a morphological body shape profile of a juvenile chimpanzee is remarkably similar to a human adult as to face, forehead and general body proportions.[68]

It is also interesting to note here that many human neotenous behavioral characteristics are also found in other juvenile animals, but are then rapidly lost when they reach the age of sexual maturity. Human is unique in that these neotenous characteristics are retained throughout life. Typical human neotenous behavioral traits are as follows:

(1) Open-mindedness and receptive to new ideas.
(2) Malleability, questing, striving, questioning and seeking.

(3) Critical testing and weighing of new ideas as well as old ones.
(4) Wide-eyed curiosity and excitement in the enjoyment of new experiences.
(5) The willingness to work hard to make sense of it all.
(6) A good sense of humor and laughter.

How often are children asked to repress such behavior and when expressed in adults they are told that they need to grow up? Clearly these neotenous traits are expressed to different degrees in different human individuals, but my guess is that in the people that age most successfully such as the super centenarians they will be found to have higher than normal levels of such neoteous traits. Many famous scientists also appear to have strong neoteous traits to the point that is a popular stereotype of a professor that still acts as a child and even likes toys of all things.

Now, the most exciting possibility here is that relatively few key regulatory genes may be involved in simultaneous control rate of development as well as the extent of neoteny during primate evolution and in particular along the ancestral descendent sequence leading to human.[69, 70] That is all these parameters may be under a central common control mechanism and could be selected evolutionary for as a group instead of individually. The very rapid evolution of these complex traits suggests this is exactly what has happened. So if it was possible to continue human evolution in the future by continuing the same process that was used so successfully in the evolutionary past, then it appears that further slowing down rate of development and increasing human neotenous characteristics would be both needed and desired. The fascinating possibility now exists that a relatively small and common set of regulatory genes control rate of development and neotenous characteristics as well as expression of the longevity determinant genes dealing with the aging effects of energy metabolism as reactive oxygen and nitrogen species. Thus, "human-directed" evolution of human along the same lines that occurred in the past and were so successful may be most desirable and not extraordinary difficult to accomplish.[3, 4, 7, 29, 49, 50]

4.3. Lifespan Energy Potential

It has been known for many years that the amount of energy consumed over a lifespan on a per body weight basis is about the same from mouse to elephant (no primates species were included except for human in these early studies).[3, 4, 6, 7, 49, 50] Since energy expenditure is a rough measure of wear and tear on an animal, these data suggested that aging is largely a wear and tear phenomena. Since the data for this correlation was largely derived in the 1930s and 1940s, I reexamined and extended it to determine how human and other primates fared as compared to other mammalian species. These data were plotted as a function specific metabolic rate (cal/g/d) versus lifespan (years).

In these comparative studies, I discovered that although the total amount of energy consumed over lifespan was similar for many different non-primate mammalian species (200 kc/g) as previous published by others, primates as a group were about twice this value (400 kc/g) and human and capuchin were again about twice that amount (800 kc/g). Within each group there were some scatter but the correlations were remarkable clear suggesting that the database was reasonably accurate. From these results, I suggest that the total energy consumed over a lifespan on a per body weight bases be defined as the species "Lifespan Energy Potential" or their LEP value and that this parameter be used in addition to lifespan to better define a species innate capacity for both an active and productive life as well as its innate longevity potential.

Now, the fact that many species have the same LEP value but different lifespan can be used to argue that the byproducts of energy metabolism such as the reactive oxygen and nitrogen species are potential causes of aging and that the data for these species was consistent with the wear and tear hypothesis of aging. But what I found that was new, was that human LEP value is about four times higher than non-primate species and about twice as high as most primate species. This interesting result suggests that humans are able to consume more oxygen per body weight over their lifespan as compared to the other primate and non-primate species. Now, how can this fact be explained? And is the answer to this question — part of the reason why humans are able to live so long?

One possibility was that mitochondria might in fact produce four times less free radicals per amount oxygen consumed or ATP produced. This possibility appeared unlikely to me at the time, however, since the energy generation efficiency of mitochondria had millions of years to evolve and recent hominid evolution occurred over a period of only several million years. This is true unless there was an advantage for leaky free radical mitochondria, and one possibility is a high mutation rate for evolution. Another more likely possibility is that perhaps human has a higher level of antioxidant defense against the related endogenous production of reactive oxygen and nitrogen species. This appeared to be a much simpler task to evolve compared to increasing the energy efficiency of mitochondria function since it required only the up-regulation of existing antioxidant genes. This prediction was soon to be tested experimentally.

I also studied the evolution of LEP value both in hominids leading to human as well as in non-human primates. As expected although specific metabolic rate varied widely from species to species, LEP value was remarkably constant and increases most dramatically only during recent hominid evolution leading to human.[3, 4, 47, 49]

4.4. Other Parameters that are Unique to Human

A literature search was undertaken to determine what other parameters might be unique in humans that could account for the unusual high human longevity potential. Some of the data found are as follows:[3, 4, 6, 7, 22, 27, 48]

(1) Serum concentration of DHEAS but not DHEA correlated positive with mammalian longevity. DHEAS is unusually high in human serum.
(2) Muscle B/A ratio of lactate dehydrogenase correlated positive with increased lifespan in many different tissues and specific areas of brain of primates. The ratio of B/A of lactate dehydrogenase is unusually high in human tissues.

However, it is important to realize that most biochemical/clinical values in human were similar to those in shorter-lived non-human primates. This includes the following three parameters that indicates they were not human longevity determinants:

(1) Body temperature.
(2) Glucose concentration.
(3) Cholesterol concentration.

In addition, it was clear that humans are not long lived because they are naturally under a food restriction like diet. *Humans are the longest lived of all primate species and they achieve this in spite of eating all the food they want all the time.* In fact, humans actually consume four times the calories over their lifespan as compared to shorter-lived species. In addition, the body temperature and glucose level of human are similar to the shorter lived species. *Thus, many of the parameters that appear to be responsible in increasing lifespan in caloric restricted animals are not the same parameters that naturally changed during the evolution of human lifespan.* The potentially important exceptions appear to be a decrease in oxidative stress and a postponement in age of sexual maturation. This fact argues against caloric restriction as a practical means to further increase human lifespan.

4.5. DNA Repair Levels as a Function of Lifespan

One of my earliest predictions to test the Longevity Determinant Gene Hypothesis was a positive correlation of DNA repair levels with increasing lifespan.[23] I was unable to get a NIH grant to test the prediction. However, a few years later, I learned that Dr. Ronald Hart undertook this experiment and confirmed the prediction.[71, 72] He found a positive correlation using fibroblast tissue culture cells derived from different mammalian species in the extent of the UV/DNA repair with their lifespan. *These experiments were therefore the first to test and confirm the predictions of the Longevity Determinant Gene Hypothesis.* Similar DNA repair studies have now been published from several different laboratories and all are in general agreement that in mammalian species a positive correlation exists for UV/DNA repair activity with species lifespan.[73]

The only problem with the Hart experiment is the so called "UV/DNA Repair Paradox" where although the human cells in culture show the greatest repair rate of UV induced damage, their sensitivity to the killing effects of UV irradiation is no different from similar cells taken from shorter lived species such as a mouse.[48] Clearly aging is not a result of UV damage and more relevant DNA repair processes that protect cells against other types of DNA damage found in major organ tissues needs to be examined. What may be happening here is that many of the DNA repair enzymes involved in repair age-related DNA damage (caused by ROS for example) of internal organs are also involved in UV/DNA repair of the skin. Thus in terms of DNA repair processes human cells are no more protected against death caused by UV irradiation, but may be more protected against other more harmful reactive oxygen species mediated damage.

4.6. Antioxidant Levels as a Function of Lifespan

4.6.1. *Superoxide Dismutase*

I next turned to testing the prediction that the antioxidant levels per Specific Metabolic Rate (SMR) in different tissues are positively correlated with lifespan. Here my main interest was in comparing human to the shorter-lived primate species.

The first antioxidant studied was Superoxide Dismutase (SOD) since this enzyme's only known role is the removal of the superoxide radical.[123] Since I was primarily interested in human and closely related non-human primate species, I established a frozen primate tissue bank for such comparative studies. Most non-human tissues came from the NIH primate centers where we had little control on ensuring freshness of tissue. Thus some variability we experienced could come from this potential problem. However, all human tissues came directly from a medical examiner's office in down-town Baltimore that consisted only of automobile accident autopsies. They were all frozen immediately and free of any known disease or pathology.

We measured SOD activity in several different tissues, but our best and most reliable results were in liver.[7, 51, 74–76, 108] The results of the SOD experiments clearly showed that human tissues had the highest level of SOD as compared to the great apes. There was also some species that had similar levels of SOD, but with a lifespan shorter than human. But since I was not seeking to determine if SOD correlated positively with lifespan this fact did not concern me. *Indeed there is absolutely no rational or hypothesis I am aware of that would predict or even suggest that the antioxidant status of a tissue taken from a given species should correlate positively to that species lifespan.* Surprisingly, some investigators apparently thought if antioxidants are important in determining aging rate then antioxidant concentration should correlate with a species lifespan. They of course found no significant correlation and then criticized my experiments for normalizing the results with specific metabolic rate (SMR),[77–85] I can find no scientific justification for their comments.

Moreover, the hypothesis I was testing was never was about antioxidants themselves acting as longevity determinants in isolation, but instead to determine if "steady state levels of oxidative related damage" is inversely related to lifespan.[48] The best test was of course to determine steady state levels of oxidative damage by direct measurement and we later did those experiments. However, an indirect method is to determine endogenous levels of the superoxide radical. This is possible by comparing the rate of endogenous production of the superoxide radical as estimated by the tissues SMR to the level of SOD activity. Thus the ratio SMR/SOD was assumed to be proportional to the endogenous free concentration of the superoxide radical that can cause damage within the cell.[48] So, the hypothesis I was testing was if the ratio SMR/SOD was proportional to the aging rate of a species or inversely proportional to their lifespan. To make this point very clear, I have indicated in previous publications that the amount of oxidative damage, now called "Oxidative Stress Status (OSS)" is under the control of several different parameters.[48, 76, 51]

The first controlling parameter is the endogenous rate of generation of the reactive oxygen species (ROS). This is clearly related to overall specific metabolic rate (SMR), but not necessarily a linear one across different species. A major source of ROS generation comes from the mitochondria that is related to both proton and electron leakage effects through the mitochondria membrane. Although it is now understood that the degree of ROS generation is under the control of a number of different parameters including fatty acid composition of the mitochondria membrane and "degree of uncoupling" controlled by specific uncoupling proteins, it is still not clear how rate of endogenous production of ROS varies with different species as a function of their lifespan.[80–83, 86–104] For example, *in vitro* measurement of the superoxide radical and hydrogen peroxide production using fragments of mitochondria parts under artificial *in vitro* incubation conditions are subject to many types of artifacts and may not be accurately reflecting true *in vivo* endogenous leakage.[77–91] However, these studies reported the remarkable result that leakage of the superoxide radical and hydrogen peroxide decrease with increase lifespan, which would be of fundamental importance if true.

But back in 1980 when our SOD comparative studies were first published it was then and still is today very reasonable to assume that the rate of ROS generation from mitochondria is proportional to the species characteristic Specific Metabolic Rate (SMR) and independent of their lifespan. That is the amount of oxygen consumed per unit time per gram weight of tissue produces the same amount of ATP and free radical leakage in different mammalian species. At that time we checked to see if the SMR of the whole animal is indeed related to the rate of oxygen consumption in the same tissue where the SOD was measured. The relation proved to be excelled as indicated in our publication.[51] *Of course if it turns out that leakage of ROS actually does decrease with increased lifespan then this result would make the positive correlation we have already found even more striking and statistically significant.*

A second controlling parameter of Oxidative Stress Status (OSS) is the concentration of antioxidants. Here, we assume that the degree of protection an antioxidant provides is based on the mass action law where the higher the tissue levels of SOD are, the greater the protection offered against the superoxide radical. These considerations indicate that the ratio of SMR per SOD would be proportional to effective free superoxide radical concentration available to generate damage.[48]

A third controlling parameter is the level of the various repair processes. Repair processes that remove oxidative damage are known for DNA, proteins and lipids. Although we knew UV repair process increased with lifespan, it was not clear then or even now if such a positive correlation exists in repair of ROS derived damage. Most importantly repair processes represent the "last resort" type of protection. For example, the repair process itself is known to create damage. Thus, I assumed that repair process would not likely decrease with age, to be in vast excess and not to increase with increasing lifespan.[48] The excellent positive correlations found that antioxidant per oxidant concentration is proportional to species lifespan supports these assumptions and suggests the possibility that DNA repair intensity may not have played a key role in the evolution of increased lifespan in primate species.

A fourth controlling parameter is the sensitivity of the target to oxidative damage and how rapidly oxidative damage could propagate through a lipid peroxidation reaction. This target sensitivity to oxidative stress was examined in a series of experiments. This was done by measuring the spontaneous rate of auto oxidation of different tissue sample suspensions as a function of the species lifespan. It was found that both tissue sensitivity to oxidation and the amount of oxidizable tissue per gram wet weight decreased with increased lifespan.[105] Similar results confirming this conclusion were later reported.[100–104, 106]

This data was interpreted as indicating that lipid membranes of cells from longer-lived species have less percentage of peroxidizable unsaturated fatty acids. Thus, we identified a new potentially important strategy to decrease oxidative stress and that is simply to decrease the sensitivity of the membranes to oxidation. However, if this increased in resistance is a result of a decrease in the unsaturated fatty acid components, then how is the fluidity of the membrane maintained and would this change in mitochondria membrane composition affect proton electron leakage rate?[105] Also since human membranes are the most resistant to auto-oxidation how would a diet high in the omega fatty acids effect oxidative stress status, membrane fluidly and lifespan?

Knowing all this we wanted to keep the experiment simple and focused only on what role SMR and SOD concentration played in determining human lifespan. If SMR is proportional to endogenous generation rate of ROS, then it follows that the ratio of SMR/SOD is proportional to effective ROS capable of producing oxidative damage. But the hypothesis being tested is if SMR/SOD is proportional to aging rate. If aging rate is inversely related to Lifespan (LS), then we have the prediction that SOD/SMR = k (LS). And this is exactly what we found. These data suggest, but do not prove of course, that the other parameters potentially effecting OSS as just reviewed may not play a dominant role in determining lifespan.[75, 107–110]

It is important to point out that it is really the outliers in our published SOD data that prove most convincing that this relation is correct.[51] For example, the data for gorilla indicating low SOD and a low SMR is consistent with large body size and low LEP. On the other hand for the lemur, its high SOD and a high SMR is consistent with its small body size and high LEP value.

In general, LEP value was found to correlate positively with SOD activity, which represents a beautiful explanation for why some species with high metabolic rate live so long. This relation is predicted by the linear relation of SOD/SMR with LS where on rearranging the variables we find SOD = k (LS)(SMR) = k LEP. This relation of SOD = kLEP was unexpected, but makes sense once you have discovered it. Thus, animals having high LEP values have unusually high SOD. High SOD would make it possible for the animals to keep the concentration of ROS low on a per amount of oxygen consumed and thus be able to consume higher amounts of energy over their lifespan.

This discovery was very exciting because for the first time I had an explanation of why the wear and tear theory worked for so many species but did not work for human and other primate species, which were the well known outliers.[7] What was discovered here is that Lifespan Energy Potential (LEP) was directly proportional to a species SOD antioxidant status. So for most non-primate species having LEP values of about 200–300 kc/g, then SOD is predicted to be similar and the rate of living hypothesis worked. But then for human with a LEP value of about 700–800 kc/g, SOD is predicted to be 2.1 to 4 times as high. We actually found SOD to be about 2.5 times as great in human as compared to mouse.

This first experiment therefore represented a great success being predicted by comparative biology and theoretical evolutionary studies that lead to a hypothesis to test and then later demonstrated experimentally to be correct. *It represents the first study to ever relate one enzyme activity (SOD) to longevity (LS) and accordingly to the maintenance of health status of human and other primates.* Thus according to our hypothesis SOD represented the first potential longevity determinant gene to be identified. These studies are of course only correlative and "Cause and Effect" studies are now justified as a follow-up. I am therefore happy to learn that later studies creating a transgenetic *Drosophila* strain having extra copies of SOD in their motor cells showed that this strain had a 30 to 40% longer lifespan. Thus, we finally have a cause and effect demonstration that SOD may indeed be a longevity determinant gene[43, 111, 112] as was predicted many years ago.

As a result of the potential significance of our original SOD/longevity study and the continued remarkable progress made in this area it is not too surprising our work has been subjected to some criticism.[77, 78, 85, 113, 114] Since this criticism has come from only two laboratories I have chosen not to respond to them in the past, since their criticism was largely in error, trivial and non-constructive in nature. I also felt that any scientists seriously interested in this area would simply review our papers first hand and decide for themselves. However, I was mistaken, since some prominent scientists in the field have been substantially influenced negatively by the publications of these two authors (for example, see Ref. 115). I will therefore briefly respond here on several of the key issues raised in these papers.

(1) The first comment is that SOD by itself is not related to lifespan when one examines all the species we studied. So, I should have concluded that SOD is not important in the control of aging rate.[77, 78, 85, 113] My answer is human does have the highest level of SOD as compared to all other shorter lifespan species that were studied and that there is a positive correlation. However, most importantly the lack of a correlation of SOD indicating it is not important in controlling aging rate makes no sense to me. I wonder what the rational might be in their thinking that SOD should correlate with lifespan. I have argued that it is the ratio of SOD/SMR that in theory should be correlated with lifespan.

(2) The above comment comes from two investigators that have since published a series of papers indicating the unimportance of antioxidants in aging as compared to the endogenous generation of ROS.[77, 78, 85] In this work, a study was published examining the SOD in pigeon, canary, toad, trout, frog, guinea pig and rat. They plotted the SOD activity in brain against the species lifespan and found no positive correlation. Instead, they found some evidence for a negative correlation in SOD and some other antioxidants. Their conclusion is that our published SOD and other antioxidant correlations with species lifespan were incorrect. But this criticism still does not make sense. I never predicted that antioxidants should correlate with lifespan, but rather lifespan should correlate inversely with a species Oxidative Stress Status (OSS). It is true however that I believe that OSS is in part determined by a species Antioxidant Status (AOS) as well as the rate of endogenous free radical production and many other factors. However, I think that plotting antioxidants activity against lifespan by itself is nonsense and cannot be interpreted without also considering endogenous ROS level the antioxidants protects against. In other words, it is the ratio of protection per degree of primary ROS generation that is predicted to be correlated to lifespan. And even here, I would also seriously hesitate to compare such diversity of animals as pigeon, canary, toads, trout and frog against one another. Many different strategies can be proposed of how oxidative stress may be controlled as I have already pointed out here and in many of my past publications. This is why I used mostly primate species with an emphasis on the great apes that are closely related to human in my comparative studies.[48]

(3) Another criticism I received is that the excellent positive correlation of SOD/ SMR with LS is really only a result of SOD being constant and the strong inverse correlation of SMA with lifespan.[83, 85, 113] This is certainly true. But let's first look at the exceptions that prove the rule and focus on the SMR issue. Although lifespan does have an inverse correlation with SMR there are many important exceptions and the most important one is human. For example, humans live much too long considering their high SMR. Human lifespan according to SMR should be shorter than gorilla, which has a lower SMR but a lifespan of only 50 years. But on recognizing that humans also have a higher SOD value than the gorilla this difference in SOD is sufficient to push lifespan up to 100 years. This result suggests why humans are able to live so long with this high SMR. It is because of their high SOD value.[74]

(4) Now, let's take on the constant SOD issue. It is true that SOD only varies about 2.5-fold where as SMR varies by about seven fold. So when many different species are taken under consideration particularly those having the same LEP values, then SMR will be the most important variable. But it is also not trivial that SOD has not changed. If for example, SOD activity decreases considerably with lifespan then no positive correlation would have been found. Thus, you cannot argue that SOD is not important simply because it does not increase with increasing lifespan of some species. We must also remember that SOD was substantially higher in human tissues. So the answer here is that antioxidants concentrations and the rate of endogenous generation of ROSs as well as many other factors are important in determining a species resultant Oxidative Stress Status (ROS). And as I had proposed many years ago it is Oxidative Stress Status that governs aging rate not antioxidants.[22, 74, 107, 110, 116] Finally, it should be recognized that the decrease in SMR with increasing lifespan should be viewed in itself to represent an important strategy to reduce oxidative stress. So, I believe that it was the combination of both the decrease in SMR and the increase in SOD that contributed to increase lifespan.[74]

(5) Still another criticism I had received was that a separate independent laboratory has never repeated the original SOD work we published.[85] This is simply not correct, for our results have not only been repeated but extended and improved by work done completely independent from our own laboratory. I am referring here to an excellent study reporting SOD activity in the brain of eleven different mammalian species having different lifespans.[52] In addition to SOD, they also measured the activity of five other enzymes not having any antioxidant activity as a control. SOD was found to have a significant positive correlation with lifespan, but the other five non-antioxidant enzymes did not. Now, the important aspect of this result is that this positive correlation was found **without** normalization of the data on a specific metabolic rate basis! Of course, the positive correlation found here still does not demonstrate the importance of SOD to lifespan. But this was not their objective. I did however normalize

their results against SMR of the species and found that the SOD/SMR ratio obtained gave a positive linear correlated with lifespan much more significantly than with SOD alone. Moreover, the slope of the line obtained is similar to what we found for the primate species. Thus, these data from another laboratory provided an important confirmation, extending our discovery to many other species and demonstrates that the correlation is special to SOD and not any random enzyme. It is of interest also that the laboratory that voiced this criticism of our work has since repeated these studies as well and confirmed our results using non-primate mammalian species.[83] Yet they never corrected their initial accusations.

4.6.2. *Dietary Antioxidants*

With the success of the SOD study we moved on to test other antioxidants for their possible significance in the evolution of human longevity. In many of these studies we used data taken from the scientific literature that were part of experiments having other research objectives. One laboratory did criticize our work using literature data instead of doing the experiments ourselves.[85] My answer is that literature data has the important advantage of not having any bias in interpretation.

However, the fact is we used literature data largely because we did not have sufficient funds to conduct the analytical measurements ourselves. In addition, there was sufficient data in the literature to justify at least a first look to determine if addition studies in our laboratory were warranted. The difficulty in using published scientific literature values is that if the results are random in nature then it is not clear if this is real or a result of bad data. However, the positive result as we had found using data from many different and independent laboratory studies is highly improbable. Also it can never be argued that we adjusted the data to fit a positive correlation. Anyone can check the data in the literature for themselves.

It is also important to point out that although it is clear that antioxidant status within cells of tissues needs to be evaluated on a per oxidant level in that tissue which that antioxidant protects against, it is not as straight forward how to evaluate antioxidant status in the serum. That is what is the initial generated oxidant level in serum? And what tissue or cells do the serum antioxidants protect? My guess is that the antioxidants status of the serum is most important to the cardiovascular system, its cells and the components within the serum as for example lipid proteins. It may be true that the steady state concentration of ROS in general in the serum is proportional to SMR of the entire organism, but this has never been demonstrated to my knowledge. Thus, when we present antioxidant data in the serum we always do so with and without normalization with SMR of the organism.

The first dietary antioxidant studied in serum was vitamin E. It is important here to also include a normalization of the vitamin E concentration against total lipid content of serum.[75, 76, 108, 117] That is, the vitamin E per cholesterol plus triglycerides. This ratio is necessary because there is an excellent positive correlation between vitamin E concentration and serum lipid concentration. The serum lipids absorb and carry most all vitamin E in the serum and serve as its major means of transport throughout the cardiovascular system. In addition, it is these lipids that primarily need the protection of the vitamin E. Thus, the ratio of vitamin E to lipid is proportional to the degree these lipids are being protected. I discovered that this ratio was clearly highest in human serum as compared with other shorter-lived mammalian species. On normalization of the data on a per SMR basis the positive correlation became even more significant.

These data suggest that vitamin E absorption is not entirely passive and that it may be subject to an activate absorption mechanism. This is because it appears unlikely that a species simply had more vitamin E in their food in proportion to their greater lifespan. We tested this idea by comparing vitamin E levels in two primate species eating the same Purina Primate Food Chow. The results showed that the longer-lived species had the higher plasma vitamin E level. There is also some evidence that selenium plays a role in the active transport of vitamin E and there is now known to be a specific receptor for the alpha tocopherol for its transport to specific tissues that is selenium dependent. There is also much evidence now for a protective role of vitamin E against a number of human diseases particularly cancer, Alzheimer's disease, and heart disease and in maintaining an active immune system. Thus, our conclusion at this stage of research indicates that vitamin E is a potentially important human longevity determinant but that the mechanism determining the high human levels needs to be identified. Finally, recent data indicate other functions for vitamin E other than an antioxidant.[123]

4.6.3. *Carotinoids*

An excellent positive correlation was found between total carotinoids level in serum and the lifespan of many different species including non-human primates and human. In these studies, humans had the highest level of total carotinoids in serum and other tissues as compared to other shorter-lived primate species.[118] On the other hand, vitamin A level did not show any correlation with lifespan. The unusually high carotinoid level in human tissues has been known for many years and gives human serum its characteristic yellow color. Some carotinoids are important as vitamin A precursors, but recently carotinoids have also been identified to be excellent antioxidants particularly as singlet oxygen scavengers.

The high level of carotinoids in human appears to be partially the result of an unusually low level of carotinoid digesting enzymes in the intestine. So perhaps a decrease in the activity of the carotinoids digestion enzymes with increased lifespan explains in part the excellent positive correlation found in carotinoids level with increased lifespan. This could be a result of changes in gene regulation selected for increasing human lifespan. Thus, the carotinoids represent a potential longevity determinant and the modifications reducing the rate of carotinoid degradation another class of longevity determinant genes.

It is of interest to note that the high levels of carotinoids in human tissues have been described as a human genetic defect since its only purpose was known only as a precursor for vitamin A. The truth may just be the opposite. The high level of carotinoids in human tissues may be an important and desirable human genetic and biochemical characteristic and responsible in part what helps promote health in humans.[118] A diet high in carotinoids content may also be the best means of obtaining the ideal amount of vitamin A without the danger of high toxic dosage as is possible when supplementing on the vitamin A and to also furnish the body with the protective carotinoids it may need as antioxidants and perhaps other functions.

4.6.4. *Vitamin C*

Vitamin C is thought to represent one of the most important protective anti-oxidants in mammalian species. Yet humans cannot synthesize vitamin C having lost this capacity early in primate evolution. On examining vitamin C levels in many different tissues and species, the general conclusion is that from mouse to man the tissues levels are so scattered that even when normalized on a per SMR basis there is no significant correlation with lifespan.[119] However, I need to be careful and limit this conclusion to serum, brain and liver tissue. There is still not sufficient data available for many other tissues to make a general statement as yet.

Although vitamin C is important for human health maintenance it appears clear that human longevity is not a result of unusually high or low tissue levels of vitamin C. Perhaps part of the explanation is the prooxidant nature of vitamin C where in the present of iron or copper ions free radicals can be generated. In addition, it is known that substantial energy is spent in the kidneys to prevent vitamin C from reaching high tissue levels. That is, there is a well-defined limit set point that prevents high blood and tissue concentrations of vitamin C. So although vitamin C is important as an antioxidant, perhaps toxic side reactions at higher tissue concentrations have reduced selective pressure for further increase in tissues concentration as selective pressure increased for longer lifespan. So in this case were the inability of human to synthesize vitamin C has been described as a genetic defect, it may actually have been a neutral mutation having little health or longevity consequence. It has also been suggested that urate may have replaced vitamin C as the antioxidant of choice for longer lived mammalian species.[119]

4.6.5. *Antioxidants having Negative Correlation with Lifespan*

Tissues level of the antioxidants glutathione, glutathione peroxidase, glutathione S transferase and catalase were determined in serum, blood and liver tissue as a function of lifespan of different mammalian species.[7, 75, 76, 107, 108, 110, 117] To my surprise, for these antioxidants a strong negative correlation with increasing lifespan was found before and after SMR normalization. Human values for these antioxidant enzymes and particularly for liver and serum were the lowest of all species examined. Similar results have been reported by other laboratories.

These results can be interpreted as follows. First, I am convinced that these data are correct since they were derived from data published from many different and independent laboratories. So it is clear that these antioxidants are less concentrated in tissues of mammalian species having increased lifespan. Now, whether a decreased concentration means a reduced role of these particular antioxidants in the evolution of increased lifespan remains to be demonstrated, but I think this is the case. The discovery of this inverse correlation is potentially very important since it is generally believed that these antioxidants are among the most important in protecting against humans against oxidative stress. So how can this inverse correlation be explained without discarding the entire hypothesis that oxidative stress plays an important role in aging?

One possible answer is lower thio-derived antioxidant levels does not necessarily mean higher oxidative stress. Could it not be that this inverse correlation occurs as part of a strategy to further lower oxidative stress? For example, the glutathione radical is very toxic and long-lived. So glutathione certainly is not an antioxidant without potential complications. In addition, the selenium in glutathione peroxide is known to be toxic and could cause problems when released during normal glutathione peroxide turnover/degradation. Finally, a similar problem may be present in the iron containing catalase where high levels of iron could be release during normal degradation of the catalase enzyme causing increase lipid peroxidation reactions and thus more lipid peroxides to be removed.

So in summary, I do not believe these negative correlations prove that all antioxidants are not important to the evolution of lifespan as has been suggested,[80, 81, 85, 90, 91] but rather this data serves to remind us that there may have been many different strategies taken to lower oxidative stress under the selective pressure to evolve longer and healthy lifespans.

A good example of such a strategy, which will be discussed in more detail later, is where we have found that with increasing lifespan in mammalian species, there is also a general decrease in the activity of the P450 cytochrome detoxification system.[120–122] Now, a characteristic of this system is the generation of hydrogen peroxide and the activation of many hydrocarbon derived mutagens. Thus, a lowering of this detoxification system may actually represent an important trade-off between the immediate benefits of the P450 detoxification system and the more long-term aging effects of a higher rate in generation of peroxides and other mutagens. But a lower rate of endogenous production of peroxides reduces the need for antioxidant enzymes to remove them, particularly if these very enzymes also had toxic effects themselves at high concentrations as the thio-derived antioxidants. So perhaps both a lowering in activity of these particular antioxidants that remove peroxides and a lowering of the P450 cytochrome system represent a strategy to actually lower oxidative stress at the cost of sacrificing the benefits of a high level P450 detoxification system. Such trade-off in evolution are common.

4.6.6. *Uric Acid*

Uric acid has both antioxidant and metal chelating properties.[123] In addition, its concentration is well known to be extraordinary high in human serum running close to saturation level. This high concentration of both serum and tissue uric acid represents an important risk factor to gout. Although uric acid is not a strong antioxidant itself, its high concentration makes it's the second highest antioxidants in a human serum sample. Thus, because uric acid is an antioxidant and represents another unusual human characteristic because of its high serum concentration, I investigated its potential role as being another longevity determinant.

Uric acid was measured in serum for a large number of primates and non-primate mammalian species.[119] An excellent linear positive concentration was found in its concentration as a function of increasing lifespan. This data was then separated into the various phyogenetic groups of primates and this further improved the correlation.

In addition to being an antioxidant and metal chelator uric acid has another interesting property. It is a neuro-stimulant similar to caffeine and other drugs having a methyl zanthine structure. Thus, uric acid is not only a potentially important longevity determinant antioxidant, but also a potential brain function enhancer. This neurostimulant property has been demonstrated by substantial data showing a positive correlation of "drive to succeed" with serum uric acid concentration in human populations.

The mechanism of how the high uric acid concentrations evolved in humans represents another example supporting alteration of gene expression in the evolution of human longevity. Much energy is use by the kidney to pump back the uric acid previously removed from the serum to maintain the high serum and tissue levels. In addition, during primate evolution, the enzyme uricase steady decreases to where in human no activity is present. So an uricase gene knockout event and kidney uric acid transport changes appear to be the mechanisms of how this potential longevity determinant evolved.

Presently most physicians consider uric acid a toxic waste product having no important biological function. Thus, even when it is only a little higher than normal in serum, physicians frequently prescribe drugs to lower its level. It has been reported however, that patients given such drugs to lower serum uric acid often feel depressed and lack energy and motivation to succeed.

It is also well known that high uric acid is associated with heart disease and that uric acid may actually help accelerate the disease. On the other hand, there is now evidence that high levels of uric acid in patients having heart disease may not be an aggravation or negative aspect of the disease but rather a potentially important adaptation.[124] Thus, new data on uric acid suggest it may be another potential human longevity determinant and that unless an individual runs a serious risk toward gout a high level of uric acid may be a desirable human characteristic to retain.

4.6.7. *Total Antioxidant Capacity*

The serum of different mammalian species contains many different antioxidants and in our comparative studies we measured only a few of these. It is very difficult to measure all the different kinds of antioxidants not only because of the high number involved and their chemical complexity but also because many are at a very low concentration. However, if we could determine the levels of the total antioxidant capacity of the serum and compare this to the major individual antioxidants that we can measure separately, then we could make some progress in estimating the presence and concentration of the antioxidants we are unable to measure. In addition, knowing total antioxidant capacity of serum could be a valuable test in its own right in evaluating if serum antioxidants are potentially important as longevity determinants.

We therefore developed a new technique to measure total antioxidant capacity of serum samples.[125–127] This procedure is called the Oxygen Radical Absorption Capacity or the ORAC assay. Using this new technique, total antioxidant capacities of serum samples were measured in different primate and non-primate species. The result showed a statistically significant linear increase in serum ORAC value with increased longevity. Most importantly it was clear that human serum had the greatest ORAC value as compared to the other shorter-lived primate species.

The major serum components accounting for antioxidant capacity are albumin, uric acid, total lipids and the dietary antioxidants vitamin C, vitamin E and the carotinoids. However, there are many other minor components such as the flavonoids. Human does not appear to have unusually high levels of albumin as compared to shorter-lived primates but does have unusually high levels of uric acid, carotinoids and vitamin E. We do not know at this time what other antioxidants such as the flavonoids and polyphenols in serum contribute to the high human ORAC value.

4.6.8. *P450 Cytochrome Detoxification System*

The activity of P450 cytochrome detoxification system and related enzymes was measured to estimate the general importance of this detoxification system to the evolution of human longevity. To our surprise, we found an inverse function in tissue concentrations of P450 activity as well as a decrease in glutathione, glutathione peroxidase and catalase as well as for glutathione *S* transferase with increased lifespan.[76, 109, 117, 120] *Thus human appears to have the least active P450 system detoxification system.* On further checking the literature, it was learned that this result should not have been surprising since it was well known in the pharmacy industry that the rate of metabolism of drugs in human is exceptionally slow as compared to the laboratory animal's mice and dogs for example.

There was also a very interesting paper reporting a decreasing ratio of the cytochromes P450 to P448 with increasing lifespan. The detoxification of the mutagen DMBA was measured using tissue culture cells taken from mammalian species having different lifespan including human.[3, 48, 75, 108, 117, 120] What was found is that P448 cytochrome is involved in mutageneous and with increasing lifespan although both P450 and P448 decreases, P448 decreases more rapidly to where none was even detectible in the human cells. These data indicate that regulatory gene change occurred during the evolution of increased lifespan to lower the potential of cells to generate mutagens related to detoxification processes. The experiments again support the general working hypothesis that alterations of regulatory gene activity is involved as a mechanism of longevity evolution in mammalian species.

These results also provided a new basis to explain why many of the antioxidant enzymes that remove hydrogen peroxide and related peroxides, particularly in the liver, are also lowest in longer-lived species. For example, it is well known that the P450 cytochrome system often produces peroxides and other ROS as by products. I therefore suggested that under evolution selection pressure for decreasing oxidative stress status, that the better trade-off was to lower the source of generation of the ROS rather than increasing thio-based and catalase enzyme antioxidant protection. That is to decrease cytochrome P450 production of peroxides by lowering the key metabolic factors such as the cytochrome P450 and P448 activities. This would be particularity beneficial if the antioxidants also had serious toxic side effects at higher concentrations as already noted.

The negative aspect of the proposed potential evolutionary trade off may be that as the result of the very low P450 cytochrome activity humans have now, there may be a more narrow range of food sources that are safe to eat because the required detoxification capacity is not present. All of this is of course only speculation based on some but certainly not enough data. Most importantly, however, is that these data represents still another demonstration of what potentially could be learned about human longevity using this comparative biochemical approach. By asking the question of "what is unique about human biology that can explain human longevity" the uniqueness we find is unusually high levels of serum uric acid and the carotinoids, unusually low levels of P450 cytochromes and the thio-based antioxidants and no correlation of vitamin C levels with longevity. In addition, it is of interest that most of these changes that occurred were a result of decrease in enzyme activities or knockout mutations. That is a knockout of an anti-longevity determinant genes. These unique human characteristics need to be evaluated in larger studies and may prove to be valuable in the design of a more appropriate diet and medical care procedures.

4.6.9. *Rate of Auto-oxidation of Tissue Homogenates*

Although the ratio of antioxidant activity per SMR was found to increase as a function of lifespan, the data indicating that the thio-based and catalase antioxidants in liver decreased with increasing lifespan was disconcerting and required further exploration. To explore this problem further we measured total antioxidant capacity in serum and found a highly positive correlation of ORAC value with increasing lifespan. Yet this was only in serum. We needed a similar test to measure net sensitivity of a whole tissue to oxidative stress. This was accomplished by measuring the spontaneous auto-oxidation rate of whole tissue homogenates.[105] In these experiments we did not need to add any endogenous generator of ROS, but simply allowed the whole tissue homogenate to spontaneous auto-oxidize. In addition, an endogenous ROS generator was not added so that we would obtain a more true indication as to the sensitivity of the tissue to both its innate ability to generate oxidants spontaneously and to resist the resulting auto-oxidation reaction.

Results indicated both the rate of autoxidation and total amount of oxidizable material per tissue weight decreased as a function of increasing lifespan.[105] These results indicated that tissues from longer-lived species were less sensitive to oxidation. This could be a result of higher levels of antioxidants and less available metal ions to catalyze lipid peroxidation reactions. However, the results also strongly indicated that tissues from longer-lived species had less oxidizeable material on a per total weight basis. Thus, we found support that longer-lived species were more protected against oxidative stress but also discovered a new mechanism of how resistance to oxidative stress may have evolved.

Since the major oxidation components we measured were products of lipid peroxidation, these data indicated that the lipid membranes from longer-lived species were less oxidizable on a per weight basis because of a less percent composition of unsaturated fatty acid. Thus, changes in the relative concentration of different membrane components, where the components themselves remain unchanged, are a result of gene regulatory changes. This result is again consistent with the longevity determinant gene hypothesis. These results were later confirmed and extended by same two laboratories that were so critical of our previous studies[79, 98, 100–104, 106, 113]

The original purpose of these autooxidation studies was to determine if longer-lived primate species and in particular human had an unusually low level of oxidative stress. The unusual low levels of liver thio and catalase antioxidants that remove peroxides in human tissues were not directly in support of this hypothesis although I have suggested a hypothesis to explain this apparent paradox. The auto-oxidation experiments however are supportive of human having the lowest oxidative stress status.

The fact that human membranes have a lower percent composition of unsaturated fatty acids raises the question of how this affects membrane fluidly which is highly dependent on essential fatty acids. In addition, because of the increased interest in adding more essential fatty acids and/or fish oil to our diet, such as the omega-3 and omega-6 fatty acids, there may now be the possibility of over dosing and increasing an individuals oxidative stress status. Experiments are urgently needed to answer these questions.

5. DNA Oxidative Damage

Previous experiments measuring antioxidant status, P450 cytochrome activity and rate of tissues auto-oxidation were done in an effort to access indirectly oxidative stress status in different primate species. However, this is no substitute for measuring directly oxidative stress status whenever possible. We therefore spent much effort measuring levels of DNA damage both in tissue DNA samples and in urine samples. We used the HPLC coupled with an ECD as well as a GC/MS instrument working with Mike Simic and Miral Dizadaroglu of NIST.[38, 48, 61, 63, 107, 110, 128–136]

In general, we found that with increasing lifespan of primate species concentration of oxidative DNA damage products decreased reaching a minimum levels in humans. This was true for both tissue DNA samples and in urine samples. The main oxidative DNA component we measured was 8-hydroxydeoxygunaosine (8Oh-dG) although we also measured thymidine glycol in tissue samples.[137] In my visit to Dr. Bruce Ames laboratory presenting a seminar on the evolution of human longevity and the potential importance of oxidative stress and many discussions thereafter, I was able to interest him to undertake similar studies using their techniques measuring oxidized nucleosides in urine sample of species having different lifespans. Their results were plotted as DNA damage versus SMR rather that lifespan. The implications are the same, however, indicating human steady state levels of oxidative damage were the lowest and short-lived species like mice and rats were the highest.

Steady state levels of oxidative DNA damage can be viewed as reflecting at least one component of the oxidative stress status of a cell. So, summarizing the results so far we have the consistent pattern that with increase lifespan in primates leading to human there is a decrease in steady state level of oxidized DNA damage, an increasing antioxidant status and a decreasing endogenous production of ROS as a result of a lower SMR and decreased cytochrome P450 activity. Whether there is also a decrease in production of ROS from mitochondria for a given amount of oxygen consumed would also be supportive but remains to be confirmed using new techniques not dependent on fragmented mitochondria.

We however discovered another very interesting result, which we felt important but have only presented preliminary results in review articles. This result was obtained when oxidative DNA damage was measured as a function of age throughout the entire lifespan. Here, we found that for most of an individual's lifespan DNA damage remained fairly constant except on being highest in infants and children in early stages of development where SMR and growth rate were also the highest and in the last 25% or so of lifespan.[61, 108, 110, 116, 117] Thus, we found a U shaped curve. Such data if confirmed could have very important implications for it suggests that the high dose of endogenous generated DNA damage occurring in early life may actually set the stage for later life age-related complications much like an acute dose of ionization radiation in young individuals would have in accelerating age like symptoms 20 years later.

We then completed the age dependent study of oxidized DNA damage for a number of different primate and non-primate mammalian species and plotted this data together as a family of curves as a function of fraction of lifespan and found that with increasing lifespan the degree of oxidized DNA damage decreased most rapidly in the early years of lifespan. These data suggest that selection pressure to decrease oxidative stress status during the evolution of increased lifespan was largely focused in the early developing years of lifespan. Thus, it may be infants and children that are most susceptible to oxidative stress and in adult life we are simply coasting into an aging status already set in place many years earlier. It is also interesting to note that these data are consistent with the pleiotropic aging hypothesis were the harmful effects of oxidative stress would have more selective pressure to be reduced at young premature ages than for the post maturation and aged individuals.

Unfortunately in these studies, we used the GC/MS method of analysis of DNA damage that disagreed with our HPLC/ECD analysis, so we never were competent enough of the data to publish it. Thus, these experiments should now be repeated using recent advanced techniques now available in DNA extraction and LC/MS/MS analysis of oxidized damage.

6. Glycation and Glycoxidation

In addition to the pleiotropic effects of oxidative stress there is another potential serious pleiotropic effect that is beneficial in youth but later contributes importantly to general aging process. This is the reactivity of the reducing sugars such as glucose reacting with proteins, nucleic acids and membranes forming adducts, cross links and then on further reaction with ROS producing an Advanced Glycated

End product (AGE products). Such glycation/glycoxidation processes can have a broad series of consequences altering proper and optimum functions of cells and tissue components and thus, contributing to the general aging process.[138–140] In general, it is well known that AGE products accumulate with age resulting in general, stiffing of tissues such as in skin, the cardiovascular system and in elastic tissues as the urine bladder and lung diaphragm. Moreover, this glycation/ glycoxidation process produces a number of toxic byproducts as the alpha oxoaldehydes (methylglyoxal) that can irreversibly destroy protein and nucleic acid function.

I was therefore interested to determine if the contributions of glycation and glycoxidation were reduced during the evolution of primate longevity.[141] In this effort, I was able to interest a colleague to collaborate with us in this study. He had developed an assay to measure a marker of this process in collagen called pentosidine. Pentosidine was measured in skin samples from three primate species and eight non-primate mammalian species. The results showed that pentosidine increased in an exponential manner with age and that the rate of this increase was inversely proportional with lifespan. Most importantly human shows the least age dependent rate of increase of this glycoxidation marker.

An important aspect of this data, however, is although the rate of accumulation of AGE products are slower with increasing lifespan this rate does not decrease to the same extent of increased lifespan. Thus, with increased lifespan the very old and long-lived human individuals carry a higher load of AGE products as well as higher general protein cross-linkage density. So old human tissue has much higher concentration of AGE products and protein cross-links than an old mouse or other shorter lived species. Therefore the problem of Glycation/ Glycoxidation has been reduced but certainly not eliminated during the evolution of increased lifespan in the primate species.

Protective and repair process are known to exist that can control the rate of accumulation of glycation/glycoxidation products. These include the Glyoxalase System, which consists of glyoxalase I and II enzymes. In addition, there are specific receptors on cells (RAGE) that detect and ingest AGE products.[142–145] Thus, during the evolution of human longevity changes in gene regulation may have occurred resulting in an increase in expression of the Glyoxalase System that may have lead to the decrease in rate of accumulation of AGE products during primate evolution. Experiments now need to be done to determine mechanism(s) that control AGE product accumulation in humans.

7. Genetic Stability, Cancer and the Dysdifferentiation Hypothesis of Aging

The existence of primary aging processes and related primary longevity determinant processes have been proposed and preliminary studies review here has supported this hypothesis. The next step is to inquire how oxidative stress causes aging and how longevity determinant genes control aging rate.

Most thinking has been that Reactive Oxygen Species (ROS), Reactive Nitrogen Species (RNS) and Glycation and Glycoxidation Reactions (GGR) cause damage that accumulates and impairs normal operation of the cell.[123] Thus, rate of accumulation of damage would be predicted to be proportional to aging rate. Yet there is little data demonstrating any physiological aging process being the result of an accumulation of either cellular or intracellular damage.

The best arguments proposed along this line of thinking have been data showing an age dependent accumulation of oxidized proteins and DNA. But this data appears to indicate that rate of accumulation of damage only becomes significant in animals that have already aged considerably such as in very old animals. For example, mitochondria do not appear to become seriously leaky in producing hydrogen peroxide or superoxide radical until the animal is already very old. The same appears to be true in the age-dependent accumulation of oxidized proteins.[146, 147] These data appear to imply that most damage appears to accumulate late in life as a result of an auto-catalytic lost of defense and repair processes. We would then have a positive feedback resulting in high rates of damage in very old aged individuals.

On the other hand, it is known that age related decline in physiological functions really begins early in life shortly after sexual maturity and continues to decline in a steady linear manner and not at an exponential rate of decline throughout life. *Thus, there appears to be no correlation in rate of accumulation of oxidative damage and rate of physiological aging.*

Lost of division potential through the shorting of telomeres does not likely seem to be a primary aging mechanism because the rate of aging appears to be the same in both dividing and non-dividing cells of an organism.[148, 149] In addition, there appears to be no correlation between the length of telomers and activity of telomerase and lifespan of different primates. For example, human telomere length appears no greater than the much shorter-lived non-human primates. Indeed, support that the finite Population Doubling Potential (PDP) of normal cell in tissues culture plays a key role in normal *in vivo* aging processes was (1) PDP decreases with increased age of the donor (2) PDP increases with increasing lifespan of the donor species and (3) cells when they reach the limits of their PDP die. Recent publications now indicate that none of these three conditions can be substantiated.

The error catastrophe of defective proteins was also a popular idea but has now been proven not to occur as expected even in *E. coli* bacteria were high degrees of error input still failed to produce a catastrophe. I had published a similar hypothesis suggesting an error catastrophe of proper gene regulation that still may be of importance.[3, 4, 7, 38, 47, 63, 110, 130, 150]

Another basic problem is that there does not appear to be sufficient amount of damage increasing with age that can in principal cause aging to the degree we know occurs. Indeed, the basic house keeping functions in the cells taken from old individuals, as DNA synthesis, protein synthesis and even cell division appeared largely intact and sound. Even the doubling potential of cells in culture taken from elderly human individuals are not significantly decreased[151] as previously reported.[152]

So how does oxidative damage cause aging? My answer is that the aging process is largely the reversal of the very differentiation and developmental processes that created the organism. I have called this process Dysdifferentiation.[38, 63, 130, 132, 133, 135, 150, 153–159] The proposal is that once the organism has reached the end of its genetically programmed stages of differentiation and development, then energy must be expended to maintain this state of differentiation. That is, the highly specialized state of differentiation of cell is not fixed or "burnt in" as one might think but instead is innately instable with the necessity of constant energy maintenance process being required to maintain its proper functional state.

Unfortunately, much needs to be learned as to mechanisms acting to stabilize and maintain proper state of differentiation.[153, 156, 160–170] However, I propose that the lifespan of an organism is directly related to the stability of its cells to maintain their proper state of differentiation. Thus the primary function of Longevity Determinant Genes is to maintain the proper differentiated state of cells. Of course, the degree and extent such maintenance processes would exist is only to the point necessary to assume sufficient postponement of aging so that death is by other causes unrelated to aging for animals living in their natural ecological nitch.

This process of age dependent lost of differentiation or dysdifferentiation occurs as a result of cells gradually losing their state of gene expression stringency resulting in general, the relaxation of gene control. This results in a gradual drifting away of cells from their optimum state of differentiation as a result of a mix alteration of improper gene expression and repression. Potential examples of dysdifferentiation are many of the age-dependent changes known to occur with age that have been defined as examples of damage rather than a change in state of differentiation. This includes lost of hormone receptors, degeneration of neurons, and the appearance of improper functional cells that are benign or cancerous.[38, 48, 61, 130, 132, 150]

One of the key arguments here in support of this hypothesis is that although oxidative damage can destroy a cell and/or cause serious impairment in function, that much less damage is required to change the differentiate state of a cell. Thus, I have proposed that the most sensitive target of a cell to oxidative damage is the genetic apparatus of the cell controlling maintenance of proper differentiation. It is well known that extraordinary small degrees of oxidative damage in a cell can switch gene express in an epigenetic-like manner. *Indeed, I believe that the same processes that were involved in changing gene expression during normal differentiation and development are the ones involved in the dysdifferentiation process. That is, no mutations are required and that the processes is largely epigenetic in nature.*

There is also much morphological data supporting that dysdifferentiation does occur. Examples are the many different types of abnormal cells that accumulate with age known as metaplasias cells, abnormal hair growth, and the age-dependent appearance of abnormal proteins. In addition, growth of normal diploid cells in tissue culture eventually reach a crisis state. After this state they frequently spontaneously transform to cancer cells. Cells taken from species having a longer lifespan are able to progress through more cell division cycles before reaching this crisis state. *Most interesting here in the remarkable stability of human cells in culture to resist this crisis state and spontaneous transformation.* These data strongly supports this hypothesis indicating that the mechanisms responsible for this stability of human cells in culture are the same mechanisms responsible for human having the longest and healthiest lifespan of primate species. This is again why I have proposed that the primary function of longevity determinant genes is the maintenance of the differentiated state of cells.

Although there is little known about stability mechanisms of differentiation, the control of oxidative stress status within a cell is likely to be involved. First, I would like to point out that oxidative damage does not have to accumulate with age in cells to cause aging or dysdifferentiation. That is, aging rate does not need to be positively correlated with rate of accumulation of oxidative damage. Instead aging rate of the organism is predicted to correlate positively with the rate of dysdifferentiation occurring which in turn is positively related to the oxidative stress status of the cell. *Thus, the probability that an alteration in gene regulation or an epigenetic event occurs is proposed to be directly related to the oxidative stress status of the cell.*

One means of how this can be understood is the concept of dwell time of an oxidative damage event. Dwell time is the length of time damage remains in the genetic apparatus before it is removed by normal repair and/or turnover processes. It is related to intensity of DNA repair. Long dwell times lead to higher rates of dysdifferentiation. This is related to intensity of endogenous production of ROS/RNS and antioxidant defense.

Thus, rate of dysdifferentiation of a cell can vary greatly by changes in the ratio of damage input to damage output. Yet in no case is it necessary for damage to accumulate with time to increase the degree of dysdifferentiation. All DNA damage is eventually repaired. And the bottom line here is that the higher the Oxidative Stress Status within the genetic apparatus the more chance there is this damage that would lead to an alteration of proper gene regulation that would lead to a dysdifferiatated cell even though this damage is repaired.

So ideally it is the steady state level of oxidative damage that needs to be low for long life and this may be more important than a slow rate of accumulation of damage. This is the reason why preventive mechanisms to reduce the initial damage of cell components together with efficient repair mechanisms to quickly remove this damage are so important.

In this model, the Oxidative Stress Status of the genetic apparatus of a cell is proportion to the probability that a dysdifferentiation event will occur. In turn, Oxidative Stress Status is proportional to aging rate of an organism and inversely proportional to lifespan. These novel concepts differ substantially with the thinking of other investigators in this field where they emphasize that oxidative damage has to accumulate with age to be important as a cause of aging. They also have argued in turn that antioxidants have to decrease with age or free radical production has to increase with age to be important to aging.[48, 108, 117, 132, 133] I disagree with this dogma.

One finial point. The amount of oxidative damage in DNA on a per base level is less in long-lived species that short-lived species. But what should be clear is that this is a direct measurement of the Oxidative Stress Status of that DNA. For example, the steady state level of oxidized DNA damage in a mouse liver tissue is much higher than for the liver tissue from human at any age. My prediction is that the low Oxidative Stress Status in human liver DNA plays a direct role in determining the rate dysdifferentiation events that will occur. And this one reason why humans live longer than mice.

We begin an investigation to test the dysdifferention hypothesis of aging by determining if improper expression of genes in cells occurred with age.[132, 133, 135, 153, 156–159, 171, 172] The first gene investigated was the hemoglobin gene.[158] Here, we found a significant increase with age of hemoglobin gene expression in brain tissue. These studies were then extended to include the c-type viruses and many other defective endogenous viruses. Results from these studies indicated significant increase with age of improper gene expression even though most improper mRNA transcribed was blocked from passing through the cell nuclear membrane. Such a selective blockage of transport of improper transcribed genes represents another potential longevity determinant mechanism protecting the proper differentiated state of the cell. We then examined the possible age dependent increase of oncogenes as *c-myc* gene and found a significant increase in expression of this gene with increasing age of mice.[130, 133, 156, 159]

Related to this work was a series of experiments we did measuring the degree of methylation of chromatin as a function of aging and aging rate. Methylation of chromatin is involved in controlling gene expression on a coarse level. A decrease in methylation frequently results in an increase in gene expression whereas an increase in methylation represses gene expression.[153] We found that as a function of age that the level of genomic 5-methyl deoxycytidine decreases significantly with age and that the rate of decrease is related to aging rate of the animal. Here, human cells were most stable in maintaining methylation and show the slowest rate of decrease in methylation status.

These studies generally support the dysdifferentiation concept. But given now that dysdifferentiation occurs, the key question is if it is enough to account for the aging process? This is really the critical question. Technology is now available to help answer that question and that is to take advantage of the microarray gene techniques to study gene expression profiles as a function of age in many different cells and tissue types. Such studies can now determine the transcription profile of tens of thousands of genes. Thus, using this new technology, it is now possible to more critically test the dysdifferentiated hypothesis of aging.

It is of interest to learn what might be the key processes that act to stabilize the differentiated state of cells. Unfortunately, surprising few papers appear to be written on this subject. Clearly oxidative stress status need to be low which is determined in part by the rate of endogenous generation of ROS, levels of antioxidants and repair intensity of nucleic acids, proteins and lipids. In addition, the genetic apparatus would be expected to be more stable in a reductive rather than an oxidative environment. Thus, oxidative/reductive regulation by the thioredoxin system may be very important. Potential stabilization factors are likely to include chromatin structure and relative concentration of the various chromatin structural proteins and related chromatin modifying enzymes. Protection against oxidative mediated damage and means of renewal and repair are also potentially important as already mentioned. Finally, means of silencing the transcription of unwanted or improper gene expression is potentially important particularly on consideration that most of the genome does not code for structural genes. But any mechanism that decreases the perturbation of the genetic apparatus of cells as oxidative damage or glycation or decreases membrane composition of unsaturated fatty acids could be important in stabilizing proper cellular differentiation.

The onset frequency of cancer increases exponentially with increasing age in all mammalian species roughly as a function of the fifth power of age. What is most interesting, however, is that rate of increase in the onset frequency of cancer in a given species is directly proportional to their innate aging rate. Thus, cancer is a major cause of death in mice by the time they reach the age of two years, but for human it is not serious until an individual reaches the age of 60 years. So it takes about 30 times the length of time to generate the same type of cancer in mice as compared to human. In fact, the argument has been made because humans have many more cells at risk than a mouse, the innate resistance of human cells to be transformed must be several order of magnitude greater.

These data suggest a close linkage of aging rate with cancer rate across all mammalian species and suggested to me that cancer and aging may have common causes and mechanism of control.[38, 63, 132, 133, 154, 171] Because cancer is also initiated by mutations and is a dysdifferentiation phenomena, then taking all of these data in consideration does strongly suggest that at least some of the causes of cancer and aging are the same. *So when longevity evolved, I have proposed that an increase resistance to cancer also evolved simultaneously because the same problems and solutions are involved.*

Now, cancer is a good example of a cellular dysdifferentiation event that is age dependent and where certain genes are expressed as a result of mutations and/epigenetic effects. Many of the mutations in genes at the initiation stage of cancer are known to be a result of oxidative damage or defective DNA repair. So antioxidants protection and DNA repair processes are known stabilizers of the differentiation state as already proposed. In this regard, I have proposed an "initiation and propagation model of aging" that is similar to the "initiation and propagation model of cancer". In this model, cancer is a special case of the general dysdifferentiation of cells that occur naturally with age. The only difference is a dysdifferentiated cancer cell grows and divides without restraint and kills the host rather quickly whereas the dysdifferentiated aged cell simply does not function as efficiently and creates inefficiency and aging processes rather than immediate death.

In this model, the aging process would consist of the initiation step followed by several propagation steps. Interesting these latter propagation steps, which is called the latent period, is known to be a function of the lifespan of the species. That is, the latent period from the initiation stage to the appearance of cancer is much shorter in short-lived species and steadily increases as the species lifespan increases. So factors controlling both sensitivity to initiation and latent period length may be key for governing overall rate of aging and lifespan. One last point here is that this dysdifferentiation model of aging suggests that most types of cancer should not be thought of as disease that can be prevented anymore than we can now prevent aging. *Instead, cancer could very well be a normal aspect of the aging process and thus, control of cancer will only come from a control of the aging process itself.*

8. Recent Support from Other Laboratories

8.1. Hominid Evolution

Since I published the original studies on the evolution of lifespan along the ancestral descendant sequence leading to *Homo sapiens* in the 1970s, many new discoveries in hominid fossils have been made providing addition estimates of brain and body size. These data were found to further confirm the original conclusion of the extraordinary rapid increase in lifespan that occurred during hominid evolution.[173–181]

Rapid evolution of male reproductive genes in the decent of man has also recently been reported.[179] There also continues to be high interest in the evolution of brain size and now, there are excellent comparative estimates of the cerebellum comparing monkeys, apes and humans.[182] In addition, excellent reviews have recently been written on the genetics and origin of the hominid species.[183]

There is also a new interest in another very close living relative to human and that is the bongos or the Pan paniscus. This great ape is closely related to the better-known chimpanzee but has a number of characteristics that make it even more human like. Thus, it will be of considerable interest to learn in future studies if the behavioral pattern of the bongos supports the proposal of learned behavior being a key driving forces in primate evolution of increased lifespan.[184, 185]

Molecular genetics continues to have an important impact on evolution studies of humans and non-human primates. For example, the sequencing of mitochondria DNA is being used not only for evolutionary lineage studies but also now in comparative biochemical functional studies.[186, 187] Much more information is also now known about mechanism of evolution including the human transposable elements[170, 188–192] and the role of epigenetics and its involvement in disease.[166] SINE-R elements are a class of retroposon elements that are derived from HERV-K endogenous retrovirus and are very active in hominoid evolution.[165, 193–198] There is now considerable evidence that during the evolution of human longevity that the retroelements and retrosequences played a major role.[199–201] Alu DNA repeats have also played an important role in human/primate evolution.[164, 202, 203] These Alu elements are involved not only in genome evolution but importantly in genome instability.[204]

There is also now additional supporting evidence that rapid speciation involved parallel and directional selection on regulatory genetic pathways, a key postulate I had proposed in the mechanism of the evolution of longevity in primate species.[199–201, 205] Another important area is phenotypic instability and gene silencing.[206, 207] Gene silencing is of increasing interest and there is some evidence that a low mutation rate is required for the evolution of large genomes where genes silencing is required for stability and lowering transcription noise level.[208] There has been some attention placed on the relation between aging, evolution and individual health span[209] and the relation of differentiation and gene regulation.[160, 166, 168, 183, 187, 208–220] Here, evolution of regulatory genetic pathways can play an important role in speciation[69, 70] and microarrays based techniques do offer new hope for understanding gene regulation from transcription profile analysis involving many thousands of genes.[14, 15, 44, 149, 161, 205, 221–226, 227]

8.2. Dysdifferentiation

The exciting news now is that the technology exists at last to measure genome wide transcription profile activity as a function of age. My first effort in the field of gerontology was to undertake a complete transcription profiling of major tissues and cell types as a function of age in the mouse very much along the same line I had used in measuring the transcription profile of ten regions of the genome of the bacterial *E. coli* as a function of cell division cycle.[22] From these and following experiments, I proposed the dysdifferentiation hypothesis of aging.[4, 27, 38, 61, 129, 130, 150] But the testing of this hypothesis required more exact, precise and most importantly sensitive assays to measure gene activity than existed when I was conducting these experiments even though I pushed the technology to its limits.[22]

This hypothesis and my initial experimental efforts in this area was for some time considered invalid in light of results of one paper indicating no changes in gene activities (then called complexity of mRNA) occurring with age.[228] This has now all changed with the recent availability of microarray analysis where transcriptional activity of thousands of genes can be determined simultaneously.[221] These new techniques, however, require skill in their use and interpretation, but the technique is improving rapidly and becoming much less expensive.[227]

Applications of this technique have now provided evidence for mitotic misregulation of genes in dividing cell fibroblast taken from humans of different age.[229] There is also now excellent data indicating that oxidative stress induces a dysdifferention processes in tissue culture cells.[230] Genome-wide transcription profiling was carried out as a function of age in *Drosophila* under normal and under oxidative stress induced by the mitochondria uncoupling agent paraquat. Results clearly showed significant gene expression changes occurring with age.[231] Some of these changes were a response to higher oxidative stress levels as a result of normal aging processes but other age-related changes were not associates with such as response.

Gene expression profiling was recently reported in skeletal muscle[225] and brain[226] in mice as function of age and caloric restriction. Here use was made of a 6347-gene microarray, which showed a rich change of transcription profile patterns as function of age and caloric restriction. Results suggest the transcription profile in the older mice is indictive of an increase oxidation stress and inflammation.

So far, these results are important for they indicate that much of the age related changes observed are not the aging processes itself, but instead a response or adaptation to the aging process. *Thus, there is still the problem here of determining if changes in gene expression are in the improper dysdifferention category or normal adaptation category.*

The importance of DNA methylation as a protective mechanism of differentiation has recently been addressed. Interest in DNA methylation is largely centered on its importance in the occurrence of cancer. Since our model suggests cancer and aging to have similar causes and control, this area is obviously of interest to us.[160, 218, 220, 232–236] Alterations in DNA methylation are widespread in cancer. An interesting new concept, however, is that it serves an adaptive role to silencing gene mutations. Yet in doing so it also shuts down needed gene function that could result in dysdiffentiation and aging.

8.3. Oxidative Stress

The first hypothesis I set out to test was the potential importance of the by-product of oxygen metabolism to be primary aging process.[48, 51] This hypothesis was tested by estimating the intensity of ROS production free to interact with cellular components. An estimate of the intensity of this ROS flux was made by measuring the ratio of antioxidant status per amount of initial endogenous generation of ROS. For example, in the case of the superoxide free radical the intensity of free superoxide available to react with cellular components and thus cause important damage affecting proper cell function was estimated by measuring the concentration of SOD in a cell per amount of initial generation of superoxide radical or SMR.

Since I had no means at that time to directly measure this initial generation of the superoxide radical, which was thought to come largely from mitochondria, I assumed that this rate would be proportional to the rate oxygen was consumed by the mitochondria. This rate of oxygen consumption was found to be proportional to the specific metabolic rate (SMR) of an animal, which is amount of oxygen consumption per day per body weight. This parameter can also be expressed on a calorie basis since there is a direct conversion of oxygen consumption to calories utilized.

Thus, I assumed that the ratio of SOD per SMR is proportional to amount of superoxide available to cause aging processes. The data generated in testing this hypothesis was positive showing that for many different primate and non-primate species that SOD per SMA was positively proportional to the species lifespan or SOD/SMR = k Lifespan. (Please see Refs. 48, 76, 117, 131 for detail equations and discussion of this issue.)

Since this paper was published in 1980, considerable progress has been made in confirming this relation of SMR being positively correlated with the rate of production of endogenous generation of ROS from mitochondria. First, the mechanism of production of Reactive Oxygen Species (ROS) and Reactive Nitrogen Species (RNS) from mitochondria has been firmed up and its general importance to aging strengthen.[93, 95–97, 99, 137, 237–239] Addition information has also been published as to the control of ROS and RNS related to proton and electron leakage.[93, 240–245] Finally and perhaps most importantly experiments have been published using *in vitro* fragmented mitochondria showing that a significant positive correlation exists between the SMR of a species and the endogenous production of ROS.[77, 80–83, 85, 89–91, 102, 113, 114, 246] This result clearly represents a very important confirmation of our findings and hypothesis.

What is surprising and very disappointing however is that the authors of this last set of papers (R. Sohol and G. Barja) interpreted their results as indicating the unimportance of antioxidants in determining the oxidative stress status of an organism. They instead emphasized that it is only the rate of endogenous generation of ROS that is important in determining aging rate.[79] This conclusion appears to be based on the fact that they did not find antioxidant status to decrease with age and that antioxidant status did not as positively related to lifespan as SMA or endogenous production of ROS. As a result, one of these authors have proposed their own separate hypothesis emphasizing this interpretation and have called it the Oxidative Stress Hypothesis of Aging.[84]

Now, certainly the rate of endogenous generations of ROS and RNS are important determinant of aging rate and that is why it was included in our analysis as being estimated as proportional to SMR. But our data already has indicated that the decrease in SMR with lifespan is an important strategy to lowering oxidative stress status. Thus, I find nothing new here in this hypothesis. Simply because they have rediscovered that SMR plays an important role in aging does not in any way indicate my hypothesis was incorrect. Instead their new data actually further supports our original findings.

In many species it is obvious that a decrease in SMR is primarily responsible for an observed increase in lifespan where antioxidant status does not significant change. But it is ridiculous to propose that antioxidants do not also play an important role particularly in species showing high LEP value as in human. Thus, it is the endogenous generation rate of ROS and antioxidants that are important and in addition to repair processes and sensitivity of targets to oxidative modification.

It has recently been shown that the rate of endogenous production of ROS and RNS are under control of nitric oxide[244, 245] and the presence of specific mitochondria uncoupling proteins.[86, 87, 92, 95, 96, 223, 241–243] Also in smaller animals that require a higher SMR to maintain a 37°C body temperature, the ratio of proton leakage to electron leakage may be important in determining the net rate of ROS production. There is even the possibility that mitochondria efficiency has improved with increase lifespan in mammalians species where there has been a decrease in the rate of ROS and RNS per rate of oxygen consumption. If this proves correct then the correlation of SOD per SMR to lifespan would be even more significant.

The spontaneous rate and degree of auto-oxidation of tissues decreases with increase lifespan.[105] These data represented an important new means in confirming the prediction that longer-lived species did evolve greater resistance to ROS related damage. These data added the new idea that not only was endogenous production rate of ROS and RNS and the endogenous level of antioxidants important to evolution of increased lifespan but that the decrease sensitivity of the target to oxidation also played an important role. I suggested that one reason for the decrease in sensitivity of tissues to lipid peroxidation might be a decrease in membrane content of unsaturated fatty acids. This suggestion has now been confirmed and extended showing that indeed, there has been a decrease in percent fatty acids in cellular and mitochondria membranes with increase of lifespan[79, 88, 98, 100–104, 106]

8.4. Redox Regulation and Oxidative Stress Homeostasis

The success of our studies indicating a general decrease in oxidative stress with increase lifespan required us to address the issue of why dietary supplement of antioxidants failed to substantially increase lifespan of experimental animals as mice and rats. The most impressively failure in this regard was from the work of Dr. Denham Harman.[247–249] His inability to demonstrate a significant increase in lifespan in normally long-lived rodent strains has been used most strongly to argue against free radicals being important as a cause of aging. Now, it is true that in animals under stress or animal strains having shorter than normal lifespan often showed an increase in life expectancy but even here lifespan is not increased significantly. Clearly if free radicals were the primary cause of aging, then feeding mice antioxidants many fold over the levels they normally would get, should increase lifespan many times over not just the few percent in the most positive examples published. My conclusion was that there appears to be a fundamental problem here with the free radical theory of aging as originally proposed.

Related to this problem it should be recognized that the well-known free radical hypothesis of aging as proposed by Dr. Harman has an equally unknown omission.[64–66] This omission is that his hypothesis never proposed a mechanism of how free radicals might be controlled **naturally** to account for the different lifespans of different species! The proposal that endogenous defense and repair processes may control the different lifespans of species was originally made by myself in 1972.[22] It was in fact this prediction that lead me to investigate if SOD and other antioxidants indeed played a role controlling aging rate.

Now that our experiments strongly support the importance of oxidative stress as an important determinant of aging rate, I needed to return back to the antioxidant supplementation experiments and try to understand why they failed to significantly increase lifespan. *The answer I arrived at is the presence of a compensation/adaptation mechanism that maintains the oxidative stress status of an animal over a wide range of dietary antioxidant intakes and other factors as exercise.* Such an oxidative stress regulatory system has never been proposed before to my knowledge. I predicted that each species had a given set point where oxidative stress would be maintained much like the set point and maintenance of body temperature. Thus, antioxidant supplements would only be of help to an organism if they were needed to correct an unusually high level of oxidative stress that could not be controlled by natural means. In addition, an important prediction of this compensation/adaptation model of oxidative stress was that lifespan could never be extended beyond the normal lifespan through dietary supplements of normal food antioxidants. This is because as long as the adaptation/compensation forces were in play a net gain in antioxidant protection beyond the genetically established set point was impossible.[108–110, 116, 117]

This compensational/adaptation model of oxidative stress regulation was based on data indicating presence of compensation processes in mice when given dietary antioxidant supplements. For example, mice on a vitamin E deficient diet have unusually high tissue levels of SOD, but when placed on a vitamin E supplement diet they had very low levels of SOD. Many other similar examples in the literature can be sited.

The oxidative stress compensation model also explains why dietary supplements of antioxidants have minimum effect in healthy long-lived mice. In Harman's experiments as well as those from other laboratories, the mistake made was not measuring the predicted effects of the antioxidant supplement diet. That is, they did not measure if the antioxidants supplements actually reduced oxidative stress. One reason why this experiment was never undertaken may have been a lack of appreciation of endogenous antioxidants as a natural control of oxidative stress in the first place. Because Harman never predicted endogenous antioxidants to play a role in determining aging rate, then how could he have proposed their role in maintaining a set point of oxidative stress controlling aging rate? My prediction is that if they did measure oxidative stress, then they would have discovered that it was not lowered in most cases and that an increase in lifespan would therefore not be expected.

The proposal of the existence of a tightly regulated maintenance of oxidative stress status has far reaching consequences in the design of experiments in testing the importance of dietary antioxidants to reduce incident of disease and extend lifespan. For example, we have found that most humans are able to maintain their set point of oxidative stress and so no matter how much additional antioxidant supplement they consumed in their diet further decrease in oxidative stress does not occur.

Thus, such compensational effects would tend to give researchers negative results for the effectiveness of dietary antioxidants supplements to disease incidence of disease. This possibility may be a basis of the difficulties clinical studies are presently having in demonstrating effectiveness of antioxidant supplements in decreasing cancer, cardiovascular and other chronic and age related disease. However, antioxidant supplements does appear to be effective in lowering an individual's oxidative stress if their initial oxidative stress is above normal or above their set point of regulation. So if individuals were initially screened for high oxidative stress and then these individuals followed with and without antioxidant supplementation, I predict that much more impressive results will be obtained in disease prevention.

One clear prediction that came from the compensation model was that to increase lifespan substantially it was necessary to bypass this natural maintenance oxidative stress. It was for this reason why I turned to the use of transgenic mice with the hope that the extra gene dosage of an antioxidant gene would produce extra antioxidants that would not be compensated against by a decrease in other antioxidants. This reasoning appears to be correct for the transgenetic thioredoxin mice because they are both longer-lived and have a greater resistance to oxidative stress.

Although I tried for some five years (1990–1995) to obtain NIA intramural funding to create a research programmed using transgenetic technology to test the longevity determinant gene hypothesis, the proposal was not considered at that time to be worth the modest funding being requested. In fact, the greatest criticism I remember was that I would probably get Downs Syndrome-like mice. Thus, the demonstration of SOD and other antioxidants as potential Longevity Determinant Genes remained to be accomplished much later by other scientists in foreign laboratories and in extramural funded NIH laboratories using the same transgenetic technology I had proposed.

8.5. Longevity Determinant Genes

The most exciting progress that has occurred most recently in the field of biogerontology in my opinion is the prediction made 30 years ago that primary aging process exists and that a class of these genes code for antioxidants appears to be true.[39–44, 111, 238, 250–252] The first phase of this research began where increased lifespan was indirectly selected for in *Drosophila*.[18, 253, 254] However, success appears to be limited where the selected longer-lived fly strains always appeared to have some corresponding disadvantage consistent with their selection of late expressed pleiotropic genes.[224, 255] For example, when such long-lived flies were mixed with short-lived flies in captivity, the short-lived flies dominated in reproduction. Also what appears to be happening here is selection for flies that have enhanced postponement in rate of development. So, we appear to have enhancement in developmental-linked and not the continuous acting classes of longevity determinant genes. Furthermore, the biochemical basis of why the flies did live longer appeared more related to the unique problems of *Drosophila* such as greater resistance to water dehydration, higher fat content or lower metabolic rate rather than antiaging mechanisms that could be translated to long-lived mammalian species.

This result for *Drosophila* is not too surprising when the life cycles of insects in general are studied. The larva stage in their life cycle is one of growth where eating is the main agenda whereas the main agenda of the adult stage life cycle is sexual reproduction and not growth. Indeed, as an extreme example, the adult Mayfly does not even have mouthparts. So, the fact that some eating occurs in the adult insects was probably a strategy to extend life a little longer to complete the sexual activity. Indeed, if one adds additional food reserves to the adult insect its lifespan will be increased as it has been for butterflies. Most cells of the adult insect are postmitoic and this fact represents another example that may serious limit large gains in lifespan of insects.

They do represent, however, good models to study the effects of oxidative stress and the potential benefits of antioxidants since their specific metabolic rate is so high. But essentially any factor that decreases metabolic rate of an insect results in some increase in lifespan. All of this suggests to me that insects are in an evolutionary dead end as far as evolution of longer lifespan is concerned in contrast to mammalian species. Thus, we must be careful not to over interpret life extension results found in *Drosophila* and too quickly move on to the next model, the mouse.

Increasing mutation rate and then selecting directly for long-lived mutants in *Drosophila* proved more successful where increased lifespan of about 35% was found. These long-lived flies also had an enhanced expression of resistance to oxidative stress and other stressors.[238, 256] Increase of lifespan due to simple point mutational changes was also discovered in yeast[257] and nematodes.[41, 44] All of these results taken collectively has created a new excitement in the field and supports the cornerstone of the longevity determinant hypothesis that a few key mutations can significantly increase lifespan. Moreover, the mechanism of action appears to be a greater resistance to a wide variety of stresses including oxidative derived stresses.

Studies now centered on the subject of identifying longevity determinant genes and determining their mechanism of action are growing in popularity in stark contrast to the period when I had first proposed their existence some 30 years ago.[48] At that time there was little respect for any scientist working in the field of biogerontology particularly with mice and of all thing to also be proposing the existence of a few longevity determinant genes. This pessimism was based largely on what was considered the obvious complexity of aging processes and the still popular pleiotropic antagonistic evolutionary hypothesis of aging proclaiming the great number of genes involved in causing aging and the impossibility of any means within reason to extend lifespan.[35-37] So not many years ago, the very idea of a relative few longevity determinant genes was considered absolutely absurd.

Now, I hear little about the hopeless and forbidding complexity of aging apparently as a result of the growing evidence that primary aging and longevity genes may indeed exist. Yet no one ever admits they were mistaken. Instead, I now hear of the genetics of human aging, the search for longevity genes, and even papers pondering the question what these genes might be from the very scientist that were most critical of these concepts only a few years ago. In addition, many recent publications in this area fails to reference the 30-year history that laid the foundation their work has been based upon.[149, 209, 219, 258-260] Nevertheless, I am very pleased that the interest and progress in this area of longevity determinant genes is progressing so rapidly.

Before leaving this section, however, it is important to note that another term for Longevity Determinant Genes was later proposed by another research group[261-264] and this term is Longevity Assurance Genes. I am not clear why this was done for it appears to represent exactly the same concept proposed in 1972 and should not be confused with the Systemic Theory of Aging proposed by George Sacher.[265] Here, George Sacher proposed the great complexity of processes governing aging rate and suggest that the brain was likely to be the organ of longevity determination. This was Sacher's stand on this issue just before the Longevity Determinant Gene Hypothesis was published in 1972. Unfortunately, the use of this alternative term since that time has served to confuse both the science and the origin of the Longevity Determinant Gene hypothesis.

Interesting studies are also now underway using centenarians as a means to identify longevity determinant genes.[14, 15, 23, 162, 163, 210–212, 266–271] There is also some data now published that plasma antioxidants are higher in centenarians particularly vitamins E and A.[272] Low unsaturated fatty acids in membranes of centenarians may also be an important contributing factor of their long life.[23, 102, 217] Interest is also returning to the MHC, HLA locus[273] where I had also indicated some Longevity Determinant Genes may be located.[25, 48, 108]

The first direct test of SOD as a longevity determinant gene was carried out by Sohal's group using a transgenetic strain of *Drosophila* carrying extra copies of the Cu, Zn SOD and catalase gene.[274, 275] It is somewhat ironic that this was the same groups most advocate in criticizing the SOD/SMR lifespan correlative studies concluding that SOD was not important and that decrease production of ROS was more important than increased SOD. Unfortunately, these experiment have received extensive criticism for not controlling the genetic background of the long-lived strain to be the same as the non-transgenetic strain. The original data coming from this laboratory was that SOD or catalase transgenic strains did not have an increase lifespan. However, when the combined SOD and catalase was achieved by forming a hybrid, then lifespan was increased by about 33%. But then questions emerged of what the contribution of the well known hybrid vigor effect might be on this new hybrid. If this experiment were done properly it could have represented an important confirmation of our hypothesis.

Fortunately, Phillips and coworkers at last did the experiment again this time by targeting the extra SOD genes into the motor neuronal tissues of *Drosophila*.[43, 111] When this was done the flies did indeed live longer by about 30% when compared to control flies now available with identical genetic background. Also importantly catalase is not required for this additional lifespan and may even be decremental in contrast to the studies reported by the Sohol's group. Increase expression of the *hsp22* genes have also been identified in long-lived *Drosophila* flies select for increased lifespan that increase resistance to oxidative stress.[112, 276]

The most exciting studies, however, are these recently reported in mice. Some effort has been made in creating transgenetic mouse strains with enhanced expression of antioxidant coding genes. Unfortunately, this effort has yet to be successful using Cu, Zn SOD and Mn SOD genes but the work in continuing (see this book[277]). However, largely by accident it was discovered by Junji Yodie's research group that a transgenetic mouse strain having extra genes for thioredoxin not only had a significantly increased resistant to a number of oxidative stress enhancing agents, but both lifespan and life expectancy was increase by 30 to 40%.[42, 250] This important discovery is now being confirmed in other laboratories and if confirmed would represent the first time an increased expression of a single gene increased lifespan substantially in a mammalian species. In the past caloric restriction was the only means known that could increase lifespan by 30 to 40% in a mammalian species. But now, by the up-regulation of a single thioredoxin antioxidant gene, a similar lifespan extension is obtained with the very important difference that now the mouse eats all it wants all the time or *ab labium*. In addition, there appears to be no potential disadvantages such as lowered body temperature, decreased immune response in early development or lower glucose levels as is present in caloric restricted mice.

An equally exciting paper reported that a mutation in the p66she adapter protein gene in mice increased resistance to a number of different types of oxidative stresses such as hydrogen peroxide and UV light and increased lifespan by 30 to 49%.[40, 251, 252] It is not yet known how this mutation increases oxidative resistance but it may involve key regulatory elements related to the p53 and p66 genes. The fundamental point here is that only one mutation was required to produce this outstanding result, which is equivalent to an increased of 30 to 40 years of healthy and productive years in the human. Thousands of mutations were not required as had been predicted by the proponents of the antagonistic pleiotropic hypothesis of aging.

9. Future Studies and Applications

Now that the longevity determinant hypothesis has received substantial support work must now focus on identifying other genes that may be equally or even more important in the recent evolution of human lifespan and move on as quickly as possible to apply these results to human. One means is to undertake a molecular/biochemical comparative study of unusually long-lived human populations to determine what unique characteristics they may have that could account for their unique longevity. This would include centenarians 80 to 120+ years old as compared to 25 and 50 years old individuals. Certainly a search for unusual expression in defense and repair mechanisms against oxidative stress and glycation/glycoxidation stress needs to be conducted but many other potential longevity determinant genes should also be screened as previous outlined.[38, 48, 61, 63, 110]

In addition to humans, it is important to obtain much more comparative data using the primates species and particularly the human/chimpanzee comparative model. Such comparative data would include physiology and disease of aging, oxidative stress profiling, hormonal profiling and risk factors to cancer and cardiovascular disease. There is now underway a genome sequencing project for the chimpanzee and the sequence information derived from this project will be extraordinarily important on comparison with human genome in identify not only genes for controlling oxidative stress, but also rate of development, neoteny and intelligence. In other words identifying all the genes that make us human.

The basic questions here are (1) what were the principle genes involved in the recent evolution of human, (2) what mechanisms occurred in changing their expression and (3) can this same evolutionary process continue but now under our control using similar mechanisms?

Transgenic technology can be used to test many different potential longevity genes. However, I believe the most important genes at this time to test in transgenetic mice are (1) Cu, Zn SOD and Mn SOD genes directed to specific tissues as the motor neurons, (2) thioredoxin/thioredoxin reductase and glutaredoxin/glutaredoxin reductase genes, (3) the genes involved in the control of glycation/glycoxidation toxicity and accumulation of AGE products (4) and the genes involved in protein repair as methionine sulfoxide reductase.

But what possible application exists in the near future that can be applied in a practical manner to increase human longevity? The obvious goal is to focus on what we know now that works and that is the thioredoxin gene families and its involvement in redox regulation as well as the possible thioredoxin and the p66she gene family. But then how can we increase the oxidative stress resistance in ourselves without genetic engineering? One possibility is the use of antioxidant memetics. Remarkable success has already been published on the use of SOD mimetics in increasing lifespan of the nematode.[39, 278–281] We now wait for antioxidant mimetics that can be taken safely orally and not only by injection so that experiments can now be conducted in mice.

Another possibility is through a deeper understanding of the redox regulation pathway. Some evidence suggests the present of key humoral regulation factors controlling whole animal oxidative stress status. *Here, it may be possible to design pharmaceutical drugs that lower the set point of oxidative stress status regulation, thus, tricking cells to increase the production of a wide range of protection/defense and repair processes by intervention of specific regulatory genetic sequences.* Such technology is now under development.[282–285] This is based on the concept that control of oxidative stress may be under relative few redox regulatory elements where only one or a few drugs may be able to up-regulate the same benefits now achieve by the thioredoxin transgenetic or p66she mouse strains.

So far the subject has been the control of aging rate and the extension of productivity and healthy years of lifespan largely through the means that naturally evolved. But what about the possibility of rejuvenation or reversing the aging process? What I am referring to here in the transformation of a senescent individual to an individual having the physical, physiological and neurological/mental characteristics of a much younger person. Although the possibilities at first appear impossible, there is one paper that published such as reversal using old gerbil rodents.[286–289] Here, the treatment of the old gerbils with the spin trap called PBN reversed age dependent memory lost to that of a younger animal.

I had proposed that the mechanism might be PBN releasing nitric oxide and thus increasing removal of the oxidized abnormal proteins. We did obtain good evidence that PBN does indeed release nitric oxide[290, 291] and that PBN in the drinking water of mice increases life expectancy but only by a few percent.[291]

However, this experiment suggest that if it were possible to decrease the rate of free radical mediated damage and/or increase the level of turnover/cell renewal/remodeling and repair processes, then, it may be possible that the organism would clean itself up and essentially be reprogrammed much like what occurs in the cloning process which is of course a clear demonstration of rejuvenation of the adult cell. So perhaps we will find evidence for similar rejuvenation processes when longevity determinant processes are turned on sufficiently high but innate aging processes remain at the same level.

10. Summary and Conclusion Highlights

(1) The success of the Longevity Determinant Gene Hypothesis has been its prediction of a few key universal primary aging processes that have been present since the evolution of the first living forms of life. These primary aging processes were countered by primary antiaging processes or longevity determinant processes. The result is that such primary aging and longevity processes still exist in all forms of life today and they control much of the nature of aging and the extent of longevity.

   We do not know yet the full scope of what all primary aging processes are, but they do include oxidative stress and glycation/glycoxidation stresses. In turn the longevity determinants are those processes that act to reduce oxidative stress and glycation/glycoxidation stress. The most important element in the Longevity Determinant Gene Hypothesis is the prediction of a few key primary genes called Longevity Determinant Genes that are held in common and control general health maintenance of all forms of life including multicellular organisms and humans. Thus, the significant extension of productive and healthy life appears possible where before it was considered impossible.

(2) The demonstration that lifespan can be significantly increased by single point mutations or by the up-regulation of a single antioxidant gene in mice strongly argues against the antagonist pleiotropic hypothesis that predicted vast genetic complexity in all the processes causing aging and thus determining lifespan.[30, 35, 36, 45] Instead, there is now compelling experimental evidence that both aging and longevity processes have co-evolved beginning from the very origin of life. Thus, the popular and standard thesis now being promoted explaining "Why We Age" although correct in part suffers by its limitation in scope by not including aging and anti-aging factors that existed from the very origin of life and further by not including an integration of the concept of evolution of longevity with the evolution of aging concept.[115, 292] As a

consequence of this limitation the proponents of the antagonistic pleiotropic hypothesis have seriously missed the prediction of primary aging processes and in turn primary anti-aging process that has now been shown to be successfully predicted by the Longevity Determinant Gene Hypothesis.[3, 4, 22, 38, 48, 61, 130]

So yes there are likely to be hundred or thousands of genes that contribute to aging according to the antagonistic pleiotropic hypothesis but these genes are developmental linked in nature and their aging effects have been effectively postponed by slowing down the rate of development. Thus, a balance approach to further lengthen lifespan is to continue both the decrease in rate of development and the up-regulation of genes protecting the dysdifferentiated stated of cells.

These comments also applies to the Disposal Soma Hypothesis of Aging, because it also does not consider the full scope of evolution of living processes and as a consequence also failed to predict primary aging and anti-aging processes. In addition, the Disposal Soma Hypothesis of Aging is not novel and consists largely of concepts earlier published in related to the Longevity Determine Genes Hypothesis.

(3) The advocates proposing and supporting a new "Oxidative Stress Hypothesis of Aging"[84] have seriously misinterpreted and consequently misguided much of the scientific community as to preexisting scientific literature content in this field and thus in the originally of their concepts and ideas. The proposal that rate of endogenous generation of ROS is primarily responsible for determining aging rate is not novel and fails to acknowledge the importance of defense and repair processes as well as sensitivity of targets to oxidative damage. Also new information as to the association of SMR to endogenous production of ROS from mitochondria represents an important contribution confirming a previous prediction made in our original SOD study. Certainly it is not proper to use this advance in an attempt to now propose a new hypothesis of aging. Instead, such new knowledge should be used to build on and refine the existing models that stand correct in basic concept and foundation. Moreover, the basis of the extended lifespan of the SOD/Catalase *Drosophila* Strain is now in question. Most importantly, however, the new proposed oxidative stress hypothesis offer no new explanation of previous data or proposal for future experiments that have not already been proposed in the existing Longevity Determinant Gene Hypothesis.

(4) Comparative analysis of different primate and non-primate species focusing on the question of "what is unique in human biology that could account for the remarkable long human lifespan" has made the following contributions:

(a) *Result*: Rate of development is proportional to aging rate in primate species.

Mechanism of change: gene regulation of development rate.

(b) *Result*: Lifespan Energy Potential (LEP) is proportional to Antioxidant Status (AOS) in primate and non-primate mammalian species with human having the highest LEP value of primate species.
   Mechanism of change: gene regulation of oxidative stress determining factors.

(c) *Result*: Serum DHEAS levels is highest in human as compared to other primates or non-primate mammalian species.
   Mechanism of change: gene regulation of DHEAS concentration.

(d) *Result*: Ratio of B/A lactate dehydrogenase is highest in many different tissues in human as compared to shorter-lived primates.
   Mechanism of change: gene regulation of relative amount of protein subunits produced.

(e) *Result*: Body Temperature, Serum Glucose Level, Cholesterol and Serum Ascorbate Levels are not scientifically different in human as compared to other primates and non-primate mammalian species.
   Mechanism of change: no change in gene regulation.

(f) *Result*: UV DNA repair levels are higher in human as compared to other primate and non-primate mammalian species.
   Mechanism of change: gene regulation of amount of repair enzyme produced.

(g) *Result*: Superoxide dismutase per rate of endogenous generation of the superoxide free radical is highest in human as compared to other primates and non-primate mammalian species.
   Mechanism of change: gene regulation of amount of SOD enzyme produced and reduction of SMR related to increase body size.

(h) *Result*: Serum levels of carotinoids are higher in human as compared to other primates and non-primate mammalian species with or without SMR normalization.
   Mechanism of change: gene regulation decreasing amount of carotinoids degrading enzyme produced.

(i) *Result*: Serum level of uric acid is higher in human as compared to other primate or non-primate mammalian species.
   Mechanism of change: gene regulation of kidney renewal processes of serum uric acid and decrease in activity and eventual lost of uricase.

(j) *Result*: Total serum antioxidant capacity (ORAC) is highest in human as compared to other primate or non-primate mammalian species.
   Mechanism of change: gene regulation of total serum antioxidant content.

(k) *Result*: Rate of tissue spontaneous autoxidation is least in human tissues as compared to other primate or non-primate mammalian species.
   Mechanism of change: gene regulation of specific saturated and unsaturated fatty acids constituents in membranes with general decrease in unsaturated fatty acids.

(l) *Result*: Oxidative Stress Status (OSS) as measured by DNA damage is least in human as compared to other primates or non-primate mammalian species.

Mechanism of change: gene regulation of all key factors controlling OSS.

(m) *Result*: Glycation/Glycoxalation rate of accumulation is least in human tissue as compared to other primates or non-primate mammalian species.

Mechanism of change: gene regulation of the glycolase I and II enzyme System and other still unknown regulatory factors.

The above results when taken collectively strongly supported the hypothesis that humans are exceptionally long-lived because of the exceptional ability of their cells to maintain proper state of differentiation for so long. This is predicated to be the result of the exceptional low oxidative stress status of human cells, the slow rate of human differentiation and development and their exceptional high degree of neoteny. All of this is a result of changes occurring in Longevity Determines Gene Regulation Patterns. These correlative studies set up the next logical stage to directly test enhanced expression of potential longevity determinant genes by transgenetic technologies, gene regulation alteration technologies and antioxidant mimetic technologies to increase lifespan.

(5) I believe a revolution in anti-aging research driven by a much brighter and optimistic future will soon become a reality that will lead to significant increases in human health, productivity and longevity. This optimism is based on the following events:

(a) The success of the transgenetic longevity determinant gene animal models showing significant extension of lifespan as a result of:

- Enhanced expression of Cu, Zn SOD genes in motor neuron tissue in transgenetic *Drosophila*.[43, 111]
- Enhanced expression of thioredoxin genes in mice that increase both resistance to oxidative stress and lifespan.[42, 250]

(b) The success of the mutation in the p66shc gene in mice that increase both resistance to oxidative stress and lifespan.[40, 251, 252]

(c) The success of a SOD mimetic pharmaceutical drug that increases the lifespan of a nematode.[39]

(d) The success of The Longevity Determinant Gene Hypothesis to predicted Longevity Determinant Genes and to have provided both direction and rational in a more universal scientific effort to better understanding of the nature of human aging and longevity and the means for its control.

Note Added in Proof

The author would like to bring to the attention to the readers of this article the following papers that are of unusually importance in the support of the thesis presented in this chapter and appeared after this paper was written:

(1) Kang, H. L., Benzer, S., Min, K. T. (2002). Life extension in *Drosophila* by feeding a drug, *Proc. Natl. Acad. Sci. USA* **22**: 838–843.

(2) Tyner, S. D. *et al.* (2002). p53 mutant mice that display early ageing-associated phenotypes, *Nature* **415**: 45–53.

(3) Puca, A. A. *et al.* (2001). A genome-wide scan for linkage to human exceptional longevity identifies a locus on chromosome 4, *Proc. Natl. Acad. Sci. USA* **98**: 10 505–10 508.

(4) Hagen, T. M. *et al.* (2002). Feeding acetyl-L-carnitine and lipoic acid to old rats significantly improves metabolic function while decreasing oxidative stress, *Proc. Natl. Acad. Sci. USA* **99**: 1870–1875.

(5) Nordberg, J. and Arner, E. S. (2001). Reactive oxygen species, antioxidants, and the mammalian thioredoxin system, *Free Radic. Biol. Med.* **31**: 1287–1312.

(6) Schafer, F. Q. and Buettner, G. R. (2001). Redox environment of a cell as viewed through the redox state of the glutathione disulfide/glutathione couple, *Free Radic. Biol. Med.* **30**: 1191–1212.

(7) Moskovitz, J., Bal-Noy, S., Williams, W. M., Requena, J., Berlett, B. S. and Stadtman, E. R. (2001). Methionine sulfoxide reductase (MSRA) is a regulator of antioxidant defense and life span in mammals, *Proc. Natl. Acad. USA*, Nov 6; **98**: 12 920–12 925.

References

1. Comfort, A. (1979). *The Biology of Senescence,* Third Edition, p. 414, Elsevier, New York.
2. Arking, R. (1998). *Biology of Aging: Observations and Principles,* Second Edition, pp. xviii, 570, Sinauer Associates, Sunderland, MA.
3. Cutler, R. G. (1978). Evolutionary biology of senescence. *In* "The Biology of Aging" (J. A. Behnke, C. E. Finch, and G. B. Moment, Eds), pp. 311–360, Plenum Press, New York.
4. Cutler, R. G. (1976). Nature of aging and life maintenance processes. *In* "Interdisciplinary Topics in Gerontology", Vol. 9 (R. G. Cutler, Ed.), pp. 83–133, S. Karger, Basel.
5. Cutler, R. G. (1978). Alterations with age in the informational storage and flow systems of the mammalian cell. *In* "Genetic Effects on Aging", Vol. 14 (D. Bergsma, D. E. Harrison, and N. W. Paul, Eds.), pp. 463–498, Alan R. Liss, Inc., New York.

6. Cutler, R. G. (1983). Species probes and aging. *In* "Intervention in the Aging Process: Basic Research, Preclinical Screening and Clinical Programs" (V. J. Cristofalo, J. Roberts, and G. Baker, Eds.), pp. 69–144, Alan R. Liss, New York.

7. Cutler, R. G. (1984). Evolutionary biology of aging and longevity in mammalian species. *In* "Aging and Cell Function" (J. E. Johnson, Ed.), pp. 1–147, Plenum Press, New York.

8. Acsadi, G. Y. and Nemeskeri, J. (1970). *History of Human Lifespan and Mortality* (K. Balas, Ed.), p. 345, Akademiai Kiado, Budapest.

9. Jeune, B. and Vaupel, J. W. (1995). Exceptional longevity: from prehistory to present. *In* "Monographs on Population Aging", Vol. 2 (B. Jeune, and J. W. Vaupel, Eds.), p. 169, Odense University Press, Campusvej, Denmark.

10. Kannisto, V. (1994). Development of oldest-old mortality, 1950–1990: evidence from 28 Developed Countries. *In* "Monographs on Population Aging", Vol. 3 (B. Jeune, and J. W. Vaupel, Eds.), p. 108, Odense University Press, Campusvej, Denmark.

11. Kannisto, V. (1996). The advancing frontier of survival: life tables for old age. *In* "Odense Monographs on Population Aging", Vol. 3 (B. Jeune, and J. W. Vaupel, Eds.), p. 135, Odense University Press, Campusvej, Denmark.

12. Thatcher, A. R., Kannisto, V. and Vaupel, J. W. (1998). The force of mortality at ages 80 to 120. *In* "Odense Monographs on Population Aging", Vol. 5 (B. Jeune, and J. W. Vaupel, Eds.), p. 124, Odense University Press, Campusvej, Denmark..

13. Jeune, B. and Andersen-Ranberg, K. (1999). What can be learned from centenarians? Ugeskr Laeger **161**: 6321–6325.

14. Yashin, A. I. *et al.* (1999). Genes, demography, and life span: the contribution of demographic data in genetic studies on aging and longevity. *Am. J. Human Genet.* **65**: 1178–1193.

15. Yashin, A. I. *et al.* (2000). Genes and longevity: lessons from studies of centenarians. *J. Gerontol. A: Biol. Sci. Med. Sci.* **55**: B319–B328.

16. Yashin, A. I. and Iachine, I. A. (1995). Genetic analysis of durations: vorrelated frailty model applied to survival of Danish twins. *Genet. Epidemiol.* **12**: 529–538.

17. Burch, P. R. J. (1969). *An Inquiry Concerning Growth, Disease and Ageing*, p. 213, University of Toronto Press, Edinburgh.

18. Arking, R. (2001). Oxidative stress and longevity determinant genes of *Drosophila*. *In* "Oxidative Stress and Aging: Advances in Basic Science, Diagnostics and Intervention" (R. G. Cutler, and H. Rodriquez, Eds.), World Scientific Singapore.

19. Kohn, R. R. (1978). *Principles of Mammalian Aging*, Second Edition, p. 240, Prentice-Hall, Inc., Englewoods, New Jersey.

20. Shock, N. W. (1984). *Normal Human Aging — The Baltimore Longitudinal Study of Aging*, p. 399, US Department of Health and Human Services, Washington D.C.

21. Cutler, R. G. (1980). Central versus peripheral aging. *In* "Aging Phenomena. Relationships Among Different Levels of Organization" (K. Oota *et al.*, Eds.), pp. 261–298, Plenum Press, New York.

22. Cutler, R. G. (1972). Transcription of reiterated DNA sequence classes throughout the lifespan of the muse. *In* "Advances in Gerontology Research", Vol. 4 (B. L. Strehler, Ed.), pp. 219–321, Academic Press, New York.

23. Franceschi, C. *et al.* (2000). Do men and women follow different trajectories to reach extreme longevity? Italian Multicenter Study on Centenarians (IMUSCE). *Aging (Milano)* **12**: 77–84.

24. Jeune, B. and Vaupel, J. W. (1999). Validation of experimental longevity. *In* "Odense Monographs on Population Aging", Vol. 6 (B. Jeune, and J. W. Vaupel, Eds.), p. 249, Odense University Press, Campusvej, Denmark.

25. Cutler, R. G. (1981). Life-span extension. *In* "Aging Biology and Behavior" (J. L. McGaugh, and S. B. Kessler, Eds.), pp. 31–76, Academic Press, New York.

26. Kohn, R. R. (1982). Cause of death in very old people. *JAMA* **247**: 2793–2797.

27. Cutler, R. G. (1974). Redundancy of information content in the genome of mammalian species as a protective mechanism determining age rate. *Mech. Aging Dev.* **2**: 381–408.

28. Cutler, R. G. (1975). Evolution of human longevity and the genetic complexity governing aging rate. *Proc. Natl. Acad. Sci. USA* **72**: 4664–4668.

29. Cutler, R. G. (1979). Evolution of human longevity: a critical overview. *Mech. Aging Dev.* **9**: 337–354.

30. Charlesworth, B. (1993). Evolutionary mechanisms of senescence. *Genetica* **91**: 11–19.

31. Charlesworth, B. and Partridge, L. (1997). Ageing: levelling of the grim reaper. *Curr. Biol.* **7**: R440–R442.

32. Charlesworth, B. (1996). Evolution of senescence: Alzheimer's disease and evolution. *Curr. Biol.* **6**: 20–22.

33. Rose, M. and Charlesworth, B. (1980). A test of evolutionary theories of senescence. *Nature* **287**: 141–142.

34. Rose, M. R. and Mueller, L. D. (2000). Ageing and immortality. *Philos. Trans. Roy. Soc. London B: Biol. Sci.* **355**: 1657–1662.

35. Wallace, D. C. (1967). The inevitability of growing old. *J. Chronic. Dis.* **20**: 475–486.

36. Wallace, D. C. (1975). A theory of the cause of aging. *Med. J. Aust.* **1**: 829–831.

37. Hamilton, W. D. (1966). The moulding of senescence by natural selection. *J. Theor. Biol.* **12**: 12–45.

38. Cutler, R. G. (1995). Longevity determinant genes, cellular dysdifferentiation and oxidative stress. *In* "Oxidative Stress and Aging" (R. G. Cutler *et al.*, Eds.), pp. 15–19, Birkhauser Press, Boston.
39. Melov, S. *et al.* (2000). Extension of life-span with superoxide dismutase/ catalase mimetics. *Science* **289**: 1567–1569.
40. Migliaccio, E. *et al.* (1999). The p66shc adaptor protein controls oxidative stress response and life span in mammals. *Nature* **402**: 309–313.
41. Finkel, T. and Holbrook, N. J. (2000). Oxidants, oxidative stress and the biology of ageing. *Nature* **408**: 239–247.
42. Yodoi, J. (1999). Redox regulation by thioredoxin family in eukaryotic life. *In* "International Symposium for Oxidative Stress, Redox Regulation and Signal Transduction; Clinical Implications", Kyodai Kaikan, Kyoto, Japan.
43. Phillips, J. P., Parkes, T. L. and Hilliker, A. J. (2000). Targeted neuronal gene expression and longevity in *Drosophila*. *Exp. Gerontol.* **35**: 1157–1164.
44. Guarente, L. and Kenyon, C. (2000). Genetic pathways that regulate ageing in model organisms. *Nature* **408**: 255–262.
45. Rose, M. R. (1991). *Evolutionary Biology of Aging*, pp. 1–221, Oxford University Press, New York.
46. Kirkwood, T. B. and Austad, S. N. (2000). Why do we age? *Nature* **408**: 233–238.
47. Cutler, R. G. (1980). Evolution of human longevity. *In* "Advances in Pathobiology, Aging, Cancer and Cell Membranes", Vol. 7 (C. Borek, D. M. Fenoglio, and D. W. King, Eds.), pp. 43–79, Thieme-Stratton, New York.
48. Cutler, R. G. (1982). Longevity is determined by specific genes: testing the hypothesis. *In* "Testing the Theories of Aging" (R. C. Adelman, and G. S. Roth, Eds.), pp. 25–114, CRC Press, Boca Raton.
49. Cutler, R. G. (1976). Evolution of longevity in primates. *J. Human Evol.* **5**: 169–202.
50. Cutler, R. G. (1979). Evolution of longevity in ungulates and carnivores. *Gerontology* **25**: 69–86.
51. Tolmasoff, J. M., Ono, T. and Cutler, R. G. (1980). Superoxide dismutase: correlation with life-span and specific metabolic rate in primate species. *Proc. Natl. Acad. Sci. USA* **77**: 2777–2781.
52. Ono, T. and Okada, S. (1984). Unique increase of superoxide dismutase level in brains of long living mammals. *Exp. Gerontol.* **19**: 349–354.
53. Kirkwood, T. B. (1977). Evolution of ageing. *Nature* **270**: 301–304.
54. Kirkwood, T. B. (1987). Immortality of the germ-line versus disposability of the soma. *Basic Life Sci.* **42**: 209–218.
55. Kirkwood, T. B. (1988). The nature and causes of ageing. *Ciba Found. Symp.* **134**: 193–207.
56. Kirkwood, T. B. (1997). The origins of human ageing. *Philos. Trans. Roy. Soc. London B: Biol. Sci.* **352**: 1765–1772.

57. Kirkwood, T. B. and Holliday, R. (1979). The evolution of ageing and longevity. *Proc. Roy. Soc. London B: Biol. Sci.* **205**: 531–546.

58. Finch, C. E. and Kirkwood, T. B. L. (2000). *Chance, Development, and Aging*, pp. 1–278, Oxford University Press, New York.

59. Finch, K. W. W. a. C. E. (1997). *Between Zeus and the Salmon: The Biodemography of Longevity*. National Research Council, Washington, D.C.

60. Finch, C. E. (1990). *Longevity, Senescence, and the Genome*, pp. 1–922, University of Chicago Press, Chicago and London.

61. Cutler, R. G. (1991). Recent progress in testing the longevity determinant and dysdifferentiation hypotheses of aging. *Arch. Gerontol. Geriatr.* **12**: 75–98.

62. Curtis, H. J. (1966). Biological mechanisms of aging. *In* "American Lecture Series", Vol. 639 (I. N. Kugelmass, Ed.), Charles C. Thomas, Springfield.

63. Cutler, R. G. (1992). Genetic stability and oxidative stress: common mechanisms in aging and cancer. *In* "Free Radicals and Aging" (I. Emerit, and B. Chance, Eds.), pp. 31–46, Birkhauser Verlag, Basel.

64. Harman, D. (1968). Free radical theory of aging: effect of free radical reaction inhibitors on the mortality rate of male LAF mice. *J. Gerontol.* **23**: 476–482.

65. Harman, D. (1969). Chemical protection against aging. *Agents Actions* **1**: 3–8.

66. Harman, D. (1969). Prolongation of life: role of free radical reactions in aging. *J. Am. Geriatr. Soc.* **17**: 721–735.

67. Harman, D. and Piette, L. H. (1966). Free radical theory of aging: free radical reactions in serum. *J. Gerontol.* **21**: 560–565.

68. Montagu, A. (1989). *Growing Young*, Second Edition, pp. 1–292, McGraw-Hill.

69. Gould, S. J. (1977). *Ontogeny and Phylogeny*, pp. ix, 501, Belknap Press of Harvard University Press, Cambridge, MA.

70. Wilson, A. C. (1976). Gene regulation in evolution. *In* "Molecular Evolution" (F. J. Ayala, Ed.), p. 277, Sinauer Associates, Inc., Sunderland, MA.

71. Hart, R. W. and Setlow, R. B. (1974). Correlation between deoxyribonucleic acid excision-repair and life-span in a number of mammalian species. *Proc. Natl. Acad. Sci. USA* **71**: 2169–2173.

72. Hart, R. W. and Setlow, R. B. (1975). DNA repair and life span of mammals. *Basic Life Sci.* 801–804.

73. Cortopassi, G. A. and Wang, E. (1996). There is substantial agreement among interspecies estimates of DNA repair activity. *Mech. Ageing Dev.* **91**: 211–218.

74. Cutler, R. G. (1983). Superoxide dismutase, longevity and specific metabolic rate. A reply. *Gerontology* **29**: 113–120.

75. Cutler, R. G. (1984). Antioxidants and longevity. *In* "Free Radicals in Molecular Biology, Aging and Disease" (D. Armstrong *et al.*, Eds.), pp. 235–266, Raven Press, New York.

76. Cutler, R. G. (1984). Antioxidants, aging and longevity. *In* "Free Radicals in Biology", Vol. VI (W. Pryor, Ed.), pp. 371–428, Academic Press, New York.

77. Perez-Campo, R. *et al.* (1993). A comparative study of free radicals in vertebrates — Part I. Antioxidant enzymes. *Comp. Biochem. Physiol.* **B105**: 749–755.

78. Lopez-Torres, M. *et al.* (1993). A comparative study of free radicals in vertebrates — Part II. Non-enzymatic antioxidants and oxidative stress. *Comp. Biochem. Physiol.* **B105**: 757–763.

79. Agarwal, S. and Sohal, R. S. (1996). Relationship between susceptibility to protein oxidation, aging, and maximum life span potential of different species. *Exp. Gerontol.* **31**: 387–392.

80. Ku, H.-H. and Sohal, R. S. (1993). Comparison of mitochondrial pro-oxidant generation and anti-oxidant defenses between rat and pigeon: possible basis of variation in longevity and metabolic potential. *Mech. Ageing Dev.* **72**: 67–76.

81. Ku, H.-H., Brunk, U. T. and Sohal, R. S. (1993). Relationship between mitochondrial superoxide and hydrogen peroxide production and longevity of mammalian species. *Free Radic. Biol. Med.* **15**: 621–627.

82. Sohal, R. S. and Brunk, U. T. (1992). Mitochondrial production of pro-oxidants and cellular senescence. *Mutat. Res.* **275**: 295–304.

83. Sohal, R. S., Sohal, B. H. and Brunk, U. T. (1990). Relationship between antioxidant defenses and longevity in different mammalian species. *Mech. Ageing Dev.* **53**: 217–227.

84. Sohal, R. S. and Weindruch, R. (1996). Oxidative stress, caloric restriction, and aging. *Science* **273**: 59–63.

85. Sohal, R. S. (1993). The free radical hypothesis of aging: an appraisal of the current status. *Aging (Milano)* **5**: 3–17.

86. Stuart, J. A. *et al.* (1999). Mitochondrial proton leak and the uncoupling proteins. *J. Bioenerg. Biomembr.* **31**: 517–525.

87. Stuart, J. A. *et al.* (2001). Mitochondrial proton leak and the uncoupling protein 1 homologues. *Biochim. Biophys. Acta* **1504**: 144–158.

88. Barja, G. (1999). Mitochondrial oxygen radical generation and leak: sites of production in states 4 and 3, organ specificity, and relation to aging and longevity. *J. Bioenerg. Biomembr.* **31**: 347–366.

89. Barja, G. (2000). The flux of free radical attack through mitochondrial DNA is related to aging rate. *Aging (Milano)* **12**: 342–355.

90. Barja, G. *et al.* (1994). A decrease of free radical production near critical targets as a cause of maximum longevity in animals. *Comp. Biochem. Physiol. Biochem. Mol. Biol.* **108**: 501–512.

91. Barja, G. *et al.* (1994). Low mitochondrial free radical production per unit O_2 consumption can explain the simultaneous presence of high longevity and high aerobic metabolic rate in birds. *Free Radic. Res.* **21**: 317–327.

92. Brand, M. D. (2000). Uncoupling to survive? The role of mitochondrial inefficiency in ageing. *Exp. Gerontol.* **35**: 811–820.
93. Brierley, E. J. *et al.* (1997). Mitochondrial involvement in the ageing process. Facts and controversies. *Mol. Cell Biochem.* **174**: 325–328.
94. Brookes, P. S. *et al.* (2001). Increased sensitivity of mitochondrial respiration to inhibition by nitric oxide in cardiac hypertrophy. *J. Mol. Cell Cardiol.* **33**: 69–82.
95. Cadenas, E. and Davies, K. J. (2000). Mitochondrial free radical generation, oxidative stress, and aging. *Free Radic. Biol. Med.* **29**: 222–230.
96. Cadenas, S. *et al.* (2000). AMP decreases the efficiency of skeletal-muscle mitochondria. *Biochem. J.* **351(Part 2)**: 307–311.
97. Demin, O. V., Kholodenko, B. N. and Skulachev, V. P. (1998). A model of $O_2 \cdot$-generation in the complex III of the electron transport chain. *Mol. Cell Biochem.* **184**: 21–33.
98. Herrero, A. and Barja, G. (1999). 8-oxo-deoxyguanosine levels in heart and brain mitochondrial and nuclear DNA of two mammals and three birds in relation to their different rates of aging. *Aging (Milano)* **11**: 294–300.
99. Liu, S. S. (1999). Cooperation of a "reactive oxygen cycle" with the Q cycle and the proton cycle in the respiratory chain — superoxide generating and cycling mechanisms in mitochondria. *J. Bioenerg. Biomembr.* **31**: 367–376.
100. Pamplona, R. *et al.* (1999). A low degree of fatty acid unsaturation leads to lower lipid peroxidation and lipoxidation-derived protein modification in heart mitochondria of the longevous pigeon than in the short-lived rat. *Mech. Ageing Dev.* **106**: 283–296.
101. Pamplona, R. *et al.* (1999). Heart fatty acid unsaturation and lipid per-oxidation, and aging rate, are lower in the canary and the parakeet than in the mouse. *Aging (Milano)* **11**: 44–49.
102. Pamplona, R. *et al.* (2000). Low fatty acid unsaturation: a mechanism for lowered lipoperoxidative modification of tissue proteins in mammalian species with long life spans. *J. Gerontol. A: Biol. Sci. Med. Sci.* **55**: B286–B291.
103. Pamplona, R. *et al.* (1998). Mitochondrial membrane peroxidizability index is inversely related to maximum life span in mammals. *J. Lipid Res.* **39**: 1989–1994.
104. Pamplona, R. *et al.* (2000). Double bond content of phospholipids and lipid peroxidation negatively correlate with maximum longevity in the heart of mammals. *Mech. Ageing Dev.* **112**: 169–183.
105. Cutler, R. G. (1985). Peroxide-producing potential of tissues: inverse correlation with longevity of mammalian species. *Proc. Natl. Acad. Sci. USA* **82**: 4798–4802.
106. Pamplona, R. *et al.* (1999). Thyroid status modulates glycoxidative and lipoxidative modification of tissue proteins. *Free Radic. Biol. Med.* **27**: 901–910.

107. Cutler, R. G. (1984). Free radicals and aging. *In* "Molecular Basis of Aging" (A. Roy and B. Chatterjee, Eds.), pp. 263–354, Academic Press, New York.

108. Cutler, R. G. (1985). Antioxidants and longevity of mammalian species. *In* "Molecular Biology of Aging" (A. D. Woodhead, A. D. Blackett, and A. Hollaender, Eds.), pp. 15–74, Plenum Press, New York.

109. Cutler, R. G. (1991). Human longevity and aging: possible role of reactive oxygen species. *In* "Annals of the New York Academy of Sciences", Vol. 621 (W. Pierpaoli, Ed.), pp. 1–28, New York Acad. Sci., New York.

110. Cutler, R. G. (1993). Oxidative stress state in aging and longevity mechanisms. *In* "Free Radicals: From Basic Science to Medicine" (G. Poli, Ed.), pp. 144–156, Birkhauser Press, Boston.

111. Parkes, T. L. *et al.* (1998). Extension of *Drosophila* lifespan by overexpression of human SOD-1 in motorneurons. *Nature Genet.* **19**: 171–174.

112. Tower, J. (1996). Aging mechanisms in fruit files. *Bioessays* **18**: 799–807.

113. Perez-Campo, R. *et al.* (1994). Longevity and antioxidant enzymes, non-enzymatic antioxidants and oxidative stress in the vertebrate lung: a comparative study. *J. Comp. Physiol.* **B163**: 682–689.

114. Lopez-Torres, M. *et al.* (1993). Maximum life span in vertebrates: relationship with liver antioxidant enzymes, glutathione system, ascorbate, urate, sensitivity to peroxidation, true malondialdehyde, *in vivo* H_2O_2, and basal and maximum aerobic capacity. *Mech. Ageing Dev.* **70**: 177–199.

115. Austad, S. N. (1997). *Why We Age. What Science is Discovering about the Body's Journey through Life*, pp. 1–244, John Wiley & Son Inc., New York.

116. Cutler, R. G., Packer, L., Bertram, J. and Mori, A. (1995). Oxidative stress and aging. *In* "Molceuclar and Cell Biology Updates" (A. A. a. L. Packer, Ed.), pp. 1–396, Birkhauser Verlag, Basel.

117. Cutler, R. G. (1991). Antioxidants and aging. *Am. J. Clin. Nutri.* **53**: 373S–379S.

118. Cutler, R. G. (1984). Carotenoids and retinol: their possible importance in determining longevity of primate species. *Proc. Natl. Acad. Sci. USA* **81**: 7627–7631.

119. Cutler, R. G. (1984). Urate and ascorbate: their possible roles as antioxidants in determining longevity of mammalian species. *Arch. Gerontol. Geriatr.* **3**: 321–348.

120. Ayala, A. and Cutler, R. G. (1991). Liver cytochrome P450 detoxification system: possible role in human aging and longevity. *In* "Liver and Aging — 1990" (K. Kitani, Ed.), pp. 337–352, Elsevier, Amsterdam.

121. Ayala, A. and Cutler, R. G. (1997). Preferential use of less toxic detoxification pathways by long-lived species. *Arch. Gerontol. Geriatr.* **24**: 87–102.

122. Ayala, A. and Cutler, R. G. (1996). The utilization of 5-hydroxyl-2-amino valeric acid as a specific marker of oxidized arginine and proline residues in proteins. *J. Free Radic. Biol. Med.* **22**: 65–80.

123. Halliwell, B. and Gutteridge, J. M. C. (1999). *Free Radicals in Biology and Medicine*, Third Edition, pp. 1–936, Clarendon Press, Oxford University Press, Oxford, New York.

124. Nieto, F. J. *et al.* (2000). Uric acid and serum antioxidant capacity: a reaction to arteriosclerosis? *Ateriosclerosis* **148**: 131–139.

125. Cao, G., Alessio, H. M. and Cutler, R. G. (1993). Oxygen radical absorbance capacity assay for antioxidants. *Free Radic. Biol. Med.* **14**: 303–311.

126. Cao, G. and Cutler, R. G. (1993). High concentrations of antioxidants may not improve defense against oxidative stress. *Arch. Gerontol. Geriatr.* **17**: 189–201.

127. Cao, G. and Cutler, R. G. (1993). New approaches for measuring plasma or serum antioxidant capacity: a methodological note. *Free Radic. Biol. Med.* **16**: 135–138.

128. Bergtold, D. S. *et al.* (1988). Urine biomarkers for oxidative DNA damage. *In* "Oxygen Radicals in Biology and Medicine" (M. G. Simic *et al.*, Eds.), pp. 483–489, Plenum Press, New York.

129. Cutler, R. G. (1976). Cross-linkage hypothesis of aging: DNA adducts in chromatic as a primary aging process. *In* "Protein and Other Adducts to DNA: Their Significance to Aging, Carcinogenesis, and Radiation Biology" (K. C. Smith, Ed.), pp. 443–493, Plenum Press, New York.

130. Cutler, R. G. (1982). The dysdifferentiative hypothesis of mammalian aging and longevity. *In* "The Aging Brain Cellular and Molecular Mechanisms of Aging in the Nervous System, Aging", Vol. 20 (E. Giacobini, G. Filogamo, and A. Vernadakis, Eds.), pp. 1–19, Raven Press, New York.

131. Cutler, R. G. (1986). Aging and oxygen radicals. *In* "Physiology of Oxygen Radicals" (A. E. Taylor, S. Matalon, and P. Ward, Eds.), pp. 251–285, Clinical Monograph Series, *Amer. Physiol. Soc.*, Bethesda, Maryland.

132. Cutler, R. G. and Semsei, I. (1989). Development, cancer and aging: possible common mechanisms of action and regulation. *J. Gerontol.* **44**: 25–34.

133. Dean, R. G., Socher, S. H. and Cutler, R. G. (1985). Dysdifferentiative nature of aging: age-dependent expression of mouse mammary tumor virus and casein genes in brain and liver tissues of the C57BL/6J mouse strain. *Arch. Gerontol. Geriatr.* **4**: 43–51.

134. Gaubatz, J. W., Prashad, N. and Cutler, R. G. (1976). Ribosomal RNA gene dosage as a function of tissue and age for mouse and human. *Biochem. Biophys. Acta* **418**: 358–375.

135. Gaubatz, J. W. and Cutler, R. G. (1990). Mouse satellite DNA is transcribed in senescent cardiac muscle. *J. Biol. Chem.* **265**: 17 753–17 758.

136. Gaubatz, J. W. and Cutler, R. G. (1978). Age-related differences in the number of ribosomal RNA genes of mouse tissues. *Gerontology* **24**: 179–207.

137. Shigenaga, M. K., Hagen, T. M. and Ames, B. N. (1994). Oxidative damage and mitochondrial decay in aging. *Proc. Natl. Acad. Sci. USA* **91**: 10 771–10 778.

138. Al-Abed, Y. *et al.* (1999). Inhibition of advanced glycation endproduct formation by acetaldehyde: role in the cardioprotective effect of ethanol. *Proc. Natl. Acad. Sci. USA* **96**: 2385–2390.

139. Asif, M. *et al.* (2000). An advanced glycation endproduct cross-link breaker can reverse age-related increases in myocardial stiffness. *Proc. Natl. Acad. Sci. USA* **97**: 2809–2813.

140. Wolffenbuttel, B. H. *et al.* (1998). Breakers of advanced glycation end products restore large artery properties in experimental diabetes. *Proc. Natl. Acad. Sci. USA* **95**: 4630–4634.

141. Sell, D. R. *et al.* (1996). Longevity and the genetic determination of collagen glycoxidation kinetics in mammalian senescence. *Proc. Natl. Acad. Sci. USA* **93**: 485–490.

142. Ranganathan, S. *et al.* (1999). Genomic sequence of human glyoxalase-I: analysis of promoter activity and its regulation. *Gene* **240**: 149–155.

143. Abordo, E. A., Minhas, H. S. and Thornalley, P. J. (1999). Accumulation of alpha-oxoaldehydes during oxidative stress: a role in cytotoxicity. *Biochem. Pharmacol.* **58**: 641–648.

144. Thornalley, P. J. (1996). Pharmacology of methylglyoxal: formation, modification of proteins and nucleic acids, and enzymatic detoxification — a role in pathogenesis and antiproliferative chemotherapy. *Gen. Pharmacol.* **27**: 565–573.

145. Thornalley, P. J. (1998). Glutathione-dependent detoxification of alpha-oxoaldehydes by the glyoxalase system: involvement in disease mechanisms and antiproliferative activity of glyoxalase I inhibitors. *Chem. Biol. Interact.* **111–112**: 137–151.

146. Stadtman, E. R. and Levine, R. L. (2000). Protein oxidation. *Ann. NY Acad. Sci.* **899**: 191–208.

147. Levine, R. L. (1998). Oxidative stress and aging. *Aging (Milano)* **10**: 151.

148. Klapper, W., Parwaresch, R. and Krupp, G. (2001). Telomere biology in human aging and aging syndromes. *Mech. Ageing Dev.* **122**: 695–712.

149. Slagboom, P. E. *et al.* (2000). Genetics of human aging. The search for genes contributing to human longevity and diseases of the old. *Ann. NY Acad. Sci.* **908**: 50–63.

150. Cutler, R. G. (1985). Dysdifferentiation and aging. *In* "Molecular Biology of Aging: Gene Stability and Gene Expression" (R. S. Sohal, L. Birnbaum, and R. G. Cutler, Eds.), pp. 307–340, Raven Press, New York.

151. Cristofalo, V. J. *et al.* (1998). Relationship between donor age and the replicative lifespan of human cells in culture: a reevaluation. *Proc. Natl. Acad. Sci. USA* **95**: 10 614–10 619.

152. Schneider, E. L. *et al.* (1981). Skin fibroblast cultures derived from members of the Baltimore longitudinal study: a new resource for studies of cellular aging. *Cytogenet. Cell Genet.* **31**: 40–46.

153. Wilson, V. L. *et al.* (1987). Genomic 5-methyldeoxycytidine decreases with age. *J. Biol. Chem.* **262**: 9948–9951.
154. Comstock, G. W. *et al.* (1997). The risk of developing lung cancer associated with antioxidants in the blood: ascorbic acid, carotenoids, alpha-tocopherol, selenium, and total peroxyl radical absorbing capacity. *Cancer Epidemiol. Biomarkers Prev.* **6**: 907–916.
155. DeWeese, T. L. *et al.* (1998). Mouse embryonic stem cells carrying one or two defective Msh2 alleles respond abnormally to oxidative stress inflicted by low-level radiation. *Proc. Natl. Acad. Sci. USA* **95**: 11 915–11 920.
156. Florine, D. L. *et al.* (1980). Regulation of endogenous murine leukemia virus-related nuclear and cytoplasmic RNA complexity in C57BL/6J mice of increasing age. *Cancer Res.* **40**: 519–523.
157. Gaubatz, J. W., Arcement, B. and Cutler, R. G. (1991). Gene expression of an endogenous retroviral-like element during murine development and aging. *Mech. Ageing Dev.* **57**: 71–85.
158. Ono, T. and Cutler, R. G. (1978). Age-dependent relaxation of gene repression: increase of endogenous murine leukemia virus-related and globin-related RNA in brain and liver of mice. *Proc. Natl. Acad. Sci. USA* **75**: 4431–4435.
159. Ono, T. *et al.* (1985). Dysdifferentiative nature of aging: age-dependent expression of MuLV and globin genes in thymus, liver and brain in the AKR mouse strain. *Gerontology* **31**: 362–372.
160. Ahuja, N. *et al.* (1998). Aging and DNA methylation in colorectal mucosa and cancer. *Cancer Res.* **58**: 5489–5494.
161. Carroll, S. B. (2000). Endless forms: the evolution of gene regulation and morphological diversity. *Cell* **101**: 577–580.
162. de Benedictis, G. *et al.* (2000). Does a retrograde response in human aging and longevity exist? *Exp. Gerontol.* **35**: 795–801.
163. de Benedictis, G. *et al.* (2000). Inherited variability of the mitochondrial genome and successful aging in humans. *Ann. NY Acad. Sci.* **908**: 208–218.
164. Szmulewicz, M. N., Novick, G. E. and Herrera, R. J. (1998). Effects of Alu insertions on gene function. *Electrophoresis* **19**: 1260–1264.
165. Tonjes, R. R., Czauderna, F. and Kurth, R. (1999). Genome-side screening, cloning, chromosomal assignment, and expression of full-length human endogenous retrovirus type K. *J. Virol.* **73**: 9187–9195.
166. Tycko, B. and Ashkenas, J. (2000). Epigenetics and its role in disease. *J. Clin. Invest.* **105**: 245–246.
167. Zs-Nagy, I., Cutler, R. G. and Semsei, I. (1988). Dysdifferentiation hypothesis of aging and cancer: a comparison with the membrane hypothesis of aging. *Ann. NY Acad. Sci.* **521**: 215–225.
168. Barton, N. H. (1996). Speciation: more than the sum of the parts. *Curr. Biol.* **6**: 1244–1246.
169. Bernstein, B. and Bernstein, H. (1991). *Aging, Sex, and DNA Repair*, p. 382, Academic Press, Inc., San Diego.

170. Deragon, J. M. and Capy, P. (2000). Impact of transposable elements on the human genome. *Ann. Med.* **32**: 264–273.
171. Semsei, I., Ma, S. and Cutler, R. G. (1989). Tissue and age specific expression of the *myc* proto-oncogene family throughout the life span of the C57BL/6J mouse strain. *Oncogene* **4**: 465–470.
172. Zs-Nagy, I. *et al.* (1993). Comparison of the lateral diffusion constant of hepatocyte membrane proteins in two wild mouse strains of considerably different longevity: FRAP studies on liver smears. *J. Gerontology* **48**: B86–B92.
173. Wood, B. and Collard, M. (1999). The human genus. *Science* **284**: 65–71.
174. Falk, D. *et al.* (2000). Early hominid brain evolution: a new look at old endocasts. *J. Human Evol.* **38**: 695–717.
175. Holden, C. (2001). Paleoanthropology. Oldest human DNA reveals Aussie oddity. *Science* **291**: 230–231.
176. Pennisi, E. (1999). From embryos and fossils, new clues to vertebrate evolution. *Science* **284**: 575–577.
177. Lockwood, C. A. and Kimbel, W. H. (1999). Endocranial capacity of early hominids. *Science* **283**: B9.
178. Wolpoff, M. H. *et al.* (2001). Modern human ancestry at the peripheries: a test of the replacement theory. *Science* **291**: 293–297.
179. Arsuaga, J. L. *et al.* (1997). Size variation in Middle Pleistocene humans. *Science* **277**: 1086–1088.
180. Conroy, G. C. *et al.* (1998). Endocranial capacity in an early hominid cranium from Sterkfontein, South Africa. *Science* **280**: 1730–1731.
181. Falk, D. (1998). Hominid brain evolution: looks can be deceiving. *Science* **280**: 1714.
182. Rilling, J. K. and Insel, T. R. (1998). Evolution of the cerebellum in primates: differences in relative volume among monkeys, apes and humans. *Brain Behav. Evol.* **52**: 308–314.
183. Ayala, F. J. and Fitch, W. M. (1997). Genetics and the origin of species: an introduction. *Proc. Natl. Acad. Sci. USA* **94**: 7691–7697.
184. Goodman, M. *et al.* (1998). Toward a phylogenetic classification of primates based on DNA evidence complemented by fossil evidence. *Mol. Phylogenet. Evol.* **9**: 585–598.
185. Kaessmann, H., Wiebe, V. and Paabo, S. (1999). Extensive nuclear DNA sequence diversity among chimpanzees. *Science* **286**: 1159–1162.
186. Garner, K. J. and Ryder, O. A. (1996). Mitochondrial DNA diversity in gorillas. *Mol. Phylogenet. Evol.* **6**: 39–48.
187. Ayala, F. J. *et al.* (1994). Molecular genetics of speciation and human origins. *Proc. Natl. Acad. Sci. USA* **91**: 6787–6794.
188. Britten, R. J. (1997). Mobile elements inserted in the distant past have taken on important functions. *Gene* **205**: 177–182.

189. Finnegan, D. J. (1989). Eukaryotic transposable elements and genome evolution. *Trends Genet.* **5**: 103–107.

190. Bennetzen, J. L. (2000). Transposable element contributions to plant gene and genome evolution. *Plant Mol. Biol.* **42**: 251–269.

191. Fedoroff, N. V. (1999). Transposable elements as a molecular evolutionary force. *Ann. NY Acad. Sci.* **870**: 251–264.

192. Smit, A. F. (1999). Interspersed repeats and other mementos of transposable elements in mammalian genomes. *Curr. Opin. Genet. Dev.* **9**: 657–663.

193. Kim, H. S. and Crow, T. J. (1999). Identification and phylogenetic analysis of novel human endogenous retroviral sequences belonging to the HERV-H family on human X and Y chromosomes. *Genes Genet. Syst.* **74**: 129–134.

194. Kim, H. S. and Crow, T. J. (2000). Phylogenetic relationships of a class of hominoid-specific retro-elements (SINE-R) on human chromosomes 7 and 17. *Ann. Human Biol.* **27**: 83–93.

195. Kim, H. S. *et al.* (2000). Phylogenetic analysis of a retroposon family as represented on the human X chromosome. *Genes Genet. Syst.* **75**: 197–202.

196. Kim, H. S. *et al.* (1999). Phylogenetic analysis of a retroposon family in African great apes. *J. Mol. Evol.* **49**: 699–702.

197. Kim, H. S. *et al.* (1999). Phylogenetic analysis of HERV-K LTR-like elements in primates: presence in some new world monkeys and evidence of recent parallel evolution in these species and in homo sapiens. *Arch. Virol.* **144**: 2035–2040.

198. Sverdlov, E. D. (2000). Retroviruses and primate evolution. *Bioessays* **22**: 161–171.

199. Johnson, N. A. and Porter, A. H. (2000). Rapid speciation via parallel, directional selection on regulatory genetic pathways. *J. Theor. Biol.* **205**: 527–542.

200. Johnson, W. E. and Coffin, J. M. (1999). Constructing primate phylogenies from ancient retrovirus sequences. *Proc. Natl. Acad. Sci. USA* **96**: 10 254–10 260.

201. Johnston, M. (1998). Gene chips: array of hope for understanding gene regulation. *Curr. Biol.* **8**: R171–R174.

202. Rowold, D. J. and Herrera, R. J. (2000). Alu elements and the human genome. *Genetica* **108**: 57–72.

203. Aleman, C. *et al.* (2000). *Cis*-acting influences on Alu RNA levels. *Nucleic Acids Res.* **28**: 4755–4761.

204. Stenger, J. E. *et al.* (2001). Biased distribution of inverted and direct Alus in the human genome: implications for insertion, exclusion, and genome stability. *Genome Res.* **11**: 12–27.

205. Hamdi, H. K. *et al.* (2000). Alu-mediated phylogenetic novelties in gene regulation and development. *J. Mol. Biol.* **299**: 931–939.

206. Comai, L. *et al.* (2000). Phenotypic instability and rapid gene silencing in newly formed arabidopsis allotetraploids. *Plant Cell* **12**: 1551–1568.

207. Kooter, J. M., Matzke, M. A. and Meyer, P. (1999). Listening to the silent genes: transgene silencing, gene regulation and pathogen control. *Trends Plant Sci.* **4**: 340–347.

208. Hurst, L. D. (1995). Evolutionary genetics. The silence of the genes. *Curr. Biol.* **5**: 459–461.

209. Cinader, B. (1989). Aging, evolution and individual health span: introduction. *Genome* **31**: 361–367.

210. Bader, G. *et al.* (1998). Apolipoprotein E polymorphism is not associated with longevity or disability in a sample of Italian octo- and non-agenarians. *Gerontology* **44**: 293–299.

211. Barzilai, N. *et al.* (2001). Offspring of centenarians have a favorable lipid profile. *J. Am. Geriatr. Soc.* **49**: 76–79.

212. Bathum, L. *et al.* (1998). Genotypes for the cytochrome P450 enzymes CYP2D6 and CYP2C19 in human longevity. Role of CYP2D6 and CYP2C19 in longevity. *Eur. J. Clin. Pharmacol.* **54**: 427–430.

213. Bladbjerg, E. M. *et al.* (1999). Longevity is independent of common variations in genes associated with cardiovascular risk. *Thromb. Haemost.* **82**: 1100–1105.

214. Gerdes, L. U. *et al.* (2000). Estimation of apolipoprotein E genotype-specific relative mortality risks from the distribution of genotypes in centenarians and middle-aged men: apolipoprotein E gene is a "frailty gene", not a "longevity gene". *Genet. Epidemiol.* **19**: 202–210.

215. Hedrick, P. W. and McDonald, J. F. (1980). Regulatory gene adaptation: an evolutionary model. *Heredity* **45**: 83–97.

216. Heijmans, B. T., Westendorp, R. G. and Slagboom, P. E. (2000). Common gene variants, mortality and extreme longevity in humans. *Exp. Gerontol.* **35**: 865–877.

217. Henon, N. *et al.* (1999). Familial versus sporadic longevity and MHC markers. *J. Biol. Reg. Homeost. Agents* **13**: 27–31.

218. Herman, J. G. and Baylin, S. B. (2000). Promoter-region hypermethylation and gene silencing in human cancer. *Curr. Topic Microbiol. Immunol.* **249**: 35–54.

219. Petropoulou, C. *et al.* (2000). Aging and longevity. A paradigm of complementation between homeostatic mechanisms and genetic control? *Ann. NY Acad. Sci.* **908**: 133–142.

220. Toyota, M. and Issa, J. P. (2000). The role of DNA hypermethylation in human neoplasia. *Electrophoresis* **21**: 329–333.

221. Burgess, J. K. and Hazelton, R. H. (2000). New developments in the analysis of gene expression. *Redox Rep.* **5**: 63–73.

222. Brown, A. R., T'so, P. O. P. and Cutler, R. G. (1991). Expression of the intracisternal *A* particle endogenous retrovirus genes over the lifetime of mouse and Syrian hamster. *Arch. Gerontol. Geriatr.* **13**: 15–30.

223. Clapham, J. C. *et al.* (2000). Mice overexpressing human uncoupling protein-3 in skeletal muscle are hyperphagic and lean. *Nature* **406**: 415–418.

224. Dudas, S. P. and Arking, R. (1995). A coordinate upregulation of antioxidant gene activities is associated with the delayed onset of senescence in a long-lived strain of *Drosophila*. *J. Gerontol. A: Biol. Sci. Med. Sci.* **50**: B117–B127.

225. Lee, C. K. *et al.* (1999). Gene expression profile of aging and its retardation by caloric restriction. *Science* **285**: 1390–1393.

226. Lee, C. K., Weindruch, R. and Prolla, T. A. (2000). Gene-expression profile of the ageing brain in mice. *Nature Genet.* **25**: 294–297.

227. Miller, R. A., Galecki, A. and Shmookler-Reis, R. J. (2001). Interpretation, design, and analysis of gene array expression experiments. *J. Gerontol. A: Biol. Sci. Med. Sci.* **56**: B52–B57.

228. Colman, P. D. *et al.* (1980). Brain poly(A)RNA during aging: stability of yield and sequence complexity in two rat strains. *J. Neurochem.* **34**: 335–345.

229. Ly, D. H. *et al.* (2000). Mitotic misregulation and human aging. *Science* **287**: 2486–2492.

230. Chenais, B. *et al.* (2000). Oxidative stress involvement in chemically induced differentiation of K562 cells. *Free Radic. Biol. Med.* **28**: 18–27.

231. Zou, S. *et al.* (2000). Genome-wide study of aging and oxidative stress response in *Drosophila melanogaster*. *Proc. Natl. Acad. Sci. USA* **97**: 13 726–13 731.

232. Robertson, K. D. and Jones, P. A. (2000). DNA methylation: past, present and future directions. *Carcinogenesis* **21**: 461–467.

233. Ripple, M. O. *et al.* (1999). Effect of antioxidants on androgen-induced AP-1 and NF-kappa B DNA-binding activity in prostate carcinoma cells. *J. Natl. Cancer Inst.* **91**: 1227–1232.

234. Toyota, M. and Issa, J. P. (1999). CpG island methylator phenotypes in aging and cancer. *Sem. Cancer Biol.* **9**: 349–357.

235. Richardson, B. and Yung, R. (1999). Role of DNA methylation in the regulation of cell function. *J. Lab. Clin. Med.* **134**: 333–340.

236. Ahuja, N. and Issa, J. P. (2000). Aging, methylation and cancer. *Histol. Histopathol.* **15**: 835–842.

237. Ross, R. E. (2000). Age-specific decrease in aerobic efficiency associated with increase in oxygen free radical production in *Drosophila melanogaster*. *J. Insect. Physiol.* **46**: 1477–1480.

238. Lin, Y. J., Seroude, L. and Benzer, S. (1998). Extended life-span and stress resistance in the *Drosophila* mutant methuselah. *Science* **282**: 943–946.

239. Miquel, J. (1998). An update on the oxygen stress-mitochondrial mutation theory of aging: genetic and evolutionary implications. *Exp. Gerontol.* **33**: 113–126.

240. Pehowich, D. J. (1999). Thyroid hormone status and membrane *n*-3 fatty acid content influence mitochondrial proton leak. *Biochim. Biophys. Acta* **1411**: 192–200.

241. Nicholls, D. G. and Budd, S. L. (2000). Mitochondria and neuronal survival. *Physiol. Rev.* **80**: 315–360.

242. Hermesh, O., Kalderon, B. and Bar-Tana, J. (1998). Mitochondria uncoupling by a long chain fatty acyl analogue. *J. Biol. Chem.* **273**: 3937–3942.

243. Gong, D. W. *et al.* (2000). Lack of obesity and normal response to fasting and thyroid hormone in mice lacking uncoupling protein-3. *J. Biol. Chem.* **275**: 16 251– 16 257.

244. Moncada, S. (2000). Nitric oxide and cell respiration: physiology and pathology. *Verh K Acad. Geneeskd Belg.* **62**: 171–179.

245. Sarkela, T. M. *et al.* (2000). The modulation of oxygen radicals production by nitric oxide in mitochondria. *J. Biol. Chem.* **5**: 5.

246. Lopez-Torres, M., Romero, M. and Barja, G. (2000). Effect of thyroid hormones on mitochondrial oxygen free radical production and DNA oxidative damage in the rat heart. *Mol. Cell Endocrinol.* **168**: 127–134.

247. Harman, D. (1996). Aging and disease: extending functional life span. *Ann. NY Acad. Sci.* **786**: 321–336.

248. Harman, D. (1994). Free-radical theory of aging. Increasing the functional life span. *Ann. NY Acad. Sci.* **717**: 1–15.

249. Harman, D. (1998). Extending functional life span. *Exp. Gerontol.* **33**: 95–112.

250. Hattori, I. *et al.* (2001). Thioredoxin-dependent redox regulation — implication in aging and neurological diseases. *In* "Oxidative Stress and Aging: Advances in Basic Science, Diagnostics and Intervention" (R. G. Cutler, and H. Rodriguez, Eds.), World Scientific Publishing Company, Singapore.

251. Giorgio, M. and Pelicci, P. G. (2001). Longevity determinent genes and oxidative stress. *In* "Oxidative Stress and Ageing: Advances in Basic Science, Diagnostics, and Intervention" (R. G. Cutler, and H. Rodriguez, Eds.), World Scientific Publishing Company, Singapore.

252. Pearson, M. *et al.* (2000). PML regulates p53 acetylation and premature senescence induced by oncogenic Ras. *Nature* **406**: 207–210.

253. Tyler, R. H. *et al.* (1993). The effect of superoxide dismutase alleles on aging in *Drosophila*. *Genetica* **91**: 143–149.

254. Arking, R. *et al.* (2000). Forward and reverse selection for longevity in *Drosophila* is characterized by alteration of antioxidant gene expression and oxidative damage patterns. *Exp. Gerontol.* **35**: 167–185.

255. Buck, S. *et al.* (2000). Extended longevity in *Drosophila* is consistently associated with a decrease in developmental viability. *J. Gerontol. A: Biol. Sci. Med. Sci.* **55**: B292–B301.

256. Rogina, B., Benzer, S. and Helfand, S. L. (1997). *Drosophila* drop-dead mutations accelerate the time course of age-related markers. *Proc. Natl. Acad. Sci. USA* **94**: 6303–6306.

257. Rothschild, H. and Jazwinski, S. M. (1998). Human longevity determinant genes. *J. LA State Med. Soc.* **150**: 272–274.

258. Miller, R. A. (1997). When will the biology of aging become useful? Future landmarks in biomedical gerontology. *J. Am. Geriatr. Soc.* **45**: 1258–1267.

259. Smith, D. W. (1995). Evolution of longevity in mammals. *Mech. Ageing Dev.* **81**: 51–60.

260. Smith, J. D. (2000). Apolipoprotein E4: an allele associated with many diseases. *Ann. Med.* **32**: 118–127.

261. Hart, R. W. and Turturro, A. (1981). Evolution and longevity-assurance processes. *Naturwissenschaften.* **68**: 552–557.

262. Hodes, R. J., McCormick, A. M. and Pruzan, M. (1996). Longevity assurance genes: how do they influence aging and life span? *J. Am. Geriatr. Soc.* **44**: 988–991.

263. Brandwagt, B. F. *et al.* (2000). A longevity assurance gene homolog of tomato mediates resistance to *Alternaria alternata f. sp. lycopersici* toxins and fumonisin B1. *Proc. Natl. Acad. Sci. USA* **97**: 4961–4966.

264. Kruse, M. *et al.* (2000). Sponge homologue to human and yeast gene encoding the longevity assurance polypeptide: differential expression in telomerase-positive and telomerase-negative cells of *Suberites domuncula*. *Mech. Ageing Dev.* **118**: 115–127.

265. Sacher, G. A. (1968). Molecular versus systemic theories on the genesis of ageing. *Exp. Gerontol.* **3**: 265–271.

266. Faure-Delanef, L. *et al.* (1998). Plasma concentration, kinetic constants, and gene polymorphism of angiotensin I-converting enzyme in centenarians. *Clin. Chem.* **44**: 2083–2087.

267. Akisaka, M. and Suzuki, M. (1998). Okinawa Longevity Study. Molecular genetic analysis of HLA genes in the very old. *Nippon Ronen Igakkai Zasshi.* **35**: 294–298.

268. Frisoni, G. B. *et al.* (2001). Longevity and the epsilon-2 allele of apolipoprotein E. The Finnish Centenarians Study. *J. Gerontol. A: Biol. Sci. Med. Sci.* **56**: M75–M78.

269. Barzilai, N. and Shuldiner, A. R. (2001). Searching for human longevity genes: the future history of gerontology in the post-genomic era. *J. Gerontol. A: Biol. Sci. Med. Sci.* **56**: M83–M87.

270. Taioli, E. *et al.* (2001). Polymorphisms of drug-metabolizing enzymes in healthy nonagenarians and centenarians: difference at GSTT1 locus. *Biochem. Biophys. Res. Commun.* **280**: 1389–1392.

271. Nemani, M. *et al.* (2000). The efficiency of genetic analysis of DNA from aged siblings to detect chromosomal regions implicated in longevity. *Mech. Ageing Dev.* **119**: 25–39.

272. Mecocci, P. *et al.* (2000). Plasma antioxidants and longevity: a study on healthy centenarians. *Free Radic. Biol. Med.* **28**: 1243–1348.

273. Caruso, C. *et al.* (2000). HLA, aging, and longevity: a critical reappraisal. *Human Immunol.* **61**: 942–949.

274. Orr, W. C. and Sohal, R. S. (1994). Extension of life-span by overexpression of superoxide dismutase and catalase in *Drosophila melanogaster*. *Science* **263**: 1128–1130.

275. Sohal, R. S. *et al.* (1995). Simultaneous overexpression of copper- and zinc-containing superoxide dismutase and catalase retards age-related oxidative damage and increases metabolic potential in *Drosophila melanogaster*. *J. Biol. Chem.* **270**: 15 671–15 674.

276. Kurapati, R. *et al.* (2000). Increased *hsp22* RNA levels in *Drosophila* lines genetically selected for increased longevity. *J. Gerontol. A: Biol. Sci. Med. Sci.* **55**: B552–B559.

277. Van Remmen, H. *et al.* (2001). Aging and oxidative stress in transgenic mice. *In* "Oxidative Stress and Aging: Advances in Basic Science and Intervention" (R. G. Culter, and H. Rodriguez, Eds.), pp. 1–54, World Scientific Publishing Company, Singapore.

278. Jung, C. *et al.* (2001). Synthetic superoxide dismutase/catalase mimetics reduce oxidative stress and prolong survival in a mouse amyotrophic lateral sclerosis model. *Neurosci. Lett.* **304**: 157–160.

279. Aston, K. *et al.* (2001). Computer-Aided Design (CAD) of Mn(II) complexes: superoxide dismutase mimetics with catalytic activity exceeding the native enzyme. *Inorg. Chem.* **40**: 1779–1789.

280. Salvemini, D. and Riley, D. P. (2000). Nonpeptidyl mimetics of superoxide dismutase in clinical therapies for diseases. *Cell Mol. Life Sci.* **57**: 1489–1492.

281. Konorev, E. A., Kennedy, M. C. and Kalyanaraman, B. (1999). Cell-permeable superoxide dismutase and glutathione peroxidase mimetics afford superior protection against doxorubicin-induced cardiotoxicity: the role of reactive oxygen and nitrogen intermediates. *Arch. Biochem. Biophys.* **368**: 421–428.

282. Segal, D. J. *et al.* (1999). Toward controlling gene expression at will: selection and design of zinc finger domains recognizing each of the 5′-GNN-3′ DNA target sequences. *Proc. Natl. Acad. Sci.* **96**: 2758–2763.

283. Gottesfeld, J. M., Turner, J. M. and Dervan, P. B. (2000). Chemical approaches to control gene expression. *Gene Expr.* **9**: 77–91.

284. Mapp, A. K. *et al.* (2000). Activation of gene expression by small molecule transcription factors. *Proc. Natl. Acad. Sci. USA* **97**: 3930–3935.

285. Beerli, R. R. *et al.* (1998). Toward controlling gene expression at will: specific regulation of the erb B-2/HER-2 promoter by using polydactyl zinc finger proteins constructed from modular building blocks. *Proc. Natl. Acad. Sci. USA* **95**: 14 628–14 633.

286. Carney, J. M. *et al.* (1991). Reversal of age-related increase in brain protein oxidation, decrease in enzyme activity, and loss in temporal and spatial memory by chronic administration of the spin-trapping compound N-tert-butyl-alpha-phenylnitrone. *Proc. Natl. Acad. Sci. USA* **88**: 3633–3636.

287. Carney, J. M. and Floyd, R. A. (1994). Brain antioxidant activity of spin traps in Mongolian gerbils. *Meth. Enzymol.* **234**: 523–526.
288. Carney, J. M. and Carney, A. M. (1994). Role of protein oxidation in aging and in age-associated neurodegenerative diseases. *Life Sci.* **55**: 2097–2103.
289. Butterfield, D. A. *et al.* (1997). Free radical oxidation of brain proteins in accelerated senescence and its modulation by N-tert-butyl-alpha-phenyl-nitrone. *Proc. Natl. Acad. Sci. USA* **94**: 674–678.
290. Chamulitrat, W. *et al.* (1993). Nitric oxide formation during light-induced decomposition of phenyl N-tert-butylnitrone. *J. Biol. Chem.* **268**: 11 520–11 527.
291. Saito, K., Yoshioka, H. and Cutler, R. G. (1998). A spin trap, N-tert-butyl-alpha-phenylnitrone extends the life span of mice. *Biosci. Biotechnol. Biochem.* **62**: 792–794.
292. Kirkwood, T. B. L. (1999). *Time of Our Lives: The Science of Human Aging*, pp. 1–277, Oxford University Press, Oxford, New York.

Chapter 65

The Genetics of Yeast Aging: Pathways and Processes

S. Michal Jazwinski

S. Michal Jazwinski • Department of Biochemistry and Molecular Biology, Louisiana State University Health Sciences Center, 1901 Perdido St., Box P7-2, New Orleans, LA 70112, USA
Tel/Fax: 504-568-4725, E-mail: sjazwi@lsuhsc.edu

1. Introduction

The yeast *Saccharomyces cerevisiae* has been a model system for the genetic and molecular analysis of aging and longevity for a decade now.[1] Its attraction in genetics and molecular biology stems from the fact that it is a unicellular eukaryote, thus facilitating sophisticated studies within a short timeframe. The use of yeast in aging research is based on the observation that individual cells divide a limited number of times, producing progeny that start the process all over.[2] In fact, this replicative life span of the individual yeast is independent of the time it takes to complete these cell divisions.[3] This and the requirement for metabolic activity to produce a daughter cell suggest that replicative life span and metabolism are closely intertwined.[4]

More recently, the use of yeast as a model to study chronological aging has been proposed.[5] Yeast cultures that have exhausted nutrients through the growth and multiplication of cells are stored in this stationary phase, and the viability of the cells is assessed over time. Cells in such cultures are characterized by very low metabolic activity, compared to those growing exponentially on which replicative life span is measured.[6] The yeast cell is the yeast organism. Thus, cell division is the reproductive activity that guarantees survival of the species. At first glance, it would appear then that chronological aging in yeast bears very little resemblance to aging in other species, in which actively metabolizing individuals capable of reproductive activity, at least for a time, are the object of study.

2. Phenomenology of Aging

There is an exponential decline in the probability of yeast cell division as a function of the number of divisions completed.[7, 8] This Gompertz mortality function breaks down at later ages, and when about 10% of the population remains, mortality rate plateaus.[9] Cell viability in stationary phase displays a much more complex behavior,[10] which suggests that periods of declining mortality alternate with ones in which mortality increases. This behavior may be due to the lysis of cells, followed by the growth of the culture until it becomes saturated again.

A multitude of phenotypic changes accompany yeast aging. These have been detailed for individual cells progressing through their replicative life span.[11] They include morphological changes, such as an increase in size, as well as physiologic changes, such as a decrease in budding rate. Altered patterns of gene expression have also been described. Given the emphasis on the population as a whole, it may not be surprising that the phenotypic characterization does not approach the same richness for chronologically aging yeast cells.[6] There are some features that may unite these seemingly disparate paradigms.[1]

3. Genes, Pathways, and Processes

The understanding of the mechanisms governing the replicative life span has expanded dramatically.[11] Currently, over twenty genes that determine yeast life span are known (Table 1). These genes encode a broad range of biochemical functions, spanning protein processing, signaling, transcriptional silencing, and transcriptional activation. This suggests that there are many mechanisms of aging. As will be seen below, there is also more than one pathway involved. Importantly, the genetic analysis of replicative life span has revealed the operation of four broad processes: metabolic control, stress resistance, gene dysregulation, and genetic stability in aging.[4]

Table 1. Yeast Longevity Genes

Process	Gene	Function
Metabolic control	RAS1	G-protein[1]
	RAS2	G-protein
	PHB1	Mitochondrial membrane-bound chaperone
	PHB2	Mitochondrial membrane-bound chaperone
	RTG2	?
	RTG3	Transcription factor subunit
Stress resistance	RAS1	G-protein
	RAS2	G-protein
	LAG1	ER-bound, GPI-anchored protein transport factor[2]
	LAC1	ER-bound, GPI-anchored protein transport factor
Gene dysregulation	RPD3	Histone deacetylase
	HDA1	Histone deacetylase
	SIR2	ADP-ribosyl transferase, histone deacetylase
	SIR4	Transcriptional silencing factor
	ZDS1	Transcriptional silencing factor
	ZDS2	Transcriptional silencing factor
	RAS2	G-protein
Genetic stability	SGS1	DNA helicase
	RAD52	DNA repair protein
	FOB1	Replication fork block protein
Other	UTH4	?
	YGL023	?
	CDC7	Protein kinase (cell cycle control)
	BUD1	G-protein (cell polarity)

[1]GTP-binding protein involved in signal transduction.
[2]Protein in endoplasmic reticulum involved in transport of glycosylphosphatidylinositol-anchored proteins.

The importance of metabolic control for longevity was first shown in studies that implicated the retrograde response in determining yeast life span.[12] The retrograde response is a pathway of interorganelle communication from the mitochondrion to the nucleus. It signals mitochondrial dysfunction and alters the expression of nuclear genes that encode a variety of cytoplasmic, mitochondrial, and peroxisomal metabolic enzymes. Induction of the retrograde response prolongs life span and makes the yeast more resistant to lethal heat stress. It results in a metabolic adjustment that involves a shift to utilization of the glyoxylate cycle to conserve carbon atoms while maintaining production of Krebs cycle intermediates, and an activation of gluconeogenesis. One can speculate that this is a compensatory response to the accumulation of mitochondrial defects with age.

The significance of stress resistance in determining replicative life span can be adduced from five sources. Mutants that were isolated for resistance to cold and starvation were shown to be longevous.[13] These mutants were also resistant to other stressors. It has been since determined that the life extension seen in these mutants is likely due to the release of transcriptional silencing complexes of Sir proteins from telomeres and silent mating type loci and their migration to the rDNA locus,[14] rather than to the enhanced resistance to stress. However, some involvement of the latter in determining life span cannot be ruled out.

A more clear-cut argument can be made for the involvement of resistance to chronic bouts of sublethal heat stress as a longevity determining factor.[15] Such chronic stress shortens life span, and this effect is more severe in the absence of the *RAS2* gene. This result seems paradoxical, because *RAS2*Δ strains are more resistant to a lethal heat shock. However, these strains do not downregulate the heat shock response rapidly, so that the yeast cells persist in a state of "shock" well after the stress is no longer operative. There is clearly a relationship between the resistance to heat and resistance to other stresses.[16] The heat shock response involves the induction of a variety of genes in addition to the classical heat shock proteins and includes proteins that protect the cell from other stressors, such as oxidative stress.[17] The induction of thermal tolerance in yeast results in a prolongation of life span.[18] This is achieved through a reduction in mortality rate, which is persistent but not permanent. Both *RAS1* and *RAS2* are necessary for this effect, as is *HSP104*. This is clearly a heritable epigenetic effect, which supports the role of stress resistance in longevity.

The resistance of yeast cells to ultraviolet radiation (UV) shows a biphasic profile with age.[19] Resistance first increases through mid-life and then plummets. This is not true of resistance to other DNA damaging agents, which shows a continuous decline with age. This suggests that it is not resistance to DNA damage itself that exhibits the biphasic pattern. The profile of UV resistance corresponds to the expression of *RAS2* during the yeast life span.[19] Because *RAS2* is required for resistance to UV,[20] this prompts the conclusion that it is the changing expression of this gene that is the source of the UV resistance profile. The deleterious events to which this *RAS2*-dependent pathway responds may involve oxidative damage to a variety of cellular components, but it does not seem to include pyrimidine dimers in DNA. Interestingly in this regard, it has been demonstrated that deficiencies in superoxide dismutases shorten the yeast replicative life span.[21, 22]

The possible role of the epigenetic inheritance of different regulatory states of chromatin was first proposed as a mechanism of yeast aging a long time ago.[23] The molecular prerequisites for such a mechanism were established when it was shown that there is a loss of transcriptional silencing of heterochromatic domains of the yeast genome with age,[24, 25] which prompted the proposal that gene dysregulation is an aging process in this organism.[4] The histone deacetylase genes *RPD3* and *HDA1* have been shown to be longevity genes in yeast, and their role in transcriptional silencing of rDNA is associated with this effect.[26] This has been followed by the implication of *Sir2*, an NAD-dependent histone deacetylase,[27] as well as proteins that affect the phosphorylation state of the silencing machinery.[28]

The instability of the rDNA repeats present as a tandem array of 100 to 200 in the yeast genome has been implicated in yeast aging. These repeats give rise to extrachromosomal rDNA circles that can amplify with age.[14] The presumptive effect of the accumulation of these circles is the enlargement and fragmentation of the nucleolus.[29] The significance of these circles in yeast aging has been opened to question recently.[12, 26, 30] The *RAD52* gene, which is involved in recombinational DNA repair, affects the yeast life span, providing evidence for a role of genetic instability in yeast aging.[31]

Only three genes have been shown to play a role in survival in the chronological life span model of yeast aging. They are *SOD1*, *SOD2*, and *RAS2*.[5] The requirement for *SOD1* and *SOD2* is perhaps not surprising. They encode the Cu, Zn and Mn superoxide dismutases, respectively. In stationary phase, yeast depend on respiration for their energy needs, rather than on fermentation. Thus, they are assaulted by oxygen radicals, which are a by-product of respiration. In addition, they do not rapidly turn over many of their cellular components, as they do when they are dividing, and therefore cellular damage can accumulate. The role of *RAS2* is more complicated. As a negative regulator of many stress responses through its effects on gene expression from the stress response regulatory element in gene promoters, the deletion of this gene actually enhances stationary phase survival. Yet, the absence of *RAS2* would not serve a stationary yeast cell well, because the gene is necessary for cells to exit from stationary phase and resume growth.[32]

4. RAS2, a Homeostatic Device in Yeast Longevity

The overexpression of *RAS2* extends replicative life span, while its deletion curtails it.[33] There is a biphasic effect of *RAS2* overexpression on yeast life span.

Past a point, further increases in Ras activity curtail life span. The life shortening effect can be largely ascribed to the stimulation of the cAMP-protein kinase A pathway, while the life prolongation to a cAMP-independent pathway.[33] This is the case for unstressed cells. In the presence of sublethal heat stress, the cAMP-protein kinase A pathway is necessary to maintain replicative life span,[15] just as it does chronological life span.[5] This already reveals the complex balancing act that is necessary for survival of the organism under constantly changing environmental conditions.

There is more to the role of *RAS2* in yeast longevity than simply the modulation of stress responses. This gene also impinges upon the retrograde response, and thus directly affects a metabolic mechanism of yeast aging.[12] *RAS2* also affects the basal level of transcriptional silencing at subtelomeric loci in yeast.[9] This gene is also important in maintenance of cell polarity during the life span[9] and in the cell division cycle.[34] *RAS2* performs the role of a homeostatic device in yeast longevity by modulating several, sometimes exclusive, pathways. It is not an integral component of these pathways, but rather a modulator, and thus it can coordinate all of them, responding to the competing demands that are placed upon the yeast cell. These competing demands are dictated by a changing environment and a fluctuating internal milieu, as well as by epigenetic exigencies.

5. Caloric Restriction in Yeast

It has recently been shown[35] that reduction of the glucose concentration in the growth medium increases yeast life span, such that the lower the glucose concentration the greater the increase (up to a point). This is also true of amino acids concentration. There is a prolongation of both mean and maximum lifespan and a retardation of the manifestations of aging. This phenomenon possesses the hallmarks of caloric restriction in mammals.

Caloric restriction operates via a separate pathway from the retrograde response in extending yeast life span.[35] The expression of the diagnostic gene for the retrograde response is not induced but rather repressed by caloric restriction. The extension of life span by caloric restriction is not suppressed by deletion of the mediators of the retrograde response *RTG2* and *RTG3*. The genetic evidence, however, suggests that there may be some overlap between the downstream longevity effectors of caloric restriction and the retrograde response. In any case, there are now clearly two metabolic mechanisms of yeast aging. Unlike the retrograde response, caloric restriction may not so much compensate for the dysfunction associated with aging, but rather to retard the development of the deficits of aging. It is not known at present whether caloric restriction in yeast enhances resistance to stress.

6. A Metabolic Continuum

The distinction between replicative life span and chronological aging may not be as stark as it seems. The largely non-dividing cells in a stationary culture and the rapidly dividing cells in a growing population are part of a metabolic continuum. The former are characterized by low metabolic activity, while in the latter it is high. The biochemistry will differ depending on the available nutrients. However, all other things being equal, it is simply the intensity of metabolic activity that differs. Thus, it may be easier to examine metabolic mechanisms of aging in the replicative life span model. In this view of a metabolic continuum, metabolic capacity is being accrued and its consequences elicited whether or not cells are dividing.[1] This would lead to a reduction of replicative life span of cells held in stationary phase. This indeed has been shown to be the case.[36] It is not, however, clear whether it is actually the period of residence in stationary phase that shortens the subsequent replicative life span, or whether it is the process of exit from stationary phase for which the cells pay a toll.

It is obvious that cells in stationary phase must have enhanced stress responses. They cannot escape stress by generating numbers through cell division. Growing populations devote more energy to growth and reproduction and less to stress resistance. This is where the *RAS2* homeostat plays such an important role, in helping to maintain the appropriate balance. However, the realities are not as straightforward, as indicated earlier. The induction of thermal tolerance extends yeast replicative life span. Perhaps what is being induced is a resistance to sublethal stresses that are normally encountered by the yeast, even under what appear to be optimal conditions. Furthermore, resistance to all stresses does not increase in stationary phase. UV resistance is higher in growing cultures.[37] From the foregoing discussion, one would expect many aging processes and mechanisms to be similar during replicative and chronological aging. However, one would also anticipate distinct aging processes and mechanisms.

7. Conclusions

The major emphasis in aging studies in yeast has been the understanding of the mechanisms underlying the limited replicative capacity of individual cells. This has led to the identification of over twenty genes that determine life span and to a delineation of the major processes that are involved in aging. The molecular mechanisms underlying these are now being elucidated. At least two metabolic mechanisms have been implicated: the retrograde response and caloric restriction. In addition, gene dysregulation, stress resistance, and genetic stability have been shown to be important in aging. Chronological aging of cells in stationary cultures has seen very limited utilization in aging studies. This model has pointed to the importance of resistance to oxygen free radicals for survival. Both models of aging are united by their operation within a metabolic continuum and by the role of the *RAS2* gene as a homeostatic device in yeast longevity. The near future will show how useful the yeast system is in providing insights into the aging of mammalian systems.

Acknowledgments

The work in the author's laboratory is supported by grants from the National Institute on Aging of the National Institutes of Health (U.S.P.H.S.).

References

1. Jazwinski, S. M. (1990). An experimental system for the molecular analysis of the aging process: the budding yeast *Saccharomyces cerevisiae*. *J. Gerontol.* **45**: B68–B74.
2. Mortimer, R. K. and Johnston, J. R. (1959). Life span of individual yeast cells. *Nature* **183**: 1751–1752.
3. Müller, I., Zimmermann, M., Becker, D. and Flömer, M. (1980). Calendar life span versus budding life span of *Saccharomyces cerevisiae*. *Mech. Ageing Dev.* **12**: 47–52.
4. Jazwinski, S. M. (1996). Longevity, genes, and aging. *Science* **273**: 54–59.
5. Longo, V. D. (1999). Mutations in signal transduction proteins increase stress resistance and longevity in yeast, nematodes, fruit flies, and mammalian neuronal cells. *Neurobiol. Aging* **20**:479–486.
6. Werner-Washburne, M., Braun, E., Johnston, G. C. and Singer, R. A. (1993). Stationary phase in the yeast *Saccharomyces cerevisiae*. *Microbiol. Rev.* **57**: 383–401.
7. Pohley, H.-J. (1987). A formal mortality analysis for populations of unicellular organisms (*Saccharomyces cerevisiae*). *Mech. Ageing Dev.* **38**: 231–243.
8. Jazwinski, S. M., Egilmez, N. K. and Chen, J. B. (1989). Replication control and cellular life span. *Exp. Gerontol.* **24**: 423–436.
9. Jazwinski, S. M., Kim, S., Lai, C.-Y. and Benguria, A. (1998). Epigenetic stratification: the role of individual change in the biological aging process. *Exp. Gerontol.* **33**: 571–580.
10. Vaupel, J. W., Carey, J. R., Christensen, K., Johnson, T. E., Yashin, A. I., Holm, N. V., Iachine, I. A., Kannisto, V., Khazaeli, A. A., Liedo, P., Longo, V. D., Zeng, Y., Manton, K. G. and Curtsinger, J. W. (1998). Biodemographic trajectories of longevity. *Science* **280**: 855–860.
11. Jazwinski, S. M. (1999). Molecular mechanisms of yeast longevity. *Trends Microbiol.* **7**: 247–252.

12. Kirchman, P. A., Kim, S., Lai, C.-Y. and Jazwinski, S. M. (1999). Interorganelle signaling is a determinant of longevity in *Saccharomyces cerevisiae*. *Genetics* **152**: 179–190.

13. Kennedy, B. K., Austriaco, N. R., Zhang, J. and Guarente, L. (1995). Mutation in the silencing gene *SIR4* can delay aging in *Saccharomyces cerevisiae*. *Cell* **80**: 485–496.

14. Sinclair, D. A., Mills, K. and Guarente, L. (1998). Molecular mechanisms of aging. *Trends Biochem. Sci.* **23**: 131–134.

15. Shama, S., Kirchman, P. A., Jiang, J. C. and Jazwinski, S. M. (1998). Role of *RAS2* in recovery from chronic stress: effect on yeast life span. *Exp. Cell Res.* **245**: 368–378.

16. Sanchez, Y., Taulien, J., Borkowich, K. A. and Lindquist, S. L. (1992). *HSP104* is required for tolerance to many forms of stress. *EMBO J.* **11**: 2357–2364.

17. Marchler, G. Schuller, C., Adam, G. and Ruis, H. (1993). A *Saccharomyces cerevisiae* UAS element controlled by protein kinase A activates transcription in response to a variety of stress conditions. *EMBO J.* **12**: 1997–2003.

18. Shama, S., Lai, C.-Y., Antoniazzi, J. M., Jiang, J. C. and Jazwinski, S. M. (1998). Heat stress-induced life span extension in yeast. *Exp. Cell Res.* **245**: 379–388.

19. Kale, S. P. and Jazwinski, S. M. (1996). Differential response to UV stress and DNA damage during the yeast replicative life span. *Dev. Genet.* **18**: 154–160.

20. Engelberg, D., Klein, C., Martinetto, H., Struhl, K. and Karin, M. (1994). The UV response involving the Ras signaling pathway and AP-1 transcription factors is conserved between yeast and mammals. *Cell* **77**: 381–390.

21. Wawryn, J., Krzepilko, A., Myszka, A. and Bilinski, T. (1999). Deficiency in superoxide dismutases shortens life span of yeast cells. *Acta Biochim. Polon.* **46**: 249–253.

22. Barker, M. G., Brimage, L. J. and Smart, K. A. (1999). Effect of Cu, Zn superoxide dismutase disruption mutation on replicative senescence in *Saccharomyces cerevisiae*. *FEMS Microbiol. Lett.* **177**: 199–204.

23. Jazwinski, S. M. (1990). Aging and senescence of the budding yeast *Saccharomyces cerevisiae*. *Mol. Microbiol.* **4**: 337–343.

24. Kim, S., Villeponteau, B. and Jazwinski, S. M. (1996). Effect of replicative age on silencing near telomeres in *Saccharomyces cerevisiae*. *Biochem. Biophys. Res. Commun.* **219**: 370–376.

25. Smeal, T., Claus, J., Kennedy, B., Cole, F. and Guarente, L. (1996). Loss of transcriptional silencing causes sterility in old mother cells of *Saccharomyces cerevisiae*. *Cell* **84**: 633–642.

26. Kim, S., Benguria, A., Lai, C.-Y. and Jazwinski, S. M. (1999). Modulation of life-span by histone deacetylase genes in *Saccharomyces cerevisiae*. *Mol. Biol. Cell* **10**: 3125–3136.

27. Imai, S., Armstrong, C. M., Kaeberlein, M. and Guarente, L. (2000). Transcriptional silencing and longevity protein *Sir2* is an NAD-dependent histone deacetylase. *Nature* **403**: 795–800.

28. Roy, N. and Runge, K. W. (2000). Two paralogs involved in transcriptional silencing that antagonistically control yeast life span. *Curr. Biol.* **10**: 111–114.

29. Sinclair, D. A., Mills, K. and Guarente, L. (1997). Accelerated aging and nucleolar fragmentation in yeast *sgs1* mutants. *Science* **277**: 1313–1316.

30. Heo, S.-J., Tatebayashi, K., Ohsugi, I., Shimamoto, A., Furuichi, Y. and Ikeda, H. (1999). Bloom's syndrome gene suppresses premature ageing caused by Sgs1 deficiency in yeast. *Genes to Cells* **4**: 619–625.

31. Park, P. U., Defossez, P. A. and Guarente, L. (1999). Effects of mutations in DNA repair genes on formation of ribosomal DNA circles and life span in *Saccharomyces cerevisiae. Mol. Cell. Biol.* **19**: 3848–3856.

32. Thevelein, J. M. (1994). Signal transduction in yeast. *Yeast* **10**: 1753–1790.

33. Sun, J., Kale, S. P., Childress, A. M., Pinswasdi, C. and Jazwinski, S. M. (1994). Divergent roles of *RAS1* and *RAS2* in yeast longevity. *J. Biol. Chem.* **269**: 18 638–18 645.

34. Baroni, M. D., Monti, P. and Alberghina, L. (1994). Repression of growth-regulated G1 cyclin expression by cyclic AMP in budding yeast. *Nature* **371**: 339–342.

35. Jiang, J. C., Jaruga, E, Repnevskaya, M. V. and Jazwinski, S. M. (2000). An intervention resembling caloric restriction prolongs life span and retards aging in yeast. *FASEB J.* 10.1096/fj.00–242fje.

36. Ashrafi, K., Sinclair, D., Gordon, J. I. and Guarente, L. (1999). Passage through stationary phase advances replicative aging in *Saccharomyces cerevisiae. Proc. Natl. Acad. Sci. USA* **96**: 9100–9105.

37. Parry, J. M., Davies, J. and Evans, W. E. (1976). The effects of "cell age" upon the lethal effects of physical and chemical mutagens in the yeast, *Saccharomyces cerevisiae. Mol. Gen. Genet.* **146**: 27–35.

Chapter 66

Oxidative Stress and Longevity Determinant Genes of Drosophila

Robert Arking

Robert Arking • Department of Biological Sciences, Wayne State University, Detroit MI 48098 USA
Tel: 313-577-2891, E-mail: rarking@biology.biosci.wayne.edu

1. Introduction

The role of oxidative stress in the aging process has been ably reviewed in recent years[1, 2] and these papers should be consulted for references and a broad and detailed review of the problem. This chapter will focus on three questions: (1) how persuasive is the evidence that specific genes conferring resistance to oxidative stress act as longevity determinant genes in *Drosophila*?, (2) how complete is our cataloging of these genes?, and (3) what does the genetic data tell us about the regulatory mechanisms and the cell biology and physiology of extended longevity in *Drosophila*?

2. Relationship between Stress and Longevity

It has been long observed that mild or non-lethal stress often has the apparently paradoxical effect of benefitting the organism by increasing its longevity.[3] Conversely, it has also been suggested that all long lived strains and mutants exhibit some form of stress resistance.[4, 5] This relationship is thought to reflect the fact that their natural environment usually exerts substantial albeit variable stresses on organisms. Evolutionary considerations of Darwinian fitness will thus impose a premium on genotypes conferring metabolic efficiency and stress resistance.[6] The magnitude of the effects of stress resistance on longevity are summarized in Table 1. Not all stressors yield the same level of response, the largest reactions being evoked by caloric restriction and by oxidative stress resistance. Since caloric restriction may well be one method of augmenting the organism's resistance to oxidative stress,[7] then the stressors with the maximum effect may be reflecting the operation of the same underlying mechanism.

Table 1. Effect of Non-Lethal Stressors on Longevity

Stressor	% Response of Experimental Animals over Controls*	Molecules Involved[†]
Cold	5%	hsp
Heat	10%	hsp & ADS
Hypergravity	12%	hsp
Physical activity	20%	ADS
Irradiation	20%	ADS & hsp
Caloric restriction	50%	ADS & others
Oxidative stress resistance	10–60%[‡]	ADS

*Data adapted from Fig. 1 of Minois (2000) with exception of last entry which is taken from articles referenced in this review.
[†]hsp = heat shock proteins; ADS = antioxidant defense system proteins.
[‡]Response noted after selection or transgenic modification of genome.

It is certainly well documented that aging involves an increase in oxidative damage in flies[8–10] and other organisms.[11] The fidelity of gene regulation may actually be well preserved in aging flies.[12] If so, then this means that the occurrence of oxidative damage in aging flies cannot be attributed to genetic dysregulation but rather must be the likely outcome of particular trajectories of gene action such that animals displaying early enhancement of oxidative stress resistance generally show delayed or lower levels of oxidative damage later in life (see below). However, it has also been noted that animals selected for extended longevity due to an enhanced resistance to oxidative stress also are characterized by a lowered developmental viability and decreased resistance to other stressors.[13, 14] Thus the enhanced resistance to one stressor, and the extended longevity flowing from that resistance, is not a free gift but is purchased at the price of a decreased fitness of the adult to other important environmental parameters. One reason why extended longevity does not appear to be common in the wild may be the inability of such organisms to thrive in variable environments characterized by multiple stressors. As a result of this brief overview, we may conclude that resistance to oxidative damage is the major, but not the only, mechanism through which the longevity determinant genes of *Drosophila* exert their effects and that its expression has effects on overall fitness.

3. Data from Candidate Gene Experiments

There are at least two general methods by which geneticists can probe the genome: candidate gene analysis and whole-genome analysis. The former method presupposes some pre-existing knowledge specifying the identity of the genes involved. As such, it allows for a rapid identification and characterization but it misses everything not included in the pre-existing knowledge and thus may give us a narrow view of reality. The latter method makes fewer assumptions regarding gene identification but are generally more expensive in both time and money, and run the risk of trading depth for breadth. Both approaches are necessary for our continued understanding of the interplay between genes and aging, and we utilize the data from both in reaching our conclusions.

3.1. CuZnSOD, MnSOD and Catalase

Early recognition of the role of oxidative stress in aging[15] led to the realization that enzymes involved in superoxide scavenging were likely candidate genes, particularly since the enzymatic activity of CuZnSOD and catalase were shown to decline in an age-specific manner in house flies.[10] It was also noted that CuZnSOD, but not catalase, expression was affected by heat shock[16] and thus might play a role in protecting the animals against the stresses of daily life. The demonstration by Phillips *et al.*[17] that CuZnSOD-null mutants of *Drosophila* had significantly decreased viability and longevity strongly suggested that this gene did in fact play an important role in modulating the rate of aging. This conclusion was later independently confirmed by Rogina *et al.*[18] using their enhancer trap-gene expression system.[19] Transgenic technology was early used to further test the role of CuZnSOD in aging. Seto *et al.*[20] used native transgenes and found that the resulting CuZnSOD over-expression (30–70%) had but a small effect (10%) on longevity and then only at elevated temperatures. A similar effect was noted by adding an ectopic bovine CuZnSOD transgene into a SOD-null mutant of *Drosophila*,[21] which increased the animals' stress resistance but had only limited success (10%) in restoring normal longevity. Attempts to use transgenes to over express CuZnSOD or catalase in normal *Drosophila* yielded essentially similar results when single transgenes were used.[22-24] However the tandem over-expression of CuZnSOD and catalase in the same animal did extend median and maximum longevity by up to 34% in some lines while simultaneously retarding oxidative damage and increasing oxidative resistance.[25, 26] Sun and Tower[27] used a binary transgenic system (FLP-OUT; see Ref. 28 for description of various transgene systems) to induce CuZnSOD overexpression in adult flies which yield proportionate increases in mean life span of up to 48%. The failure of CuZnSOD alone to induce life span extension in these different experiments has never been fully explained but it is likely to involve differences in genetic background between the strains used (see Ref. 28). Subsequent analysis[29] showed that the absence of the CuZnSOD gene has a number of important pleiotropic effects: (1) adult sensitivity to paraquat, (2) male sterility, (3) female semi-sterility, (4) adult hyperoxia sensitivity, (5) larval radiation sensitivity, (6) developmental sensitivity to glutathione depletion, and (7) adult life span reduction. Thus, both under- and over-expression of the CuZnSOD gene leads to significant decreases in fitness.[13, 14, 29]

All of the above mutational or transgenic alterations of gene expression ostensibly affect all tissues of the organism at all stages. But there is much information showing that most genes have characteristic tissue and stage-specific expression patterns (e.g. Ref. 30). Thus it was important when Parkes *et al.*[31] showed that a GAL4-UAS transgene expression system which selectively targeted CuZnSOD expression to the adult motor neuron was capable both of (a), restoring the normal adult life span of CuZnSOD-null mutants and (b), extending by 40% the adult life span of an otherwise wildtype fly. Over-expression of CuZnSOD in the adult central nervous system, adult muscle or larval body has no effect on adult longevity.[32] This suggests that oxidative stress within the adult motor neuron may be the rate limiting factor in fly longevity, a topic we shall return to below.

MnSOD is the mitochondrial version of superoxide dismutase. Given the crucial role of mitochondria in energy metabolism and ROS generation, it seemed logical that this gene product would likely play an important role in modulating the life span. This assumption is borne out by the recent data. Sun *et al.*[33] used the FLP-OUT system to induce the over-expression of MnSOD I adult flies and reported that MnSOD showed increases in expression of upto 75%. This yielded a 33% increase in mean life span and a 37% increase in maximum life span. Phillips and his colleagues[34] have also used transgenes to over-express the MnSOD gene in wild type animals and find that they obtain lifespan extensions of greater than 60% They also note that the over-expressed MnSOD shows mediocre rescue of the CuZnSOD-null mutant, thus establishing that the two enzymes operate in functionally different compartments. However, Mockett *et al.*[35] reported no difference in the life spans of their transgenic over-expressed MnSOD lines and the controls, even though the transgenic animals did significantly over express MnSOD mRNA, protein and enzyme activity. The transgenic animals released slightly less peroxide from their mitochondria and exhibited slightly less oxidative damage than did their controls. The difference between these two sets of experiments is puzzling and has been attributed to a genetic background in the genetically marked *yellow white* animals used by Mockett *et al.*[35, 36] such that the control MnSOD levels are already at an optimum and overexpression would have no functional effect. This is certainly consistent with the conclusions reached in Sec. 2 above and with the fact that the *yellow* gene, for example, is known to interact with at least 17 other genes and may well have more global effects on gene expression.[37] A variation of this explanation may reside in the fact that SOD is known to have a pro-oxidant effect due to a decreased steady state level of superoxide resulting in altered reaction kinetics and the oxidation of the target molecule it was supposed to protect.[38] The degree of oxidation may depend on the genetic background. This means that solutions worked out on one set of genotypes may not be applicable to all other genotypes. Some genotypes will likely be recalcitrant to the proffered explanation and will require customized interventions.

In contrast to the SOD data, acatalesemic mutants of *Drosophila* are essentially normal when reared under standard conditions as long as they have at least 3% of the normal catalase expression level,[39] a finding consistent with the transgene work[23, 27] and which suggests that catalase is not normally a rate limiting factor in longevity.

3.2. Other Genes

Given the complexity of the oxidative stress resistance process, it is inevitable that other genes are involved. Glutathionine is perhaps the most abundant low molecular weight antioxidant present in the animal and represents a potential clue to other candidate genes. Mockett *et al.*[40] over expressed the glutathione reductase gene in transgenic *yellow white Drosophila* and obtained upto 100%

overexpression of the enzyme. Longevity was significantly enhanced under hyperoxic conditions but not under normoxic conditions, suggesting that glutathione reductase may not be a rate-limiting factor in anti-aging defenses under normal conditions but may well be one when the level of oxidative stress is elevated.

Another set of candidate genes are those which regulate the expression of the antioxidant structural genes. Two different mutant searches have identified such genes. In the first, Lin *et al.*[41] did P-element mutagenesis of the third chromosome and screened for long life (relative to the *white* control strain) at 29°C. One homozygous mutant, named *methuselah (mth)*, lived upto 35% longer and was more resistant to paraquat, starvation and high temperature. The *mth* gene appears to code for a transmembrane G protein-coupled receptor presumably involved in the regulation of stress response genes. This last point needs to be critically tested. The second search also used P-element mutagenesis but focused on the second chromosome and identified seven mutants which significantly affected paraquat resistance, CuZnSOD and catalase activity, oxidative damage levels and longevity (from −40% to +20%; Ref. 42). The *trans*-acting mutants formed two groups, one of which acted as if they were normally positive regulators of CuZnSOD and catalase in wildtype animals and the other of which acted as if they were normally negative regulators. The molecular identity of these mutants is now being determined. In summary, the point is that there is an apparent complex web of regulatory genes on chromosomes 2 and 3 which regulate the antioxidant structural genes both singly and coordinately. In the long term, these represent attractive targets for interventions into the aging process.

It is useful not to restrict our focus to only antioxidant genes. For example, Leffelaar and Grigliatti[43] isolated several conditional (i.e. temperature sensitive) DNA repair mutants *(mus)* which acted so as to significantly decrease adult longevity relative to controls when raised under restrictive conditions. These and other repair genes likely play an important role in longevity determination (see Sec. 7).

4. Data from Genome-Wide Analysis

Selection experiments represent one form of a genome-wide screen for longevity determinant genes. We shall discuss two independent projects which each generated long lived strains which have been extensively analyzed. The Wayne State (WSU) lines[44, 45] consist of non-selected normal lived strains (R) which served as the progenitor for indirectly selected long lived (L, 2L) strains and short-lived (2E) strains. Each of these longevity classes is itself composed of a number (2–4) of replicate sister lines selected in parallel. Early work suggested that oxidative stress resistance was a major factor in the extended longevity and so a candidate gene approach was adopted on the L strains.[46] The important points in the present context are as follows:

(1) The long lived La strain depends on the early and coordinate up-regulation of CuZnSOD and MnSOD for the expression of its extended longevity.[9]
(2) The antioxidant enzyme activity is inversely correlated with the levels of oxidative damage.
(3) All long lived L strains are highly resistant to paraquat.[47]
(4) The expression of the antioxidant structural genes is under the control of positive and negative trans-acting regulatory genes.[42]
(5) The mean daily metabolic rate is unchanged, consequently the metabolic potential of the long-lived La strain is increased.[48, 49]
(6) The La strain animals exhibit either an unaltered or a decreased resistance to other stressors relative to their controls.[13, 47, 50]
(7) La and Lb sister strains with identical longevity phenotypes have significantly different patterns of antioxidant gene expression.[51]

The WSU L strains are, to my knowledge, the only strains which have been analyzed both by a candidate gene analysis (above) and by genome-wide QTL mapping. J. W. Curtsinger and his colleagues[52] found four QTLs in the (L × R) recombinant inbred strains that accounted for almost all of the selection response. They are located on chromosomes 2 and 3. The major QTLs for both paraquat resistance and longevity are coincident with each other and are centered over the locus of the CuZnSOD gene. Both types of analysis suggest that the WSU L strains live long because of a specific up-regulation of antioxidant defense genes.

However, RAPD analysis of the L and R strains showed that there also exist loci in these strains with either a sex-specific or a sex-shared effect on longevity.[53] Furthermore, Mackay and her colleagues[54] showed in other strains that different environments evoke different patterns of QTL loci with effects on longevity. This is reminiscent of the glutathione reductase data.[40] The genetic effects we observe in the controlled conditions of the lab may only be a subset of the genetic effects potentially present in a natural population.

The University of California-Irvine (UCI) lines were developed by Rose[55] using a different wild type progenitor stock but the same indirect selection protocols as used for the WSU lines. Interestingly enough, these lines appear to have an overlapping but different set of traits associated with the extended longevity of the selected O lines relative to their progenitor B lines, the most important of which in the present context are:

(1) The O line animals are significantly more resistant to environmental stresses (starvation, dessication, etc.) than the B line controls and have significantly increased lipid and glycogen content.[56]

(2) Continued selection for starvation and/or dessication resistance also led to continued increase in longevity.[57]

This suggested that there might in fact be two different genetic mechanisms underlying the extended longevity in the WSU and UCI lines. However, this is probably not the case for two reasons. First, using selection for starvation resistance on another set of strains failed to yield a correlated response of increased longevity,[58] suggesting that the correlation between the two may be strain specific and thus not yield a robust indication of possible candidate genes. Second, it has recently been demonstrated that the UCI O lines are significantly more resistant to paraquat than are their B line progenitors and controls.[59] Resistance to oxidative stress thus appears to be a robust correlated response to indirect selection for long life and should be very informative in identifying the underlying genetic mechanism(s). The information from my lab and that of J. W. Curtsinger's cited above supports this statement.

5. Data Implicating the Role of Mitochondria

Over-expression of MnSOD has been shown to have large effects on longevity (see above). However, there exist data implicating other aspects of the role of mitochondria (mt) in the aging process. It is currently believed that the mitochondria are the major source of ROS responsible for endogenous oxidative damage. Genes affecting this process might well be important presumptive longevity determinant genes. Schwarze *et al.*[60] showed that there are significant changes in mitochondrial function which can be attributed to oxidative stress. Their data suggests that an initial blocking of cytochrome c oxidase (COX) activity, perhaps due to oxidative damage to the inner membrane, leads to an impaired electron flow and increased ROS formation, followed by a reduction in mtRNA abundance and decreased COX activity. This is consistent with the finding that mitochondrial proteins such as aconitase and adenine nucleotide translocatase (an integral protein of the inner membrane) are preferentially oxidized during aging.[61]

Genetic data to support this scenario has been provided by Driver and Tawadros.[62] Using the WSU strains, crosses were made so as to result in flies having the Ra nuclear genes combined with the La, Lb or Rb mt genomes. Longevity measurements showed that strains with the Ra nuclear genome and either the La or Lb mt genomes lived significantly longer than did flies with the Ra nuclear genome and either a Ra or Rb mt genome. Interestingly enough, the effect of the mt genomes on longevity is diet dependent. As expected, there is an inverse correlation between the level of ROS production in young animals and longevity. Clearly, understanding the genetic and functional differences between the L and R mitochondria should extend our knowledge of the longevity determinant genes.

6. Discussion

6.1. Quality of the Evidence

We posed three questions in the Introduction which we will now answer. With regard to the first question, the evidence discussed above showing that certain specific genes confer oxidative resistance to the organism and act as longevity determinant genes is really quite good. With regard to the completeness of our catalog, the data is concentrated in a few favorite genes and thus our knowledge is very patchy and incomplete. The candidate gene approach has identified circa 5 genes which have been analyzed in any depth, while the genome-wide studies have shown that many more genes remain to be identified and characterized. Therefore one necessary approach is to assay isogenic and inbred lines with the DNA microarray technology so as to determine the gene sets involved in the aging processes of short-, normal- and long-lived strains and mutants. Table 2 summarizes the recent data showing the global gene expression patterns observed in post-mitotic tissues of the mouse.

The effects of the calorically restricted delayed aging can be interpreted consistent with the hypothesis presented below, but it is also clear that there is

Table 2. Observed Changes in Gene Expression of Post-Mitotic Tissues of the Mouse During Normal and Delayed Aging*

Experiment	Normal Aging	CR* Delayed Aging
Effect of CR[†] on mouse muscle	↑ Stress response ↑ Neuronal injury ↓ Energy metabolism	↑ Biosynthesis ↑ Protein turnover ↑ Energy metabolism ↓ Macromolecular damage
Effect of CR[†] on mouse brain	↑ Stress response ↑ Inflammatory response ↓ Protein turnover ↓ Growth factors	↓ Stress response Better immune modulation ↓ Protein synthesis ↑ Growth factors ↑ DNA synthesis

*Data for top panel taken from Lee, C.-K., Kloop, R. G., Weindruch R. and Prolla, T. A. (1999). Gene expression profile of aging and its retardation by caloric restriction. *Science* **285**: 1390–1393; and the data for the bottom panel was taken from Lee, C.-K., Weindruch, R. and Prolla, T. A. (2000). Gene-expression profile of the ageing brain in mice. *Nature Genet.* **25**: 294–297.
[†]Caloric restriction.

a complicated interplay between genes affecting ROS scavenging, protein turnover, energy metabolism, stress response and repair processes. We need to know the expression patterns of the comparable genes in *Drosophila* before we can adequately answer the third question posed above.

6.2. Implications of Tissue-Specific Age Failure

The striking tissue specificity of CuZnSOD to the motor neurons[31] needs to be understood. In a discussion of their work, Parkes *et al.*[32] used the concept of tissue-specific age-related failure thresholds to explain their result. According to this line of reasoning, the motor neuron would have the lowest such threshold in the fly's body; that is why reducing its level of oxidative stress improves the longevity of the organism. Presumably the transgenic fly died because some other tissue reached its age-related failure threshold. One wonders if there is a limit to the number of transgenic "patches" one can effectively apply to an organism. Parkes *et al.*[32] raise an alternative view of this concept; namely that the CuZnSOD transgene might have altered the ROS-mediated signal transduction and neuroendocrine signaling pathways and extended longevity in this indirect manner. A deep understanding of the aging process and the genes involved requires an understanding of these matters.

7. Conclusion and Summary Hypothesis

The major longevity determinant genes of *Drosophila* are mostly involved in resistance to oxidative stress. This is fully compatible with the situation in mammals.[63] There are three aspects of the conserved cellular response to oxidative stress:

(1) avoidance of the generation of ROS (involving but not limited to genes such as CuZnSOD, MnSOD and catalase);
(2) scavenging of ROS once produced (involving but not limited to thioredoxin, glutaredoxin, glutathione synthetase, glutathione reductase, and others);
(3) repair of ROS induced damage once inflicted (involving but not limited to DNA repair genes, heat shock proteins, proteosomes and others).

The genes included in Categories A and B can be viewed as encompassing a *strategy of prevention* of oxidative damage. The genes included in Category C can be viewed as encompassing a *strategy of repair* of oxidative damage. Based on all the above data, we now propose a hypothesis consistent with evolutionary and molecular constraints, as follows:

(a) Long lived animals have a global gene expression pattern which is geared to the scavenging of ROS and thus to the *prevention* of oxidative damage (responses A and B, prevention strategy).

(b) Normal lived animals have a global gene expression pattern which is geared to the *repair* of oxidative damage caused by ROS (response C, repair strategy).

(c) Short lived animals have a global gene expression pattern which is geared to neither repair nor to prevention of oxidative damage (null response, suicide strategy).

(d) Changes in the expression of the antioxidant defense system genes initiate the gene cascades necessary to change global gene expression so as to permit the animal to shift into a prevention mode and thus delay the onset of senescence.

The longevity determinant genes are those that act within the above framework. This hypothesis leads to a broader framework in which we can identify genes affecting oxidative stress and its consequences in different environments.

Acknowledgments

I thank Prof. Craig N. Giroux for his insightful discussions on oxidative stress resistance processes.

References

1. Martin, G. M., Johnson, T. E. and Austad, S. (1996). Genetic analysis of ageing: role of oxidative damage and environmental stresses. *Nature Genet.* **13**: 25–34.

2. Allen, R. G. (1998). Oxidative stress and superoxide dismutase in development, aging and gene regulation. *Age* **21**: 47–76.

3. Minois, N. (2000). Longevity and aging: beneficial effects of exposure to mild stress. *Biogerontology* **1**: 15–29.

4. Parsons, P. A. (1995). Inherited stress resistance and longevity: a stress theory of ageing. *Heredity* **75**: 216–221.

5. Johnson, T. E., Lithgow, G. J. and Murakami, S. (1996). Hypothesis: interventions that increase the response to stress offer the potential for effective life prolongation and increased health. *J. Gerontol. Biol. Sci.* **51**: B392–B395.

6. Parsons, P. A. (1997). Success in mating: a coordinated approach to fitness through genotypes incorporating genes for stress resistance and heterozygous advantage under stress. *Behav. Genet.* **27**: 75–81.

7. Yu, B. P. (1999). Calorie restriction: a potent mechanistic solution to the oxygen paradox. *In* "The Paradoxes of Longevity" (J. M. Robine *et al.* Eds.), Springer, Berlin.

8. Oudes, A. J., Herr, C. M., Olsen, Y. and Fleming, J. E. (1998). Age-dependent accumulation of advanced glycation end-products in adult *Drosophila melanogaster*. *Mech. Ageing Dev.* **100**: 221–229.

9. Arking, R., Burde, V., Graves, K., Hari, R., Feldman, E., Zeevi, A., Soliman, S., Saraiya, A., Buck, S., Vettraino, J., Sathrasala, K., Wehr, N. and Levine, R. L. (2000). Forward and reverse selection for longevity in *Drosophila* is characterized by alteration of antioxidant gene expression and oxidative damage patterns. *Exp. Gerontol.* **35**: 167–185.

10. Sohal, R. J., Ferman, K. J., Allen, R. G. and Cohen, N. R. (1983). Effects of age on oxygen consumption, superoxide dismutase, catalase, glutathione, inorganic peroxide and chloroform-soluble antioxidant in the adult housefly, *Musca domestica*. *Mech. Ageing Dev.* **24**: 185–195.

11. Stadtman, E. (1992). Protein oxidation and aging. *Science* **257**: 1220–1224.

12. Rogina, B., Vaupel, J. W., Partridge, L. and Helfand, S. L. (1998). Regulation of gene expression is preserved in aging *Drosophila melanogaster*. *Curr. Biol.* **8**: 475–478.

13. Kuether, K. and Arking, R. (1999). *Drosophila* selected for extended longevity are more sensitive to heat shock. *Age* **22**: 175–180.

14. Buck, S., Vettriano, J. and Arking, R. (2000). Extended longevity in *Drosophila* is consistently associated with a decrease in developmental viability. *J. Gerontol. Biol. Sci.* **55A**: B292–B301.

15. Harman, D. (1956). Aging, a theory based on free radicals and radiation chemistry. *J. Gerontol.* **11**: 298–300.

16. Fleming, J. E., Niedzwiecki, A. and Reveillaud, I. (1994). Stress induced expression of Cu, Zn superoxide dismutase and catalase in senescent *Drosophila melanogaster*. In "New Strategies in Prevention and Therapy: Biological Oxidants and Antioxidants" (L. Packer, and E. Cardenas, Eds.), pp. 181–192, Hippokates Verlag, Stuttgart, Germany.

17. Phillips, J. P., Campbell, Michaud, Charbonneau and Hilliker, A. J. (1989). Null mutations of Cu, Zn superoxide dismutase in *Drosophila* confers hypersensitivity to paraquat and reduced longevity. *Proc. Natl. Acad. Proc. USA* **86**: 2761–2765.

18. Rogina B. and Helfand, S. L. (2000). Cu, Zn superoxide dismutase deficiency accelerates the time course of an age-related marker in *Drosophila melanogaster*. *Biogerontology* **1**: 163–169.

19. Rogina, B. and Helfand, S. L. (1996). Timing of expression of a gene in the adult *Drosophila* is regulated by mechanisms independent of temperature and metabolic rate. *Genetics* **143**: 1643–1651.

20. Seto, N. O. L., Hayashi, S. and Tener, G. M. (1990). Over-expression of Cu, Zn superoxide dismutase in *Drosophila* does not affect life span. *Proc. Natl. Acad. Sci. USA* **87**: 4270–4274.

21. Reveillaud, I, Phillips, J., Duyf, B., Hilliker, A., Kongpachith, A. and Fleming, J. E. (1994). Phenotypic rescue by a bovine transgene in a Cu, Zn superoxide

dismutase-null mutant of *Drosophila melanogaster. Mol. Cell Biol.* **14**: 1302–1307.

22. Orr, W. C. and Sohal, R. S. (1992). The effects of catalase gene overexpression on life span and resistance to oxidative stress in transgenic *Drosophila melanogaster. Arch. Biochem. Biophys.* **297**: 35–41.

23. Orr, W. C. and Sohal, R. S. (1993). Effects of Cu, Zn superoxide dismutase overexpression of life span and resistance to oxidative stress in transgenic *Drosophila melanogaster. Arch. Biochem. Biophys.* **301**: 34–40.

24. Griswold, C. M., Mathews, A. L., Bewley, K. E. and Mahaffey, J. W. (1993). Molecular characterization of rescue of acatalesemic mutants of *Drosophila melanogaster. Genetics* **134**: 731–788.

25. Orr, W. C. and Sohal, R. S. (1994). Extension of life-span by over-expression of superoxide dismutase and catalase in *Drosophila melanogaster. Science* **263**: 1128–1130.

26. Sohal, R. S., Agarwal, A. and Orr, W. C. (1995). Simultaneous over-expression of copper- and zinc-containing superoxide dismutase and catalase retards age-related oxidative damage and increases metabolic potential in *Drosophila melanogaster. J. Biol. Chem.* **270**: 15 671–15 674.

27. Sun, J. and Tower, J. (1999). FLP recombinase-mediated induction of Cu, Zn superoxide dismutase transgene expression can extend the life span of adult *Drosophila melanogaster* flies. *Mol. Cell. Biol.* **19**: 216–228.

28. Tower, J. (2000). Transgenic methods for increasing *Drosophila* life span. *Mech. Ageing Dev.* **in press**.

29. Parkes, T. L., Kirby, K., Phillips, J. P. and Hilliker, A. J. (1998a). Transgenic analysis of the cSOD-null phenotypic syndrome in *Drosophila. Genome* **41**: 642–651.

30. Klichko, V. I., Radyuk, S.V. and Orr, W. C. (1999). CuZnSOD promoter-driven expression in the *Drosophila* central nervous system. *Neurobiol. Aging* **20**: 557–543.

31. Parkes, T. L., Elia, A. J., Dickinson, D., Hilliker, A. J., Phillips, J. P. and Boulianne, G. L. (1998). Extension of *Drosophila* lifespan by over-expression of human SOD-1 in motorneurons. *Nature Genet.* **19**: 171–174.

32. Parkes, T. L., Hilliker, A. J. and Phillips, J. P. (1999). Motorneurons, reactive oxygen and life span in *Drosophila. Neurobiol. Aging* **20**: 531–535.

33. Sun, J., Folk, D., Bradley, T. and Tower, J. (2000). Induced over-expression of mitochondrial Mn superoxide dismutase extends the life span of adult *Drosophila melanogaster* with decreasing O_2 consumption. 41st Annual Drosophila Research Conference, March. Abstract #116, p. A41.

34. Phillips, J. P. (2000). Personal communication, 1 August.

35. Mockett, R. J., Orr, W. C., Rahmandar, J. J., Benes, J. J., Radyuk, S. V., Klichko, V. I. and Sohal, R. S. (1999). Over-expression of Mn-containing superoxide dismutase in transgenic *Drosophila melanogaster. Arch. Biochem. Biophys.* **371**: 260–269.

36. Sohal, R. S., Mockett, R. J. and Orr, W. C. (2000). Current issues concerning the role of oxidative stress in aging: a perspective. *Results Problems Cell Diff.* **29**: 45–66.

37. Flybase=http://flybase.bio.indiana.edu

38. Offer, T., Russo, A. and Samuni, A. (2000). The pro-oxidative activity of SOD and nitroxide SOD mimics. *FASEB J.* **14**: 1215–1223.

39. Mackay, W. J. and Bewley, G. C. (1989). The genetics of catalase in *Drosophila melanogaster*: isolation and characterization of acatalasemic mutants. *Genetics* **122**: 643–652.

40. Mockett, R. J., Sohal, R. S. and Orr, W. C. (1999b). Over-expression of glutathione reductase extends survival in transgenic *Drosophila melanogaster* under hyperoxia and not under normoxia. *FASEB J.* **13**: 1733–1742.

41. Lin, Y.-J., Seroude, L. and Benzer, S. (1998). Extended life-span and stress resistance in the *Drosophila* mutant *methuselah*. *Science* **282**: 943–946.

42. Arking, R. (2001). Gene expression and regulation in the extended longevity phenotypes of *Drosophila*. *In* "Healthy Aging for Functional Longevity: Molecular and Cellular Interactions in Senescence". *Ann. NY Acad. Sci.* **in press**.

43. Leffelaar, D. and Grigliatti, T. A. (1984). A mutation in *Drosophila* that appears to accelerate aging. *Dev. Genet.* **4**: 199–210.

44. Luckinbill, L. S., Arking, R., Clare, M. J., Cirocco, W. C. and Buck, S. (1984). Selection for delayed senescence in *Drosophila melanogaster*. *Evolution* **38**: 996–1004.

45. Arking, R. (1987). Successful selection for increased longevity in *Drosophila*: analysis of the survival data and presentation of a hypothesis on the genetic regulation of longevity. *Exp. Gerontol.* **22**: 199–220.

46. Dudas, S. P. and Arking, R. (1995). A coordinate up-regulation of antioxidant gene activities is associated with the delayed onset of senescence in a long-lived strain of *Drosophila*. *J. Gerontol. Biol. Sci.* **50A**: B117–B127.

47. Force, A. G., Staples, T., Soliman, S. and Arking, R. (1995). Comparative biochemical and stress analysis of genetically selected *Drosophila* strains with different longevities. *Dev. Genet.* **17**: 340–351.

48. Arking, R., Buck, S., Wells, R. A. and Pretzlaff, R. (1988). Metabolic rates in genetically based long-lived strains of *Drosophila*. *Exp. Gerontol.* **23**: 59–76.

49. Ross, R. E. (2000). Age-specific decrease in aerobic efficiency associated with increase in oxygen free radical production in *Drosophila melanogaster*. *J. Insect Physiol.* **46**: 1477–1480.

50. Vettraino, J., Buck, S. and Arking, R. Direct selection for paraquat resistance in *Drosophila* involves reciprocal effects on the P450 system and the antioxidant defense system but does not extend longevity. **Submitted**.

51. Arking, R, Burde, V., Graves, K., Hari, R., Feldman, E., Zeevi, A., Soliman, S., Saraiya, A., Buck, S., Vettraino, J. and Sathrasala, K. (2000). Identical longevity

phenotypes are characterized by different patterns of gene expression and oxidative damage. *Exp. Gerontol.* **35**: 353–373.

52. Curtsinger, J. W. (2000). Personal communication, 14 July.

53. Curtsinger, J. W., Fukui, H. H., Resler, A. S., Kelly, K. and Khazaeli, A. A. (1998). Genetic analysis of extended life span in *Drosophila melanogaster*. I. RAPD screen for genetic divergence between selected and control lines. *Genetica* **104**: 21–32

54. Nudzhin, S. V., Pasyukova, E. G., Dilda, C. L., Zeng, Z.-B. and Mackay, T. F. C. (1997). Sex specific quantitative trait loci affecting longevity in *Drosophila melanogaster. Proc. Natl. Acad. Sci. USA* **94**: 9734–9739.

55. Rose, M. R. (1984). Laboratory evolution of postponed senescence in *Drosophila melanogaster. Evolution* **38**: 1004–1010.

56. Service, P. M. (1987). Physiological mechanisms of increased stress resistance in *Drosophila melanogaster* selected for postponed senescence. *Physiol. Zool.* **60**: 321–326.

57. Rose, M. R., Vu, L. N., Park, S. U. and Graves, J. L., Jr. (1992). Selection on stress resistance increases longevity in *Drosophila melanogaster. Exp. Gerontol.* **27**: 241–250.

58. Harshman, L. G., Moore, K. M., Sty, M. A. and Magwire, M. M. (1999). Stress resistance and longevity in selected lines of *Drosophila melanogaster. Neurobiol. Aging* **20**: 521–529.

59. Harshman, L. G. and Haberer, B. A. (2000). Oxidative stress resistance: a robust correlated response to selection in extended longevity lines of *Drosophila melanogaster. J. Gerontol. Biol. Sci.* **in press**.

60. Schwarze, S. R., Weindruch, R. and Aiken, J. (1998). Oxidative stress and aging reduce COX I RNA and cytochrome c oxidase activity in *Drosophila. Free Radic. Biol. Med.* **25**: 740–747.

61. Yan, L. J. and Sohal, R. S. (1998). Mitochondrial adenine nucleotide translocase is modified oxidatively during aging. *Proc. Natl. Acad. Sci. USA* **95**: 12 896–12 901.

62. Driver, C. and Tawadros, N. (2000). Cytoplasmic genomes that confer additional longevity in *Drosophila melanogaster. Biogerontology,* **in press**.

63. Kapahi, P., Boulton, M. E. and Kirkwood, T. B. L. (1999). Positive correlation between mammalian life span and cellular resistance to stress. *Free Radic. Biol. Med.* **26**: 495–500.

SECTION 11

Longevity Determinant Genes and Oxidative Stress

Chapter 67

Toward a Unified Theory of Aging — What Mammals Can Learn from Worms and Other Ephemeral Creatures

Robert J. Shmookler Reis

Robert J. Shmookler Reis • Departments of Geriatrics, Medicine, Toxicology, and Biochemistry & Molecular Biology, University of Arkansas for Medical Sciences, and Central Arkansas Veterans Health Care System, 4300 West 7th Street, Research 151, Little Rock, AR 72205
Tel: 501-257-5560, E-mail: ReisRobertJS@exchange.uams.edu

1. Summary

Genetic approaches have recently made enormous strides toward defining the mechanisms by which longevity is determined in several "lower" organisms. The most dramatic advances have come from studies of the nematode *C. elegans,* with important contributions also from budding yeast (*Saccharomyces cerevisiae*) and the fruitfly *Drosophila melanogaster.* Remarkably, after decades of apparently discordant evidence from these and other model systems, a consistent picture is emerging, which appears to also fit the more limited data on mammals, from rodents to humans. Many of the disparate phenomena associated with aging and with differences in life span among species are beginning to fall into place under the broad canopy of metabolically generated free radical stress and its sequelae, and the defenses and adaptations that have evolved to deal with them.

2. Theories of Aging

Many theories have been proposed to account for the pervasive occurrence of senescence; these can be grouped into two broad categories, involving programmed *versus* stochastic mechanisms.

2.1. Programmed Aging Theories

The idea that aging might follow a genetic program, analogous to development, is difficult to reconcile with the absence of extreme-longevity variants in most species, which would be expected to arise readily from disruption of such a program. The nematode *C. elegans* has recently emerged as a notable exception, although life extension appears to be a fortuitous side-effect of mutation to a program regulating development. Several temperature-sensitive mutations, to genes within a developmental-arrest pathway, greatly increase the longevity (by 1.5- to 4-fold) of adults that had matured at lower, permissive temperatures.[1-3] This "dauer" pathway, which shows remarkable structural and functional homology to the insulin/IGF-1 receptor pathway of mammals, also affects fertility and metabolism, especially of lipids.[3-6] The longevity extension arising from certain dauer-constitutive mutations may thus also support other, non-programmatic theories of aging (see below).

Recent data on telomere shortening as a function of cell proliferation has thrust *telomere attrition* into the spotlight as a possible mechanism for "programmed aging". Proliferation of diploid, untransformed mammalian cells in culture is normally limited to a fixed number of divisions characteristic of the species, cell type, and (perhaps) donor age.[7, 8] As a function of accrued cell divisions in such diploid cells, either *in vitro* or *in vivo,* the ends of chromosomes — comprising short-repeat DNA arrays called telomeres — become progessively shorter.[9–12] Although other factors are involved, it now appears that this limitation in several human cell strains can be obviated by transgenic over-production of the telomerase reverse transcriptase protein,[13] part of the telomerase ribonucleoprotein responsible for extension of telomere ends. Moreover, inhibition of the RNA component of telomerase can terminate the otherwise unlimited proliferation of transformed, "immortal" cell lines.[14, 15] While these observations are of great interest and potential utility in developing new cancer therapies and means for tissue replacement, their relevance to *in vivo* aging is less clear since even predominantly post-mitotic animals (such as insect and nematode adults) undergo senescence, and species such as mammals, comprising many tissues which retain proliferative capacity, do not generally exhaust that capacity during the course of normal aging. *Mus musculus* with a homozygous-defective telomerase reverse transcriptase (*TRT*) gene do not have curtailed life spans until the sixth knockout generation, or the fourth generation after backcrossing into a C57BL background with shorter telomeres.[16, 17] The accelerated mortality then seen may reflect premature senescence, or more plausibly attrition of specific highly proliferative tissues.[17, 18]

2.2. Stochastic Error Accumulation

Aging could arise as a consequence of random genetic or epigenetic errors that accrue over time, resulting in impaired protein synthesis and a deterioration of cellular function. One member of this family of theories, *error catastrophe,* postulates that such errors occurring in the enzymes (or their genes) which govern DNA, RNA or protein synthesis, could engender positive feedback loops leading to exponential increases in synthetic errors.[19, 20] This specific prediction has been found to be invalid in several animal model systems undergoing normal aging.[21] *Free radical damage* is another, likelier source of stochastic errors, in which metabolic byproducts of oxidative metabolism result in irreversible cellular damage (see next section). Cells normally generate free radicals such as peroxide, superoxide, and hydroxyl radicals, roughly in proportion to metabolic rate. Aging in a number of animal species is characterized by a progressive decline in the levels of antioxidant defenses, responsible for inactivating these metabolites.[22, 23] Although dietary supplementation with exogenous antioxidants has not generally been successful in preventing or delaying the declines of senescence, *caloric restriction* — typically by 25–35% of *ad libitum* intake, while maintaining nutritional sufficiency — is the one intervention that reliably extends life-span in a wide variety of taxa, including yeast, nematodes, and mammals.[24–29] While no direct evidence demonstrates that this strategy is effective in humans, early (premortality) indices of aging are significantly retarded in calorically-restricted rhesus monkeys.[25, 30] A plausible mechanism for extension of longevity with caloric restriction is through the attenuation of metabolic rate and hence of free radical generation. The accompanying ~ 1°C reduction in core body temperature, and diminution of peak metabolic output,[30, 31] could have adverse consequences in less-benign environments than the laboratory setting. Specifically, reduced core body temperature might slow the myriad of biochemical reactions for which rates are temperature dependent, and thus in itself may marginally impair muscle strength, reaction time, etc. for young animals. In mammals, there is a dual risk with respect to infections, since the combined effects of hypometabolism and hypothermia could compromise both immune function and the production of fever — a protective elevation of body temperature to levels deleterious to bacterial and viral growth. With aging, however, calorically-restricted animals suffer less immune deterioration,[32, 33] and thus reap a net benefit.

A more explicit trade-off between longevity and reproduction — termed *"antagonistic pleiotropy"* — has been suggested as a rationale for the evolution of senescence, and has received some support from experimental data on relatively short-lived, high-fecundity animals. In contrast, species which have attenuated fertility for other reasons (e.g. limited prey or other resources), and which either have prolonged fertile periods or which care for their young, might have undergone selection for increased life-span. Longevity is of variable advantage in varying environments, but any selective advantage vanishes if predation prevents individuals from attaining senescence. Life span thus may not be strictly fixed by evolution but may instead be subject to random mutations (since the force of selection is so weak at late age), or allowed to respond to levels of predation, environmental stress, and food availability. Arguments such as these are sometimes jointly elevated to the status of an "evolutionary theory of aging"[34] — but could be regarded more accurately as a set of quite separate hypotheses and conjectures dealing with specific aspects of senescence viewed from the perspective of an evolutionary biologist.[35–37]

3. Oxidative Damage and Senescence

Early indications that oxidative damage might be limiting to longevity came from biochemical arguments and interventions,[38, 39] physiological and biochemical studies of oxidation damage and defenses, primarily in aging *Drosophila* and other insects,[22, 23, 40] and from inter-species comparisons.[41, 42] Moreover, many laboratories have reported age-dependent loss of functional mitochondria, and of mitochondrial DNA integrity,[26, 43–46] consistent with the generation of free radicals as byproducts of mitochondrial metabolism. These arguments were not widely regarded as compelling, and by their nature — comparing species of varying longevity, or comparing young to aged adults of any species — *cannot* discern causes from effects. Recently, however, a number of advances employing genetic and reverse-genetic approaches have greatly strengthened the inference of oxidative-damage processes in a variety of aging systems.

3.1. Evidence from Insects

In Drosophila, overexpression of both Cu, Zn SOD and catalase in the same transgenic strain can extend longevity by 10–15%, although neither transgene alone is sufficient.[40, 47] Targeting Cu, Zn SOD to *Drosophila* motorneurons proved far more effective than global expression for extending life span, with increases as large as 40%,[48, 49] implying both a pivotal role of the nervous system in insect longevity, and some countervailing effects exerted in other tissues or early in development. Insertional mutation to the gene for sodium dicarboxylate cotransporter, a key membrane transporter of Krebs cycle intermediates, was reported to produce "a near doubling" of mean life span — although only data supporting a 1.5- to 1.6-fold increase were shown.[50] By either measure, however, impairment of this metabolic "gatekeeper" produced the largest mutational increase in *Drosophila* longevity yet documented.

3.2. Evidence from Worms

The characterization of the *Age-1* mutation in *C. elegans,* the first mutation shown to markedly (~ 1.6-fold) extend metazoan longevity, led to the realization that these long-lived mutant worms have increased late-life expression of catalase and Cu, Zn SOD[51, 52] as well as upregulation of MnSOD, mitochondrial superoxide dismutase encoded by *sod-3*.[53] Once cloned, *age-1* was found to encode the catalytic subunit of a phosphatidylinositol-3-OH kinase, or PI3K.[3] This gene participates in one of the two dauer-pathway branches,[54–56] which is strikingly similar to the signaling pathways activated by insulin and IGF-1 in mammals.[6, 57] The *daf-2* gene encodes a member of the insulin/IGF-1 receptor family,[58] and appears to phosphorylate both itself and the regulatory subunit of PI3K. The ligand for this DAF-2 receptor-kinase remained unknown until the recent cloning of a gene coding for a 95-residue insulin-like precursor polypeptide, required for normal longevity.[59] The precursor is processed to a single peptide, like IGF-1, but with a predicted structure similar to insulin.[59] The AGE-1 PI3 kinase is opposed by a phosphatase encoded by *daf-18*, a homologue of the PTEN tumor suppressor whose substrate is phosphoinositol-3,4,5-triphosphate, the product of AGE-1 kinase.[5, 60] Downstream components of the signaling cascade include a kinase (encoded by *pdk-1*) conveying the PI3K signal to the AKT-1 and AKT-2 kinases.[6]

The *daf-2* branch of the dauer pathway produces all of its effects — on longevity, developmental arrest, and metabolism — through *daf-16*, encoding a transcription factor of the Fork-head or "winged helix" family.[4, 61] Thus, in double mutants, abrogation of *daf-16* blocks longevity effects and all other tested manifestations of *daf-2* and *age-1* mutants.[1, 2, 55, 56, 62] Other *daf-2* and *age-1* mutant phenotypes include relative resistance to UV, heat, and oxidative stresses,[63, 64] as well as constitutive formation of dauer larvae, whose regulation is regarded as the primary function of this pathway. Remarkably, neuron-specific expression of transgenes for *age-1* or *daf-2* restores a wild-type life span to long-lived mutants defective for the same gene.[65] This implies that, in this pathway, longevity is either limited by metabolic damage to neurons or is regulated by neuroendocrine signaling. In contrast, some metabolic effects of *age-1* and *daf-2* mutants (affecting lipid utilization) are rescued primarily by expression specific to muscle,[65] and perhaps other tissues not assessed. These results are consistent with an earlier implication of sensory neuron function in *C. elegans* longevity[66] and are stunningly reminiscent of the *Drosophila* experiments in which longevity was most benefited by motorneuron-specific overexpression of human SOD-1.[48, 49]

A second genetic circuit implicated in nematode longevity is the "clock" pathway. Mutations to any of four *C. elegans* genes (*clk-1*, *clk-2*, *clk-3*, and *gro-1*) lead to dysregulation of the rates of development, motility, egg laying, food ingestion, and defecation. The *clk-1* gene has been cloned, and found to encode a homolog of the yeast *COQ7* gene.[67] *COQ7* is required for synthesis of coenzyme Q or ubiquinone, a prenylated benzoquinone lipid that functions both as a metabolic regulator in mitochondria and as a membrane antioxidant elsewhere.[68–70] Alleles of *clk-1* display slowing and increased variation in the timing of embryonic and larval development, and also of adult behaviors such as pharyngeal pumping (food intake) and defecation.[69, 71] Mean total life span was also extended 14–23% for the *qm30* mutation, and by 7–40% for the *e2519* mutation, to *clk-1*,[69, 71] although most of the increase could be attributed to extended development time.[72] A COQ-deficient diet also extends adult life span by 25–67%.[73]

Neither *daf-2*, *age-1*, nor *clk-1* mutations extend longevity in the absence of a functional cytosolic catalase encoded by *ctl-1*,[74] implying that antioxidant effects must mediate all of their life-extension phenotypes. It is also intriguing that a chemical mimetic of both SOD and catalase extends the life span of wild-type and *mev-1* mutant worms.[75] *Mev-1* encodes a subunit of succinate dehydrogenase cytochrome b,[76] and when mutated renders worms short-lived and oxygen sensitive, presumably by heightened superoxide production.[77] Metabolic activity has been compared between worms of normal and extended life span, by measuring oxygen consumption, "metabolic potential" and ATP levels,[78] or carbon dioxide production.[79] Although these measures did not all produce unambiguous results, a few firm conclusions emerge (reviewed in Refs. 73 and 80). It appears that *daf-2(e1370)*, at least in the context of a second mutation to *clk-1*, acts throughout adult life to decrease metabolic potential (luminescence elicited from lucigenin, a parameter of uncertain physiological import), while nevertheless *increasing* respiration rate and ATP levels, interpreted as "apparent uncoupling of energy production and consumption".[78] It may not be necessary to invoke uncoupling, however, since oxygen consumption and ATP levels were similarly elevated in a *daf-2;clk-1* double mutant, relative to wild-type or *clk-1* worms, when rescaled for metabolic time.[73] Van Voorhies and Ward[79] confirmed the increase in oxygen consumption by *daf-2* adults, but paradoxically found a halving of their carbon dioxide release — leading the authors to conclude that these long-lived worms have a slowed metabolism. Detection of generated CO_2 is certainly more sensitive than monitoring of oxygen consumed, because the CO_2 accrues above a zero background. Nonetheless, this may not be an ideal measure by which to compare the metabolic rates of wild-type and dauer-mutant worms. The dauer larva retains the glyoxylate shunt in place of the citric acid cycle that supplants it during normal development, thereby avoiding the loss of two CO_2 molecules per cycle. It is quite possible that *daf-2* and *age-1* adults may include, among their dauer-like features, partial reversion to this conservative dauer metabolism. Increased use of the glyoxylate cycle in *daf-2* adults would account for their paradoxical production of twofold *less* carbon dioxide than wild-type, in the face of 1.6-fold *more* oxygen consumption.[79]

3.3. Beyond Oxygen: The Nitrogen Connection

It is worth noting that oxidative damage is not the only dire consequence of metabolism. Nitrogen metabolism can generate reactive nitrogen species including nitrites, nitrates, S-nitrosoglutathione (GSNO) and peroxynitrite (ONOO⁻), which are generally cytotoxic and highly neurotoxic.[81–84] Aging nematodes, however, suffer yet a second assault from nitrogen metabolism: ammonia, the main form of nitrogenous waste excreted by most aquatic and a few terrestrial animals, is itself quite toxic,[85] and in particular is highly neurotoxic.[86–89] We assessed the ammonia production by aging *C. elegans* adults, and found a marked age-specific decline in all strains, of constant log-linear slope, superimposed on marked genotype-specific variation in the initial level.[73] Although all tested long-lived strains displayed very low ammonia production, many clock mutant strains (in particular *clk-2*) appeared to have greatly reduced ammonia for their near-normal or modestly increased longevities. The effect of clock-pathway mutations on life span, however, may have been attenuated as a consequence of suboptimal dietary coenzyme Q or its benzoquinone precursors.[90]

3.4. Still to be Assigned

There are several additional mutations that increase longevity in *C. elegans*, for which the mechanism is not presently known to directly involve metabolic rate or antioxidant defenses. These include a mutation to an unknown gene termed *age-2*,[91] and two mutations for which the "primary" defect is in spermatogenesis: *spe-26*[63, 92] and *spe-10*,[93] each accompanied by modest but significant effects on longevity and stress resistance. Finally, some mutations screened initially for thermotolerance appear to have longevity as a correlated phenotype,[63] as discussed in the next section. Substitution of an associated but more easily scored trait in place of life span, such as resistance to a specific stress, spares effort in screening but to some extent preordains the types of genes to be discovered.

4. Stress Response and Aging

A number of studies have noted that life-extending mutations — in particular *age-1* and *daf-2* in *C. elegans*, but also *methuselah* in *Drosophila* — often also confer resistance to environmental stresses such as oxidative agents, heat shock, UV irradiation, starvation, or desiccation.[39, 53, 63, 64, 94] Such observations could be construed as evidence for mutational slowing of an underlying aging process, and thus improved resistance to all stresses.[63, 64] An alternative explanation is that poikilotherms normally vary their metabolic rate with temperature and food supply, and may thus be adapted to enter a hypometabolic state that enhances both stress resistance and the activities of antioxidant enzymes.[23] We propose a third interpretation, not necessarily limited to poikilotherms: that oxygen- and nitrogen-radical reactions underlie not only senescence but also the damage incurred by many stressors, *all* of which are kept in abeyance by any genetic or environmental means of reducing the total metabolic load. For example, ultraviolet irradiation induces reactive oxygen species via the mitochondria,[45] whereas heat shock is known to involve oxidative damage in yeast[95] — perhaps by impairing antioxidant defenses while occupying protein chaperones with products of acute denaturation. In *C. elegans*, starvation appears to induce many of the same molecular responses as senescence,[96] and the common factor could be lipid oxidation elicited by starvation.[97] Moreover, heat shock, desiccation, UV, and many other stresses induce metallothioneins, metal-binding proteins that scavenge free radicals and protect against stress-induced damage.[98] Class III metallothioneins are specific to the central nervous system of mammals, offsetting its heightened vulnerability to those stressors and to reactive nitrogen species.[83, 99]

5. An Emerging Synthesis — With Gaps

Only a decade ago, theories of aging abounded — each supported by evidence specific to one or a few organisms — leading to the proposal of a (not-so-)unifying hypothesis: that each biological system has its own "weakest links" defining the nature of senescent deterioration for that system.[100, 101] Today, in contrast, the evidence appears to be converging on a single mechanism to account for many disparate aging phenomena. As argued above, oxidative damage can account for many of the manifestations of aging — including DNA damage, the accumulation of somatic mutations, loss of membrane fluidity, mitochondrial impairment, and the modification, cross-linking, and irreversible denaturation of proteins. The widespread phenomenon of life extension by caloric restriction implies that metabolically generated byproducts may limit longevity in a remarkable range of organisms — including yeast,[27–29] *C. elegans*,[24] and mammals.[26, 30]

5.1. Repair — The End is Near!

Telomere attrition remains an apparent intrusion into this harmony, since the proposed and widely accepted mechanism of telomere loss is failure to replicate the ends of linear chromosomes.[102] Yet even here, oxidative damage could perhaps play an instrumental role. The most direct evidence connecting telomere attrition to metabolic control comes from budding yeast, in which the mother cells can be monitored and undergo cellular senescence, attenuated by a sort of caloric restriction.[28, 29] Guarente and coworkers[27] demonstrated elegantly that the effect of either nutritional or genetic dietary attenuation depends strictly on the integrity of the telomeric silencer SIR2 and of NAD synthesis. In cultured human cells, reduction of telomere length occurs at about 50–100 base pairs per cell,[10, 11] substantially more rapidly than predicted from failure to resynthesize primase sites.[103] The difference has been attributed to exonuclease erosion of telomere ends,[103] or the introduction of DNA cross-links and nicks in telomeres, which may arise primarily through oxidative damage.[104] Although it could be argued that telomeres limit the replicative capacity of untransformed, somatic cells, and that this is rarely a factor in organismic senescence, there is some evidence to the contrary in highly replicative tissues, and particularly in immune senescence (see below). It is useful to distinguish between "cellular aging" — the limit to cell proliferation,[7] due largely to telomere attrition — and organismic senescence, which occurs even in predominantly postmitotic animals and in nondividing tissues of mammals.[105] Instances of cellular aging *in vivo* comprise a subset of the changes characteristic of whole-animal senescence, and arise from the limits to division potential of precursors for gametes, T cells, fibroblasts, intestinal epithelial cells, etc. It is hardly surprising that these tissues are precisely those affected in later generations of telomerase-knockout mice: depleted gonads, compromised immune response, failure of wound healing, and sparse intestinal lining.[17, 18]

Other aging phenomena that are *not* entirely explicable through oxidative damage, and which may be system- or species-specific in occurrence, include self-replicating DNA molecules in clonally propagating fungi (reviewed in Ref. 106), the generation of ribosomal DNA circles in yeast[107] and of extrachromosomal DNA circles in cultured human cells,[108] ectopic or "leaky" gene expression *in vivo* and *in vitro*,[109-111] DNA methylation changes observed in mammalian tissues and cells,[112-117] age-dependent activation of transposable elements and retroviruses (reviewed in Ref. 106), and the accumulation of site-specific mitochondrial DNA deletions.[118-121]

6. Positional Mapping of Natural Gene Variations Controlling Longevity

Much of the evidence strongly implicating metabolic generation of oxygen radicals derives, as discussed above, from a handful of mutagen-induced, single-gene lesions that extend life span.[91, 122-124] Within this small sample, the prevalence of mutations impacting metabolic rate may signify the pivotal role of metabolism in the etiology of aging, or may simply reflect the labyrinthine complexity of such pathways and the relative ease of their disruption. The fact remains that we really do not yet know the full extent of any longevity-affecting pathway, nor what other genes and pathways might also impact life span. It is widely held that longevity is "tuned" by evolution in each species, a view strongly supported by the quite recent (in evolutionary terms) extension of human life span to twice that of our closest primate relatives — but the natural genetic variation, that made such modulation possible, remains unknown. That information can only come from genetic studies designed to map loci based on function. Maximum-likelihood positions are calculated for genetic determinants of a trait such as life span, by in effect seeking peaks of correlation between genotype and phenotype among segregating cross progeny. This has been accomplished, with most of the population variance in life span explained by a modest number of polymorphic loci, in both *C. elegans*[125-129] and *Drosophila*.[130-132] From several such crosses, the full set of genes with similarly marked effects on life span can be enumerated; in *C. elegans*, they number between 13 and a few dozen.[127] Positional cloning of these genes is only a matter of time.

It is likely that quite distinct, although perhaps overlapping, sets of genes will be revealed by mutational screens *versus* gene mapping. Because the extant complement of genes evolved over millions of years, under selection for compatibility with the rest of the genome and the range of environments encountered, the vast majority of new mutations are expected to be deleterious. Thus, finding a random mutation that increases longevity is comparable to hitting your radio with a hammer and suddenly picking up Radio Tokyo — it would never happen if prior function were optimized for survival. The observation of such "chance improvements" strongly implies that the genes in question also serve other roles, which *are* impaired by the mutations, and which are more important than longevity from evolution's perspective. If a gene had actually been optimized for another purpose — such as to enhance survival in harsh environments, or early reproduction — a random mutation should be neutral or harmful to that function but might, incidentally, allow increased longevity in benign circumstances. Genes discovered by mapping answer to different constraints. They must exist as polymorphisms in a species, and be dimorphic between the parents in a cross, since monomorphic loci cannot contribute to genetic variance. Alleles will be generally be fully functional gene copies, with only exceptional and partial loss-of-function alleles being tolerated. This is because, whether or not a polymorphism is balanced, each of the alleles has undergone the scrutiny of natural selection and is adapted to some environmental and genetic context. Mapping, unlike mutational screens, is not limited to genes of pleiotropic effect — quite possibly a small subset of all genes affecting longevity.

We have used the mapping of longevity quantitative trait loci (QTLs, genes conjointly producing a scalar effect on a trait) to localize genes strongly influencing longevity in nematode populations, without bias by our preconceptions as to their nature. Thus far, four different interstrain crosses have yielded a total of 13 significant loci, most of them confirmed by identification in more than one cross (Refs. 125–127 and unpublished data). Four of these QTLs have been isolated in a constant genetic background by repeated back-crossing, and retain the expected effects on median and maximal life span (10–25%). Each of these back-crossed QTLs has resistance to one or more stresses (heat, UV, H_2O_2 or paraquat) as an associated phenotype, colocalizing even in fine-mapped recombinant congenic lines. Consistent with this, several of the longevity QTLs colocalize with loci mapped for resistance to UV, heat shock, or hydrogen peroxide — but none confers resistance to all three stresses (Kang *et al.* submitted).

7. Conclusions and Prospects

The phenomenon of aging is universal among metazoa, or very nearly so,[101, 105] and we appear now to be approaching a consensus that senescence is so widespread precisely because it arises from common causes, intrinsically coupled to multicellular life. In brief, several aspects of living (eating, metabolizing, reproducing) are actually quite bad for you, and if done to excess will kill you — or at least hasten death. What can be done? We already know (or, in the case of primates, strongly suspect) that limitation of food intake is the one means most certain to extend life span. "Compliance" is likely to loom large here, although this would seem to be an area in which self-determination is completely appropriate: some can live an extended, though spartan, life of caloric restriction and exercise, while others may prefer to burn their candle at a faster but more enjoyable pace.

It may be possible, however, to "eat your cake and *not* have had it", at least metabolically speaking. With the elucidation of regulatory pathways governing both metabolic activity and antioxidant defenses, some reprieve from the Grim Reaper may be afforded by pharmacological simulation of genetic lesions that inhibit metabolic pathways or that contribute to antioxidant defenses. We should be cautious, however, in extrapolating from invertebrates to humans. It is remarkable enough that genes and pathways remain conserved to the extent they do; predictions of what life extension can be attained in humans are obviously without basis at present. A reasonable objective, and one quite likely to be attainable, would be to help those dealt a short-lived hand of genetic cards to attain the same life span and health span as the longest-lived among us.

Acknowledgments

Supported in part by grants from the National Institute on Aging (R01-AG091413 and P01-AG13918) and the Department of Veterans Affairs (Merit Review and REAP).

References

1. Kenyon, C., Chang, J., Gensch, E., Rudner, A. and Tabtiang, R. (1993). A *Caenorhabditis elegans* mutant that lives twice as long as wild type. *Nature* **366**: 461–464.
2. Larsen, P. L., Albert, P. S. and Riddle, D. L. (1995). Genes that regulate both development and longevity in *Caenorhabditis elegans*. *Genetics* **139**: 1567–1583.
3. Morris, J. Z., Tissenbaum, H. A. and Ruvkun, G. (1996). A phosphatidyli-nositol-3-OH kinase family member regulating longevity and diapause in *Caenorhabditis elegans*. *Nature* **382**: 536–539.
4. Ogg, S., Paradis, S., Gottlieb, S., Patterson, G. I., Lee, L., Tissenbaum, H. A. and Ruvkun, G. (1997). The Fork head transcription factor DAF-16 transduces insulin-like metabolic and longevity signals in *Caenorhabditis elegans*. *Nature* **389**: 994–999.
5. Ogg, S. and Ruvkun, G. (1998). The *Caenorhabditis elegans* PTEN homolog, DAF-18, acts in the insulin receptor-like metabolic signaling pathway. *Mol. Cell* **2**: 887–893.
6. Paradis, S., Ailion, M., Toker, A., Thomas, J. H. and Ruvkun, G. (1999). A PDK1 homolog is necessary and sufficient to transduce AGE-1 PI3 kinase signals that regulate diapause in *Caenorhabditis elegans*. *Genes Dev.* **13**: 1438–1452.
7. Hayflick, L. (1965). The limited in vitro lifespan of human diploid cell strains. *Exp. Cell Res.* **37**: 614–636.

8. Cristofalo, V. J., Allen, R. G., Pignolo, R. J., Martin, B. G. and Beck, J. C. (1998). Relationship between donor age and the replicative lifespan of human cells in culture: a reevaluation. *Proc. Natl. Acad. Sci. USA* **95**: 10 614–10 619.

9. Hastie, N. D., Dempster, M., Dunlop, M. G., Thompson, A. M., Green, D. K. and Allshire, R. C. (1990). Telomere reduction in human colorectal carcinoma and with ageing. *Nature* **346**: 866–868.

10. Harley, C. B., Futcher, A. B. and Greider, C. W. (1990). Telomeres shorten during ageing of human fibroblasts. *Nature* **345**: 458–460.

11. Harley, C. B. (1997). Human ageing and telomeres. *Ciba Found. Symp.* **211**: 129–139.

12. Son, N. H., Murray, S., Yanovski, J., Hodes, R. J. and Weng, N. (2000). Lineage-specific telomere shortening and unaltered capacity for telomerase expression in human T and B lymphocytes with age. *J. Immunol.* **165**: 1191–1196.

13. Bodnar, A. G., Ouellette, M., Frolkis, M., Holt, S. E., Chiu, C. P., Morin, G. B., Harley, C. B., Shay, J. W., Lichtsteiner, S. and Wright, W. E. (1998). Extension of life-span by introduction of telomerase into normal human cells. *Science* **279**: 349–352.

14. Feng, J., Funk, W. D., Wang, S. S., Weinrich, S. L., Avilion, A. A., Chiu, C. P., Adams, R. R., Chang, E., Allsopp, R. C. and Yu, J. (1995). The RNA component of human telomerase. *Science* **269**: 1236–1241.

15. Shammas, M. A., Simmons, C. G., Corey, D. R. and Shmookler Reis, R. J. (1999). Telomerase inhibition by peptide nucleic acids reverses "immortality" of transformed human cells. *Oncogene* **18**: 6191–6200.

16. Rudolph, K. L., Chang, S., Lee, H. W., Blasco, M., Gottlieb, G. J., Greider, C. and DePinho, R. A. (1999). Longevity, stress response, and cancer in aging telomerase-deficient mice. *Cell* **96**: 701–712.

17. Herrera, E., Samper, E., Martin-Caballero, J., Flores, J. M., Lee, H. W. and Blasco, M. A. (1999). Disease states associated with telomerase deficiency appear earlier in mice with short telomeres. *EMBO J.* **18**: 2950–2960.

18. Lee, H. W., Blasco, M. A., Gottlieb, G. J., Horner, J. W., Greider, C. W. and DePinho, R. A. (1998). Essential role of mouse telomerase in highly proliferative organs. *Nature* **392**: 569–574.

19. Orgel, L. E. (1963). The maintenance of the accuracy of protein synthesis and its relevance to ageing: a correction. *Proc. Natl. Acad. Sci. USA* **49**: 412–517.

20. Orgel, L. E. (1970). The maintenance of the accuracy of protein synthesis and its relevance to ageing: a correction. *Proc. Natl. Acad. Sci. USA* **67**: 1476.

21. Shmookler Reis, R. J. (1976). Enzyme fidelity and metazoan aging. *Interdisc. Top. Gerontol.* **10**: 11–23.

22. Agarwal, S. and Sohal, R. S. (1994). Aging and protein oxidative damage. *Mech. Ageing Dev.* **75**: 11–19.

23. Sohal, R. S., Mockett, R. J. and Orr, W. C. (2000). Current issues concerning the role of oxidative stress in aging: a perspective. *Results. Prob. Cell Diff.* **29**: 45–66.

24. Klass, M. R. (1977). Aging in the nematode *Caenorhabditis elegans*: major biological and environmental factors influencing life span. *Mech. Ageing Dev.* **6**: 413–429.

25. Weindruch, R. (1996). The retardation of aging by caloric restriction: studies in rodents and primates. *Toxicol. Pathol.* **24**: 742–745.

26. Ramsey, J. J., Harper, M. and Weindruch, R. (2000). Restriction of energy intake, energy expenditure, and aging. *Free Radic. Biol. Med.* **29**: 946–968.

27. Lin, S. J., Defossez, P. A. and Guarente, L. (2000). Requirement of NAD and SIR2 for life-span extension by calorie restriction in *Saccharomyces cerevisiae*. *Science* **289**: 2126–2128.

28. Jiang, J. C., Jaruga, E., Repnevskaya, M. V. and Jazwinski, S. M. (2000). An intervention resembling caloric restriction prolongs life span and retards aging in yeast. *FASEB J.* **14**: 2135–2137.

29. Jazwinski, S. M. (2000). Metabolic control and ageing. *Trends. Genet.* **16**: 506–511.

30. Wanagat, J., Allison, D. B. and Weindruch, R. (1999). Caloric intake and aging: mechanisms in rodents and a study in nonhuman primates. *Toxicol. Sci.* **52**: 35–40.

31. Lane, M. A., Baer, D. J., Rumpler, W. V., Weindruch, R., Ingram, D. K., Tilmont, E. M., Cutler, R. G. and Roth, G. S. (1996). Calorie restriction lowers body temperature in rhesus monkeys, consistent with a postulated anti-aging mechanism in rodents. *Proc. Natl. Acad. Sci. USA* **93**: 4159–4164.

32. Miller, R. A. (1996). The aging immune system: primer and prospectus. *Science* **273**: 70–74.

33. Chen, J., Astle, C. M. and Harrison, D. E. (1998). Delayed immune aging in diet-restricted B6CBAT6 F1 mice is associated with preservation of naive T cells. *J. Gerontol. A: Biol. Sci. Med. Sci.* **53**: B330–B337.

34. Rose, M. R. (1991). *Evolutionary Biology of Aging*, Oxford University Press, New York.

35. Medawar, P. B. (1952). *An Unsolved Problem in Biology*, H. K. Lewis, London.

36. Maynard Smith, J. (1963). Temperature and the rate of ageing in poikilotherms. *Nature* **199**: 400–402.

37. Comfort, A. (1970). Biological theories of aging. *Human Dev.* **13**: 127–139.

38. Harman, D. (1968). Free radical theory of aging: effect of free radical reaction inhibitors on the mortality rate of male LAF mice. *J. Gerontol.* **23**: 476–482.

39. Harman, D. (1998). Aging and oxidative stress. *J. Int. Fed. Clin. Chem.* **10**: 24–27.

40. Sohal, R. S., Agarwal, A., Agarwal, S. and Orr, W. C. (1995). Simultaneous overexpression of copper- and zinc-containing superoxide dismutase and catalase retards age-related oxidative damage and increases metabolic potential in *Drosophila melanogaster. J. Biol. Chem.* **270**: 15 671–15 674.
41. Cutler, R. G. (1985). Peroxide-producing potential of tissues: inverse correlation with longevity of mammalian species. *Proc. Natl. Acad. Sci. USA* **82**: 4798–4802.
42. Cutler, R. G. (1991). Antioxidants and aging. *Am. J. Clin. Nutri.* **53**: 373S–379S.
43. Wallace, D. C., Brown, M. D., Melov, S., Graham, B. and Lott, M. (1998). Mitochondrial biology, degenerative diseases and aging. *Biofactors* **7**: 187–190.
44. Melov, S. (2000). Mitochondrial oxidative stress. Physiologic consequences and potential for a role in aging. *Ann. NY Acad. Sci.* **908**: 219–225.
45. Gniadecki, R., Thorn, T., Vicanova, J., Petersen, A. and Wulf, H. C. (2000). Role of mitochondria in ultraviolet-induced oxidative stress. *J. Cell Biochem.* **80**: 216–222.
46. Cadenas, E. and Davies, K. J. (2000). Mitochondrial free radical generation, oxidative stress, and aging. *Free Radic. Biol. Med.* **29**: 222–230.
47. Orr, W. C. and Sohal, R. S. (1994). Extension of life-span by overexpression of superoxide dismutase and catalase in *Drosophila melanogaster. Science* **263**: 1128–1130.
48. Parkes, T. L., Elia, A. J., Dickinson, D., Hilliker, A. J., Phillips, J. P. and Boulianne, G. L. (1998). Extension of *Drosophila* lifespan by overexpression of human SOD-1 in motorneurons. *Nature Genet.* **19**: 171–174.
49. Phillips, J. P., Parkes, T. L. and Hilliker, A. J. (2000). Targeted neuronal gene expression and longevity in Drosophila. *Exp. Gerontol.* **35**: 1157–1164.
50. Rogina, B., Reenan, R. A., Nilsen, S. P. and Helfand, S. L. (2000). Extended life-span conferred by cotransporter gene mutations in *Drosophila. Science* **290**: 2137–2140.
51. Vanfleteren, J. R. (1993). Oxidative stress and ageing in *Caenorhabditis elegans. Biochem. J.* **292**: 605–608.
52. Larsen, P. L. (1993). Aging and resistance to oxidative damage in *Caenorhabditis elegans. Proc. Natl. Acad. Sci. USA* **90**: 8905–8909.
53. Honda, Y. and Honda, S. (1999). The daf-2 gene network for longevity regulates oxidative stress resistance and Mn-superoxide dismutase gene expression in *Caenorhabditis elegans. FASEB J.* **13**: 1385–1393.
54. Vowels, J. J. and Thomas, J. H. (1992). Genetic analysis of chemosensory control of dauer formation in *Caenorhabditis elegans. Genetics* **130**: 105–123.
55. Gottlieb, S. and Ruvkun, G. (1994). daf-2, daf-16 and daf-23: genetically interacting genes controlling Dauer formation in *Caenorhabditis elegans. Genetics* **137**: 107–120.

56. Dorman, J. B., Albinder, B., Shroyer, T. and Kenyon, C. (1995). The age-1 and daf-2 genes function in a common pathway to control the lifespan of *Caenorhabditis elegans*. *Genetics* **141**: 1399–1406.

57. Tissenbaum, H. A. and Ruvkun, G. (1998). An insulin-like signaling pathway affects both longevity and reproduction in *Caenorhabditis elegans*. *Genetics* **148**: 703–717.

58. Kimura, K. D., Tissenbaum, H. A., Liu, Y. and Ruvkun, G. (1997). daf-2, an insulin receptor-like gene that regulates longevity and diapause in *Caenorhabditis elegans*. *Science* **277**: 942–946.

59. Kawano, T., Ito, Y., Ishiguro, M., Takuwa, K., Nakajima, T. and Kimura, Y. (2000). Molecular cloning and characterization of a new insulin/IGF-like peptide of the nematode *Caenorhabditis elegans*. *Biochem. Biophys. Res. Commun.* **273**: 431–436.

60. Mihaylova, V. T., Borland, C. Z., Manjarrez, L., Stern, M. J. and Sun, H. (1999). The PTEN tumor suppressor homolog in *Caenorhabditis elegans* regulates longevity and dauer formation in an insulin receptor-like signaling pathway. *Proc. Natl. Acad. Sci. USA* **96**: 7427–7432.

61. Lin, K., Dorman, J. B., Rodan, A. and Kenyon, C. (1997). daf-16: an HNF-3/forkhead family member that can function to double the life-span of *Caenorhabditis elegans*. *Science* **278**: 1319–1322.

62. Gems, D., Sutton, A. J., Sundermeyer, M. L., Albert, P. S., King, K. V., Edgley, M. L., Larsen, P. L. and Riddle, D. L. (1998). Two pleiotropic classes of daf-2 mutation affect larval arrest, adult behavior, reproduction and longevity in *Caenorhabditis elegans*. *Genetics* **150**: 129–155.

63. Lithgow, G. J., White, T. M., Melov, S. and Johnson, T. E. (1995). Thermotolerance and extended life-span conferred by single-gene mutations and induced by thermal stress. *Proc. Natl. Acad. Sci. USA* **92**: 7540–7544.

64. Murakami, S. and Johnson, T. E. (1996). A genetic pathway conferring life extension and resistance to UV stress in *Caenorhabditis elegans*. *Genetics* **143**: 1207–1218.

65. Wolkow, C. A., Kimura, K. D., Lee, M. S. and Ruvkun, G. (2000). Regulation of *Caenorhabditis elegans* life-span by insulinlike signaling in the nervous system. *Science* **290**: 147–150.

66. Apfeld, J. and Kenyon, C. (1999). Regulation of lifespan by sensory perception in *Caenorhabditis elegans*. *Nature* **402**: 804–809.

67. Ewbank, J. J., Barnes, T. M., Lakowski, B., Lussier, M., Bussey, H. and Hekimi, S. (1997). Structural and functional conservation of the *Caenorhabditis elegans* timing gene clk-1. *Science* **275**: 980–983.

68. Vajo, Z., King, L. M., Jonassen, T., Wilkin, D. J., Ho, N., Munnich, A., Clarke, C. F. and Francomano, C. A. (1999). Conservation of the *Caenorhabditis elegans* timing gene clk-1 from yeast to human: a gene required for ubiquinone biosynthesis with potential implications for aging. *Mamm. Genome* **10**: 1000–1004.

69. Lakowski, B. and Hekimi, S. (1996). Determination of life-span in *Caenorhabditis elegans* by four clock genes. *Science* **272**: 1010–1013.

70. Lakowski, B. and Hekimi, S. (1998). The genetics of caloric restriction in *Caenorhabditis elegans*. *Proc. Natl. Acad. Sci. USA* **95**: 13 091–13 096.

71. Felkai, S., Ewbank, J. J., Lemieux, J., Labbe, J. C., Brown, G. G. and Hekimi, S. (1999). CLK-1 controls respiration, behavior and aging in the nematode *Caenorhabditis elegans*. *EMBO J.* **18**: 1783–1792.

72. Thaden, J. J. and Shmookler Reis, R. J. (2000). Ammonia, respiration, and longevity in nematodes. *Age* **23**: 75–84.

73. Larsen, P. L. and Clarke, C. F. (2002). Extension of life-span in *Caenorhabditis elegans* by a diet lacking coenzyme. *Science* **295**: 120–123.

74. Taub, J., Lau, J. F., Ma, C., Hahn, J. H., Hoque, R., Rothblatt, J. and Chalfie, M. (1999). A cytosolic catalase is needed to extend adult lifespan in *Caenorhabditis elegans* daf-C and clk-1 mutants. *Nature* **399**: 162–166.

75. Melov, S., Ravenscroft, J., Malik, S., Gill, M. S., Walker, D. W., Clayton, P. E., Wallace, D. C., Malfroy, B., Doctrow, S. R. and Lithgow, G. J. (2000). Extension of life-span with superoxide dismutase/catalase mimetics. *Science* **289**: 1567–1569.

76. Ishii, N., Fujii, M., Hartman, P. S., Tsuda, M., Yasuda, K., Senoo-Matsuda, N., Yanase, S., Ayusawa, D. and Suzuki, K. (1998). A mutation in succinate dehydrogenase cytochrome b causes oxidative stress and ageing in nematodes. *Nature* **394**: 694–697.

77. Honda, S., Ishii, N., Suzuki, K. and Matsuo, M. (1993). Oxygen-dependent perturbation of life span and aging rate in the nematode. *J. Gerontol.* **48**: B57–B61.

78. Braeckman, B. P., Houthoofd, K., De Vreese, A. and Vanfleteren, J. R. (1999). Apparent uncoupling of energy production and consumption in long-lived Clk mutants of *Caenorhabditis elegans*. *Curr. Biol.* **9**: 493–496.

79. Van Voorhies, W. A. and Ward, S. (1999). Genetic and environmental conditions that increase longevity in *Caenorhabditis elegans* decrease metabolic rate. *Proc. Natl. Acad. Sci. USA* **96**: 11 399–11 403.

80. Gems, D. (1999). Nematode ageing: putting metabolic theories to the test. *Curr. Biol.* **9**: R614–R616.

81. Chiueh, C. C. (1999). Neuroprotective properties of nitric oxide. *Ann. NY Acad. Sci.* **890**: 301–311.

82. Floyd, R. A. (1999). Antioxidants, oxidative stress, and degenerative neurological disorders. *Proc. Soc. Exp. Biol. Med.* **222**: 236–245.

83. Cai, L., Klein, J. B. and Kang, Y. J. (2000). Metallothionein inhibits peroxynitrite-induced DNA and lipoprotein damage. *J. Biol. Chem.* **275**: 38 957–38 960.

84. Schopfer, F., Riobo, N., Carreras, M. C., Alvarez, B., Radi, R., Boveris, A., Cadenas, E. and Poderoso, J. J. (2000). Oxidation of ubiquinol by peroxynitrite:

implications for protection of mitochondria against nitrosative damage. *Biochem. J.* **349**: 35–42.

85. Wright, P. A. (1995). Nitrogen excretion: three end products, many physiological roles. *J. Exp. Biol.* **198**: 273–281.

86. Dolinska, M., Hilgier, W. and Albrecht, J. (1996). Ammonia stimulates glutamine uptake to the cerebral non-synaptic mitochondria of the rat. *Neurosci. Lett.* **213**: 45–48.

87. Willard-Mack, C. L., Koehler, R. C., Hirata, T., Cork, L. C., Takahashi, H., Traystman, R. J. and Brusilow, S. W. (1996). Inhibition of glutamine synthetase reduces ammonia-induced astrocyte swelling in rat. *Neuroscience* **71**: 589–599.

88. Kosenko, E., Felipo, V., Montoliu, C., Grisolia, S. and Kaminsky, Y. (1996). Effects of acute hyperammonemia in vivo on oxidative metabolism in nonsynaptic rat brain mitochondria. *Metab. Brain Dis.* **12**: 69–82.

89. Kosenko, E., Kaminski, Y., Lopata, O., Muravyov, N. and Felipo, V. (1999). Blocking NMDA receptors prevents the oxidative stress induced by acute ammonia intoxication. *Free Radic. Biol. Med.* **26**: 1369–1374.

90. Jonassen, T., Larsen, P. L. and Clarke, C. F. (2001). A dietary source of coenzyme Q is essential for growth of long-lived *Caenorhabditis elegans* clk-1 mutants. *Proc. Natl. Acad. Sci. USA* **65** [epub ahead of print].

91. Yang, Y. and Wilson, D. L. (1999). Characterization of a life-extending mutation in age-2, a new aging gene in *Caenorhabditis elegans*. *J. Gerontol. A: Biol. Sci. Med. Sci.* **54**: B137–B142.

92. Van Voorhies, W. A. (1992). Production of sperm reduces nematode lifespan. *Nature* **360**: 456–458.

93. Cypser, J. R. and Johnson, T. E. (1999). The spe-10 mutant has longer life and increased stress resistance. *Neurobiol. Aging* **20**: 503–512.

94. Lin, Y. J., Seroude, L. and Benzer, S. (1998). Extended life-span and stress resistance in the *Drosophila* mutant methuselah. *Science* **282**: 943–946.

95. Davidson, J. F., Whyte, B., Bissinger, P. H. and Schiestl, R. H. (1996). Oxidative stress is involved in heat-induced cell death in *Saccharomyces cerevisiae*. *Proc. Natl. Acad. Sci. USA* **93**: 5116–5121.

96. Cherkasova, V., Ayyadevara, S., Egilmez, N. and Shmookler Reis, R. J. (2000). Diverse *Caenorhabditis elegans* genes that are upregulated in dauer larvae also show elevated transcript levels in long-lived, aged, or starved adults. *J. Mol. Biol.* **300**: 433–448.

97. Murakami, R., Tanaka, A. and Nakamura, H. (1997). The effect of starvation on brain carnitine concentration in neonatal rats. *J. Pediatr. Gastroenterol. Nutri.* **25**: 385–387.

98. Ghoshal, K. and Jacob, S. T. (2000). Regulation of metallothionein gene expression. *Prog. Nucleic. Acid. Res. Mol. Biol.* **66**: 357–384.

99. Erickson, J. C., Hollopeter, G., Thomas, S. A., Froelick, G. J. and Palmiter, R. D. (1997). Disruption of the metallothionein-III gene in mice: analysis of

brain zinc, behavior, and neuron vulnerability to metals, aging, and seizures. *J. Neurosci.* **17**: 1271–1281.

100. Shmookler Reis, R. J. (1989). Model systems for aging research: syncretic concepts and diversity of mechanisms. *Genome* **31**: 406–412.

101. Shmookler Reis, R. J. (1990). *Review of Biological Research In aging* (M. Rothstein, Ed.), pp. 293–313, A. R. Liss, New York.

102. McEachern, M. J., Krauskopf, A. and Blackburn, E. H. (2000). Telomeres and their control. *Ann. Rev. Genet.* **34**: 331–358.

103. Zakian, V. A. (1997). Life and cancer without telomerase. *Cell* **91**: 1–3.

104. von Zglinicki, T. (2000). Role of oxidative stress in telomere length regulation and replicative senescence. *Ann. NY Acad. Sci.* **908**: 99–110.

105. Finch, C. E. (1990). *Longevity, Senescence, and the Genome*, University of Chicago Press, Chicago.

106. Nikitin, A. G. and Shmookler Reis, R. J. (1997). Role of transposable elements in age-related genomic instability. *Genet. Res.* **69**: 183–195.

107. Park, P. U., Defossez, P. A. and Guarente, L. (1999). Effects of mutations in DNA repair genes on formation of ribosomal DNA circles and life span in *Saccharomyces cerevisiae. Mol. Cell Biol.* **19**: 3848–3856.

108. Riabowol, K., Shmookler Reis, R. J. and Goldstein, S. (1985). Interspersed repetitive and tandemly repetitive sequences are differentially represented in extrachromosomal covalently closed circular DNA of human diploid fibroblasts. *Nucleic. Acids. Res.* **13**: 5563–5584.

109. Ono, T., Dean, R. G., Chattopadhyay, S. K. and Cutler, R. G. (1985). Dysdifferentiative nature of aging: age-dependent expression of MuLV and globin genes in thymus, liver and brain in the AKR mouse strain. *Gerontology* **31**: 362–372.

110. Wareham, K. A., Lyon, M. F., Glenister, P. H. and Williams, E. D. (1987). Age related reactivation of an X-linked gene. *Nature* **327**: 725–727.

111. Goldstein, S., Jones, R. A., Hardin, J. W., Braunstein, G. D. and Shmookler Reis, R. J. (1990). Expression of alpha- and beta-human chorionic gonadotropin subunits in cultured human cells. *In Vitro Cell Dev. Biol.* **26**: 857–864.

112. Shmookler Reis, R. J. and Goldstein, S. (1982). Interclonal variation in methylation patterns for expressed and non-expressed genes. *Nucleic. Acids. Res.* **10**: 4293–4304.

113. Shmookler Reis, R. J. and Goldstein, S. (1982). Variability of DNA methylation patterns during serial passage of human diploid fibroblasts. *Proc. Natl. Acad. Sci. USA* **79**: 3949–3953.

114. Shmookler Reis, R. J., Finn, G. K., Smith, K. and Goldstein, S. (1990). Clonal variation in gene methylation: c-H-ras and alpha-hCG regions vary independently in human fibroblast lineages. *Mutat. Res.* **237**: 45–57.

115. Issa, J. P., Ottaviano, Y. L., Celano, P., Hamilton, S. R., Davidson, N. E. and Baylin, S. B. (1994). Methylation of the oestrogen receptor CpG island links ageing and neoplasia in human colon. *Nature Genet.* **7**: 536–540.

116. Cross, S. H. and Bird, A. P. (1995). CpG islands and genes. *Curr. Opin. Genet. Dev.* **5**: 309–314.

117. Issa, J. P. (2000). CpG-island methylation in aging and cancer. *Curr. Top. Microbiol. Immunol.* **249**: 101–118.

118. Melov, S., Lithgow, G. J., Fischer, D. R., Tedesco, P. M. and Johnson, T. E. (1995). Increased frequency of deletions in the mitochondrial genome with age of *Caenorhabditis elegans*. *Nucleic. Acids. Res.* **23**: 1419–1425.

119. Melov, S., Schneider, J. A., Coskun, P. E., Bennett, D. A. and Wallace, D. C. (1999). Mitochondrial DNA rearrangements in aging human brain and in situ PCR of mtDNA. *Neurobiol. Aging* **20**: 565–571.

120. Ozawa, T. (1998). Mitochondrial DNA mutations and age. *Ann. NY Acad. Sci.* **854**: 128–154.

121. Cortopassi, G. A. and Wong, A. (1999). Mitochondria in organismal aging and degeneration. *Biochim. Biophys. Acta* **1410**: 183–193.

122. Riddle, D. L., Swanson, M. M. and Albert, P. S. (1981). Interacting genes in nematode dauer larva formation. *Nature* **290**: 668–671.

123. Klass, M. R. (1983). A method for the isolation of longevity mutants in the nematode *Caenorhabditis elegans* and initial results. *Mech. Ageing Dev.* **22**: 279–286.

124. Duhon, S. A., Murakami, S. and Johnson, T. E. (1996). Direct isolation of longevity mutants in the nematode *Caenorhabditis elegans*. *Dev. Genet.* **18**: 144–153.

125. Ebert, R. H., Cherkasova, V. A., Dennis, R. A., Wu, J. H., Ruggles, S., Perrin, T. E. and Shmookler Reis, R. J. (1993). Longevity-determining genes in *Caenorhabditis elegans*: chromosomal mapping of multiple noninteractive loci. *Genetics* **135**: 1003–1010.

126. Ebert, R. H., Shammas, M. A., Sohal, B. H., Sohal, R. S., Egilmez, N. K., Ruggles, S. and Shmookler Reis, R. J. (1996). Defining genes that govern longevity in *Caenorhabditis elegans*. *Dev. Genet.* **18**: 131–143.

127. Ayyadevara, S., Ayyadevera, R., Hou, S., Thaden, J. J. and Shmookler Reis, R. J. (2001). Genetic mapping of quantitative trait loci governing longevity of *Caenorhabditis elegans* in recombinant-inbred progeny of a Bergerac-BO x RC301 interstrain cross. *Genetics* **157**: 655–666.

128. Shook, D. R. and Johnson, T. E. (1999). Quantitative trait loci affecting survival and fertility-related traits in *Caenorhabditis elegans* show genotype-environment interactions, pleiotropy and epistasis. *Genetics* **153**: 1233–1243.

129. Shook, D. R., Brooks, A. and Johnson, T. E. (1996). Mapping quantitative trait loci affecting life history traits in the nematode *Caenorhabditis elegans*. *Genetics* **142**: 801–817.

130. Pasyukova, E. G., Vieira, C. and Mackay, T. F. (2000). Deficiency mapping of quantitative trait loci affecting longevity in *Drosophila melanogaster*. *Genetics* **156**: 1129–1146.
131. Vieira, C., Pasyukova, E. G., Zeng, Z. B., Hackett, J. B., Lyman, R. F. and Mackay, T. F. (2000). Genotype-environment interaction for quantitative trait loci affecting life span in *Drosophila melanogaster*. *Genetics* **154**: 213–227.
132. Nuzhdin, S. V., Pasyukova, E. G., Dilda, C. L., Zeng, Z. B. and Mackay, T. F. (1997). Sex-specific quantitative trait loci affecting longevity in *Drosophila melanogaster*. *Proc. Natl. Acad. Sci. USA* **94**: 9734–9739.

Chapter 68

Mouse Models and Longevity

Marco Giorgio and Pier Giuseppe Pelicci

Keywords: Mouse models, aging, genes.

Marco Giorgio and **Pier Giuseppe Pelicci** • Istituto Europeo di Oncologia, Via Ripamonti 435, 20141 Milano, Italy; IFOM (Firc Institute of Molecular Oncology) Via Serio 21, 20139 Milan-Italy.

1. Summary

In this review, we will briefly summarize and discuss the mouse model systems that are available for investigating the genetics of aging. Our interest in using mice as a model system to study aging stems from their evolutionarily closeness to humans. This obvious consideration, however, might bias interpretation of the results. The criteria that are used to define the aging process are frequently based on our knowledge of the process of aging in humans (body functioning, reproductive activity, frequency of disease, lifespan, etc.). Mechanisms of aging and associated phenotypes are, however, frequently species specific. For example, the rate of metabolism, lethargic strategies, reproductive rate, etc. all affect aging and may change the patterns of aging in different mammalian species. Interpretation of results obtained from studies on mouse aging should, therefore, be critically evaluated when transposed to humans.

2. The Aging Process in Mice

The majority of the mouse strains currently used in research laboratories were derived at the beginning of the 20th Century from the wild *Mus musculus,* subspecies *domesticus* and *musculus*. The lifespan of wild *Mus musculus* is 1–1.5 years, as observed in open field Northern European populations, protected from predators, supplied *ad libitum* with standard laboratory mouse diet and kept under controlled temperature. The lifespan of the different laboratory mouse inbred strains, instead, varies from 1.5–3 years.[1-4] Differences in aging and lifespan might result from the high degree of homozygosity, originated by inbreeding, of the laboratory mouse as compared with the *wild-type* (wt) population. The relevance of inbreeding effects on polygenic controlled phenotypes has been recently demonstrated for reproductive capability.[5] Although little is known about the effects of homozigosity on aging and lifespan traits, one should be, therefore, careful in interpreting results of genetic investigations carried out on inbred-derived mouse strains, particularly if mouse models are used to infer conclusions about the process of aging in humans.

Apart from these limitations, the usage of laboratory mice is a powerful tool to help the study of genetics of mammalian aging.[6-10] Indeed, mouse models represent the only experimental system now available to approach the genetic effects of aging in complex biological process as diverse as, for example, memory functionality, cancer incidence, and sensory capability. Furthermore, the phenotypic traits of aging in mice and humans are similar for the majority of tissues (vessels, skin, nervous system, reproductive apparatus, immune system, etc.).

3. Mouse Models of Longevity

Lifespan in mammals is strongly affected by the aging rate: all the genetic diseases that accelerate the aging rate provoke a lifespan reduction. In this vision, longevity represents the synthesis of the phenotypic expression of allelic forms that delay aging processes.

The only known spontaneous mutations that confer prolonged mouse lifespan are the dwarf mutations, e.g. non-allelic mutants of Ames (df) and Snell (dw), two genes which function in the same pathway of pituitary gland ontogeny regulation. Mice homozygous for either the df or dw mutations have impaired GH, prolactin, and TSH secretion and develop severe proportional dwarfism, hypothyroidism, and infertility. Interestingly, both mutants have increased lifespan of about 50 percent.[11-13] Lower body temperature and reduced oxidative stress, due to the absence of hormones, are considered the mechanism responsible for the prolonged lifespan of these mutants.[14]

On the other hand, there are spontaneous mutants selected for shortened lifespan, such as the senescence-accelerated mouse, SAMP.[15] The mean lifespan of SAMP inbred strains is 9.7 months. They show an accelerated senescence with earlier appearance of age-associated disorders like senile amyloidosis, brain atrophy, hearing impairment, and osteopenia. The biochemical defect of the SAMP mutant is not yet identified. Mitochondrial respiratory alterations, together with reduced repair enzyme activity on oxidative-stress-induced mutations imply the involvement of abnormal mitochondrial function,[16, 17] suggesting also that in these short life-span mouse mutants, the altered response to oxidative stress is a plausible cause of the mutated phenotype.[18, 19]

The effects of environmental factors on mouse lifespan have been largely investigated and food intake has been demonstrated to represent an important non-genetic determinant of longevity. Long-term 40% caloric restriction induces a significant (approximately 30%) extension of lifespan in both mice and rats, as compared to *ad libitum* fed control animals. Food-restricted rodents also display a delayed development of a broad spectrum of age-associated diseases, including cancer.[20-22] The reduction of oxidative damage due to modification of hormonal levels and metabolism has been suggested as part of the mechanism underlying the anti-aging effects of caloric restriction.[23-25] Noteworthy, caloric restriction has also been shown to lengthen lifespan in primates.[26]

4. Murine Reverse Genetics and Aging Research

During the last two decades the interest in mouse genetics increased further due to the introduction of the transgenic technology. A number of mouse mutants have been generated which represent important model systems for the genetics of aging or age-associated diseases. For the latter, many mouse strains are now available, whose description is beyond the scope of this brief review. As an example, transgenization of different mutant forms of the amyloid precursor or of the *presenilin-2* proteins correlates with functional, cognitive, and pathological defects in the mouse brain, which resembles the symptoms of the Alzheimer's Disease.[27, 28]

Models of accelerated aging are Klotho and telomerase-deficient mice. The Klotho gene encodes for a cell surface protein with glycosidase activity, whose physiological function is not clear. Mice carrying a homozygous disruption of the Klotho locus exhibit multiple pathological conditions resembling human aging and shortened lifespan.[29, 30]

Telomerase deficiency is also associated with a shortened lifespan. However, the contribution of telomerase to life-extension in mammals has yet to be investigated. As an example, there is no correlation between telomere length and maximal lifespan among a wide variety of species[31–33] and even within different strains species of mice.

Transgenic mouse models in aging research have also been used to evaluate the accumulation of DNA mutations during lifespan. Using the Lac Z coding sequence as transgene, the frequency of somatic mutations in various tissues and in different genetic backgrounds has been measured.[34]

Until recently, the only transgenic mice mutant available with increased longevity was the AlphaMUPA, where the urokinase-type plasminogen activator (uPA), an extracellular protease implicated in tissue remodeling, is overexpressed in the brain. These mice have low serum concentrations of corticosteroids, low body temperature, and learning defects. Spontaneous food-intake is reduced by 20% and their lifespan increased by approximately 20 percent. It is hypothesized that caloric restriction due to an eating alteration is the mechanism responsible for the longevity phenotype observed in this transgenic mutant.[35, 36]

Accumulation of oxidative damage produced by Reactive Oxygen Species (ROS) is considered the most plausible proximal mechanism of aging.[37–43] Indeed, in *Drosophila*, over-expression of both the wild-type Cu, Zn Superoxide Dismutase (SOD) scavenging enzyme and its mutated form FALS (responsible for the Familial Amyotrophic Lateral Sclerosis) in the motoneurons of transgenic mutants provokes prolonged lifespan of about 40 percent.[44] In mice, instead, transgenization of normal SOD induces severe neuronal and immunological abnormalities and shortened lifespan[45–47] while transgenization of FALS SOD results in defects similar to those observed in FALS.[48] The different results obtained in flies and in mice might reflect metabolic differences among tissues or species and underscore the intrinsic difficulty in extrapolating results across distant species.

Recently, we have bred knockout mice deficient in the protein p66[ShcA]. To our surprise, their lifespan increased 35 percent with respect to the weight (WT).[49]

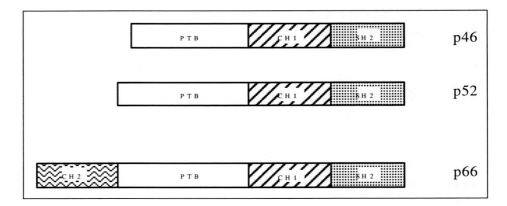

Fig. 1. Modular organization of ShcA proteins.

5. The p66^{ShcA} Protein

Three ShcA isoforms of 52 (p52^{shcA}), 46 (p46^{shcA}), and 66 (p66^{shcA}) kDa have been characterized.[50, 51] They share an aminoterminal SH2 domain, an adjacent glycine/proline rich region (CH1) and a carboxyterminal Phospho Tyrosine Binding domain (PTB). The p66^{shcA} isoform is characterized by an additional amino-terminal CH region, CH2[51] (see Fig. 1).

The SH2-containing ShcA proteins are cytoplasmic substrates of activated tyrosine kinases (TK) and have been implicated in the transmission of activation signals from TKs to Ras proteins.[50, 52] Upon phosphorylation, ShcA proteins form stable complexes with cellular tyrosine-phosphorylated polypeptides, including RTKs and receptors devoid of intrinsic TK activity. These interactions are mediated by the ShcA SH2 and/or PTB domains.[50, 53]

P52^{shcA} and p46^{shcA} proteins are phosphorylated by all receptor TKs (RTKs) tested to date, including the EGF receptor (EGFR),[50] the platelet derived growth-factor receptor,[54] the hepatocyte growth-factor receptor,[55] the erbB-2 receptor,[56] the insulin receptor,[57] the fibroblast growth-factor receptor,[58] and the nerve-growth-factor receptor.[59] P52^{shcA} and p46^{shcA} proteins are also involved in signaling from cytoplasmic TKs, since they are constitutively phosphorylated in cells that express activated Lck, Src, Fps, or Sea.[53, 60–63] In addition, they are rapidly phosphorylated on tyrosine after ligand stimulation of surface receptors that have no intrinsic TK activity, but are thought to signal by recruiting and activating cytoplasmic TKs (e.g. IL-2, erythropoietin, G-CSF, GM-CSF, B- and T-cell receptors, CD4, CD8).[64–68]

Phosphorylated ShcA proteins form stable complexes with the Grb2 adaptor protein, through direct binding of the Grb2 SH2 domain to the major ShcA tyrosine-phosphorylation sites (Tyr 239, 240, and 317).[69] Grb2 is constitutively complexed with SOS, a ubiquitously expressed Ras guanine nucleotide exchange factor for

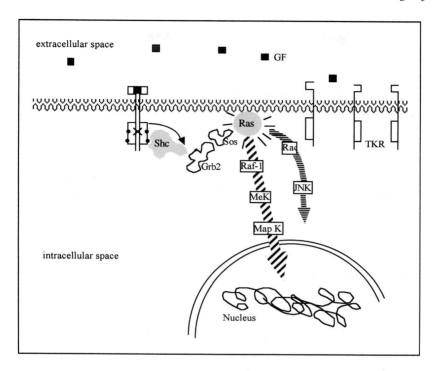

Fig. 2. ShcA signal transduction pathway.

Ras.[70–73] Recruitment of the Grb2/SOS complex results in the membrane relocalization of SOS, an event considered sufficient to induce Ras activation,[74] thus suggesting that ShcA proteins are involved in the regulation of Ras (Fig. 2).

Consistently with this model, over-expression of $p52^{shcA}/p46^{shcA}$ enhances EGF- or GM-CSF induced MAP kinase activation and *fos* promoter activity.[51, 68] At variance, $p66^{shcA}$ over-expression has no effect on MAP kinase activation and inhibits *fos* activity, though this is not the consequence of the inability of $p66^{shcA}$ to become phosphorylated following EGF stimulation and to form stable complexes with activated EGFR or Grb2.[51] Therefore, $p66^{shcA}$ could, unlike $p52^{shcA}/p46^{shcA}$, be part of a signaling pathway that inhibits growth factor-regulated genes, like *fos*. This positive-negative signal balance of the various Shc isoforms might be regulated by growth-factor induced tyrosine-phosphorylation of $p52^{shcA}/p46^{shcA}$ and the intracellular levels of $p66^{shcA}$ expression.

Little is known about the mechanisms that regulate expression of the three Shc isoforms *in vivo*. Different regulatory mechanisms might control the expression of the two Shc transcripts in different cell types, since $p52^{shcA}/p46^{shcA}$ are found in every cell type with invariant reciprocal relationships, whereas levels of $p66^{shcA}$ expression varies from cell type to cell type and are absent in others. $P66^{shcA}$, for example, is not expressed in hemopoietic cells.[50, 51]

We have recently demonstrated that $p66^{shcA}$ is involved in the signal transduction pathways activated by environmental stresses, as shown by the fact that $p66^{shcA}$ becomes serine-phosphoryled in cells treated with UV or H_2O_2. We took advantage of Mouse primary Embryonic Fibroblast (MEFs) derived from mice carrying a targeted mutation of the Shc locus to analyze the effects of $p66^{shcA}$ overexpression and $p66^{shcA}$ ablation on the cellular response to H_2O_2. The mutation disrupts the exon encoding the p66 CH2 region, without affecting the $p52^{shcA}/p46^{shcA}$ coding sequences.

WT MEFs were sensitive to H_2O_2 treatment, displaying more than 70% cell death after 24-h exposure to 400 µM H_2O_2. Overexpression of $p66^{shcA}$ increased WT MEFs susceptibility to H_2O_2. In contrast, $p66^{shcA}-/-$ MEF cells were more resistant to the same dose of H_2O_2 and more than 70% of the cells survived after 24 h of treatment. Expression of the $p66^{shcA}$ cDNA into $p66^{shcA}-/-$ cells restored a normal response to H_2O_2.[49] We also examined the ability of $p66^{shcA}-/-$ mice to resist oxidative stress *in vivo* using the paraquat resistance protocol. $P66^{shcA}-/-$ mice survived paraquat intoxication significantly better than the weight.[49]

6. The Ablation of $p66^{ShcA}$ Increases Lifespan in Mice

The increased resistance of $p66^{shcA}-/-$ MEFs and mice prompted us to investigate the effect of this mutation on lifespan. Thirty-seven mice born in August 1996 from $p66^{shcA}+/-$ heterozygous parents were not sacrificed and maintained under identical conditions. They consisted of 14 wt, 8 $p66^{ShcA}+/-$, and 15 $p66^{shcA}-/-$ mice. After 28 months of observation, all the wt animals had died (median survival of 25.37 ± 0.63 months), while 3 of the 8 heterozygous (37%) and 11 of the 15 homozygous (73%) were still alive. The remaining 3 $p66^{shcA}+/-$ died after additional two months (median survival of 27.40 ± 2.819 months in the $p66^{shc}+/-$ mice). Two $p66^{shcA}-/-$ mice also died after two months; the remaining nine are still alive (lifespan longer than 31 months). The comparison of survival curves obtained using the Kaplan and Meier method showed a highly significant difference between the three groups (log-rank $p = 0.0002$). Cumulative survival did not differ significantly between wild type and heterozygous ($p = ns$ [0.057]). The cumulative survival in the $p66^{shcA}-/-$ group was 71.4% ($p < 0.01$ versus $p66^{shcA}+/-$ and wt) (see Fig. 3). It appears, therefore, that homozygous mutation of $p66^{shcA}$ correlates with prolonged survival in mice.[49]

7. Mammalian Aging Inducer Genes

Enhanced resistance to environmental stress (starvation, heat shock, UV, paraquat-induced oxidative stress, ethanol) correlates well with prolonged lifespan in invertebrates.[75–82] A positive correlation between lifespan and cellular resistance

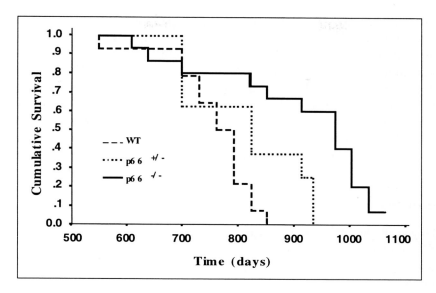

Fig. 3. Survival curves of WT, p66^{shcA}+/−, and p66^{shcA}−/− mice.

to stress also has been documented among different mammalian species.[83] However, whether the effects of the p66^{shcA} oxidative-stress response and lifespan are causally linked remains to be established.

Phenotypic analysis of the p66^{shcA}−/− mice has not revealed, till now, significant abnormalities. However, the possibility remains that the peculiar stabling conditions of laboratory animals prevents the expression of phenotypic traits which would be negatively selected under wild type conditions. Evaluation of p66^{shcA}−/− mice under conditions of stress is ongoing.

In summary, the physiological function of p66^{shcA} remains an open question. If we assume that the p66^{shcA} knock-out mice have no defective phenotype, we may then speculate that the existence of such genes, as p66^{shcA}, are evolutionary selected to "regulate" lifespan. These genes might act as "aging inducers" to reduce the lifespan of a given species in order to achieve maximum fitness of the *population*.[84–87] Then, the possibility of identifying lifespan-regulating genes in mammals will be important in improving aging-associated disorders in humans.

8. Conclusions

Mouse transgenic models will be extremely powerful to help investigate aging processes and the genetics of longevity in mammals. Precautions have to be taken before extrapolating discoveries from murine models to humans, especially when considering such a complex phenotype as "lifespan". We found that mice deficient in p66^{shcA} live longer, therefore suggesting a role for p66^{shcA} in determining lifespan. Even if a huge amount of work has to be done to define entirely the mechanism of p66^{shcA} action, the discoveries of such a target gene is a starting point to approach knowledge regarding the genetics of longevity in mammals. The application of high-throughput technologies for gene expression analysis, like microarray chips, in the available mouse models of longevity might, for example, allow the creation of a catalog of lifespan-affecting longevity genes.

References

1. Hogan, B., Beddington, R., Costantini, F. and Lacy, E. (1994). *Manipulating the Mouse Embryo*, Cold Spring Harbor Laboratory Press.
2. Rugh, R. (1990). *The Mouse: Its Reproduction and Development*, Oxford University Press.
3. Festing, M. (1979). *Inbred Strains in Biomedical Research*, Oxford University Press.
4. Goodrick, C. L. (1975). Life-span and the inheritance of longevity of inbred mice. *J. Gerontol.* **30**(3): 257–263.
5. Meagher, S., Penn, D. J. and Potts, D. K. (2000). Male-male competition magnifies inbreeding depression in wild house mice. *Proc. Natl. Acad. Sci. USA* **9**(7): 3324–3329.
6. Miller, R. A., Austad, S., Burke, D., Chrisp, C., Dysko, R., Galecki, A. and Jackson, A. (1999). Monnier exotic mice as models for aging research: polemic and prospectus. *Neurobiol. Aging* **20**(2): 217–231.
7. Miller, R. A. and Nadon, N. L. *Principles of Animal Use for Gerontological Research.*
8. Nebert, D. W., Brown, D. D., Towne, D. W. and Eisen, H. J. (1984). Association of fertility, fitness and longevity with the murine Ah locus among (C57BL/6N) (C3H/HeN) recombinant inbred lines. *Biol. Reprod.* **30**(2): 363–373; *J. Gerontol. A: Biol. Sci. Med. Sci.* 2000 March; **55**(3): B117–B123.
9. Wolf, N. S., Li, Y., Pendergrass, W., Schmeider, C. and Turturro, A. (2000). Normal mouse and rat strains as models for age-related cataract and the effect of caloric restriction on its development. *Exp. Eye Res.* **70**(5): 683–692.
10. Harrison, D. E. and Roderick, T. H. (1997). Selection for maximum longevity in mice. *Exp. Gerontol.* **32**(1/2): 65–78.
11. Brown-Borg, H. M., Borg, K. E., Meliska, C. J. and Bartke, A. (1996). Dwarf mice and the aging process. *Nature* **384**(6604): 33.
12. Gage, P. J., Lossie, A. C., Scarlett, L. M., Lloyd, R. V. and Camper, S. A. (1995). Ames dwarf mice exhibit somatotrope commitment but lack growth hormone-releasing factor response. *Endocrinology* **136**(3): 1161–1167.

13. Bartke, A., Brown-Borg, H. M., Bode, A. M., Carlson, J., Hunter, W. S. and Bronson, R. T. (1998). Does growth hormone prevent or accelerate aging? *Exp. Gerontol.* **33**(7/8): 675–687.

14. Hunter, W. S., Croson, W. B., Bartke, A., Gentry, M. V. and Meliska, C. J. (1999). Low body temperature in long-lived Ames dwarf mice at rest and during stress. *Physiol. Behav.* **67**(3): 433–437.

15. Takeda, T., Hosokawa, M., Takeshita, S., Irino, M., Higuchi, K., Matsushita, T., Tomita, Y., Yasuhira, K., Hamamoto, H., Shimizu, K., Ishii, M. and Yamamuro, T. (1981). A new murine model of accelerated senescence. *Mech. Aging Dev.* **17**(2): 183–194.

16. Nakahara, H., Kanno, T., Inai, Y., Utsumi, K., Hiramatsu, M., Mori, A. and Packer, L. (1998). Mitochondrial dysfunction in the senescence accelerated mouse (SAM). *Free Radic. Biol. Med.* **24**(1): 85–92.

17. Choi, J. Y., Kim, H. S., Kang, H. K., Lee, D. W., Choi, E. M. and Chung, M. H. (1999). Thermolabile 8-hydroxyguanine DNA glycosylase with low activity in senescence-accelerated mice due to a single-base mutation. *Free Radic. Biol. Med.* **27**(7/8): 848–854.

18. Higuchi, K., Hosokawa, M. and Takeda, T. (1999). Senescence-accelerated mouse. *Meth. Enzymol.* **309**: 674–686.

19. Mori, A., Utsumi, K., Liu, J. and Hosokawa, M. (1998). Oxidative damage in the senescence-accelerated mouse. *Ann. NY Acad. Sci.* **854**: 239–250.

20. Weindruch, R., Walford, R. L., Fligiel, S. and Guthrie, D. (1986). The retardation of aging in mice by dietary restriction: longevity, cancer, immunity and lifetime energy intake. *J. Nutri.* **116**(4): 641–654.

21. Pugh, T. D., Oberley, T. D. and Weindruch, R. (1999). Dietary intervention at middle age: caloric restriction but not dehydroepiandrosterone sulfate increases lifespan and lifetime cancer incidence in mice. *Cancer Res.* **59**(7): 1642–1648.

22. Weindruch, R. and Walford, R. L. (1982). Dietary restriction in mice beginning at 1 year of age: effect on life-span and spontaneous cancer incidence. *Science* **215**(4538): 1415–1418.

23. Sohal, R. S. and Weindruch, R. (1996). Oxidative stress, caloric restriction, and aging. *Science* **273**(5271): 59–63.

24. Merry, B. J. (2000). Calorie restriction and age-related oxidative stress. *Ann. NY Acad. Sci.* **908**: 180–198.

25. Masoro, E. J. (2000). Caloric restriction and aging: an update. *Exp. Gerontol.* **35**(3): 299–305.

26. Couzin, J. (1998). Low-calorie diets may slow monkeys' aging. *Science* **282**(5391): 1018.

27. Chapman, P. F., White, G. L., Jones, M. W., Cooper-Blacketer, D., Marshall, V. J., Irizarry, M., Younkin, L., Good, M. A., Bliss, T. V., Hyman, B. T., Younkin, S. G. and Hsiao, K. K. (1999). Impaired synaptic plasticity and learning in aged amyloid precursor protein transgenic mice. *Nat. Neurosci.* **2**(3): 271–276.

28. Oyama, F., Sawamura, N., Kobayashi, K., Morishima-Kawashima, M., Kuramochi, T., Ito, M., Tomita, T., Maruyama, K., Saido, T. C., Iwatsubo, T., Capell, A., Walter, J., Grunberg, J., Ueyama, Y., Haass, C. and Ihara, Y. (1998). Mutant presenilin 2 transgenic mouse: effect on an age-dependent increase of amyloid beta-protein 42 in the brain. *J. Neurochem.* **71**(1): 313–322.

29. Kuro-o, M., Matsumura, Y., Aizawa, H., Kawaguchi, H., Suga, T., Utsugi, T., Ohyama, Y., Kurabayashi, M., Kaname, T., Kume, E., Iwasaki, H., Iida, A., Shiraki-Iida, T., Nishikawa, S., Nagai, R. and Nabeshima, Y. I. (1997). Mutation of the mouse klotho gene leads to a syndrome resembling aging. *Nature* **390**(6655): 45–51.

30. Roush, W. (1997). Fast-forward aging in a mutant mouse? *Science* **278**(5340): 1013.

31. Rudolph, K. L., Chang, S., Lee, H. W., Blasco, M., Gottlieb, G. J., Greider, C. and DePinho, R. A. (1999). Longevity, stress response, and cancer in aging telomerase-deficient mice. *Cell* **96**(5): 701–712.

32. Wyllie, F. S., Jones, C. J., Skinner, J. W., Haughton, M. F., Wallis, C., Wynford-Thomas, D., Faragher, R. G. and Kipling, D. (2000). Telomerase prevents the accelerated cell aging of Werner syndrome fibroblasts. *Nature Genet.* **24**(1): 16–17.

33. Goyns, M. H. and Lavery, W. L. (2000). Telomerase and mammalian aging: a critical appraisal. *Mech. Aging Dev.* **114**(2): 69–77.

34. Vijg, J., Dolle, M. E., Martus, H. J. and Boerrigter, M. E. (1997). Transgenic mouse models for studying mutations in vivo: applications in aging research. *Mech. Aging Dev.* **99**(3): 257–271.

35. Miskin, R., Masos, T., Yahav, S., Shinder, D. and Globerson, A. (1999). AlphaMUPA mice: a transgenic model for increased lifespan. *Neurobiol. Aging* **20**(5): 555–564.

36. Meiri, N., Masos, T., Rosenblum, K., Miskin, R. and Dudai, Y. (1994). Overexpression of urokinase-type plasminogen activator in transgenic mice is correlated with impaired learning. *Proc. Natl. Acad. Sci. USA* **91**(8): 3196–2000.

37. Jazwinski, S. M. (1996). Longevity, genes, and aging. *Science* **273**(5271): 54–59.

38. Sohal, R. S. and Weindruch, R. (1996). Oxidative stress, caloric restriction, and aging. *Science* **273**(5271): 59–63.

39. Lithgow, G. J. and Kirkwood, T. B. (1996). Mechanisms and evolution of aging. *Science* **273**(5271): 80.

40. Berlett, B. S. and Stadtman, E. R. (1997). Protein oxidation in aging, disease, and oxidative stress. *J. Biol. Chem.* **272**(33): 20 313–20 316.

41. Beckman, K. B. and Ames, B. N. (1998). The free radical theory of aging matures. *Physiol. Rev.* **78**(2): 547–581.

42. Lenaz, G. (1998). Role of mitochondria in oxidative stress and aging. *Biochim. Biophys. Acta* **1366**(1/2): 53–67.

43. Johnson, F. B., Sinclair, D. A. and Guarente, L. (1999). Molecular biology of aging. *Cell* **96**(2): 291–302.

44. Elia, A. J., Parkes, T. L., Kirby, K., St. George-Hyslop, P., Boulianne, G. L., Phillips, J. P. and Hilliker, A. J. (1999). Expression of human FALS SOD in motorneurons of *Drosophila*. *Free Radic. Biol. Med.* **26**(9/10): 1332–1338.

45. Huang, T. T., Carlson, E. J., Gillespie, A. M., Shi, Y. and Epstein, C. J. (2000). Ubiquitous overexpression of Cu, Zn superoxide dismutase does not extend lifespan in mice. *J. Gerontol. A: Biol. Sci. Med. Sci.* **55**(1): B5–B9.

46. Nabarra, B., Casanova, M., Paris, D., Paly, E., Toyoma, K., Ceballos, I. and London, J. (1997). Premature thymic involution, observed at the ultrastructural level, in two lineages of human SOD-1 transgenic mice. *Mech. Aging Dev.* **96**(1/3): 59–73.

47. Tu, P. H., Raju, P., Robinson, K. A., Gurney, M. E., Trojanowski, J. Q. and Lee, V. M. (1996). Transgenic mice carrying a human mutant superoxide dismutase transgene develop neuronal cytoskeletal pathology resembling human amyotrophic lateral sclerosis lesions. *Proc. Natl. Acad. Sci. USA* **93**(7): 3155–3160.

48. Liu, R., Althaus, J. S., Ellerbrock, B. R., Becker, D.A. and Gurney, M. E. (1998). Enhanced oxygen radical production in a transgenic mouse model of familial amyotrophic lateral sclerosis. *Ann. Neurol.* **44**(5): 763–770.

49. Migliaccio, E., Giorgio, M., Mele, S., Pelicci, G., Reboldi, P., Pandolfi, P. P., Lanfrancone, L. and Pelicci, P. G. (1999). The p66shc adaptor protein controls oxidative stress response and lifespan in mammals. *Nature* **402**(6759): 309–313.

50. Pelicci, G., Lanfrancone, L., Grignani, F., McGlade, J., Cavallo, F., Forni, G., Nicoletti, I., Grignani, F., Pawson, T. and Pelicci, P. G. (1992). A novel transforming protein (SHC) with an SH2 domain is implicated in mitogenic signal transduction. *Cell* **70**(1): 93–104.

51. Migliaccio, E., Mele, S., Salcini, A. E., Pelicci, G., Lai, K. M., Superti-Furga, G., Pawson, T., Di Fiore, P. P., Lanfrancone, L. and Pelicci, P. G. (1997). Opposite effects of the p52shc/p46shc and p66shc splicing isoforms on the EGF receptor-MAP kinase-fos signalling pathway. *EMBO J.* **16**(4): 706–716.

52. Bonfini, L., Migliaccio, E., Pelicci, G., Lanfrancone, L. and Pelicci. P. G. (1996). Not all Shc's roads lead to Ras. *Trends Biochem. Sci.* **21**(7): 257–261.

53. Pelicci, G., Lanfrancone, L., Salcini, A. E., Romano, A., Mele, S., Grazia-Borrello, M., Segatto, O., Di Fiore, P. P. and Pelicci, P. G. (1995). Constitutive phosphorylation of Shc proteins in human tumors. *Oncogene* **11**(5): 899–907.

54. Yokote, K., Mori, S., Hansen, K., McGlade, J., Pawson, T., Heldin, C. H. and Claesson-Welsh, L. (1994). Direct interaction between Shc and the platelet-derived growth factor beta-receptor. *J. Biol. Chem.* **269**(21): 15 337–15 343.

55. Pelicci, G., Giordano, S., Zhen, Z., Salcini, A. E., Lanfrancone, L., Bardelli, A., Panayotou, G., Waterfield, M. D., Ponzetto, C., Pelicci, P. G. and Comoglio,

P. M. (1995). The SH2 adaptor protein Shc mediates transduction of the motogenic signal triggered by SF/HGF. *Oncogene* **11**: 899–907.

56. Segatto, O., Pelicci, G., Giuli, S., Digiesi, G., Di Fiore, P. P., McGlade, J., Pawson, T. and Pelicci, P. G. (1993). Shc products are substrates of erbB-2 kinase. *Oncogene* **8**: 2105–2112.

57. Skolnik, E. Y., Lee, C.-H., Batzer, A., Vicentini, L. M., Zhou, M., Daly, M., Myers, M. J., Backer, J. M., Ullrich, A., White, M. F. and Schlessinger, J. (1993). The SH2\SH3 domain-containing protein Grb2 interacts with tyrosine-phosphorylated IRS1 and Shc: implications for insulin control of ras signalling. *EMBO J.* **12**: 1929–1936.

58. Vainikka, S., Joukov, V., Wenstrom, S., Bergman, M., Pelicci, G. and Alitalo, K. (1994). Signal transduction by fibroblast growth factor receptor-4 (FGFR-4): comparison with FGFR-1. *J. Biol. Chem.* **269**: 18 320–18 326.

59. Borrello, M. G., Pelicci, G., Arighi, E., De Filippis, L., Greco, A., Bongarzone, I., Rizzetti, M. G., Pelicci, P. G. and Pierotti, M. (1994). The oncogenic versions of the RET and TRK tyrosine kinases bind SHC and GRB2 adaptor proteins. *Oncogene* **9**: 1661–1668.

60. MacGlade, J., Cheng, A., Pelicci, G., Pelicci, P. G. and Pawson, T. (1992). SHC proteins are phosphorylated and regulated by the v-Src and v-fps protein tyrosine kinases. *Proc. Natl. Acad. Sci.* **89**: 8869–8873.

61. Crowe, A., McGlade, J., Pawson, T. and Hayman. M. J. (1994). Phosphorylation of the SHC proteins on tyrosine correlates with the transformation of fibroblasts and erythroblasts by the v-sea tyrosine kinase. *Oncogene* **9**: 537–544.

62. Dilworth, S., Brewster, C., Jones, M., Lanfrancone, L., Pelicci, G. and Pelicci, P. G. (1994). Transformation by polyoma virus middle T-antigen requires the binding and tyrosine phosphorylation of Shc. *Nature* **367**: 87–89.

63. Baldari, C. T., Pelicci, G., Di Somma, M. M., Milia, E., Giuli, S., Pelicci, P. G. and Telford, J. L. (1995). Inhibition of CD4/p56lck signaling by a dominant negative mutant of the Shc adaptor protein. *Oncogene* **10**(6): 1141–1147.

64. Burns, L. A., Karnitz, L. M., Sutor, S. L. and Abraham, R. T. (1993). Interleukin-2-induced tyrosine phosphorylation of p52^{shc} in T lymphocytes. *J. Biol. Chem.* **268**: 17 659–17 661.

65. Damen, J. E., Liu, L., Kutler, R. B. and Krystal, G. (1993). Erythropoietin stimulates the tyrosine phosphorylation of Shc and its association with Grb2 and a 145-Kd tyrosine phosphorylated protein. *Blood* **82**: 2296–2303.

66. Ravichandran, K. S., Lee, K. K., Songyang, Z., Cantley, L. C., Burn, P. and Burakoff, S. J. (1993). Interaction of Shc with the z chain of the T cell receptor upon T cell activation. *Science* **262**: 902–905.

67. Lanfrancone, L., Pelicci, G., Brizzi, M. F., Casciari, C., Giuli, S., Pegoraro, L., Pawson, T. and Pelicci, P. G. (1995). Overexpression of Shc proteins potentiates the proliferative response to the granulocyte-macrophage colony-stimulating factor and recruitment of Grb2/SOS and Grb2/p140 complexes to the β receptor subunit. *Oncogene* **10**: 907–917.

68. Matsuguchi, T., Salgia, R., Hallek, M., Eder, M., Druker, B., Ernst and Griffin, J. (1994). Shc phosphorylation in myeloid cells is regulated by GM-CSF, IL-3 and steel factor and is constitutively increased by p210[BCR/ABL]. *J. Biol. Chem.* **269**: 5016–5021.

69. Salcini, A. E, McGlade, J., Pelicci, G., Nicoletti, I., Pawson, T. and Pelicci, P. G. (1994). Formation of Shc-Grb2 complexes is necessary to induce neoplastic transformation by overexpression of Shc proteins. *Oncogene* **9**: 2827–2836.

70. Batzer, L. N., Daly, R., Yajnik, V., Skolnik, E., Chardin, P., Bar-sagi, D., Margolis, B. and Schlessinger, J. (1993). Guanine-nucleotide-releasing factor hSos1 binds to Grb2 and links receptor tyrosine kinases to Ras signalling. *Nature* **363**: 85–88.

71. Buday, L. and Downward, J. (1993). Epidermal growth factor regulates p21[ras] through the formation of a complex of receptor, Grb2 adapter protein, and Sos nucleotide exchange factor. *Cell* **73**: 611–620.

72. Chardin, P., Camonis, J., Gale, W. L., Van Aelst, L., Schlessinger, J., Wigler, M. H. and Bar-sagi, D. (1993). Human Sos1: a guanine nucleotide exchange factor for Ras that binds to Grb2. *Science* **260**: 1338–1343.

73. Egan, S. E., Giddings, B. W., Brooks, M. W., Buday, L., Sizeland, A. M. and Weinberg, R. (1993). Association of Sos Ras exchange protein with Grb2 is implicated in tyrosine kinase signal transduction and transformation. *Nature* **363**: 45–51.

74. Aronheim, A., Engelberg, D., Li, N., Al-Alawi, N., Schlessinger, J. and Karin, M. (1994). Membrane targeting of the nucleotide exchange factor SOS is sufficient for activating the Ras signaling pathway. *Cell* **78**: 949–961.

75. Sun, J., Childress, A. M., Pinswasdl, C. and Jazwinski, S. (1994). Divergent roles of *RAS1* and *RAS2* in yeast longevity. *J. Biol. Chem.* **269**: 18 638–18 645.

76. Kennedy, B. K., Austriaco, N. R., Zhang, J. and Guarente, L. (1995). Mutation in the silencing gene *SIR4* can delay aging in *Saccharomyces cerevisiae*. *Cell* **80**: 485–496.

77. Murakami, S. and Johnson, T. E. (1996). A genetic pathway conferring life extension and resistance to UV stress in *Caenorhabditis elegans*. *Genetics* **143**: 1207–1218.

78. Larsen, P. L., Albert, P. S. and Riddle, D. L. (1995). Genes that regulate both development and longevity in *Caenorhabditis elegans*. *Genetics* **139**: 1576–1583.

79. Service, P. M., Hutchinson, E. W., MacKinley, M. D. and Rose, M. R. (1985). Resistance to environmental stress in *Drosophila melanogaster* selected for postponed senescence. *Physiol. Zool.* **58**: 380–389.

80. Lin, Y. J., Seroude, L. and Benzer, S. (1998). Extended life-span and stress resistance in the *Drosophila* mutant *methuselah*. *Science* **282**: 943–946.

81. Ishi, N., Fujii, M., Hartman, P. S., Tsuda, M., Yasuda, K., Senoo-Matsuda, N., Yanase, S., Ayusawa, D. and Suzuki, K. (1998). A mutation in sucinate dehydrogenase cytochrome *b* causes oxidative stress and aging in nematodes. *Nature* **394**: 694–697.

82. Orr, W. C. and Sohal, R. S. (1994). Extension of life-span by overexpression of superoxide dismutase and catalase in *Drosophila melanoganster*. *Science* **263**: 1128–1130.

83. Kapahi, P., Boulton, M. E. and Kirkwood, T. B. (1999). Positive correlation between mammalian lifespan and cellular resistance to stress. *Free Radic. Biol. Med.* **26**(5/6): 495–500.

84. Weismann, A. Ueber die Dauer des Lebens. Fisher. Jena.

85. Rose, M. and Charlesworth, B. (1980). A test of evolutionary theories of senescence. *Nature* **287**(5778): 141–142.

86. Hughes, K. A. and Charlesworth, B. (1994). A genetic analysis of senescence in *Drosophila*. *Nature* **367**(6458): 64–66.

87. Skulachev, V. P. (1997). Aging is a specific biological function rather than the result of a disorder in complex living systems: biochemical evidence in support of Weismann's hypothesis. *Biochemistry (Mosc)* **62**(11): 1191–1195.

Chapter 69

Oxidative Stress, DNA Repair and Centenarian Survival

Richard Marcotte and Eugenia Wang*

Richard Marcotte • The Bloomfield Center for Research in Aging, Lady Davis Institute for Medical Research, The Sir Mortimer B. Davis-Jewish General Hospital and Department of Medicine, McGill University, Montréal, Québec, Canada
Eugenia Wang • Department of Biochemistry and Molecular Biology, University of Louisville School of Medicine, Louisville, Kentucky 40292
*Corresponding author.
Tel: 502/852-2556, -5217, E-mail: Eugenia.Wang@Louisville.edu

1. Introduction

Death is the ultimate destiny of living organisms. As organisms age, in general (seen from the viewpoint of average age) there is an inverse relationship between attainable maximum life span and morbidity: the more an organism ages, the more likely it will die.[1,2] (However, in many organisms the mortality curve plateaus in extreme age; in humans this occurs around age 95.) In the case of human populations, medical advances have reduced the likelihood of dying in infancy, or in large epidemics of infectious diseases, so that life expectancy has doubled in the last century. Advances in medicine have also contributed to increased life expectancy by developing better tools to treat diseases of aging, such as cardiovascular disorders, cancers, neurodegenerative diseases, osteoporosis, and others. Genes involved in these diseases are being discovered at an extraordinary pace, possibly leading to future therapies so that elderly people can enjoy an even healthier extended post-reproductive life. Identifying genes which are dysregulated in diseases of aging may provide cues to "true" longevity genes; proper regulation of these genes may provide healthier years up to a hundred years old. But what exactly are "true" longevity genes? Do they prevent age-dependent diseases, or extend life beyond curing the diseases? Both notions are likely; they clearly serve "mortality avoidance", by both supporting survival and delaying death.

We do not know that the milestone that Jeanne Calment set by reaching the age of 122 years is the maximum age attainable by a human, but it raises some interesting questions as to the implication of lifestyle and environment upon extreme longevity (Ms. Calment smoked up to 100 years of age).[3] The balance of genetic *versus* environmental contributions to aging is difficult to evaluate in man, but is thought to be in the range of 30 : 70 respectively, as deduced from twin studies.[4,5] Siblings of centenarians have a four-fold increased chance of living to ninety years old compared to the general population,[6] suggesting that a strong familial component may indeed impact greatly on extreme longevity, such as that of centenarians.

Genetic studies from other organisms have amassed evidence for reactive oxygen species (ROS) and metabolism as key mediators in extending longevity. In the nematode *Caenorhabditis elegans*, mutations that lead to an extended life span affect the *daf-2* and *age-1* genes, homologues of the human insulin receptor and the p110 subunit of phosphatidylinositol 3-kinase, as well as clk-1, a protein homologue to the yeast CAT5 protein, which is involved in the synthesis of coenzyme Q, a component of the mitochondrial electron transport chain.[7,8] Moreover, in nematodes, a mutation in succinate dehydrogenase cytochrome *b* decreases life span by 60%.[9] In *Drosophila melanogaster*, a mutation in the gene *methuselah*, which bears high homology to the G protein-coupled transmembrane receptor, extends life span by 35% over the parental line.[10] Moreover, this mutant shows resistance to oxidative stress, high temperature, and starvation. Other evidence for ROS implication in aging is the fact that genes involved in free radical detoxification, such as Cu, Zn superoxide dismutase, catalase, and glutathione reductase, extend life span when over-expressed in *Drosophila*.[11, 12] Oxidative stress also seems to be involved in mammalian aging; in a recent report, Migliaccio *et al.* reported a 30% increase in the life span of mice mutated for the adapter protein p66[shc].[13] Surprisingly, p66[shc] functions in the insulin pathway just as the *Caenorhabditis elegans daf* mutants, suggesting a common evolutionary pathway for longevity. However, even though these organisms sometime turn out to be useful models for understanding aging-related phenomena that also occur in man, human aging is a complex interplay of genetic and environmental factors, only some of which are shared with lower organisms. Since humans are one of the only species (if not the only one) that spend more than half of their life post-reproductively, the issue of fitness beyond childbearing age is inescapable. Centenarians are the eldest of the elderly, and genes that push back the known age limit set by Mme. Calment are likely to emerge in the coming years. So far, reports of genetic studies aiming for longevity genes are correlative; these studies are reviewed in the following sections.

2. The Mitochondrial Fountain of Youth

Mitochondria, thought to have arisen from a symbiotic association a few million years ago, possess their own genome, which codes for 13 polypeptides along with two ribosomal messages and 22 tRNAs. These proteins are all components of the mitochondrial respiratory chain, responsible for producing adenosine triphosphate (ATP) through oxidative phosphorylation, which all cells require to function properly. Through a set of successive translocations, electrons are transferred from different electron carriers onto molecular oxygen to produce water. This process is not totally efficient, since molecular oxygen often escapes the transport chain without being completely reduced to water, in the form of oxygen radicals and hydrogen peroxide. These reactive oxygen species (ROS) have a very detrimental effect on cells if they are not rapidly detoxified; they cause DNA damage, protein oxidation, and lipid peroxidation, affecting the general homeostatic status of affected cells.[14, 15] Since mitochondrial DNA is closer to the source of the free radicals than nuclear DNA, it is more vulnerable to cumulative damage during the aging process, and it has recently been shown that oxidative damage seems to increase with age.[16]

An increasing number of reports have demonstrated the presence of small and large deletions, insertions, oxidative and alkylation derivatives of nucleotides, in mitochondrial DNA of variously aged tissues from different individuals. These damages to mitochondrial, but not to nuclear, DNA are inversely proportional to the maximum life span.[17, 18] Using PCR, an accumulation of 4977 mitochondrial base pair (bp) deletions was detected in muscle and liver of 30-year old individuals, while the same deletions appear only after 60 years in the testis, and even later in brain and heart, where they increase rapidly after 75 years of age.[19] This observation is not an artifact, as long-extended PCR of total mitochondrial DNA genome failed to yield any full-length product in a subsequent study.[20] However, these deletions are present in only a small percentage of mitochondria in affected cells, which argues against any role in cell deterioration, and suggests that accumulation with age is a stochastic process. On the other hand, point mutations were discovered in the 1000 bp mitochondrial DNA replication region.[21] These mutations occur with very high frequency, and increase with age in the same individuals. This may lead to a selective advantage as mutated mitochondria replicate at a faster rate than non-affected mitochondria, subverting the normal behavior to a more detrimental one. However, these mutations in the mitochondrial control region for replication have so far not been shown to alter the normal functioning of the mitochondria or the affected cells.

A few studies demonstrate some age-associated changes in mitochondrial function, mainly in complex II + III of the respiratory chain, which is mostly coded by the nuclear genome. Complex I contains more subunits encoded by the mitochondrial genome, but it has so far not been shown to be impaired in aged individuals. Even though these studies suggest the involvement of mitochondrial DNA rearrangements in the aetiology of aging, their accumulation may be mainly correlative; any actual role these changes may play in the aging phenotype is still debatable. Although mutated mitochondrial DNA is found to some extent in neurodegenerative diseases, we do not know that it is causally linked to the actual phenotype. Moreover, if we hypothesize that these deletions should have an actual impact on the longevity of the affected individuals, centenarians should have a more stable mitochondrial genome. Based on this notion, all described mutations, whether deletions, insertions, or point mutations, should impair normal mitochondrial function in aged people; therefore, centenarians should possess a superior homeostatic status, where either the accumulation of these defects occurs at a slower rate or not at all, or they have inherited genes whose actions counteract the deleterious effects of the mutations.

This hypothesis leads to the question of whether centenarians inherit protective genes against mitochondrial degeneration. In one study, the entire mitochondrial DNA from 11 Japanese centenarians was sequenced by automated sequencing.[22] Three nucleotide substitutions causing amino acid replacement were more frequent in centenarians than in a control group. These three substitutions were linked because only rarely is one of them found alone in an individual. One of the variations, a C-to-A transversion at nucleotide position 5178 (Mt5178A) within the NADH dehydrogenase subunit 2 gene, was further screened for in 37 centenarians and 252 healthy blood donors. Mt5178A was found in 62% of the centenarians, compared to 45% of the blood donors. So far, no functional studies have tested whether this transversion affects the activity of NADH dehydrogenase subunit 2. Paradoxically, no difference in activity of NADH dehydrogenase, succinate dehydrogenase, or cytochrome *c* oxidase was seen in fibroblast cell lines established from old donors (94 years old).[23] In a second study, using ancestrally associated polymorphism combinations (defined as haplogroups), 212 centenarians and 275 control donors from Northern Italy were screened for differences in distribution of these haplogroups.[24] Strikingly, the J haplogroup increased from 2% to 23% between male controls and male centenarians. On the other hand, the U haplogroup went from 23.5% in male controls to 4% in male centenarians. While the same patterns were not statistically significant in females, the trends also showed an increase and a decrease in the J and U haplogroups respectively. As the authors noted, this is not the first study where gender differences are seen in longevity association studies. Gender differences are not unexpected, since in general females constitute 3/4 of all centenarians, suggesting the existence of gender-dependent life-extending genes. Interestingly, these results show a marked geographic dependence, since centenarians from Southern Italy did not exhibit the same predisposition for these haplogroups. Moreover, the Mt5178A transversion found in Japanese centenarians is located in the M haplogroup, which is not found in Italian centenarians.

The difference in haplotype distribution between Japanese and Italian centenarians evokes the old questions of the environmental impact on gene action, and how many genes regulate longevity in man. Do ten thousand genes each exert a weak effect, or do a few genes, 10 or so, have a strong effect on life span? From emerging results from *Caenorhabditis elegans* and fruit flies, it seems unlikely that thousands of genes act synergistically to promote extreme longevity, but more likely that a few master genes regulate essential processes, balancing the equilibrium between damage and repair, and allowing life extension by overcoming the age-related accumulation of insults. Furthermore, it is certain that the few (in the dozens) master genes are classed into two groups: those shared by all centenarians ("pan-human longevity assurance genes" — hLAGs), and those exhibited by only certain cultural, ethnic or geographically unique subgroups ("endemic hLAGs").

3. Centenarians Overcome Degeneration of the DNA Repair Process

Why do mutations accumulate in the mitochondrial genome and the nuclear DNA? There are two main possibilities: (1) genomic damage increases with age, and/or (2) the DNA repair capacity decreases with age. The mutational rate for mitochondrial DNA is estimated to be 8–20 fold higher than that of nuclear DNA.[15, 18] It was long argued that mitochondria did not possess DNA repair capabilities; this is now known to be untrue.[25] Mitochondria clearly possess a number of DNA repair mechanisms that are as efficient as nuclear DNA repair pathways. However, so far no nucleotide excision repair pathway has been described in mitochondria. This explains why mitochondria cannot remove DNA adducts as efficiently as nuclear DNA. The major support for the notion that DNA repair plays a role in the aetiology of aging is the so-called "premature aging syndromes". The RecQ family of DNA helicases is implicated in at least three human premature aging genetic disorders and/or cancer predisposition, including Werner's syndrome (WS), Bloom's syndrome (BS), and Rothmund-Thomson syndrome (RTS).[26] These syndromes, along with Cockayne syndrome and Hutchinson-Gilford progeria, all share to a certain extent some clinical features with normal aging: all these progeroid syndromes have in common a disrupted gene that is implicated in DNA repair and/or transcription. The best-characterized protein among these syndromes, the Werner protein (WRN), is a helicase and 3' to 5' exonuclease.[26] It interacts with a variety of proteins, including PCNA, topoisomerase I, DNA polymerase delta, Ku protein, replication protein A (RPA), and tumor suppressor p53.[26-29] Disruption of these interactions may lead to increased DNA damage, supposedly mimicking the accumulative nature of DNA damage in normal aging. Werner's syndrome exhibits an enhanced frequency of deletions and rearrangements, while Bloom's syndrome displays an elevated frequency of sister chromatid exchange. Moreover, there is some evidence for an increase in ROS generation in progeroid fibroblast cell lines, coupled with a decrease in detoxifying enzymes. Results from progeroid syndromes are hard to reconcile with normal aging, since the amount of DNA damage found in the former is much higher. In lymphocytes, however, the DNA repair capacity of aged individuals, 75 to 80 years old, is not down-regulated, but similar to that of a group of 35 to 39 year olds.[30] Expression levels of two proteins involved in DNA repair response are either unchanged or increased in lymphocytes from nonagenarians; DNA binding activity of Ku protein is marginally affected. Interestingly, poly(ADP-ribose) polymerase is increased 1.6 fold in centenarians.[31, 32] Since this enzyme represents an immediate response to oxidative DNA damage, it is tempting to speculate that enhanced oxidative DNA damage response is of survival value, because of greater counteracting power to combat oxidative stress in centenarian lymphocytes, and that this is in fact an adaptive response to keep unwanted damage from accumulating.

Similar studies from dermal fibroblasts also find decreased DNA repair with normal aging. Goukassian *et al.*[33] demonstrated a decrease in the removal of UV-induced pyrimidine pyrimidone photoproducts and thymine dimers. This apparent decrease is associated with down-regulation of several proteins involved in nucleotide excision repair: ERCC3, PCNA, RPA, XPA, and p53 mRNA and protein levels were decreased when comparing a 21–37 year donor age group with a 62–88 year group. This result is in part confirmed by cDNA array screening with mRNA from actively dividing fibroblasts from young, middle age, and old humans: PCNA is decreased 5.7 fold, while other proteins not characterized in the previous study are also decreased.[34] Since no reports yet describe DNA repair activity in centenarian fibroblasts, we do not know whether their behavior differs from that of their younger (octogenarian) counterparts. Nevertheless, we suggest that centenarians constitute a categorically separate group; either they do not exhibit the age-dependent decline in DNA repair, or they actually maintain robust DNA repair machinery, analogous to individuals in their 30s or 40s.

4. Tumor Suppressors or "Longevity Promoters"

For centenarians to live to such a respectable age, they must have avoided the traps set by many environmental and genetic cues throughout life. In fact, cancer susceptibility increases with age up to 90 years old, and then decreases by half at 100 years old.[35] Centenarians seem to have better defense mechanisms against detrimental agents to which individuals are exposed in daily situations. Since siblings of centenarians have an increased chance of living to a very honorable age,[6] the likelihood of inheriting two non-mutated alleles of a "good" gene is higher than inheriting either one allele or both alleles of a bad or mutated gene. Therefore, centenarians probably do not inherit a germline mutation in a known tumor suppressor, so that like their siblings they will not acquire a second hit in their somatic cells, leading to tumor development; the absence of mutations in genes that prevent cancer development increases the likelihood of living to an extreme age. Tumor suppressor genes have been discovered at an extraordinary pace in the last 10 years; over 30 are described so far, ranging from transcriptional regulators to mismatch repair gene and apoptosis regulators.[36]

The well-known p53 is involved in many cellular processes, such as cell cycle arrest, apoptosis, and DNA repair. It is mutated in about 50% of all cancers, making it the most frequently mutated tumor suppressor. There are common variations, polymorphisms, in the DNA sequence of p53. One of them, a C to G transversion leading to an arginine-proline change at codon 72, has been thoroughly studied in various cancers, with no real consensus as to whether it renders cells more susceptible to malignant growth (Refs. 37–39, and references therein). But if indeed one amino acid gives a selective advantage, it should select for centenarians, who lack the deleterious polymorphism. This was the starting belief for Bonafè *et al.*[37, 39] in studying the distribution of codon 72 in centenarians. However, they found no statistical differences between the alleles, as both were present at similar levels in a control group and centenarians. Since about 20% of all individuals are affected by cancer, about 50% of all cancers have a mutated p53, and the selective advantage from lacking the single mutant p53 may be only 1.5–2 fold, significant differences would be attainable only with a sample size over 4000 centenarians, as argued by Sun *et al.*[38] Therefore, since the centenarian sample size is small (in the hundreds for any given population), it is difficult to determine whether Bonafè's results are due to the lack of a sufficient cohort for a statistically meaningful study.

More than 10 polymorphisms have been described in p53 alone, and there are polymorphisms in other tumor suppressors as well. Taken as a group, lacking these deleterious polymorphisms could have a net beneficial effect on the quality of life of centenarians, and possibly explain their decreased incidence of lethal cancer. Interestingly, BARD1, a BRCA-1 associated RING domain protein, is decreased several fold in aged fibroblasts. Knowing the role of BRCA-1 in oxidative damage repair, it is tempting to speculate that this decreased expression may play a role in skin deterioration with age. Moreover, ATM and ATR, two DNA-damage sensing proteins (ATM is associated with a genetic disorder, ataxia telangiectasia, characterized by increased genomic instability and cancer susceptibility), are also down-regulated in aged fibroblasts. Recently, ATM was found to have a protective role against oxidative stress,[40] suggesting again that decreased ATM expression could be in part responsible for increased DNA damage in aged cells. Thus, an imbalance in any tumor suppressive pathway could have a huge effect on cell and even organ homeostasis, as these tumor suppressors regulate important cellular processes, and sometimes even other tumor suppressors, e.g. ATM and p53. It will be interesting to evaluate polymorphisms in these DNA repair genes as candidate markers for the distinctive genetic status of centenarians. It is tempting to suggest that extremely long-lived individuals carry "good" polymorphisms; it is already evident that mortality rates plateau before reaching the extremes of old age, and this observation may indicate an anomaly in the seemingly monotonic decline in age-dependent molecular operations.

5. Conclusion

There are two main theories about the complexity of the aetiology of aging. First, the rate of aging is a very complex phenomenon, influenced by small hits in thousands of genes, leading to the gradual deterioration of vital processes as people age. This is the main belief governing the somatic mutation theory of aging. Second, in spite of the complex aetiology of aging, a few key regulating proteins may act on common pathways to determine the rate of aging. This "aging determinant gene hypothesis" is strengthened by the existence of premature aging syndromes, where only one gene is mutated to give a phenotype reminiscent of aging. However, even here the phenotypes are not totally identical to the normal aging process, suggesting the disruption of other genes for a complete aging aetiology. The above sections describe genes regulating processes likely to affect the rate of aging. Some are correlated with extreme longevity, as their decline in expression is lacking in centenarians as compared with the general population. Others are involved in overcoming diseases of aging, therefore

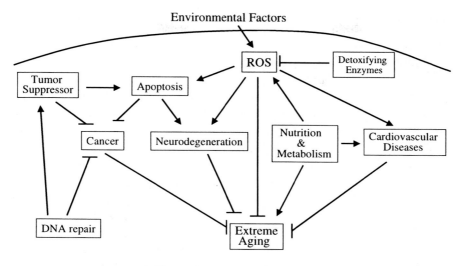

Fig. 1. Marcotte and Wang.

permitting centenarians to escape from ambushes on the path to death. A summary is depicted in Fig. 1. At first glance, Fig. 1 immediately argues for the hypothesis that the sum of a large number of genes contributes to the rate of aging. However, there are several interactions among different classes of functional genes. Therefore, one gene in particular may affect several important processes that lead to an increased rate of aging, or to more uniform aging, thought to be the key to survive 100 years. For example, the Werner protein is a DNA helicase involved in DNA repair, recombination, and transcription. It can certainly be classified as a tumor suppressor, since its absence leads to genomic instability. It associates with p53, which is involved in apoptosis, ROS generation, and cell cycle arrest. Thus a defect in one gene can lead to a cascade of signal transduction, detrimental to the organism as a whole, as in Werner's syndrome, where a small set of 10–20 master genes may regulate a complicated phenotype such as aging. Undoubtedly, new genetic approaches and cDNA array technology will allow the discovery of such master genes, and lead to the understanding of key genetic mechanisms which allow centenarians to escape life-threatening old-age perils and live happily to 100 years and beyond.

Acknowledgments

We are indebted to Mr. Emmanuel Petroulakis and Mr. Alan Bloch for proofreading this review. This work was supported by a grant to EW from the Defense Advanced Research Project Agency (DARPA) of the US Department of Defense.

References

1. Stearns, S. C., Ackermann, M., Doebeli, M. and Kaiser, M. (2000). Experimental evolution of aging, growth, and reproduction in fruitflies. *Proc. Natl. Acad. Sci. USA* **97**: 3309–3313.

2. Martin, G. M., Austad, S. N. and Johnson, T. E. (1996). Genetic analysis of aging: role of oxidative damage and environmental stresses. *Nature Genet.* **13**: 25–34.

3. Sulman, P. (2000). Design for living. *Sci. Am.* **11**: 18-21.

4. Andersen-Ranberg, K., Christensen, K., Jeune, B., Sytthe, A., Vasegaard, L. and Vaupel, J. W. (1999). Declining physical abilities with age: a cross-sectional study of older twins and centenarians in Denmark. *Age Aging* **28**: 373–377.

5. Herskind, A. M., McGue, M., Holm, N. V., Sorensen, T. I., Harvald, B. and Vaupel, J. W. (1996). The heritability of human longevity: a population-based study of 2872 Danish twin pairs born 1870-1900. *Human Genet.* **97**: 319–323.

6. Perls, T. T., Bubrick, E., Wager, C. G., Vijg, J. and Kruglyak, L. (1998). Siblings of centenarians live longer. *Lancet* **351**: 1560–1561.

7. Wood, W. B. (1998). Aging of *Caenorhabditis elegans*: Mosaics and mechanisms. *Cell* **95**: 147–150.

8. Hekimi, S., Lakowski, B., Barnes, T. M. and Ewbank, J. J. (1998). Molecular genetics of life span in *Caenorhabditis elegans*: how much does it teach us? *Trends Genet.* **14**: 14–20.

9. Ishii, N., Fujii, M., Hartman, P. S., Tsuda, M., Yasuda, K., Senoo-Matsuda, N., Yanase, S., Ayusawa, D. and Suzuki, K. (1998). A mutation in succinate dehydrogenase cytochrome *b* causes oxidative stress and aging in nematodes. *Nature* **394**: 694–696.

10. Lin, Y.-J., Seroude, L. and Benzer, S. (1998). Extended life-span and stress resistance in the *Drosophila* mutant *methuselah*. *Science* **282**: 943–946.

11. Sun, J. and Tower, J. (1999). FLP recombinase-mediated induction of Cu, Zn superoxide dismutase transgene expression can extend the life span of adult *Drosophila melanogaster* flies. *Mol. Cell. Biol.* **19**: 216–28.

12. Mockett, R. J., Sohal, R. S. and Orr, W. C. (1999). Overexpression of glutathione reductase extends survival in transgenic *Drosophila melanogaster* under hyperoxia but not normoxia. *FASEB J.* **13**: 1733–1742.

13. Migliaccio, E., Giorgio, M., Mele, S., Pelicci, G., Reboldi, P., Pandolfi, P. P., Lanfrancone, L. and Pelicci, P. G. (1999). The p66[sch] adaptor protein controls oxidative stress response and life span in mammals. *Nature* **402**: 309–313.

14. Berlett, B. S. and Stadtman, E. R. (1997). Protein oxidation in aging, disease, and oxidative stress. *J. Biol. Chem.* **272**: 20 313–20 316.

15. Beckman, K. B. and Ames, B. N. (1997). Oxidative decay of DNA. *J. Biol. Chem.* **272**: 19 633–19 636.

16. Beckman, K. B. and Ames, B. N. (1998). The free radical theory of aging matures. *Physiol. Rev.* **78**: 547–581.

17. Hayashi, J., Otha, S., Kagawa, Y., Kondo, H., Kaneda, H., Yonekawa, H., Takai, D. and Miyabayashi, S. (1994). Nuclear but not mitochondrial genome involvement in human age-related mitochondrial dysfunction. Functional integrity of mitochondrial DNA from aged subjects. *J. Biol. Chem.* **269**: 6878–6883.

18. Barja, G. and Herrero, A. (2000). Oxidative damage to mitochondrial DNA is inversely related to maximum life span in the heart and brain of mammals. *FASEB J.* **14**: 312–318.

19. Lee, H. C., Pang, C. Y., Hsu, H. S. and Wei, Y. H. (1994). Differential accumulation of 4977 bp deletion in mitochondrial DNA of various tissues in human aging. *Biochim. Biophys. Acta* **1226**: 37–43.

20. Melov, S., Shoffner, J. M., Kaufman, A. and Wallace, D. C. (1995). Marked increase in the number and variety of mitochondrial DNA rearrangements in aging human skeletal muscle. *Nucleic Acids Res.* **23**: 4122–4126.

21. Michikawa, Y., Mazzucchelli, F., Bresolin, N., Scarlato, G. and Attardi, G. (1999). Aging-dependent large accumulation of point mutations in the human mtDNA control region for replication. *Science* **286**: 774–779.

22. Tanaka, M., Gong, J. S., Zhang, J., Yoneda, M. and Yagi, K. (1998). Mitochondrial genotype associated with longevity. *Lancet* **351**: 185–186.

23. Allen, R. G., Keogh, B. P., Tresini, M., Gerhard, G. S., Volker, C., Pignolo, R. J., Horton, J. and Cristofalo, V. J. (1997). Development and age-associated differences in electron transport potential and consequences for oxidant generation. *J. Biol. Chem.* **272**: 24 805–24 812.

24. De Benedictis, G., Rose, G., Carrieri, G., De Luca, M., Falcone, E., Passarino, G., Bonafè, M., Monti, D., Baggio, G., Bertolini, S., Mari, D., Mattace, R. and Franceschi, C. (1999). Mitochondrial DNA inherited variants are associated with successful aging and longevity in humans. *FASEB J.* **13**: 1532–1536.

25. Croteau, D. L., Stierum, R. H. and Bohr, V. A. (1999). Mitochondrial DNA repair pathways. *Mutat. Res.* **434**: 137–148.

26. Karow, J. K., Wu, L. and Hickson, I. D. (2000). RecQ family helicases: roles in cancer and aging. *Curr. Opin. Genet. Dev.* **10**: 32–38.

27. Cooper, M. P., Machwe, A., Orren, D. K., Brosh, R. M., Ramsden, D. and Bohr, V. A. (2000). Ku complex interacts with and stimulates the Werner protein. *Genes Dev.* **14**: 907–912.

28. Lebel, M., Spillare, E. A., Harris, C. C. and Leder, P. (1999). The Werner syndrome gene product co-purifies with the DNA replication complex and interacts with PCNA and topoisomerase I. *J. Biol. Chem.* **274**: 37 795–37 799.

29. Kamath-Loeb, A. S., Johansson, E., Burgers, P. M. and Loeb, L. A. (2000). Functional interaction between the Werner Syndrome protein and DNA polymerase delta. *Proc. Natl. Acad. Sci. USA* **97**: 4603–4608.

30. King, C. M., Bristow-Craig, H. E., Gillespie, E. S. and Barnett, Y. A. (1997). In vivo antioxidant status, DNA damage, mutation and DNA repair capacity

in cultured lymphocytes from healthy 75- to 80-year-old humans. *Mutat. Res.* **377**: 137–147.

31. Muiras, M. L., Muller, M., Schachter, F. and Burkle, A. (1998). Increased poly(ADP-ribose) polymerase activity in lymphoblastoid cell lines from centenarians. *J. Mol. Med.* **76**: 346–354.

32. Frasca, D., Barattini, P., Goso, C., Pucci, S., Rizzo, G., Bartolini, C., Costanzo, M., Errani, A., Guidi, L., Antico, L., Tricerri, A. and Doria, G. (1998). Cell proliferation and ku protein expression in aging humans. *Mech. Aging Dev.* **100**: 197–208.

33. Goukassian, D., Gad, F., Yaar, M., Eller, M. S., Nehal, U. S. and Gilchrest, B. A. (2000). Mechanisms and implications of the age-associated decrease in DNA repair capacity. *FASEB J.* **14**: 1325–1334.

34. Ly, D. H., Lockhart, D. J., Lerner, R. A. and Schultz, P. G. (2000). Mitotic misregulation and human aging. *Science* **287**: 2486–2492.

35. Stanta, G., Campagner, L., Cavallieri, F. and Giarelli, L. (1997). Cancer of the oldest old. What we learned from autopsy studies. *Clin. Geriatr. Med.* **13**: 55–68.

36. Macleod, K. (2000). Tumor suppressor genes. *Curr. Opin. Genet. Dev.* **10**: 81–93.

37. Bonafè, M., Olivieri, F., Mari, D., Baggio, G., Mattace, R., Sansoni, P., De Benedictis, G., De Luca, M., Bertolini, S., Barbu, C., Monti, D. and Franceschi, C. (1999). P53 variants predisposing to cancer are present in healthy centenarians. *Am. J. Human Genet.* **64**: 292–295.

38. Sun, Y., Keshava, C., Sharp, D. S., Weston, A. and McCanlies, E. C. (1999). DNA sequence variants of P53: cancer and aging. *Am. J. Human Genet.* **65**: 1779–1782.

39. Bonafè, M., Olivieri, F., Mari, D., Baggio, G., Mattace, R., Berardelli, M., Sansoni, P., De Benedictis, G., De Luca, M., Marchegiani, F., Cavallone, L., Cardelli, M., Giovagnetti, S., Ferrucci, L., Amadio, L., Lisa, R., Tucci, M. G., Troiano, L., Pini, G., Gueresi, P., Morellini, M., Sorbi, S., Passeri, G., Barbi, C., Valensin, S., Monti, D., Deiana, L., Pes, G. M., Carru, C. and Franceschi, C. (1999). P53 codon 72 polymorphism and longevity: additional data on centenarians from continental Italy and Sardinia. *Am. J. Human Genet.* **65**: 1782–1785.

40. Takao, N., Li, Y. and Yamamoto, K.-I. (2000). Protective roles for ATM in cellular response to oxidative stress. *FEBS Lett.* **472**: 133–136.

SECTION 12

Non-Invasive Assessment of Oxidative Stress and Therapeutic Intervention

Chapter 70

Strategies for Controlling Oxidative Stress: Protection Against Peroxynitrite and Hydroperoxides by Selenoproteins and Selenoorganic Compounds

Helmut Sies* and Gavin E. Arteel

Keywords: Ebselen, GSH peroxidase, peroxynitrite, plasma, oxidative stress.

Helmut Sies • Institut für Physiologische Chemie I, Heinrich-Heine-Universität, Postfach 101007, D-40001, Düsseldorf, Germany
Gavin E. Arteel • Laboratory of Hepatobiology and Toxicology, Department of Pharmacology, CB #7365 Mary Ellen Jones Building, University of North Carolina at Chapel Hill, Chapel Hill, NC 27599-7365, USA

*Corresponding Author.
Tel: +49-211-811-2707, E-mail: helmut.sies@uni-duesseldorf.de

1. Introduction

The balance between prooxidants and antioxidants is critical for survival and functioning of aerobic organisms. An imbalance favoring prooxidants and/or disfavoring antioxidants, potentially leading to damage, has been called *oxidative stress*.[1, 2] During aging, the oxidant/antioxidant balance is shifted toward oxidative stress. While oxidants can directly damage tissues, oxidant reactions can also initiate or alter cellular signaling cascades that can serve to amplify the oxidant's effect. The effect of both of these pathways on aging are reviewed elsewhere in this book. However, it is clear that maintaining or supplementing the antioxidant defense of an organism can beneficial effects in attenuating the progression of aging. The physiological and pharmacological strategies for antioxidant defense are organized in three categories: *prevention*, *interception*, and *repair* (see Ref. 3). Most antioxidants work at the level of interception. The main point of *intercepting* a damaging species, once formed, is to exclude it from further activity. Further, interception often leads transfer of the prooxidant away from more sensitive compartments of the cell. The purpose of this review to discuss some of the general biochemistry of the selenium and glutathione systems in relation to antioxidant defense, and to and cite some recent work with this system in the specific context of protection against peroxynitrite.

2. Glutathione Peroxidase and Oxidant Interception Reactions

Using radiolabeled selenium, ~30 selenium-containing proteins have been identified,[4] with known functions for about 12; all of which contain selenocysteine as the selenoaminoacid. Of the functionally-identified selenoproteins, there are four different glutathione peroxidases, including a special one found in the gastrointestinal tract (see recent review, Ref. 5), the "classical" GPx of bovine erythrocytes, discovered by Mills,[6] shown to be a selenoprotein,[7, 8] and another GSH peroxidase that acts on peroxidized phospholipids in biological membranes (PHGPx).[9] As the name implies, all of these enzymes work to reduce organic hydroperoxides (ROOH). Although ROOH species are generally not potent oxidants themselves, they are both the product and precursor to other more reactive oxidizing species. Therefore, tight control of the levels of ROOH species is critical for the antioxidant defense of the cell.

The enzymatic catalysis reaction cycle of glutathione peroxidase (GPx) is thought to proceed in three main steps, involving the enzyme-bound seleno-cysteine, which is present as the selenol (R–SeH). In the first step (reaction (1), the organic hydroperoxide (ROOH) reacts to yield the selenenic acid [R–SeOH] and the corresponding alcohol (ROH) in a two-electron reaction. The following two sequential one-electron steps consist of the reduction by thiols, e.g. GSH. Reaction (2) gives the selenosulfide and water, and reaction (3) regenerates the selenol and GSSG. The overall reaction is summarized in reaction (4).

$$R-SeH + ROOH \longrightarrow R-SeOH + ROH \tag{1}$$

$$R-SeOH + GSH \longrightarrow R-Se-SG + H_2O \tag{2}$$

$$R-Se-SG + GSH \longrightarrow R-SeH + GSSG \tag{3}$$

$$ROOH + 2GSH \xrightarrow{\text{GPx}} ROH + H_2O + GSSG. \tag{4}$$

In reactions (2) and (3), GSH serves more as a ancillary reductant than as a direct antioxidant *per se*. In the presence of physiological concentrations of GSH, reactions (2) and (3) are sufficiently rapid as to not be rate-limiting.[10] Therefore, the rate constant for the entire catalytic reaction (k_{app}) is the same as the bimolecular rate constant driving reaction (1), and has been coined a "tert-uni ping-pong" mechanism.

3. Peroxynitrite, Oxidative Stress and Aging

Peroxynitrite ($ONOO^-$) is produced by the diffusion-limited reaction of nitric oxide and superoxide anion. Peroxynitrite is stable, but upon protonation to peroxynitrous acid (pK_a 6.8), it decays to nitrate with a rate constant of 1.3 s^{-1} at 25°C. Peroxynitrous acid is highly reactive, yielding oxidizing and nitrating species (see Ref. 11). At the level of the whole organism, the reactive chemistry of peroxynitrite can be considered beneficial, because of its cytotoxicity to bacteria[12] or other invading organisms. Inflammatory cells, such as macrophages and neutrophils, produce large amounts of both nitric oxide and superoxide, which in turn rapidly form peroxynitrite; however, overproduction of peroxynitrite can lead to normal tissue damage *via* its oxidation and nitration reactions (see Ref. 13, for review) and may contribute to aging.[14]

Strategies of peroxynitrite interception requires that the intercepting molecule is available at the corresponding site near the target to be protected, or near the source of peroxynitrite generation. Further, the reaction product formed between peroxynitrite and the intercepting molecule should be recyclable in order to permit the maintenance of a catalytic defense line. For assessing the detoxifying capacity of a given compound with a prooxidant, it is useful to consider the rate constant of this reaction. Table 1 lists the rate constants for some selenium-containing compounds and proteins as well as the concentration required to inhibit the oxidation of dihydrorhodamine 123 by peroxynitrite by 50%, which is a relative index of the bimolecular reaction rate constant.

Table 1. Interception of Peroxynitrite by Selenium-Containing compounds and Proteins

Compound	[a]Rate Constant $(M^{-1} s^{-1})$	[b]Dihydrorhodamine-123 Oxidation Half-Maximal Inhibitory Concentration (μM)
Small Molecules		
Ebselen	2.0×10^6 (17)	0.15 (61)
Ebsulfur	—	15 (61)
2-(Methylseleno)benzanilide	1.2×10^4 (62)	0.8 (24)
Ebselen selenoxide	—	100 (24)
Selenomethionine	2.4×10^3 (20)	0.3 (61)
Methionine	1.8×10^2 (63)	20 (61)
Selenocystine	—	2.5 (61)
Cystine	—	$> 10^3$ (61)
Sodium selenite	—	$> 10^4$ (61)
Glutathione	5.8×10^2 (64)	12 (24)
Proteins		
PHGPx	—	0.05 (61)
Glutathione peroxidase (GPx)	8.0×10^6 (25)	0.2 (24)
Carboxymethylated GPx	—	0.1 (24)
Oxidized GPx	7.4×10^5 (25)	—
Albumin	5.6×10^3 (65)	7 (24)

[a]Second-order rate constants are for pH 7.4, 25°C.
[b]Dihydrorhodamine 123 (0.5 μM), DTPA (0.1 mM), and peroxynitrite (0.1 μM).

4. Protection Against Peroxynitrite by Organoselenium Compounds

4.1. Ebselen

The organoselenium compound, 2-phenyl-1,2-benzisoselenazole-3(2H)-one (ebselen), has antioxidant properties. This compound has been extensively studied as a glutathione peroxidase mimic (see Ref. 15 for review). It was of considerable interest to note that ebselen also reacts with peroxynitrite.[16, 17] The compound acts catalytically in reducing peroxynitrite to nitrite in a first step, followed by subsequent reduction of the resultant selenoxide back to ebselen by reducing equivalents (e.g. GSH), analogous to reactions (1)–(4):

$$ONOO^- + 2GSH \longrightarrow NO_2^- + H_2O + GSSG. \tag{5}$$

In addition to glutathione serving as a reducing supply, the mammalian seleno-protein thioredoxin reductase can also reduce ebselen selenoxide at the expense of NADPH (Ref. 18; see below). Thus, ebselen can act both as a catalyst reducing hydroperoxides and as a catalyst reducing peroxynitrite (see Ref. 19).

4.2. Selenomethionine

Selenomethionine is oxidized to the selenoxide by peroxynitrite with a second-order rate constant approximately 100-fold higher than for the reaction of methionine with peroxynitrite[20] and selenomethionine residues on proteins may help protect against peroxynitrite. However, recent work suggested that albumin from selenium-supplemented individuals did not apparently protect against bolus addition of peroxynitrite relative to albumin isolated from individuals prior to selenium supplementation, although selenomethionine levels in albumin were higher after supplementation.[21]

5. Protection Against Peroxynitrite by Selenoproteins

The major selenium-containing aminoacids in the body are selenocysteine and selenomethionine. The incorporation of selenium into selenocysteine is specific and mediated at the ribosomal level, where the specialized selenocysenyl-tRNA recognizes the UGA codon in the SECIS secondary structure of the mRNA (see Ref. 22, for review). Conversely, the incorporation of selenomethionine into proteins is non-specific and directly related to dietary intake of selenium/ selenomethionine.[23]

5.1. Glutathione Peroxidase

The selenocysteine-containing glutathione peroxidase (GPx) can act as a peroxynitrite reductase [see reaction (5)], preventing oxidation and nitration reactions caused by peroxynitrite.[24] Increases in nitrite during exposure to peroxynitrite were observed with GPx,[24] indicating two-electron reduction of peroxynitrite; however, the nitrite yield was less than complete ($\sim 50\%$). The second-order rate constant for the reaction of glutathione peroxidase (tetrameric) with peroxynitrite is 8.0×10^6 M^{-1} s^{-1}.[25] While there is no net loss of GPx activity when GPx is maintained in the reduced state by supplying reductants,[24, 25] GPx is inactivated in the absence of GSH[26] or upon exposure to nitric oxide donors.[27] Comparing the second-order rate constants (Table 1) with their respective intracellular concentrations, it is likely that GPx outcompetes thiols for the direct reaction with peroxynitrite. The estimated rate of peroxynitrite reduction by GPx within the cell is estimated to be 16 s^{-1} (see Ref. 13). Increasing the level of selenoproteins (e.g. GPx 14-fold) by selenium supplementation attenuated mitogen-activated protein kinase (P38, JNK1/2 and ERK1/2) activation by peroxynitrite in cultured WB-F344 rat liver cells.[28] Thus, the reaction of GPx with peroxynitrite is considered a biologically efficient detoxication pathway *in vivo*. In view of the recently reported evidence for peroxynitrite as a signaling molecule in flow-dependent activation of JNK,[29] the GPx reaction would also be modulatory in some circumstances.

5.2. Selenoprotein P

Selenoprotein P in human plasma also protects against peroxynitrite,[30] suggesting that it may serve as a protectant against peroxynitrite in human blood. The heparin-binding domains of selenoprotein P enable surface coating of cellular membranes (e.g. endothelial cells; Ref. 31). This effect could serve two purposes *in vivo*: (1) as a means of concentrating selenoprotein P on cell surfaces for targeted defense against oxidants; (2) reducing equivalents could be transferred to bound oxidized selenoprotein P from the cell layer, allowing for maintenance of a defense line. Recent work with surface plasmon resonance has indicated that heparin has two binding sites for selenoprotein P, one with a binding constant in the low nM range and the other in the mid nM range.[32]

5.3. Thioredoxin Reductase

Thioredoxin reductase, coupled with thioredoxin and NADPH, is an efficient general protein disulfide reductase (see Ref. 33 for review). Mammalian thioredoxin reductase is a selenoprotein[34] and has a much broader substrate specificity than its *Escherichia coli* counterpart, including a number of organoselenium compounds (e.g. selenocystine[35]). Mammalian thioredoxin reductase can function in the reduction of peroxynitrite by selenocysteine or ebselen,[18] maintaining these compounds in a catalytic cycle at the expense of NADPH. Upon administration, extracellular ebselen is present in human plasma as an albumin complex.[36] Thioredoxin reductase, coupled with thioredoxin and NADPH, can also reduce the selenenyl sulfide complex of BSA-ebselen and release free ebselen.[18]

6. Potential Protection Against Aging by Pharmacologic/ Nutritive Supplementation with Organoselenium Compounds

6.1. Ebselen and Other GPx Mimics

In addition to being an antioxidant, ebselen exhibits antiinflammatory actions in a number of experimental models, which may or may not be attributed to its antioxidant effects.[15, 37, 38] Ebselen has also been shown to be protective in other *in vivo* models involving oxidative stress, such as heart ischemia-reperfusion injury,[39] ozone-induced pulmonary inflammation,[40] focal ischemic stroke,[41] and alcoholic liver injury.[42] Indeed, the compound has been developed for clinical use

(Phase III), as recently reviewed.[43] The indications addressed were ischemic stroke.[19, 44, 45] Unlike inorganic selenium and selenomethionine, ebselen is not toxic to mammals, most likely because it contains bound selenium. It should be mentioned that under pathophysiological conditions other reactivities of organoselenium compounds can also play important roles as well. Indeed, GPx mimics have been described as *in vivo* immune response modifiers.[46] For example, ICAM-1 and VCAM-1 expression induced by TNF-α were found to be inhibited,[47] as were TNF-α and neutrophil-induced endothelial alterations[48] and TNF-mediated apoptosis[49] by GPx mimics.

6.2. Selenium Supplementation

The concept of selenium as an essential mammalian micronutrient dates back to the 1950s;[50] prior to this study, the toxicity of selenium was the major source of research interest and is reviewed elsewhere.[23, 51] The use of nutritive supplementation in cases of regional selenium deficiency or as a pharmacologic supplementation as a cancer preventative has received recent attention.[52-54] Further, deficits in micronutrients, such as selenium, occur often in the aged population,[55] and may play a role neurodegeneration disorders associated with aging (e.g. Parkinson's disease and multiple sclerosis).[56-60] Therefore, selenium supplementation may be a protective therapy under these conditions.

Acknowledgments

Support by the Deutsche Forschungsgemeinschaft, SFB 503, Project B1, and by the National Foundation for Cancer Research (NFCR), Bethesda, M.D., is gratefully acknowledged. Helmut Sies is a Fellow of the NFCR.

References

1. Sies, H. (1985). Oxidative stress: introductory remarks. *In* "Oxidative Stress" (H. Sies, Ed.), pp. 1–8, Academic Press, London.
2. Sies, H. (1986). Biochemistry of oxidative stress. *Angew. Chem. Int. Ed. England* **25**: 1058–1071.
3. Sies, H. (1993). Strategies of antioxidant defense. *Eur. J. Biochem.* **215**: 213–219.
4. Behne, D., Weiss-Nowak, C., Kalcklosch, M., Westphal, C., Gessner, H. and Kyriakopoulos, A. (1994). Application of nuclear analytical methods in the investigation and identification of new selenoproteins. *Biol. Trace Elem. Res.* **43–45**: 287–297.

5. Brigelius-Flohé, R. (1999). Tissue-specific functions of individual glutathione peroxidases. *Free Radic. Biol. Med.* **27**: 951–965.

6. Mills, G. C. (1957). Hemoglobin catabolism. I. Glutathione peroxidase, an erythrocyte enzyme which protects hemoglobin from oxidase breakdown. *J. Biol. Chem.* **229**: 189–197.

7. Flohé, L., Günzler, W. A. and Schock, H. H. (1973). Glutathione peroxidase: a selenoenzyme. *FEBS Lett.* **32**: 132–134.

8. Rotruck, J. T., Pope, A. L., Ganther, H. E., Swanson, A. B., Hafeman, D. G. and Hoekstra, W. G. (1973). Selenium: biochemical role as a component of glutathione peroxidase. *Science* **179**: 588–590.

9. Ursini, F., Maiorino, M. and Gregolin, C. (1985). The selenoenzyme phospholipid hydroperoxide glutathione peroxidase. *Biochim. Biophys. Acta* **839**: 62–70.

10. Flohé, L., Loschen, G., Günzler, W. A. and Eichele, E. (1972). Glutathione peroxidase, V. The kinetic mechanism. *Hoppe Seylers. Z. Physiol. Chem.* **353**: 987–999.

11. Beckman, J. S. (1996). The physiological and pathophysiological chemistry of nitric oxide. *In* "Nitric Oxide: Principles and Actions" (J. Lancaster, Ed.), pp. 1–82, Academic Press, San Diego, California.

12. Zhu, L., Gunn, C. and Beckman, J. S. (1992). Bactericidal activity of peroxynitrite. *Arch. Biochem. Biophys.* **298**: 452–457.

13. Arteel, G. E., Briviba, K. and Sies, H. (1999). Protection against peroxynitrite. *FEBS Lett.* **445**: 226–230.

14. Stadtman, E. R. and Berlett, B. S. (1998). Reactive oxygen-mediated protein oxidation in aging and disease. *Drug Metab Rev.* **30**: 225–243.

15. Sies, H. (1993). Ebselen, a selenoorganic compound as glutathione peroxidase mimic. *Free Radic. Biol. Med.* **14**: 313–323.

16. Masumoto, H. and Sies, H. (1996). The reaction of ebselen with peroxynitrite. *Chem. Res. Toxic.* **9**: 262–267.

17. Masumoto, H., Kissner, R., Koppenol, W. H. and Sies, H. (1996). Kinetic study of the reaction of ebselen with peroxynitrite. *FEBS Lett.* **398**: 179–182.

18. Arteel, G. E., Briviba, K. and Sies, H. (1999). Function of thioredoxin reductase as a peroxynitrite reductase using selenocystine or ebselen. *Chem. Res. Toxicol.* **12**: 264–269.

19. Sies, H. and Masumoto, H. (1997). Ebselen as a glutathione peroxidase mimic and as a scavenger of peroxynitrite. *Adv. Pharmacol.* **38**: 229–246.

20. Padmaja, S., Squadrito, G. L., Lemercier, J. N., Cueto, R. and Pryor, W. A. (1996). Rapid oxidation of DL-selenomethionine by peroxynitrite. *Free Radic. Biol. Med.* **21**: 317–322.

21. Hondal, R. J., Motley, A. K., Hill, K. E. and Burk, R. F. (1999). Failure of selenomethionine residues in albumin and immunoglobulin G to protect against peroxynitrite. *Arch. Biochem. Biophys.* **371**: 29–34.

22. Stadtman, T. C. (1996). Selenocysteine. *Ann. Rev. Biochem.* **65**: 83–100.
23. Schrauzer, G. N. (2000). Selenomethionine: a review of its nutritional significance, metabolism and toxicity. *J. Nutr.* **130**: 1653–1656.
24. Sies, H., Sharov, V. S., Klotz, L. O. and Briviba, K. (1997). Glutathione peroxidase protects against peroxynitrite-mediated oxidations. A new function for selenoproteins as peroxynitrite reductase. *J. Biol. Chem.* **272**: 27 812–27 817.
25. Briviba, K., Kissner, R., Koppenol, W. H. and Sies, H. (1998). Kinetic study of the reaction of glutathione peroxidase with peroxynitrite. *Chem. Res. Toxicol.* **11**: 1398–1401.
26. Padmaja, S., Squadrito, G. L. and Pryor, W. A. (1998). Inactivation of glutathione peroxidase by peroxynitrite. *Arch. Biochem. Biophys.* **349**: 1–6.
27. Asahi, M., Fujii, J., Suzuki, K., Seo, H. G., Kuzuya, T., Hori, M., Tada, M., Fujii, S. and Taniguchi, N. (1995). Inactivation of glutathione peroxidase by nitric oxide. Implication for cytotoxicity. *J. Biol. Chem.* **270**: 21 035–21 039.
28. Schieke, S. M., Briviba, K., Klotz, L. O. and Sies, H. (1999). Activation pattern of mitogen-activated protein kinases elicited by peroxynitrite: attenuation by selenite supplementation. *FEBS Lett.* **448**: 301–303.
29. Go, Y. M., Patel, R. P., Maland, M. C., Park, H., Beckman, J. S., Darley-Usmar, V. M. and Jo, H. (1999). Evidence for peroxynitrite as a signaling molecule in flow-dependent activation of c-Jun NH(2)-terminal kinase. *Am. J. Physiol.* **277**: H1647–H1653.
30. Arteel, G. E., Mostert, V., Oubrahim, H., Briviba, K., Abel, J. and Sies, H. (1998). Protection by selenoprotein P in human plasma against peroxynitrite-mediated oxidation and nitration. *Biol. Chem.* **379**: 1201–1205.
31. Wilson, D. S. and Tappel, A. L. (1993). Binding of plasma selenoprotein P to cell membranes. *J. Inorg. Biochem.* **51**: 707–714.
32. Arteel, G. E., Franken, S., Kappler, J. and Sies, H. (2000). Binding of selenoprotein P to heparin: characterization with surface plasmon resonance. *Biol. Chem.* **381**: 265–268.
33. Bjornstedt, M., Kumar, S., Bjorkhem, L., Spyrou, G. and Holmgren, A. (1997). Selenium and the thioredoxin and glutaredoxin systems. *Biomed. Env. Sci.* **10**: 271–279.
34. Tamura, T. and Stadtman, T. C. (1996). A new selenoprotein from human lung adenocarcinoma cells: purification, properties, and thioredoxin reductase activity. *Proc. Natl. Acad. Sci. USA* **93**: 1006–1011.
35. Bjornstedt, M., Hamberg, M., Kumar, S., Xue, J. and Holmgren, A. (1995). Human thioredoxin reductase directly reduces lipid hydroperoxides by NADPH and selenocystine strongly stimulates the reaction via catalytically generated selenols. *J. Biol. Chem.* **270**: 11 761–11 764.
36. Wagner, G., Schuch, G., Akerboom, T. P. and Sies, H. (1994). Transport of ebselen in plasma and its transfer to binding sites in the hepatocyte. *Biochem. Pharmacol.* **48**: 1137–1144.

37. Parnham, M. J., Leyck, S., Kuhl, P., Schalkwijk, J. and van den Berg, W. B. (1987). Ebselen: a new approach to the inhibition of peroxide-dependent inflammation. *Int. J. Tissue React.* **9**: 45–50.

38. Schewe, T. (1995). Molecular actions of ebselen — an antiinflammatory antioxidant. *Gen. Pharmacol.* **26**: 1153–1169.

39. Maulik, N. and Yoshida, T. (2000). Oxidative stress developed during open heart surgery induces apoptosis: reduction of apoptotic cell death by ebselen, a glutathione peroxidase mimic. *J. Cardiovasc. Pharmacol.* **36**: 601–608.

40. Ishii, Y., Hashimoto, K., Hirano, K., Morishima, Y., Mochizuki, M., Masuyama, K., Nomura, A., Sakamoto, T., Uchida, Y., Sagai, M. and Sekizawa, K. (2000). Ebselen decreases ozone-induced pulmonary inflammation in rats. *Lung* **178**: 225–234.

41. Gladilin, S., Bidmon, H. J., Divanach, A., Arteel, G. E., Witte, O. W., Zilles, K. and Sies, H. (2000). Ebselen lowers plasma interleukin-6 levels and glial heme oxygenase-1 expression after focal photothrombotic brain ischemia. *Arch. Biochem. Biophys.* **380**: 237–242.

42. Kono H., Arteel G. E., Rusyn I., Sies H. and Thurman R. G. (2000) Ebselen prevents early alcohol-induced liver injury in rats. *Free Radic. Biol. Med.* **in press**.

43. Parnham, M. J. and Sies, H. (2000). Ebselen: prospective therapy for cerebral ischaemia. *Exp. Opin. Invest. Drugs* **9**: 607–619.

44. Yamaguchi, T., Sano, K., Takakura, K., Saito, I., Shinohara, Y., Asano, T. and Yasuhara, H. (1998). Ebselen in acute ischemic stroke: a placebo-controlled, double-blind clinical trial. Ebselen Study Group. *Stroke* **29**: 12–17.

45. Ogawa, A., Yoshimoto, T., Kikuchi, H., Sano, K., Saito, I., Yamaguchi, T. and Yasuhara, H. (1999). Ebselen in acute middle cerebral artery occlusion: a placebo-controlled, double-blind clinical trial. *Cerebrovasc. Dis.* **9**: 112–118.

46. Wendel, A., Kuesters, S. and Tiegs, G. (1997). Ebselen — an in vivo immune response modifier. *Biomed. Env. Sci.* **10**: 253–259.

47. d'Alessio, P., Moutet, M., Coudrier, E., Darquenne, S. and Chaudiere, J. (1998). ICAM-1 and VCAM-1 expression induced by TNF-α are inhibited by a glutathione peroxidase mimic. *Free Radic. Biol. Med.* **24**: 979–987.

48. Moutet, M., d'Alessio, P., Malette, P., Devaux, V. and Chaudiere, J. (1998). Glutathione peroxidase mimics prevent TNF-alpha and neutrophil-induced endothelial alterations. *Free Radic. Biol. Med.* **25**: 270–281.

49. Tiegs, G., Kusters, S., Kunstle, G., Hentze, H., Kiemer, A. K. and Wendel, A. (1998). Ebselen protects mice against T cell-dependent, TNF-mediated apoptotic liver injury. *J. Pharmacol. Exp. Ther.* **287**: 1098–1104.

50. Schwarz, K. and Foltz, C. M. (1957). Selenium as an intergral part of factor-3 against dietary necrotic liver degeneration. *J. Am. Chem. Soc.* **79**: 3292–3293.

51. Buell, D. N. (1983). Potential hazards of selenium as a chemopreventive agent. *Sem. Oncol.* **10**: 311–321.

52. Beck, M. A. and Levander, O. A. (1998). Dietary oxidative stress and the potentiation of viral infection. *Ann. Rev. Nutr.* **18**: 93–116.

53. Ganther, H. E. (1999). Selenium metabolism, selenoproteins and mechanisms of cancer prevention: complexities with thioredoxin reductase. *Carcinogenesis* **20**: 1657–1666.

54. Rayman, M. P. (2000). The importance of selenium to human health. *Lancet* **356**: 233–241.

55. Lesourd, B. M. (1997). Nutrition and immunity in the elderly: modification of immune responses with nutritional treatments. *Am. J. Clin. Nutr.* **66**: 478S–484S.

56. Chazot, G. and Broussolle, E. (1993). Alterations in trace elements during brain aging and in Alzheimer's dementia. *Prog. Clin. Biol. Res.* **380**: 269–281.

57. Emard, J. F., Thouez, J. P. and Gauvreau, D. (1995). Neurodegenerative diseases and risk factors: a literature review. *Soc. Sci. Med.* **40**: 847–858.

58. Johannsen, P., Velander, G., Mai, J., Thorling, E. B. and Dupont, E. (1991). Glutathione peroxidase in early and advanced Parkinson's disease. *J. Neurol. Neurosurg. Psychiatry* **54**: 679–682.

59. Clausen, J., Jensen, G. E. and Nielsen, S. A. (1988). Selenium in chronic neurologic diseases. Multiple sclerosis and Batten's disease. *Biol. Trace Elem. Res.* **15**: 179–203.

60. Johnson, S. (2000). The possible role of gradual accumulation of copper, cadmium, lead and iron and gradual depletion of zinc, magnesium, selenium, vitamins B2, B6, D, and E and essential fatty acids in multiple sclerosis. *Med. Hypotheses* **55**: 239–241.

61. Briviba, K., Roussyn, I., Sharov, V. S. and Sies, H. (1996). Attenuation of oxidation and nitration reactions of peroxynitrite by selenomethionine, selenocystine and ebselen. *Biochem. J.* **319**: 13–15.

62. Masumoto, H. and Sies, H. (1996). The reaction of 2-(methylseleno)benzanilide with peroxynitrite. *Chem. Res. Toxicol.* **9**: 1057–1062.

63. Pryor, W. A., Jin, X. and Squadrito, G. L. (1994). One- and two-electron oxidations of methionine by peroxynitrite. *Proc. Natl. Acad. Sci. USA* **91**: 11 173–11 177.

64. Lee, J. L., Hunt, J. A. and Groves, J. T. (1997). Rapid decomposition of peroxynitrite by manganese poryphyrin-antioxidant redox couples. *Bioorg. Med. Chem. Lett.* **7**: 2913–2918.

65. Radi, R., Beckman, J. S., Bush, K. M. and Freeman, B. A. (1991). Peroxynitrite oxidation of sulfhydryls. The cytotoxic potential of superoxide and nitric oxide. *J. Biol. Chem.* **266**: 4244–4250.

Chapter 71

Salen Manganese Complexes, Combined Superoxide Dismutase/Catalase Mimetics, Demonstrate Potential for Treating Neurodegenerative and Other Age-Associated Diseases

Susan R. Doctrow*, Christy Adinolfi, Michel Baudry, Karl Huffman, Bernard Malfroy, Catherine Bucay Marcus, Simon Melov, Kevin Pong, Yongqi Rong, Janet L. Smart and Georges Tocco

Susan R. Doctrow, Christy Adinolfi, Karl Huffman, Bernard Malfroy, Catherine Bucay Marcus, Janet L. Smart and Georges Tocco • Eukarion, Inc., Bedford, MA 01730
Simon Melov • The Buck Institute for Age Research, Novato, CA 94949
Michel Baudry, Kevin Pong and Yongqi Rong • Neuroscience Program, University of Southern CA, Los Angeles, CA 90089

*Corresponding Author, Eukarion, Inc., 6F Alfred Circle, Bedford, MA 01730
Tel: 781-275-0424, Ext: 12, E-mail: s.doctrow@eukarion.com

1. Summary

Reactive oxygen species (ROS) are implicated in numerous pathological processes, including inflammatory, ischemic, and neurodegenerative diseases. Salen manganese complexes are low molecular weight synthetic compounds that have both superoxide dismutase and catalase activities, thus mimicking two enzymes involved in normal antioxidant defense. The ability to catalytically scavenge multiple ROS, including superoxide and hydrogen peroxide, makes salen manganese complexes potentially valuable for the broad range of diseases in which oxidative stress is implicated. The compounds protect cultured cells, including primary neurons, from several types of oxidative insults. Furthermore, they are protective in a variety of *in vivo* models for ROS-associated diseases, including rodent models for stroke and excitotoxic neuronal death and for neurodegeneration resulting from mitochondrial oxidative stress. Recently, the compounds have also been shown to extend the lifespan of a mulicellular organism, the nematode *Caenorhabditis elegans*. In general, the results obtained with the salen manganese complexes support the concept that use of a multifunctional catalytic antioxidant may serve as a very effective therapeutic approach applicable to a broad range of diseases and, possibly, to the enhancement of health during the aging process.

2. Introduction: Reactive Oxygen Species and Disease

Oxygen, although essential for aerobic life, can be converted to highly reactive species, including superoxide ion, hydrogen peroxide, and hydroxyl radical, known collectively as reactive oxygen species (ROS). Increased ROS formation under pathological conditions is believed to cause cellular damage through chemical interactions with proteins, lipids, and DNA (reviewed in Ref. 1). In inflammatory states, ROS can be formed via a number of pathways, including the NADPH oxidase system in activated polymorphonuclear neutrophils and monocyte/ macrophages.[2, 3] ROS have also been implicated as key mediators of tissue injury after ischemia and reperfusion.[4, 5] Endogenous ROS scavengers such as the enzymes superoxide dismutases (SOD) and catalase, which catalytically destroy superoxide ion and hydrogen peroxide, respectively, provide some degree of tissue protection. Also, several molecules synthesized endogenously or acquired through diet, such as ascorbate (vitamin C), alpha-tocopherol (vitamin E), and glutathione have antioxidant properties. Presumably, however, when ROS generation exceeds the capacity of endogenous scavengers to neutralize them, tissues become vulnerable to damage, a condition often referred to as "oxidative stress".[6] Consequently, antioxidant therapies are being investigated for a wide variety of disorders in which oxidative stress is believed to play a significant role in the associated pathophysiologies (reviewed in Ref. 7).

3. Synthetic Superoxide Dismutase Mimetics as Potential Therapeutics

Various studies have investigated the clinical efficacy of SOD. Locally-injected SOD has, for example, shown some promise in rheumatoid arthritis[1, 8] and radiation protection.[8, 9] However, in many other instances, the results of clinical and other *in vivo* studies have been disappointing because, as a protein, SOD has a number of delivery and stability shortcomings. Efforts to improve the pharmacokinetic properties of SOD have included encapsulation into liposomes or conjugation to high molecular weight polyethylene glycol. Because of the equivocal results obtained with SOD, there has been an interest in developing synthetic SOD mimetics that should have more favorable pharmaceutical properties such as, for example, cost-effectiveness, stability, and oral availability, than a protein such as SOD. The first low molecular weight SOD mimetics to be described were copper complexes that showed anti-tumor promoting activity in mice.[10, 11] More recently, several manganese complexes purported to mimic manganese-SOD have been described. The complex formed between manganese and the chelator desferrioxamine has SOD activity[12] and has shown some protective activity against kainate-induced neurotoxicity in the rat.[13] Salen manganese complexes,[14] the focus of the present review, and porphyrin-manganese complexes[15, 16] have been reported to have SOD activity as well as more favorable stability properties than the earlier complexes. The porphyrin-manganese complexes have shown biological efficacy, for example, in an acute lung injury model.[17] A class of manganese macrocyclic ligand complexes with SOD activity has also been described[18] with later generation analogs showing enhanced catalytic activity and efficacy in inflammatory models[19] and in a septic shock model.[20]

4. Salen Manganese Complexes are Catalytic Scavengers of Multiple Types of Reactive Oxygen Species, Providing Advantages Over Superoxide Dismutase

As mentioned above, we have previously reported that synthetic salen-manganese complexes have SOD activity[14] and subsequently reported that these compounds have catalase activity as well, catalytically converting hydrogen peroxide to oxygen.[21, 22] In this regard, the complexes may mimic properties of bacterial manganese-containing catalases, so called "pseudocatalases".[1, 23] Consistent with their catalase activity, the compounds can also carry out per-oxidative reactions, an alternative means of scavenging hydrogen peroxide in the presence of a suitable substrate.[22] The manganese porphyrins also exhibit catalase/peroxidase activities, as well as the ability to scavenge peroxynitrite.[16, 24] Evidence indicates that this latter property may also be shared by the salen manganese complexes (Tocco, Adinolfi, and Doctrow, unpublished data; Refs. 25 and 26). The macrocyclic manganese complexes do not show activity against hydrogen peroxide or other ROS, and have hence been described as highly selective SOD mimetics useful for probing the role of superoxide in biological processes.[19]

This review will focus only on the salen manganese complexes, which we believe have several properties that might facilitate their potential usefulness as therapeutic agents. First, as low molecular weight synthetic molecules they have the potential advantages mentioned above with regard to cost-effectiveness, pharmaceutical formulation and delivery. Second, they act catalytically, presumably enhancing their efficiency over noncatalytic ROS scavengers such as vitamin E. Third, their ability to destroy superoxide anion, hydrogen peroxide and potentially other ROS should enhance their ability to protect tissues in various disease states involving the production of multiple ROS species. Furthermore, while the SOD enzymes very efficiently destroy superoxide, in doing so they produce hydrogen peroxide, a toxic and longer-lived ROS. Thus, in certain circumstances, SOD might itself be damaging, while compounds with both SOD and catalase activities, such as the salen manganese complexes, would not. For example, in rat brain slices subjected to anoxia-reoxygenation, bovine Cu, Zn SOD exacerbated synaptic damage[22] while EUK-8, a prototype salen manganese complex was protective.[22, 27] In another example, EUK-8 attenuated permeability increases in Caco-2 intestinal cell monolayers subjected to lactic acidosis, which induces a cellular oxidative stress as evidenced by iron delocalization, lipid peroxidation and protein oxidation. In this system, Cu, Zn SOD was either ineffective or deleterious, while bovine liver catalase, alone or in combination with the SOD, was protective to the same degree as EUK-8.[28]

5. EUK-8, a Prototype Salen Manganese Complex with Efficacy in a Broad Range of Disease Models

The salen manganese complexes have been found to be protective in a variety of models for ROS-associated disease. The earliest pharmacology studies (reviewed in Ref. 22) used a prototype salen manganese complex, EUK-8, whose SOD activity was first described by Baudry *et al.*[14] In a stringent porcine endotoxemia model for the Adult Respiratory Distress Syndrome, EUK-8 infused intravenously prevented acute lung injury, as assessed by several functional parameters including dynamic lung compliance and arterial oxygenation, and abrogated lung lipid peroxidation and edema.[21, 29] In murine experimental allergic encephalomyelitis (EAE), a model for multiple sclerosis, EUK-8 administered intraperitoneally prevented the development of severe paralysis.[30] In another model for chronic autoimmune disease, the *mrl/lpr* mouse, which develops a syndrome similar to human systemic lupus erythematosus and dies of glomerulonephrosis,[31] EUK-8 treatment prolonged survival (Fig. 1) and delayed the onset of proteinuria (data not shown). EUK-8 protected the perfused iron-overloaded rat heart from ischemia-reperfusion injury[32, 33] and, as discussed above, preserved synaptic

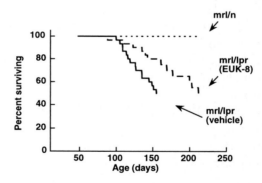

Fig. 1. Effect of EUK-8 treatment on survival of *mrl/lpr* mice. *Mrl/lpr* mice, which develop chronic autoimmune disease and *mrl/n* mice, a nonautoimmune control strain[31] were treated three times weekly with EUK-8 (1 mg) intraperitoneally, beginning at the age of 8 weeks. The dose in adult mice corresponded to approximately 25 mg/kg. Survival data were collected for each group until median lifespan was reached. Animals surviving beyond median lifespan were censored. Group sizes were 30 for mrl/lpr mice and 16 for *mrl/n* mice. There were no deaths in either group of *mrl/n* mice (vehicle and EUK-8 treated groups are pooled in the figure). EUK-8 did not affect lymphoproliferation in the mrl/lpr mice, as assessed by spleen weights. Furthermore, there were no differences in weight gain among the four groups, and no histopathological abnormalities attributed to chronic EUK-8 treatment. The enhancement of survival of mrl/lpr mice treated with EUK-8 as compared to vehicle was statistically significant (Log-Rank and Wilcoxon tests, $p < 0.05$).

function in rat hippocampal slices subjected to anoxia-reoxygenation.[27] In organotypic hippocampal slice cultures, EUK-8 decreased oxidative stress and inhibited cell death caused by beta-amyloid peptides.[34] This review will now focus on more recent pharmacological studies, with particular emphasis on neurodegenerative diseases and aging.

6. EUK-134, an EUK-8 Analog with Improved Catalase and Cytoprotective Activities *In Vitro*

Structural modifications of EUK-8 have led to the development of salen manganese complexes with improved properties. The structures of EUK-8 and EUK-134, another salen manganese complex, are shown in Fig. 2. These compounds exhibit equivalent SOD activities,[35] but EUK-134 is a more active catalase [Fig. 3(A)]. EUK-134 completely protects cultured human fibroblasts from hydrogen peroxide-induced toxicity, while EUK-8 is nearly ineffective [Fig. 3(B)]. In contrast, both compounds effectively protect PC12 cells from toxicity elicited by sin-1, with EUK-134 being more potent [Fig. 3(C)]. Since sin-1 generates superoxide and nitric oxide,[36] the cytoprotective activity of EUK-8 and EUK-134 may be related to their SOD activities. However, it may also involve direct interactions with peroxynitrite, the product of these two ROS.[37] In support of the latter mechanism, both compounds protect human umbilical vein endothelial cells from peroxynitrite-induced toxicity.[25] It is clear that the relative activities of these and other salen manganese complexes in various disease models will depend upon numerous factors, including which ROS cause the pathology of interest. Similar to their ability to protect cell lines in models illustrated here, the salen manganese complexes also protect primary neuronal cultures against various forms of oxidative injury. They protect rat cortical neurons against hydrogen peroxide and sin-1 toxicity[26] and rat primary dopaminergic neurons against toxicity by MPTP and 6-hydroxydopamine, two agents believed to mimic the neuropathology of Parkinson's disease *in vivo*.[38] In the latter study,

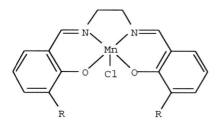

EUK-8: R = H
EUK-134: R = OCH$_3$

Fig. 2. Structures of EUK-8 and EUK-134. (Reprinted from Ref. 35 with permission from the publisher.)

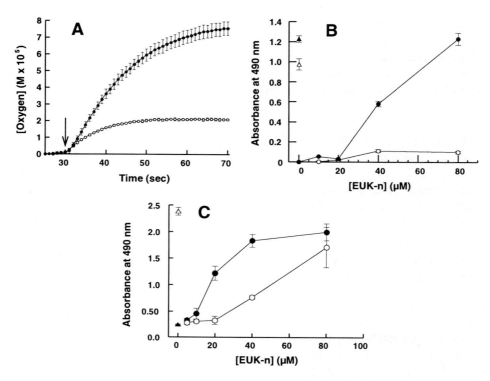

Fig. 3. Comparative activities of EUK-8 (open circles) and EUK-134 (solid circles). (A) Catalase activity, assayed by monitoring the conversion of hydrogen peroxide (10 mM, added at arrow) to oxygen, using a polarographic oxygen electrode as described previously.[21] EUK-8 or EUK-134 was present at 10 μM. (Reprinted from Ref. 35 with permission from the publisher.) (B) Cytoprotective activity against hydrogen peroxide, generated using glucose and glucose oxidase. Human skin fibroblasts were incubated for 18 h with: medium without glucose oxidase (open triangle); medium with glucose oxidase and 5 μg/ml bovine liver catalase (solid triangle); medium with glucose oxidase and EUK-8 or EUK-134, as indicated. Cell viability was assessed using the XTT assay. Further experimental details are given in Ref. 35. (Reprinted from Ref. 35, with permission from the publisher.) (C) Cytoprotective activity against sin-1 toxicity. PC12 cells were incubated with no additions (open triangle), 1 mM sin-1 (closed triangle) or 1 mM sin-1 plus EUK-8 or EUK-134, as indicated (open, solid circles, respectively). Cell viability was assessed after 18 h as described for (B).

both MPP[+] and 6-hydrogydopamine increased tyrosine nitration of tyrosine hydroxylase in the dopaminergic neuron cultures. This protein tyrosine nitration, a form of cellular injury attributed to peroxynitrite,[39] was completely prevented by EUK-134 (Fig. 4). Taken together, these observations support the potential value of salen manganese complexes in treating neurological disorders, as has been further confirmed by several recent *in vivo* studies, summarized below.

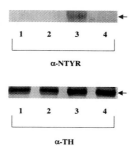

Fig. 4. Tyrosine nitration of tyrosine hydroxylase in neurons treated with MPP⁺. Primary cultures of rat dopaminergic neurons were pre-treated with EUK-134 (0.5 µM) for 24 h and with the neurotoxin MPP⁺ (10 µM) for 3 hours. Cells were lysed, extracted, and processed for immunoprecipitation of tyrosine hydroxylase and Western blot analysis for 3-nitrotyrosine (A) or tyrosine hydroxylase (B). Experimental details are given in Ref. 38. Lanes 1, control; 2, EUK-134; 3, MPP⁺ 4, MPP⁺ and EUK-134. Similar results were obtained in neurons treated with 6-hydroxydopamine.[38] (Reprinted from Ref. 38, with permission from the publisher.)

7. Salen Manganese Complexes are Neuroprotective *In Vivo* in Models of Stroke and Seizure-Induced Excitotoxicity

Salen manganese complexes are neuroprotective in rodent stroke models involving focal ischemia induced by permanent arterial occlusion. In such experimental models, extensive brain infarction occurs within several hours of the occlusion. Besides the small core necrotic brain region whose blood flow is irreversibly disrupted by the occlusion, most of the infarct volume consists of a "penumbral" region, in which cell death is believed to result from excitotoxic and inflammatory processes and, more recently, apoptotic processes as well.[40] This penumbral infarction is presumed to be amenable to therapeutic intervention with neuroprotective agents capable of interfering with these injury cascades.[41] In inflammation, activated leukocytes secrete damaging mediators including superoxide and other ROS.[42,43] There is also evidence that stimulation with excitatory amino acids causes superoxide production in neurons.[44,45] Thus, there is reason to believe that an ROS scavenging compound would have neuroprotective efficacy in a focal brain ischemia model. In one study,[35] EUK-8 and EUK-134 were administered to rats 3 h after permanent occlusion of the left middle cerebral artery. Both compounds reduced brain infarction, based on an assessment of brain infarct volumes 21 h after occlusion. As shown in Fig. 5(A), EUK-134 was more efficacious than EUK-8. The protection by EUK-134 persisted, even when infarcts were measured 72 h after occlusion [Fig. 5(B)], suggesting that the compound prevented, rather than delayed, the formation of the infarct.

More recently, the effects of EUK-134 were examined in a another type of *in vivo* model for excitotoxic neuronal death. This model involved systemic

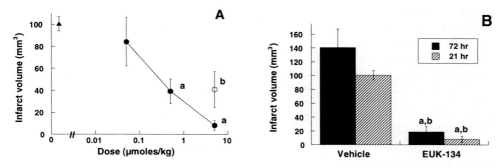

Fig. 5. Activity of EUK-8 and EUK-134 in a rat stroke model. Rats were subjected to permanent middle cerebral artery occlusion (MCAo), treated intravenously with EUK-8, EUK-134 or saline vehicle, and brain infarcts were analyzed as described previously.[35] Statistical analysis was by ANOVA followed by Dunnett's *t* test. (A) Brain infarct volumes, measured at 21 h post-MCAo in rats receiving vehicle (triangle), EUK-8 (open circle) or EUK-134 (solid circles). Statistical significance: a, $p < 0.01$; b, $p < 0.05$. (B) Brain infarct volumes, measured at 21 or 72 h post MCAo in rats treated with vehicle or EUK-134 (5 µm/kg), as indicated. Statistical significance: a, versus vehicle-treated group killed at 21 h ($p < 0.01$); b, versus vehicle treated group killed at 72 h ($p < 0.01$). The two vehicle-treated groups did not differ from one another, nor did the two EUK-134-treated groups ($p > 0.05$). (Both figures reprinted from Ref. 35, with permission from the publisher.)

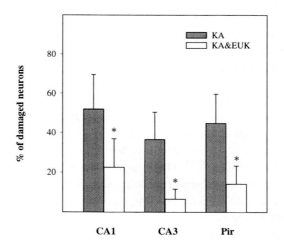

Fig. 6. Effects of EUK-134 on neuronal death in rats treated with kainic acid (KA). Rats were treated systemically with KA (10 mg/kg) and also received injections of saline vehicle or EUK-134 (5 mg/kg) as described previously.[48] Brain sections, obtained 24 h after KA administration, were processed for hematoxylin/eosin staining and analyzed to assess neuronal viability in hippocampus (CA1 and CA3 regions) and piriform cortex. Experimental details are given in Ref. 48. (*, $p < 0.01$.)

administration of kainic acid (KA), a glutamate analog, to rats. Under these conditions, KA causes repeated episodes of seizure activity lasting 4–6 h, resulting in death of neurons in limbic structures by what is believed to be an excitotoxic injury.[46, 47] EUK-134 did not affect seizure activity in KA-treated rats, but prevented the resulting neuronal death, while inhibiting several indices of oxidative stress.[48] Figure 6 shows damaged neurons, assessed by hematoxylin/eosin staining, in the hippocampus (CA1 and CA3) and piriform cortex of KA-treated rats with and without EUK-134. In all three regions, EUK-134 treatment significantly reduced the percent of damaged neurons. Two transcriptional factors known to be responsive to oxidative and other cellular stresses, AP-1 and NF-kappa B, were activated at 16 h after KA administration (Fig. 7). Activation of both transcriptional

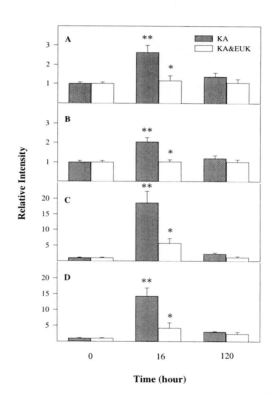

Time (hour)

Fig. 7. Effects of EUK-134 on transcriptional factor activation in KA-treated rats. Rats were treated as described for Fig. 6, but brain region samples were obtained at 16 and 120 h after KA administration and active levels of NF-κB and AP-1 were determined by gel shift assay and image analysis, as described in more detail in Ref. 48. The DNA-binding activities of NF-kappa B (A and B) and AP-1 (C and D) were quantified in piriform cortex (A and C) and hippocampus (B and D). Data are expressed as ratios between values found in control and treated animals and represent the means ($\pm$ sd) of 6 animals. Statistical significance: **, KA versus control treatment; * versus KA versus KA plus EUK-134, $p < 0.05$ (ANOVA followed by Tukey's test).

CA1   CA3

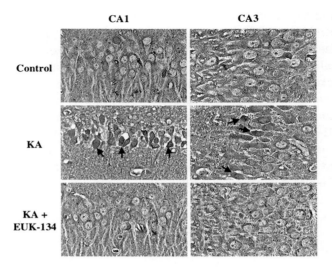

Control

KA

KA +
EUK-134

Fig. 8. Effect of EUK-134 on protein nitration in KA-treated rats. Rats were treated as described for Fig. 6, but were killed 8 h after KA treatment and brains were processed for immunohistochemistry for 3-nitrotyrosine. Further experimental details are given in Ref. 48. The figure shows nitrotyrosine staining in hippocampal regions CA1 and CA3 from rats treated as indicated. (Reprinted from Ref. 48, with permission from the publisher.)

factors was significantly reduced in rats treated with EUK-134. This study also examined protein tyrosine nitration, an indication of peroxynitrite-mediated injury, within hippocampal regions (CA1 and CA3) by immunochemical staining for 3-nitrotyrosine (Fig. 8). Relative to controls, rats treated with KA showed positive staining indicative of protein tyrosine nitration in neurons. In rats treated with KA and EUK-134, this staining was almost completely absent, indicating that EUK-134 prevented protein tyrosine nitration *in vivo*. As discussed above, this inhibition may be due to the prevention of peroxynitrite formation by the SOD activity of EUK-134, and/or by direct scavenging of peroxynitrite.

8. Salen Manganese Complexes Rescue a Spongiform Neurodegenerative Disorder Associated with Mitochondrial Oxidative Stress in Mice Lacking MnSOD

The two studies presented above illustrate the efficacy of salen manganese complexes in neuronal injury models where a role for ROS has been hypothesized. More recently, the compounds have been tested in a genetic mouse model designed to promote increased oxidative stress, specifically, to the mitochondria. Mice lacking Cu, Zn SOD, the cytosolic form of SOD, have a normal lifespan and exhibit a relatively mild phenotype, including, for example, exacerbated brain injury in a stroke model.[49] In contrast, mice lacking MnSOD, the mitochondrial form of SOD (SOD2), exhibit an extremely severe phenotype. For example, on a CD1 background, lack of SOD2 results in death within the first week of life, apparently of a dilated cardiomyopathy.[50] These observations are particularly compelling in consideration of the fact that Cu, Zn SOD is the predominant form of the enzyme, accounting for well over 90% of cellular SOD levels. Thus, it seems apparent that oxidative stress to the mitochondrion, the major site of ROS generation in the cell, has profound consequences.

Previously, the strain of mice lacking SOD2 on a CD1 background (*sod2−/−* mice) were treated with a Mn-containing porphyrin known as MnTBAP, which has been reported to have SOD activity as well as other ROS-scavenging properties.[16] Treatment of the *sod2−/−* mice with MnTBAP was found to double their lifespan (mean ± sd of 8 ± 4 and 16 ± 6 for untreated and MnTBAP treated mice, respectively).[51] Furthermore, the treated mice did not develop dilated cardiomyopathy, or other phenotypes associated with mitochondrial oxidative stress, such as hepatic lipid accumulation, indicating that MnTBAP rescued these manifestations of the *sod2−/−* phenotype. However, the treated mice developed a fatal neurological phenotype not previously seen in this genetic background. Essentially, this involved a profound movement disorder and widespread spongiform brain pathology, with severe vacuolization in several regions including cortex and brainstem.[51] Since MnTBAP was known not to cross the blood-brain barrier, these data were interpreted as indicating that rescue of the peripheral neonatal causes of death in the *sod2−/−* mice by the porphyrin enabled unmasking of the neurological phenotype, which takes longer to develop. In support of this interpretation, the small percentages of untreated *sod2−/−* that live beyond 2 weeks also develop indications of the neurological syndrome (S. Melov, unpublished data).

Since the salen manganese complexes have shown efficacy in animal models involving neurological pathologies, we therefore investigated the ability of these compounds to rescue the neurological phenotype in *sod2−/−* mice.[55] *Sod2−/−* mice were treated daily with EUK-8 or EUK-134 from postnatal day 3 until their death. As shown in Fig. 9, EUK-8 treatment prolonged survival of *sod2−/−* mice beyond that seen with MnTBAP treatment . This effect was dose-dependent, with 1 mg/kg EUK-8 exhibiting a lesser increase in survival compared to that seen with 30 mg/kg.[55] Mice treated with the higher dose of EUK-8 also showed no evidence of the movement disorder and no evidence of spongiform pathology in their brains at 18 days of age (Fig. 10). Further, *sod2−/−* mice treated with either EUK-8 (or EUK-134) at the higher dose for longer than 5 weeks showed no evidence of spongiform changes at either the ultrastructural or light microscopic level at any age. The cause of death in EUK-8 treated mice is not yet known. It may involve another unmasked neurological phenotype but this remains to be

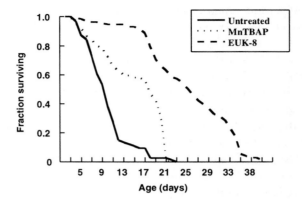

Fig. 9. Survival of *sod2−/−* mice treated with catalytic antioxidants. *Sod2−/−* mice[50] were untreated, or were treated daily, beginning at postnatal day 3, with intraperitoneal injections of MnTBAP (5 mg/kg) or EUK-8 (30 mg/kg). The results of Kaplan Meier survival analysis are shown. Differences in survival between each pair of groups were highly significant (Wilcoxon and Log Rank tests, $p < 0.05$). Group numbers were 134, 193, and 136 for Untreated, MnTBAP-treated, and EUK-8 treated groups, respectively. EUK-134 ($n = 19$) was as effective as EUK-8 at 30 mg/kg, while EUK-8 (1 mg/kg, $n = 47$) had a lesser effect, and EUK-134 (1 mg/kg, $n = 11$) had no significant effect (data not shown).

Fig. 10. Spongiform neural pathology in *sod2−/−* mice treated with catalytic antioxidants. *Sod2−/−* mice, treated as described for Fig. 9, were killed at 18 days and their brains processed for histopathology. The figure shows the cerebral cortex (hematoxylin/eosin staining) of representative mice treated with A, MnTBAP (5 mg/kg), B, EUK-8 (1 mg/kg) and C, EUK-8 (30 mg/kg).

characterized. EUK-134 was also effective but, in contrast to the observations with the rat stroke model,[35] it was no more potent or efficacious than EUK-8.[55] A possible explanation for this finding is that SOD activity, being equivalent for the two compounds, is more important than catalase activity for this animal model. While not the only possible interpretation, it is a reasonable one based on the nature of the genetic defect in the *sod2−/−* mice, namely, a specific deficit in superoxide scavenging activity.

9. Salen Manganese Complexes Extend Lifespan of a Multicellular Organism

Overall, the salen manganese complexes show efficacy in numerous animal models, including those for pathologies that, in humans, are age-associated. This certainly raises the question of whether the compounds would interfere with the aging process itself. Recently, we demonstrated that treatment with salen manganese complexes extends the lifespan of a multicellular organism, the nematode *Caenorhabditis elegans* (*C. elegans*), which is frequently used as a model for research in aging. In this study,[52] wildtype *C. elegans* were cultured in liquid medium containing EUK-134 or EUK-8. One such experiment is shown in [Fig. 11(A)]. EUK-134 treatment caused an increase in mean lifespan averaging about 44% over multiple experiments, as summarized in detail elsewhere.[52] Comparable results were obtained with EUK-8, with no dose-dependence observed for either compound (0.05 to 5 mM in the culture medium). Follow-up studies are

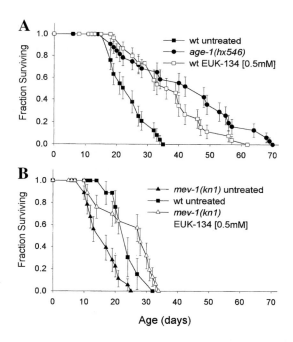

Fig. 11. Lifespan extension in *C. elegans* by EUK-134. *C. elegans* were maintained in liquid culture in the presence of, where indicated, EUK-134. Experimental details are described in the original study.[52] Reprinted from Ref. 52, with permission from the publisher. (A) Survival of *C elegans* wildtype strain, with and without EUK-134 (0.5 mM), and for comparison, the longer-lived *C. elegans age-1*(hx546) mutant strain.[54] (B) Survival of *C. elegans mev-1* mutant with or without EUK-134 (0.5 mM) and, for comparison, *C. elegans* wildtype strain.

underway to investigate the mode of EUK-134 or EUK-8 ingestion in this model, the dose-dependence and mechanism of action of the compounds, and other key issues. However, the hypothesized mechanism of action, involving ROS scavenging, is supported by an experiment utilizing the *mev-1* (kn1) nematode. This strain has a mutation in the cytochrome b subunit of complex II of the electron transport chain, resulting in an elevated accumulation of oxidative damage during aging, an increased sensitivity to oxygen, and a significantly shorter lifespan as compared to the wildtype strain.[53] Treatment of the *mev 1* strain with 0.5 mM EUK-134 extended lifespan by over 60% [Fig. 11(B)]. This finding is consistent with the hypothesis that EUK-134 treatment causes a decrease in chronic oxidative stress in the *mev-1* mutant, essentially, normalizing its lifespan to approximate that of the wildtype.

10. Conclusions

Salen manganese complexes have the ability to scavenge multiple ROS and, hence, are potentially useful in treating a variety of pathological conditions associated with oxidative stress. Consistent with this proposal, the compounds are protective in numerous experimental disease models, including the excitotoxic and degenerative neurological pathologies summarized in this review. Further, their ability to increase lifespan of a multicellular organism, *C. elegans*, suggests that the compounds might attenuate multiple degenerative processes that occur during normal aging. The applicability of these findings to mammalian aging remains to be determined but the protectiveness of the compounds in several rodent models for oxidative pathologies is promising. In general, the results obtained with the salen manganese complexes support the concept that use of a multifunctional catalytic antioxidant may serve as a very effective therapeutic approach applicable to a broad range of diseases and possibly to the enhancement of health during the aging process.

Acknowledgments

We acknowledge support from the NIH for research on salen manganese complexes as SOD/catalase mimetics, in particular the National Institute for Allergy and Infectious Disease for support of studies in autoimmune mice (R43 AI36763 to S.R.D. and Eukarion), the National Institute of General Medical Sciences for support of studies in stroke (R44 GM57770 to S.R.D. and Eukarion), and the National Cancer Institute (R44 CA83575 to S.R.D. and Eukarion) for support of initial preclinical development of a salen manganese complex. Studies in *sod2−/−* mice were supported by NIH grant AG18679 (to S.M.). We also acknowledge Dr. Douglas Wallace (Center for Molecular Medicine, Emory University, Atlanta, GA), in whose laboratory studies were first conducted, by S.M., in *sod2−/−* mice and *C. elegans*, and Dr. Gordon Lithgow (University of Manchester, UK), whose laboratory was responsible for the *C. elegans* lifespan studies reviewed here and originally published by Melov *et al.*[52]

References

1. Halliwell, B. and Gutteridge, J. M. C. (1989). *Free Radicals in Biology and Medicine*, Second Edition, 543 pages, Clarendon Press, Oxford.
2. Klebanoff, S. J. (1988). Phagocytic cells: products of oxygen metabolism. *In* "Inflammation: Basic Principles and Clinical Correlates" (I. M. G. J. I. Gallin, and R. Snyderman, eds.), pp. 391–444, Raven Press Ltd., New York.
3. Anderson, B. O., Brown, J. M. and Harken, A. H. (1991). Mechanisms of neutrophil-mediated tissue injury. *J. Surg. Res.* **51**: 170–179.
4. McCord, J. M. (1987). Oxygen-derived radicals: a link between reperfusion injury and inflammation. *Fed. Proc.* **46**: 2402-2406.
5. Granger, D. N. (1988). Role of xanthine oxidase and granulocytes in ischemia-reperfusion injury. *Am. J. Physiol.* **255**: H1269–H1275.
6. Sies, H. (1986). Biochemistry of oxidative stress. *Angewandte Chemie.* **25**: 1058–1071.
7. Rice-Evans, C. A. and Diplock, A. T. (1993). Current status of antioxidant therapy. *Free Radic. Biol. Med.* **15**: 77–96.
8. Flohe, L. (1988). Superoxide dismutase for therapeutic use: clinical experience, dead ends, and hopes. *Mol. Cell Biochem.* **84**: 123–131.
9. Sanchiz, F., Milla, A., Artola, N., Julia, J. C., Moya, L. M., Pedro, A. and Vila, A. (1996). Prevention of radiation-induced cystitis by orgotein: a randomized study. *Anticancer Res.* **16**: 2025–2028.
10. Kensler, T. W., Bush, D. M. and Kozumbo, W. J. (1983). Inhibition of tumor promotion by a biomimetic SOD. *Science* **221**: 75–77.
11. Yamamoto, S., Nakadate, T., Aizu, E. and Kato, R. (1990). Anti-tumor promoting action of pthalic acid mono-n-butyl ester cupric salt. *Carcinogenesis* **11**: 749–754.
12. Darr, D., Zarilla, K. A. and Fridovich, I. (1987). A mimic of superoxide dismutase activity based upon desferrioxamine B and manganese (IV). *Arch. Biochem. Biophys.* **258**: 351–355.
13. Bruce, A. J., Najm, I., Malfroy, B. and Baudry, M. (1992). Effects of desferrioxamine/manganese complex, a superoxide dismutase mimic, on kainate-induced pathology in rat brain. *Neurodegeneration* **1**: 265–271.
14. Baudry, M., Etienne, S., Bruce, A., Palucki, M., Jacobsen, E. and Malfroy, B. (1993). Salen-manganese complexes are superoxide dismutase-mimics. *Biochem. Biophys. Res. Commun.* **192**: 964–968.

15. Faulkner, K. M., Liochev, S. I. and Fridovich, I. (1994). Stable Mn(III) porphyrins mimic superoxide dismutase in vitro and substitute for it in vivo. *J. Biol. Chem.* **269**: 23 471–23 476.

16. Patel, M. and Day, B. J. (1999). Metalloporphyrin class of therapeutic catalytic antioxidants. *Trends Pharmacol. Sci.* **20**: 359–364.

17. Day, B. J. and Crapo, J. D. (1996). A metalloporphyrin superoxide dismutase mimetic protects against paraquat-induced lung injury in vivo. *Toxicol. Appl. Pharmacol.* **140**: 94–100.

18. Riley, D. P. and Weiss, R. H. (1994). Manganese macrocyclic ligand complexes as mimics of SOD. *J. Am. Chem. Soc.* **116**: 387–388.

19. Salvemini, D. *et al.* (2000). A nonpeptidyl mimic of superoxide dismutase with therapeutic activity in rats. *Science* **286**: 304–306.

20. Macarthur, H., Westfall, T. C., Riley, D. P., Misko, T. P. and Salvemini, D. (2000). Inactivation of catacholamines by superoxide gives new insights on the pathogenesis of septic shock. *Proc. Natl. Acad. Sci. USA* **97**: 9753–9758.

21. Gonzalez, P. K., Zhuang, J., Doctrow, S. R., Malfroy, B., Benson, P. F., Menconi, M. J. and Fink, M. P. (1995). EUK-8, a synthetic superoxide dismutase and catalase mimetic, ameliorates acute lung injury in endotoxemic swine. *J. Pharmacol. Exp. Ther.* **275**: 798–806.

22. Doctrow, S. R., Huffman, K., Marcus, C. B., Musleh, W., Bruce, A., Baudry, M. and Malfroy, B. (1997). Salen-manganese complexes: combined superoxide dismutase/catalase mimics with broad pharmacological efficacy. *In* "Advances in Pharmacology, 'Antioxidants in Disease Mechanisms and Therapeutic Strategies' " (H. Sies, Ed.), pp. 247–270, Academic Press, New York.

23. Dismukes, G. C. (1993). Polynuclear manganese enzymes. *In* "Bioinorganic Catalysis" (J. Reediik, Ed.), pp. 317–346, Marcel Dekker, Inc., New York.

24. Groves, J. T. (1999). Peroxynitrite: reactive, invasive, and enigmatic. *Curr. Opin. Chem. Biol.* **3**: 226–235.

25. RayChaudhury, A., Frischer, H. and Malik, A. B. (1996). Antiproliferative and cytotoxic effects of nitric oxide in endothelial cells. *In* "Biology of Nitric Oxide" (S. Moncada *et al.*, Ed.), Portland Press, London.

26. Anderson, I., Adinolfi, C., Doctrow, S., Huffman, K., Joy, K., Malfroy, B., Soden, P., Rupniak, H. T. and Barnes, J. C. (2000). Oxidative signalling and inflammatory pathways in Alzheimer's disease. *In* "Neuronal Signal Transduction in Alzheimer's Disease" (C. O'Neill, and B. Anderton, Eds.), Portland Press, London, **in press**.

27. Musleh, W., Bruce, A., Malfroy, B. and Baudry, M. (1994). Effects of EUK-8, a synthetic catalytic superoxide scavenger, on hypoxia- and acidosis-induced damage in hippocampal slices. *Neuropharmacology* **33**: 929–934.

28. Gonzalez, P. K., Doctrow, S. R. and Fink, M. P. (1997). Role of oxidant stress and iron delocalization in acidosis-induced intestinal epithelial hyperpermeability. *Shock* **8**: 108–114.

29. Gonzalez, P. K., Zhuang, J., Doctrow, S. R., Malfroy, B., Smith, M., Menconi, M. J. and Fink, M. P. (1995). Delayed treatment with EUK-8, a novel synthetic superoxide dismutase (SOD) and catalase (CAT) mimetic, ameliorates acute lung injury in endotoxemic pigs. *Surg. Forum* **46**: 72–73.

30. Malfroy, B., Doctrow, S. R., Orr, P. L., Tocco, G., Fedoseyeva, E. V. and Benichou, G. (1997). Prevention and suppression of autoimmune encephalomyelitis by EUK-8, a synthetic catalytic scavenger of oxygen reactive metabolites. *Cell. Immunol.* **177**: 62–68.

31. Theofilopoulos, A. N. and Dixon, F. J. (1985). Murine models of systemic lupus erythematosus. *Adv. Immunol.* **37**: 269–391.

32. Pucheu, S., Boucher, F., Sulpice, T., Tresallet, N., Bonhomme, Y., Malfroy, B. and deLeiris, J. (1996). EUK-8, a synthetic catalytic scavenger of reactive oxygen species, protects isolated iron-overloaded rat heart from functional and structural damage induced by ischemia/reperfusion. *Cardiovasc. Drugs Ther.* **10**: 331–339.

33. Pucheu, S., Boucher, F., Malfroy, B. and DeLeiris, J. (1995). Protective effect of the superoxide scavenger EUK-8 against ultrastructural alterations induced by ischemia and reperfusion in isolated rat hearts. *Nutrition (Suppl.)* **11**: 582–584.

34. Bruce, A. J., Malfroy, B. and Baudry, M. (1996). Beta-amyloid toxicity in organotypic hippocampal cultures: protection by EUK-8, a synthetic catalytic free radical scavenger. *Proc. Natl. Acad. Sci. USA* **93**: 2312–2316.

35. Baker, K., Bucay-Marcus, C., Huffman, C., Kruk, H., Malfroy, B. and Doctrow, S. R. (1998). Synthetic combined superoxide dismutase/catalase mimics are protective as a delayed treatment in a rat stroke model: a key role for reactive oxygen species in ischemic brain injury. *J. Pharmacol. Exp. Ther.* **284**: 215–221.

36. Holm, P., Kankaanranta, H., Metsa-Ketela and Moilanen, E. (1998). Radical releasing properties of nitric oxide donors GEA 3162, SIN-1, and S-nitroso-N-acetylpenicillamine. *Eur. J. Pharmacol.* **346**: 97–102.

37. Squadrito, G. L. and Pryor, W. A. (1998). Oxidative chemistry of nitric oxide: the roles of superoxide, peroxynitrite, and carbon dioxide. *Free Radic. Biol. Med.* **25**: 392–403.

38. Pong, K., Doctrow, S. R. and Baudry, M. (2000). Prevention of 1-methyl-4-phenylpyridinium and 6-hydroxydopamine-induced nitration of tyrosine hydroxylase and neurotoxicity by EUK-134, a superoxide dismutase and catalase mimetic, in cultured dopaminergic neurons. *Brain Res.* **881**: 182–189.

39. Ischiropoulos, H. and Al-Mehdi, A. B. (1995). Peroxynitrite-mediated oxidative protein modifications. *FEBS Lett.* **364**: 279–282.

40. Mattson, M. P., Culmsee, C. and Yu, Z. F. (2000). Apoptotic and antiapoptotic mechanisms in stroke. *Cell Tissue Res.* **301**: 173–187.

41. Ziven, J. A. and Choi, D. W. (1991). Stroke Therapy. *Sci. Am.* **July**: 56–63.

42. Matsuo, Y., Onodera, H., Shiga, Y., Nakamura, M., Ninomiya, M., Kihara, T. and Kogure, K. (1994). Correlation between myeloperoxidase-quantified

neutrophil accumulation and ischemic brain injury in the rat. Effects of neutrophil depletion. *Stroke* **25**: 1469–1475.

43. Clark, R. K., Lee, E. V., Fish, C. J., White, R. F., Price, W. J., Jonak, Z. L., Feuerstein, C. Z. and Barone, F. C. (1993). Development of tissue damage, inflammation and resolution following stroke: an immunohistochemical and quantitative planimetric study. *Brain Res. Bull.* **31**: 565–572.

44. Lafon-Cazal, M., Pietri, S., Culcasi, M. and Bockaert, J. (1993). NMDA-dependent superoxide production and neurotoxicity. *Nature* **364**: 535–537.

45. Dugan, L. L., Sensi, S. L., Canzoniero, L. M. T., Handran, S. D., Rothman, S. M., Lin, T. S., Goldberg, M. P. and Choi, D. W. (1995). Mitochondrial production of reactive oxygen species in cortical neurons following exposure to d-methyl-d-aspartate. *Neuron.* **15**: 6377–6388.

46. Bruce, A. and Baudry, M. (1995). Oxygen free radicals in rat limbic structures after kainate-induced seizures. *Free Radic. Biol. Med.* **18**: 993–1002.

47. Rong, Y. and Baudry, M. (1996). Seizure activity results in a rapid induction of nuclear factor-kappa B in adult, but not juvenile rat limbic structures. *J. Neurochem.* **67**: 662–668.

48. Rong, Y., Doctrow, S. R., Tocco, G. and Baudry, M. (1999). EUK-134, a synthetic superoxide dismutase and catalase mimetic, prevents oxidative stress and attenuates kainate-induced neuropathology. *Proc. Natl. Acad. Sci. USA* **96**: 9897–9902.

49. Kondo, T., Reaume, A. G., Huang, T. T., Carlson, E., Murakami, K., Chen, S. F., Hoffman, E. K., Scott, R. W., Epstein, C. J. and Chan, P. H. (1997). Reduction of Cu, Zn superoxide dismutase activity exacerbates neuronal cell injury and edema formation after transient focal cerebral ischemia. *J. Neurosci.* **17**: 4180–4189.

50. Li, Y. *et al.* (1995). Dilated cardiomyopathy and neonatal lethality in mutant mice lacking manganese superoxide dismutase. *Nature Genet.* **11**: 376–381.

51. Melov, S., Schneider, J. A., Day, B. J., Hinerfeld, D., Coskun, P., Mirra, S. S., Crapo, J. D. and Wallace, D. C. (1998). A novel neurological phenotype in mice lacking mitochondrial manganese superoxide dismutase. *Nature Genet.* **18**: 159–163.

52. Melov, S., Ravenscroft, J., Malik, S., Gill, M. S., Walker, D. S., Clayton, P. E., Wallace, D. C., Malfroy, B., Doctrow, S. R. and Lithgow, G. J. (2000). Extension of lifespan with superoxide dismutase/catalase mimetics. *Science* **289**: 1567–1569.

53. Ishii, N., Fujii, M., Hartman, P. S., Tsuda, M., Yasuda, K., Senoo-Matsuda, N., Yanase, S., Ayusawa, D. and Suzuki, K. (1998). A mutation in succinate de-hydrogenase cytochrome b causes oxidative stress and ageing in nematodes. *Nature* **394**: 697–697.

54. Johnson, T. E. (1990). Increased life-span of age-1 mutants in *Caenorhabditis elegans* and lower Gompartz rate of aging. *Science* **249**: 908–912.

55. Melov, S., Doctrow, S. R., Schneider, J. A., Haberson, J., Patel, M., Coskun, P. E., Huffman, K., Wallace, D. C. and Malfroy, B. (2001). Lifespan extension and rescue of *spongiform encephalopathy* in superoxide dismutase 2 nullizygous mice treated with superoxide dismutase-catalase mimetics. *J. Neurosci.* **21**: 8348–8353.

Chapter 72

Glutathione Peroxidase Mimics

Timothy C. Rodell*

Timothy C. Rodell, M.D. • OXIS International, Inc., 6040 N. Cutter Circle, Suite 317, Portland, OR 97217-3935

*Current address: RxKinetix, Inc., 1172 Century Drive, Suite 260, Louisville, CO 80027 Tel: (303) 926-1900, E-mail: tcrodell@aol.com

1. Summary

Glutathione peroxidase (GPx) is a selenium containing antioxidant enzyme that reduces hydroperoxides in the presence of glutathione (GSH). Its biological activity appears to be not only related to its antioxidant activity but also to its effect on lowering intracellular peroxide concentrations, resulting in reduction in transcription factor activation and production of down stream mediators. Several small molecule catalytic mimics of GPx have been developed and tested initially in animals and man. Of these, ebselen and BXT-51072 are the best characterized and have shown initial encouraging results in several human diseases.

2. Role of Glutathione Peroxidase (GPx) and Rationale for Therapeutic Use of GPx Mimics

Glutathione peroxidases are a family of enzymes, of which at least four have been described, which contain selenium at the active site, and which catalyze the reduction of inorganic (hydrogen peroxide) and organic hydroperoxides using GSH as a cofactor. The intracellular form of GPx, along with catalase, are the primary routes of reduction of intracellular hydrogen peroxide. Because catalase does not reduce organic peroxides, GPx is the primary route of reduction of organic hydroperoxides, including lipid hydroperoxides produced by the reaction of oxygen radicals with phospholipids in cell membranes.[1, 2]

Intracellular peroxide concentration appears to be a primary mediator of activation of the transcription factor NF-κB therefore, GPx activity may be a major regulator of transcription factor activation and subsequent inflammatory and apoptotic events. A recent study by Arrigo, *et al.* supports this concept.[3]

While a number of antioxidant approaches to the modification of disease have been, or are being attempted, including augmentation of superoxide dismutase (SOD) activity either with the enzyme SOD itself or SOD mimetics, as well as augmentation of thiol concentrations with GSH and N-acetylcysteine, the augmentation of GPx activity represents an attractive therapeutic option for a number of reasons. Among these include the fact that augmenting SOD activity could result in increased peroxide concentrations, which may explain its apparent bell shaped dose response curve in some systems.[4] In addition, antioxidant supplementation with large proteins may not target the critical intracellular compartment. Stoichometric small molecule antioxidants such as spin traps may require high concentrations in order to be effective. Finally, recent data suggesting that GPx mimics may also reduce peroxynitrite make it likely that GPx mimics may in fact affect most, if not all of the biologically relevant reactive oxygen species.

These considerations support the concept that augmentation of GPx activity with a low molecular weight, catalytic molecule that penetrates cells would be a rational therapeutic alternative for the treatment of conditions associated with oxidative stress.

3. Ebselen

The first synthetic small molecule described with GPx-like activity was ebselen (Fig. 1). While a number of other activities have been ascribed to ebselen, including direct, non-catalytic antioxidant activity[5] and reduction in neutrophil derived superoxide production through inhibition of NADPH oxidase,[6] its predominant biological effect is believed to be mediated by its GPx-like activity. This activity was originally described in 1984 and shown to be dependent on the presence of selenium just as in native GPx.[5] Like native GPx, ebselen reduces organic hydroperoxides, oxidizing GSH as a cofactor to glutathione disulfide (GSSG). Ebselen has been studied in numerous biological systems and has been demonstrated to have broad antioxidant and cytoprotective activity.[5] It is also the first synthetic GPx mimic to be studied in humans.

The majority of clinical work with ebselen has focused on acute diseases of the central nervous system. A small, double-blind, placebo controlled trial in patients with middle cerebral artery occlusion suggested an improved outcome in patients treated with ebselen, however, statistical significance in this study was only seen in a subset of the overall intent-to-treat analysis.[7] In patients with subarachnoid hemorrhage, there was also a suggestion of improved outcome as measured by Glasgow Outcome Scale in the subset of enrolled patients who displayed delayed neurologic deficits, however, again in the intent-to-treat analysis, there was no difference seen between ebselen treated and placebo treated patients.[8] More encouraging were the results of a trial in ischemic stroke in which patients who were treated with ebselen displayed significantly improved Glasgow Outcome Scale scores at one month when compared to placebo. This effect approached but did not reach statistical significance at 3 months.[9] These data, taken together, strongly suggest a role for ebselen in the treatment of acute ischemic neurologic disease.

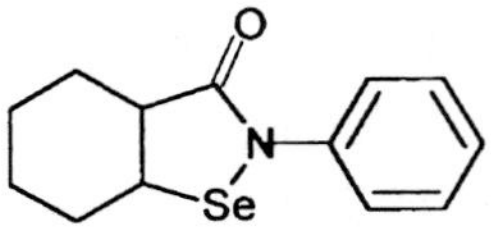

Fig. 1. Structure of Ebselen.

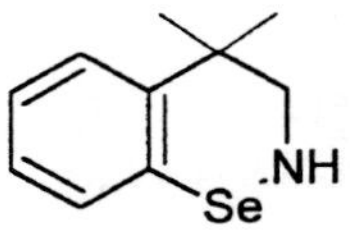

Fig. 2. Structure of BXT-51072.

4. BXT-51072

BXT-51072 (Fig. 2) is a member of a large family of organo-selenium compounds first reported by Chaudiere and colleagues in 1988.[10] Like ebselen, BXT-51072 catalyzes the reduction of inorganic and organic hydroperoxides using glutathione as a co-factor.

Based on rate of consumption of GSH, BXT-51072 has approximately fourfold greater GPx activity than ebselen, however, it demonstrates a greater biological potency increase over ebselen than would be predicted from relative GPx activities alone.[11]

BXT-51072 and related compounds protect human umbilical vein endothelial cells (HUVECs) from hydroperoxide induced cytotoxicity, as well as blocking neutrophil adhesion induced by tumor necrosis factor α (TNF-α), P-selectin (Fig. 3)

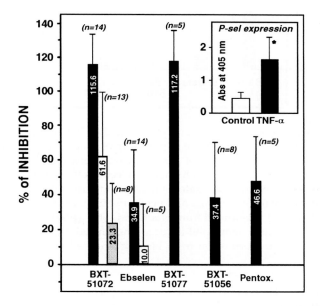

Fig. 3. Effect of GPx-mimics and pentoxifylline on the TNF-α induced expression of P-selectin by HUVEC. These cells were exposed or nonexposed to 1 ng/ml TNF-α for 3.5 h and then fixed. The expression of P-selectin was determined by primary binding with a mouse monoclonal antibody to human P-selectin followed by secondary binding with an alkaline phosphatase-conjugated rabbit antimouse IgG (see insert for $n = 14$ independent experiments). GPx mimics were tested at 10 µM (black bar), 4 µM (white bar) or 1 µM (gray bar) by pretreating HUVEC for 1 h and further cotreating them during TNF-α activation. Pentoxifylline was tested at 50 µM following the same treatment. The effect of each compound is expressed as a percentage of inhibition of the TNF-α induced expression of P-selectin after substraction of the control value. *$p < 0.001$. (Reprinted from Moutet, M., *et al.* (1998). Glutathione peroxidase mimics prevent TNF-α and neutrophil-induced endothelial alterations. *Free Radic. Biol. Med.* **25**(3): 270–281, with permission from Elsevier Science.[11])

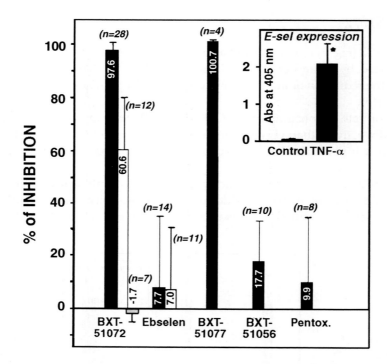

Fig. 4. Effect of GPx-mimics and pentoxifylline on the TNF-α induced expression of E-selectin by HUVEC. These cells were exposed or nonexposed to 1 ng/ml TNF-α for 3.5 h and then fixed. The expression of E-selectin was determined by primary binding with a mouse monoclonal antibody to human E-selectin followed by secondary binding with an alkaline phosphatase-conjugated rabbit antimouse IgG (see insert for $n = 28$ independent experiments). GPx mimics and pentoxifylline were tested as described in the legend of Fig. 3. *$p < 0.001$. (Reprinted from Moutet, M., *et al.* (1998). Glutathione peroxidase mimics prevent TNF-α and neutrophil-induced endothelial alterations. *Free Radic. Biol. Med.* **25**(3): 270–281, with permission from Elsevier Science.[11])

and E-selectin (Fig. 4) expression and blocking TNF-α induced production of interleukin-8 (IL-8) (Fig. 5)[11] and IL-6.[12] BXT-51072 also blocks the TNF-α mediated expression of VCAM and ICAM.[13] In these systems, ebselen is relatively inactive.

The above described effects on inflammatory mediator production are presumed to be mediated by the effects of BXT-51072 on up-regulation of the transcription factor NF-κB. In cultured rat alveolar macrophages, BXT-51072 inhibits the up-regulation of NF-κB as measured by gel shift with an IC$_{50}$ of 16 nM. In HUVECs, 10 μM BXT-51072 shows a 50% reduction in the nuclear translocation of the p65 subunit of NF-κB in response to TNF-α.[12]

In vivo, oral administration of BXT-51072 has been shown to improve survival, block hepatocellular damage and block TNF-α production in galactosamine sensitized mice challenged with endotoxin.[14]

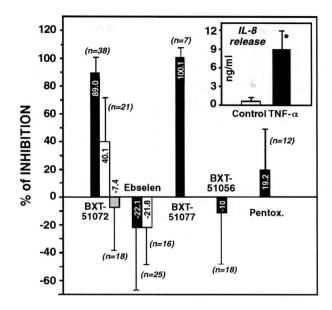

Fig. 5. Effect of GPx-mimics and pentoxifylline on the TNF-α induced release of IL-8 by HUVEC. These cells were exposed or nonexposed to 1 ng/ml TNF-α for 3.5 hours. Culture media were collected and Il-8 release was determined using commercial ELISA kits (see insert for $n = 27$ independent experiments). GPx mimics and pentoxifylline were tested as described in the legend of Fig. 3. *$p < 0.001$. (Reprinted from Moutet, M., *et al.* (1998). Glutathione peroxidase mimics prevent TNF-α and neutrophil-induced endothelial alterations. *Free Radic. Biol. Med.* **25**(3): 270–281, with permission from Elsevier Science.[11])

In a porcine model of restenosis, twice daily oral administration of BXT-51072 resulted in a significant decrease in endothelial proliferation following angioplasty induced coronary artery injury, combined with a non-significant trend toward a decrease in smooth muscle proliferation and concentric narrowing. Animals treated with BXT-51072 in this model also demonstrated preservation of endothelially mediated coronary vascular reactivity and decreased production of superoxide in the coronary artery wall.[15]

In a murine model of inflammatory bowel disease, in which colitis was induced by feeding dextran in drinking water, BXT-51072 improved clinical manifestations of colitis including diarrhea, bleeding and weight loss when administered prophylactically or following establishment of colitis.[16]

4.1. Clinical Studies

In Phase I human clinical trials, BXT-51072 was demonstrated to be rapidly absorbed orally and generally well tolerated.

In a recently completed Phase II trial in ulcerative colitis, the results of which have not yet been fully reported, patients treated with BXT-51072 demonstrated a reduction in clinical manifestations as demonstrated by a reduction in colitis activity index (CAI) which was accompanied by a reduction in visual sigmoidoscopy score and histopathologic score as well as a reduction in markers of oxidative stress in colon biopsies.[17] Because this was an open label, non-placebo controlled trial, these initial encouraging results will have to be confirmed in a larger, better controlled study.

5. Conclusion

Small molecule mimics of glutathione peroxidase appear to have similar biological effects to the native enzyme and have demonstrated encouraging results in pre-clinical and clinical scenarios. Both ebselen and BXT-51072 have demonstrated initial suggestions of efficacy in man, which will require further confirmation in additional clinical studies. The broad *in vitro* and *in vivo* anti-inflammatory and anti-apoptotic activity of molecules such as BXT-51072 and, to a lesser extent, ebselen, which appears to be mediated at least partially by their effects on transcription factor activation, lends strong support to the central role of oxidative processes in inflammatory and apoptotic diseases, and supports a potentially profound role for small molecule antioxidants such as these in the treatment of a wide range of conditions.

References

1. Ursini, F., Maiorino, M., Brigelius-Flohé, R., Aumann, K. D., Roveri, A., Schomburg, D. and Flohé, L. (1995). Diversity of glutathione peroxidases. *Meth. Enzymol.* **252**: 38–53.
2. Diplock, A. T. (1994). Antioxidants and free radical scavengers. *In* "Free Radical Damage and Its Control" (C. A. Rice-Evans, and R. H. Burdon, Eds.), pp. 113–130, Elsevier Science B.V., Amsterdam.
3. Kretz-Remy, C., Mehlen, P., Mirault, M. E. and Arrigo, A. P. (1996). Inhibition of IκB-α phosphorylation and degradation and subsequent NF-κB activation by glutathione peroxidase overexpression. **133**(5): 1083–1093.
4. J. McCord., personal communication.
5. Noguchi, N. and Niki, E. (1997). Antioxidant properties of ebselen. *In* "Handbook of Synthetic Antioxidants", pp. 285–303, Marcel Dekker, Inc., New York.
6. Cotgreave, I., Duddy, St., Kass, G., Thompson, D. and Moldeus, P. (1989). Studies on the anti-inflammatory activity of ebselen. *Biochem. Pharmacol.* **38**(4): 649–656.

7. Ogawa, A., Yoshimoto, T., Kikuchi, H., Sano, K., Saito, I., Yamaguchi, T. and Yasuhara, H. (1999). Ebselen in acute middle cerebral artery occlusion: a placebo-controlled, double-blind clinical trial. *Cerebrovasc. Dis.* **9**(2): 112–118.
8. Saito, I, Asano, T., Sano, K. , Takakura, K., Abe, H., Yoshimoto,T., Kikuchi, H., Ohta, T. and Ishibashi, S. (1998). Neuroprotective effect of an antioxidant, ebselen, in patients with delayed neurological deficits after aneurysmal subarachnoid hemorrhage. *Neurosurgery* **42**(2): 269–277.
9. Yamaguchi, T., Sano, K., Takakura, K., Saito, I., Shinohara, Y., Asano, T. and Yasuhara, H. (1998). Ebselen in acute ischemic stroke: a plecebo-controlled, double-blind clinical trial. Ebselen Study Group. *Stroke* **29**(1): 12–17.
10. Chaudiere, J., Yadan, J. C., Erdelmeier, I., Tailhan-Lomont, C. and Moutet, M. (1994). Design of new selenium-containing mimics of glutathione peroxidase. *In* "Oxidative Processes and Antioxidants" (R. Paoletti, Ed.), pp. 165–184, Raven Press Ltd., New York.
11. Moutet, M., D'Alessio, P., Malette, P., Devaux, V. and Chaudiere, J. (1998). Glutathione peroxidase mimics prevent TNF-α and neutrophil-induced endothelial alterations. *Free Radic. Biol. Med.* **25**(3): 270–281.
12. Moutet, M., personal communication.
13. D'Alessio, P., Moutet, M., Coudrier, E., Darquenne, S. and Chaudiere, J. (1998). ICAM-1 and VCAM-1 expression induced by TNF-α are inhibited by a glutathione peroxidase mimic. *Free Radic. Biol. Med.* **24**(6): 979–987.
14. Moutet, M., Giroud, C., Henry, N., Monnier, D. and Chaudiere, J. Protective effect of the glutathione peroxidase mimic BXT-51072 in a murine model of acute hepatitis. Abstract, conference proceedings "Oxidative Stress and Redox Regulation: Cellular Signaling, AIDS, Cancer and Other diseases".
15. Robinson, K. A., personal communication.
16. Ziemniak, J., Moutet, M., Yadan, J. C. and Mir, N. (1997). The activity of a novel glutathione peroxidase mimic, BXT-51072, in dextran sulfate induced inflammatory bowel disease. Abstract, Annual Meeting of American Gastroenterological Association and American Association for the Study of Liver Diseases.
17. Zhang, Y., Rodell, T., Murphy, T., Robinson, C., Banan, A., Choudhary, S. and Keshavarzian, A. (2000). Orally administered glutathione peroxidase-mimetic (BXT-51072) is a potent anti-oxidant: protection against oxidation and nitration of inflamed colonic mucosa in patients with ulcerative colitis (UC). *Gastroenterology* **118**(4): A589.

Chapter 73

Nutritional Approaches to Reducing Oxidative Stress

Bryant Villeponteau

Keywords: Aging, age-related disease, cancer, cardiovascular disease, nutritional supplements, mitochondria, clinical trials, free radicals, antioxidants, oxidative stress.

Bryant Villeponteau • Health Span Sciences, 4953 Smith Canyon Ct., San Diego, CA 92130
Tel: 858-794-8578, E-mail: bvillepo@san.rr.com

1. Summary

Life expectancy has increased dramatically in developed countries in the last hundred years. Although infectious diseases were the biggest killers a hundred years ago, the age-related diseases now account for most morbidity and mortality in the developed world. To increase life expectancy in the next hundred years, medical science will have to deal more effectively with the age-related diseases and with the aging process itself. While anti-aging drugs and genetic methods offer much hope for the longer-term, evidence is accumulating that dietary changes and nutritional supplements may help reduce oxidative free radicals and thereby slow the aging process and the progression of the age-related diseases. The existing data, including both animal and clinical results, suggest that human life expectancy may indeed be significantly increased in the near term by optimizing diet and using nutritional supplements.

2. The Role of Oxidative Stress in Aging

Although first proposed in 1956,[1] the free radical hypothesis of aging had rather limited scientific support until the last decade, when results from many studies indicated that free-radical damage accumulates with age and contributes to age-related degenerative diseases such as Alzheimer's, arthritis, cardiovascular disease, cancer, and osteoporosis.[2-5] Free radicals are highly unstable molecules with at least one unpaired electron. In an attempt to achieve a stable electronic structure, free radicals react in destructive ways with the surrounding DNA, lipid membranes, and cellular proteins. According to this model, non-repaired free radical damage accumulates to cause or promote both aging and the age-related degenerative diseases.

Although non-oxidative processes can generate free radicals, the vast bulk of free radicals are generated by oxidative metabolism.[2, 6] Indeed, oxidative stress is often used as another name for free radical damage. In eukaryotic cells, most oxidative metabolism is carried out in mitochondria, so that mitochondria are the primary generators of free radicals.[7-10] It is for this reason that mitochondrial DNA and membranes are especially vulnerable to the accumulation of free radical damage.[11] Some of the mitochondrial free radicals leak out and cause damage to genomic DNA, cellular membranes, structural proteins, and cell enzymes. Moreover, as the mitochondria accumulate damages by the oxidative free radicals, they leak out proportionally more free radicals.[8, 12] Eventually, the cell can repair less and less of the free radical damage, which leads to the accumulation of genomic defects and damaged cellular membranes and proteins throughout the cell. While there is no universally agreed paradigm for the causes of aging, the accumulation of free radical damage provides a model of aging and degenerative disease that has enjoyed widespread support by many scientists based on its explanatory power and its supporting data.[13–17]

While an exhaustive review of the data supporting the role of oxidative stress in aging and disease is beyond the scope of this review, it is interesting to look at the potential role of oxidative stress in cases where lifespan has been lengthened experimentally. Many studies have demonstrated that the aging process can be slowed in yeast, worms, and flies using genetics[14, 15] or in mammals using dietary restriction.[18] While the ultimate causes of the increased lifespan can be debated, one can argue that reduction in oxidative stress, either by reducing free radical production or by providing greater antioxidant protection, is a common factor underlying the increased longevity in both the genetic experiments and the dietary restricted mammals. For example, the *C. elegans* longevity mutants all appear to have reduced metabolic rates[19] and increased levels of antioxidant protection lengthens *Drosophila* lifespan.[20, 21] Likewise, long-lived dietary restricted mammals have altered metabolism and lower levels of oxidative stress.[18, 22] The contribution of oxidative stress to aging has also been supported by the many studies pointing to a significant role of oxidative stress in promoting all of the major age-related diseases.[1–5, 13, 17]

3. Reducing Oxidative Stress — Lowering Exposure to Environmental Oxidants

There are at least three ways to reduce oxidative stress: (1) Lower exposure to environmental oxidants; (2) Lower the *in vivo* generation of oxidative free radicals; and (3) Increase the *in vivo* levels of antioxidants. The environmental oxidants have long been recognized as a threat to human heatlh. The environmental air pollutants that cause oxidative stress such as ozone, hydroxides, and peroxides, have been implicated in promoting cancer and other degenerative diseases in polluted cities worldwide. Most environmental oxidants originate in the incomplete burning of petroleum fuel in automobiles, trucks, and buses as well as in fossil fuel used in electrical power plants. More efficient engine designs that generate less oxidative pollutants and catalytic converters that neutralize most of the oxidants that are produced have greatly lowered the environmental oxidants in many cities of the developed world. Thus, lowering one's exposure to smog reduces the exposure to harmful environmental oxidants, but air pollution is not a major problem in most cases.[23]

Tobacco smoking is a much more important environmental source of oxidants due to the incomplete burning (oxidation) of the tobacco leaf.[23] Smoking increases oxidative stress levels and is known to increase greatly the risk of various types of cancer, heart disease, and stroke. Even second-hand smoke is now thought to be a health risk. Reducing one's exposure to tobacco smoke is the best way to reduce environmental oxidative stress.

4. Reducing Oxidative Stress — Lower the *In Vivo* Generation of Oxidative Free Radicals

The great bulk of all oxidative free radicals are generated inside the cell rather than coming from the environment.[2, 6, 7, 10] Thus, inhibiting free radical production is one avenue for reducing oxidative stress. Since most cellular free radicals are generated by mitochondria as by-products of energy production and respiration,[7, 9, 24] stabilizing mitochondria structure and energy efficiency is an obvious target for lowering production of oxidative free radicals. This is analogous to making a more fuel-efficient and cleaner-burning gasoline engine that generates far less oxidative pollutants.

The omega-3 fatty acids EPA and DHA are available nutritional supplements that appear to stabilize mitochondrial membranes.[25–27] Omega-3 consumption is typically deficient in developed western countries, so omega-3 supplementation may help stabilize mitochondrial structure and reduce free radical production and leakage.[28] As to the energy efficiency of mitochondria, it is known that magnesium is required for many of the oxidation-reduction steps in mitochondria and that magnesium deficiency enhances free radical production.[29, 30] Since magnesium is often deficient in many individuals,[31–33] supplementation with at least the minimum daily requirement of magnesium (400 mg) should have value in optimizing mitochondrial efficiency and minimizing free radical leakage.

5. Reducing Oxidative Stress — Increase the *In Vivo* Levels of Antioxidant Defenses

Antioxidants are a diverse class of molecules that have the ability to neutralize many of the free radicals that are taken in from the environment or are generated internally. Antioxidants are widely distributed in nature, but are concentrated mostly in plants. The commercially available antioxidant supplements, which include β-carotene, vitamins C and E, are well known nutritional products that are consumed by hundreds of millions of people worldwide. Despite this worldwide acceptance, it has sometimes been difficult to demonstrate beneficial results using these nutritional antioxidants.[34–36] One major problem is that antioxidants typically exhibit both pro-oxidant as well as antioxidant potential.[36–38] Thus, the beneficial capacity of an antioxidant to neutralize free radicals must be balanced against its inherent capacity to become a harmful pro-oxidant. This dual nature of antioxidants is observed in the typical U-shaped dose-reaction curve for many antioxidants, where a beneficial effect is seen at low doses, but at higher doses the antioxidant plateaus or is even harmful. In practice, then, the pro-oxidant effects of antioxidants can often partially or wholly cancel out the benefit of free-radical neutralization by the antioxidant. To get the best results with most antioxidants, dosage must be carefully optimized, so as to maximize the beneficial antioxidant activity while minimizing the harmful pro-oxidant potential.

The above discussion suggests that the dual nature of antioxidants is a major limitation to the success of antioxidant therapies in reducing oxidative stress. In an attempt to identify antioxidants that get around this pro-oxidant limitation, I first looked at nutritional antioxidants that had high antioxidant activity per μg. I predicted that high potency antioxidants might have lower pro-oxidant activity. In the first high-potency antioxidant supplement, I developed ThioMax as a proprietary mixture of α-lipoic acid and *n*-propyl gallate. These two nutritional antioxidants are respectively about 50-fold and 100-fold more biologically potent than vitamin E[39] and thus might be expected to have low pro-oxidant activity. This expectation of low pro-oxidant activity was given greater credence by the available data describing the strong beneficial effects of each of these antioxidants. For example, α-lipoic acid has been reported to protect against cataracts, diabetes, and neuro-degeneration.[40–43] The food preservative n-propyl gallate has been shown to extend the lifespan of flies[44, 45] and to protect brain, liver, and erythrocytes.[39, 46–48] As discussed below, our clinical trial with ThioMax suggest that it does reduce oxidative stress by significant amounts.

In a later attempt to identify antioxidants with low pro-oxidant activity, potential catalytic antioxidants were surveyed. Superoxide Dismutase (SOD) and Catalase are good examples of enzymes that regulate the oxidation-reduction of transitions metals like zinc and manganese and thus can be rapidly recycled after reducing a free-radical oxidant. Adding more SOD and Catalase enzymes in genetic experiments has extended the lifespan of flies,[20] indicating that catalytic antioxidants have a powerful ability to extend lifespan. Since genetic approaches currently have severe limitations in humans and the catalytic enzymes themselves would be degraded in the digestive track, I looked for small-molecule SOD and Catalase mimics that were more stable than the peptide enzymes. Manganese-binding synthetic molecules have been described that act as small molecule mimics of Superoxide Dismutase (SOD) and Catalase.[49–52] I reasoned that some plant antioxidants may also have this SOD and Catalase mimic activity. Noting that the plant tetramer to hexamer gallate-ester proanthocyanidin polymers tightly binds transition metals like manganese and zinc, I hypothesized that gallate-ester proanthocyanidin polymers may bind metals like manganese so as to promote manganese-mediated catalytic oxidation-reduction. Preliminary experimental and modeling work suggested that the tetramer to hexamer gallate-ester proanthocyanidin polymers in association with manganese (the combined proprietary product is named Thiogen[TM]) might mimic the catalytic antioxidant enzymes manganese SOD and manganese Catalase (Villeponteau, unpublished). A clinical trial with this nutritional supplement demonstrated that Thiogen is very effective in reducing oxidative stress (see below). All of the Thiogen results to date are fully consistent with its proposed catalytic nature. Since Thiogen contains a natural plant antioxidant extract that does not require FDA approval, more research to test the effectiveness of this potential catalytic antioxidant is clearly warranted.

6. Clinical Trials on Nutritional Supplements that May Reduce Oxidative Stress

Based on the assumption that oxidative stress plays an important role in the aging process, our major goal was to identify an oral formulation that maximally reduced oxidative stress. We first addressed the question of whether the first formulations should be drugs or nutritional supplements. Any anti-aging drug would likely take 8 to 12 years for FDA approval and would have added safety concerns in that anti-aging drugs would likely be taken by healthy people for a very long period of time. In contrast, nutritional supplements are expected to have fewer side effects, can be developed and tested quickly, and do not require FDA approval. Based on these considerations, we focused on developing nutritional supplements that would effectively lower oxidative stress.

Our first attempt at a formulation that might effectively reduce oxidative stress was YouthGuard® (Table 1, column 1). YouthGuard® contained the antioxidant ThioMax (a proprietary mix of α-lipoic acid and *n*-propyl gallate) and mitochondrial-stabilizing components containing magnesium and EPA/DHA omega-3 extracts. Because calcium can compete with magnesium absorption in the gut, the calcium to magnesium ratio was maintained at less than one in the formulation. YouthGuard® also contained a full complement of essential vitamins and most minerals.

YouthGuard® was tested against a placebo in a double blind clinical trial for a period of six weeks.[53] Urine samples taken at the beginning and end of the trial were tested for oxidative stress using standard antibody assays for DNA damage, isoprostanes, and lipid peroxides. The results of these three assays were normalized and then averaged to obtain the total oxidative stress value for each volunteer's sample. Oxidative stress was found to be significantly reduced by 27% with YouthGuard® supplementation ($P = 0.019$), whereas oxidative stress in the placebo actually drifter upward somewhat.[53] Thus, YouthGuard® supplementation did significantly reduce oxidative stress, although our results to date do not tell us which ingredients in YouthGuard® had the biggest effects on oxidative stress.

In a second clinical trial,[53] we tested the proprietary plant extract (Thiogen) containing tetramer to hexamer proanthocyanidin gallate-ester polymers and manganese against 400 IU of vitamin E and a commercial grape-seed extract with mostly lower subunit polymers of proanthocyanidin (Table 1, columns 2–4). The Thiogen-containing formulation (column 4) dramatically reduced oxidative stress

Table 1. The Composition of Four Supplements Tested in Human Clinical Trials

	YouthGuard®	Centrum-Vit. E	Vi-Mix-Grape	Vi-Mix-Thiogen
β-carotine	5000 IU			
α-lipoic acid	50 mg			
Biotin		30 mcg		
Boron		150 mcg		
Calcium	450 mg	162 mg	360 mg	360 mg
Chloride		72 mg		
Chromium	150 mcg	65 mcg	100 mcg	100 mcg
Citrus bioflavonoids	50 mg			
Copper	2 mg	2 mg	0.8 mg	0.8 mg
Folic acid	900 mcg	400 mcg	900 mcg	900 mcg
Grape seed extract*			80 mg	80 mg
Iodine	100 mcg	150 mcg	80 mcg	80 mcg
Iron		18 mg		
Lecithin	100 mg			
Magnesium	500 mg	100 mg	400 mg	400 mg
Manganese	5 mg	3.5 mg	3 mg	3 mg
Molybdenum	100 mcg	160 mcg	50 mcg	50 mcg
NAC	50 mg			
Niacinamide	150 mg	20 mg	100 mg	100 mg
Nickel		5 mcg		
n-propyl gallate	10 mg			
PABA	10 mg			
Pantathenic acid	50 mg	10 mg		
Phosphorus		109 mg		
Potassium		80 mg		
Proanthocyanidin**				120 mg
Riboflavin	20 mg	1.7 mg	10 mg	10 mg
Selenium	50 mcg	20 mcg		
Silicon		2 mg		
Thiamin	20 mg	1.5 mg	9 mg	9 mg
Tin		10 mcg		
Vanadium		10 mcg		
Vitamin A		5000 IU	3000 IU	3000 IU
Vitamin B12	200 mcg	6 mcg	200 mcg	200 mcg
Vitamin B6	30 mg	2 mg	12 mg	12 mg
Vitamin C	200 mg	60 mg	200 mg	200 mg
Vitamin D	400 IU	400 IU	400 IU	400 IU
Vitamin E		400 IU/30 IU	30 IU	30 IU
Vitamin K		25 mcg		
Zinc	30 mg	15 mg	30 mg	30 mg

*95% grape seed extract with low levels of proanthocyanidin gallic ester oligomers over tetramers.
**High levels of tetramer to hexamer proanthocyanidin gallic esters.

by 38% ($P = 0.025$). In contrast, 400 IU of vitamin E (column 2) had a non-significant 15% reduction ($P = 0.25$) while the grape seed proanthocyanidins (column 3) may have actually increased oxidative stress somewhat ($P = 0.117$). In this case, we know that the other components (vitamins and minerals) in the formulation were not a major determining factor in reducing oxidative stress.

7. Extending Lifespan by Lowering Insulin and Glucose Levels

The *C. elegans* and *Drosophila* genetic work on longevity has providing important clues to the aging process in these animals and has led to efforts to generalize these results to mammals. For example, the *C. elegans* work has pointed to the possible role of insulin-like growth factor receptors and metabolism in worm lifespan determination.[14, 54–56] Dietary restriction studies in mammals have also pointed to a role for the insulin pathway and general metabolism in longevity, as insulin and glucose levels fall in long-lived dietary restricted mammals and metabolism is apparently altered.[57, 58] Moreover, human diabetics with altered insulin metabolism appear to suffer accelerated aging and are at a higher risk of developing many of the age-related diseases.[55, 59] Thus, data from very different fields argue that lowering insulin and glucose levels and altering metabolism might promote longevity.

In the case of Type-II diabetes, dietary changes such as lowering carbohydrate consumption are often recommended to bring down glucose and insulin levels. Moreover, clinical trials with chromium[60–64] or α-lipoic acid[65–68] have indicated that these nutritional supplements may be beneficial in modulating insulin levels or glucose levels. For the overweight non-diabetic person, reducing total carbohydrate intake and/or consuming carbohydrates with a lower glycemic index are sometimes recommended to lower insulin and circulating glucose levels. These same dietary changes and nutritional supplements may be used by normal healthy individuals to reduce insulin and glucose levels. The available data, though indirect, suggest that low glycemic foods, chromium, and α-lipoic acid may extend healthy life span through their effects on lowering insulin and glucose levels.

8. Dietary Changes and Nutritional Supplements that Counteract Age-Related Diseases

There are many dietary changes and nutritional supplements that have been recommended for treating various age-related diseases. Any dietary supplement that prevents or delays the progression of one or more age-related disease has the potential to extend life expectancy, whether or not it alters the aging process *per se*.

Multi-component nutritional products such as YouthGuard® have many other ingredients that may slow the progression of one or more age-related diseases via pathways other than oxidative stress. For example, high blood homocysteine levels promote atherosclerosis plaques and clogged arteries.[69] Folic acid, vitamins B6 and B12 have been shown to reduce homocysteine levels in the blood, and appear to lower cardiovascular risk.[70] In addition, low folate levels have been linked to enhanced DNA damage and increased cancer risks.[71-74] Another common deficiency, magnesium, may be useful for treating high blood pressure,[32, 75, 76] cardiovascular diseases,[33, 77-80] Alzheimer's,[81-83] and osteoporosis.[84] Moreover, the omega-3 fatty acids appear to reduce the risks of cardiovascular disease[85-88] and cancer.[89, 90] Finally, deodorized garlic extracts,[91-93] help the body fight both viral and bacterial infections. Chronic inflammation due to persistent infections is thought to play a role in many of the age-related diseases.

9. Conclusions

While the ultimate causes of aging are still contested, most investigators believe that oxidative stress plays a role in the process. Oxidative stress has also been strongly linked to the age-related diseases. We have shown that oxidative stress can be reduced by 27% to 38% using high potency or catalytic nutritional antioxidants with low pro-oxidant activity. Since oxidative damage apparently increases with age,[7, 10, 94, 95] nutritional supplements that reduce oxidative stress may provide effective agents to slow aging and the progression of age-related diseases. More work needs to be done on these supplements to show that the observed reduction in oxidative stress can be translated into longer lifespans in worms and mice. As important, nutritional supplements that dramatically lower oxidative stress need to be tested in human clinical trails for their ability to prevent or slow the progression of the individual age-related diseases linked to oxidative stress.

References

1. Harman, D. (1988). Free radicals in aging. *Mol. Cell Biochem.* **84**: 155.
2. Ames, B. N., Shigenaga, M. K. and Hagen, T. M. (1993). Oxidants, antioxidants, and the degenerative diseases of aging. *Proc. Natl. Acad. Sci. USA* **90**: 7915.
3. Marchioli, R. (1999). Antioxidant vitamins and prevention of cardiovascular disease: laboratory, epidemiological and clinical trial data [see comments]. *Pharmacol. Res.* **40**: 227.
4. Ambrosone, C. B., Freudenheim, J. L., Thompson, P. A., Bowman, E., Vena, J. E., Marshall, J. R., Graham, S., Laughlin, R., Nemoto, T. and Shields, P. G.

(1999). Manganese superoxide dismutase (MnSOD) genetic polymorphisms, dietary antioxidants, and risk of breast cancer. *Cancer Res.* **59**: 602.

5. Christen, Y. (2000). Oxidative stress and Alzheimer disease. *Am. J. Clin. Nutri.* **71**: S621.

6. Ames, B. N. (1995). Understanding the causes of aging and cancer. *Microbiologia* **11**: 305.

7. Ames, B. N., Shigenaga, M. K. and Hagen, T. M. (1995). Mitochondrial decay in aging. *Biochim. Biophys. Acta* **1271**: 165.

8. Shigenaga, M. K., Hagen, T. M. and Ames, B. N. (1994). Oxidative damage and mitochondrial decay in aging. *Proc. Natl. Acad. Sci. USA* **91**: 10 771.

9. Perez-Campo, R., Lopez-Torres, M., Cadenas, S., Rojas, C. and Barja, G. (1998). The rate of free radical production as a determinant of the rate of aging: evidence from the comparative approach. *J. Comp. Physiol.* **B168**: 149.

10. Sastre, J., Pallardo, F. V., Garcia de la Asuncion, J. and Vina, J. (2000). Mitochondria, oxidative stress and aging. *Free Radic. Res.* **32**: 189.

11. Barja, G. and Herrero, A. (2000). Oxidative damage to mitochondrial DNA is inversely related to maximum life span in the heart and brain of mammals. *FASEB J.* **14**: 312.

12. Hagen, T. M., Yowe, D. L., Bartholomew, J. C., Wehr, C. M., Do, K. L., Park, J. Y. and Ames, B. N. (1997). Mitochondrial decay in hepatocytes from old rats: membrane potential declines, heterogeneity and oxidants increase. *Proc. Natl. Acad. Sci. USA* **94**: 3064.

13. Ames, B. N. and Shigenaga, M. K. (1992). Oxidants are a major contributor to aging. *Ann. NY Acad. Sci.* **663**: 85.

14. Jazwinski, S. M. (1996). Longevity, genes, and aging. *Science* **273**: 54.

15. Campisi, J. (1997). The biology of replicative senescence. *Eur. J. Cancer* **33**: 703.

16. Osiewacz, H. D. (1997). Genetic regulation of aging. *J. Mol. Med.* **75**: 715.

17. Beckman, K. B. and Ames, B. N. (1998). The free radical theory of aging matures. *Physiol. Rev.* **78**: 547.

18. Roth, G. S., Ingram, D. K. and Lane, M. A. (1999). Calorie restriction in primates: will it work and how will we know? *J. Am. Geriatr. Soc.* **47**: 896.

19. Van Voorhies, W. A. and Ward, S. (1999). Genetic and environmental conditions that increase longevity in *Caenorhabditis elegans* decrease metabolic rate. *Proc. Natl. Acad. Sci. USA* **96**: 11 399.

20. Parkes, T. L., Elia, A. J., Dickinson, D., Hilliker, A. J., Phillips, J. P. and Boulianne, G. L. (1998). Extension of *Drosophila* lifespan by overexpression of human SOD1 in motorneurons [comment] [see comments]. *Nature Genet.* **19**: 171.

21. Arking, R., Burde, V., Graves, K., Hari, R., Feldman, E., Zeevi, A., Soliman, S., Saraiya, A., Buck, S., Vettraino, J., Sathrasala, K., Wehr, N. and Levine, R. L. (2000). Forward and reverse selection for longevity in *Drosophila* is

characterized by alteration of antioxidant gene expression and oxidative damage patterns. *Exp. Gerontol.* **35**: 167.

22. Greenberg, J. A. and Boozer, C. N. (2000). Metabolic mass, metabolic rate, caloric restriction, and aging in male Fisher 344 rats. *Mech. Aging Dev.* **113**: 37.

23. Ames, B. N. and Gold, L. S. (1998). The causes and prevention of cancer: the role of environment. *Biotherapy* **11**: 205.

24. Barja, G. (1998). Mitochondrial free radical production and aging in mammals and birds. *Ann. NY Acad. Sci.* **854**: 224.

25. Demaison, L., Sergiel, J. P., Moreau, D. and Grynberg, A. (1994). Influence of the phospholipid n-6/n-3 polyunsaturated fatty acid ratio on the mitochondrial oxidative metabolism before and after myocardial ischemia. *Biochim. Biophys. Acta* **1227**: 53.

26. Oudart, H., Groscolas, R., Calgari, C., Nibbelink, M., Leray, C., Le Maho, Y. and Malan, A. (1997). Brown fat thermogenesis in rats fed high-fat diets enriched with n-3 polyunsaturated fatty acids. *Int. J. Obes. Relat. Metab. Disorder* **21**: 955.

27. Pehowich, D. J. (1999). Thyroid hormone status and membrane n-3 fatty acid content influence mitochondrial proton leak. *Biochim. Biophys. Acta* **1411**: 192.

28. Fernandes, G. (1994). Dietary lipids and risk of autoimmune disease. *Clin. Immunol. Immunopathol.* **72**: 193.

29. Rock, E., Astier, C., Lab, C., Vignon, X., Gueux, E., Motta, C. and Rayssiguier, Y. (1995). Dietary magnesium deficiency in rats enhances free radical production in skeletal muscle. *J. Nutri.* **125**: 1205.

30. Bada, V., Kucharska, J., Gvozdjakova, A., Herichova, I. and Gvozdjak, J. (1996). The cytoprotective effect of magnesium in global myocardial ischemia. *Bratisl Lek Listy* **97**: 587.

31. Costello, R. B. and Moser-Veillon, P. B. (1992). A review of magnesium intake in the elderly. A cause for concern? *Magnes. Res.* **5**: 61.

32. Kawano, Y., Matsuoka, H., Takishita, S. and Omae, T. (1998). Effects of magnesium supplementation in hypertensive patients: assessment by office, home, and ambulatory blood pressures. *Hypertension* **32**: 260.

33. Yang, C. Y. (1998). Calcium and magnesium in drinking water and risk of death from cerebrovascular disease. *Stroke* **29**: 411.

34. Duthie, G. G. and Bellizzi, M. C. (1999). Effects of antioxidants on vascular health. *Brit. Med. Bull.* **55**: 568.

35. Gaziano, J. M. (1999). Antioxidant vitamins and cardiovascular disease. *Proc. Assoc. Am. Physicians.* **111**: 2.

36. van der Pols, J. C. (1999). A possible role for vitamin C in age-related cataract. *Proc. Nutri. Soc.* **58**: 295.

37. Medina-Navarro, R., Duran-Reyes, G. and Hicks, J. J. (1999). Pro-oxidating properties of melatonin in the in vitro interaction with the singlet oxygen. *Endocr. Res.* **25**: 263.

38. Stocker, R. (1999). The ambivalence of vitamin E in atherogenesis. *Trends Biochem. Sci.* **24**: 219.

39. Behl, C., Davis, J. B., Lesley, R. and Schubert, D. (1994). Hydrogen peroxide mediates amyloid beta protein toxicity. *Cell* **77**: 817.

40. Packer, L., Witt, E. H. and Tritschler, H. J. (1995). alpha-Lipoic acid as a biological antioxidant. *Free Radic. Biol. Med.* **19**: 227.

41. Packer, L., Tritschler, H. J. and Wessel, K. (1997). Neuroprotection by the metabolic antioxidant alpha-lipoic acid. *Free Radic. Biol. Med.* **22**: 359.

42. Ames, B. N. (1998). Micronutrients prevent cancer and delay aging. *Toxicol. Lett.* **102–103**: 5.

43. Hagen, T. M., Ingersoll, R. T., Lykkesfeldt, J., Liu, J., Wehr, C. M., Vinarsky, V., Bartholomew, J. C. and Ames, A. B. (1999). (R)-alpha-lipoic acid-supplemented old rats have improved mitochondrial function, decreased oxidative damage, and increased metabolic rate. *FASEB J.* **13**: 411.

44. Bains, J. S., Sharma, S. P. and Garg, S. K. (1992). Effect of propyl gallate feeding on glutathione content in ageing *Zaprionus paravittiger (Diptera)*. *Gerontology* **38**: 192.

45. Bains, J. S., Kakkar, R. and Sharma, S. P. (1996). Gender specific alterations in antioxidant status of aging *Zaprionus paravittiger* fed on propyl gallate. *Biochem. Mol. Biol. Int.* **40**: 731.

46. Wu, T. W., Fung, K. P., Zeng, L. H., Wu, J. and Nakamura, H. (1994). Propyl gallate as a hepatoprotector in vitro and in vivo. *Biochem. Pharmacol.* **48**: 419.

47. Bhatnagar, A. (1995). Attenuation of reperfusion injury by the antioxidant n-propyl gallate. *J. Cardiovasc. Pharmacol.* **26**: 343.

48. Wu, J., Sugiyama, H., Zeng, L. H., Mickle, D. and Wu, T. W. (1998). Evidence of Trolox and some gallates as synergistic protectors of erythrocytes against peroxyl radicals. *Biochem. Cell Biol.* **76**: 661.

49. Samuni, A., Mitchell, J. B., DeGraff, W., Krishna, C. M., Samuni, U. and Russo, A. (1991). Nitroxide SOD-mimics: modes of action. *Free Radic. Res. Commun.* **12–13**: 187.

50. Musleh, W., Bruce, A., Malfroy, B. and Baudry, M. (1994). Effects of EUK-8, a synthetic catalytic superoxide scavenger, on hypoxia- and acidosis-induced damage in hippocampal slices. *Neuropharmacology* **33**: 929.

51. Pucheu, S., Boucher, F., Sulpice, T., Tresallet, N., Bonhomme, Y., Malfroy, B. and de Leiris, J. (1996). EUK-8 a synthetic catalytic scavenger of reactive oxygen species protects isolated iron-overloaded rat heart from functional and structural damage induced by ischemia/reperfusion. *Cardiovasc. Drugs Ther.* **10**: 331.

52. Baker, K., Marcus, C. B., Huffman, K., Kruk, H., Malfroy, B. and Doctrow, S. R. (1998). Synthetic combined superoxide dismutase/catalase mimetics are protective as a delayed treatment in a rat stroke model: a key role for reactive oxygen species in ischemic brain injury. *J. Pharmacol. Exp. Ther.* **284**: 215.

53. Villeponteau, B., Cockrell, R. and Feng, J. (2000). Nutraceutical interventions may delay aging and the age-related diseases. *Exp. Gerontol.* **in press.**
54. Vanfleteren, J. R. and Braeckman, B. P. (1999). Mechanisms of life span determination in *Caenorhabditis elegans. Neurobiol. Aging* **20**: 487.
55. Thomas, J. H. and Inoue, T. (1998). Methuselah meets diabetes. *Bioessays* **20**: 113.
56. Tissenbaum, H. A. and Ruvkun, G. (1998). An insulin-like signaling pathway affects both longevity and reproduction in *Caenorhabditis elegans. Genetics* **148**: 703.
57. Kemnitz, J. W., Roecker, E. B., Weindruch, R., Elson, D. F., Baum, S. T. and Bergman, R. N. (1994). Dietary restriction increases insulin sensitivity and lowers blood glucose in rhesus monkeys. *Am. J. Physiol.* **266**: E540.
58. Bodkin, N. L., Ortmeyer, H. K. and Hansen, B. C. (1995). Long-term dietary restriction in older-aged rhesus monkeys: effects on insulin resistance. *J. Gerontol. A: Biol. Sci. Med. Sci.* **50**: B142.
59. Schleicher, E. D., Wagner, E. and Nerlich, A. G. (1997). Increased accumulation of the glycoxidation product N(epsilon)-(carboxymethyl)lysine in human tissues in diabetes and aging. *J. Clin. Invest.* **99**: 457.
60. Cobo, J. M. and Castineira, M. (1997). Oxidative stress, mitochondrial respiration, and glycemic control: clues from chronic supplementation with Cr3+ or As3+ to male Wistar rats. *Nutrition* **13**: 965.
61. Fox, G. N. and Sabovic, Z. (1998). Chromium picolinate supplementation for diabetes mellitus. *J. Family Pract.* **46**: 83.
62. Guan, X., Matte, J. J., Ku, P. K., Snow, J. L., Burton, J. L. and Trottier, N. L. (2000). High chromium yeast supplementation improves glucose tolerance in pigs by decreasing hepatic extraction of insulin. *J. Nutri.* **130**: 1274.
63. Hellerstein, M. K. (1998). Is chromium supplementation effective in managing Type-II diabetes? *Nutri. Rev.* **56**: 302.
64. Morris, B. W., MacNeil, S., Hardisty, C. A., Heller, S., Burgin, C. and Gray, T. A. (1999). Chromium homeostasis in patients with Type-II (NIDDM) diabetes. *J. Trace Elem. Med. Biol.* **13**: 57.
65. Rudich, A., Tirosh, A., Potashnik, R., Khamaisi, M. and Bashan, N. (1999). Lipoic acid protects against oxidative stress induced impairment in insulin stimulation of protein kinase B and glucose transport in 3T3-L1 adipocytes. *Diabetologia* **42**: 949.
66. Konrad, T., Vicini, P., Kusterer, K., Hoflich, A., Assadkhani, A., Bohles, H. J., Sewell, A., Tritschler, H. J., Cobelli, C. and Usadel, K. H. (1999). alpha-Lipoic acid treatment decreases serum lactate and pyruvate concentrations and improves glucose effectiveness in lean and obese patients with Type-2 diabetes. *Diabetes Care* **22**: 280.
67. Khamaisi, M., Potashnik, R., Tirosh, A., Demshchak, E., Rudich, A., Tritschler, H., Wessel, K. and Bashan, N. (1997). Lipoic acid reduces glycemia

and increases muscle GLUT4 content in streptozotocin-diabetic rats. *Metabolism* **46**: 763.

68. Jacob, S., Ruus, P., Hermann, R., Tritschler, H. J., Maerker, E., Renn, W., Augustin, H. J., Dietze, G. J. and Rett, K. (1999). Oral administration of RAC-alpha-lipoic acid modulates insulin sensitivity in patients with Type-2 diabetes mellitus: a placebo-controlled pilot trial. *Free Radic. Biol. Med.* **27**: 309.

69. Mujumdar, V. S., Hayden, M. R. and Tyagi, S. C. (2000). Homocyst(e)ine induces calcium second messenger in vascular smooth muscle cells. *J. Cell Physiol.* **183**: 28.

70. Bunout, D., Garrido, A., Suazo, M., Kauffman, R., Venegas, P., de la Maza, P., Petermann, M. and Hirsch, S. (2000). Effects of supplementation with folic acid and antioxidant vitamins on homocysteine levels and LDL oxidation in coronary patients. *Nutrition* **16**: 107.

71. Blount, B. C. and Ames, B. N. (1995). DNA damage in folate deficiency. *Baillieres Clin. Haematol.* **8**: 461.

72. Blount, B. C., Mack, M. M., Wehr, C. M., MacGregor, J. T., Hiatt, R. A., Wang, G., Wickramasinghe, S. N., Everson, R. B. and Ames, B. N. (1997). Folate deficiency causes uracil misincorporation into human DNA and chromosome breakage: implications for cancer and neuronal damage. *Proc. Natl. Acad. Sci. USA* **94**: 3290.

73. Ames, B. N. (1999). Micronutrient deficiencies. A major cause of DNA damage. *Ann. NY Acad. Sci.* **889**: 87.

74. Kim, Y. I. (1999). Folate and cancer prevention: a new medical application of folate beyond hyperhomocysteinemia and neural tube defects. *Nutri. Rev.* **57**: 314.

75. Witlin, A. G. and Sibai, B. M. (1998). Magnesium sulfate therapy in preeclampsia and eclampsia. *Obstet. Gynecol.* **92**: 883.

76. Touyz, R. M. and Milne, F. J. (1999). Magnesium supplementation attenuates, but does not prevent, development of hypertension in spontaneously hypertensive rats. *Am. J. Hypertens* **12**: 757.

77. Ascherio, A., Rimm, E. B., Hernan, M. A., Giovannucci, E. L., Kawachi, I., Stampfer, M. J. and Willett, W. C. (1998). Intake of potassium, magnesium, calcium, and fiber and risk of stroke among US men. *Circulation* **98**: 1198.

78. Kronon, M. T., Allen, B. S., Hernan, J., Halldorsson, A. O., Rahman, S., Buckberg, G. D., Wang, T. and Ilbawi, M. N. (1999). Superiority of magnesium cardioplegia in neonatal myocardial protection. *Ann. Thorac. Surg.* **68**: 2285.

79. Rubenowitz, E., Axelsson, G. and Rylander, R. (1999). Magnesium and calcium in drinking water and death from acute myocardial infarction in women. *Epidemiology* **10**: 31.

80. Crippa, G., Sverzellati, E., Giorgi-Pierfranceschi, M. and Carrara, G. C. (1999). Magnesium and cardiovascular drugs: interactions and therapeutic role. *Ann. Ital. Med. Int.* **14**: 40.

81. Glick, J. L. (1990). Dementias: the role of magnesium deficiency and an hypothesis concerning the pathogenesis of Alzheimer's disease [published erratum appears in *Med, Hypotheses* 1990 December; **33**(4): preceding 301]. *Med. Hypotheses* **31**: 211.

82. Durlach, J. (1990). Magnesium depletion and pathogenesis of Alzheimer's disease. *Magnes. Res.* **3**: 217.

83. Durlach, J., Bac, P., Durlach, V., Durlach, A., Bara, M. and Guiet-Bara, A. (1997). Are age-related neurodegenerative diseases linked with various types of magnesium depletion? *Magnes. Res.* **10**: 339.

84. Toba, Y., Kajita, Y., Masuyama, R., Takada, Y., Suzuki, K. and Aoe, S. (2000). Dietary magnesium supplementation affects bone metabolism and dynamic strength of bone in ovariectomized rats. *J. Nutri.* **130**: 216.

85. Hashimoto, M., Shinozuka, K., Shahdat, H. M., Kwon, Y. M., Tanabe, Y., Kunitomo, M. and Masumura, S. (1998). Antihypertensive effect of all-cis-5,8,11,14,17-icosapentaenoate of aged rats is associated with an increase in the release of ATP from the caudal artery. *J. Vasc. Res.* **35**: 55.

86. Aguila, M. B. and Mandarim-de-Lacerda, C. A. (1999). Numerical density of cardiac myocytes in aged rats fed a cholesterol-rich diet and a canola oil diet (n-3 fatty acid rich). *Virchows. Arch.* **434**: 451.

87. von Schacky, C. (2000). n-3 fatty acids and the prevention of coronary atherosclerosis. *Am. J. Clin. Nutri.* **71**: S224.

88. Siscovick, D. S., Raghunathan, T., King, I., Weinmann, S., Bovbjerg, V. E., Kushi, L., Cobb, L. A., Copass, M. K., Psaty, B. M., Lemaitre, R., Retzlaff, B. and Knopp, R. H. (2000). Dietary intake of long-chain n-3 polyunsaturated fatty acids and the risk of primary cardiac arrest. *Am. J. Clin. Nutri.* **71**: S208.

89. Anti, M., Marra, G., Armelao, F., Bartoli, G. M., Ficarelli, R., Percesepe, A., De Vitis, I., Maria, G., Sofo, L., Rapaccini, G. L. *et al.* (1992). Effect of omega-3 fatty acids on rectal mucosal cell proliferation in subjects at risk for colon cancer [see comments]. *Gastroenterology* **103**: 883.

90. Bougnoux, P. (1999). n-3 polyunsaturated fatty acids and cancer. *Curr. Opin. Clin. Nutri. Metab. Care* **2**: 121.

91. Cellini, L., Di Campli, E., Masulli, M., Di Bartolomeo, S. and Allocati, N. (1996). Inhibition of *Helicobacter pylori* by garlic extract (Allium sativum). *FEMS. Immunol. Med. Microbiol.* **13**: 273.

92. Chung, J. G., Chen, G. W., Wu, L. T., Chang, H. L., Lin, J. G., Yeh, C. C. and Wang, T. F. (1998). Effects of garlic compounds diallyl sulfide and diallyl disulfide on arylamine N-acetyltransferase activity in strains of *Helicobacter pylori* from peptic ulcer patients. *Am. J. Chin. Med.* **26**: 353.

93. Gail, M. H., You, W. C., Chang, Y. S., Zhang, L., Blot, W. J., Brown, L. M., Groves, F. D., Heinrich, J. P., Hu, J., Jin, M. L., Li, J. Y., Liu, W. D., Ma, J. L., Mark, S. D., Rabkin, C. S., Fraumeni, J. F., Jr. and Xu, G. W. (1998). Factorial trial of three interventions to reduce the progression of precancerous gastric

lesions in Shandong, China: design issues and initial data. *Contr. Clin. Trials* **19**: 352.

94. Lu, C. Y., Lee, H. C., Fahn, H. J. and Wei, Y. H. (1999). Oxidative damage elicited by imbalance of free radical scavenging enzymes is associated with large-scale mtDNA deletions in aging human skin. *Mutat. Res.* **423**: 11.

95. Kowald, A. and Kirkwood, T. B. (2000). Accumulation of defective mitochondria through delayed degradation of damaged organelles and its possible role in the ageing of post-mitotic and dividing cells. *J. Theor. Biol.* **202**: 145.

Chapter 74

UV-DNA Repair Enzymes in Liposomes

Daniel Yarosh

Daniel Yarosh • Applied Genetics Inc. Dermatics, 205 Buffalo Avenue, Freeport, New York 11520
Tel: (516)-868-9026, E-mail: DanYarosh@AGIDERM.COM

1. Introduction: What is Aging?

When Henry Ford was perfecting the mass assembly of the Model T, he sent his senior engineers out to the junkyards across the country to examine the remains of his discarded cars. The engineers reported back that the Model T's they saw were in bad shape: the tires were bald, the upholstery was ripped, the engine blocks were cracked and the axels bent. However, they noted, the shock absorbers on most of the junked cars were intact and still functional. Henry Ford immediately fired the supervisor in charge of the shock absorbers, and ordered that the new supervisor change vendors for this part. He explained that it was a terrible waste of money to equip his cars with high quality shock absorbers that outlived the rest of the equipment.

Henry Ford understood a principle of aging: it is a waste of resources to maintain one organ that will outlast the rest of the body. The best strategy for maximizing an animal's resources is to have all the body's organs deteriorate at about the same rate. The time that this takes is called "longevity" and there is little doubt that an organism's genes dictate its length. In fruit flies, for example, the age at death is strongly related to the age at which reproduction begins, suggesting that programmed events early in life trigger a program that controls lifespan.[1] The first part of aging, therefore, is longevity, or the maximum lifespan, and is under genetic control.

Industrial nations have been reaching closer and closer to this maximum lifespan, particularly with improvements in public health in the 19th century and medical advances in the 20th century. Today, 13% of their population is over 65 but in 30 years this will grow to 23%. The increase in the average lifespan has been achieved by controlling the effect of environmental forces, such as pollution and infectious disease. This second part of aging, therefore, is largely not under genetic control but is determined by environmental exposure, and is something we can alter, both as individuals and as societies.

What is commonly call "aging", therefore, has a ceiling that is controlled by genes (longevity), and the steps by which the ceiling is reached that is controlled by the environment (aging), as shown in Fig. 1.

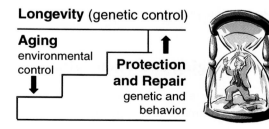

Fig. 1

Efforts toward protection from the environment and repair of any damage have increased the average lifespan and inched closed to the limits imposed by longevity. An understanding of the molecular biology behind each of these processes will make climbing those stairs more enjoyable, or less painful, and actually breaking through the ceiling.

2. The Molecular Genetics of Longevity: Telomerase and DNA Repair

We have only just begun to investigate the molecular genetics of longevity. The best understood system was first discovered in the last decade as one controlled by the enzyme "telomerase". Telomerase is a polymerase that fills in the frayed ends of chromosomes when they are replicated, and it is essential to cell division. As cells progress through their lineage by dividing, they gradually lose expression of the telomerase gene until they no longer are able to fill in the chromosome ends and become unable to divide.[2] This is the molecular explanation for the "Hayflick phenomenon", that is, that cells have a limited lifespan in culture that is not dependent on time, but on the number of cell divisions. Recently, gene therapy in mice was used to restore telomerase activity in liver, and the aging process leading to cirrhosis was reversed.[3] The genetic control of telomerase expression, therefore, is one way in which genes coordinately control longevity in all dividing cells. Telomerase cannot explain all of longevity, but it may be a model to understand how other gene systems determine the maximum lifespan.

Even with functional telomerase cells die from environmental factors. Just living at 37°C destroys about 10 000 DNA bases per genome per day.[4] During oxygen metabolism for aerobic energy about 2% of the oxygen is inevitably converted into free radicals.[5] Sunlight is a serious carcinogen that humans are exposed to every day. In addition, infections and autoimmune diseases result from defects in our immune system, often triggered by infectious agents.

Defense mechanisms are available to cope with these challenges. Among the most widely studied are DNA repair systems to replace damaged DNA bases and antioxidants to soak up free radicals. Damaged cells are eliminated by apoptosis and non-functional proteins are turned-over by proteases.

3. The Spiral of Life: Aging and Photoaging

These processes comprise a Spiral of Life: normal tissue is damaged by environmental assault. This damage is sometimes overt but often subclinical and without any visible symptoms. An inflammatory response ensues in which defense mechanisms are engaged. This results in repair of the damage, but the tissue is

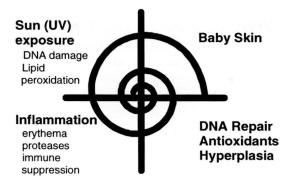

Fig. 2

never quite restored to its original state. Over the decades, repetitive cycles of damage, inflammation and repair lead to aging.

Perhaps the best understood example is Photoaging, or the effect of sun on skin (Fig. 2). Here the molecular biology is well understood. The ultraviolet component of the solar light spectrum is the most damaging to skin. UV rays are absorbed by DNA, leading to photoproducts that inactive the functions of DNA. In addition, UV produces oxygen radicals that oxidize lipids and proteins. The skin then springs into action with an inflammatory response, characterized by erythema (sunburn), the release of proteases and cytokines. Infectious agents that may try to take advantage of this compromised situation are sought out and destroyed. A temporary period of immune suppression follows. During that time repair systems are activated that excise the DNA lesions and replace the damaged DNA. Antioxidant systems absorb the oxygen radicals, and the damaged lipids are cleaved, digested and replaced. Irreparably damaged cells are discarded by apoptosis and hyperplasia fills in the missing tissue.

Unfortunately the damaged components are not always completely repaired. Mutations characteristic of UV exposure are commonly found in the DNA of sun exposed skin. In addition, in signaling the damage and marshalling the defense, irrevocable losses are sustained to the skin's infrastructure. Intercellular signals, such as cytokines, activate proteases that chew on the collagen support and dilate blood vessels. They also suppress some aspects of the immune system that allow mutant cells to grow.

Following any one battle with UV rays, these changes are virtually undetectable. However, chronic sun exposure in the face of normal DNA repair still results in persistent damage that can be detected in the DNA of skin. The continual rounds of damage, inflammation and repair over the course of 60 years of sun exposure, even with subclinical damage, finally accumulate changes that we recognize as aged skin.

Perhaps this model of photoaging applies to other organ systems. The continual exposure of lungs to dust, smoke and infectious agents, the gastrointestinal tract to natural poisons in food and more infectious agents, the cardiac and circulatory systems to oxidized lipids in the blood, leads to repeated cycles of subclinical damage, inflammation and repair that over time reduces and finally destroys each of these organ systems in the process of aging. It should not be surprising, therefore, as we have conquered the more common acute causes of death, such as small pox and polio, a myriad of chronic disease of each major organ system now replace them. Cancer in each of the major organ systems, of course, is characterized by long latency and is the ultimate expression of the gradual loss of defensive tools. These are the manifestations of aging that make the steps toward the ceiling of longevity unpleasant or even intolerable.

4. UV-DNA Repair Enzymes in Liposomes

4.1. Engineering DNA Repair in Skin

The skin is among the first targets of anti-aging therapy since the outward manifestations of the Spiral of Life are first seen here. The aging mechanisms are better understood than in any other organ system, and tools are available to enhance DNA repair. The approach at AGI Dermatics is to purify DNA repair enzymes that repair damaged caused by solar UV, and to encapsulate them in liposomes, or microscopic lipid vesicles. These pH-sensitive liposomes are specially engineered to cross the stratum corneum, localize in the epidermis, and deliver the DNA repair enzyme into the keratinocytes.[6]

The cyclobutane pyrimidine dimer (CPD), formed in DNA by fusing adjacent pyrimidine bases in a single strand, has been the substrate for therapy, since these are formed most frequently after UV exposure and have turned out to be responsible for many of the biological effects of UV. Of course CPD's in DNA interfere with replication and result in mutations. But persistent CPD's also trigger a response in keratinocytes that produce antigen-specific immune suppression,[7] and this immunosuppression is essential to the development of tumors.

In 1975, Tanaka, Sekiguchi and Okada[8] demonstrated that the bacteriophage T4 endonuclease V could initiate excision repair in human cells in culture. This endonuclease specifically recognizes cyclobutane pyrimidine dimers (CPD) in DNA, the most common lesion caused by sunlight in DNA. It initiates base excision repair by cutting the single-strand of DNA at the site of the CPD. Enzymes of the host cell then follow to remove the lesion and resynthesize undamaged single-strand DNA using the opposite strand as a template. We therefore encapsulated T4 endonuclease V into liposomes and these are termed "T4N5 Liposomes".

A similar enzyme is photolyase, which absorbs visible light and uses the energy to specifically cleave CPD to restore the original pyrimidine bases.[9] Photolyase injected into human cells and then exposed to photoreactivating light increase the repair of UV-induced DNA damage.[10] Photolyase encapsulated in liposomes are termed "Photosomes®".

4.2. Repair of DNA Damage

These enzymes repair CPD's in human cells after application in liposomes. T4N5 liposomes deliver the enzyme within 60 min, and repair is initiated immediately. By 6 h almost half the CPD's have been removed.[11] The delivery of this liposomal enzyme in a hydrogel lotion and applied to UV-irradiated human volunteers increased the repair of CPD's in the skin compared to untreated controls.[12]

Photosomes also are delivered to cells by liposomes and when exposed to visible/UV-A light the photolyase reverses up to 50% of the CPD's.[13, 14] Photosomes in a hydrogel lotion applied to UV-irradiated skin also removed 50% of the CPD's, particularly in the basal layer.[15] These studies demonstrate that these engineered liposomes deliver enzymes into cells of the skin.

4.3. Protection from UV Immunosuppression

Unrepaired lesions in DNA set the stage for activation of oncogenes or inactivation of tumor suppressor genes. However, UV can also promote the growth of tumors.[16] This is accomplished by suppression of the immune surveillance system through the release of immunosuppressive cytokines that generate T-suppressor cells.[7] Both T4N5 liposomes and Photosomes repair of CPD's and reduce UV-induced immunosuppression.[14, 15, 17] The result of this removal of DNA damage and prevention of immunosuppression is that DNA repair liposomes reduced the incidence of skin cancer in UV-irradiated mice and reduced the tumor yield by 50%.[18, 19]

4.4. Clinical Testing

T4N5 liposome lotion has been tested in human subjects over the last 7 years. Fifteen volunteers with a history of skin cancer were UV irradiated and treated with four doses of T4N5 liposome lotion before biopsy and measurement of CPD. Sites treated with T4N5 liposomes showed increased repair compared to untreated sites.[12] Twelve XP patients were irradiated with UV and treated once with T4N5 liposome lotion. After 6 h, a biopsy was taken to measure CPD. The treated sites showed 20% fewer CPD per cell than untreated control sites.[20] And, as mentioned above, Photosomes reduced UV-induced immunosuppression in human volunteers.[15]

In the most comprehensive study of precancerous lesions ever undertaken in the DNA repair deficiency disease xeroderma pigmentosum (XP), 30 patients were recruited who had a history of skin cancer or AK. All lesions were removed prior to the beginning of the trial. By random and blind selection, 20 patients applied T4N5 liposome lotion to their skin daily, and 10 patients applied a placebo control lotion. New AKs and skin cancers were scored every 3 months for one year. The results showed that XP patients using T4N5 liposome lotion had their rate of new AKs reduced by 68% compared to placebo controls.[21]

These studies suggest that it is feasible and indeed practical to remove lesions in DNA and slow the progression of skin aging leading to cancer.

5. Commercial Challenges and Opportunities in Anti-Aging Medicine

There are several challenges to be met. The new drugs that come out of anti-aging research will be preventative medicines, and will face a regulatory system that favors treatment programs over prevention programs. This makes their approval more uncertain than a therapeutic drug, and discourages the investment of resources by biotech companies and the capital markets that will be necessary to make significant strides. When the drugs are approved, they will face a system of insurance reimbursement and third party payment that also favors therapeutic over preventative drugs. The countervailing force is the overwhelming desire of the public to have anti-aging drugs.

6. Conclusion

We are entering an era when molecular biology can offer answers to the genetic limits of longevity and the environmental contribution to aging. We have much more to learn about the molecular controls on life span, and we have precious few tools to manipulate these mechanisms. Our knowledge is much greater about the environmental influences on aging. Here we are just entering the phase of application of molecular biology to prevention of aging.

Acknowledgments

I would like to thank Drs. Barbara Gilchrest, John Voorhees and Margaret Kripke for their contributions that have inspired me in the development of these ideas.

References

1. Sgro, C. and Partridge, L. (1999). A delayed wave of death from reproduction in *Drosophila*. *Science* **286**: 2521–2524.
2. Kruk, P., Rampino, N. and Bohr, V. (1995). DNA damage and repair in telomeres: relation to aging. *Proc. Natl. Acad. Sci. USA* **92**: 258–262.
3. Rudolph, K., Chang, S., Millard, M., Schreiber-Agus, N. and DePhinho, R. (2000). Inhibition of experimental liver cirrhosis in mice by telomerase gene delivery. *Science* **287**: 1253–1258.
4. Lindahl, T. and Nyberg, B. (1972). Rate of depurination of native deoxyribonucleic acid. *Biochemistry* **11**: 3610–3618.
5. Darr, D. and Fridovich, I. (1994). Free radicals in cutaneous biology. *J. Invest. Dermatol.* **102**: 671–675.
6. Yarosh, D., Bucana, C., Cox, P., Alas, L., Kibitel, J. and Kripke, M. (1994). Localization of liposomes containing a DNA repair enzyme in murine skin. *J. Invest. Dermatol.* **103**: 461–468.
7. Nishigori, C., Yarosh, D., Donawho, C. and Kripke, M. (1996). The immune system in ultraviolet carcinogenesis. *J. Invest. Dermatol. Symp. Proc.* **1**: 143–146.
8. Tanaka, K., Sekiguchi, M. and Okada, Y. (1975). Restoration of ultraviolet-induced unscheduled DNA synthesis of xeroderma pigmentosum cells by the concomitant treatment with bacteriophage T4 endonuclease V and HVJ (Sendai virus). *Proc. Natl. Acad. Sci. USA* **72**: 4071–4075.
9. Yasui, A. and Eker, A. (1995). DNA photolyases. *In* "DNA Damage and Repair" (J. Nickoloff, and M. Hoekstra, Eds.), pp. 9–32, Humana Press Inc., Totowa, New Jersey.
10. Rosa, L, Vermeulen, W., Bergen Henegouwen, J. *et al.* (1990). Effects of microinjected photoreactivating enzyme on thymine dimer removal and DNA repair synthesis in normal human and xeroderma pigmentosum fibroblasts. *Cancer Res.* **50**: 1905–1910.
11. Yarosh, D., O'Connor, A., Alas, L., Potten, C. and Wolf, P. (1999). Photoprotection by topical DNA repair enzymes: molecular correlates of clinical studies. *Photochem. Photobiol.* **69**: 136–140.
12. Wolf, P., Maier, H., Mullegger, R. *et al.* (2000). Topical treatment with liposomes containing T4 endonuclease V protects human skin in vivo from ultraviolet-induced upregulation of interleukin-10 and tumor necrosis factor-α. *J. Invest. Dermatol.* **114**: 149–156 (XP clinical trial summary).
13. Yarosh, D. and Klein, J. (1994). Biological consequences of repair of UV-induced DNA damage. *Trends Photochem. Photobiol.* **3**: 175–181.
14. Vink, A., Moodycliffe, A., Shreedhar, V. *et al.* (1997). The inhibition of antigen-presenting activity of dendritic cells resulting from UV irradiation of murine skin is restored by *in vitro* photorepair of cyclobutane pyrimidine dimers. *Proc. Natl. Acad. Sci. USA* **94**: 5255–5260.

15. Stege, H., Roza, L., Vink, A. *et al.* (2000). Enzyme plus light therapy to repair DNA damage in ultraviolet-B-irradiated human skin. *Proc. Natl. Acad. Sci. USA* **97**: 1790–1795.

16. Yarosh, D. and Kripke, M. (1996). DNA repair and cytokines in antimutagenesis and anticarcinogenesis. *Mutat. Res.* **350**: 255–260.

17. Kripke, M., Cox, P., Alas, L. and Yarosh, D. (1992). Pyrimidine dimers in DNA initiate systemic immunosuppression in UV-irradiated mice. *Proc. Natl. Acad. Sci. USA* **89**: 7516–7520.

18. Yarosh, D., Alas, L., Yee, V., Oberyszyn, A., Kibitel, J., Mitchell, D., Rosenstein, R., Spinowitz, A. and Citron, M. (1992). Pyrimidine dimer removal enhanced by DNA repair liposomes reduces the incidence of UVR skin cancer in mice. *Cancer Res.* **52**: 4227–4231.

19. Bito, T., Ueda, M., Nagano, T., Fujii, S. and Ichihashi, M. (1995). Reduction of ultraviolet-induced skin cancer in mice by topical application of DNA excision repair enzymes. *Photoderm. Photoimm. Photomed.* **11**: 9–13.

20. Yarosh, D., Klein, J., Kibitel, J. *et al.* (1996). Enzyme therapy of xeroderma pigmentosum: safety and efficacy testing of T4N5 liposome lotion containing a prokaryotic DNA repair enzyme. *Photoderm. Photoimm. Photomed.* **12**: 122–130.

21. Yarosh, D., Klein, J., O'Connor, A., Hawk, J. and Rafal, E. (2001). Effects of topically applied T4 endonuclease V in liposomes on skin cancer in xeroderma pigmentosum: a randomized study. *Lancet* **357**: 926–929.

Chapter 75

Mitochondrial Dysfunction in Aging and Disease: Development of Therapeutic Strategies

James A. Dykens, Amy K. Carroll, Amy C. Wright and Beth Fleck

Keywords: Electron transport, respiration, oxidative phosphorylation, superoxide, nitric oxide, permeability transition, cyclosporin A, bongkrekic acid, adenine nucleotide translocase, MELAS, LHON, MERRF, apoptosis.

James A. Dykens, Amy K. Carroll, Amy C. Wright and **Beth Fleck** • MitoKor, 11494 Sorrento Valley Rd, San Diego, CA USA

*Corresponding Author.
Tel: 858-509-5617

1. Summary

Mitochondrial dysfunction is known to cause several dozen well-defined diseases, and is increasingly implicated in models of constitutive aging, as well as in pathology ranging from chronic neurodegenerative diseases to acute myocardial infarction. Mitochondria fail via a number of mechanisms, including irreversible permeability transition, increased free radical production, BAX/Bcl-2 interactions, and Ca^{2+} overload, with each mechanism offering opportunities, as well as pitfalls, for possible therapeutic agents. A novel, cell-based, fluorescence resonance energy transfer (FRET) assay for mitochondrial membrane potential *in situ* has been developed, and the resulting high throughput compound discovery program holds promise for development of mitochondrially-targeted therapies.

2. Introduction

On-going discoveries of novel deletions and single nucleotide polymorphisms in mitochondrial DNA are fostering the paradigm shift that is afoot in the field of mitochondrial biology, a trend destined to continue given that mitochondria are the only metazoan organelle that contain a completely functional, extra-nuclear genome (Refs. 1–3; see chapter by Aubrey de Grey). However, the current renaissance in the field of mitochondrial physiology is atypical in that it is not being driven wholly by the techniques of molecular biology. Instead, reinvigorated investigations into mitochondrial bioenergetics, ultrastructure, electron transport and coupling of electron transport and phosphorylation (oxidative phosphorylation; OXPHOS) have been prompted in large measure by the realization that mitochondria occupy a unique central role in determining cell viability, senescence and death. Mitochondria are not only the key proximate determinants of cellular energetic and oxidative status, thermogenesis, and Ca^{2+} buffering capacity, but they also integrate the various pro- and anti-apoptotic signals into actions that will ultimately either induce, or repress, apoptosis.

Given such pervasive and crucial roles, it should not be surprising that mitochondrial dysfunction figures prominently in constitutive processes such as aging, chronic neurodegenerative diseases, and acute pathology associated with transient ischemia and reperfusion. Since the initial description of the first mitochondrial disease by Rolf Luft in 1959, dozens of diseases with a mitochondrial etiology have been described, including mitochondrial encephalopathy, lactic acidosis, and strokelike episodes syndrome (MELAS) and myoclonus epilepsy with ragged-red fibers (MERRF), many of them associated with known deletions or other defects in mitochondrial DNA (mtDNA).[4-7] Despite the strides made in characterizing the symptomology and etiology of these ailments, the continuing paucity of therapeutic agents remains a major hurdle, and thus an opportunity for drug development.

Mitochondrial impairment has also been increasingly implicated in more prevalent sporadic degenerative diseases. For example, impairment of complex IV in the electron transport system (cytochrome c oxidase) is found in systemic and neuronal tissues from Alzheimer's disease patients, whereas specific impairment of complex I is detected in tissues from Parkinson's disease patients.[8-10] Indeed, our insight into the molecular etiology of a host of neurodegenerative diseases has been fundamentally altered by studies implicating mitochondrial dysfunction in Ca^{2+} dyshomeostasis, free radical production and neuronal death.[11-15] Similarly, our understanding of the normal aging process has been substantially illuminated by genetic and bioenergetic studies implicating mitochondrial senescence and associated radical generation and energetic decline in gradual cellular deterioration (Refs. 16–20; see chapter this volume by Kirkwood).

The primary bioenergetic function of mitochondria is the complete oxidation of glucose, amino acids, and fats to CO_2 and water coupled with generation of ATP via OXPHOS to yield 36 moles of ATP per mole of glucose, 18-fold more than anaerobic glycolysis. This process is entirely dependent on both the impermeability of the inner mitochondrial membrane, where the proteins that catalyze the electron transfer reactions of respiration are embedded, and their catalytic function, which is also dependent upon their orientation and lateral mobility within that membrane. The sequential reduction/oxidation reactions of the respiratory system are coupled stochiometrically to translocation of protons from the matrix to the inner membrane space, resulting in a ~ 220 mV electrochemical potential (matrix negative). The majority of this (180 mV) is due to ion potential and is termed $\Delta\Psi_m$, with the remainder due to a pH gradient (ΔpH). Peter Mitchell was awarded a 1978 Nobel Prize for his work showing that the potential energy of this gradient is coupled to ADP phosphorylation via activity of F_0-F_1 ATP synthtase, the molecular mechanisms of which have only been recently elucidated.[21-22] In any event, establishment and maintenance of $\Delta\Psi_m$ is directly dependent not only on the electron transfer system (ETS), but also on the impermeability of the inner membrane. Loss of inner membrane impermeability, with complete mitochondrial failure, can occur via several mechanisms, but the well-known ability of agents such cyclosporin A and bongkrekic acid to moderate this collapse of $\Delta\Psi_m$ suggests that at least some of these mechanisms are amenable to therapeutic intervention.[23]

3. Opportunities for Mitochondrial Therapeutics

The obvious ramification of the recent studies implicating mitochondrial dysfunction in pathology in chronic neurodegenerative diseases,[3, 10, 11, 24–27] and in acute ischemic pathology associated with stroke and myocardial infarction[12, 13, 28, 29] is the opportunity for development of therapeutic strategies aimed at improving (or salvaging) mitochondrial function under these circumstances. This may be especially true because proper mitochondrial function depends not only on structural integrity of the inner membrane and correct quaternary structure of the respiratory complexes, but also on the integration of microanatomy with exceedingly complex biochemical machinery. This machinery includes: (1) the redox centers of the electron transport system, some of which are composed of over 40 separate protein subunits encoded by both the mitochondrial and nuclear genomes, (2) the enzymes of β-oxidation and the Kreb's cycle, (3) the cornucopia of specialized trans-membrane transport mechanisms designed to import, export, and target metabolites and proteins across the outer, and more impermeable, inner mitochondrial membranes, including most notably the ATP synthase, (4) the machinery responsible for replication, repair, transcription and translation of the mitochondrial genome (16.5 kB in humans, encoding 13 subunits essential for electron transport system), and (5) the mechanisms for growth and replication (*de novo* mitochondrial formation does not occur).

Such bewildering complexity, plus the diversity of pathology in which mitochondrial dysfunction is implicated, may prove the point that failure rates increase as a function of complexity. However, a more optimistic view is that such complexity offers a myriad of interventional opportunities. Moreover, because mitochondrial dysfunction elicits so many dire consequences, even modest success at improving mitochondrial function is likely to yield substantial benefits.

An obvious direction for development of mitochondrially directed therapies is the prevention (or induction) of apoptosis. A number of apoptogens (such as TNF-α) induce translocation of cytosolic members of the BAX and Bcl-2 family of proteins (among others) to the mitochondrial membranes where they likely form hetero- and homo-multimers to create pores that breach the essential impermeability of the membranes.[30, 31] If the balance of pro-apoptotic proteins recruited to the mitochondrion exceeds the anti-apoptotic capacity, the organelle loses its potential and swells as water enters along its concentration gradient. In so doing, cytochrome c and apoptosis induction factor (AIF), a FAD containing oxidoreductase with high homology to ascorbate reductase, are expelled from the mitochondrial intermembrane space into the cytosol.[32] Once in the cytoplasm, these proteins activate caspases and induce nuclear chromatin condensation and fragmentation, respectively, as part of the apoptotic cascade. The initial models of cytochrome c and AIF release were based on mitochondrial swelling with consequent rupture of outer membrane and non-selective efflux. However, there is now evidence from several labs that under some conditions, selective cytochrome c release occurs in the absence of swelling without coincident release of other comparable sized intermembrane proteins, like adenylate kinase.[33, 34] Such data likely reflect more precise, and multi-factorial, regulation of apoptosis than has previously been suspected, and underscores the opportunities for development of novel interventional techniques.

3.1. Mitochondrial Permeability Transition and Radical Production

Much research on mitochondrially-based therapeutics has focused on several sporadic neurodegenerative diseases and excitotoxicity, primarily because the evidence for oxidative pathology and mitochondrial involvement is more fully developed in these areas.[11] However, examination of acute cellular and mitochondrial pathology associated with ischemia and reoxygenation of aerobically poised tissues such as myocardium or kidney, and chronic mitochondrial impairment associated with aging, reveals similar mechanisms of mitochondrial failure (albeit of different magnitudes and timing). Although superficially disparate, these circumstances share in common several functional parallels with mitochondrial dysfunction in neurodegenerative disease, such as reductions in the ATP/ADP ratio that can arise when ATP utilization outpaces production, or when availability of O_2 as the terminal electron acceptor in the mitochondrial electron transfer system is inadequate due to hypoxia, or when genetic or radical-induced defects in electron transfer components chronically impair OXPHOS.[11, 35]

The convergence of free radical pathophysiology, degenerative disease (including the decline of pancreatic B-cells in diabetes), and aging has also been fostered by evidence linking mitochondrial radical production and bioenergetic physiology as _synergistic_ contributors to cell senescence and death.[17, 28, 36-40] For example, free radical exposure greatly potentiates, by lowering the levels of Ca^{2+} required, the catastrophic loss of inner membrane impermeability and consequent permeability transition (PT), that entails an irreversible collapse of mitochondrial membrane potential.[28, 34, 36-38] Despite its disastrous nature, PT is moderated by cyclosporin A and bongkrekic acid, providing encouragement that small molecule development strategies targeting PT hold promise.

Controversy surrounding the true nature of PT remains, primarily because none of the extant models completely accounts for all the data. Nevertheless, a likely cause for PT is an irreversible conformational shift in the adenine nucleotide translocase (ANT), a transmembrane carrier that uses $\Delta\Psi_m$ to exchange ATP for ADP across the inner membrane.[36, 41, 42] One of the most abundant proteins in the mitochondrion, ANT is found at contact points between the outer and inner mitochondrial membranes where it is usually associated with a number of other proteins, including cyclophilin D (the binding site of cyclosporin A), creatine kinase, and voltage dependent anion channel (VDAC), among others. ANT binds both bongkrekic acid and atractyloside (a thistle toxin) in the adenylate binding site with high affinity and specificity. Both agents block adenylate exchange, and both lead to mitochondrial failure and cell death, reiterating the essential role mitochondrial ATP plays in cell viability. Interestingly, for full catalytic capacity ANT requires small amounts of cardiolipin, a membrane lipid found almost exclusively within the inner mitochondrial membrane, suggesting the possibility of allosteric modulation as a strategy for compound development with fewer potentially deleterious pitfalls than agents binding at the adenylate site.

Regardless of the real nature of PT, $\Delta\Psi_m$ collapse obviously abrogates ATP production, and also causes mitochondrial swelling as water enters along its osmotic gradient. Importantly, PT greatly accelerates free radical production from mitochondria.[24] When provided with adequate cofactors, oxidizable substrates and O_2, free radical production from well-coupled mitochondria isolated from rat cerebral cortex and cerebellum is below the levels of detection of electron paramagnetic resonance spin-trapping techniques. However, immediately upon exposure to μM Ca^{2+} concentrations, spin adducts with hyperfine splitting characteristics consistent with both carbon- and oxygen-centered radicals are readily detected under identical conditions. Moreover, radical exposure (H_2O_2 plus Fe-EDTA) substantially increases subsequent radical production almost 10-fold, suggesting that oxidative stress arising from one mitochondrion likely oxidatively imperils neighboring organelles.[24, 25]

In contrast to the potentiation of PT induced by radical exposure, high ATP-to-ADP ratios and even the mere presence of adenylates, serve to protect against Ca^{2+}-mediated induction of PT.[28, 34–38, 41, 42] Indeed, mitochondria exposed to physiologically relevant concentrations (mM) of ATP tolerate exposures to 8-fold higher Ca^{2+} concentrations before undergoing PT. As such, susceptibility to PT is a result of an equilibrium between oxidative and energetic factors, suggesting that slightly shifting this equilibrium might be a more readily attainable therapeutic goal than complete blockade of PT, which may well have undesirable consequences from the stand-point of adenylate exchange.

Not unexpectedly, given the declines in mitochondrial function and corresponding increases in oxidative markers associated with normal aging, mitochondria in tissues from aged animals show increased susceptibility to PT.[43, 44] Although such observations are intriguing, it remains an unjustified inference to assume that small molecules capable of improving mitochondrial stability and function will be efficacious as anti-aging treatments; aging is a sum of many parts, some of which are relatively independent of mitochondrial status. Nevertheless, it seems unlikely that improving the energetic status of senescent tissues will prove injurious, as long as the concomitant oxidative component of mitochondrial function can be simultaneously moderated. Indeed, it has already been shown that small molecule superoxide dismutase mimetics with catalase activity can extend lifespan in *C. elegans*.[18]

3.2. Antioxidants as Mitochondrial Therapeutics

From a thermodynamic perspective, the half-cell reduction potentials indicate that any of mitochondrial electron transport components can autoxidize to univalently reduce oxygen to superoxide radical ($O_2^{\cdot-}$). However, it appears that between 75–90% of the $O_2^{\cdot-}$ produced by mitochondria arises from autoxidation of the semiquinone radical form of CoQ, with the remainder largely due to autoxidation of Complex I (Refs. 11 and 28; see Beckman this volume). Mitochondrial radical production is in direct proportion to the rate of respiration, and ambient P_{O2}; the more aerobically poised and energetically demanding the cell, the higher the rate of radical production.[45] An oft-quoted approximation is that between 1–6% of the respired O_2 is univalently reduced to $O_2^{\cdot -}$. However most of these estimates were generated using excessive substrate concentrations,[46] and under ambient atmospheric partial pressure of oxygen (P_{O2} in air $\simeq$ 159 Torr) which is about 4-fold higher than levels prevailing in most human tissues (40 Torr), both of which would artifactually inflate rates of mitochondrial radical production. Nevertheless, calculations based on biologically relevant oxidative markers, such as nitrotyrosine formation and oxidation of DNA, combined with respiratory rates, indicate that estimates of univalent reduction to yield $O_2^{\cdot -}$ of between 1–5% total respiration are not unreasonable. Regardless of ambient radical production, it is clear not only that mitochondrial radical production is exacerbated when the components of the electron transport system are in a more fully reduced state, as occurs under conditions of diminished O_2 availability such as during ischemia, but also that excessive Ca^{2+} loading, with consequent permeability transition, greatly accelerates radical production.[24, 25, 47]

Although oxidative stress potentiates PT, enthusiasm for development of antioxidants as therapeutic agents needs to be tempered by a clear-eyed appraisal of some of the variables. In assessing antioxidant efficacy, it bears re-emphasis that the appropriate antioxidant (e.g. having a redox potential that will permit reaction with the radical at hand) must also be appropriately situated if it is to function; cytosolic superoxide dismutase will do little to moderate intramitochondrial $O_2^{\cdot -}$ production, and intramembranous vitamin E will do little to forestall cytosolic radical reactivity. This is especially so because the reaction between $O_2^{\cdot -}$ and SOD is about 6-fold slower than the competing reaction of $O_2^{\cdot -}$ with NO; when NO is available, formation of peroxynitrite anion will be preferred over dismutation, even in the presence of what appears to be ample capacity to remove $O_2^{\cdot -}$. As detailed elsewhere in this volume (see chapter by Beckman), the biochemistry of nitric oxide (NO) *in vivo* is complicated by the intracellular environment where thiol reductants, metals and a plethora of electron acceptors foster formation of a diverse family of reactive species, such as nitrosonium (NO^+), nitric oxide radical ($NO^\cdot$), and nitroxyl anion (NO^-), each of which can undergo subsequent reactions.[48]

Numerous transition metals capable of catalyzing radical and free radical cascades are essential for life, yet even minor perturbations in metal metabolism profoundly alter the cellular and mitochondrial oxidative milieu that can have both short term effects, such as plasma membrane peroxidation with loss of integrity that yields necrosis, and long term effects such as inactivation of key regulatory enzymes, and mutation of nDNA and mtDNA.[49]

The upshot of such considerations is that, although the data are compelling that oxidative load increases in a host of pathologies and with age,[20, 36, 50, 51] and that mitochondrial dysfunction and radical production increase with age, antioxidants designed to repress a specific radical species based on extrapolation from its *in vitro* behavior may be misleading *in vivo* where spontaneous interconversions and compartmentalization are the rule. As a result, effective antioxidant therapies will have to operate under the simultaneous constraints of: (1) functioning catalytically and in concert with endogenous antioxidants, as opposed to mass action radical scavengers that do not recycle, (2) possessing reaction kinetics sufficient to out-compete endogenous radical reactions, (3) obtaining appropriate localization in order to moderate the dominant radical reaction, (4) achieving long term efficacy without impairment of constitutive metal and oxidative metabolism.

Despite the limitations of such constraints, several antioxidant development programs have achieved sufficient pre-clinical and clinical efficacy to warrant continued pursuit. Interestingly, a disparity between modest *in vitro*, versus excellent *in vivo*, efficacy for many of these compounds implies a circuitous mechanism(s) of action. For example, several small molecule radical scavengers under development (including SOD mimetics) likely show efficacy in animal stroke models not by repressing neuronal radical production and hence forestalling neuronal death, but rather by moderating the upregulation of vascular endothelial proteins that effect adhesion and recruitment of circulating leukocytes.[52] A likely scenario is that these compounds reduce infarct size and repress neuronal apoptosis by diminishing leukocyte recruitment, and hence preventing the massive radical production resulting from activation of neutrophils, lymphocytes and other pro-inflammatory cells. In this light, these compounds are indeed acting as antioxidants, but not in the manner assumed. (As an aside, this supposition is readily tested by assessing efficacy of these compounds in animals rendered neutropenic via anti-PMN antibodies.) Moreover, recent data showing increased longevity in *C. elegans* treated with synthetic SOD/catalase mimetics not only supports the hypothesis that aging is an expression of chronic oxidative load,[18] but also provides encouragement that antioxidant strategies hold promise.

3.3. Functional Screening Based on Mitochondrial Membrane Potential

It is evident that dyshomeostasis of both Ca^{2+} and oxidative status is associated with mitochondrial dysfunction, and that both Ca^{2+} and oxidative regulation are progressively impaired in normal aging and in a host of pathologies.[53, 54] For example, comparable Ca^{2+} dyshomeostasis precedes neuronal death following exposure to dicarboxylic excitotoxins,[15, 55] and after transient hypoxia in both neurons and cardiac myocytes.[56] Although mitochondrial Ca^{2+} sequestration via the uniporter is the major Ca^{2+} buffering system in most cells, the data also indicate a substantial role for Ca^{2+} mismanagement by endoplasmic reticulum, independent of adenylate status, that increases with age and in some pathologies.[57, 58]

Clearly free cytosolic Ca^{2+} has a number of direct physiological effects, such as activation of phospholipase A_2, activation of proteases, and alterations in numerous metabolic pathways including Kreb's cycle enzymes. However, depending on oxidative status and the magnitude of Ca^{2+} load, even slightly immoderate increases in cytosolic free Ca^{2+} can perturb oxidative phosphorylation, and destroy mitochondrial function via induction of PT.[59, 60]

Assessment of mitochondrial membrane potential ($\Delta\Psi_m$) provides the single most comprehensive measure of mitochondrial bioenergetic function, primarily because it directly depends upon proper integration of the entire suite of diverse metabolic pathways that converge at the mitochondria.[29, 61] Although $\Delta\Psi_m$ is only one of several processes that establish and maintain the protonmotive force (Δp) across the mitochondrial inner membrane, it is the major component of that force, and it is the predominant driving force responsible for mitochondrial Ca^{2+} uptake, ATP generation, and radical production.[62] However, a major bottleneck for development of small molecule therapeutics designed to improve mitochondrial stability has been the paucity of techniques available to assess $\Delta\Psi_m$ in a high through-put format.

Previous assessment of $\Delta\Psi_m$ *in situ* has been done using membrane-permeant cationic dyes that partition between the cytoplasm and the mitochondrial matrix according to Nernstian dictates of concentration and potential. Although useful, such potentiometric dyes have several inherent limitations, most notably a lack of specificity for mitochondrial potential, as opposed to the potential present at the surrounding plasma membrane. Moreover, care must be taken to use these dyes at concentrations where they neither deplete $\Delta\Psi_m$, nor self-quench and thereby obscure the kinetics and magnitude of responses.[62] Such low dye concentrations are not amenable to most of technologies available for high-throughput screening of compound libraries that are the basis for contemporary drug discovery programs.

To circumvent this bottleneck, we have developed a novel assay for $\Delta\Psi_m$ based on fluorescence resonance energy transfer (FRET) between two dyes that must be co-localized to the mitochondria. Because the efficiency of energy transfer declines as a function of the intramolecular separation to the sixth power (r^6), the assay reports only those interactions within the mitochondria. In so doing, the FRET assay avoids the confounding variable of plasma membrane potential that has previously made it difficult to isolate $\Delta\Psi_m$ in intact cells in high throughput modes.

In the MitoKor FRET assay for $\Delta\Psi_m$, the excitation dye is nonyl acridine orange (NAO), a stain for cardiolipin (diphosphatidyl glycerol) that is distributed almost exclusively (> 99%) in the mitochondrial inner membrane.[63] The second dye is tetramethylrhodamine (TMRE), a potentiometric dye taken up into the mitochondrial matrix in proportion to $\Delta\Psi_m$ and the imposed concentration gradients.[64] Once in the matrix, TMRE quenches the emission of NAO, FRET between these two dyes therefore only occurs when $\Delta\Psi_m$ exists and both dyes are present. It is the specificity of NAO staining for cardiolipin, combined with the absolute prerequisite for close proximity of two dyes, that allows this FRET assay to report $\Delta\Psi_m$ unconfounded by dye associated with the plasma membrane potential.

Importantly, FRET does not occur when non-mitochondrial fluorescent dyes are substituted. For example, no FRET with TMRE is apparent when NAO is replaced by calcein or carboxyfluorescein, both of which have suitably overlapping excitation spectra, but both of which remain in the cytosol. Likewise, no FRET with NAO is detectable when TMR is replaced by the cytoplasmic dye carboxy SNAFL, despite it having appropriate excitation spectrum for FRET. Such findings are not surprising in light of the stringent requirement for close proximity in order for FRET to occur.[65]

Uptake and retention of the potentiometric dye depends on $\Delta\Psi_m$, so that the extent of TMR quenching of NAO directly reflects the magnitude of $\Delta\Psi_m$. Conversely, dissipation of $\Delta\Psi_m$ results in efflux of the potentiometric dye from

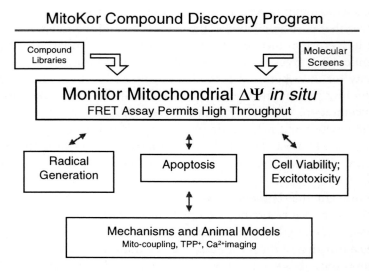

Fig. 1. High-throughput Compound Discovery Program Based on $\Delta\Psi_m$ — Assessment of $\Delta\Psi_m$ in high throughput mode permits compound screening against a variety of triggers to induce mitochondrial collapse. Initial screens have been for agents that moderate Ca^{2+}-mediated loss of $\Delta\Psi_m$, with Ca^{2+} loading via ionophores or via direct addition of Ca^{2+} to digitonin-permeabilized cells. Prevention of mitochondrial PT in the FRET assay predicts consequent reductions in apoptosis (determined as caspase-3 activation and annexin binding), improvements in cellular viability and adenylate status, and moderation of neuronal excitotoxicity, all of which serve as secondary, corroborative assays. Mechanistic tertiary assays entail classical polarography, assessment of $\Delta\Psi_m$ and Ca^{2+} using TPP and ion-selective electrodes, respectively, mitochondrial swelling and radical generation, among others.

the mitochondrion with a corresponding loss of proximity, and hence dequenching of NAO; in this assay, loss $\Delta\Psi_m$ is detected as an increase in the NAO signal (Fig. 1).

Collapse of $\Delta\Psi_m$, and permeability transition, can be induced in intact cells using Ca^{2+} ionophores, such as ionomycin and A-23187, or more directly by adding Ca^{2+} to cells where the plasma membrane has been permeabilized by low concentrations of digitonin. In the presence of oxidizable substrates $\Delta\Psi_m$ shows a transient decline upon addition of low Ca^{2+} concentrations, followed by complete recovery. Exposure to additional Ca^{2+} under these conditions yields irreversible collapse of $\Delta\Psi_m$ that, because it is moderated by cyclosporin A, reflects PT.[65, 66]

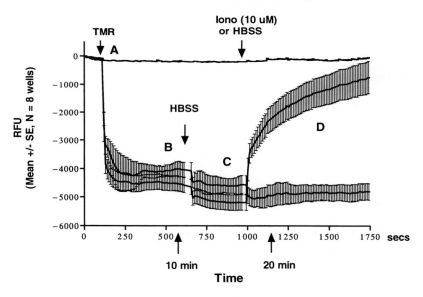

Fig. 2. FRET based HTS assay for $\Delta\Psi_m$ — Collapse of $\Delta\Psi_m$ in intact neuroblastoma cells (plated @ 60K/ well, 24 h prior to assay) is induced in a dose-dependent manner by Ca^{2+} load via the ionophore ionomycin.[63] Cells are stained with the cardiolipin stain nonyl acridine orange (NAO, 85 nM) for 5 min prior to initiation of assay. Quenching of the initial NAO signal (A) begins immediately upon addition of potentiometric dye tetramethylrhodamine (TMRE; final concentration = 150 nM) that is sequestered by the mitochondria in direct proportion to $\Delta\Psi_m$. Steady state of quenching is obtained within 1–2 min (B), and monitoring after addition of compound (C) can reveal potential effects of compound under scrutiny. Addition of ionomycin (5 μM) initiates a Ca^{2+}-mediated collapse of $\Delta\Psi_m$ (D) that is moderated by cyclosporin A and bongkrekic acid, indicating that it is due to permeability transition. In this assay protocol, each well can be normalized to the amount of initial quenching, greatly reducing artifactual variability due to cell plating and dye loading. CCCP (0.5 μM) also collapses $\Delta\Psi_m$, while oligomycin induces hyperpolarization.[63]

The FRET assay for $\Delta\Psi_m$ is currently serving as a gate-keeping primary screen in MitoKor's compound discovery and development program, with a series of secondary assays in place designed to corroborate independently the data from the $\Delta\Psi_m$ assay (Fig. 2). For example, agents that effectively moderate Ca^{2+}-mediated $\Delta\Psi_m$ collapse should also correspondingly moderate cytochrome c release and caspase 3 activation. Similarly, compounds that moderate $\Delta\Psi_m$ collapse induced by combined radical exposure and Ca^{2+} load, would be expected to moderate cytotoxicity in a model where these stressors are used to kill cells. The output of such a screening and structure-activity relationship optimization program will be compounds that can preserve (or improve) $\Delta\Psi_m$ under conditions of combined oxidative stress and Ca^{2+} dyshomeostasis that are relevant to a host of both chronic and acute pathologies. Given the similarities in mitochondrial dysfunction that underlie many of these pathologies and those associated with senescence, such compounds may also prove efficacious in moderating the progression of normal aging.

Acknowledgments

For the sake of brevity, most citations are reviews summarizing large areas of research. The author apologizes to the many scientists who have contributed invaluably to these fields, but whose work could not be specifically acknowledged. I thank M.D. Dykens for editorial assistance.

References

1. Ozawa, T. (1999). Mitochondrial genome mutation in cell death and aging. *J. Bioenerg. Biomembr.* **31**: 377–390.
2. Dykens, J. A., Moos, W. H. and Davis, R. E. (1999). An introduction to mitochondrial genetics and physiology. *Drug Dev. Res.* **46**: 2–13.
3. Moos, W. H., Dykens, J. A. and Davis, R. E. (1999). Mitochondrial biology and neurodegenerative diseases. *Pharm. News* **6**: 15–29.
4. Suomalainen, A. (1997) Mitochondrial DNA and disease. *Ann. Med.* **29**: 235–246.
5. Graff, C., Clayton, C. and Larsson, N. G. (1999). Mitochondrial medicine — recent advances. *J. Int. Med.* **246**: 11–23.
6. Simon, D. K. and Johns, D. R. (1999). Mitochondrial disorders: clinical and genetic features. *Ann. Rev. Med.* **50**: 111–127.
7. Wallace, D. C. (1999). Mitochondrial diseases in man and mouse. *Science* **283**: 1482–1488.
8. Cassarino, D. S. and Bennett, J. P., Jr. (1999). An evaluation of the role of mitochondria in neurodegenerative diseases: mitochondrial mutations and

oxidative pathology, protective nuclear responses, and cell death in neurodegeneration. *Brain Res. Brain Res. Rev.* **29**: 1–25.

9. Schapira, A. H. (1999). Mitochondrial involvement in Parkinson's disease, Huntington's disease, hereditary spastic paraplegia and Friedreich's ataxia. *Biochem. Biophys. Acta* **1410**: 159–170.

10. Beal, M. F. (2000). Energetics in the pathogenesis of neurodegenerative diseases. *Trends Neurosci.* **23**: 298–304.

11. Beal, M. F., Bodis-Wollner, I. and Howell, N. (1997). *Neurodegenerative Diseases: Mitochondria and Free Radicals in Pathogenesis,* John Wiley & Sons.

12. Reynolds, I. J. (1999). Mitochondrial membrane potential and the permeability transition in excitotoxicity. *Ann. NY Acad. Sci.* **893**: 33–41.

13. Siesjo, B. K., Elmer, E., Janelidze, S., Keep, M., Kristian, T., Ouyang, Y. B. and Uchino, H. (1999). Role and mechanisms of secondary mitochondrial failure. *Acta Neurochir. Suppl. (Wien)* **73**: 7–13.

14. Lipton, P. (1999). Ischemic cell death in brain neurons. *Physiol. Rev.* **79**: 1431–1568.

15. Stout, A. K., Raphael, H. M., Kanterewicz, B. I., Klann, E. and Reynolds, I. J. (1998). Glutamate-induced neuron death requires mitochondrial calcium uptake. *Nat. Neurosci.* **1**: 366–373.

16. Kirkwood, T. B. and Kowald, A. (1997). Network theory of aging. *Exp. Gerontol.* **32**: 395–399.

17. Cortopassi, G. A. and Wong, A. (2000). Mitochondria in aging and degeneration. *Biochim. Biophys. Acta* **1410**: 183–193.

18. Melov, S., Ravenscroft, J., Malik, S., Gill, M. S., Walker, D. W., Clayton, P. E., Wallace, D. C., Malfroy, B., Doctrow, S. R. and Lithgow, G. J. (2000). Extension of life-span with superoxide dismutase and catalase mimetics. *Science* **289**: 1567–1569.

19. Melov, S. (2000). Mitochondrial oxidative stress. Physiological consequences and potential role in aging. *Ann. NY Acad. Sci.* **908**: 219–225.

20. Samson, F. E. and Nelson, S. R. (2000). The aging brain, metals and oxygen free radicals. *Cell Mol. Biol. (Noisy-le-grand)* **46**: 699–707.

21. Groth, G. (2000). Molecular models of the structural arrangement of subunits and the mechanism of proton translocation in the membrane domain of F(1)F(0) ATP synthase. *Biochim. Biophys. Acta* **1458**: 417–427.

22. Nicholls, D. G. and Ferguson, S. J. (1992). *Bioenergetics 2,* p. 255, Academic Press, London.

23. Leventhal, L., Sortwell, C. E., Hanbury, R., Collier, T. J., Kordower, J. H. and Palfi, S. (2000). Cyclosporion A protects striatal neurons in vitro and in vivo from 3-nitropropionic acid toxicity. *J. Comp. Neurol.* **425**: 471–478.

24. Dykens, J. A. (1994). Isolated cerebellar and cerebral mitochondria produce free radicals when exposed to elevated Ca^{2+} and Na^+: implications for neurodegeneration. *J. Neurochem.* **63**: 584–591.

25. Dykens, J. A. (1995). Mitochondrial radical production and mechanisms of oxidative excitotoxicity. *In* "The Oxygen Paradox" (K. J. A. Davies, and F. Ursini, Eds.), pp. 453–467, Cleup Press (University of Padova).

26. Shults, C. W., Haas, R. H. and Beal, M. F. (1999). A possible role of coenzyme Q10 in the etiology and treatment of Parkinson's disease. *Biofactors* **9**: 267–272.

27. Fiskum, G., Murphy, A. N. and Beal, M. F. (1999). Mitochondria in neurogeneration: acute ischemia and chronic neurodegenerative diseases. *J. Cereb. Blood Flow Metab.* **19**: 351–369.

28. Dykens, J. A. (1997). Mitochondrial free radical production and the etiology of neurodegenerative disease. *In* "Neurodegenerative Diseases: Mitochondria and Free Radicals in Pathogenesis" (M. F. Beal, I. Bodis-Wollner, and N. Howell, Eds.), pp. 29–55, John Wiley & Sons, New York.

29. Dykens, J. A. (1999). Free radicals and mitochondrial dysfunction in excitotoxicity and neurodegenerative diseases. *In* "Cell Death and Diseases of the Nervous System" (V. E. Koliatos, and R. R. Ratan, Eds.), pp. 45–68, Humana Press, NJ.

30. Kroemer, G. and Reed, J. C. (2000). Mitochondrial control of cell death. *Nat. Med.* **6**: 513–519.

31. Teng, C. S. (2000). Protooncogenes as mediators of apoptosis. *Int. Rev. Cytol.* **197**: 137–202.

32. Daugus, E., Nochy, D., Ravagnan, L., Loeffler, M., Susin, S. A., Zamzami, N. and Kroemer, G. (2000). Apoptosis-inducing factor (AIF): a ubiquitous mitochondrial oxidoreductase involved in apoptosis. *FEBS Lett.* **476**: 118–123.

33. Andreyev, A. Y., Fahy, B. and Fiskum, G. (1998). Cytochrome c release from brain mitochondria is independent of the mitochondrial permeability transition. *FEBS Lett.* **439**: 373–376.

34. Bernardi, P., Colonna, R., Costantini, P., Eriksson, O., Fontaine, E., Ichas, F., Massari, S., Nicolli, A., Petronilli, V. and Scorrano, L. (1998). The mitochondrial permeability transition. *Biofactors* **8**: 273–281.

35. Benzi, G. and Moretti, A. (1995). Age and peroxidative stress-related modifications of the cerebral enzymatic activities linked to mitochondria and the glutathione system. *Free Radic. Biol. Med.* **19**: 77–101.

36. Bernardi, P., Scorrano, L., Colonna, R., Petronilli, V. and Di Lisa, F. (1999). Mitochondria and cell death. Mechanistic aspects and methodological issues. *Eur. J. Biochem.* **264**: 687–701.

37. Sastre, J., Pallardo, F. V., Garcia de la Asuncion, J. and Vina, J. (2000). Mitochondria, oxidative stress and aging. *Free Radic. Res.* **32**: 189–198.

38. Halestrap, A. P., Doran, E., Gillespie, J. P. and O'Toole, A. (2000). Mitochondria and cell death. *Biochem. Soc. Trans.* **28**: 170–177.

39. Wollheim, C. B. (2000). Beta-cell mitochondria in the regulation of insulin secretion; a new culprit in Type-II diabetes. *Diabetologia* **43**: 265–277.

40. Anderson, C. M. (1999). Mitochondrial dysfunction in diabetes mellitus. *Drug Dev. Res.* **46**: 67–79.

41. Crompton, M. (1999). The mitochondrial permeability transition pore and its role in cell death. *Biochem. J.* **341**: 233–249.

42. Fontaine, E. and Bernardi, P. (1999). Progress on the mitochondrial permeability transition pore: regulation by complex I and ubiquinone analogs. *J. Bioenerg. Biomembr.* **31**: 335–345.

43. Miro, O., Casademont, J., Casals, E. Perea, M., Urbano-Marques, A., Rustin, P. and Cardellach, F. (2000). Aging is associated with increased lipid peroxidation in human hearts, but not with mitochondrial respiratory chain enzyme defects. *Cardiovasc. Res.* **47**: 624–631.

44. Mather, M. and Rottenberg, H. (2000). Aging enhances the activation of the permeability transition pore in mitochondria. *Biochem. Biophys. Res. Commun.* **273**: 603–608.

45. Barja, G. and Herrero, A. (2000). Oxidative damage to mitochondrial DNA is inversely related to maximum life span in the heart and brain of mammals. *FASEB J.* **14**: 312–318.

46. Beckman, K. B. and Ames, B. N. (1998). Mitochondrial aging: open questions. *Ann. NY Acad. Sci.* **854**: 118–127.

47. Lemasters, J. J., Qian, T., Bradham, C. A., Brenner, D. A., Cascio, W. E., Trost, L. C., Nishimura, Y., Nieminen, A. L. and Herman, B. (1999). Mitochondrial dysfunction in the pathogenesis of necrotic and apoptotic cell death. *J. Bioenerg. Biomembr.* **31**: 305–319.

48. Boveris, A., Costa, L. E., Poderoso, J. J., Carreras, M. C. and Cadenas, E. (2000). Regulation of mitochondrial respiration by oxygen and nitric oxide. *Ann. NY Acad. Sci.* **899**: 121–135.

49. Halliwell, B. (1999). Oxygen and nitrogen are pro-carcinogens. Damage to DNA by reactive oxygen, chlorine and nitrogen species: measurement, mechanism and the effects of nutrition. *Mutat. Res.* **443**: 37–52.

50. Sohal, R. S. and Sohal, B. H. (1991). Hydrogen peroxide release by mitochondria increases during aging. *Mech. Ageing Dev.* **7**: 187–202.

51. Mecocci, P., MacGarvey, U., Kaufman, A. E., Koontz, D., Shoffner, J. M., Wallace, D. C. and Beal, M. F. (1993). Oxidative damage to mitochondrial DNA shows marked age-dependent increases in human brain. *Ann. Neurol.* **34**: 609–616.

52. Kosonen, O., Kankaanranta, H., Uotila, J. and Moilanen, E. (2000). Inhibition by nitric oxide-releasing compounds of E-selectin expression in and neutrophil adhesion to human endothelial cells. *Eur. J. Pharmacol.* **394**: 149–156.

53. Buchholz, J., Tsai, H., Foucart, S. and Duckles, S. P. (1996). Advancing age alters intracellular calcium buffering in rat adrenergic nerves. *Neurobiol. Aging* **17**: 885–892.

54. Pascale, A. and Etcheberrigaray, R. (1999). Calcium alterations in Alzheimer's disease: pathophysiology, models and therapeutic opportunities. *Pharmacol. Res.* **39**: 81–88.
55. Murphy, S. N., Thayer, S. A. and Miller, R. J. (1987). The effects of excitatory amino acids on intracellular calcium in single mouse striatal neurons in vitro. *J. Neurosci.* **7**: 4145–4158.
56. Siegmund, B., Schluter, K.-D. and Piper, H. M. (1993). Calcium and the oxygen paradox. *Cardiovasc. Res.* **27**: 1778–1783.
57. Paschen, W. and Doutheil, J. (1999). Disturbances in the functioning of endoplasmic reticulum: a key mechanism underlying neuronal cell injury? *J. Cereb. Blood Flow Metab.* **19**: 1–18.
58. Mattson, M. P., LaFerla, F. M., Chan, S. L., Leissring, M. A., Shepel, P. N. and Geiger, J. D. (2000). Calcium signaling in the ER: its role in neuronal plasticity and neurodegenerative disorders. *Trends Neurosci.* **23**: 222–229.
59. Murphy, A. N., Fiskum, G. and Beal, M. F. (1999). Mitochondrial ion neurodegeneration: bioenergetic function in cell life and death. *J. Cereb. Blood Flow Metab.* **19**: 231–245.
60. Skulachev, V. P. (1999). Mitochondrial physiology and pathology; concepts of programmed death of organelles, cells and organisms. *Mol. Aspects Med.* **20**: 139–184.
61. Nicholls, D. G. and Budd, S. L. (1998). Neuronal excitotoxicity: the role of mitochondria. *Biofactors* **8**: 287–299.
62. Nicholls, D. G. and Ward, M. W. (2000). Mitochondrial membrane potential and neuronal glutamate excitotoxicity: mortality and millivolts. *TINS* **23**: 166–174.
63. Maftah, A., Petit, J. M., Ratinaud, M. H. and Julien, R. (1989). 10-N nonylacridine orange: a fluorescent probe which stains mitochondria independently of their energetic state. *Biochem. Biophys. Res. Commun.* **164**: 185–190.
64. Mason, W. T. (1993). *Fluorescent and Luminescent Probes for Biological Activity*, Academic Press.
65. Dykens, J. A. and Stout, A. K. (2001). Fluorescent dyes and assessment of mitochondrial membrane potential in FRET modes. *In* "Methods in Cell Biology" (E. Schon, and L. Pons, Eds.), Academic Press, **in press**.
66. Li, P. A., Kristian, T., He, Q. P. and Siesjo, B. K. (2000). Cyclosporin A enhances survival, ameliorates brain damage, and prevents secondary mitochondrial dysfunction after a 30 minute period of transient cerebral ischemia. *Exp. Neurol.* **165**: 153–163.

Chapter 76

Nitric Oxide-Blocking Therapeutics

Ching-San (Monte) Lai[†] and Norman K. Orida[*,‡]

Ching-San (Monte) Lai and Norman K. Orida • Medinox, Inc., 11575 Sorrento Valley Road, Suite 201, San Diego, CA 92121 USA
[†]Dr. Lai is President and CEO of Medinox, Inc.
[‡]Dr. Orida is Vice President of Business Development of Medinox, Inc.
[*]Corresponding Author.
Tel: 858-793-4820, E-mail: nkorida@medinox.com, E-mail: cslai@medinox.com

1. Summary

Dithiocarbamates are small molecules that bind both nitric oxide (NO) and free iron. With the ability to scavenge these two very important pro-inflammatory mediators, dithiocarbamates have broad therapeutic potential for treating oxidative stress-related human diseases and disorders. Here, we describe NOX-100, a proprietary dithiocarbamate, under development at Medinox as a new NO-blocking agent.

2. NO and NO-Regulating Therapeutics

Just over a decade ago, NO was merely thought of as an environmental pollutant in automobile exhaust and city smog. This view changed in 1987 when the role of NO as a signaling molecule in the body was elucidated.[1,2] Since 1987, more than 45 000 publications have illustrated the important roles of NO in health and disease, and revealed a beneficial/harmful duality for the molecule. On one hand, low levels of NO production are essential for homeostatic functions such as blood pressure regulation, memory, and host defense. On the other hand, excessive NO production is a key factor in the pathology of septic shock, ischemic stroke, allograft rejection, diabetes, neurodegenerative diseases, and many other inflammatory conditions. These observations stimulated much research on the development of NO-regulating therapeutics.

Earlier on, much work was devoted to identifying inhibitors of the nitric oxide synthases (NOS), the enzymes that synthesize NO from L-arginine. NO is produced in the body by three different isoforms of the NOS enzyme. Two isoforms are constitutively active and Ca^{2+}-dependent. Neuronal NOS (nNOS) is found in neurons and plays a role in memory and pain perception. Endothelial NOS (eNOS) is found in vascular endothelial cells and plays an active role in vasodilation and blood pressure regulation. The third isoform, inducible NOS (iNOS), is expressed in macrophages, astrocytes, microglial cells, neutrophils and many other tissues, and is induced in response to stimulation by inflammatory cytokines. NO production by iNOS is 1000-fold greater than any constitutive NOS isoform and contributes mainly to the pathogenesis of many inflammatory conditions.

Although NOS inhibitors have shown proof of concept in preclinical studies, attempts to develop the compounds as human therapeutics have not yet been successful. Non-specific NOS inhibitors, such as L-arginine analogs, have potential problems because they not only affect NO production from iNOS but also production from other isoforms required for vital homeostatic functions such as blood pressure regulation.[3,4] This problem was revealed in Glaxo Wellcome's recent Phase III septic shock trial for L-NMMA (L-N-monomethyl arginine), which was terminated early.[5]

Lately, iNOS has become a preferred target for therapeutic intervention. Despite an intense effort, no iNOS inhibitor has yet been developed as a therapeutic. There are potential pitfalls in developing iNOS inhibitors as well. For example, iNOS is expressed constitutively in a number of tissues including kidney, ileum, retina, uterus, and platelets, suggesting a basal function of the enzyme in these tissues.[6] Furthermore, iNOS is crucial for protection against lethal tuberculosis infection.[7]

As an alternative to NOS inhibitors, chemical scavengers can be used to remove NO directly from biological systems. A wide variety of compounds, ranging from small molecular weight "spin traps" to large proteins, have been shown to be potential NO-blockers.

3. Dithiocarbamate Molecules as NO-Blocking Agents

Because of its short-lived nature, NO is difficult to detect and the development of low molecular weight dithiocarbamate spin traps, which capture NO in a metal chelator complex, have greatly facilitated the study of NO in biological systems using electron paramagnetic resonance (EPR) spectroscopy.[8–12]

Dithiocarbamates bind NO in a two-step process as shown in Fig. 1. First, two dithiocarbamate molecules (NOX) combine with Fe^{2+} to form a complex [(NOX)$_2$-Fe], which is held together via the planar coordination of the Fe^{2+} by

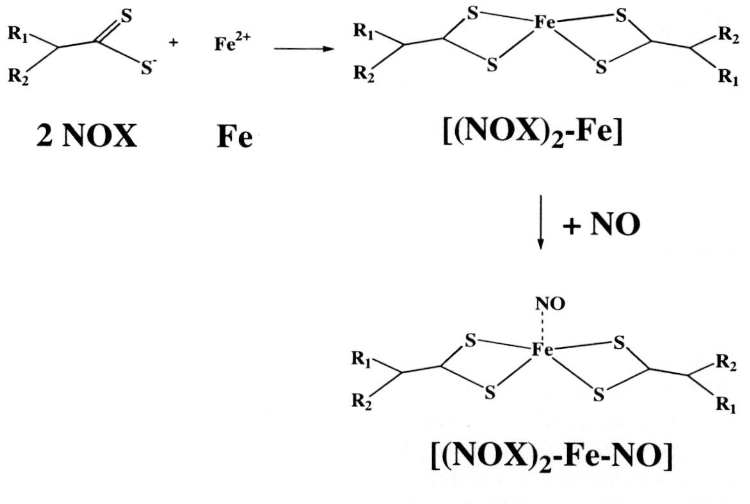

Fig. 1. Schematic illustration of how a representative dithiocarbamate (NOX) binds NO in a two-step process. First, two dithiocarbamate molecules combine with ferrous iron to form a [(NOX)$_2$-Fe] metal chelator complex. Next, the [(NOX)$_2$-Fe] complex binds a molecule of NO. Once in the [(NOX)$_2$-Fe-NO] complex, NO is neutralized and prevented from participating in additional reactions.

two dithiocarbamates. Next, the [(NOX)$_2$-Fe] complex tightly binds to NO, forming an NO-containing complex, [(NOX)$_2$-Fe-NO].[13] The stoichiometry of the [(NOX)$_2$-Fe-NO] complex was determined by isotope substitution studies using ^{17}O and ^{15}N and confirmed that the complex contained one Fe and a single NO molecule.[14]

The rate constant of the reaction of NO with [(NOX)$_2$-Fe] is estimated to be approximately 1×10^8 M^{-1}s^{-1} and the equilibrium constant between Fe^{2+} and dithiocarbamates is approximately 10^4 M^{-1}.[15] These properties indicate that the [(NOX)$_2$-Fe] complex has a high affinity for NO, which is greater than the affinity of hemoglobin for NO (approximately $2–5 \times 10^7$ M^{-1}s^{-1}).[16] Conversely, the comparatively low equilibrium constant of dithiocarbamates for ferrous iron[15] suggests that dithiocarbamates will not remove metal ions from metalloenzymes, which may account for their excellent safety profile in animal studies (Medinox, Inc., unpublished studies).

The binding of NO to the [(NOX)$_2$-Fe] complex has been studied *in vitro* and *in vivo* using electron paramagnetic resonance (EPR) spectroscopy. [(NOX)$_2$-Fe] interacts with NO to form the stable and water-soluble [(NOX)$_2$-Fe-NO] complex, whose characteristic three-line spectrum can readily be detected by EPR spectroscopy.[17, 18] The three-line spectrum has also been observed during experimental sepsis in living mice, which is evidence of NO formation *in vivo* [Fig. 2(a)]. In the presence of the nitric oxide synthase inhibitor, L-N-monomethyl-arginine (L-NMMA), the spectrum amplitude was reduced considerably, consistent with the notion that NO trapped by the [(NOX)$_2$-Fe] complex was produced via the NO synthase pathway [Fig. 2(b)].

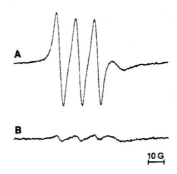

Fig. 2. (A) EPR spectrum of a [(NOX)$_2$-Fe-NO] complex in a urine sample obtained from a mouse 6 h after bacterial lipopolysaccharide (LPS) challenge. The spectrum shows the characteristic — line spectrum of the [(NOX)$_2$-Fe-NO] complex. In this study, a dithiocarbamate was injected intravenously after the induction of excessive NO production by LPS and the [(NOX)$_2$-Fe-NO] complex was recovered in the urine. If the nitric oxide synthase inhibitor, L-NMMA, was injected after LPS challenge, much less signal was observed, (B) showing that nitric oxide synthase activity was required to produce the observed EPR signal. (Reprinted from Komarov *et al.* (1995). Detection of nitric oxide production in mice by spin-trapping electron paramagnetic resonance spectroscopy. *Biochemica et Biophysica Acta* **1272**: 29–36, with permission from Elsevier Science.[10])

4. Dual Anti-Inflammatory Activities of Dithiocarbamates

As described above, NO binds to [(NOX)$_2$-Fe] but not to the dithiocarbamate molecule itself. The effectiveness of dithiocarbamates as NO-blocking agents therefore depends on the presence of free Fe^{2+}. Iron is a powerful oxidant, and is injurious to tissues because it promotes free radical-mediated oxidative damage.[19] Thus, most iron is sequestered and stored in intracellular ferritin *in vivo*.[20] When attacked by superoxide anion radical and/or NO, ferritin releases Fe^{2+}, thus increasing the low molecular-weight iron pool.[21, 22] Free iron levels are known to increase considerably during tissue injury[23, 24] and can contribute to free-radical-induced oxidative stress.[25]

NOX-100 is a proprietary dithiocarbamate under development at Medinox as a critical care therapeutic. Like other dithiocarbamates, NOX-100 must bind free Fe^{2+} first in order to bind NO. We believe that the release of iron at sites of damaged or inflamed tissues localizes the formation of [(NOX)$_2$-Fe] complexes and targets NOX-100's dual NO- and iron-binding activities to sites of inflammation. By simultaneously removing both deleterious free iron and excessive NO at sites of inflammation, NOX-100 is a dually effective anti-inflammatory agent.

NO is a potent vasodilator and removing too much of it could lead to hypertension. However, we have not observed this to occur in animal studies — even after administering extremely high doses of NOX-100. The reason that NOX-100 is incapable of removing all NO is due to differences in the kinetics and distribution between NOX-100 and NO. NO, is a small, uncharged diatomic gas of 30 daltons that diffuses isotropically and rapidly across cell membranes and tissues.[16, 26] Conversely, NOX-100 is a charged molecule of approximately 300 daltons that diffuses much more slowly and unevenly. The 20–30% volume distribution of NOX-100 confirms that the compound is primarily distributed in the circulation and interstitial space (Medinox, Inc., unpublished studies). The differential distribution renders NOX-100 incapable of scavenging all NO, particularly within tissues such as smooth muscle where NO exerts its physiological effect.

5. Selected Animal Studies of NOX-100

The safety and efficacy profiles of NOX-100 as an NO-blocking agent have been evaluated in numerous preclinical studies. Here, we present a few selected animal studies as examples to illustrate the potential medical applications of NOX-100 in the field of critical care medicine.

5.1. Restoration of Blood Pressure in Septic Shock

The induction of excessive NO production by bacterial endotoxin is a well-characterized event of septic shock.[27] Excessively produced NO from iNOS leads

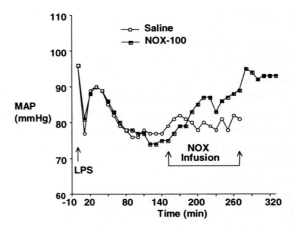

Fig. 3. NOX-100 restores blood pressure during experimental septic shock in the rat. Following challenge with LPS at time zero, rats were administered a bolus of NOX-100 or an equal volume of saline intravenously 150 min later. The NOX-100 treated animals continued to receive NOX-100 as an infusion (0.1 mmoles/kg/h) for an additional 2 hours. Control animals received an equivalent infusion volume of saline. Data represent averages from 16 animals each in the NOX-100 and saline treated groups. (Unpublished studies from Medinox, Inc.)

to extreme vasodilation and loss of vascular tone, which results in systemic hypotension. Systemic hypotension causes insufficient tissue perfusion, which in turn, leads to multiple organ failure and death.

Figure 3 shows the effect of NOX-100 in a rat model in which septic shock was induced by lipopolysaccharide (LPS). Two hours after LPS challenge, the blood pressure falls precipitously and remains low. This hypotensive condition is caused by the induction of iNOS and the synthesis of large amounts of the powerful vasodilator, NO. The administration of NOX-100 in saline at 2 h after LPS challenge restored blood pressure to normotensive levels. Note that NOX-100 did not produce hypertension, indicating that the compound did not remove all of the NO. In contrast, administration of the equivalent volume of saline alone failed to restore blood pressure.

5.2. Prevention of Bacterial Translocation During Endotoxemia

Bacterial translocation (BT), which is defined as the migration of microbes across the intestinal barrier, is central to the pathogenesis of gut-origin sepsis.[28] The compromise of intestinal barrier function, which can arise during endotoxemia,[29] leads to BT and is correlated with excessive NO production.[30] BT is associated with cellular damage caused by peroxynitrite, which is formed by the reaction of NO and superoxide.[31] Peroxynitrite damages epithelial cells by increasing transepithelial permeability and activating the poly (ADP-ribose) synthethase, thereby depleting cellular ATP.[32]

Using a rat model, Dickinson, *et al.*[33] demonstrated that NOX-100 has a profound effect in preventing BT. In saline-treated controls, bacteria translocated from the gut to the mesenteric lymph nodes in 56% of the animals compared to only 15% in the NOX-100 treated group. Most strikingly, none of the NOX-100-treated animals had positive blood cultures, which were observed in 28% of the saline treated control animals. This indicated that NOX-100 completely blocked the passage of viable bacteria from the intestine to the circulation. The study also revealed that NOX-100 treatment reduced the number of apoptotic nuclei and amount of nitrotyrosine immunoreactivity, indicating that by removing NO, peroxynitrite formation and subsequent peroxynitrite-mediated tissue damage could be prevented. These results have direct implications for NOX-100 in the prevention of bacterial translocation following surgical procedures.

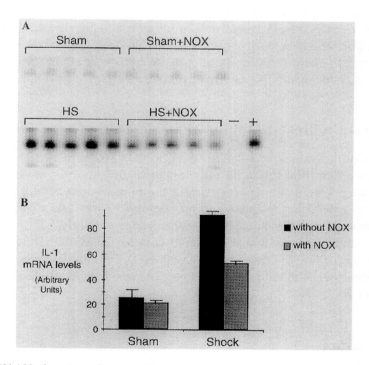

Fig. 4. NOX-100 decreases hepatic IL-1β mRNA transcription in hemorrhagic shock. Liver samples were collected 24 h after hemorrhage. Total RNA was isolated, and IL-1β mRNA transcription was determined by RT-PCR. PCR products were separated on a 10% polyacrylamide gel, developed with the use of a PhosphorImager (A) and quantified by scanning densitometry (B). (Reprinted from Menenzes *et al.* (1999). A novel nitric oxide scavenger decreases liver injury and improves survival after hemorrhagic shock. *Am. J. Physiol.* **277** (*Gastrointest. Liver Physiol.* 40): G144–G151, with permission from The American Physiological Society.[37])

5.3. Reduction of Inflammatory Gene Expression During Hemorrhagic Shock

Severe hemorrhage results in an oxidative stress-related inflammatory response that includes the upregulation of iNOS expression.[34] Previously, NOS inhibition was shown to reduce tissue injury, suggesting that excessive NO contributes to the pathophysiology of hemorrhagic shock.[35] Therefore, a reduction in the pro-inflammatory response with an NO-blocker such as NOX-100 might have a beneficial effect in hemorrhagic shock.

Harbrecht, *et al.*[36] showed that a selective iNOS inhibitor reduced transcription pro-inflammatory factor activation and cytokine production in the rat lung during hemorrhagic shock. Using the same rat model, this same group recently evaluated whether removal of excessive NO with NOX-100 could also have a therapeutic benefit during hemorrhagic shock.[37] Their study revealed that NOX-100 profoundly improved the physiological condition of the animals when it was infused during the post-hemorrhage resuscitation. NOX-100 restored blood pressure, decreased liver injury, decreased hepatic neutrophil infiltration and greatly improved survival after 24 hours. Moreover, at the gene level, NOX-100 treatment reduced the levels of TNF-α and IL-1β mRNA (Fig. 4). Infusion with the equivalent volume of saline produced no beneficial effect on blood pressure or gene expression.

6. Clinical Studies of Dithiocarbamate Nitric Oxide-Blockers

Medinox, Inc. completed a safety study of NOX-100 in hemodialysis patients in December 2000. The results in this dose-escalating study revealed that there were no drug related side effects even in the highest dose cohort (50 mg/kg). NOX-100 is now being evaluated in an ongoing Phase I/IIa clinical study in septic shock. The multicenter study, which began in July 2001, is currently obtaining safety and preliminary efficacy data for NOX-100 while testing the concept that blocking the excessive NO with NOX-100 could alleviate the severe hypotension associated with this condition. The trial is scheduled for completion in Summer 2002.

7. Conclusion

The discovery of NO as a signaling molecule opened the door for NO-blocking drugs that might treat the multitude of human diseases and disorders that involve excessive NO production. A *sine qua non* and primary obstacle is the need to eliminate excessive, pathology-causing NO while sparing the small amounts that are essential for vital functions. Our preclinical studies showed clearly that NOX-100 is a safe and effective NO-blocking agent. Clinical studies are now in progress to evaluate the safety, tolerability and efficacy of NOX-100 in human critical care indications. As a new technology for eliminating excessive NO, Medinox's NO-blockers represent a promising approach for addressing the beneficial/harmful duality of this key gas molecule.

References

1. Ignarro, L. J., Buta, G. M., Wood, K. S., Bryns, R. E. and Chaudhuri, G. (1987). Endothelium-derived relaxing factor produced and released from artery and vein is nitric oxide. *Proc. Natl. Acad. Sci. USA* **84**: 9265–9269.
2. Palmer, R. B., Ferrige, A. G. and Moncada, S. (1987). Nitric oxide release accounts for the biological activity of endothelium-derived relaxing factor. *Nature* **327**: 524–526.
3. Robertson, F. M, Offner, P. J., Ciceri, D. P., Becker, W. K. and Pruitt, B. A. (1994). Detrimental hemodynamic effects of nitric oxide synthase inhibition in septic shock. *Arch. Surg.* **129**: 149–156.
4. Cobb, J. P. and Danner, R. L. (1996). Nitric oxide and septic shock. *JAMA* **275**: 1192–1196.
5. Glaxo Wellcome plc press release. April 24, 1998.
6. Kone, B. C. (1997). Nitric oxide in renal health and disease. *Am. J. Kidney Dis.* **30**: 311–333.
7. MacMicking, J. D., North, R. J., LaCourse, R., Mudgett, J. S., Shah, S. K. and Nathan, C. F. (1997). Identification of NOS2 as a protective locus against tuberculosis. *Proc. Natl. Acad. Sci. USA* **94**: 5423–5428.
8. Kubrina, L. N., Caldwell, W. S., Mordvintcev, P. I., Malenkova, I. V. and Vanin, A. F. (1992). EPR evidence for nitric oxide production from guanido nitrogens of L-arginine in animal tissues in vivo. *Biochim. Biophys. Acta.* **1099**: 223–237.
9. Komarov, A., Mattson, D., Jones, M. M., Singh, P. K. and Lai, C.-S. (1993). In vivo spin trapping of nitric oxide in mice. *Biochim. Biophys. Res. Commun.* **195**: 1191–1198.
10. Komarov, A. M. and Lai, C. S. (1995). Detection of nitric oxide production in mice by spin-trapping electron paramagnetic resonance spectroscopy. *Biochim. Biophys. Acta* **1272**: 29–36.
11. Lancaster, J. R. J., Langrehr, J. M., Bergonia, H. A., Murase, N., Simmons, R. L. and Hoffman, R. A. (1992). EPR detection of heme and nonheme iron-containing protein nitrosylation by nitric oxide during rejection of rat heart allograft. *J. Biol. Chem.* **267**: 10 994–10 998.
12. Komarov, A. M., Kramer, J. H., Mak, I. T. and Weglicki, W. B. (1997). EPR detection of endogenous nitric oxide in postischemic heart using lipid and aqueous-soluble dithiocarbamate-iron complexes. *Mol. Cell Biochem.* **175**: 91–97.

13. Gibson, J. F. (1962). Unpaired electron in nitroso-bis(dimethyldithiocarbamato) iron(II). *Nature* **196**: 64.

14. Kotake, Y., Tanigawa. M. and Ueno, I. (1996). Spin trapping isotopically-labelled nitric oxide produced from [^{15}N]L-arginine and [^{17}O]dioxygen by activated macrophages using a water soluble Fe^{2+}-dithiocarbamate spin trap. *Free Radic. Res.* **23**: 287–295.

15. Paschenko, S. V., Khramtsov, V. V., Skatchkov, M. P., Plyusnin, V. F. and Bassenge, E. (1996). EPR and laser flash photolysis studies of the reaction of nitric oxide with water soluble NO trap Fe(II)-proline-dithiocarbamate complex. *Biochem. Biophys. Res. Commun.* **225**: 577–584.

16. Vaughn, M. W., Kuo, L. and Liao, J. C. (1998). Effective diffusion distance of nitric oxide in the microcirculation. *Am. J. Physiol.* **274**: H1705–H1714.

17. Mordvintcev, P., Mulsch, A., Busse, R. and Vanin, A. (1991). On-line detection of nitric oxide formation in liquid aqueous phase by electron paramagnetic resonance spectroscopy. *Anal. Biochem.* **199**: 142–146.

18. Lai, C. S. and Komarov, A. M. (1994). Spin trapping of nitric oxide produced in vivo in septic-shock mice. *FEBS Lett.* **345**: 120–124.

19. Chevion, M, Berenshtein, E. and Zhu, B. Z. (1999). The role of transition metal ions in free radical-medicated damage. *In* "Reactive Oxygen Species in Biological Systems" (D. L. Gilbert, and C. A. Colton, Eds.), pp. 103–132, Plenum, New York.

20. Guyton, A. C. and Hall, J. E. (1996). *Textbook of Medical Physiology*, pp. 430–431, W.B. Saunders Co., Philadelphia.

21. Biemond, P., Swaak, A. J., Beindorff, C. M. and Koster, J. F. (1986). Superoxide-dependent and superoxide-independent mechanisms of iron mobilization from ferritin by xanthine oxidase. Implications for oxygen-free-radical-induced tissue destruction during ischaemia and inflammation. *Biochem. J.* **239**: 169–173.

22. Biemond, P., van Eijk, H. G., Swaak, A. J. and Koster, J. F. (1984). Iron mobilization from ferritin by superoxide derived from stimulated poly-morphonuclear leukocytes. Possible mechanism in inflammation diseases. *J. Clin. Invest.* **73**: 1576–1579.

23. Chevion, M., Jiang, Y., Har-El, R., Berenshtein, E., Uretzky, G. and Kitrossky, N. (1993). Copper and iron are mobilized following myocardial ischemia: possible criteria for tissue injury. *Proc. Natl. Acad. Sci. USA* **90**: 1102–1106.

24. Ambrosio, G., Zweier, J. L., Jacobus, W. E., Weisfeldt, M. L. and Flaherty, J. T. (1987). Improvement of postischemic myocardial function and metabolism induced by administration of deferoxamine at the time of reflow: the role of iron in the pathogenesis of reperfusion injury. *Lab. Invest.* **76**: 906–915.

25. McCord, J. M. (1998). Iron, free radicals, and oxidative injury. *Sem. Hematol.* **35**: 5–12.

26. Lancaster, J. R. J. (1994). Simulation of the diffusion and reaction of endogenously produced nitric oxide. *Proc. Natl. Acad. Sci. USA* **91**: 8137–8141.

27. Thiemermann, C. (1990). Nitric oxide and septic shock. *Gen. Pharmac.* **29**: 159–166.

28. Alexander, J. W., Boyce, S. T., Babcock, G. F., Gianotti, L., Peck, M. D., Dunn, D. L., Types, T., Childress, C. P. and Ash, S. K. (1990). The process of microbial translocation. *Ann. Surg.* **212**: 496–510.

29. Deitch, E. A., Berg, R. D. and Specian, R. (1987). Endotoxin promotes the translocation of bacteria from the gut. *Arch. Surg.* **122**: 185–190.

30. Sorrels, D. L., Friend, C., Koltuksuz, U., Courcoulas, A., Boyle, P., Garrett, M. S., Watkins, A., Rowe, M. I. and Ford, H. R. (1996). Inhibition of nitric oxide with aminoguanidine reduces bacterial translocation after endotoxin challenge. *Arch. Surg.* **131**: 1155–1163.

31. Beckman, J. S. and Koppenol, W. H. (1996). Nitric oxide, superoxide, and peroxynitrite: the good, the bad and the ugly. *Am. J. Physiol.* **271 (Cell Physiol. 40)**: C1424–C1437.

32. Kennedy, M., Denengerg, A. G., Szabo, C. and Salzman, A. (1998). Poly (ADP-ribose) synthetase activation mediates increased permeability induced by peroxynitrite in CaCo-2BB3 cells. *Gastroenterology* **114**: 510–518.

33. Dickinson, E., Tuncer, R., Nadler, E., Boyle, P., Alber, S., Watkins, S. and Ford, H. (1999). NOX, a novel nitric oxide scavenger, reduces bacterial translocation in rats after endotoxin challenge. *Am. J. Physiol.* **277 (Gastrointest. Liver Physiol. 40)**: G1281–G1287.

34. Shenkar, R., Coulson, W. F. and Abraham, E. (1994). Hemorrhage and resuscitation induce alterations in cytokine expression and the development of acute lung injury. *Am. J. Respir. Cell Mol. Biol.* **10**: 290–297.

35. Yao, Y. M., Bahrami, S., Leightfried, G., Redl, H. and Schlag, G. (1996). Significance of NO in hemorrhage-induced hemodynamic alterations, organ injury, and mortality in rats. *Am. J. Physiol.* **270 (Heart Circ. Physiol. 34)**: H973–H979.

36. Harbrecht, B. G., Wu, B., Watkins, S. C., Billiar, T. R. and Peitzman, A. B. (1997). Inhibition of nitric oxide synthesis during severe shock but not following resuscitation increases hepatic injury and neutrophil accumulation in hemorrhaged rats. *Shock* **8**: 415–421.

37. Menenzes, J., Hierholzer, C., Watkins, S. C., Lyons, V., Peitzman, A. B., Billiar, T. R., Tweardy, D. J. and Harbrecht, B. G. (1999). A novel nitric oxide scavenger decreases liver injury and improves survival after hemorrhagic shock. *Am. J. Physiol.* **277 (Gastrointest. Liver Physiol. 40)**: G144–G151.

Chapter 77

Nitric Oxide and Mammalian Aging: "Surplus" and "Deficit" as the Two Faces of a Pharmaceutical Currency for Nitric Oxide-Modulator Drugs

David R. Janero

David R. Janero • NitroMed, Inc., 12 Oak Park Drive, Bedford, Massachusetts 01730 USA
Tel: 781-685-9749, E-mail: djanero@nitromed.com

1. Nitric Oxide (NO) Physiological Chemistry Defines the Therapeutic Potential of NO Modulation

In less than two decades since its pharmacological identification in living tissue, nitric oxide (NO) has become recognized across species as a ubiquitous natural biomediator essential to mammalian homeostasis and health.[1] Laboratory and clinical investigations continue to extend and refine our understanding of NO's chemical reactivity and versatile physiology in both health and disease.[2-4] This growing knowledge base has fostered rational pharmaceutical attempts to modulate selectively tissue NO status for therapeutic benefit.[3, 5] Although within the last decade controlled inhalation of NO gas has been used against certain bronchopulmonary disorders,[3, 5] therapeutic NO modulation reaches back to at least 1879 when a NO-producing compound, glyceryl trinitrate (nitroglycerine), became an accepted cardiovascular drug for relief of chest pain (angina pectoris).[6]

One attribute of NO's physiological chemistry impacts greatly upon the potential therapeutic utility of NO modulation: NO's "double-edged", "two-faced", or "Janus-faced" character. That is, NO [and nitrogen species derived from the NO radical[3, 4]] can exert both positive physiological (i.e., protective, regulatory) and adverse pathological (i.e., injurious, toxic) effects.[7] Perhaps for this reason, cellular NO production/activity is controlled by the NO synthase (NOS) enzyme family, signal transduction pathways, and biomolecules that trap or inactivate NO.[2, 3, 7] None of these systems is obligatory or fail-safe.[8] Thus, the low basal level of NO constitutively produced by the vascular endothelial NOS isoform (eNOS) maintains a vasodilator tone essential to blood pressure and regulation of blood flow.[9] Cytokine or endotoxin induction of the inducible NOS isoform (iNOS) protects tissue from invading pathogens by transiently potentiating local NO formation to cytotoxic, antiseptic levels.[10] But sustained, systemic iNOS induction can compromise vascular tone and incite the organ failure associated with septic shock.[11]

That NO's physiological chemistry encompasses "the good, the bad, and the ugly"[12] is a fundamental consideration when attempting to alter tissue NO tone for safe and effective therapeutic benefit. This is so because, within a pharmaceutical context, NO's "double-edged" character takes on additional meanings. NO has the potential to shift between disease-causing and disease-fighting roles, depending upon a myriad of factors often peripheral to NO's molecular properties, such as target-tissue environment.[2, 7, 12] Furthermore, the intrinsic nature of a specific biological response to NO (regulatory, protective, or deleterious) need not define or limit its ultimate therapeutic impact in the organism. For example, NO's suppression of net cell replication helps combat proliferative vascular syndromes, whether by inciting reversible cytostasis, programmed cell death (apoptosis), or toxic cell necrosis.[13]

Table 1. Some Age-Related Diseases/Conditions Thought to Have a NO-Related Pathogenic Component

Disease State or Condition	NO Determinant[a]	Reference
Cardiovascular system		
Restenosis	Deficit	(13)
Atherosclerosis	Deficit or surplus	(14, 15)
Hypertension	Deficit	(16)
Coronary heart disease	Deficit	(17)
Heart failure	Deficit	(18)
Angiogenesis	Deficit	(19)
Stroke	Surplus	(20)
Nervous system		
Alzheimer's disease	Deficit or surplus	(21, 22)
Huntington's disease	Deficit or surplus	(23, 24)
Parkinson's disease	Deficit or surplus	(23, 25)
Cognitive impairment	Deficit or surplus	(26, 27)
Glaucoma/retinal degeneration	Surplus	(28)
Retinitis/uveitis	Surplus	(28)
Neurogenic pain	Surplus	(29)
Epilepsy	Deficit	(30)
Presbyacusis (hearing loss)	Surplus	(31)
Urogenital system		
Erectile dysfunction	Deficit	(32)
Glomerular kidney dysfunction	Surplus	(33)
Benign prosthetic hyperplasia	Deficit	(34)
Reproductive system		
Menopause onset	Deficit	(35)
Musculoskeletal system		
Arthritis	Surplus	(36)
Reduced skeletal muscle force	Deficit or surplus	(37, 38)
Immune system		
Immunity/susceptibility to infection	Deficit	(39)
Gastrointestinal system		
Anti-inflammatory drug-induced damage	Deficit	(40)
Metabolic/nutritional		
Diabetes	Deficit or surplus	(41, 42)
Insulin resistance	Deficit	(43)
Early satiation/anorexia	Deficit	(44)
Miscellaneous		
Tumorigenesis	Deficit	(45)

[a]NO deficit or NO surplus specified as putative pathogenic component.

2. NO Modulation as a Therapeutic Intervention in Age-Related Diseases

Another manifestation of NO's "double-edged" character emerges from evidence that both NO excess and NO deficiency (either quantitative or functional) contribute to the etiopathology of several conditions, particularly in the aging population[13-45] (Table 1). The debilitating, if not life-threatening, nature of many of these diseases underscores their important unmet medical needs. In some cases, compelling evidence exists implicating both NO excess and NO deficiency in their pathogenesis, perhaps reflecting an age-related imbalance between NO-protective/reparative and NO-damaging activities.

Regardless of the specific malady, the information presented in Table 1 suggests that "NO reduction" and "NO augmentation" are rational and useful routes of targeted NO-modulation therapy. Indeed, both are being actively pursued by the pharmaceutical industry.[3, 46] Agents that would trap or neutralize NO and selective inhibitors of the high-output iNOS isoform are routinely sanctioned for treatment of diseases to which excess NO might contribute.[46, 47] The main difficulty associated with commercializing such agents has been to design new, proprietary chemical entities having excellent bioavailability while targeting only the disease-causing NO surplus and not the constitutive, low-level NO production vital to organ function.

In comparison to NO reduction, NO augmentation might represent the more commercially accessible NO-modulation modality. This concept is supported by NO's low intrinsic toxicity,[48] the extant technology for quantifying accurately NO and NO surrogate dosing,[49] and the available means of delivering and targeting specific amounts of not only NO gas, but also a diversity of NO sources (NOS genes, NO donors).[3, 5, 13, 20, 40] From a practical drug discovery and development standpoint, the variegated molecular approaches toward NO-donor design and synthesis would assist lead selection and optimization.[3, 13] The overall challenge to NO-augmentation therapy is to harness the bioactivity of supplemental NO for efficient salutary impact without inducing adverse side effects, such as systemic hypotension.

3. "NO-Enhanced" Medicines: An Approach to NO Augmentation

To meet this challenge, NitroMed, Inc., has formulated a diverse technology platform for NO-augmentation therapy. One aspect of this platform features proprietary molecules of the S-nitrosothiol (thionitrite) chemical class bearing a characteristic $-S-NO$ moiety as a NO-donor functionality. Small-molecule and protein S-nitrosothiols are tissue constituents, potential intermediates of NO metabolism, and pharmacologically active NO surrogates.[50, 51] S-Nitrosothiols have

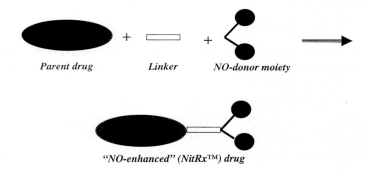

Fig. 1. Schematic illustration of a NitroMed approach toward NO-enhanced medicines. A parent drug is covalently modified by attaching to it a NO-donor moiety (e.g., S-nitrosothiol) with an inert synthetic linker. The resulting proprietary new chemical entity, a "NO-enhanced" NitRx™ drug, expresses the pharmacology of both the parent drug and NO. To this intent, it is envisioned that NO formation and linker cleavage *in vivo* would generate the parent drug with an unaltered profile of action.

been administered alone for supplementation of therapeutically active NO[52] or in combination with existing medicines to enhance their activity and/or safety and extend their product life cycle while leaving their pharmacology unaltered. To the latter intent, an existing drug is either co-administered with an S-nitrosothiol NO donor[53] or derivatized ["S-nitrosated"[50]] to a new chemical entity covalently linked to an S-nitrosothiol functionality[54] (Fig. 1). Applications of these approaches and allied technologies originating from NitroMed, Inc., and others have recently been detailed.[3] Specific illustrations of NitroMed's drug-enhancement technology with S-nitrosothiol NO donors are now presented as applied to age-related urogenital and gastrointestinal disorders.

4. NO-Enhanced Medicines for Male Erectile Dysfunction (MED)

MED is the inability to attain or sustain penile rigidity for satisfactory sexual performance.[32] MED is a common chronobiological disorder: e.g., in the Massachusetts Male Aging Study, 39% of men aged 40 and 67% of men aged 70 had MED to some degree, and 5% of men aged 40 and 15% of men aged 70 were completely impotent.[55] Endogenous NO production from endothelium and nitrergic nerves plays a fundamental role in the relaxation of penile corpus cavernosum and arteriolar smooth muscle responsible for the erectile response.[32] Although of diverse etiology, MED reflecting organic impotence can be caused by local NO production/release insufficient for the smooth muscle relaxation required to allow penile engorgement with blood and a consequent erection.[32] Locally injected α-adrenergic receptor antagonists to counteract smooth muscle contraction are accepted for the clinical management of organic impotence when combined with vasodilator drugs (most effectively, prostaglandin E$_1$).[56] Sildenafil citrate (Viagra™; Pfizer, Inc.), a phosphodiesterase subtype-5 inhibitor that attenuates degradation of NO's second messenger, cGMP, has also entered the MED drug armamentarium.[57] Only limited success has been obtained with these agents: 30–50% of men with organic impotence fail to respond to either prostaglandin E$_1$ or sildenafil treatment.[57, 58] Pain induction by prostaglandin E$_1$[56] and potential cardiovascular risks of sildenafil[59] are further shortcomings.

Two novel approaches for MED therapy based upon the concept of NO-enhancement have been pursued by NitroMed, Inc. In the first, a new class of molecules, S-nitrosated α-adrenergic receptor antagonists, was synthesized by linking a NO-donor S-nitrosothiol functionality to an α-adrenergic antagonist by an inert, organic-ester tether (Fig. 2). Their NO-donor properties in relaxing human penile smooth muscle, their α-adrenergic receptor antagonism, and their ability to induce erection in laboratory animals without systemic side-effects after acute administration by local injection suggest that S-nitrosated analogs of yohimbine and moxisylyte may be useful drugs for MED.[60] Therapeutic NO targeting would be enhanced by the local co-delivery of a source of an essential mediator of penile erection, NO, and an agent that inhibits the predominant endogenous pathway

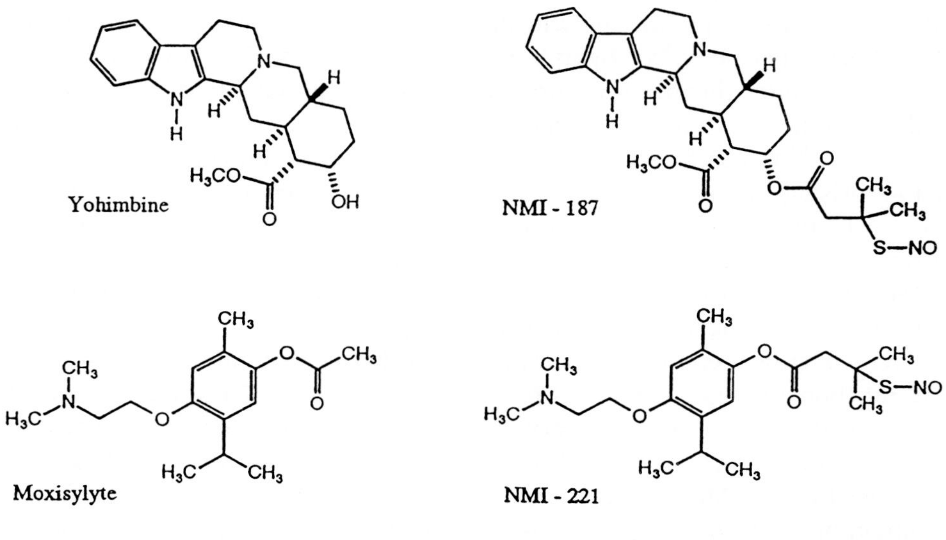

Fig. 2. Chemical structures of the α-adrenergic receptor antagonists yohimbine and moxisylyte and their respective, NO-enhanced analogs, NMI-187 and NMI-221, as synthesized by NitroMed, Inc.[60]

suppressing penile erection, an α-adrenergic antagonist, in one bi-functional molecule.

A potential for combination MED therapy based upon NO enhancement of prostaglandin E_1 action arises from demonstration that prostaglandin E_1 interacts synergistically with the S-nitrosothiol NO donor S-nitrosoglutathione (GSNO) to relax penile smooth muscle.[61] Furthermore, GSNO consistently relaxed human penile smooth muscle whether or not it responded to prostaglandin E_1.[61] These data suggest that the combination of prostaglandin E_1 and GSNO could potentially decrease the number of MED patients refractory to prostaglandin E_1 and reduce associated penile pain by lowering the prostaglandin E_1 dose required for satisfactory clinical response.

5. NO-Enhanced Medicines for Drug-Induced Gastrointestinal (GI) Damage

In many countries, nonsteroidal anti-inflammatory drugs (NSAIDs) constitute the most frequently taken over-the-counter and prescription medicines, reflecting their potent anti-inflammatory, analgesic, and antipyretic properties. Chronic NSAID use increases with age, primarily for symptomatic relief from musculoskeletal conditions such as osteoarthritis and as prophylaxis against cardiovascular disease.[40] A large proportion of people 65 years of age and older holds a current or recent NSAID prescription, the prevalence of NSAID use in this subpopulation accounting for about 90% of all NSAID prescriptions.[62] Their effectiveness and popularity notwithstanding, NSAIDs display a well-recognized side effect, GI toxicity.[40, 62] Symptomology of NSAID toxicity ranges from the relatively benign (heartburn, pain, dyspepsia) to the GI bleeding, ulceration, and perforation accounting for most of the disability and death from chronic NSAID use.[63] Given these epidemiological and population considerations, it is not surprising that the GI side effects of NSAIDs are age-related and represent the most important and common adverse drug effects reported to regulatory authorities.[62, 63] Selective inhibitors of the cyclooxygenase (COX) isoform induced in the setting of inflammation, COX-2, have improved the NSAID safety profile, but not without attendant liabilities and long-term safety questions.[64]

NO is a critical endogenous GI protectant that helps maintain the structural and functional integrity of the GI tract and repair injured GI tissue.[65] Thus, NO supplementation is considered a prime therapeutic modality for reducing NSAID-induced GI toxicity.[40] As with MED (above), two general approaches have been explored by NitroMed, Inc., to enhance the therapeutic and safety profiles of common NSAIDs. In the first, acute oral co-administration of the S-nitrosothiol NO-donor GSNO with the NSAID piroxicam reduced the extent of piroxicam's GI toxicity without altering its pharmacokinetic or analgesic properties (Fig. 3).[53] Glutathione alone was not GI-protective against piroxicam-induced stomach

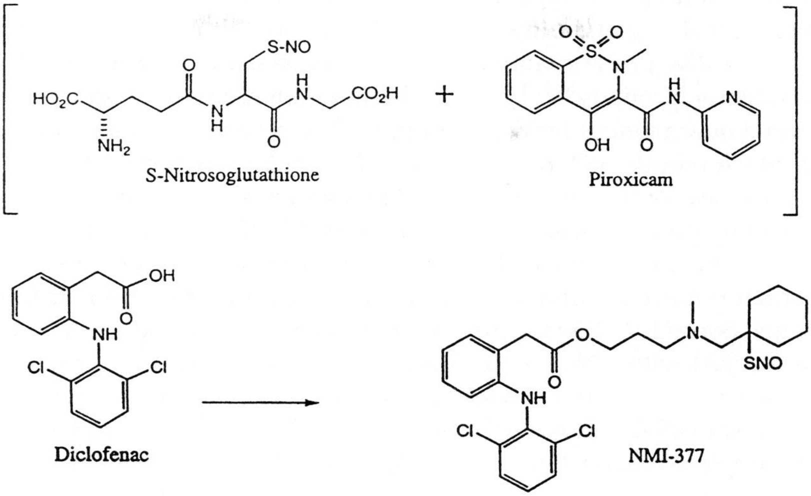

Fig. 3. Structure outline illustrating two NitroMed applications of nitrosothiol NO donors to reduce NSAID-induced GI toxicity. (A) Co-administration of the NO donor S-nitrosoglutathone (GSNO) and the NSAID piroxicam.[53] (B) Derivatization of the NSAID diclofenac with an organic linker bearing a nitrosothiol NO-donor moiety to produce the new chemical entity NMI-377, an NO-enhanced NSAID.[54]

erosions, implicating GSNO's *S*-nitrosothiol (i.e., NO-donor) group in the salutary response.[53]

Alternatively, NitroMed has covalently linked the *S*-nitrosothiol functionality onto known NSAIDs by *S*-nitrosation of the parent drug, as exemplified by several *S*–NO-diclofenac derivatives[54] (Fig. 3). In acute studies following single oral administration, many *S*–NO-diclofenac compounds displayed analgesic and anti-inflammatory activities at least equivalent to the parent, non-*S*-nitrosated NSAID, but caused significantly less GI irritation than did an equivalent diclofenac dose.[54] Such data imply that derivatization of an NSAID with an *S*-nitrosothiol NO donor improves the safety profile of this class of drugs by endowing the parent NSAID with GI-sparing properties.

6. Considerations Regarding the Development and Clinical Application of NO-Modulator Drugs

The central role of NO as a mediator of cell (patho) physiology and increasing evidence implicating altered NO status in several diseases have generated great interest in modulating exogenously tissue NO for symptomatic benefit or therapeutic gain.[3, 5, 7, 40, 46] Many disease states at least theoretically amenable to NO modulation have a chronobiological component (Table 1). Thus far, anti-anginal organic nitrates and NO gas for bronchopulmonary indications represent the prime examples of NO-based therapy by exploiting NO's potent vasodilator property.[6, 7, 9, 13] The pharmaceutical industry continues to explore several new avenues for discovering and developing efficacious NO-modulatory agents for both NO reduction and NO augmentation.[3] Dietary/nutritional approaches for NO modulation have also been espoused that may have life-extending properties, at least in the mouse.[66–68] While aimed at combating human illness, the therapeutically-oriented pharmaceutical efforts rest largely upon laboratory data, for only very recently have clinical trials been announced in the United States for NO-modulatory drugs such as performance-enhanced NSAIDs derivatized with a NO-donor functionality.[3] A contemporary report relating the GI protection offered by NO-donor-NSAID conjugates to specific inhibition of an important cellular enzyme[69] brings targeted molecular resolution to therapeutic NO supplementation. Further data along these lines could help identify "receptors" or "transducers" of NO activity whose direct modulation might allow disease-specific, and even personalized (patient-specific), NO-based therapy with minimal risk.

As our appreciation of the NO's physiological chemistry increases, the list of maladies potentially amenable to NO-modulation therapy should grow. In turn, refinement of current approaches toward tissue NO modulation and their utilization as commercially viable technology platforms with which to address additional diseases will most certainly depend upon detailing the mechanistic pharmacology of NO-modulating agents. Therein lies an operational dichotomy that may simply be another manifestation of NO's "Janus-faced" character: the need to know much more about NO's scope of action while seeking simultaneously to extract medical benefit out of our extant limited knowledge of NO biology.

Acknowledgments

The author thanks W.A. Pryor for suggesting that he contribute to this forum; D. Spooner for illustration assistance; and R. Earl and J. Schroeder for structure drawings.

References

1. Fleming, I. and Busse, R. (1999). NO: the primary EDRF. *J. Mol. Cell. Cardiol.* **31**: 5–14.
2. Patel, R. P., Mc Andrew, J., Sellak, H., White, C. R., Jo, H., Freeman, B. A. and Darley Usmar, V. M. (1999). Biological aspects of reactive nitrogen species. *Biochim. Biophys. Acta* **1411**: 385–400.

3. Janero, D. R. (2000). Nitric oxide (NO)-related pharmaceuticals: contemporary approaches to therapeutic NO modulation. *Free Radic. Biol. Med.* **28**: 1495–1506.
4. Beckman, J. S. (2002). Nitric oxide, peroxynitrite and ageing. *In* "Oxidative Stress and Aging: Diagnostics, Intervention and Longevity" (R. G. Cutler, and H. Rodriguez, Eds.), (Chap. 4, this volume.)
5. Hou, Y. C., Janczuk, A. and Wang, P. G. (1999). Current trends in the development of nitric oxide donors. *Curr. Pharm. Design* **5**: 417–441.
6. Parrat, J. R. (1979). Nitroglycerine — the first one hundred years: new facts about an old drug. *J. Pharm. Pharmacol.* **31**: 801–809.
7. Wink, D. A. and Mitchell, J. B. (1998). Chemical biology of nitric oxide: insights into regulatory, cytotoxic, and cytoprotective mechanisms of nitric oxide. *Free Radic. Biol. Med.* **25**: 434–456.
8. Young, D. V., Serebrynanik, D., Janero, D. R. and Tam, S. W. (2000). Suppression of proliferation of human coronary artery cells by the nitric oxide donor, S-nitrosoglutathione, is cGMP independent. *Mol. Cell Biol. Res. Commun.* **4**: 32–36.
9. Lyons, D. (1997). Impairment and restoration of nitric oxide-dependent vasodilation in cardiovascular disease. *Int. J. Cardiol.* **62**: S101–S109.
10. Beck, K. F., Eberhardt, W., Frank, S., Huwiller, A., Messmer, U. K., Mühl, H. and Pfeilschifter, J. (1999). Inducible nitric oxide synthase: role in cellular signaling. *J. Exp. Biol.* **202**: 645–653.
11. Symeonides, S. and Balk, R. A. (1999). Nitric oxide in the pathogenesis of sepsis. *Infect. Dis. Clin. North Am.* **13**: 449–463.
12. Beckman, J. S. and Koppenol, W. H. (1996). Nitric oxide, superoxide, and peroxynitrite: the good, the bad, and the ugly. *Am. J. Physiol.* **271**: C1424–C1437.
13. Janero, D. R. and Ewing, J. F. (2000). Nirtic oxide and post-angioplasty restenosis: pathological correlates and therapeutic potential. *Free Radic. Biol. Med.* **29**: 1199–1221.
14. Shimokawa, H. (1999). Primary endothelial dysfunction: atherosclerosis. *J. Mol. Cell. Cardiol.* **31**: 23–37.
15. Kojda, G. and Harrison, D. (1999). Interactions between NO and reactive oxygen species: pathophysiological importance in atherosclerosis, hypertension, diabetes, and heart failure. *Cardiovasc. Res.* **43**: 562–571.
16. Taddei, S. and Salvetti, A. (1996). Pathogenic factors in hypertension: endothelial factors. *Clin. Exp. Hypertens.* **18**: 323–335.
17. Lüscher, T. F., Tanner, F. C., Tschudi, M. R. and Noll, G. (1993). Endothelial dysfunction in coronary artery disease. *Ann. Rev. Med.* **44**: 395–418.
18. Paulus, W. J. (1996) Paracrine coronary endothelial modulation of diastolic left ventricular function in man: implications for diastolic heart failure. *J. Cardivasc. Fail.* **2**: S155–S164.

19. Rivard, A., Fabre, J. E., Silver, M., Chen, D., Murohara, T., Kearney, M., Magner, M., Asahara, T. and Isner, J. M. (1999). Age-dependent impairment of angiogenesis. *Circulation* **99**: 111–120.

20. Floyd, R. A. and Hensley, K. (2000). Nitrone inhibition of age-associated oxidative damage. *Ann. NY Acad. Sci.* **899**: 222–237.

21. Thomas, T. (2000). Monoamine oxidase-B inhibitors in the treatment of Alzheimer's disease. *Neurobiol. Aging* **21**: 343–348.

22. De la Monte, S. M., Lu, B. X., Sohn, Y. K., Etienne, D., Kraft, J., Ganju, N. and Wands, J. R. (2000). Aberrant expression of nitric oxide synthase III in Alzheimer's disease: relevance to cerebral vasculopathy and neuro-degeneration. *Neurobiol. Aging* **21**: 309–319.

23. Meyer, R. C., Spangler, E. L., Kametani, H. and Ingram, D. K. (1998). Age-associated memory impairment. Assessing the role of nitric oxide. *Ann. NY Acad. Sci.* **854**: 307–311.

24. Molina, J. A., Jiménez Jiménez, F. J., Ortí Pareja, M. and Navarro, J. A. (1998). The role of nitric oxide in neurodegeneration. Potential for pharmacological intervention. *Drugs Aging* **12**: 251–259.

25. Joseph, J. A., Villalobos Molina R., Denisova, N., Erat, S., Jimenez, N. and Strain, J. (1996). Increased sensitivity to oxidative stress and the loss of muscarinic receptor responsiveness in senescence. *Ann. NY Acad. Sci.* **786**: 112–119.

26. Ingram, D. K., Spangler, E. L., Meyer, R. C. and London, E. D. (1998). Learning in a 14-unit T-maze is impaired in rats following systemic treatment with N-omega-nitro-L-arginine. *Eur. J. Pharmacol.* **341**: 1–9.

27. Vernet, D., Bonavera, J. J., Swerdloff, R. S., Gonzalez Cadavids, N. F. and Wang, C. (1998). Spontaneous expression of inducible nitric oxide synthase in the hypothalamus and other brain regions of aging rats. *Endocrinology* **139**: 3254–3261.

28. Becquet, F., Courtois, Y. and Goureau, O. (1997). Nitric oxide in the eye: multifaceted roles and diverse outcomes. *Surv. Opthalmol.* **42**: 71–82.

29. Karlsten, R. and Gordh, T. (1997). How do drugs relieve neurogenic pain? *Drugs Aging* **11**: 398–412.

30. Pereira de Vasconcelos, A., Marescaux, C. and Nehlig, A. (1998). Age-dependent regulation of seizure activity by nitric oxide in the developing rat. *Brain Res. Dev. Brain Res.* **107**: 315–319.

31. Reuss, S., Schaeffer, D. F., Laages, M. H. and Riemann, R. (2000). Evidence for increased nitric oxide production in the auditory brain stem of the aged dwarf hamster (Phodopus sungorus): an NADPH-diaphorase histochemical study. *Mech. Aging Dev.* **112**: 125–134.

32. Melman, A. and Gingell, J. C. (1999). The epidemiology and pathophysiology of erectile dysfunction. *J. Urol.* **161**: 5–11.

33. Reckelhoff, J. F., Hennington, B. S., Kanji, V., Racusen, L. C., Schmidt, A. M., Yan, S. D., Morrow, J., Roberts, L. J. and Salahadueen, A. K. (1999). Chronic

aminoguanidine attenuates renal dysfunction and injury in aging rats. *Am. J. Hypertens.* **12**: 492–498.

34. Crone, J. K., Burnett, A. L., Chamness, S. L., Strandberg, J. D. and Chang, T. S. (1998). Neuronal nitric oxide synthase in the canine prostate: aging, sex steroid, and pathology correlations. *J. Androl.* **19**: 358–364.

35. Tempfer, C., Moreno, R. M., O'Brien, W. E. and Gregg, A. R. (2000). Genetic contributions of the endothelial nitric oxide synthase gene to ovulation and menopause in a mouse model. *Fertil. Steril.* **73**: 1025–1031.

36. Jang, D. and Murrell, G. A. C. (1998). Nitric oxide in arthritis. *Free Radic. Biol. Med.* **24**: 1511–1519.

37. Richmonds, C. R., Boonyapisit, K., Kusner, L. L. and Kaminski, H. J. (1999). Nitric oxide synthase in aging rat skeletal muscle. *Mech. Aging Dev.* **109**: 177–189.

38. Viner, R. I., Ferrington, D. A., Williams, T. D., Bigelow, D. J. and Schöneich. C. (1999). Protein modification during biological aging: selective tyrosine nitration of the SERCA2a isoform of the sarcoplasmic reticulum Ca^{2+}-ATPase in skeletal muscle. *Biochem. J.* **340**: 657–669.

39. Koike, E., Kobayashi, T., Mochitate, K. and Murakami, M. (1999). Effect of aging on nitric oxide production by rat alveolar macrophages. *Exp. Gerentol.* **34**: 889–894.

40. Bandarage, U. K. and Janero, D. R. (2001). Nitric oxide-releasing nonsteroidal anti-inflammatory drugs: novel gastrointestinal-sparing drugs. *Mini Rev. Med. Chem.* **1**: S7–S70.

41. Horning, M. L., Morrison, P. J., Banga, J. D., Stroes, E. S. and Rabelink, T. J. (1998). Nitric oxide availability in diabetes mellitus. *Diabetes Metab. Rev.* **14**: 241–249.

42. Rothe, H. and Kolb, H. (1999). Startegies of protection from nitric oxide toxicity in islet inflammation. *J. Mol. Med.* **77**: 40–44.

43. Meneilly, G. S., Battistini, B. and Floras, J. S. (2000). Lack of effect of sodium nitroprusside on insulin-mediated blood flow and glucose disposal in the elderly. *Metabolism* **49**: 373–378.

44. Morley, J. E. (1997). Anorexia of aging: physiologic and pathologic. *Am. J. Clin. Nutri.* **66**: 760–773.

45. Khare, V., Sodhi, A. and Singh, S. M. (1999). Age-dependent alterations in the tumoricidal functions of tumor-associated macrophages. *Tumour Biol.* **20**: 30–43.

46. Lai, C.-S. and Orida, N. K. (2002). Nitric oxide blocking therapeutics. *In* "Oxidative Stress and Aging: Diagnostics, Intervention and Longevity" (R. G. Cutler, and H. Rodriguez, Eds.), (Chap. 76, this volume.)

47. Babu, B. R. and Griffith, O. W. (1998). Design of isoform-selective inhibitors of nitric oxide synthase. *Curr. Opin. Chem. Biol.* **2**: 491–500.

48. Gordge, M. P. (1998). How toxic is nitric oxide? *Exp. Nephrol.* **6**: 12–16.

49. Ewing, J. F. and Janero, D. R. (1998). Specific *S*-nitrosothiol (thionitrite) quantification as solution nitrite after vanadium (III) reduction and ozone-chemiluminescent detection. *Free Radic. Biol. Med.* **25**: 621–628.

50. Hogg, N. (2000). Biological chemistry and clinical potential of *S*-nitrosothiols. *Free Radic. Biol. Med.* **28**: 1478–1486.

51. Ewing, J. F., Young, D. V., Janero, D. R., Garvey, D. S. and Grinnell, T. A. (1997). Nitrosylated bovine serum albumin derivatives as pharmacologically active nitric oxide congeners. *J. Pharmacol. Exp. Ther.* **283**: 947–954.

52. De Belder, A. J., MacAllister, R., Radomski, M. W., Moncada, S. and Vallance, P. J. (1994). Effects of *S*-nitrosoglutathione in the human forearm circulation: evidence for selective inhibition of platelet activation. *Cardiovasc. Res.* **28**: 691–694.

53. Tam, S. W., Saha, J. K., Garvey, D. S., Schroeder, J. D., Shelekhin, T. E., Janero, D. R., Chen, L., Glavin, A. and Letts, L. G. (2000). Nitrosothiol-based NO-donors inhibit the gastrointestinal mucosal damaging actions of NSAIDs. *Inflammopharmacology* **8**: 81–88.

54. Bandarage, U. K., Chen, L., Fang, X., Garvey, D. S., Glavin, A., Janero, D. R., Letts, L. G., Mercer, G. J., Saha, J. K., Schroeder, J. D., Shumway, M. J. and Tam, S. W. (2000). Nitrosothiol esters of diclofenac: synthesis and pharmacological characterization as gastrointestinal-sparing prodrugs. *J. Med. Chem.* **43**: 4005–4016.

55. Feldman, H. A., Goldstein, I., Hatzichristou, D. G., Krane, R. J. and McKinlay, J. B. (1994). Impotence and its medical and psychosocial correlates: results of the Massachusetts male aging study. *J. Urol.* **151**: 143–180.

56. Buvat, J., Lemaire, A. and Herbaut Buvat, M. (1996). Intracavernous pharmacotherapy: comparison of moxisylyte and prostaglandin E_1. *Int. J. Impot. Res.* **8**: 41–46.

57. Boolell, M., Allen, M. J., Ballard, S. A., Gepi-Attee, S., Muirhead, G. J., Naylor, A. M., Osterloh, I. H. and Gingell, C. (1996). Sildenafil: an orally active Type-5 cyclic GMP-specific phsophodiesterase inhibitor for the treatment of penile erectile dysfunction. *Int. J. Impot. Res.* **8**: 47–52.

58. Porst, H. (1996). The rationale for prostaglandin E_1 in erectile failure: a survey of worldwide experience. *J. Urol.* **155**: 802–815.

59. Kloner, R. A. (2000). Cardiovascular risk and sildenafil. *Am. J. Cardiol.* **86**: 57F–61F.

60. Sáenz de Tejada, I., Garvey, D. S., Schroeder, J. D., Shelekhin, T., Letts, L. G., Fernández, A., Cuevas, B., Gabancho, S., Martínez, V., Angulo, J., Trocha, M., Marek, P., Cuevas, P. and Tam, S. W. (1999). Design and evaluation of nitrosylated α-adrenergic receptor antagonists as potential agents for the treatment of impotence. *J. Pharmacol. Exp. Ther.* **290**: 121–128.

61. Angulo, J., Cuevas, P., Moncada, I., Martin-Morales, A., Allona, A., Fernandez, A., Gabancho, S., Ney, P. and Sáenz de Tejada, I. (2000). Rationale

for the combination of PGE_1 and *S*-nitrosoglutathione to induce relaxation of human penile smooth muscle. *J. Pharmacol. Exp. Ther.* **295**: 586–593.

62. Tenenbaum, J. (1999). The epidemiology of nonsteroidal anti-inflammatory drugs. *Can. J. Gastroenterol.* **13**: 119–122.

63. Langman, M. J. (1999). Risks of anti-inflammatory drug-associated damage. *Inflamm. Res.* **48**: 236–238.

64. Lichtenstein, D. R. and Wolfe, M. M. (2000). COX-2-selecive NSAIDs. New and improved? *JAMA* **284**: 1297–1299.

65. Kawano, S. and Tsuji, S. (2000). Role of mucosal blood flow: a conceptual review in gastric mucosal injury and protection. *J. Gasteroenterol. Hepatol.* **15**: D1–D6.

66. McCarty, M. F. (1998). Vascular nitric oxide may lessen Alzheimer's risk. *Med. Hypotheses* **51**: 465–476.

67. Saito, K., Yoshioka, H. and Cutler, R. G. (1998). A spin trap, N-tert-butyl-alpha-phenylnitrone, extends the life span of mice. *Biosci. Biotechnol. Biochem.* **62**: 792–794.

68. Janero, D. R. (2001). Nutritional aspects of nitric oxide: human health implications and therapeutic opportunities. *Nutrition* **17**: 896–903.

69. Fiorucci, S., Antonelli, E., Santucci, L., Morelli, O., Miglietti, M., Federici, B., Mannucci, R., Del Soldato, P. and Morelli, A. (1999). Gastrointestinal safety of nitric oxide-derived aspirin is related to inhibition of ICE-like cysteine proteases in rats. *Gastroenterology* **116**: 1089–1106.

Chapter 78

Genechips for Age/Stress-Related Gene Expression and Analysis

Chris Seidel[†] and Ralph M. Sinibaldi[*,‡]

Chris Seidel and **Ralph M. Sinibaldi** • Operon Technologies, 1000 Atlantic Ave, Alameda CA

Current Addresses:
[†]Department of Molecular and Cell Biology, UC Berkeley, Berkeley CA, 94720
[‡]Genospectra Inc., 46540, Fremont Blvd., Fremont, CA, 94536

*Corresponding Author.
Tel: 510-865-8644, E-mail: seidel@phageT4.org, E-mail: rsinibaldi@genospectra.com

1. Introduction

Oxidative stress and aging encompass a wide variety of cellular phenomena ranging from simple repair of damaged DNA to initiating programmed cell death through apoptosis. DNA microarrays can be used to visualize the behavior of all genes as the cell responds to various kinds and amounts of oxidative stress. In this way one has the opportunity to dissect and unravel the myriad processes taking place and identify the regulatory elements involved.

2. Method for Measuring Oxidative Stress and Age-Related Genes

Oxygen radicals and other active oxygen species arise from a number of mechanisms in biology.[1] Cells have evolved complex and versatile mechanisms for dealing with the various kinds of damage that can occur in the presence of such redox-active oxygen species (ROS), and for regulating the redox environment within the cell. Incomplete reduction of oxygen to water during respiration can give rise to ROS such as superoxide (O^-), H_2O_2, and OH. Radiolysis of H_2O is also an abundant source of active oxygen species.[2] These molecules cause cytotoxic lesions in the form of damage to virtually all cellular components. DNA undergoes approximately 100 different modifications in response to free radicals, affecting all components of the molecule.[2-4] The DNA repair mechanisms for some of these lesions are unknown. Proteins and lipids are also well characterized targets of active oxygen species. The response of *Escherichia coli* to oxidative stress reveals at least two distinct sets of genes for dealing separately with superoxide and H_2O_2. Groups of genes involved in a coordinated response to some stimulus are often referred to as "stimulons". Each of the oxidative stress stimulons in *Escherichia coli* consists of dozens of genes, and only a handful of each set are understood in any detail. Each set contains a subset of genes that share a common regulator. Two-dimensional (2-D) gel electrophoresis of cells exposed to peroxide reveals induction of 30–40 polypeptides, nine of which fall under the control of one regulator, oxyR.[5,6] A similar situation exists for superoxide in which 30–40 proteins are induced, nine of which form what is known as the soxR regulon.[7,8] In neither case has the regulon been fully characterized and it is thought that other regulons have yet to be discovered.[9]

In eukaryotes the situation becomes even more complex. Recently Godon *et al.*[10] defined a "peroxide stimulon" in the budding yeast *Saccharomyces cerevisiae* consisting of 167 proteins whose expression is affected by a factor of at least 1.5 fold when examined by 2-D gel analysis of cellular extracts from cells exposed to low levels of H_2O_2. The peroxide stimulon is likely to consist of many separate pathways responding en masse to the peroxide insult. It is not known what genes are responding to common regulatory elements, and which are responding to the integration of various combinations of regulatory elements. It is known that yeast can mount different responses to oxidative stress depending on the nature of the stress. As in *Escherichia coli*, it has been shown that yeast have distinct responses to H_2O_2 and superoxide.[11, 12] However, many of the genes involved have not been identified.

While the response to oxidative stress is complicated, involving the up and down regulation of hundreds of proteins, effective methods have recently been developed for unravelling the complex response of the cell to virtually any stimuli. DNA microarrays allow the parallel analysis of every gene in an organism at any point in time. The ability to examine the relative state of every RNA in an organism allows the researcher an unprecedented amount of information from which to decipher the cellular response to a situation, and serves as a perfect compliment to other methods of large scale parallel analysis, such as 2-D gel electrophoresis. DNA microarrays consist of DNA sequences corresponding to a gene or parts of a gene spotted down or synthesized onto a surface in an ordered array. Each location on the array is a unique sequence capable of hybridizing to a labeled target sequence present in a complex mixture of sequences representing the sum total of RNAs extracted from a cell or a group of cells. After hybridization and washing away of excess probe, the array is scanned and the presence of bound molecules is detected and quantified. In this way, the relative abundance of hundreds or thousands of genes can be assayed at a time, in parallel.

3. DNA Microarrays

Two basic types of DNA microarrays are currently in use. The first type involves oligo synthesis on the surface of a chip.[13] These arrays are produced by Affymetrix. Oligos of length 20 to 25 bases are synthesized on a surface using a combinatorial stepwise photolithograpic process. The resulting surface contains a high density grid of features to which hybridization of a labeled sample can occur. The technique benefits from from the basic fact that at the current time it is possible to synthesize more than 100 000 features per square centimeter, and this number continues to increase. Given the size of most genomes this allows several 25-mer stretches to represent each gene in an organism, and also allows for a matching set of control features, containing a single base mismatch. By comparing direct match to mismatch features for several regions of a gene one can get high quality quantitative information for expression of a particular gene, or exons of a gene. The main disadvantage to this technique is the cost, and the resources required to make the photolithographic mask required to make the chip. It is also necessary to know in advance what sequence will be constructed on the chip. A promising variation on this method involving oligo synthesis on a surface using high density ink jet printer technology is now available.

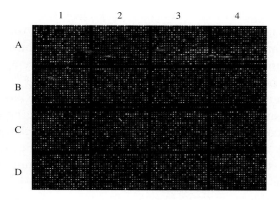

Fig. 1. Fluorescence image of a yeast genome DNA micro-array from the experiment described in Table 1. cDNA from cells exposed to 1 mM H_2O_2 for 20 min. was labeled with Cy5 (red) while cDNA from mock treated cells was labeled with Cy3 (green). The array is hybridized and scanned at 550 nm and 650 nm. The false color overlay indicates the relative amounts of cDNA in each population. Dark gray (Red) spots indicate transcripts with higher abundance in H_2O_2 exposed cells, whereas light gray (green) spots denote transcripts with higher abundance in control cells. Very light gray (yellow) spots indicate roughly equal abundance in each population.

The other kind of DNA microarray that currently enjoys the broadest range of use is the spotted DNA microarray.[14] This method was developed at Stanford University and popularized by Pat Brown and Joe DeRisi.[15] Because of its simplicity, and accessibility of resources, it is possible for researchers to build their own DNA chips if they so choose. This method relies on glass slides which have been coated with poly-L-lysine. Double stranded DNA, typically in the form of PCR products, is then robotically spotted down onto the surface from microtiter plates. The negatively charged DNA sticks to the positively charged poly-L-lysine surface. We have recently shown that long oligos work just as well, if not better than PCR products as hybridization probes for robotically spotted DNA microarrays.[16, 35] These arrays typically make use of a two color labeling scheme to compare the global gene expression patterns of two populations of cells. For instance, to examine which genes are induced or repressed by oxidative stress, one could compare control cells to cells exposed to hydrogen peroxide. The first step is to perform the experiment and isolate RNA from each population of cells. Each sample is then independently labeled by reverse transcription and incorporation of a unique fluorescent dye. After labeling, the samples are mixed together and allowed to hybridize to the array. Upon scanning, the relative fluorescence of each dye at every spot on the array is compared. In this way one can determine the relative amount of expression for every gene between the two samples. An assumption of this technique is that on the whole, transcription of most genes will be unchanged, whereas individual genes or groups of genes can change dramatically. The array contains controls to normalize for differences in the amount of input RNA. This technique has been used successfully to examine and unveil an extraordinary range of regulatory space in yeast from cells undergoing heat shock,[17] diauxic shift,[18] the cell cycle,[19] to the response of the genome to mutation and exposure to compounds.[20] It has also revealed an unprecedented way of viewing, classifying, and understanding human cancers.[21]

Arrays produced using 70-mer oligos combine the best characteristics of both approaches. They are specific and they are easy to use. In organisms, such as *Saccharomyces cerevisiae*, the designed oligonucleotide sequences are BLASTed against the whole genome to ensure that there is no cross-hybridization owing to sequence relatedness or gene overlap. In humans, the designed sequences can be checked against known genes to minimize cross-hybridization, or if one wishes to make a mammalian-general array, sequences can be chosen that cross-hybridize to humans, mice, and so on. We employ a rational design to select 70mers (or DNA sequences) that are sequence optimized and normalized for hybridization temperature. The 70mers are designed to be complementary to sequences near the 3' ends of genes and have no significant secondary structure. In addition, because of the uniform size, we are able to print the oligos at a normalized concentration so that every spot contains a consistent amount of nucleic acid. Hybridization-normalized DNA sequences can be designed to be $\pm 3°C$ of each other, thus assuring consistent hybridization for all the DNA sequences on the array. Sequence optimization is a design process in which we minimize cross-hybridization and overlapping gene hybridization by choosing sequences in the gene that do not crosshybridize with other genes. The oligonucleotides can also be designed to detect alternatively spliced genes. One criticism of the oligonucleotide approach for the analysis of gene expression is that it is not very sensitive and multiple oligo-nucleotides to cover a gene must be used. This is true with shorter oligonucleotides (15- to 20mer), but the 70mers yield good sensitivity and offer the great specificity of shorter sequences. We compared the sensitivity of the 35-, 50-, 70-, and 90mer for detecting highly expressed genes and genes expressed at moderate or low levels in yeast, and the 70mer exhibited the best characteristics of specificity and sensitivity. We are confident that 70mer oligo-based arrays offer an accurate and cost efficient way to monitor gene expression.

4. Gene Expression of Cells Exposed to Hydrogen Peroxide

Godon *et al.* used 2 dimensional gel analysis of extracts from cells exposed to 0.2, 0.4, and 0.8 mM H_2O_2 for 15 minutes to define the hydrogen peroxide stimulon. To get a glimpse at the genomic transcriptional response of cells to H_2O_2, we exposed cells to 1 mM H_2O_2, harvested the RNA, and analyzed the resulting

Table 1. Selected Peroxide Stimulon Proteins Compared to Microarray Data

Gene	ORF	H_2O_2 Stimulation index	$H_2O_2/$ Control 20 min.	$H_2O_2/$ Control 30 min.	Description
Antioxidant scavenging/defense					
CCP1	YKR066C	6	6.0	4.4	Cytochrome c peroxidase
CTT1	YGR088W	14.7	1.4	1.1	Catalase T
GLR1	YPL091W	2.1	1.3	1.3	Glutathione reductase
SOD1	YJR104C	4.3	2.6	1.2	Copper/zinc superoxide dismutase
SOD2	YHR008C	5.9	2.5	2.1	Manganese superoxide dismutase
TRR1	YDR353W	12.2	10.0	8.2	Thioredoxin reductase/ NADPH-dependent
TRX1	YLR043C	11.5	0.79	0.76	Thioredoxin 1
TRX2	YGR209C	11.5	4.5	4.7	Thioredoxin 2
TSA1	YML028W	5.9	8.7	10.9	Thiol-specific antioxidant
GRX1	YCL035C	5	0.74	0.61	Putative glutaredoxin
	YDR453C	15	3.6	3.7	Similar to Tsa1p
AHP1	YLR109W	3.1	1.9	1.6	alkyl hydroperoxide reductase
GRE2	YOL151W	5.9	8.6	8.0	
CTA1	YDR256C		6.7	7.6	Catalase A
LYS7	YMR038C		2.1	1.9	Cu chaperone for Sod1
GRX2/TTR1	YDR513W		1.6	1.3	Glutaredoxin
Amino acid mzetabolism					
ARO4	YBR249C	3.1	0.52	0.31	2-Dehydro-3-deoxyphospho- heptonate aldolase
CPA2	YJR109C	2.1	1.3	1.1	Carbamoylphosphate synthase
GDH1	YOR375C	0.36	1.3	1.4	Glutamate dehydrogenase HIS4
	YCL030C	2.3	0.54	0.37	Hydrolyase
ILV2	YMR108W	0.31	0.78	0.73	Acetolactate synthase
ILV3	YJR016C	0.6	0.85	0.84	Dihydroxyacid dehydratase
ILV5	YLR355C	0.15	0.7	0.66	Ketol-acid reductoisomerase
SHMT2	YLR058C	0.44	0.9	0.84	Serine hydroxymethyl- transferase
Arginine metabolism					
ARG1	YOL058W	3.1	1.1	1.5	Arginosuccinate synthetase
CAR2	YLR438W		1.5	2.7	Ornithine aminotransferase
CPA1	YOR303W		2.1	3.2	Carbamoyl phosphate synthetase
CAR1	YPL111W		2.8	3.7	Arginase
Lysine metabolism					
LYS20	YDL182W	1.8	4.1	3.5	Homocitrate synthase
LYS9	YNR050C	0.55	3.7	4.0	Saccharopine dehydrogenase
LYS21	YDL131W		3.3	2.1	Homocitrate synthase
LYS4	YDR234W		2.2	2.7	Homoaconitase

Table 1 (*Continued*)

Gene	ORF	H_2O_2 Stimulation index	$H_2O_2/$ Control 20 min.	$H_2O_2/$ Control 30 min.	Description
LYS2	YBR115C		2.0	2.0	Alpha aminoadipate reductase
LYS1	YIR034C		4.3	4.3	Saccharopine dehydrogenase
FRE1	YLR214W		3.9	2.8	Fe Homeostasis, Fe/Cu reductase
ISU2	YOR226C		6.8	5.9	Fe Homeostasis
ISU1	YPL135W		3.5	2.3	Fe Homeostasis
FRE7	YOL152W		8.6	8.6	
TCA cycle					
LPD1	YFL018C	0.45	0.72	0.62	Dihydrolipoamide dehydrogenase
MDH1	YKL085W	0.6	0.57	0.43	Malate dehydrogenase, mitochondrial
PDB1	YBR221C	0.34	1.4	1.3	Pyruvate dehydrogenase complex
SDH3	YKL141W		2.4	2.1	Succinate dehydrogenase complex
PCK1	YKR097W		3.4	4.3	Phosphoenolpyruvate carboxykinase
ACO1	YLR304C		2.8	1.8	Aconitase
IDH2	YOR136W		3.6	2.4	Isocitrate dehydrogenase
CIT3	YPR001W		2.7	2.5	Citrate synthase
Miscellaneous					
RAD55	YDR076W		2.3	4.1	DNA repair and recombination
XRS2	YDR369C		6.8	6.2	DNA repair and recombination
RAD51	YER095W		1.8	2.0	DNA repair and recombination
FDH1	YOR388C		23.7	42.6	Formate Dehydrogenase
	YPL275W		14.0	22.6	Formate Dehydrogenase
	YPL276W		6.9	14.9	Formate Dehydrogenase
SFA1	YDL168W		2.8	2.6	Formaldehyde metabolism
AAD14	YNL331C		3.4	4.4	Putative aryl alcohol reductase
AAD15	YOL165C		3.6	3.5	Putative aryl alcohol reductase
AAD3	YCR107W		2.9	3.1	Putative aryl alcohol reductase
AAD6	YFL056C		4.6	6.2	Putative aryl alcohol reductase
	YFL057C		4.1	4.6	Putative aryl alcohol reductase
TAH18	YPR048W		4.2	4.6	Oxidreductase
	YIR035C		3.3	4.7	Oxidoreductase
	YDR453C		3.6	3.7	Oxidoreductase, similar to TSA1p
	YML131W		5.0	5.1	Oxidoreductase
ROX1	YPR065W		4.0	3.7	O2 dep transcription

Table 1 (*Continued*)

Gene	ORF	H_2O_2 Stimulation index	$H_2O_2/$ Control 20 min.	$H_2O_2/$ Control 30 min.	Description
CYC1	YJR048W		4.4	3.0	Oxidative phosphorylation
SDL1	YIL167W		6.9	2.8	Serine dehydratase/ gluconeogenesis
MIG2	YGL209W		3.0	1.2	Glucose repression
GTT2	YLL060C		3.4	3.0	Glutathione transferase

cDNA with DNA microarrays. We compared the results to those found by Godon *et al.* and have listed a small portion of the results in Table 1. Genes with antioxidant scavenging/defense properties showed the same basic profile transcriptionally as was found by 2-D gel analysis of proteins. We saw up-regulation of cytochrome c peroxidase (CCP1), and both superoxide dismutases — SOD1 and SOD2. Other oxidant scavenging enzymes that stood out on the arrays and agree with the 2-D gel analysis are thioredoxin II (TRX2) and thioredoxin reductase (TRR1), as well as thioperoxidase (TSA1) and the related protein YDR453C. We were able to extend this category by the addition of peroxisomal catalase A (CTA1), thioltransferase (TTR1), and LYS7, a protein which acts as a Cu chaperone to SOD1. CTA1 and TTR1 have been shown previously to be induced by oxidative stress.[22-24] Thioredoxins I and II were unable to be distinguished by 2-D gel electrophoresis and are listed together as being induced by 11.5 fold. While TRX2 is known to be induced by oxidative stress,[25, 26] little is known regarding TRX1 expression. We detected unique expression profiles for each of these genes, and found that while TRX2 was significantly up-regulated relative to control cells, TRX1 was not induced. TRX1 has many cellular roles including stress response, amino acid metabolism, and membrane trafficking,[27, 28] and is required along with TRX2 for the antioxidant activity of the cytosolic peroxidase AHP1.[29] What role differential regulation of TRX1 and TRX2 may play in the cell is unclear.

Two notable exceptions for which we did not see up-regulation in response to H_2O_2 are the cytosolic catalase T (CTT1), and Glutaredoxin (GRX1). CTT1 is known to be induced under conditions of oxidative stress,[30] and shares Yap1 and Skn7 dependent regulatory behavior with other genes of the stimulon, such as CCP1 and TSA1.[31] *Saccharomyces cerevisiae* contains at least two glutaredoxin genes, GRX1 and GRX2 (aka TRR1). While both are up-regulated during oxidative stress in response to hydrogen peroxide,[32] mutants in each differ in their sensitivity to oxidative stress. GRX1 has been found to be important for response to menadione whereas GRX2 mutants are hyper sensitive to H_2O_2.[33] We saw no up-regulation of GRX1 and only a slight increase in GRX2 expression.

The slight discrepancies between the microarray data and the 2-D gel analysis can be due to many factors. Two-dimensional gels examine the final protein products of a gene, whereas microarrays examine the products of transcription. While it is often the case that transcription is tightly linked to protein expression, the extent of post-transcriptional regulation for many genes of the peroxide stimulon has not been examined. Indeed microarrays can be employed to examine post-transcriptional effects by isolating polysomes and seeing what RNAs are being actively translated.[34] Furthermore, the response of many genes in the peroxide stimulon changes due to seemingly small differences in peroxide concentration (Godon *et al.*). Our experiments used 1 mM peroxide for 20 and 30 minutes whereas Godon used 0.2 to 0.8 mM for 15 minutes. The kinetics of the peroxide stress response divides genes into distinct classes with most genes peaking at 15 minutes after exposure and returning to pre-stress levels after one hour, while other genes show a relatively delayed response and peak after one hour (Godon *et al.*).

Using DNA microarrays one can rapidly measure a kinetic profile of expression for every gene across a time course. Since the typical microarray experiment gives a readout for the transcriptional state of every gene in the organism, the data sets are very sensitive to certain kinds of artifacts. For instance, cells can undergo cold shock while sitting in a cold centrifuge rotor, which may not be the intent of the experimenter. Differences in time, cell density, or solvent exposure can lead to complicated nonspecific effects. For this reason it is important to treat samples to be compared, as identically as possible. Developing methods for the rapid processing of cells will help the consistency of the data. For example, harvesting cells by filtering takes seconds whereas centrifugation takes several minutes. Lastly, the cells we used for this experiment are mutant for the transcription elongation factor TFIIS. While there are no genes in yeast that are known to be regulated by TFIIS, further experiments will determine if it makes a contribution to the expression of any genes in the peroxide stimulon. In many cases we were able to extend the categories of genes found by Godon *et al.* For instance several members of the pathways for arginine and lysine biosynthesis were shown to be induced. Proteins involved in translation were repressed in general, in agreement with the data of Godon *et al.* Dozens of genes were found to be significantly induced that were not found by 2-D gel in the peroxide stimulon, including proteins involved in iron homeostasis, DNA repair, and many oxidoreductases.

5. Concluding Remarks

Since the announcement of the completion of sequencing of the human genome in the year 2000, researchers have the freedom to build and use microarrays of their own design which span all human genes. DNA microarrays are the perfect tool for unraveling the complex regulatory network of genes involved in the response to oxidative stress and aging. Microrarrays allow the study of hundreds or thousands of genes per experiment. Parallel analysis of this scale is required to see the coordinated expression of pathways that unfold in response to oxidative stress.

References

1. Fridovich, I. (1978). The biology of oxygen radicals. *Science* **201**: 875–880.
2. von Sonntag, C. (1987). *The Chemical Basis of Radiation Biology*, Taylor & Francis, London.
3. Dizdaroglu, M. (1992). Oxidative damage to DNA in mammalian chromatin. *Mutat. Res.* **275**: 331–342.
4. Imlay, J. A. and Linn, S. (1988). DNA damage and oxygen radical toxicity. *Science* **240**: 1302–1309.
5. Christman, M. F., Morgan, R. W., Jacobson, F. S. and Ames, B. N. (1985). Positive control of a regulon for defenses against oxidative stress and some heat shock proteins in *Salmonella typhimurium*. *Cell* **41**: 753–762.
6. Greenberg, J. T. and Demple, B. (1989). A global response induced in *Eschericia coli* by redox–cycling agents overlaps with that induced by peroxide stress. *J. Bacteriol.* **171**: 3933–3939.
7. Greenberg, J. T., Monach, P. A., Chou, J. H., Josephy, P. D. and Demple, B. (1990). Positive control of a global antioxidant defense regulon activated by superoxide generating agents in *Eschericia coli*. *Proc. Natl. Acad. Sci. USA* **87**(16): 6181–6185.
8. Tsaneva, I. R. and Weiss, B. (1990). SoxR, a locus governing a superoxide response regulon in *Escherichia coli* K-12. *J. Bacteriol.* **172**: 4197–4205.
9. Demple, B. (1991). Regulation of bacterial oxidative stress genes. *Ann. Rev. Genet.* **25**: 315–337.
10. Godon, C., Lagniel G., Lee, J., Buhler, J.-M., Kieffer, S., Perrot, M., Boucherie, H., Toledano, M. B. and Labarre, J. (1998). The H_2O_2 Stimulon in *Saccharomyces cerevisiae*. *J. Biol. Chem.* **273**(34): 22 480–22 489.
11. Jamieson, D. J. (1992). *Saccaromyces cerevisiae* has distinct adaptive responses to both hydrogen peroxide and menadione. *J. Bacteriol.* **174**(20): 501–507.
12. Flattery-O'Brien, J., Collinson, L. P. and Daws, I. W. (1993). *Saccaromyces cerevisiae* has an inducible response to menadione which differs from that to hydrogen peroxide. *J. Gen. Microbiol.* **139**: 501–507.
13. Chee, M., Yang, R., Hubbell, E., Berno, A., Huang, X. C., Stern, D., Winkler, J., Lockhart, D. J., Morris, M. S. and Fodor, S. P. (1996). Accessing genetic information with high-density DNA arrays. *Science* **274**: 610–614.

14. Schena, M., Shalon, D., Davis, R. W. and Brown, P. O. (1995). Quantitative monitoring of gene expression patterns with a complementary DNA micro-array. *Science* **270**: 467–470.

15. DeRisi, J. L. and Brown, P. O. The MGuide — the molecular biologists complete guide for assembling your own microarrayer. http://cmgm.stanford.edu/pbrown/mguide/

16. Sinibaldi, R. M., O'connell, C., Seidel and Rodriquez, H. (2001). Gene expression analysis on medium-density oligonucleotide arrays. *In* "Methods in Molecular Biology" (J. B. Rampal, Ed.), Vol. 170, Humana Press.

17. Lashkari, D. A., DeRisi, J. L., McCusker, J. H. Y., Namath, A. F., Gentile, C., Hwang, S. Y., Brown, P. O. and Davis, R. W. (1997). Yeast microarrays for genome wide parallel genetic and gene expression analysis. *Proc. Natl. Acad. Sci. USA* **94**(24): 13 057–13 062.

18. DeRisi, J. L., Vishwanath, I. R. and Brown, P. O. (1997). Exploring the metabolic and genetic control of gene expression on a genomic scale. *Science* **278**: 680–686.

19. Spellman, P. T., Sherlock, G., Zhang, M. Q., Iyer, V. R., Anders, K., Eisen, M. B., Brown, P. O., Botstein, D. and Futcher, B. (1998). Comprehensive identification of cell cycle-regulated genes of the yeast *Saccharomyces cerevisiae* by microarray hybridization. *Mol. Biol. Cell* **9**(12): 3273–3297.

20. Hughes, R. H. *et al.* (2000). Functional discovery via a compendium of expression profiles. *Cell* **102**: 109–126.

21. Ross, T. D. *et al.* (2000). Systematic variation in gene expression patterns in human cancer cell lines. *Nature Gen.* **24**: 227–235.

22. Lee, J., Romeo, A. and Kosman, D. J. (1996). Transcriptional remodeling and G1 arrest in dioxygen stress in *Saccharomyces cerevisiae. J. Biol. Chem.* **271**(40): 24 885–24 893.

23. Luikenhuis, S., Perrone, G., Dawes, I. W. and Grant, C. M. (1998). The yeast *Saccharomyces cerevisiae* contains two glutaredoxin cenes that are required for protection against reactive oxygen species. *Mol. Biol. Cell* **9**: 1081–1091.

24. Grant, C. M., Luikenhuis, S., Beckhouse, A., Soderbergh, M. and Dawes, I. W. (2000). Differential regulation of glutaredoxin gene expression in response to stress conditions in the yeast *Saccharomyces cerevisiae. Biochim. Biophys. Acta* **1490**: 33–42.

25. Morgan, B. A., Banks, G. R., Toone, W. M., Raitt, D., Kuge, S. and Johnston L. H. (1997). The Skn7 response regulator controls gene expression in the oxidative stress response of the budding yeast *Saccharomyces cerevisiae. EMBO J.* **16**(5): 1035–1044.

26. Lee, J., Godon, C., Lagniel, G., Spector, D., Garin, J., Labarre, J. and Toledano, M. B. (1999). Yap1 and Skn7 control two specialized oxidative stress response regulons in yeast. *J. Biol. Chem.* **274**(23): 16 040–16 046.

27. Slekar, K. H., Kosman, D. J. and Culotta, V. C. (1996). The yeast copper/zinc superoxide dismutase and the pentose phosphate pathway play overlapping roles in oxidative stress protection. *J. Biol. Chem.* **271**: 28 831–28 836.

28. Xu, Z. and Wickner, W. (1996). Thioredoxin is required for vacuole inheritance in *Saccharomyces cerevisiae*. *J. Cell Biol.* **132**(5): 787–794.

29. Lee, J., Spector, D., Godon, C., Labarre, J. and Toledano, M. B. (1999). A new antioxidant with alkyl hydroperoxide defense properties in yeast. *J. Biol. Chem.* **274**(8): 4537–4544.

30. Jungmann, J., Reins, H. A., Lee, J., Romeo, A., Hassett, R., Kosman, D. and Jentsch, S. (1993). MAC1, a nuclear regulatory protein related to Cu-dependent transcription factors is involved in Cu/Fe utilization and stress resistance in yeast. *EMBO J.* **12**(13): 5051–5056.

31. Lee, J., Godon, C., Lagniel, G., Garini, J., Labarref, J. and Toledano, B. (1999). Yap1 and Skn7 control two specialized oxidative stress response regulons in yeast. *J. Biol. Chem.* **274**(23): 16 040–16 046.

32. Luikenhuis, S., Perrone, G., Dawes, I. W. and Grant, C. M. (1998). The yeast *Saccharomyces cerevisiae* contains two glutaredoxin genes that are required for protection against reactive oxygen species. *Mol. Biol. Cell* **9**: 1081–1091.

33. Rodriguez-manzaneque, M. T., Ros, J., Cabiscol, E., Sorribas, A. and Herrero, E. (1999). Grx5 glutaredoxin plays a central role in protection against protein oxidative damage in Saccharomyces cerevisiae. *Mol. Cell. Biol.* **19**(12): 8180–8190.

34. Johannes, G., Carter, M. S., Eisen, M. B., Brown, P. O. and Sarnow, P. (1999). Identification of eukaryotic mRNAs that are translated at reduced cap binding complex eIF4F concentrations using a cDNA microarray. *Proc. Natl. Acad. Sci. USA* **96**(23): 13 118–13 123.

35. Chen, M., Ten Bosch, J., Beckman, K., Salijoughi, S., Seidel, C., Tuason, N., Larka, L., Lam, H., Sinibaldi, R. and Saul, R. (1999). Covalent attachment of sequence optimized PCR products and long oligos for DNA microarrays. *Microbial Comparative Genomics* **4**(2): 116.

SECTION 13

Future Trends in Health Care

Future Trends in Health Care

Chapter 79

Preventive Medicine and Health Care Reimbursement: A Necessary but Not Sufficient Condition for Wider Access to Preventive Health Services

Tali Arik

Tali Arik • 504 w. 36th St, Hays, K5 67601
Tel: (785) 625-3072, E-mail: t.arik@worldnet.att.net

1. Introduction

It is widely assumed that broadening healthcare coverage and reimbursement is a necessary and sufficient condition for making healthcare services more widely and possibly universally available. There is great attraction in this thesis. There is a large, incontrovertible body of evidence that preventive medicine works. Early detection of a variety of pathologic conditions is important to appropriately instituting measures which then decrease the morbidity and mortality attendant to those conditions. All medical diagnostic and therapeutic measures have a cost. Therefore, it appears reasonable to assume and often is assumed that if cost was removed from the equation, services proven to be of diagnostic and therapeutic value would be used by all. It will be shown that such is not the case. Even if healthcare reimbursement was not an issue, there are many barriers to the widespread application of preventive medicine. Better, cheaper, more widely available healthcare reimbursement will not remove most of these barriers. Consideration of all the issues involved may prompt more creative, thoughtful debates among healthcare providers, policy advocates, and the public than those that prevail today, characterized by their near-obsession with reimbursement issues and even proposals regarding nationalized, universal healthcare reimbursement. Only then will the prospects brighten for making proven preventive healthcare measures more widely available, much less those for which the evidence is less convincing or about which there remains question, debate, and honest difference of opinion. To the extent that the terms of the debate shift to include consideration of issues other than healthcare reimbursement, an element of realism will have entered the necessary debate regarding the many and serious barriers to the wider utilization of preventive medical services.

Preventive medicine works. Its proper practice decreases morbidity and mortality, including disease-specific and all-cause mortality. It is the primary reason for the extraordinary gains in life expectancy enjoyed by millions of individuals since 1900. The theme of the O2SA conference and the book of which this is a chapter is that by understanding primary mechanisms of diseases-in this case oxidative stress — we can better apply the basic science of oxidative stress to diagnostics and therapeutics with a view to improving human health. This is our goal, as scientists, researchers, physicians, healthcare providers and policy makers. These improvements would result from primary and secondary prevention, and from the application of actual therapeutics once pathology has been diagnosed. Some of the interventions may improve quality of life; others may facilitate reaching maximal lifespan. Some may accomplish both. Given these goals, it is reasonable to ask what barriers exist to the implementation of preventive medical strategies and technologies. Further, should there exist barriers, it should be asked what steps might be taken to broaden the application of such valuable strategies and technologies, making them available to all that may benefit.

2. Consequences of Increasing Human Life Span

The socioeconomic and political implications of increasing the human life span and optimizing the functional human life span are explored in the chapter by Rodriguez and Banks. We are already experiencing the consequences of these improvements, which are real and now documented. It is widely accepted that the average human life span has steadily increased since 1900 in the United States for all population subsets (data from the Bureau of the Census and the National Center for Health Statistics). It is also clearly documented that this improvement in lifespan is not increasing the burden of disabled elderly, burdened with suffering and disease, though writers continue to put forth this incorrect view. Declines in the prevalence of chronic disability have been observed up to age 95, based on data from the 1982, 1984, 1989, and 1994 National Long Term Care Surveys.[1] Manton's review of these data "suggest that there have been statistically significant and biologically important changes in the age rate of loss of biological fitness in the United States population associated with increases in life expectancy above age 65". Vita *et al.* rebut the view that greater longevity may lead to a greater burden of disability. Their longitudinal study published in the April 9, 1999 *New England Journal of Medicine* showed that "persons with better health habits survive longer [and that] disability is postponed and compressed into fewer years at the end of life".[2] Richard J. Hodes, M.D., Director of the National Institute on Aging reiterated the finding of long-term declines in rates of disability in older Americans since 1982. He stated, "... there are approximately 1.4 million fewer Americans disabled today than there would have been if the rates of disability observed in 1982 had remained constant".[3]

No intervention(s) can attenuate the decline in individual organ systems that begin in the third decade. The absence of such therapies does not negate the fact that the likelihood of achieving robust "old age" (after 65) is far greater now than ever before in societies such as ours in which advanced medical care is available.[4] As an example, successful treatment of an otherwise fatal myocardial infarction in a previously healthy and fully functional 85 year-old is now the expected outcome for that individual in this era of technologically advanced cardiovascular care, accompanied by the heightened expectations such care has created for patients and the public. After successful treatment that individual may now reach the age of 91, which is the average additional life expectancy of one who reaches age 85. The view that "old age" is a sequence of age-related diseases culminating in infirmity and death is no longer correct, and is the exception, not the rule. It is simply not correct that "infirmity remains the lot of those older than 80, however much the media may dote on the 90 year-old marathon runner".[5] They dote with good reason based on solid epidemiological evidence. For all practical purposes, currently available and applied medical technology IS postponing aging and its consequences. For those reaching ages 65, 75, 85, 90, and 100, the current additional expected life expectancies would bring them to ages 82, 86, 91, 94, and 102, respectively.[6] Even as long ago as 1983, it was shown that only 35% of those over age 85 were impaired in any activity required for daily living, and only 20% lived in a nursing home.[7] The number of centenarians in the United States is now greater than 50 000, more than three times that in 1980. They are the fastest growing age group in the world. Perls' research, published as the New England Centenarian Study, was discussed in his book, *Living to 100: Lessons in Living to Your Maximum Potential at Any Age*.[8] He and his co-authors concluded, "... people live to 100, not by living in spite of disease, but by avoiding or delaying it as long as possible. We have replaced the saying 'The older you get, the sicker you get' with the more accurate observation: "The older you get, the healthier you've been".

Many authors have explored the broad theme of prevention and its implications for healthcare reimbursement and healthcare costs, as in William B. Schwartz, MD's recent book, *Life Without Disease: The Pursuit of Medical Utopia*.[9] His review of the evidence led him to predict that improvements in diet, lifestyle, environment, and advances in molecular and genetic medicine will help us approach a "medical nirvana" absent of disease, accompanied by a fall in healthcare costs. Before that would happen, the inexorable restraining forces of healthcare cost containment would come into play. When molecular and genetic therapies are initially available they are unquestionably expensive. That is the experience to date. Denham Harman's writings echo these themes, from the perspective of the Free Radical Theory of Aging (FRTA). He cites work in the FRTA as applied to mice, in which the average life expectancy at birth (ALE-B) was increased, whereas normal life span was little changed. The net result was an increase in the functional lifespan which shortened the duration of senescence and increased the number of older animals. He believes the US population since 1960 mimics this pattern, citing a variety of demographic data supporting this hypothesis. Harman then points out the correlation of this pattern with various documented trends in antioxidant use. He states, "The increases in percentage of elderly people in the population since 1960 and the declining incidence of chronic disability among them, decreases in cancer mortality since 1991, and continuing declines in cardiovascular disease are in accord with the beneficial effects expected from the growing use of antioxidant supplements since the 1960s on disease and life span (e.g. coronary artery disease, cancer, and life span) as well as the growing publicity about the ability of fruits and vegetables to decrease disease incidence by depressing free radical damage".[10]

What preventive measures have contributed to improvements in life span and the decreased rate of development of disability in the elderly? It is accepted that what are generally known as "public health measures" have contributed most to this favorable outcome. Improvements in sanitation and the quality of food and water, widespread availability of antibiotics after their discovery, and universal vaccinations have been the result of government and community-sponsored efforts, not having anything to do with healthcare reimbursement.[11] What other preventive measures appear to have favorably influenced morbidity and mortality? To the extent that they have favorably influenced it, what is the evidence that availability of healthcare reimbursement has facilitated the implementation and application of these measures? Does the availability of healthcare reimbursement indeed facilitate and promote preventive healthcare? The answers to these questions will bear directly on the value of making healthcare reimbursement programs applicable to diagnostic and therapeutic interventions regarding oxidative stress and aging. It is assumed by the author that there is evidence that diagnostic and therapeutic interventions based on oxidative stress do bear positively on aging and/or age-related diseases, and may maximize or even extend the human lifespan. The evidence for this view is large and is beyond the scope of this review, though some of it will be presented in this chapter. A wealth of this evidence has been presented in the chapters of this text.

3. Technologies for Age-Related Diseases

An overview of various technologies (the word will be used broadly to encompass both diagnostic methods and treatments — pharmacological, preventive, and otherwise) for the detection, prevention, and amelioration of several so-called "age-related diseases" will be presented, with reference to these technologies' positive effects on morbidity and mortality. This will place into perspective the wide range of valuable technologies now available to positively impact the ability to reach maximum predicted lifespan, and optimize functional lifespan. It will be the technologies that are available now that will be the focus of questions regarding healthcare reimbursement. Specifically, would the wider availability of healthcare reimbursement improve the utilization of these technologies? Further, is there presently any evidence that such availability actually has a positive impact on outcomes, regardless of our expectations or what logic might suggest?

In cardiovascular disease, the treatment of lipid abnormalities reduces cardiovascular mortality, all-cause mortality, and morbidity. With respect to cardiovascular disease, reaching the "Holy Grail" of maximizing life span and optimizing functional lifespan may be said to be a possibility. There are many reviews summarizing the data that in aggregate are convincing and not considered assailable. One excellent review is "The Proving of the Lipid Hypothesis" by Gilbert Thompson.[12] He succinctly described the impact of the landmark trials now known widely by their acronyms 4S and WOSCOPS, which convinced even the skeptical, foremost among them Dr. Michael Oliver. In an article published in *The Lancet*, Thompson went on to state that "Statins prevent coronary heart disease".[13] These two clinical endpoint trials were followed by three other trials, CARE, AFCAPS, and LIPID, all showing a reduction in cardiovascular events and no increase in non-cardiovascular mortality in statin-treated individuals. As stated by the renowned William Roberts, "The statin drugs are to atherosclerosis what penicillin was to infectious disease".[14] Thompson concludes with an allusion to healthcare reimbursement issues, stating, "However, all too often they are not prescribed, often because of budgetary constraints". As we shall see, this is far from the only issue affecting their use. For other reviews with similar emphasis, see, "Cholesterol and Coronary Heart Disease: The 21st Century", by Scott M. Grundy,[15] the editorial, "Death Rates From Coronary Disease-Progress and a Puzzling Paradox" by Daniel Levy, M.D. and Thomas J. Thom, of the NHLBI,[16] "Shattuck Lecture-Cardiovascular Medicine at the Turn of the Millennium: Triumphs, Concerns, and Opportunities", by Eugene Braunwald, M.D.,[17] and "Achievements in Public Health, 1900–1999: Decline in Deaths from Heart Disease and Stroke — United States, 1900–1999", from the Centers for Disease Control.[18] Finally, a meta-analysis of 28 randomized controlled trials including 49 477 patients with pharmacological intervention and 56 636 controls concluded that statins reduce risk for fatal and nonfatal stroke, coronary heart disease mortality, and all-cause mortality.[19]

Analysis of 84 129 women participating in the Nurses' Health Study showed that women in a cohort meeting criteria defining low cardiac risk (3% of the study population) had a relative risk of coronary events of only 0.17 compared with all other women in the study. Eighty-two percent of all coronary events in the study could be attributed to lack of adherence to the low-risk pattern defined in the study. The low-risk criteria included guidelines involving diet, exercise, and abstinence from smoking.[20] What role might healthcare reimbursement have in prompting adherence to guidelines such as these, shown to confer low-risk status?

4. Preventive Medicine, Cardiovascular Disease, and Screening

Again, there is a wealth of data demonstrating the roles of preventive medication and lifestyle factors in decreasing cardiovascular morbidity and mortality. Entire texts are devoted solely to this subtopic of cardiology.[21] The large numbers of people able to be studied, and clearly definable and measurable endpoints easily lends cardiovascular medicine to cost-effectiveness analyses, of which hundreds have been published in the literature. Prototypical are studies such as those by Goldman which examined costs, effectiveness, and cost-effectiveness for a hypothetical intervention in 10 000 patients for 5 years,[22] by McGehee *et al.* which assessed benefits and costs associated with diet therapy for dyslipidemia,[23] by Sikand *et al.* which assessed the cost savings generated by avoiding hypolipidemic medications as a consequence of dietary intervention,[24] and striking data from the Framingham Heart Study and the Nutritional Health and Nutrition Examination Survey (NHANES). This latter analysis yielded the "estimate that reducing saturated fat intake by 1% to 3% would prevent 32 000 CHD events and yield a combined savings in medical expenditures and lost earnings of $4.1 billion over 10 years (estimates in 1993 US dollars).[25] The 1998 Supplement to The American Journal of Cardiology, titled "A Symposium: LCAS in Context", chaired by Antonio M. Gotto, Jr., M.D. was entirely devoted to an examination of cost-effectiveness issues in lipid-lowering.[26] "The Future of Cardiology: Utilization and Costs of Care" discusses a variety of scenarios with respect to heart disease costs, including the impact of improvements in the prevention of cardiovascular disease.[27] These are the types of data gathered and conclusions generated from studies done regarding cost-effectiveness.

The applicability of such analyses is emphasized by the widespread prevalence of atherosclerosis, declared by Margaret Winker, M.D. in the January 6, 1999 *JAMA* to be "The Emerging Epidemic of Atherosclerosis".[28] That year, a series of articles published in the *JAMA* highlighted the prevalence of cardiovascular disease risk factors among adolescents and young adults,[29] and the universal presence of atherosclerosis in subjects as young as age 15.[30] The authors concluded their paper forcefully, stating "Atherosclerosis begins in youth. Fatty streaks and clinically significant raised lesions increase rapidly in prevalence and extent during the 15- to 34-year age span. Primary prevention of atherosclerosis, as contrasted with primary prevention of clinically manifest atherosclerotic disease, must begin in childhood or adolescence". A separate study of 204 persons ages 2 to 39 concluded similarly.[31] The accompanying editorial, "When Should Heart Disease Prevention Begin?", by J. Michael Gaziano, M.D., MPH of Harvard Medical School concluded that "... considerably more attention must be paid to developing more effective strategies for reducing risk factors in children and young adults ... there are divergent opinions and substantial remaining uncertainty about the effectiveness and wisdom of more intensive screening and treatment of elevated blood pressure and cholesterol levels in children and young adults. This uncertainty should not detract from the clear case for redoubling efforts to promote population-wide good health practices in children and young adults that may help to decrease the risks of future cardiovascular disease".[32] What might be the role of healthcare reimbursement in better achieving these goals?

This lengthy review of the issues regarding cardiovascular disease, its risk factors, its prevention, and the intense study of cost-effectiveness issues serves as a paradigm for similar discussions regarding other major "age-related diseases", including cancer, diabetes, cognitive decline, and osteoporosis. These will not be treated in the same detail in this review. The emphasis on cardiovascular diseases and their risk factors is appropriate and well placed. The largest expenditure of healthcare dollars is on behalf of them. The greatest morbidity, mortality, and losses in production are due to them.[33] Analyses of gains in life expectancy clearly demonstrate that attention to only four cardiovascular risk factors, including reducing diastolic blood pressure to normal, reducing cholesterol to 200 mg/dl, reducing weight to an ideal level, and quitting cigarette smoking, will generate more gain in life expectancy for men and women than all other preventive interventions combined in populations at elevated risk.[34] Again, it must be asked how healthcare reimbursement can influence positively these four risk factors, as an example.

According to Brawley and Kramer, cigarette smoking is the most avoidable risk factor for cancer and cardiovascular disease.[35] They state that "Cessation and avoidance of smoking have the potential to save and extend more lives than any other public health activity. It is estimated that 400 000 Americans die prematurely every year because of cigarette smoking". Is there a healthcare reimbursement solution to the problem of smoking?

Screening programs for cancer are widely promoted and have even become politicized. For example, it is a point of contention in the 2000 New York Senatorial race as to whether Mr. Lazio or Ms. Clinton has done more to promote breast cancer screening. The debate regarding breast cancer screening is no longer rational and scientific, having "become so contentious that effective communication and rational discussion on this topic have been compromised ... the debate is less about facts than it is about perceptions and values. There is disagreement about how to fairly describe facts about risk and how to avoid misperceptions that may distort assessment of risk. Other sources of disagreement concern the potential harms of screening, the relative roles of physicians and patients in decision making, and how to factor cost into screening decisions. The entire decision making process has also been charged by single-issue advocacy groups and a kind of gender rivalry ... Lessons from this debate may apply to other medical problems that have small degrees of risk and whose management is strongly debated".[36] Breast cancer screening has been shown to save lives in women over the age of 50.[37] It has not been shown to be the case for women aged 40–49.[38, 39] Even were it so, its effectiveness as a preventive measure in women at average risk is less than one-tenth that of the benefit conferred in preventing cardiovascular disease by quitting smoking, or one fourth that of having a Pap smear.[40]

The only screening tests shown to decrease cancer mortality other than mammography are the Pap smear,[41] sigmoidoscopy for people over age 50,[42] and the PSA for prostate cancer (but only in a single European study, data presented by Boyle and Bartsch at the meeting of the American Urological Society, May 2000, and submitted to *The Lancet*). These procedures are reimbursed to one degree or another by insurance. This has allowed the study of the relationships between their utilization and the availability or lack of healthcare reimbursement of them. Two recent studies demonstrated a significant improvement in the detection of colon cancer using colonoscopy. In a *New England*

Journal of Medicine editorial, Dr. Daniel Podolsky discussed the reports. He stated, "although the studies ... fall short of proving that life expectancy is increased by performing colonoscopic screening of persons 50 years of age or older who are at average risk for colorectal cancer, such an extrapolation of the data is virtually irresistible ... These two new reports reinforce the growing suspicion among physicians that in recommending flexible sigmoidoscopy to screen persons for colorectal cancer, we are promoting a suboptimal approach. The failure of insurance companies to cover the costs of colonoscopic screening is no longer tenable".[43] Dr. Podolsky concludes his editorial, stating, "I believe it is time for both government and private insurers to provide coverage for colonoscopic screening for all persons 50 years of age or older who are at average risk for colon cancer. As many people have pointed out, relying on flexible sigmoidoscopy is as clinically logical as performing mammography of one breast to screen women for breast cancer". Is this good science, or reminiscent of the breast cancer screening debates? Again, many issues bear on the subject of healthcare reimbursement, including the scientific evidence, the advisability from a public health standpoint, cost-effectiveness, and availability of presumably finite healthcare dollars and resources. The recent studies regarding the early detection of colorectal cancer are similar to studies reporting the early detection of lung cancer using spiral CT.[44] Whether such cancers are indolent and "pseudodisease", or actually threatening and in need of resection is the subject of scholarly debate.[45] Another study revealed that "An intense regimen of screening for lung cancer did nothing to reduce mortality from the disease in a large cohort of smokers followed up for 20–25 years", reported by Pamela Marcus, Ph.D., at the annual meeting of the American Society of Preventive Oncology.[46] "Annual transvaginal ultrasound screening detected ovarian cancer at an early stage in a study of more than 14 000 women at high risk for the disease. But it's still too soon for widespread adoption of annual ultrasound screening for this purpose. Larger studies are needed to justify the expense of the screen and further refine its use so that women do not undergo unnecessary surgery based on misleading results, Dr. John R. van Nagell, the study's lead author, cautioned in an interview".[47] In the same news report of the findings presented at the annual meeting of the Society of Gynecologic Oncologists, Dr. Beth Y. Karlan, Director of Gynecologic Oncology at Cedars-Sinai in Los Angeles, commented, "Although these data are encouraging, it cannot be assumed that ultrasound screening for ovarian cancer saves lives because the study lacks statistical power to measure the effects of screening on mortality ... The ability of this intervention to alter the natural history of ovarian cancer remains unproven".

Clearly, such findings and reports heighten the public's anxiety and demand for healthcare reimbursement for screening studies for cancer, without regard to the evidence that such studies influence mortality or outcomes. Such demands and anxieties further complicate the debates regarding healthcare reimbursement.

The front page of the August 25–27 *USA Today* proclaimed, "The Inside story … for about \$700, you can have your body scanned to spot potential problems …" accompanied by a color photograph of a scanner. Quoting from the story, Dr. Harvey Eisenberg, "a controversial radiologist, uses a CT scanner at his HealthView center for Preventive Medicine here [Newport Beach, California] to peer inside healthy people's bodies in order to tell them what unsuspected health problems lurk there. HealthView, one of the first such centers in the nation, initially attracted the wealthy, worried well. But as insurance companies pay for more of these scans, and the prices have come down, there are more average-income people in the waiting room. Critics say these unwarranted tests waste money, cause false alarm and put a strain on the rest of the health system. But business is booming". This story highlights the polarized views regarding the use of screening tests.

   Two separate, peer-reviewed publications recently made similar criticisms regarding the use of Electron Beam CT (EBCT) scanning. EBCT has been seriously researched and is the subject of many scientific papers in the peer-reviewed medical literature. Unfortunately, it has become heavily commercialized and is promoted directly to consumers as a "Heart Test" to detect early, asymptomatic coronary atherosclerosis. A nine-member panel of experts representing the American College of Cardiology and the American Heart Association recommended the technique not be used to diagnose obstructive coronary artery disease. They published their findings in an "Expert Consensus Document" in the July issue of the *Journal of the American College of Cardiology*, and the July 4 issue of *Circulation*. The low test specificity of EBCT was cited as its main drawback, resulting "in additional expense and unnecessary testing to rule out a diagnosis of CAD".[48] From the "Expert Consensus Document" is the statement, "… the published literature does not clearly define which asymptomatic people require or will benefit from EBCT. Additional appropriately designed studies of EBCT for this purpose are strongly encouraged. In the setting of this degree of uncertainty, EBCT screening should not be made available to the general public without a physician's request".[49] The total number of all CT scans done in the US has been rising at the rate of 1 million a year, with a large part of that growth coming from "seemingly healthy Americans willing to fork over hundreds of dollars of their own money for it" (*Wall Street Journal*, March 23, 2000).

   These are thoughtful admonitions from experts who have considered carefully all the available evidence and who are competent to judge that evidence. Such advice does not sit well with a generation of Baby-Boomers whose members research health matters themselves and come to their own conclusions, even if they are unfamiliar with Bayesian theory, let alone definitions of test sensitivity, specificity, and positive and negative predictive power. A Reuters's report from August 11, 2000 states that 98 million Americans are now using the Internet to find healthcare information, twice the number of a year ago. Of 1001 adults surveyed in May and June of 2000, 86% reported using the Internet to get information on healthcare or specific diseases, compared with 71% in 1999. *Managed Care News* reported in May 1999 that the type of the health information most commonly retrieved from the Internet is that pertaining to a specific disease or medical condition (48%), followed by educational information (34%), dietary information (32%), and medication and drug information (29%). Yet, there are legitimate concerns as expressed by Dr. Thomas Reardon, president of the American Medical Association. In an article published in *The New York Times* March 6, 2000, he stated, "... we do support patients having access to good, reliable, information" ... the problem is that the information on the Internet varies from sound to irresponsible ..."Right now, anybody can put anything on the Internet". The medical association put out the statement, "The AMA is greatly concerned that a substantial proportion of information on the Internet might be inaccurate, erroneous, misleading, or fraudulent and thereby pose a threat to the public". By making demands of insurers that they are not qualified to make, the "empowered Baby-Boomers" add further pressure to medical inflation and divert healthcare resources from more appropriate use, and diminish dollars available for healthcare reimbursement of more soundly established diagnostics and therapeutics. These decisions are often based on information they derive from the Internet. For many "Boomers", the Internet is their primary or even sole source of healthcare information.

5. Baby-Boomers Leading the Charge for Health Care Reimbursement

The "Boomers" reliance on the Internet for medical information mirrors their acceptance and popularization of Alternative and Complementary Medicine (ACM), and is consistent with their suspicion — and even rejection — of scientifically established medical authority. These issues are explored in a thorough review titled "The Persuasive Appeal of Alternative Medicine" accompanied by an editorial, "Weighing the Alternatives: Lessons from the Paradoxes of Alternative Medicine".[50] Bearing on the issue of healthcare reimbursement is the fact, astonishing to many, that more is spent out-of pocket on alternative health care than out-of -pocket for conventional care. Hence, it is not surprising that those persons with larger incomes are the primary users (and advocates) of ACM.

Despite their higher disposable incomes, the "Boomers" have led a charge for healthcare reimbursement for ACM. Again, the "worried well Baby-Boomers" are a driving force, distorting the priorities for appropriate healthcare reimbursement. In 1999 was published a systematic literature review from MEDLINE of all articles in alternative medicine that were clinical trials. The total annual number of articles listed in MEDLINE rose to a peak of 400 000 a year by 1996, whereas the number indexed under alternative medicine peaked at 1500 articles per year in 1986, and has remained stable since. By 1996, approximately 10% of these alternative medicine articles were of the clinical trial type. The authors concluded "the number of clinical trials is still low, offering little comfort for those demanding evidence of efficacy from well-conducted, randomized controlled trials".[51] Another review titled "Evidence-based medicine and complementary medicine [CM]", concluded that undertaking and implementing an evidence-based approach in CM "would require a greater commitment to science, critical thinking, and evidence-based medicine than is currently found in the CM community".[52] Why would healthcare reimbursement offer coverage for therapies for which there is little evidence, if any, or at least less evidence than for other therapies, other than for business reasons?[53] In fact, an ongoing three-year survey of insurers revealed that market demand was the primary reason insurers offered healthcare reimbursement for ACM services.[54] Primary reasons for not covering ACM services included lack of research on efficacy and lack of standards of practice. It is ironic that these "therapies" are often covered when the aforementioned colonoscopies or statin therapies are not, for which there is strong evidence of diagnostic and therapeutic value, respectively.

6. Barriers to the Implementation of Preventive Diagnostics and Therapeutics

Aggressive preventive therapies as demonstrated in the Diabetes Control and Complications Trial (DCCT) lead to significantly less retinopathy and nephropathy.[55] A PET scan (cost approximately $1300, not covered by insurance) "is the most accurate way to diagnose Alzheimer's disease.[56] The National Osteoporosis Foundation has issued recommendations for the prevention and treatment of osteoporosis.[57] All of these are candidates for healthcare reimbursement, and have a good to excellent evidence base on which the justification for reimbursement could solidly rest.

Given its status as epidemic, some consideration should be given to obesity, though it is not often thought of as an "age-related disease" — though given its increasing incidence with age, it justifiably could be considered so.[58] Given that "Most obesity treatment methods are ineffective over the long term", how is this major public health problem impacted by healthcare reimbursement, if at all?[59]

These types of medical problems-cardiovascular diseases, cancer, osteoporosis, cognitive decline, diabetes, and obesity-lend themselves to study with respect to the impact of healthcare reimbursement on their respective diagnostics and treatments. A MEDLINE search conducted with the search defined by the terms "preventive", "screening", and "insurance reimbursement" and restricted to citations in English with abstracts published from 1998 to the present yielded 209 citations. Most of the citations were of studies concerned with these medical problems. Having healthcare reimbursement was demonstrated to be predictive of having screening mammography in nine studies.[60-66] One demonstrated a similar correlation with screening for prostate cancer.[67]

A larger number of studies demonstrated that a lack of healthcare reimbursement predicted decreased utilization of preventive services. A detailed survey of 49 604 respondents aged 55 to 64 years, published in 1999, concluded that surveyed individuals in this age group "with health insurance coverage are much more likely than those without coverage to have a routine checkup every 2 years, to have their blood pressure checked, and to have a regular source of care. Insured women are more likely to undergo a Papanicolaou test, clinical breast examination, and mammograms than uninsured women".[68] NHANES III, conducted from 1988 to 1994, showed that uninsured women were more likely to be current smokers, sedentary, overweight and to consume less fiber, vitamin C, folate, calcium, and potassium than insured women. Compared with insured women, uninsured women were less likely to have had their blood pressure checked during the previous 6 months, to have had their cholesterol level checked, and to be aware of hypercholesterolemia. Insured women (24.9%) were three times more likely to use estrogen replacement therapy than uninsured women (7.9%). NHANES III data collected from 1988 to 1994 suggest that women without health insurance have a worse cardiovascular disease risk factor profile and use healthcare services less frequently than women with health insurance.[69] In a study of 2118 diabetics older than 18 years, those without healthcare reimbursement were at high risk for underuse of preventive care. In fact, only 3% of insulin users and 1% of non-users met all five of the American Diabetes Association standards for care in the previous year.[70] A study of skin cancer screening showed that there was a "significant association of noncompliance to seek recommended follow-up care with lack of health insurance ...".[71] In a study of late-stage cervical cancer, uninsured women were more likely to be diagnosed at a late stage.[72]

Few studies have evaluated utilization of multiple preventive services by a specific population relative to the availability of healthcare reimbursement. One examined the three cancer screening practices of Pap smear, mammography, and clinical breast examination. It concluded that having health insurance was a predictor of having mammography.[73] One "studied all patients with incident cases of melanoma or colorectal, breast, or prostate cancer in Florida in 1994 for whom the stage at diagnosis and insurance status was known". It found that "persons lacking health insurance and persons insured by Medicaid are more likely to be diagnosed with late stage cancer at diverse sites, and efforts to improve access to cancer-screening services are warranted for these groups".[74] In a comparison study of low-income uninsured and insured women, a survey was done regarding whether they were counseled by a healthcare provider regarding any of seven types of preventive health behaviors in the previous 12 months in which they received care. "Uninsured women were less than half as likely to receive counseling on three or more preventive topics as were mothers on Medicaid".[75]

In the age of managed care, it has been asked whether the type of insurance coverage affects the provision of preventive healthcare services. One study examined the prevalence of physician advice for four preventive health behaviors for cardiovascular disease in the context of type of insurance and regular source of care. Although the study cited data that "ninety-seven percent of physicians believe it their responsibility to give advice in order to modify patient behavior to reduce risk factors ... the actual provision of preventive services by physicians is relatively rare".[76] Physician characteristics predictive of physician counseling include the physician's own personal behavior. Some studies suggest that primary care physicians are more likely to provide preventive services than some specialists, with the exception that counseling by a specialist to reduce risk factors is directly related to the subspecialty's specific priorities for counseling. Some studies show a gender bias in regard to screening, with female physicians more likely to recommend mammograms and Pap smears. The California study showed that "the single most important factor associated with the receipt of preventive care" was regular source of care. The study concluded that regular source of care was the most important factor in providing preventive care advice. It warned, "managed care plans that do not ensure continuity of care and protect the patient-physician relationship may have a detrimental effect on physician advice for prevention".

The California Behavioral Risk Factor Surveillance Surveys, which collected data in 1989 and 1990, compared Pap smear, mammogram, fecal occult blood test, and proctoscopic exam rates for adults with three different types of insurance coverage — group HMO, IPA HMO — with standard indemnity plan.[77] Rates of screening were greatest for those enrolled in group HMO plans. Screening was less likely for those with the other two plans, which had equal screening rates. Those with no healthcare reimbursement were screened the least. Of interest, the strongest predictor of having any screening was having a regular healthcare provider, as in the aforementioned study. Another study found that one of the few differences between HMO's and non-HMO's was the higher use of preventive services by HMO enrollees.[78] Another comparison study of similar insurance plans that examined three measures of preventive services found no differences in the rates of provision of the measured preventive services.[79] A 1999 survey conducted by the Kaiser Family Foundation and the Harvard University School of Public Health surveyed physician and nurse experiences with managed care. There was an overwhelmingly negative response regarding most aspects of managed care, due to perceived barriers to care they believed were created and perpetuated by the plans themselves. Nevertheless, 70% of the physicians and about 50% of the nurses surveyed said managed care had increased the use of practice guidelines and disease management protocols. According to 45% of the physicians and 42% of the nurses, managed care increased the likelihood that patients would get preventive services. The plans were said by 33% of physicians to have helped them encourage a patient to practice better health.[80] It would appear from this review of the published evidence, though the studies are few, that availability of healthcare reimbursement, particularly HMO type plans, improves access to preventive services such as screening. Are changes in healthcare reimbursement therefore the answer to the problem of insufficient practice and provision of preventive medicine, from both diagnostic and therapeutic perspectives?

In the discussion to follow, it will become clear that a multitude of factors other than healthcare reimbursement impact the provision of preventive medical services, in addition to those alluded to previously. It will be clear that many of these have little or nothing to do with modes of healthcare reimbursement. This suggests difficulty in implementing even highly desirable, proven preventive health measures throughout the eligible candidate population. An overarching proposal such as the universal availability of appropriate healthcare reimbursement, perhaps emphasizing or initially being limited to preventive healthcare services, may appear an attractive, potential solution. Such a proposal will be seen to fall short of the mark and is not a realistic approach to improving the general health of the population. There are many other barriers to preventive care that are not addressed by the universal availability of healthcare reimbursement. Even when coverage is available to subsets of the population (HMO enrollees, for example), the preventive services available are infrequently used. A 1998 survey of California health plans showed that only 2–3% of enrollees participated in health promotion programs sponsored by the plans, regardless of their type.[81] In fact, only 35–45% of the HMO's, and no PPO's or indemnity plans assessed the impact of health promotion programs on health risks and behaviors, health status, or health costs. This led the authors of that survey to the grim conclusion that "For the majority of California's PPO and indemnity plans, health promotion is not an integral part of their business. For the majority of HMO's, health promotion programs are offered primarily as a marketing vehicle".

What about clinical practice guidelines, of which there are now hundreds published in the peer-reviewed medical literature? At times they have been viewed as a near panacea, with the hope that they would bring uniformity to the diagnosis and treatment of patients in accord with the best evidence available. Experience with guidelines tempers any enthusiasm they may engender. Despite the energy put into developing guidelines and the promise their application may hold, their passive dissemination (mailings, publications, CME) is ineffective in increasing their use, unless accompanied by reminders or financial incentives.[82] Of the hundreds of issued clinical guidelines, those of the National Cholesterol Education Program (NCEP) are likely the most widely disseminated and well known, and which have the strongest evidence base. Yet studies have shown that NCEP guidelines have done little to alter physician behavior in assessing risk, counseling, and treating even patients at high risk for coronary events.[83] Physicians were found to comply poorly with these highly publicized guidelines, the values of which are little disputed. How poorly do they comply? In one 4 state survey of 45 medical practices, 33% of patients with documented cardiovascular disease (CVD) were not screened with lipid panels, 45% were not receiving dietary counseling, 67% were not receiving medication for elevated levels of cholesterol, and only 14% with CVD had achieved target lipid levels.[84] The authors concluded, among other things, that "It is clear in this study and nationwide there is a significant gap in the primary care screening and treatment of cholesterol disorders in patients with CVD". They discuss the numerous reasons for the consistent failure of physicians to practice according to the widely accepted and endorsed NCEP guidelines. They offer a number of suggestions for improvement, but none depend on or would be facilitated by any changes in current modes of healthcare reimbursement.

In a survey of physician preventive practices in diabetics, important oversights in care were attributed to "oversight" by the offending physicians.[85] From poorly managed hypertension,[86] to a failure to recommend cancer screening even when physicians agree with the guidelines,[87] to multiple studies documenting underutilization of antilipid therapies[88] and aspirin[89] in the treatment and prevention of CVD, to inadequate diagnosis and treatment of osteoporosis, in which less than 2% of women with risk factors undergo bone densitometry and no more than 20% take any bone-sparing or protective medications,[90] physicians consistently fail to practice "state-of-the-art" preventive medicine, for which peer-reviewed guidelines exist. Continuing medical education (CME) is not the answer.[91] Physician's failure to follow guidelines was the subject of a comprehensive review published in *JAMA* in 1999, in which a MEDLINE search produced 76 articles that included 120 different surveys investigating 293 potential barriers to physician adherence to guidelines.[92] Failure to adequately inform patients prior to diagnosing and treating them is a common and significant barrier. In one large study of informed decision making, only 17.2% of patients were completely informed when the medical decision to be made was "basic". The percentage dropped to only 0.5% when the decision was "complex".[93] Physicians' problem-oriented as opposed to prevention-oriented approach to patients, neutral to absent feedback on the their prevention efforts, time constraints, lack of incentives (including financial), lack of training, and lack of perceived legitimacy of preventive diagnostics and therapeutics are other important and common barriers.[94] Few of these barriers other than lack of financial incentive to practice preventive medicine are likely to be favorably impacted by changes in existing modes of healthcare reimbursement.

Numerous patient barriers to preventive medicine exist and must be acknowledged. Gender, communication issues[95] and "health literacy"[96] negatively impact the provision of preventive medical services, but are unlikely to be ameliorated by changes in healthcare reimbursement. Patients fail to comply with recommendations for screenings. As discussed above, this is one aspect in which having healthcare reimbursement appears to have a demonstrable positive effect. They are not compliant with recommended therapies for a wide variety of reasons, but the healthcare reimbursement of recommended therapeutics is known to be a critical aspect of patient compliance, especially as the cost of the regimen increases. Patient education,[97] motivation, and social, ethnic, and cultural factors all affect patients' acceptance of preventive diagnostics and therapeutics.[98] Confusion engendered in the minds of patients who do not understand the nuances of debate between physicians with legitimate differences of opinion also adversely affect patients' acceptance of recommendations. Can we expect women to unquestioningly continue their screening mammography when the press reports uncritically the results of a study published in *The Lancet* titled, "Is screening for breast cancer with mammography justifiable?" which concluded that "... there is no reliable evidence that screening decreases breast cancer mortality"?[99] Are antioxidants truly useful in treating or preventing CVD?[100] The manner in which physicians debate these issues and the way those legitimate, scientific debates are reported to the public affect significantly their attitudes toward health recommendations. The public should be able to participate in meaningful debates based on evidence-based medicine to help direct priorities to interventions most likely to improve public health. The public should be able to understand the type of data displayed in the review, "The Need for Perspective in Evidence-Based Medicine" which tabulates the projected outcomes for a variety of diagnostic and therapeutic interventions.[101] This would provide the basis for intelligent, rational debate about allocation of healthcare resources and reimbursement. Data of the type discussed in the review, "Lifetime Costs for Preventive Medical Services: A Model" are similarly important, and should be part of the public's healthcare knowledge base to facilitate appropriate decision-making.[102]

Declines in public awareness about key health issues such as hypertension have been documented.[103] This can occur as other, more "popular" health information "crowds out" an older, less fashionable but critically important message that may bear constant repetition.[104] Health plans that report their data publicly have been shown to provide preventive healthcare services to higher percentages of their beneficiaries than plans not reporting their data.[105]

Clearly, examination of national trends in CVD morbidity and mortality do not support increases in total healthcare expenditures. The United States has by far the highest per capita healthcare expenditure of any country. Yet, CVD mortality is lower in 18 of 33 industrialized countries than in the US. In 15 of 33 industrialized countries, the decline in the CVD death rate is faster than in the

US.[106] In a 1990 Harvard survey, the life expectancy of US women ranked 16th out of 26 developed nations. Men ranked 18th. Instead of spending more on healthcare, the discussions should focus on the wiser allocation of current spending.

7. Conclusion

It can be stated that preventive diagnostics and therapeutics are realities in medicine. They are available and effective, with a strong, and in some cases unassailable, evidence base. An unfortunate reality is that their application throughout the population is uneven, even when their value and efficacy is proven and beyond doubt. It is tempting to think that the wider availability of healthcare reimbursement will significantly and positively impact the provision and utilization of preventive medical technologies. The data available from studies of healthcare reimbursement suggests this might be the case. However, a comprehensive review demonstrates that there are numerous barriers to the widespread provision and utilization of preventive medical technologies, most of which are not sensitive, related, or amenable to modification by simply adjusting or modifying the prevailing modes of healthcare reimbursement. Healthcare reimbursement is a necessary condition for wider availability of preventive medical services. Given that it is one of many necessary conditions, it cannot be a sufficient condition.

Acknowledgments

I am grateful to Ms. Candra Faulkner for her technical assistance in formatting and preparing this document for submission, and to Ms. Karen Fanning, Library Associate of the Verde Valley Medical Center in Cottonwood, Arizona for assisting me in the research for this submission.

References

1. Manton, Kenneth, G., Corder, L. and Stallard, E. (1997). Chronic disability trends in elderly United States populations: 1982–1994. *Proc. Natl. Acad. Sci. USA* **94**: 2593–2598.
2. Vita, A. J., Terry, R. B., Hubert, H. B. and Fries, J. F. (1998). Aging, health risks, and cumulative disability. *New England J. Med.* **338**: 1035–1041.
3. Glaser, V. P. (1999). Institute profile: National Institute on Aging. *J. Anti-Aging Med.* **2**: 319–324.
4. Resnick, N. M. (1998). Geriatric medicine. *In* "Harrison's Principles of Internal Medicine" (A. S. Fauci, Ed.), pp. 37–46, McGraw-Hill, New York.

5. Rose, M. R. (1999). Can human aging be postponed? *Scientific Am.* **281**: 106–111.
6. Resnick, N. M. (1998). Geriatric medicine. *In* "Harrison's Principles of Internal Medicine" (A. S. Fauci, Ed.), pp. 37–46, McGraw-Hill, New York.
7. Resnick, N. M. (1998). Geriatric medicine. *In* "Harrison's Principles of Internal Medicine" (A. S. Fauci, Ed.), pp. 37–46, McGraw-Hill, New York.
8. Perls, T. T., Silver, M. H., Lauerman, J. F. and Hutter-Silver, M. (1999). *In* "Living to 100: Lessons in Living to Your Maximum Potential at Any Age", Basic Books, New York.
9. Schwartz, W. B. (1999). *In* "Life Without Disease: The Pursuit of Medical Utopia", University of California Press, Berkeley.
10. Harman, D. (1999). Aging: minimizing free radical damage. *J. Anti-Aging Med.* **2**: 15–36.
11. Ernster, V. L. and Colford, J. M., Jr. (1998). Host and disease: influence of demographic and socioeconomic factors*. *In* "Harrison's Principles of Internal Medicine" (A. S. Fauci, Ed.), pp. 37–46, McGraw-Hill, New York.
12. Thompson, G. R. (1999). The proving of the lipid hypothesis. *Curr. Opin. Lipid.* **10**: 201–205.
13. Oliver, M. (1991). Might treatment of hypercholesterolemia increase non-cardiac mortality? *Lancet* **337**: 1529–1531.
14. Roberts, W. C. (1996). The underused miracle drugs: the statin drugs are to atherosclerosis what penicillin was to infectious disease. *Am. J. Cardiol.* **78**: 377–378.
15. Grundy, S. M. (1997). Cholesterol and coronary heart disease: the 21st century. *Arch. Int. Med.* **157**: 1177–1183.
16. Levy, D. and Thom, T. J. (1998). Death rates from coronary disease-progress and a puzzling paradox. *New England J. Med.* **339**: 915–917.
17. Braunwald, E. (1997). Shattuck lecture-cardiovascular medicine at the turn of the millennium: triumphs, concerns, and opportunities. *New England J. Med.* **337**: 1361–1369.
18. Centers for Disease Control. (1999). Achievements in public health, 1900–1999: decline in deaths from heart disease and stroke — United States, 1900–1999. *MMWR* **48**: 649–656.
19. Bucher, H. C., Griffith, L. E. and Guyatt, G. H. (1998). Effect of HMGcoA reductase inhibitors on stroke. A meta-analysis of randomized, controlled trials. *Ann. Int. Med.* **128**: 89–95.
20. Stampfer, M. J., Hu, F. B., Manson, J. E., Rimm, E. B. and Willett, W. C. (2000). Primary prevention of coronary heart disease in women through diet and exercise. *New England J. Med.* **343**: 16–22.
21. Wong, N. D., Black, H. R. and Gardin, J. M., Eds. (2000). *Preventive Cardiology*, McGraw-Hill. New York.
22. Goldman, L. (1998). Cost awareness in medicine. *In* "Harrison's Principles of Internal Medicine" (A. S. Fauci, Ed.), pp. 49–52, McGraw-Hill, New York.

23. McGehee, M. M., Johnson, E. Q. and Rasmussen, H. M. (1995). Benefits and costs of medical nutrition therapy by registered dieticians for patients with hypercholesterolemia. *J. Am. Diet. Assoc.* **95**: 1041–1043.

24. Sikand, G., Kashyap, M. L. and Yang, I. (1998). Beneficial outcome and cost savings of medical nutrition therapy in hyper-cholesterolemia. *J. Am. Diet. Assoc.* **98**: 889–894.

25. Oster, G. and Thompson, D. (1996). Estimated effects of reducing dietary saturated fat intake on the incidence and costs of coronary heart disease in the United States. *J. Am. Diet. Assoc.* **96**: 127–131.

26. Gotto, A. M., Ed. (1998). A symposium: LCAS in context. *Am. J. Cardiol.* **82**(6A): 2M–4M.

27. Steinwachs, D. M., Collins-Nakai, R. L., Cohn, L. H., Garson, A., Jr. and Wolk, M. J. (2000). The future of cardiology: utilization and costs of care. *J. Am. Coll. Cardiol.* **35**: 1092–1099.

28. Winker, M. A. (1999). The emerging epidemic of atherosclerosis: call for papers. *JAMA* **281**: 84–85.

29. Winkleby, M. A., Robinson, T. N., Sundquist, J. and Kraemer, H. C. (1999). Ethnic variation in cardiovascular disease risk factors among children and young adults: findings from the Third National Health and Nutrition Examination Survey, 1988–1994. *JAMA* **281**: 1006–1013.

30. Strong, J. P., Malcom, G. T., McMahan, C. A., Tracy, R. E., Newman, W. P., Herderick, G. E. and Cornhill, J. P. (1999). Prevalence and extent of atherosclerosis in adolescents and young adults: implications for prevention from the pathobiological determinants of atherosclerosis in youth study. *JAMA* **281**: 727–735.

31. Berenson, G. S., Srinivasan, S. R., Bao, W., Newman, W. P., Tracy, R. T. and Wattigney, M. S. (1998). Association between multiple cardiovascular risk factors and atherosclerosis in children and young adults. *New England J. Med.* **343**: 16–22.

32. Gaziano, J. M. (1998). When should heart disease prevention begin? *New England J. Med.* **338**: 1690–1690.

33. Burke, G. L. and Bell, R. A. (2000). Trends in cardiovascular disease: incidence and risk factors. *In* "Preventive Cardiology" (N. D. Wong, H. R. Black, and J. M. Gardin, Eds.), pp. 21–46, McGraw-Hill, New York.

34. Wright, J. W. and Weinstein, M. C. (1998). Gains in life expectancy from medical interventions-standardizing data on outcomes. *New England J. Med.* **339**: 380–386.

35. Brawley, O. W. and Kramer, B. S. (1998). Prevention and early detection of cancer. *In* "Harrison's Principles of Internal Medicine" (A. S. Fauci, Ed.), pp. 499–505, McGraw-Hill, New York.

36. Ransohoff, D. F. and Harris, R. P. (1997). Lessons from the mammography screening controversy: can we improve the debate? *Ann. Int. Med.* **127**: 1029–1034.

37. Brawley, O. W. and Kramer, B. S. (1998). Prevention and early detection of cancer. *In* "Harrison's Principles of Internal Medicine" (A. S. Fauci, Ed.), pp. 499–505, McGraw-Hill, New York.
38. Brawley, O. W. and Kramer, B. S. (1998). Prevention and early detection of cancer. *In* "Harrison's Principles of Internal Medicine" (A. S. Fauci, Ed.), pp. 499–505, McGraw-Hill, New York.
39. Salzman, P., Kerlikowske, K. and Phillips, K. Cost-effectiveness of extending screening mammography guidelines to include women 40 to 49 years of age. *Ann. Int. Med.* **127**: 955–965.
40. Wright, J. W. and Weinstein, M. C. (1998). Gains in life expectancy from medical interventions-standardizing data on outcomes. *New England J. Med.* **339**: 380–386.
41. Brawley, O. W. and Kramer, B. S. (1998). Prevention and early detection of cancer. *In* "Harrison's Principles of Internal Medicine" (A. S. Fauci, Ed.), pp. 499–505, McGraw-Hill, New York.
42. Brawley, O. W. and Kramer, B. S. (1998). Prevention and early detection of cancer. *In* "Harrison's Principles of Internal Medicine" (A. S. Fauci, Ed.), pp. 499–505, McGraw-Hill, New York.
43. Podolsky, D. K. (2000). Going the distance — the case for true colorectal-cancer screening. *New England J. Med.* **343**: 207–208.
44. Zoler, M. L. (2000). Spiral CT detects lung cancers at early stage. *Int. Med. News.* July 1, 2000.
45. Moon, M. A. (2000). Imaging reveals indolent cancers. *Int. Med. News.* May 1, 2000.
46. Moon, M. A. (2000). Mortality rate unaffected by intensive lung cancer screenings. *Int. Med. News.* May 1, 2000.
47. Boschert, S. (2000) Ultrasound effectively detects early-stage ovarian cancer. *Int. Med. News.* March 15, 2000.
48. Expert Consensus Group (2000). New document explores potential and limitations of EBCT. *Cardiology.* July, 2000
49. O'Rourke, R. A. *et al.* (2000). American College of Cardiology/American Heart Association Expert Consensus Document on electron-beam computed tomography for the diagnosis and prognosis of coronary artery disease. *J. Am. Coll. Cardiol.* **36**: 326–340.
50. Kaptchuk, T. J. and Eisenberg, D. M. (1998). The persuasive appeal of alternative medicine. *Ann. Int. Med.* **129**: 1061–1070.
51. Barnes, J., Abbot, N. C., Harkness, E. F. and Ernst, E. (1999). Articles on complementary medicine in the mainstream medical literature. *Arch. Int. Med.* **159**: 1721–1725.
52. Vickers, A. (1998). Evidence-based medicine and complementary medicine. *Evidence-Based Medicine.* November/December 1998.
53. Eisenberg, D. (1997). Advising patients who seek alternative medical therapies. *Ann. Int. Med.* **127**: 61–69.

54. Pelletier, K. R., Astin, J. A. and Haskell, W. L. (1999). Current trends in the integration and reimbursement of complementary medicine by managed care organizations (MCOs) and insurance providers: 1998 update and cohort analysis. *Am. J. Health Promotion* **14**: 125–133.

55. The Diabetes Control and Complications Trial/Epidemiology of Diabetes Interventions and Complications Research Group. (2000). Retinopathy and nephropathy in patients with Type-1 diabetes four years after a trial of intensive therapy. *New England J. Med.* **342**: 381–389.

56. Zoler, M. L. (2000). PET accurate enough to diagnose Alzheimer's. *Int. Med. News.* July 15, 2000.

57. Heinemann, D. F. (2000). Osteoporosis: an overview of the National Osteoporosis Foundation clinical practice guide. *Geriatrics* **55**: 31–36.

58. MedscapeWire. (2000). Leading Medical Institutions Join Forces to Battle Obesity Epidemic. MedscapeWire. February 28, 2000.

59. Douketis, J. D., Feightner, J. W., Attia, J., Feldman, W. F. and the Canadian Task Force on Preventive Care. (1999). Periodic health examination, 1999 update: 1. Detection, prevention, and treatment of obesity. *CMAJ* **160**: 513–525.

60. May, E. S., Kiefe, C. I., Funkhouser, E. and Fouad, M. N. (1999). Compliance with mammography guidelines: physician recommendation and patient adherence. *Prev. Med.* **28**: 386–394.

61. Bush, R. A. and Langer, R. D. (1998). The effects of insurance coverage and ethnicity on mammography utilization in a postmenopausal population. *Western J. Med.* **168**: 236–240.

62. Maxwell, A. E., Bastani, R. and Warda, U. S. (1998). Mammography utilization and related attitudes among Korean-American women. *Women Health* **27**: 89–107.

63. Rankow, E. J. and Tessaro, I. (1998). Mammography and risk factors for breast cancer in lesbian and bisexual women. *Am. J. Health Behav.* **22**: 403–410.

64. Laws, M. B. and Mayo, S. J. (1998). The Latina Breast Cancer Control Study, year one: factors predicting screening mammography by Latina women in Massachusetts. *J. Commun. Health* **23**: 251–267.

65. Zambrana, R. E., Breen, N., Fox, S. A. and Gutierrez-Mohamed, M. L. (1999). Use of cancer screening practices by Hispanic women: analyses by subgroup. *Prev. Med.* **29**: 466–477.

66. Centers for Disease Control. (1998). Self-reported use of mammography and insurance status among women aged > or = 40 years — United States, 1991–1992 and 1996–1997. *MMWR* **47**: 825–830.

67. Eisen, S. A. *et al.* (1999). Sociodemographic and health status characteristics with prostate cancer screening in a national cohort of middle-aged male veterans. *Urology* **53**: 516–522.

68. Powell-Greiner, E., Bolen, J. and Bland, S. (1999). Health care coverage and use of preventive services among the near elderly in the United States. *Am. J. Public Health* **89**: 882–886.

69. Ford, E. S., Will, J. C., De Proost Ford, M. A. and Mokdad, A. H. (1998). Health insurance status and cardiovascular disease risk factors among 50–64-year-old US women: findings from the Third National Health and Nutrition Examination Survey. *J. Women Health* **7**: 997–1006.

70. Beckles, G. L. A. *et al.* (1998). Population-based assessment of the level of care among adults with diabetes in the US. *Diabetes Care* **21**: 1432–1438.

71. Jonna, B. P., Delfino, R. J., Newman, W. G. and Tope, W. D. (1998). Positive predictive value for presumptive diagnosis of skin cancer and compliance with follow-up among patients attending a community screening program. *Prev. Med.* **27**: 611–616.

72. Ferrante, J. M., Gonzalez, E. C., Roetzheim, R. G., Pal, N. and Woodard, L. (2000). Clinical and demographic predictors of late-stage cervical cancer. *Arch. Family Med.* **9**: 439–445.

73. Zambrana, R. E., Breen, N., Fox, S. A. and Gutierrez-Mohamed, M. L. (1999). Use of cancer screening practices by Hispanic women: analyses by subgroup. *Prev. Med.* **29**: 466–477.

74. Roetzheim, R. G. *et al.* (1999). Effects of health insurance and race on early detection of cancer. *J. Natl. Cancer Inst.* **91**: 1409–1415.

75. Nickel, J. T. *et al.* (1998). Preventive health counseling reported by uninsured women with limited access to care. *J. Health Care Poor Underserved* **9**: 293–308.

76. Gentry, D., Longo, D. R., Housemann, R. A., Loiterstein, D. and Brownson, R. C. (1999). Prevalence and correlates of physician advice for prevention: impact of type of insurance and regular source of care. *J. Health Care Finance* **26**: 78–97.

77. Gordon, N. P., Rundall, T. G. and Parker, L. (1998). Type of health care coverage and the likelihood of being screened for cancer. *Med. Care* **38**: 636–645.

78. Reschovsky, J. D. and Kemper, P. (2000). Do HMO's make a difference? Introduction. *Inquiry* **36**: 374–377.

79. Reschovsky, J. D. and Kemper, P. and Tu, H. (2000). Does type of health insurance affect health care use and assessments of care among the privately insured? *Health Serv. Res.* **35**: 221–237.

80. Landers, S. J. (1999). Study highlights managed care conflicts. *Am. Med. News.* August 16, 1999.

81. Schauffler, H. H. and Chapman, S. A. (1998). Health promotion and managed care: surveys of California's health plans and population. *Am. J. Prev. Med.* **14**: 161–167.

82. Gifford, D. R. *et al.* (1999). Improving adherence to dementia guidelines through education and opinion leaders. *Ann. Int. Med.* **131**: 237–246.

83. Frolkis, J. P. (1999). Do physicians follow NCEP guidelines? *Cardiol. Rev.* **16**: 11–14.

84. McBride, P. *et al.* (1998). Primary care practice adherence to National Cholesterol Education Program guidelines for patients with coronary heart disease. *Arch. Int. Med.* **158**: 1238–1244.

85. (1999). *Arch. Int. Med.* **159**: 294–302.

86. Berlowitz, D. R. *et al.* (1998). Inadequate management of blood pressure in a hypertensive population. *New England J. Med.* **339**: 1957–1963.

87. Malin, J. L. *et al.* (2000). Organizational systems used by California capitated medical groups and independent practice associations to increase cancer screening. *Cancer* **88**: 2824–2831.

88. Aronow, W. A. (1999). Treatment of hypercholesterolemia in older persons with coronary artery disease. *Clin. Geriatrics* **7**: 93–99.

89. Xhignesse, M. *et al.* (1999). Antiplatetet and lipid-lowering therapies for the secondary prevention of cardiovascular disease: are we doing enough? *Can. J. Cardiol.* **15**: 185–189.

90. Boschert, S. (2000). Few women get bone density scans. *Int. Med. News.* July 1, 2000

91. Davis, D. (1999). Impact of formal continuing medical education: do conferences, workshops, rounds, and other traditional continuing education activities change physician behavior or health outcomes? *JAMA* **282**: 867–874.

92. Cabana, M. D. *et al.* (1999). Why don't physicians follow clinical practice guidelines? A framework for improvement. *JAMA* **282**: 1458–1465.

93. Braddock, C. H. *et al.* (1999). Informed decision making in outpatient practice: time to get back to basics. *JAMA* **282**: 2313–2320.

94. Cabana, M. D. *et al.* (1999). Why don't physicians follow clinical practice guidelines? A framework for improvement. *JAMA* **282**: 1458–1465.

95. Northrup, L. M. (1999). Communication aids patient compliance. *Physician Practice Options*, May 15, 1999.

96. Gazmararian, J. A. *et al.* (1999). Health literacy among Medicare enrollees in a managed care organization. *JAMA* **281**: 545–557.

97. Haggerty, J. *et al.* (1999). Screening mammography referral rates for women ages 50 to 69 years by recently-licensed family physicians: physician and practice environment correlates. *Prev. Med.* **29**: 391–404.

98. Cabana, M. D. *et al.* (1999). Why don't physicians follow clinical practice guidelines? A framework for improvement. *JAMA* **282**: 1458–1465.

99. Gotzsche, P. C. and Olsen, O. (2000). Is screening for breast cancer with mammography justifiable? *Lancet* **355**: 129–181.

100. Ewy, G. A. (1999). Antioxidant therapy for coronary artery disease: don't paint the walls without treating the termites! *Arch. Int. Med.* **159**: 294–302.

101. Woolf, S. H. (1999). The need for perspective in evidence-based medicine. *JAMA* **282**: 2358–2365.
102. Filak, A. T. *et al.* (1999). Lifetime costs for preventive medical services: a model. *J. Family Pract.* **48**: 706–710.
103. MedscapeWire (2000). Low hypertension awareness among Americans with high risk for complications. *MedscapeWire*. March 13, 2000.
104. MedscapeWire. (1999). Decline in awareness and treatment of high blood pressure could pose a serious public health threat. *MedscapeWire*. August 19, 1999.
105. Prager, L. O. (1999). Spotlight spurs improvement. *Am. Med. News.* August 16, 1999.
106. Braunwald, E. (1997). Shattuck lecture-cardiovascular medicine at the turn of the millennium: triumphs, concerns, and opportunities. *New England J. Med.* **337**: 1361–1369.

Chapter 80

Tissue Engineering

Hidetomi Terai, Alexander Sevy and Joseph P. Vacanti*

Hidetomi Terai, Alexander Sevy and **Joseph P. Vacanti** • Department of Surgery, Massachusetts General Hospital, 55 Fruit Street, Warren 1157, Boston, MA02114

*Corresponding Author.
Tel: (617) 724 1725, E-mail: jvacanti@partners.org

1. Introduction

In the last half of the 20th century, innovative drugs, surgical procedures, and medical devices have contributed dramatically to human health care. Medical advances have been shown to save and improve the quality of human life for the young and especially the elderly population, particularly in the field of transplantation.

While these advances bring us wider possible therapeutic choices, they also present new problems such as the critical shortage of donor organs, the necessity for life-long immunosuppressive drugs and of course, the high transplantation expenses including perioperative intensive care. For example, the number of patients awaiting a liver transplant increased from 616 in 1988 to 14 000 in 1999, while the number of patients who have undergone a liver transplant only increased from 1713 to 4700 during the same period.[1] This data shows a discrepancy between the available and needed donor organs and suggests that the gap will be much wider in the coming years. Furthermore, with the fastest-growing segment of the United States population comprised of those 65 years old and older, the demand for organs will increase over time.

Selective cell transplantation was proposed by investigators instead of entire organ transplantation to alleviate donor organ shortages. This approach utilizes cultured and expanded cells obtained from small amounts of donor tissue. There are also lower risks and expenses associated with major surgical procedures and protracted hospitalizations. Furthermore, when autologus cells are used, it may also be possible to avoid the need for immunosuppression. However, the theory behind cell transplantation produced no sufficiently viable results. It was not until the new concept of "tissue engineering" was applied that cell transplantation could potentially serve as a future alternative to whole organ transplantation.

2. Tissue Engineering

Tissue engineering is an interdisciplinary field that applies the principles and methods of engineering and the life sciences toward the development of biological substitutes aimed at the creation, restoration, preservation, or improvement of damaged tissues and organs due to injury, disease, genetic abnormality, aging, or aging-related diseases. Tissue engineering will play a significant role in longevity medicine as the world's population matures and the demand for replacement tissue increases. In our investigations, synthetic biodegradable polymer scaffolds have been mainly applied as delivery vehicles for cells to make new living tissues (Fig. 1). This approach is based on the following observations: (1) Every tissue undergoes remodeling. (2) Isolated cells tend to reform into the appropriate tissue structure under appropriate experimental conditions. (3) Although isolated cells have the capacity to form the appropriate tissue structure, they do so only

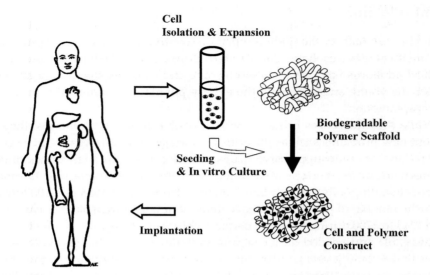

Fig. 1. Schematic illustration of the tissue engineering methods using biodegradable polymers.

to a limited degree when placed in a suspension into tissues. (4) Tissue cannot be implanted in large volumes without sufficient nutrient diffusion or oxygen supply from capillaries.[2] Thus, polymer scaffolds have been designed to provide enough surface area and volume as a template to guide cell organization and growth, while also allowing the diffusion of nutrients and oxygen to the transplanted cells. Eventually the polymer scaffolds degrade *in vivo* through hydrolysis after serving their function, which avoids a long-term foreign body reaction.

Investigators have attempted to engineer every type of tissue by employing not only these devices, but also other materials suitable for the growth of individual cells or tissues such as skin, nerve, esophagus, intestine, bladder ... etc.[3–7] Although there are differences in the process of constructing new functional tissue, the basic concept is similar for all the tissues. This chapter will discuss advances in the fields and recent achievements of engineered tissues in our laboratory.

3. Tissue Engineering of Liver

The liver plays a critical role in a broad spectrum of physiological functions including metabolism, storage, synthesis and detoxification. The only established curative treatment for patients suffering from end-stage or metabolic liver diseases is liver transplantation. Isolated cell transplantation has been proposed as a strategy to replace lost functions in a variety of organs, and the focus for end-stage liver disease has been on hepatocyte transplantation. Hepatocytes are liver parenchymal cells that perform most of liver's functions and make up 80% of the cytoplasmic mass in the liver.[8] For successful hepatocyte transplantation, these basic characteristics should be incorporated: (1) Hepatocytes are anchorage-dependent cells and require an insoluble extra cellular matrix (ECM) for survival, reorganization, proliferation, and function. (2) Hepatocytes are highly metabolic cells that require rapid access to a steady supply of oxygen and nutrients. (3) Hepatocytes exhibit tremendous regenerative capacity with hepatotrophic stimulation *in vivo*.[9, 10] Various methods and implantation sites have been investigated for hepatocyte transplantation based on this insight.[11-23] Direct injections were performed as a free cell graft by using recipients' stromal tissue as an extra cellular matrix, however, there were no sufficient results exhibiting long-term liver support. Microcarriers, developed as an alternative method to direct injection, can present ECM features that affect hepatocyte function and survival.[24-26] Furthermore, microcarriers can facilitate the identification and analysis of transplanted cells. Hepatocytes encapsulated in biocompatible membranes were also used for cell transplantation and suggest the possibility of xeno-, or allo-graft cell transplantation because this structure provides protection from the immune system.[27, 28] In both the cases of microcarrier beads and encapsulation, better results were obtained than with direct injection methods. While these methods have the potential to support liver function temporarily, they are insufficient for the replacement of an orthotopic liver transplantation. The aims of current studies are for larger mass transplantation and longer survival periods.

As scaffolding for cells, our laboratory has chosen synthetic biodegradable fabricated polymer using non-woven polyglycolic acid (PGA) fibers (Fig. 2).[29, 30] They have a highly porous structure that allows for the diffusion of oxygen and nutrients to implanted cells and provides enough space for hepatocyte reorganization and neovascularization from surrounding tissue. Furthermore, these biomaterials degrade *in vivo* by hydrolysis into lactic acid and glycolic acid, which are then incorporated into the tricarboxylic acid cycle and harmlessly excreted.[31] They essentially disappear after serving as a template that guides cell organization and growth, leaving behind only the new natural tissue. Additionally, they provide a large surface area for cells, allowing greater mass transplantation than previous methods. Using this method, we demonstrated the survival of transplanted hepatocytes in the small intestine mesentery. Improvements were shown in the engraftments and survival of transplanted hepatocytes that received hepatotrophic stimulation after a portcaval shunt and partial hepatectomy, and those that were seeded on prevascularized polymer devices. The implanted hepatocytes also exhibited a partial correction of single-enzyme liver defects.[32]

Although small masses of hepatocytes have been implanted successfully and have exhibited their specific functions, it is still difficult to transplant a mass of hepatocytes that is adequate to fully compensate for defects in liver function. The

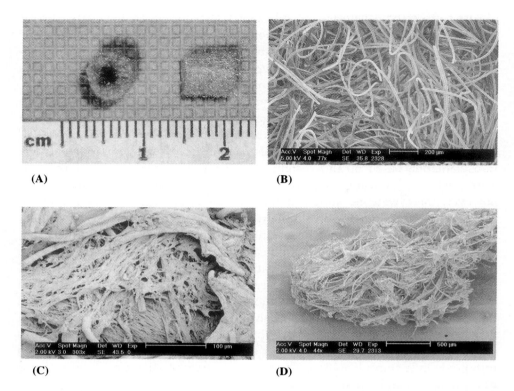

(A)   (B)   (C)   (D)

Fig. 2. (a) Biodegradable polymer tube (Non-woven polyglycolic acid (PGA) fibers). (b) Scanning electron photomicrographs of a PGA scaffold. (c) PGA filled with cells (rat osteoblasts) and rich with extracellular matrix. (d) Cell-polymer constructs before implantation.

recipient site must immediately provide transplanted cells with oxygen and nutrients; however, most cells inside the polymer cannot survive long enough for neovascularization into the engraftment. The problem of sufficient vascularization must be solved in order to fabricate any kind of three-dimensional, solid, and normally highly vascularized organ with tissue-engineering. To overcome this problem, many approaches have been attempted. Kim *et al.* used biodegradable polymer tubes as cell vehicles and implanted them directly into vascular conduits perfused with portal blood. This method is considered to have advantages not only for a steady blood supply, but also for a blood supply that includes the rich hepatotrophic factors present in portal blood.[33] The other approach is to develop a polymer device with an integrated vascular network that provides access to the blood almost immediately after implantation. Kim *et al.* used a special three-dimensional biodegradable polymer to evaluate the survival and function of hepatocytes under static and flow conditions.[34] The polymer scaffolds, which are constructed at MIT, have an intrinsic network of interconnected channels fabricated with a three-dimensional printing (3DP) technique.[35] This technique enables the design and fabrication of complex, three-dimensional, synthetic, porous biodegradable polymer scaffolds in any shape or size with a high degree of macroarchitectual and microarchitectual complexity. As another approach to establish vascular networks, we have focused on constructing capillary networks *in vitro*, which have the potential to solve the critical problems of nutrient and gas exchange for most tissue engineered constructs.[36]

4. Tissue Engineering of Cardiovascular Tissue

Since Voorhees developed the first fabric graft using Vinyon N and treated aortic aneurysms with these grafts in 1952, many researchers have been engaged in developing more ideal materials for vascular grafts.[37] Currently, polyethylene terephthalate (Dacron) and expanded poly-tetrafluoroethylene (ePTFE) are the most widely used prosthetic vascular grafts. In 1960, Starr, Edwards, and Harken performed the first successful surgical implantation of a mechanical heart valve.[38, 39] Currently available prosthetic heart valves include mechanical prosthetic valves, homografts and xenografts. These prostheses each show good clinical results, however these substitutes still have well-known limitations including the inability to grow or remodel, the risk of thrombogenesis or infection, and foreign body reactions. One of the main targets of tissue engineering is to construct prostheses that would overcome such problems as rejection, coagulation, and growth to allow function similar to the original tissues.

For vascular grafts, the blood-contacting surface and the biomechanical properties of the graft become important variables in determining patency especially in a small caliber vessel replacement. With this in mind, biohybrid grafts have been developed by using a variety of biological substances and synthetic grafts. Proteins, anticoagulants and antibiotics have been bonded to the surface of Dacron or ePTFE and can modulate the acute and chronic reactions of recipient.[40–42] When Herring reported that endothelial cell transplantation onto a graft surface could enhance graft survival in 1978, researchers started to use cell transplantation for biological grafts.[43] In 1986, Weinberg and Bell made a completely biological blood vessel by using animal collagen gels combined with bovine endothelial cells, fibroblasts, and smooth muscle cells.[44] However, these constructs failed to show the burst strength and mechanical properties required for *in vivo* implantation even when they had been reinforced with Dacron meshes before implantation. Wesolowski and Ruderman reported that a slowly absorbable vascular graft could induce a host regenerative process producing a "new" functional artery.[45, 46] In 1979, Bowald reported that an entirely bioresorbable graft using Vicryl (a copolymer of polyglycolide and polylactide) sheets resulted in aneurysmal dilation and rupture.[47, 48] Greisler performed grafts using woven polyglycolic acid (PGA) in a rabbit model, and he detected the endothelialization in the graft and the replacement of PGA with cellular matrix.[49, 50] These basic concepts, cell transplantation and bioresorbable devices, became the basis of tissue-engineered vascular grafts. Shinoka *et al.* replaced a 2-cm pulmonary artery segment in lambs with a PGA tubular construct seeded with autologous arterial or venous cells expanded *in vitro*. These grafts showed patency after 11 and 24 weeks. Endothelial lining and extracellular matrix, including collagen and elastic fibers were also observed in these constructs. However, the same constructs implanted in the systemic circulation resulted in aneurysm formation.[51] Shum-Tim *et al.* used a PGA-polyhydroxyalkanoate (PHA) co-polymer, which has a much longer degradation time than PGA, and successfully replaced the abdominal aorta with this scaffold.[52] Niklason *et al.* succeeded in making small tissue-engineered vessels with cultured bovine smooth muscle cells and endothelial cells. The construct had burst strength greater than 2000 mmHg and showed contractile responses to pharmacological agents. They also showed excellent patency in the femoral artery position after 28 days of implantation.[53] Tissue engineered heart valves have also been successfully implanted in lamb models.[54, 55]

Tissue engineered heart valves and vessels can solve oxygen and nutrient supply problems. Careful observation of mechanical properties and long term study are still needed, however these tissues have great promise for numerous current clinical applications as well as in the future.

5. Tissue Engineering of Cartilage

The principal function of articular cartilage is variable load bearing in functional activities through a range of motion. Reconstruction of damaged articular cartilage is one of the most common therapies for locomotive tissue, because severe damage to cartilage is traditionally thought to be irreversible and an eventual secondary cause of osteoarthritis after several decades of damage.[56] An estimated 1 million patients per year need therapy for damaged cartilage due to injury, arthritis, congenital abnormalities, or trauma.[2] Cartilage is composed of a matrix and a single cell type called chondrocytes, which are not vascularized like other tissues because they do not need a high oxygen supply. That is why cartilage has been the most successful tissue for tissue engineering. In fact, clinical use of autologous chondrocytes for human knee cartilage has already been reported and is available on commercial basis.[57, 58]

The first published data that describes cartilage repair by isolated cell transplantation is seen in the report by Chesterman and Smith in 1968.[59] Since this report was published, there have been attempts to transplant cartilage utilizing several different materials and procedures. One major concern with tissue engineered cartilage is the delivery and anchorage of cells to the donor site. Periosteal flaps, fibrin-based glues, demineralized bone, collagen gel, biodegradable, and non-biodegradable synthetic polymers have been used as templates for this purpose.[60–70] However, in the majority of studies, the repair tissue was found to be of variable morphological quality, ranging from fibrous to hyaline cartilage with biomechanical properties that were subnormal and tended to deteriorate over time.

We have utilized biodegradable synthetic polymers as templates for chondrocyte implantation. Biodegradable synthetic polymers can be generated in many shapes.[71] Other structural cartilage tissues like trachea, ear, and nose are also in development.[72–74] Furthermore, these polymers enable the production of complex structural tissues by allowing the combination of different cells that are seeded on different parts of the polymers. Isogai made tissue-engineered phalanges by combining osteocytes, chondrocytes, tenocytes and periosteum using biodegradable polymers. Cell interactions are important in creating more durable articular cartilage and have been demonstrated between transplanted osteocytes and chondrocytes leading to better joint construction.[75]

6. Conclusion

With the continued critical scarcity of donor tissues and organs, tissue engineering offers tremendous potential for alleviating the limitations of current therapy, however, many important issues remain to be solved. For example, there is not yet a system for supplying blood to large transplanted masses of fabricated visceral organs such as the liver or kidney. The establishment of such a system will be essential for the next era of tissue engineering.[36] However, establishing improved methods for cell source and cell expansion remains a problem, for which advances in stem cell biology may provide essential solutions in the future.[76, 77] Further advances in the area of biomaterials and chemical engineering may provide better products that can directly modify cell growth, proliferation, and differentiation with molecular biological substances.[78, 79]

Tissue engineering is a relatively new therapeutic approach for human health, but interdisciplinary cooperation between biology, engineering, and medicine can substantially improve human life by providing artificial organs for the vast need that exists in our society.

References

1. Data source: United Network for Organ Sharing;
   http://www.unos.org/frame_Default.asp
2. Langer, R. and Vacanti, J. P. (1993). Tissue engineering. *Science* **260**: 920–926.

3. LaFrance, M. L. and Armstrong, D. W. (1999). Novel living skin replacement biotherapy approach for wounded skin tissues. *Tissue Eng.* **5**: 153–170.

4. Hadlock, T., Sundback, C., Hunter, D., Cheney, M. and Vacanti, J. P. (2000). A polymer foam conduit seeded with schwann cells promotes guided peripheral nerve regeneration. *Tissue Eng.* **6**: 119–127.

5. Sato, M., Ando, N., Ozawa, S., Miki, H. and Kitajima, M. (1994). An artificial esophagus consisting of cultured human esophageal epithelial cells, polyglycolic acid mesh, and collagen. *Asaio J.* **40**: M389–M392.

6. Kaihara, S., Kim, S. S., Kim, B. S., Mooney, D., Tanaka, K. and Vacanti, J. P. (2000). Long-term follow-up of tissue-engineered intestine after anastomosis to native small bowel. *Transplantation* **69**: 1927–1932.

7. Atala, A., Freeman, M. R. and Vacanti, J. P. (1993). Implantation in vivo and retrieval of artificial structures consisting of rabbit and human urothelium and human bladder muscle. *J. Urol.* **150**: 608–612.

8. Jones. A. L. (1990). Anatomy of the normal liver. *In* "Hepatology. A Textbook of Liver Disease " (D. Zakim, and T. D. Boyer, Eds.), pp. 3–29, W.B. Saunders Co., PA.

9. Kaufmann, P. M., Sano, K., Uyama, S., Takeda, T. and Vacanti, J. P. (1994). Heterotopic hepatocyte transplantation: assessing the impact of hepatotrophic stimulation. *Transplant Proc.* **26**: 2240–2241.

10. Sano, K., Cusick, R. A. and Lee, H., *et al.* (1996). Regenerative signals for heterotopic hepatocyte transplantation. *Trasplant Proc.* **28**: 1859–1860.

11. Baumgartner, D., LaPlante-O'Neill, P. M., Sutherland, D. E. and Najarian, J. S. (1983). Effects of i ntrasplenic injection of hepatocytes, hepatocyte fragments and hepatocyte culture supernatants on d-galactosamine-induced liver failure in rats. *Eur. Surg. Res.* **15**: 129–135.

12. Vroemen, J. P., Buurman, W. A., Heirwegh, K. P., van der Linden, C. J. and Kootstra, G. (1986). Hepatocyte transplantation for enzyme deficiency disease in congenic rats. *Transplantation* **42**: 130–135.

13. Kasai, S., Sawa, M., Kondoh, K., Ebata, H. and Mito, M. (1987). Intrasplenic hepatocyte transplantation in mammals. *Transplant Proc.* **19**: 992–994.

14. Then, P., Sandbichler, P., Erhart, R., Dietze, O., Klima, G. and Vogel, W. (1991). Hepatocyte transplantation into the lung for treatment of acute hepatic failure in the rat. *Transplant Proc.* **23**: 892–893.

15. Zhang, H., Miescher-Clemens, E., Drugas, G., Lee, S. M. and Colombani, P. (1992). Intrahepatic hepatocyte transplantation following subtotal hepatectomy in the recipient: a possible model in the treatment of hepatic enzyme deficiency. *J. Pediatr. Surg.* **27**: 312–315.

16. Makowka, L., Rotstein, L. E., Falk, R. E., Falk, J. A., Nossal, N. A. and Langer, B. (1980). Allogeneic and xenogeneic hepatocyte transplantation in experimental hepatic failure. *Transplantation* **30**: 429–435.

17. Johnson, L. B., Aiken, J., Mooney, D., Schloo, B. L., Griffith-Cima, L. and Langer, R. (1994). The mesentery as a laminated vascular bed for hepatocyte transplantation. *Cell Transplant* **3**: 273–281.
18. Ricordi, C., Lacy, P. E., Callery, M. P., Park, P. W. and Flye, M. W. (1989). Trophic factors from pancreatic islets in combined hepatocyte-islet allografts enhance hepatocellular survival. *Surgery* **105**: 218–223.
19. Gupta, S., Aragona, E., Vemuru, R. P., Bhargava, K. K., Burk, R. D. and Chowdhury, J. R. (1991). Permanent engraftment and function of hepatocytes delivered to the liver: implications for gene therapy and liver repopulation. *Hepatology* **14**: 144–149.
20. Jaffe, V., Darby, H., Selden, C. and Hodgson, H. J. (1988). The growth of transplanted liver cells within the pancreas. *Transplantation* **45**: 497–498.
21. Vroemen, J. P., Buurman, W. A., van der Linden, C. J., Visser, R., Heirwegh, K. P. and Kootstra, G. (1988). Transplantation of isolated hepatocytes into the pancreas. *Eur. Surg. Res.* **20**: 1–11.
22. Selden, C., Gupta, S., Johnstone, R. and Hodgson, H. J. (1984). The pulmonary vascular bed as a site for implantation of isolated liver cells in inbred rats. *Transplantation* **38**: 81–83.
23. Sandbichler, P., Then, P., Vogel, W., Erhart, R., Dietze, O. and Philadelphy, H. (1992). Hepatocellular transplantation into the lung for temporary support of acute liver failure in the rat. *Gastroenterology* **102**: 605–609.
24. Demetriou, A. A., Reisner, A., Sanchez, J., Levenson, S. M., Moscioni, A. D. and Chowdhury, J. R. (1988). Transplantation of microcarrier-attached hepatocytes into 90% partially hepatectomized rats. *Hepatology* **8**: 1006–1009.
25. Demetriou, A. A., Levenson, S. M., Novikoff, P. M., Novikoff, A. B., Chowdhury, N. R. and Whiting, J. (1986). Survival, organization, and function of microcarrier-attached hepatocytes transplanted in rats. *Proc. Natl. Acad. Sci. USA* **83**: 7475–7479.
26. Demetriou, A. A., Whiting, J. F., Feldman, D., Levenson, S. M., Chowdhury, N. R. and Moscioni, A. D. (1986). Replacement of liver function in rats by transplantation of microcarrier-attached hepatocytes. *Science* **233**: 1190–1192.
27. Dixit, V., Darvasi, R., Arthur, M., Brezina, M., Lewin, K. and Gitnick, G. (1990). Restoration of liver function in Gunn rats without immunosuppression using transplanted microencapsulated hepatocytes. *Hepatology* **12**: 1342–1349.
28. Dixit, V. and Gitnick, G. (1995). Transplantation of microencapsulated hepatocytes for liver function replacement. *J. Biomat. Sci. Polym. Ed.* **7**: 343–357.
29. Vacanti, J. P., Morse, M. A., Saltzman, W. M., Domb, A. J., Perez-Atayde, A. and Langer, R. (1988). Selective cell transplantation using bioabsorbable artificial polymers as matrices. *J. Pediatr. Surg.* **23**: 3–9.
30. Uyama, S., Kaufmann, P. M., Takeda, T. and Vacanti, J. P. (1993). Delivery of whole liver-equivalent hepatocyte mass using polymer devices and hepatotrophic stimulation. *Transplantation* **55**: 932–935.

31. Athanasiou, K. A., Niederauer, G. G. and Agrawal, C. M. (1996). Sterilization, toxicity, biocompatibility and clinical applications of polylactic acid/polyglycolic acid copolymers. *Biomaterials* **17**: 93–102.

32. Takeda, T., Kim, T. H., Lee, S. K., Langer, R. and Vacanti, J. P. (1995). Hepatocyte transplantation in biodegradable polymer scaffolds using the Dalmatian dog model of hyperuricosuria. *Transplant Proc.* **27**: 635–636.

33. Kim, S. S., Kaihara, S., Benvenuto, M. S., Kim, B. S., Mooney, D. J. and Vacanti, J. P. (1999). Small intestinal submucosa as a small-caliber venous graft: a novel model for hepatocyte transplantation on synthetic biodegradable polymer scaffolds with direct access to the portal venous system. *J. Pediatr. Surg.* **34**: 124–128.

34. Kim, S. S., Utsunomiya, H., Koski, J. A., Wu, B. M., Cima, M. J. and Sohn, J. (1998). Survival and function of hepatocytes on a novel three-dimensional synthetic biodegradable polymer scaffold with an intrinsic network of channels. *Ann. Surg.* **228**: 8–13.

35. Sachs, E. M., Cima, M. J., Williams, P., Brancazio, D. and Cornie, J. (1992). Three dimensional printing. *J. Eng. Ind.* **114**: 481–488.

36. Kaihara, S., Borenstein, J., Koka, R., Lalan, S., Ochoa, E. R. and Ravens, M. (2000). Silicon micromachining to tissue engineer branched vascular channels for liver fabrication. *Tissue Eng.* **6**: 105–117.

37. Voorhees, A. B. J., Jarerzki, A. and Blakemore, A. (1952). The use of tubes constructed of Vinyon N cloth in bridging arterial defects. *Ann. Surg.* **135**: 332.

38. Starr, A. and Edwards, M. L. (1961). Mitral replacement: clinical experience with a ball-valve prosthesis. *Ann. Surg.* **154**: 726–740.

39. Harken, D. E., Soroff, H. S., Taylor, W. J., Lefemine, A. A., Gupta, S. K. and Lunzer, S. (1960). Partial and complete prostheses in aortic insufficiency. *J. Thorac. Cardiovasc. Surg.* **40**: 744–762.

40. Eberhart, R. C., Munro, M. S., Williams, G. B., Kulkarni, P. V., Shannon, W. A., Jr. and Brink, B. E. (1987). Albumin adsorption and retention on C18-alkyl-derivatized polyurethane vascular grafts. *Artif. Organs* **11**: 375–382.

41. Park, K. D., Okano, T., Nojiri, C. and Kim, S. W. (1988). Heparin immobilization onto segmented polyurethane-urea surfaces — effect of hydrophilic spacers. *J. Biomed. Mat. Res.* **22**: 977–992.

42. Goeau-Brissonniere, O., Mercier, F., Nicolas, M. H., Bacourt, F., Coggia, M. and Lebrault, C. (1994). Treatment of vascular graft infection by in situ replacement with a rifampin-bonded gelatin-sealed Dacron graft. *J. Vasc. Surg.* **19**: 739–741.

43. Herring, M., Gardner, A. and Glover, J. (1978). A single-staged technique for seeding vascular grafts with autogenous endothelium. *Surgery* **84**: 498–504.

44. Weinberg, C. B. and Bell, E. (1986). A blood vessel model constructed from collagen and cultured vascular cells. *Science* **231**: 397–400.

45. Wesolowski, S. A., Fries, C. C., Domingo, R. T., Liebig, W. J. and Sawyer, P. N. (1963). The compound prosthetic vascular graft. A pathologic survey. *Surgery* **53**: 19.

46. Ruderman, R. J., Hegyeli, A. F., Hattler, B. G. and Leonard, F. (1972). A partially biodegradable vascular prosthesis. *Trans. Am. Soc. Artif. Int. Organs* **18**: 30–37.

47. Bowald, S., Busch, C. and Eriksson, I. (1979). Arterial regeneration following polyglactin 910 suture mesh grafting. *Surgery* **86**: 722–729.

48. Bowald, S., Busch, C. and Eriksson, I. (1980). Absorbable material in vascular prostheses: a new device. *Acta Chir. Scand.* **146**: 391–395.

49. Greisler, H. P. (1982). Arterial regeneration over absorbable prostheses. *Arch. Surg.* **117**: 1425–1431.

50. Greisler, H. P., Kim, D. U., Price, J. B. and Voorhees, A. B., Jr. (1985). Arterial regenerative activity after prosthetic implantation. *Arch. Surg.* **120**: 315–323.

51. Shinoka, T., Shum-Tim, D., Ma, P. X., Tanel, R. E., Isogai, N. and Langer, R. (1998). Creation of viable pulmonary artery autografts through tissue engineering. *J. Thorac. Cardiovasc. Surg.* **115**: 536–545.

52. Shum-Tim, D., Stock, U., Hrkach, J., Shinoka, T., Lien, J. and Moses, M. A. (1999). Tissue engineering of autologous aorta using a new biodegradable polymer. *Ann. Thorac. Surg.* **68**: 2298–2304.

53. Niklason, L. E., Gao, J. and Abbott, W. M. (1999). Functional arteries grown in vitro. *Science* **284**: 489–493.

54. Shinoka, T., Ma, P. X., Shum-Tim, D., Breuer, C. K., Cusick, R. A. and Zund, G. (1996). Tissue-engineered heart valves. Autologous valve leaflet replacement study in a lamb model. *Circulation* **94**: II164–II168.

55. Stock, U. A., Nagashima, M., Khalil, P. N., Nollert, G. D., Herden, T. and Sperling, J. S. (2000). Tissue-engineered valved conduits in the pulmonary circulation. *J. Thorac. Cardiovasc. Surg.* **119**: 732–740.

56. Fassbender, H. G. (1987 November 20). Role of chondrocytes in the development of osteoarthritis. *Am. J. Med.* **83**: 17–24.

57. Mandelbaum, B. R., Browne, J. E., Fu, F., Micheli, L., Mosely, J. B., Jr. and Erggelet, C. (1998). Articular cartilage lesions of the knee. *Am. J. Sports Med.* **26**: 853–861.

58. Richardson, J. B., Caterson, B., Evans, E. H., Ashton, B. A. and Roberts, S. (1999). Repair of human articular cartilage after implantation of autologous chondrocytes. *J. Bone Joint Surg. [Br]* **81**: 1064–1068.

59. Chesterman, P. J. and Smith, A. U. (1968). Homotransplantation of articular cartilage and isolated chondrocytes. An experimental study in rabbits. *J. Bone Joint Surg. [Br]* **50**: 184–197.

60. Green, W. T., Jr. (1977). Articular cartilage repair. Behavior of rabbit chondrocytes during tissue culture and subsequent allografting. *Clin. Orthop.* 237–250.

61. Speer, D. P., Chvapil, M., Volz, R. G. and Holmes, M. D. (1979). Enhancement of healing in osteochondral defects by collagen sponge implants. *Clin. Orthop.* 326–335.

62. Wakitani, S., Kimura, T., Hirooka, A., Ochi, T., Yoneda, M. and Yasui, N. (1989). Repair of rabbit articular surfaces with allograft chondrocytes embedded in collagen gel. *J. Bone Joint Surg. [Br]* **71**: 74–80.

63. Klompmaker, J., Jansen, H. W., Veth, R. P., Nielsen, H. K., de Groot, J. H. and Pennings, A. J. (1992). Porous polymer implants for repair of full-thickness defects of articular cartilage: an experimental study in rabbit and dog. *Biomaterials* **13**: 625–634.

64. Mow, V. C., Ratcliffe, A., Rosenwasser, M. P. and Buckwalter, J. A. (1991). Experimental studies on repair of large osteochondral defects at a high weight bearing area of the knee joint: a tissue engineering study. *J. Biomech. Eng.* **113**: 198–207.

65. O'Driscoll, S. W., Keeley, F. W. and Salter, R. B. (1986). The chondrogenic potential of free autogenous periosteal grafts for biological resurfacing of major full-thickness defects in joint surfaces under the influence of continuous passive motion. An experimental investigation in the rabbit. *J. Bone Joint Surg. [Am]* **68**: 1017–1035.

66. O'Driscoll, S. W. and Salter, R. B. (1984). The induction of neochondrogenesis in free intra-articular periosteal autografts under the influence of continuous passive motion. An experimental investigation in the rabbit. *J. Bone Joint Surg. [Am]* **66**: 1248–1257.

67. von Schroeder, H. P., Kwan, M., Amiel, D. and Coutts, R. D. (1992). The use of polylactic acid matrix and periosteal grafts for the reconstruction of rabbit knee articular defects. *J. Biomed. Mat. Res.* **25**: 329–339.

68. Vacanti, C. A., Langer, R., Schloo, B. and Vacanti, J. P. (1991). Synthetic polymers seeded with chondrocytes provide a template for new cartilage formation. *Plast. Reconstr. Surg.* **88**: 753–759.

69. Messner, K. and Gillquist, J. (1993). Synthetic implants for the repair of osteochondral defects of the medial femoral condyle: a biomechanical and histological evaluation in the rabbit knee. *Biomaterials* **14**: 513–521.

70. Freed, L. E., Grande, D. A., Lingbin, Z., Emmanual, J., Marquis, J. C. and Langer, R. (1994). Joint resurfacing using allograft chondrocytes and synthetic biodegradable polymer scaffolds. *J. Biomed. Mat. Res.* **28**: 891–899.

71. Kim, W. S., Vacanti, J. P., Cima, L., Mooney, D., Upton, J. and Puelacher, W. C. (1994). Cartilage engineered in predetermined shapes employing cell transplantation on synthetic biodegradable polymers. *Plast. Reconstr. Surg.* **94**: 233–237.

72. Vacanti, C. A., Paige, K. T., Kim, W. S., Sakata, J., Upton, J. and Vacanti, J. P. (1994). Experimental tracheal replacement using tissue-engineered cartilage. *J. Pediatr. Surg.* **29**: 201–205.

73. Hadlock, T. A., Vacanti, J. P. and Cheney, M. L. (1997). Tissue engineering for auricular reconstruction. *Facial Plast. Surg. Clin. North Am.* **5**: 311–317.

74. Shastri, V. P., Martin, I. and Langer, R. (2000). Macroporous polymer foams by hydrocarbon templating. *Proc. Natl. Acad. Sci. USA* **97**: 1970–1975.

75. Isogai, N., Landis, W., Kim, T. H., Gerstenfeld, L. C., Upton, J. and Vacanti, J. P. (1999). Formation of phalanges and small joints by tissue-engineering. *J. Bone Joint Surg. [Am]* **81**: 306–316.

76. Ohgushi, H. and Caplan, A. I. (1999). Stem cell technology and bioceramics: from cell to gene engineering. *J. Biomed. Mat. Res.* **48**: 913–927.

77. Revel, M. (2000). Ongoing research on mammalian cloning and embryo stem cell technologies: bioethics of their potential medical applications. *Israel Med. Assoc. J.* **2(Suppl)**: 8–14.

78. Pierschbacher, M. D. and Ruoslahti, E. (1984). Cell attachment activity of fibronectin can be duplicated by small synthetic fragments of the molecule. *Nature* **309**: 30–33.

79. Vandorpe, J., Schacht, E., Dunn, S., Hawley, A., Stolnik, S. and Davis, S. S. (1997). Long circulating biodegradable poly(phosphazene) nanoparticles surface modified with poly(phosphazene)-poly(ethylene oxide) copolymer. *Biomaterials* **18**: 1147–1152.

Chapter 81

Database Analysis of Human Intervention Studies (The Kronos Longitudinal Study of Aging and Anti-aging Interventions)

Christopher B. Heward

Christopher B. Heward • The Kronos Group, 4455 East Camelback Road, Suite 135, Phoenix, AZ 85018
Tel: (602) 778-7482, E-mail: heward@thekronosgroup.com

1. Introduction

There is still much uncertainty and debate about the biology of aging. Although there is no consensus about the fundamental mechanisms responsible for it, most scientists share the practical view that aging is a universal, intrinsic and deleterious process that manifests itself as a decline in functional capacity over time.[1,2] Like pornography, we all recognize aging when we see it. The problem, for scientists, is how to measure it.

The Baltimore Longitudinal Study of Aging (BLSA), initiated in 1958, is America's oldest continuing scientific examination of human aging.[3,4] Its purpose is to determine those age-associated changes and processes that can be attributed to aging, *per se*, rather than to the diseases that sometimes accompany aging. To accomplish this, BLSA participants are thoroughly tested at 2-year intervals in order to reveal changes in the functioning of a variety of important biochemical and physiological systems. These include various vital organs, the immune system, metabolism, the endocrine system (hormone levels), mental abilities, etc. As a result of this sustained effort, the BLSA database has become nothing less than a national treasure, in terms of its contribution to our understanding of human aging.

Perhaps the most important conclusion emerging from the BLSA data is the finding that there is no single, chronological timetable of human aging.[5] People do not all age at the same rate and, even within one individual, different tissues and organs often age at different rates. This suggests that aging involves a variety of distinct processes or determining factors. These factors can be grouped into two general categories — genetic (hereditary) and environmental, but the variety in the pattern of their expression in different individuals is enormous.

Such individual differences probably explain the failure of simple models of aging, such as the Gompertz equation,[6] to accurately describe the pattern of human mortality at late ages. Gompertz showed that the age-specific mortality rate increases exponentially as a function of age. For humans, between the ages of 30 and 90, mortality rate doubles about every eight years. Thereafter, the rate of increase in mortality, for those lucky few that manage to live so long, slows down.[7] It is important to understand exactly why this slowing occurs because it may point the way toward interventions that will slow it even more or slow it at an earlier age.

One approach toward developing a better model of aging and its relationship to mortality is multivariate analysis using quantitative biomarker measurements of specific physiological variables that reflect functional capacity.[8] Such a model, based upon the appropriate physiological variables may be validated by its ability to predict late age-specific mortality. Identifying which variables are important will greatly facilitate the development of targeted interventions designed to delay or even reverse age-related loss of function and reduce mortality. This is the ultimate goal of the Kronos Longitudinal Study of Aging and Anti-aging Interventions (KLSAAI).

2. Physiological Reserve Capacity

Natural selection favors individuals that become reproductively successful. This is usually achieved by developing greater capacity in vital systems that enable them to better escape predation, disease, accidents, and environmental extremes. In mammals, selective pressure diminishes after reproductive success because there is, apparently, no procreative advantage to greater longevity. In the absence of predation, longevity is determined by the level of physiological reserve that remains after reproductive maturity. Excess physiological reserve capacity evolves as a secondary consequence of a natural selection process based upon reproductive competence, not longevity. Longevity is not a primary evolutionary goal. Thus, physiological reserve capacity is not conserved over time. Molecular damage occurs at a rate greater than molecular repair. The result is declining functionality and increasing vulnerability to accidents and disease (i.e. aging).

The most straightforward way to gauge the decline in physiological function associated with biological aging is to measure the functional reserves of the cells, tissues, organ systems, or organisms under study. This is the approach taken by the BLSA. Figure 1 shows a hypothetical example of several indicators of physiological function, measured longitudinally, for a typical individual in the BLSA. It charts this individual's functional capacity for several measures of physiological function over his lifetime. Although there is considerable heterogeneity in the rate of decline of the many measures of physiological function as we age, the general trend can be expressed as a total body average that reflects an individual's lifespan potential (Fig. 2). This is useful in graphically expressing an idealized pattern of physiological functional capacity throughout a

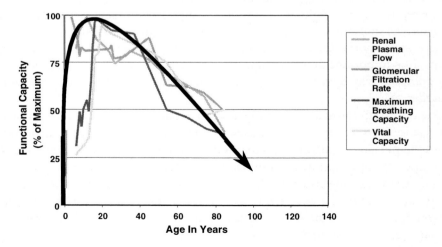

Fig. 1

Lifespan Potential

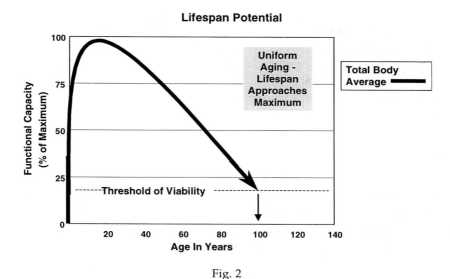

Fig. 2

typical human life span. The graph, of course, incorporates periods of both development and aging.

The first phase of life is characterized by development and maturation. It begins at birth and ends when the individual becomes fully mature (i.e. completely functional, both mentally and physically). Depending upon the individual, this usually happens between the ages of 20 to 30 and is punctuated by a general peak in functional capacity.

This peak in functional capacity marks the beginning of the second phase of life. It is characterized by degeneration and functional decline (aging). The early years of this phase are associated with a high degree of sexual (reproductive) activity, excellent health, intellectual prowess, physical strength, and general vigor. This is followed by "middle age", more obviously associated with a progressive, degenerative decline in function. Libido wanes and sexual activity becomes less frequent. Women go through menopause and men begin to experience the first signs of a gradual and constant decline of testosterone. There is a general loss of physical strength due to a loss of muscle mass and an increase in body fat. Flexibility and mobility wane. Wrinkled skin, aching joints, graying hair, slower reflexes, and reading glasses declare to the world that, "the bloom is off the rose". In addition, insomnia often becomes a problem — minor, at first, but increasing with age. Minor health problems begin to occur more frequently and last longer. As time goes by, there is increasing physiological stress, complete with hormone imbalances and deficiencies, resulting in a negative nitrogen balance, loss of lean muscle mass and increasing fat mass. General health, functional capacity, and vitality are all in decline. The rate of decline is highly variable from person to person, but nobody escapes it. Finally, for everyone, as the progressive degeneration continues and accelerates, symptoms of bodily imbalance reach clinical thresholds and one or more diseases are diagnosed. Immune function continues to decline and susceptibility to infections increases. Medical intervention, at this point, is usually aimed at relieving the symptoms of disease rather than eliminating their causes. Youthful health and vitality become faint memories and, finally, the battle for life, itself, is lost. This is the inexorable pattern of human aging.

3. Clinical Interventions

The above-described pattern of a typical human life span is normal, but it is, obviously, not healthy. The concomitant relationship between health and functional capacity is undeniable — loss of one is loss of the other. This fact alone belies the notion of a healthy 100-year old man. Every 100-year old man has health problems, most of which are age-related. Even so, many people would be happy if they made it to the age of 100. The vast majority of us die prematurely, at a much younger age, usually from age-related diseases (Fig. 3).

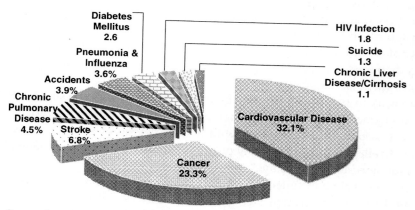

Major Causes of Death in the USA

Diabetes Mellitus 2.6

Pneumonia & Influenza 3.6%

Accidents 3.9%

Chronic Pulmonary Disease 4.5%

Stroke 6.8%

Cancer 23.3%

HIV Infection 1.8

Suicide 1.3

Chronic Liver Disease/Cirrhosis 1.1

Cardiovascular Disease 32.1%

Source: Guyer B, Strobino DM, Ventura SJ et al. Annual Summary of Vital Statistics - 1995

Fig. 3

3.1. Remedial Medicine — Diagnose and Treat

One of the reasons for this stems from the fact that we do not age uniformly. Some organs age more rapidly than others. Different people age differently as a direct result of the interaction between genetics and environment. Figure 4 shows an example of uneven aging. In this case, it is the patient's cardiovascular system that is declining prematurely (perhaps due to environmental factors, such as a

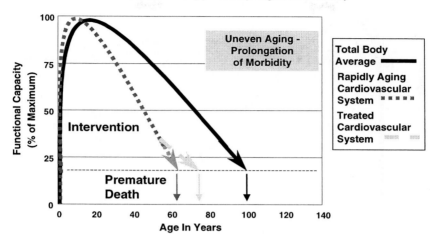

Fig. 4

poor diet). This decline in functional capacity threatens to end this patient's life well before the age of 100. Cardiovascular disease is a common cause of death in the United States today. The conventional approach is to wait until symptoms of actual disease develop before taking action. Only then is the problem diagnosed and only when treatment becomes "medically necessary" is intervention begun. If treatment is successful, then life is briefly prolonged, albeit at a relatively low level of functional capacity. Perhaps the greatest shortcoming of the current health care delivery system in the United States today is that it focuses almost exclusively upon treating existing disease, rather than preventing its onset.

3.2. Preventive Medicine — Assess and Intervene

Without becoming mired in the debate about whether aging itself is a disease, it is clear that the natural consequences of aging cry out for targeted early interventions designed to prevent the onset of age-related diseases. The KLSAAI is aimed at studying the practice of high technology preventive medicine. It is focused on keeping people healthy as they age through comprehensive clinical assessment and early intervention. It employs high technology assessments that help elucidate a patient's functional status and suggest targeted interventions that may help to preserve youthful functional capacity. Various disease risk factors and indicators of functional capacity are measured objectively in order to determine efficacy.

As depicted in Fig. 5, the goal is to identify those processes and organ systems that are declining prematurely and, where possible, intervene in ways designed to return them to more youthful (healthy) norms. If successful, this will result not

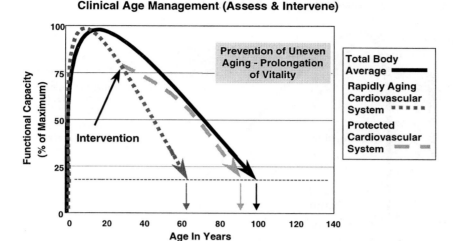

Fig. 5

only in prolongation of the patient's life, but prolongation of the period of his life span associated with a high level of physiological function, free of age-related diseases and symptoms.

4. General Clinical Guidelines

The KLSAAI is a long term, scientifically directed and evidence-based, clinical intervention program designed to move the latest scientific developments in preventive medicine out of the research laboratory and into the clinic. The clinical guidelines used in the study are designed to improve each patient's health and functionality.

Each patient's intervention program is unique and customized specifically to meet his or her individual needs. This customization is based upon the clinical research protocol outlined below.

5. Step 1 — Baseline Data

The first step in the program is to obtain a comprehensive and thorough understanding of each patient's current physiological status. This is accomplished through a three-stage process of information and data acquisition on each patient.

Stage one begins with the patient completing the Kronos Patient Intake Questionnaire (KPIQ). This process is interactive and unlike any that the patient is likely to have experienced before. The KPIQ provides the patient's complete medical background, including his/her family's medical history and it is usually filled out via the Internet in the comfort of the patient's home. The form is flexible and exhaustive in scope, but easy to complete because the patient is presented only with those questions that are pertinent to him/her, in terms of gender, age, and medical history. In addition, the form does not need to be completed all at one time. The information collected becomes an important part of the patient's medical record and it is included in the Kronos database. It is also used in the development of the patient's intervention protocol.

Stage two consists of an extensive "hands on" physical examination by a licensed and specially trained physician. During this examination, the physician will take the patient's vital signs and determine his general health status. In addition, the patient will be examined to screen for diseases and conditions that are best discovered by direct examination. Cancer (prostate, breast, skin, etc.), cardiac arrhythmia, pulmonary disease, and hypertension, are just a few examples of such health problems.

The physical exam also includes the measurement of the patient's physiological status as determined through a series of biomarker-like tests. These tests are designed to assess functional capacity. They include more than 50 physiological tests measuring such things as basal metabolic rate, reaction time, hearing, vision, static balance, short term memory, flexibility, strength, skin elasticity and UV damage, bone mineral density, body composition (by DEXA), cardiovascular function (by 15-lead exercise ECG with full metabolic cart), and pulmonary capacity. Information obtained from this initial physical examination becomes a part of each patient's permanent medical record and the Kronos database. It, along with their laboratory results, is used to help develop their custom intervention program.

Stage three involves obtaining a complete "biochemical picture" of each Kronos patient. To accomplish this, a thorough battery of blood and urine tests is given. This may include more than 150 different biochemical values. In order to achieve the maximum reliability in our laboratory test results, Kronos has designed and fully equipped a, state of the art, analytical laboratory, unlike any other anywhere in the world. Kronos' goal was to create a clinical reference laboratory built to the highest research laboratory standards. It recently became the first outside laboratory to be certified by Johns Hopkins University.

The laboratory tests measure a broad spectrum of biochemical parameters (Tables 1–8). The testing includes, a complete cardiovascular panel, routine liver and kidney function tests, an extensive hormone panel assessing specific aspects of endocrine function, a standard 24-item chemistry panel, complete blood count with micro, a metals panel that measures the most biologically important macro and trace metals, a broad-spectrum oxidative stress profile, and targeted cancer screening tests. In addition, a number of the tests are designed to help identify specific, sub-clinical, often asymptomatic, health conditions that may need immediate prophylactic intervention.

Table 1. Cardiovascular Profile

Assay	Reference Range	Units
Apolipoprotein A1	94–178	mg/dL
Apolipoprotein B	52–163	mg/dL
Ascorbate (vitamin C)	5–28	ug/mL
CRP-high sensitivity	0.000–0.500	mg/dL
Cholesterol	150–200	mg/dL
Coenzyme Q10**	0.5–1.2	ug/mL
Ferritin	24–360	ng/mL
Folic acid	3.1–17.5	ng/mL
Hemoglobin A1C	4.3–6.7	%
HDL	65–100	mg/dL
Homocysteine	5.0–9.0	umol/L
Iron	50–170	ug/dL
Iron bind % sat	13–45	%
Iron bind cap — available (AIBC)	130–375	ug/dL
Iron bind cap — total (TIBC)	228–428	ug/dL
LDL (direct)	50–100	mg/dL
LP(a)	0–11	mg/dL
ORAC (total)**	3000–5500	uM
Tocopherol, alpha**	6.0–16.5	ug/mL
Tocopherol, delta**	0.05–0.25	ug/mL
Tocopherol, gamma**	0.6–5.0	ug/mL
Triglycerides	40–150	mg/dL
Vitamin B12	210-911	pg/mL

**For investigational use only. Assay not approved for diagnostic purposes.

Table 2. Liver and Kidney Function Profile

Assay	Reference Range	Units
Liver function indicators		
Proteins		
Total protein	6.2–8.0	g/dL
Albumin	3.4–5.5	g/dL
Globulin	2.0–3.5	g/dL
Bilirubin, total	0.0–1.5	mg/dL
Bilirubin, direct	0.0–0.65	mg/dL
Bilirubin, indirect	0.00–0.85	mg/dL
Enzymes		
SGOT (AST)	10–40	U/L
SGPT (ALT)	10–55	U/L
GGTP	1–94	U/L
LDH	100–190	U/L
Alkaline phosphatase	0–135	U/L
Renal function indicators		
Bun	8–25	mg/dL
Creatinine (urine)	0.6–1.5	mg/dL
Creatinine (24 h urine)		
Creatinine (serum)		

**For investigational use only. Assay not approved for diagnostic purposes.

Table 3. Endocrine Function Profile

Assay	Reference Range	Units
Pancreatic hormones		
Insulin	6–27	ulU/ml
Adrenal hormones		
DHEA S	0.26–2.0	ug/mL
Cortisol	8.7–22.4	ug/dL
Sex hormones		
Dihydrotestosterone	300–850	pg/mL
Estradiol	10–50	pg/mL
Estriol (uncong)	0–1.7	ng/mL
Estrone	25–150	pg/mL
Progesterone	0.12–0.8	ng/mL
SHBG	15–100	nmol/L
Testosterone	1.2–8.7	ng/mL
Free testost index	14.8–94.8	T/SHBG
Thyroid hormones		
T3 (total)	0.7–1.9	ng/mL
T4 (total)	6.4–11.7	ug/dL
T3 (free)	2.4–6.4	pg/mL
T4 (free)	0.6–1.8	ng/dL
TSH	0.34–5.6	ulU/mL
Pituitary hormones		
IGF-1 (growth hormones)	37–224	ng/mL
TSH	0.34–5.6	ulU/mL

**For investigational use only. Assay not approved for diagnostic purposes.

Table 4. Chemistry Profile 24

Assay	Reference Range	Units
Albumin	3.4–5.5	g/dL
Alkaline phosphatase	0–135	U/L
Bilirubin, direct	0.0–0.65	mg/dL
Bilirubin, indirect	0.00–0.85	mg/dL
Bilirubin, total	0.0–1.5	mg/dL
Bun	8–25	mg/dL
Calcium	8.0–10.5	mg/dL
Chloride	98–106	mEq/L
Cholesterol	150–200	mg/dL
Creatinine	0.6–1.5	mg/dL
GGTP	1–94	U/L
Globulin	2.0–3.5	g/dL
Glucose	70–110	mg/dL
Iron	50–170	ug/dL
LDH	100–190	U/L
Magnesium	1.8–2.5	mg/dL
Phosphorus	2.5–4.8	mg/dL
Potassium	3.5–5.0	mEq/L
SGOT (AST)	10–40	U/L
SGPT (ALT)	10–55	U/L
Sodium	135–145	mEq/L
Total protein	6.2–8.0	g/dL
Triglycerides	40–150	mg/dL
Uric acid	3.6–8.5	mg/dL

**For investigational use only. Assay not approved for diagnostic purposes.

Table 5. Complete Blood Count (CBC)

Assay	Reference Range	Units
WBC	4.0–11.0	K/uL
RBC	4.30–6.00	M/uL
Hemoglobin	13.0–18.0	g/dL
Hematocrit	40.0–53.0	%
MCV	78.0–100.0	fL
MCH	27.0–31.0	Pg
MCHC	31.0–37.0	g/dL
RDW	11.5–14.5	%
Platelet count	130–450	K/uL
MPV	6.5–11.6	fL
Neutrophils %	40–85	%
Lymphocytes %	15–50	%
Monocytes	3–15	%
Eosinophils %	0–7	%
Basophils %	0–2	%
Neutrophils, ABS	1.6–9.4	K/uL
Lymphocytes, ABS	0.6–5.5	K/uL
Monocytes, ABS	0.1–1.6	K/uL
Eosinophils, ABS	0.0–0.8	K/uL
Basophils, ABS	0.0–0.2	K/uL

**For investigational use only. Assay not approved for diagnostic purposes.

Table 6. Trace Metal Profile

Assay	Reference Range	Units
Essential macrominerals		
Calcium	8.0–10.5	mg/dL
Iron	50–170	ug/dL
Magnesium	1.8–2.5	mg/dL
Sodium	135–145	mEq/L
Potassium	3.5–5.0	mEq/L
Chloride	98–106	mEq/L
Essential trace metals		
Chromium	0.12–2.10	ug/L
Cobalt	0.1–1.4	ug/L
Copper	700–1400	ug/L
Manganese	0.5–5.1	ug/L
Molybdenum	0.1–3.0	ug/L
Selenium	85–230	ug/L
Zinc	530–2910	ug/L
Toxic trace metals		
Aluminum	1–20	ug/L
Antimony	0.00–0.71	ug/L
Arsenic	1.7–15.4	ug/L
Barium	0–80	ug/L
Beryllium	0.0–0.1	ug/L
Cadmium	0.0–2.20	ug/L
Lead	0–23	ug/L
Mercury	0.5–5.8	ug/L
Nickel	0.6–7.5	ug/L
Strontium	13–80	ug/L
Thallium	0.0–1.0	ug/L
Tin	0.0–8.2	ug/L

**For investigational use only. Assay not approved for diagnostic purposes.

Table 7. Oxidative Stress Profile

Assay	Reference Range	Units
Albumin	3.4–5.5	g/dL
Ascorbate (vitamin C)	5–28	ug/mL
Bilirubin (direct)	0.0–0.65	mg/dL
Bilirubin (total)	0.0–1.5	mg/dL
Carotene, alpha**	20–400	ng/mL
Carotene, beta**	50–710	ng/mL
Ceruloplasmin	25–45	mg/dL
Coenzyme Q10**	0.5–1.2	ug/mL
Cryptoxanthin, beta**	5–200	ng/mL
Ferritin	24–360	ng/mL
Glucose	70–110	mg/dL
Glutathione	20–100	uM
Hemoglobin A1C	4.3–6.7	%
Peroxides AQ UR**	5.0–38.0	uM
Peroxides AQ SER**	0.5–5.0	uM
Iron	50–170	ug/dL
Iron bind % Sat	13–45	%
Iron bind cap (AIBC)	130–375	ug/dL
Iron bind cap (TIBC)	228–428	ug/dL
Lutein**	40–600	ng/mL
Orac (total)**	3000–5500	uM
Retinol**	400–1300	ng/mL
Retinyl palmitate**	10–190	ng/mL
Selenium	85–230	ug/L
Thiols**	100–240	UM
Tocopherol, alpha**	6.0–16.5	ug/mL
Tocopherol, delta**	0.05–0.25	ug/mL
Tocopherol, gamma**	0.6–5.0	ug/mL
Uric acid	3.6–8.5	mg/dL
Zeaxanthin**	10–150	ng/mL

**For investigational use only. Assay not approved for diagnostic purposes.

Table 8. Cancer Risk Profile

Assay	Reference Range	Units
PSA (total)	0.0–4.0	ng/mL

Through this three-stage process, insight into each patient's functional status and reserve capacity is achieved. This process establishes each patient's individual baseline data, which is important for several reasons:

First — The patient's baseline information is necessary to develop his customized intervention program. All interventions are based on each patient's unique physiological and biochemical requirements.

Second — The baseline data provide a starting point from which to measure a patient's progress over the years. This progress is measured through additional testing, repeated periodically, in a manner similar to the BLSA.

Third — The initial baseline test results will be used to create a cross-sectional database to establish age-adjusted norms for each test. This will become possible only when sufficient numbers of individuals, of each sex and across the entire age spectrum, have been tested. These data will function as a cross-sectional control, indicating the normal rate of aging prior to intervention. Although, from a rigorous scientific standpoint, this is not ideal, from a practical standpoint, given the clinical setting, it is the best we can do. Still, given sufficient numbers of patients, these cross-sectional data will be useful in the statistical analysis that will eventually validate (or invalidate) the efficacy of the interventions.

6. Step 2 — Interventions

Since customized compounding is essential to the development of individualized patient treatment protocols, Kronos also has a compounding pharmacy. This provides the highest possible level of control over the quality of the products prescribed to the patient. The pharmacy has established a state-of-the-art quality control verification system to insure the quality of all of its products. The pharmaceutical staff has over 71 years of combined experience in all forms of compounding and delivery systems, including topical creams, injectibles, sub-linguals, suppositories, and encapsulation. The ability to produce virtually any preparation that a physician can prescribe gives the Kronos physician great flexibility in designing each patient's custom intervention program. A computerized follow-up system facilitates our ability to monitor compliance as well. In general, the interventions will take the following forms:

6.1. Dietary Supplementation

Some scientists maintain that Americans get enough vitamins from a typical "well balanced" diet and that additional supplementation is a waste of money; however, many do not share this view. The US Government's recommended daily allowance (RDA) is not designed for optimum health and longevity. It is clearly not optimum for many vitamins (e.g. B complex, C, E, etc.) and minerals (e.g. zinc, selenium, etc.) and it ignores other important supplements entirely (e.g. co-enzyme Q-10, beta-carotene, etc.). Therefore, a customized regimen of dietary supplements is developed for all patients based upon their biochemical test results. Supplementation regimens may include vitamins, minerals, herbs, specially derived nutrients, nootropics, and pharmaceuticals, if needed.

One of the most common targets of intervention that is addressed by supplementation is oxidative stress. Reactive oxygen species (ROS) are produced by all cells that use oxygen as products of essential, energy-producing processes (aerobic metabolism) and, also, to a lesser extent, by ionizing radiation (X-rays, cosmic rays, etc.) and certain toxic chemicals in the environment.[9–16] ROS damage cell components (DNA, proteins, and lipids) and thereby weaken the functional integrity of tissues. They are known to have important causative roles in a variety of age-related diseases, such as cardiovascular disease, cancer, and Alzheimer's disease. It has been hypothesized that an individual's steady state level of ROS mediated damage (oxidative stress) is associated with the incidence of age related disease in that individual.[17–19] The lower the rate of damage, the higher the probability of a long and healthy life. By measuring various markers of oxidative damage together with known protection factors (endogenous and exogenous antioxidants), it is possible to gage an individual's level of oxidative stress. In many cases, an individual exhibiting a high level of oxidative stress will show significant improvement after a period of targeted supplementation, proper exercise, and moderate lifestyle changes. An example of how this might occur is shown in Fig. 6.

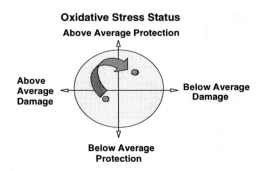

Fig. 6

6.2. Hormone Replacement

One of the most promising interventions designed to achieve significant longevity and youthful function involves maintaining optimum homeostatic balance using bio-identical hormone replacement (or supplementation) therapy. In principle, the basic approach is simple. The hormones in which one is deficient are determined and these hormones are replaced using bio-identical hormones in amounts sufficient to return blood levels to youthful norms. Hormone replacement therapy is a long-term proposition, possibly continuing for the rest of a patient's life. Thus, great care must be taken to insure that optimum blood levels of the hormones are achieved and maintained. Figure 7 shows how urinary estrogen level may be used to monitor bio-identical estrogen replacement therapy in a post-menopausal woman.

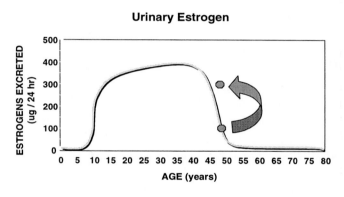

Fig. 7

6.3. Diet (Optimum Nutrition)

Dietary intervention is notoriously difficult to control in a clinical environment. Thus, dietary intervention will be limited to an educational program designed to inform patients about foods that are not consistent with optimum health. The dietary guidelines include what specific foods to eat and not to eat, how to prepare them, and a simple explanation of the role of calories in aging. They encourage consumption of the minimum number of calories required to obtain all of the vital nutrients that are necessary for optimum health. The goal is to build an understanding of how nutrition and energy intake affect aging in order to help make better nutrition a daily habit of each Kronos patient.

In a nutshell, the dietary guidelines can be summed up in three words, "NO EMPTY CALORIES". The diet promoted at Kronos contains a wide variety of nutritious fruits, green and colorful vegetables, nuts, and other nutrient dense foods, including lean meat, fish, and poultry. Thus, the diet consists of foods rich in protein, good-quality essential fats, and antioxidants. It limits intake of saturated fats and "trans-" fatty acids. It restricts the intake of refined carbohydrates such as grains, sugars, and concentrated starches. If followed, the diet will produce better nutrition with fewer calories.

6.4. Exercise

The benefits of exercise in achieving optimum health and longevity are extremely well documented.[20-25] The exercise guidelines emphasize the correct way to start an exercise program, what elements of cardiovascular exercise are important, and why resistance training should be included. A basic exercise program, designed to fit his individual needs, is recommended to each patient. Each patient is encouraged to seek the advice of a certified personal trainer or other exercise professional.

6.5. Pharmaceuticals

The role of pharmaceuticals in delaying the onset of age-related disease is a limited but important one. As a general rule, Kronos prefers to utilize natural interventions to correct homeostatic imbalances, if at all possible. In the absence of a natural therapy, pharmaceuticals may be useful in treating early symptoms of age-related diseases not otherwise subject to control. Certain age-related degenerative processes cannot be addressed using supplements and/or natural hormones. Thus, pharmaceutical therapy may be more appropriate for correcting and/or preventing age-related changes in these instances.

7. Step 3 — Ongoing Monitoring

Once admitted into the program, patients are continually monitored for physiological and biochemical health status. First and foremost, it must be verified that the initial intervention protocol is working. Therefore, as part of this monitoring process, each patient's progress is charted. When appropriate, a trend line is established for each parameter that is changing significantly. This will provide an indication of how well the patient is doing, so that they will not have to wait for years to observe the results of their efforts. Timely feedback on the efficacy of a patient's intervention program is essential to maintaining the motivation necessary for long-term adherence.

Information from the follow-up testing is used to improve and refine each patient's intervention program. In addition, the results from all measurements, on every patient, will be combined in a computer database system. Ongoing analysis of this database will allow the determination of which interventions work well and which do not. This information will then be used to devise new and improved intervention regimens.

8. Conclusion

Since 1889, when Brown-Sequard unsuccessfully attempted to rejuvenate failing testicular function in human subjects,[26] scientists have sought interventions designed to combat the aging process. In recent years, with a better understanding of the biochemical and neuroendocrine determinants of aging, anti-aging research is on a more solid foundation.[27-29] Most current studies are concerned with elucidating regulatory mechanisms of aging, but they also have practical implications.[30, 31]

Today, by carefully applying technology developed during the past twenty years, it may be possible to significantly delay the onset of many age-related diseases. The KLSAAI is a practical, custom tailored, medically safe and scientifically sound program specifically designed for this purpose. Currently, the program is limited to the types of interventions described above (i.e. supplements, hormones, diet, exercise, drugs, and monitoring). Scientifically validating that targeted interventions can significantly retard the onset of age-related diseases and prolong functional lifespan will go a long way toward capturing the imagination of the masses and stimulating more aggressive research into the development of new and better technologies.

In the future, however, better methods for measuring the consequences of aging will be developed, including immune function testing, cognitive function testing, cancer screening tests, and genetic testing. These tests will, in turn, lead to improved techniques for maintaining youthful function in all areas. New technologies, such as tissue grafting, stem cell therapy, "age-breaker" drugs, SOD and Catalase mimetics, improved hormonal therapies, and genetic engineering are all on the horizon. Each year a patient gains, by preserving his health and prolonging his life, improves his chances of being around to enjoy the benefits of the tremendous rejuvenation technologies that will, almost certainly, become available in the future. Because they represent an ideal group on which to conduct clinical trials, it is likely that patients enrolled in the KLSAAI will play a significant role as volunteer research subjects in clinical trials aimed at facilitating the transfer of these technologies from the laboratory to the clinic. The long-term value of the KLSAAI patient database will be incalculable in this regard.

Acknowledgments

The author would like to thank Debbie Werner and Carol Jackson for their helpful contributions in the preparation of this manuscript.

References

1. Balin, A. K. (Ed.) (1994). *Practical Handbook of Human Biologic Age Determination*m, CRC Press, Inc.
2. Strehler, B. (1982). *Time, Cells, and Aging*, Academic Press, New York.
3. No Author. (1989). *Older and Wiser: The Baltimore Longitudinal Study of Aging*, NIH Publication No. 89-2797. US Government Printing Office, Washington, DC.
4. Shock, N. W., Greulich, R. C., Andres, R., Arenberg, D., Costa, P. T., Jr., Lakatta, E. G. and Tobin, J. D. (1984). *Normal Human Aging: The Baltimore Longitudinal Study of Aging*, NIH Publication No. 84-2450. US Government Printing Office, Washington, DC.
5. Hayflick, L. (1994). *How and Why We Age*, Balantine Books, New York.
6. Gompertz, B. (1825). On the nature of the function expressive of the law of human mortality and on a new mode of determining the value of life contingencies. *Philos. Trans. Royal Soc.* London **115**: 513–585.
7. Ham, R. G. and Veomett, M. J. (1980). *Mechanisms of Development*, C. V. Mosby, St. Louis.
8. Manton, K. G., Woodbury, M. A. and Stallard, E. (1995). Sex differences in human mortality and aging at late ages: the effect of mortality selection and state dynamics. *Gerontologist* **35**(5): 597–608.
9. Cutler, R. G. (1979). Evolution of human longevity: a critical overview. *Mech. Aging Dev.* **9**: 337–354.
10. Cutler, R. G. (1982). The dysdifferentiative hypothesis of mammalian aging and longevity. *In* "The Aging Brain Cellular and Molecular Mechanisms of Aging in the Nervous System, Aging" (E. Giacobini, G. Filogamo, and A. Vernadakis, Eds.), Vol. 20, pp. 1–19, Raven Press, New York.
11. Cutler, R. G. (1982). Longevity is determined by specific genes: testing the hypothesis. *In* "Testing the Theories of Aging" (R. Adelman, and G. Roth, Eds.), pp. 25–114, CRC Press, Boca Raton, FLA.
12. Cutler, R. G. (1984). Evolutionary biology of aging and longevity. *In* "Aging and Cell Structure" (J. E. Johnson, Ed.), Vol. 2, pp. 371–428, Plenum Press, New York.
13. Orr, W. C. and Sohal, R. S. (1994). Extension of life span by overexpression of superoxide dismutase and catalase in *Drosophila melanogaster*. *Science* **263**: 1128–1130.
14. Tolmasoff *et al.* (1980). Superoxide dismutase: correlation with life span and specific metabolic rate in primate species. *Proc. Natl. Acad. Sci. USA* **77**: 2777–2781.
15. Cutler, R. G. and Semsei, I. (1980). Development, cancer and aging: possible common mechanisms of action and regulation. *J. Gerontol.* **44**: 25–34.
16. Cutler, R. G. (1985). Peroxide-producing potential of tissues: correlation with the longevity of mammalian species. *Proc. Natl. Acad. Sci. USA* **82**: 4798–4802.

17. Cutler, R. G. (1991). Antioxidants and aging. *Am. L. Clin. Nutri.* **53**: S373–S379.
18. Cutler, R. G. (1984c). Antioxidants, aging and longevity. *In* "Free Radicals in Biology" (W. Pryor, Ed.), Vol. VI, pp. 371–428, Academic Press, New York.
19. Cutler, R. G. (1985a). Urate and ascorbate: their possible role as antioxidants in determining longevity of mammalian species. *Arch. Gerontol. Geriatrics* **3**: 321-348.
20. Jones, T. F. and Eaton, C. B. (1995). Exercise prescription. *Am. Family Physician* **52**(2): 543–550, 553–555.
21. Butler, R. N., Davis, R., Lewis, C. B., Nelson, M. E. and Strauss, E. (1998). Physical fitness: benefits of exercise for the older patient. 2. *Geriatrics* **53**(10): 46, 49–52, 61–62.
22. Coll, R., Izquierdo, J. and Salto, G. (1991). Benefit of physical exercise in medicine. *Ann. Med. Int.* **8**(2): 101.
23. No author. (1998). American College of Sports Medicine Position Stand. Exercise and physical activity for older adults. *Med. Sci. Sports Exerc.* **30**(6): 992–1008.
24. Schilke, J. M. (1991). Slowing the aging process with physical activity. *J. Gerontol. Nurs.* **17**(6): 4–8.
25. Samitz, G. (1998). Physical activity for decreasing cardiovascular mortality and total mortality. A public health perspective. *Wien. Klin. Wochenschr.* **18, 110**(17): 589–596.
26. Brown-Sequard, C. E. (1889). Des effects produits chez l'homme par des injections sous-cutanees d'um liquide retire des testicules frais de cobayes et de dheins. *Computes Rend. Soc. Biol.* **41**: 415–422.
27. Butler, R. N., Fossel, M., Pan, C. X., Rothman, D. and Rothman, S. M. (2000). Anti-aging medicine. What makes it different from geriatrics? *Geriatrics* **55**(6): 36, 39–43.
28. Butler, R. N. (2000). "Anti-aging" elixirs. *Geriatrics* **55**(6): 3–4.
29. Yu, B. P. (1999). Approaches to anti-aging intervention: the promises and the uncertainties. *Mech. Ageing Dev.* **111**(2/3): 73–87.
30. Rattan, S. I. (1998). Is gene therapy for aging possible? *Indian J. Exp. Biol.* **36**(3): 233–236.
31. Holloszy, J. O. (2000). The biology of aging. *Mayo. Clin. Proc.* **75(Suppl.)**: S3–S8, discussion S8–S9.

Chapter 82

Funding Trends and Possibilities in Oxidative Stress and Aging Research

Huber R. Warner and Bradley Wise

Huber R. Warner • Biology of Aging Program, National Institute on Aging, Bethesda, MD 20892
Tel: 301/496-4996, E-mail: warnerh@nia.nih.gov
Bradley Wise • Neuroscience and Neuropsychology of Aging Program, National Institute on Aging, Bethesda, MD 20892
Tel: 301/496-9350, E-mail: wiseb@nia.nih.gov

1. Introduction

Oxidative stress has long been thought to be a causal factor in a variety of acute and chronic conditions including stroke, cardiovascular disease, neuro-degeneration, cancer, cataracts, etc. This article documents in limited detail the financial commitments of the various Institutes and Centers at the National Institutes of Health for funding extramural research to understand how oxidative stress is causally related to the onset and development of these pathological conditions. Age is a common risk factor for the onset of all of these conditions, so the role of the National Institute on Aging is particularly relevant in the battle to alleviate and delay the symptoms of human disease.

2. Funding Sources

When I was asked to participate in this publication and the subsequent conference in April 2001, I wondered what I could contribute that would be interesting to these audiences. I decided that it might be useful to provide a summary of what the National Institutes of Health (NIH) spends to support extramural research on oxidative stress. Such a summary is shown in Table 1. These data come from a search of the NIH database of funded grants using only the key words "oxidative stress"; not included in this analysis are cooperative agreements ("U" mechanisms) which would include clinical studies and clinical trials. Table 1 includes only those Institutes and Centers (ICs) which spent at least

Table 1. NIH Funding for Extramural Research on Oxidative Stress in FY 1999

IC	R01, R03, R15, R21, R37	Program Projects	Centers	K Awards	Fellowships	SBIR	Total
NHLBI	$24 023	$7160	$3154	$2094	$218		$36 649
NIA	$8102	$7564	$191	$363			$16 220
NINDS	$8334	$1653	$2002	$625	$174	$337	$13 125
NCI	$7746	$4705				$595	$13 046
NIGMS	$4228	$159	$1623		$43	$777	$6830
NIEHS	$4844	$603	$330		$78		$5855
NIDDK	$3701	$937	$133	$239	$42	$366	$5418
NEI	$4328			$133			$4461
NIAAA	$2493		$774	$150	$19		$3436
NIAID	$2257				$83		$2340
NCCAM	$569		$1494				$2063
NICHD	$626	$1103					$1729
NIMH	$256	$1211			$126		$1593
Totals	$70 938	$25 095	$8207	$3604	$783	$2075	$110 072

$1 million for new and non-competing grants in fiscal year 1999 (FY 1999). ICs spending between $100 000 and $1 million are not shown in Table 1, and include the National Center for Research Resources ($849 K), the National Institute of Arthritis and Musculoskeletal and Skin Diseases ($769 K), the National Institute on Drug Abuse ($872 K), the National Institute of Dental and Craniofacial Research ($486 K) and the National Institute on Deafness and Other Communication Disorders ($408 K). All other ICs not shown in Table 1 spent $100 000 or less for research on oxidative stress in FY 1999.

It is perhaps not unexpected that the IC most committed to funding research on oxidative stress is the National Heart Lung and Blood Institute. An entire issue of Free Radical Biology and Medicine was recently devoted to the role of oxidation in atherosclerosis.[1] Stanley Hazen organized a forum on this subject and documented the recent explosive growth of interest in the oxidation of low density lipoprotein and its role in atherosclerosis, as the number of peer-reviewed publications published per year on this subject escalated from near zero in 1985 to more than 300 in 1999.[2] Even so, he acknowledged that "the critical question, is LDL oxidation important in atherosclerosis, is still unanswered", the major problem being the difficulty in distinguishing between correlative and causal associations.

The other institutes providing at least $10 million for funding extramural research for oxidative stress in FY 1999 were the National Institute on Aging, the National Institute on Neurological Disorders and Stroke, and the National Cancer Institute (Table 1). It was especially surprising to me that the NIA ranked as high as it did. This is even more apparent when the data are recalculated as a percentage of the total budget for each institute (Table 2). When the data are normalized in this way it is clear that research on oxidative stress is a relatively high priority for the NIA extramural funding. While 2.7% of the total budget may not seem like a large commitment it must be noted that the total budget figures in Table 2 include funds for both the intramural research programs and operating expenses. Further breakdown of the NIA funding reveals that all of this research on oxidative

Table 2. Percentage of Total Budget Spent for Extramural Research on Oxidative Stress

IC	Funding for Oxidative Stress	Total Budget	%
NIA	$16 220	$599 741	2.7
NHLBI	$36 649	$1 781 389	2.1
NIEHS	$5 855	$388 228	1.51
NINDS	$13 125	$898 521	1.46
NEI	$4 461	$396 634	1.12
NIGMS	$6 830	$1 196 798	0.57
NIDDK	$5 418	$996 189	0.54
NCI	$13 046	$2 900 435	0.45

stress is funded by either the Biology of Aging Program or the Neuroscience and Neuropsychology of Aging Program. The majority (68%) is funded by the Biology of Aging Program, and the remaining 32% is focused on neuroscience research.

The above analysis reflects two assumptions about research on aging. The first is what appears by now to be general agreement that oxidative stress plays a significant role in aging and age-related pathology as predicted almost fifty years ago by Denham Harman.[3] The evidence supporting this point of view remains equivocal, as most of it is correlative.[4] Much effort has been expended to show that biomarkers of oxidative damage increase with age, or that levels of antioxidants decline with age, but definitive results have been hard to obtain. Even the impact of supplementation with vitamin E on risk of age-related diseases appears to remain an open question. A robust increase in longevity of a cohort of individuals or reduction in the rate of aging (whatever that means) in response to intervention with a proven antioxidant would strengthen the case, but such results are few. So far, a known spin-trap compound has been shown to extend the life span of a strain of senescence-accelerated mice,[5] an SOD/catalase mimetic has been shown to extend the life span of *Caenorhabditis elegans*,[6] and over-expression of human Cu, Zn superoxide dismutase in fruit fly motorneurons extends their life span.[7] Caloric restriction, which extends both average and maximum life span in every system in which it has been adequately tested, also reduces oxidative stress, but it is not clear whether there is a causal relationship between this reduction and life span extension. Such results are encouraging, but the picture remains far from complete.

The second assumption, which is becoming more generally accepted, is that the loss of neurons in various regions of the brain which accompanies neuro-degenerative diseases appears to be at least partially the consequence of oxidative stress-induced apoptosis.[8] Both direct and indirect evidence has been obtained implicating oxidative stress, mitochondrial dysfunction, and apoptosis in a variety of neurodegenerative diseases, including Alzheimer's disease, Parkinson's disease, amyotrophic lateral sclerosis (ALS) and Friedriech's ataxia. In an NIA-supported study, vitamin E treatment delayed the progression of Alzheimer's disease in moderately impaired patients by about six months.[9] In animal studies, withdrawal of neurotrophic support to neurons is associated with increases in cellular oxidative stress, leading eventually to neuronal death,[10] while consuming a diet rich in antioxidants prevented and reversed the deleterious effects of aging on neuronal and cognitive behavioral function in aged rats.[11] Although much evidence points to a role for oxidative stress in age-related neurodegeneration, the story remains unfinished, as indicated earlier.

Recognizing this, NIA staff have issued several program announcements (PA) and requests for applications (RFA) in the past ten years, soliciting applications on oxidative stress and its implications. These solicitations include:

PA-93-017   Oxidative damage, antioxidant defense and aging (NIA only).

PA-96-058   Mechanisms of cell death and injury in neurodegenerative disease (with NINDS, NIEHS, NIMH).
http://grants.nih.gov/grants/guide/pa-files/PA-96-058.html

PA-99-034   Drug discovery for the treatment of Alzheimer's disease (with NIMH).
http://grants.nih.gov/grants/guide/pa-files/PAS-99-034.html

PA-99-054   Xenobiotics and cell death/injury in neurodegenerative disease (with NIEHS, NINDS, NIMH).
http://grants.nih.gov/grants/guide/pa-files/PAS-99-054.html

PA-00-081   Aging, oxidative stress and cell death (NIA only).
http://grants.nih.gov/grants/guide/pa-files/PA-00-081.html

RFA: AG-01-002   Molecular and neural mechanisms underlying the effects of caloric restriction on health and longevity (NIA only).
http://grants.nih.gov/grants/guide/rfa-files/RFA-AG-01-002.html

These announcements highlight a variety of general topics of interest to the NIA, including: measuring oxidative stress/validating biomarkers; age-related changes in relevant enzymes; role of mitochondrial function and dysfunction in creating oxidative stress and/or inducing apoptosis; the development of relevant animal model systems, especially transgenic mice; development and characterization of dietary, pharmacological or genetic interventions; identification and characterization of single nucleotide polymorphisms in relevant genes.

Some of the major general questions about oxidative stress and aging remaining to be answered are:

❖ Is oxidative damage to DNA, proteins and lipids causally related to aging, and if so, how? How does this damage trigger apoptosis?

❖ Is normal mitochondrial metabolism the major source of reactive oxygen species, leading to apoptosis of cells in post-mitotic tissues such as brain and muscle? What other sources are there, and how are all of these affected by aging?

❖ What factors determine the cell and tissue specificity of oxidative stress-induced pathology, e.g. why is neurons loss in ALS specific for motorneurons?

❖ Development of better animal models for studying oxidative stress-induced pathology, including genetically altered mice.

❖ Development and testing of intervention strategies, including gene therapy, and possibly cell replacement therapy.

3. Conclusion

Although traditionally there has been considerable interest in trying to "prove" the free radical theory of aging, this is an irrelevant goal for the NIA. NIA staff recognize that there are many factors which contribute to what we call aging, and our priorities include funding research to identify: (1) how oxidative damage contributes to age-related changes, (2) which of these age-related changes are causally related to aging pathology, and (3) whether interventions can be developed to prevent, reverse, or at least retard age-related oxidative stress-induced pathology. Thus, NIA grant programs in the Biology of Aging and the Neuroscience and Neuropsychology of Aging Programs include these objectives among the many other priorities of the NIA.

Table 1. The figures in Table 1 indicate total dollars (direct and indirect) spent in Fiscal Year 1999 by various institutes of the National Institutes of Health, sorted by grant mechanisms. R01, R03, R15, R21 and R37 are research project grants, with most funding being for R01s, the standard grant mechanism. Multicomponent awards include Program Projects (P01) and Centers. K Awards include a variety of career awards, with most of the funds being used for salaries. SBIR includes both Small Business Innovation Research and Small Business Technology Transfer Awards. The figures were obtained from the NIH database on funded grants, using only the key words "oxidative stress". All figures are in thousands of dollars.

Abbreviations are: NHLBI — National Heart Lung and Blood Institute; NIA — National Institute on Aging; NINDS — National Institute of Neurological Disorders and Stroke; NIGMS — National Institute of General Medical Sciences; NIEHS — National Institute of Environmental Health Sciences; NCI — National Cancer Institute; NIDDK — National Institute of Diabetes and Digestive and Kidney Diseases; NEI — National Eye Institute; NIAAA — National Institute on Alcohol Abuse and Alcoholism; NIAID — National Institute of Allergy and Infectious Disease; NCCAM — National Center for Complementary Medicine; NICHD — National Institute of Child Health and Human Development, and NIMH — National Institute of Mental Health.

Table 2. See Legend to Table 1 for explanation of abbreviations. All figures are in thousands of dollars, and in total dollars (direct and indirect) spent in Fiscal Year 1999. The figures were obtained from the NIH database on funded grants, using only the key words "oxidative stress". Total budgets were obtained at the website *http://www.od.nih.gov/ofm/budget/00conference.stm*

References

1. Forum: role of oxidation in atherosclerosis. (2000). *Free Radic. Biol. Med.* **28**: 1681–1826.
2. Hazen, S. (2000). Oxidation and aging. *Free Radic. Biol. Med.* **28**: 1683–1684.
3. Harman, D. (1956). Aging: a theory based on free radical and radiation chemistry. *J. Gerontol.* **11**: 298–300.

4. Warner, H. R. and Starke-Reed, P. (1997). Oxidative stress and aging. *In* "Oxygen, Gene Expression and Cellular Function" (L. B. Clerch, and D. J. Massaro, Eds.), pp. 139–149, Marcel Dekker, Inc., New York.

5. Edamatsu, R., Mori, A. and Packer, L. (1995). The spin-trap N-tert-alpha-phenyl butylnitrone prolongs the life span of the senescence accelerated mouse. *Biochem. Biophys. Res. Commun.* **211**: 847–849.

6. Melov, S., Ravenscroft, J., Malik, S., Gill, M. S., Walker, D. W., Clayton, P. E., Wallace, D. C., Malfroy, B., Doctrow, S. R. and Lithgow, G. J. (2000). Extension of life-span with superoxide dismutase/catase mimetics. *Science* **289**: 1567–1568.

7. Parkes, T. L., Elia, A. J., Dickson, D., Hilliker, A. J., Phillips, J. P. and Boulianne, G. L. (1998). Extension of *Drosophila* lifespan by over-expression of human SOD-1 in motorneurons. *Nature Genet.* **19**: 171–174.

8. Warner, H. R., Hodes, R. J. and Pocinki, K. (1997). What does cell death have to do with aging? *J. Am. Geriatr. Soc.* **45**: 1140–1146.

9. Sano, M., Ernesto, C., Thomas, R. G., Klauber, M. R., Schafer, K., Grundman, M., Woodbury, P., Growdon, J., Cotman, C. W., Pfeiffer, E., Schneider, L. S. and Thal, L. J. (1997). Controlled trial of selegiline, alpha-tocopherol, or both as treatment for Alzheimer's disease. *New England J. Med.* **336**: 1216–1222.

10. Pucha, G. V., Deshmukh, M. and Johnson, E. M. (1999). BAX translocation is a critical event in neuronal apoptosis: regulation by neuroprotectants, BCL-2, and caspases. *J. Neurosci.* **19**: 7476–7485.

11. Joseph, J. J., Shukitt-Hale, B., Denisova, N. A., Bielinski, D., Martin, A., McEwen, J. J. and Bickford, P. C. (1999). Reversals of age-related declines in neuronal signal transduction, cognitive, and motor behavioral deficits with blueberry, spinach, or strawberry dietary supplementation. *J. Neurosci.* **19**: 8114–8121.

SECTION 14

Human Life Extension

Chapter 83

The Ethics of Longevity: Should We Extend the Healthy Maximum Lifespan?

Michael Fossel*,†

Michael Fossel • Box 630, Ada, Michigan 49301
Voice/Fax: 616-676-4099, E-mail: shigoto@earthlink.net
*Clinical Professor of Medicine, Michigan State University
†Editor-in-Chief, Journal of Anti-Aging Medicine

1. Introduction

As we will soon extend the healthy human lifespan well into the several century range,[1-3] it is appropriate to consider the ethical implications of doing so. If we can extend the healthy human lifespan, is it ethical? Too often overlooked through ignorance, the obverse question is equally important, viz. if we can extend the healthy lifespan and chose not to do so, is *that* ethical? Both questions deserve exploration and play a role in assessing the ethics of our responsibility and of our choices.

2. Defining the Issue

Is an extension of the human lifespan — and the ethical questions which it engenders — novel? Historical data shows that the human lifespan has increased substantially over the past two centuries,[4] suggesting this is not a fresh ethical issue, yet we treat it here as though it is without any historic parallel.

The inconsistency — and the reason for it being an unprecedented inflection in human culture — accrues from the substantial difference between the concepts of the mean and the maximum lifespan. This difference is conceptual, but has profound and parallel differences in our clinical ability to achieve either increase, as well as for their drastically different cultural implications. The consequent ethical issues are both ethically unexplored and crucial to defining who we are as ethical human beings.

The maximum lifespan, despite myth and exaggerated claims to the contrary, has never been altered. This is the lifespan that you might expect if you had the best available genes, the least traumatic environment, the least exposure to infectious agents, an optimal food supply, clean water, fresh air, and unbelievably good luck. In any generation, well less that one in several hundred thousand human beings have ever come even close to attaining this — almost apocryphal — lifespan. Living a maximum lifespan is extraordinarily rare, and we have never been capable of altering or extending it. We have been totally unable to alter the fundamental processes of aging that underlie and mandate this limit. Estimated as 120 years, the maximum human lifespan is a biologically fixed limitation of the human genome. It is dictated by the genes peculiar to each species and is not currently modifiable by dietary supplements, exercises, hormones, or any known drug.

In distinct contrast to the maximum lifespan, the mean (or expected) lifespan is eminently modifiable. Interventions include adequate and balanced nutritional intake, avoidance of infectious agents (e.g. clean water and immunizations), avoidance of trauma (e.g. use of seat belts), and similar good (and timeless) advice. Largely as a result of such interventions (particularly those aimed at obstetrics and childhood diseases), the mean lifespan in most developed nations has increased three-fold — from 25 to 75 years — over the past two centuries. None of these approaches affect the maximum lifespan.

A subtle exception to this otherwise inviolable rule is caloric restriction. An optimal nutritional intake is quite different from having all the food (balanced diet or not) that one desires. In every vertebrate species yet studied dietary restriction increases the "maximum" lifespan by an average of 40%.[5] Such animals are fed approximately 70% of an *ad libitum* diet, all nutritional needs (vitamins, co-factors, minerals, amino acids, essential lipids, etc.) are met completely, and only total calories are limited severely. In this regard, maximum lifespan (at least by one definition) is elastic. Are there any additional such interventions which affect the maximum lifespan? Currently, the answer is no.

Selective breeding can extend the maximum lifespan over several generations, but is not applicable to human clinical intervention,[6] though it proves that our genes determine fundamental aging and the maximum lifespan. Direct genetic intervention supports the same point. In a number of species, alteration of one to three genes results in extensive elongation (6–8 fold) of the maximum lifespan without any apparent costs to fecundity, health, or competitiveness.[7] Current work on cell senescence suggests that clinical interventions affecting the underlying pathology of age-related diseases may be quite effective clinically.[8, 9] In short, we may soon be able to prevent age-related diseases and extend our lifetime health.[10] Such approaches may allow us to alter not merely the mean lifespan, which has recurrent historic precedent, but the maximum lifespan, which does not.[11]

3. The Ethics of Caring, the Ethics of Choice

Previous increases in the mean lifespan were gradual and did not alter the *pace* of the human lifespan. You might die young, but you aged normally. As we increased the mean lifespan, we "added back" a substantial number of children and young adults into our population by preventing early deaths. We altered neither the rate of aging nor the maximum attainable lifespan of the survivors. You might not die young, but you still aged at exactly the same rate as your ancestors had throughout history.

Altering the maximum human lifespan, on the other hand, will not only prevent personal suffering, pain, and disease, but will have profound, pervasive, and largely unpredictable social consequences.[12] The increase will be sudden, occurring over years rather than centuries. It will affect nearly everyone, not merely those who died young. It will substantially slow aging and significantly extend the lifespan in a novel manner. One hundred year olds in 1700, though uncommon, looked much as 100 year-olds do now. As we alter the rate of aging (and the maximum lifespan), however a 100-year-old might be indistinguishable from a 30-year-old. This is an unprecedented revolution in human health. Moreover, treatments which extend the human lifespan beyond the current maximum can only do so by preventing age-related diseases and by improving age-related health in a fundamental way.[13] Biology precludes extending the lifespan unless we achieve better (and longer) health. If we have lifespans of 200, it will only be because the 100 year old has the physiological health of a young adult. To extent life requires that we extend health. Focusing solely on lifespan *per se* is specious and disingenuous, for the actual effect will be upon human health, with the concomitant side effect being an extended lifespan. We can improve health without extending life, but we cannot extend life (past maximum lifespan) without improving health. The two are linked biologically and inseparable ethically. To forego an extension in lifespan is to be willing to condone disease we could otherwise cure.[14]

The social consequences of such a revolution are unforeseeable. We can envision effects of an extended lifespan on family structure, child-raising, gender roles, age roles, the economy, our culture, our aspirations, and all that makes us human.[15] The novelty and potential impact of extending lifespan therefore begs ethical evaluation. The question is not simply one of the ethical benefits and costs of extending the human lifespan, but equally, the benefits and costs of not doing so. Legally restricting such technology is not the same as never having invented the technology at all. To the contrary, whether we extend the lifespan or not, we still remain faced with our active ethical responsibility.

4. Scylla and Charybdis

Currently, lacking an ability to extend the healthy lifespan, we bear no ethical responsibility for the deaths and quotidian suffering. Ethical responsibility hinges on the ability to intervene. Once we can intervene, however, we acquire immediate and irrevocable ethical liability. If we can improve health and thereby increase lifespan (with any foreseeable social costs), we are responsible for the social impact; if we choose not to treat disease or increase the lifespan, we are then equally responsible for the suffering of those we have refused to help (and all these foreseeable social costs). If we avoid social disruption by restricting technology, we are then responsible for the suffering of our fellow human beings who are not treated. An initially simple ethical question takes on a complex mantle, demanding more careful soul searching than most might be willing to engage in. The ethical question is not whether a longer *lifespan* is good or bad, but which cultural *outcome* is better. We might avoid the foggy Scylla of overpopulation, but only by crashing into the rocky Charybdis of a culture that refuses to treat suffering. Whichever we choose, we bear the ethical responsibility for the social consequences.

If we choose not to permit lifespan extension, the option may appear to be misleadingly passive ("let us just leave things as they are now"), but our choice is active. Closing our eyes and denying culpability implicates us in a criminal stupidity, as well as the unavoidable ethical effects of our decision. We shoulder the responsibility willy-nilly once the technology (and therefore the choice) becomes possible. We cannot legitimately claim that we are "not making a decision" or "letting things be the way they always were". Things will never be as they were; there is no avoiding the ethical responsibility by pretending that we live in the past, powerless to help those around us.

And if we enact legal restrictions on treatments that can profoundly alleviate human suffering, they will likely be circumvented to save lives, and in that very act will cost us all the more dearly. Laws that encourage lawlessness serve only to further destroy the social fabric. If we restrict a personal medical treatment, even for the best of social reasons, we play a dangerous game, one that cannot be whitewashed as reflecting the social virtues of a simpler past. Rather it not only epitomizes an apparent lack of care, but encourages further social carelessness in others. Equally, however, we cannot close our eyes and simply accept that lifespan extension "will happen anyway", pretending that our ethical decision, if powerless, is likewise unimportant. In so doing, we share ethically in the outcome, countenanced by our silence and our passive acceptance.

While the likely outcome will be to lessen individual human suffering from cancer, arthritis, heart disease, and Alzheimer's dementia, there is also the unknown social outcome. What will happen to violence, freedom, hopes, fear, dreams, salvation, and human culture? By treating human disease and thereby extending the healthy human lifespan, do we advance civilization or destroy it? While we cannot predict accurately, we are yet responsible for the choice.

5. Suffering

Consider a physician with a suffering patient and an effective medicine. The physician must ethically balance both side effects and the patient's current suffering. Side effects might appear trivial in significant disease (e.g. hair loss in a patient who will otherwise die of cancer), but unacceptable in treating a minor problem (e.g. hair loss in mild acne). An ethical physician makes a careful judgment, balancing both the actual disease and potential side effects. The quality of this judgment, not its outcome, determines the quality of the ethics. If the judgment is prudent (assessing and balancing risks), but the patient dies through completely unforeseeable side effects, the physician is still competent and ethical. But if the physician ignored quite foreseeable risks and blithely treated the patient anyway, then even if the patient (by good luck) survives, the physician is still incompetent (both clinically and ethically). Our ethics pivot on the quality (the wisdom, intelligence, and personal caring) of our decisions and not on the desirability of the actual outcome. Ethical behavior requires that we honestly and diligently assess the potential outcomes of our actions. Failure is not culpable; failure to try is ethical betrayal.

In treating aging diseases, we must take into account both the patient's suffering and the risks of side effects. If we extend the human lifespan, we must take account of both the suffering we prevent and the social outcomes of doing so. If we do not extend the human lifespan, we must take into account the suffering we are unwilling to prevent and the social outcomes of doing so.

What are the outcomes that hang in the balance? At a first glance, the medical issues appear relatively simple and the social issues muddy. To an extent, we are already aware of the current medical outcomes: the diseases we currently live with. Unfortunately, those most familiar with such outcomes (the aged) are not always those most involved in discussions such as these (often the young and middle-aged). The latter have less personal (and perhaps an intellectually-pretentious) knowledge of the concrete tragedies of aging. Few middle-aged ethicists understand constant pain, the loss of memory and function in dementia, urinary incontinence, shortness of breath, or dependency. Several years ago, a Danish bioethicist argued against *in vitro* fertilization until she suffered an infection and required it in order to bear a child, then became an advocate (and a regrettably silent one) of exactly the same technology. How often are our ethical decisions made in ignorance?

The beneficial (and apparently obvious) medical results may also be unexpected. The vague notion that we will "prevent age-related disease and suffering" is an emotional rallying point, but does it approximate reality? The technology might merely postpone such diseases and suffering. This would still be a benefit (would you rather have a heart attack tomorrow or in twenty years?). But what if you defer Alzheimer's dementia for a century, but equally extend the period of mindless dependence for a century? Fortunately, the biology of the matter suggests that this would not occur,[16] though the question deserves an honest appraisal, and bears strongly on the ethics involved. We cannot extend lifespan by extending only the period of senescence and suffering. As discussed above, to extend the lifespan, we must necessarily extend and improve human health.

6. Social Implications

The social outcomes are far more puzzling and error-prone. The errors come from two sources: the complexity (indeed, impossibility) of social projection and the emotional tendency to ignore and misrepresent the other side of any difficult ethical argument. Let us sketch out the possible social costs and benefits of extending the human lifespan.

The all-too-obvious risks are overpopulation,[17] environmental stress, and a consequent loss of quality-of-life. Although the population increase is difficult to project precisely (particularly within specific ethnic groups as opposed to overall increase, or as exact rate-of-increase as opposed to a final asymptote), a general rule-of-thumb serves remarkably well in approximating the total effect: the population increases proportionately to the mean lifespan increase. If we double the mean lifespan, we double the final population.[18] This assumes no change in birth rate (a shaky assumption) and no linked increases in causes of death (e.g. more frequent viral plagues or increased violence within a denser population). The environmental stresses might be proportional to population, although such stresses are more likely related to poverty and lack of education, each of which might also be strongly affected (in either direction) by a longer lifespan and higher population. The quality of life is generally assumed to be linked to both population density and environmental stress, although an optimal population density (by any of several possible measures) is remarkably difficult to agree upon.[19] Quality of life is again perhaps more strongly correlated with a strong economy and education than with most other quantifiable factors. The truth is that while population is very likely to increase, we not only have very little ability to predict where it might stabilize, we have only emotional arguments or complex and hazy intellectual arguments to offer in predicting the outcome for the quality of life or the environment. This is not to denigrate these concerns, but rather to highlight their uncertainty. It is not that the outcome is rosy, so much as unpredictable. While it is easy to suggest that a longer lifespan may have significant social and environmental costs, honesty and logic make it impossible to construct, *a priori*, a tightly constructed, unimpeachable proof of the position. We might reasonably anticipate social problems, but not firmly predict them. We can reasonably agree that extending the human lifespan will entail significant social changes and (since change itself is difficult) social stress. Interesting times are inherently strenuous to those living through them.

Equally, however, the extension of the healthy lifespan may have social benefits. Consider the unprecedented effects of having older individuals who are still vigorous, healthy, and productive. Such individuals not only have much to offer the world economically, but culturally. Daily we bury those with educational degrees, with a lifetime of experience and, perhaps, with wisdom. Surely, extending the human lifespan might make us more civilized.[20] We would have longer perspectives, more experience, and more productive energy to apply to the problems of the world and our fellow human beings. And if living for two hundred years precludes retirement, why not continue doing a job you love or chose a new career, one made richer by your breadth of experience?

Such arguments — concerning both the costs and the benefits of extending the lifespan — are rarefied, insubstantial, even elusive. They lack an immediacy and tangibility that allows for careful judgment. Nonetheless, their lack of data and numerical predictability, their difficult subjectivity, and their dependence upon guesswork do not preclude their being right. If we knew for certain that we might finally achieve (or destroy) true human civilization, we could make better ethical decisions. The reality is that we can only make informed guesses. The reader is referred to reviews focusing on the economics[21] and overall social, cultural, and ethical implications of extending lifespan.[22]

7. Unintended Consequences

Many discussions of lifespan extension stop here and exhibit a curious blind spot. They assume (without being explicit) that the social outcome of not permitting lifespan extension is equivalent to not developing lifespan extension. In other words, we could simply pass a law, make lifespan extension illegal and we will remain as we are now. This is an intellectually criminal error. Consider an analogous medical example.

Assume that we have a five-year old patient dying of pneumonia in 1900 when we have no antibiotics and (essentially) no therapy for the disease. Compare this to the same five-year old child in 2000, when we have many available antibiotics and other available medical care. Consider the starkly contrasting ethical implications of being unable to treat the child in 1900 (when we have no choice) with intentionally choosing not to treat the child in 2000 (when we could save her life). These two ethical situations stand in austere contrast to one another. Allowing a child to die through inaction carries an ethical revulsion; submitting to the inevitable does not. There is a vast ethical divide between a world populated by people who want to help (but are incapable of doing so), and a world populated by people (who are quite capable of helping but) completely unwilling to aid others who are suffering. Intentional restriction of therapy is not the same as mere ignorance of therapy. Ethics is fundamentally based on helping others and requires clear and overwhelming risks to justify the intentional infliction of suffering through our inaction.

A culture in which physicians were eminently capable of treating you and your family, but society actively prevented such care is an unethical culture. Since age-related diseases and lifespan are both the direct outcome of the fundamental processes of aging, this would mean no novel (and inextricably related) therapies would be permitted for the basic diseases of aging: arthritis, cancer, heart disease, strokes, or Alzheimer's dementia. The distaste aroused by a physician who lets a child die must be magnified and considered before we chose a culture in which all of us stand back and allow, even encourage, preventable human suffering. Comparisons to the Nazi horrors documented in the Nuremberg Trials are neither irrelevant nor easy to bear in this context. We risk losing our humanity if we make naively intellectual choices based on one-sided social calculations of the risks inherent in extending lifespan. The social cost of having a society which condones intentional suffering is far from negligible, but nor does it necessarily trump all other considerations. An assessment of the ethics of extending the healthy human lifespan must consider both the costs of refusing to treat suffering and equally the costs and benefits of longer lives.

The ethics of longevity are potent, relevant, and pressing. Sometime within the next decade we will demonstrate that we can alter the aging process and, at least theoretically, extend the healthy human lifespan. Consider what we will have done. We will not merely have extended the lifespan, but will have increased and extended our individual health over that longer lifespan. Biology requires that we cannot extend the lifespan without increasing health and preventing disease. The future will not bring a longer life, but a longer life without the fear and suffering of cancer, Alzheimer's dementia, heart disease, osteoarthritis, immune senescence, and a host of other problems. Aging is not a "pure" clinical phenomenon, but expresses itself in such diseases.[23, 24]

8. Summary

The ethics of altering aging and extending the healthy human lifespan are frequently caricatured. The simplistic dichotomy is that of extended lifespan versus increased world population and social unrest. A more honest appraisal, demonstrating respect for our ethics, sees that the question is really one of the costs and benefits incurred if we treat the diseases of aging. On the one hand, the outcome will be an extended healthy lifespan, with unpredictable social costs or inestimable cultural benefits. On the other hand, if we restrict such treatment, we demonstrate a lack of personal caring for the tragedies of others and we encourage potential societal disruption and lawlessness, both cutting deeply into the forces that drive our ethics in the first place. Do we take uncertain but likely social risks, or do we knowingly ignore the suffering of our fellow man and thereby incur other, perhaps even more daunting, social costs? The difficulty of such a decision and the wisdom, honesty, and attention we put into facing it, are the measure of an ethical decision. Although the social impact of treating age-related disease may be unpredictable and profound (whether detriments or benefits), ultimately it is the human suffering that we prevent, postpone, or cure that is the more tangible measure of human ethics. In the most fundamental analysis, it is our willingness to help others that is the most profound and only true measure of both our ethics and our humanity.

References

1. Fossel, M. (1996). *Reversing Human Aging*, William Morrow and Company, New York.
2. Bova, B. (1998). *Immortality: How Science is Extending Your Life Span — and Changing the World*, Avon Books, New York.
3. Banks, D. A. and Fossel, M. (1997). Telomeres, cancer, and aging; altering the human lifespan. *JAMA* **278**(16): 1345–1348.

4. Smith, W. E. (1993). *Human Longevity*, pp. 11–19, Oxford University Press, New York.
5. Lane, M. A., Black, A., Ingram, D. K. and Roth, G. S. (1998). Calorie restriction in nonhuman primates: implications for age-related disease risk. *J. Anti-Aging Med.* **1**: 315–326.
6. Fossel. (1996). Op Cit, pp. 121–122.
7. Dorman, J. B., Albinder, B., Shroyer, T. and Kenyon, C. (1995). The age-1 and daf-2 genes function in a common pathway to control the lifespan of *Caenorhabditis elegans*. *Genetics* **141**: 1399–1406.
8. Fossel, M. (2000). Cell senescence and human aging: a review of the theory. *In Vivo* **14**: 29–34.
9. Fossel, M. (2000). The role of cell senescence in human aging. *In* "Endocrinology of Aging" (B. Bercu, and R. Walker, Eds.), Springer-Verlag, New York.
10. Fossel, M. (due for publication 2002). *Cell Senescence in Human Aging and Disease*, Oxford University Press, New York.
11. Fossel, M. (1998). Telomerase and the aging cell; implications for human health. *JAMA* **279**: 1732–1735.
12. Fossel. (1996). Op Cit.
13. Fossel, M. (2000). The role of telomerase in age-related degenerative disease and cancer. *In* "Interorganellar Signaling in Age-Related Disease" (M. P. Mattson, Ed.), JAI Press, **in press**.
14. Fossel. (2002). Op Cit.
15. Fossel. (1996). Op Cit, pp. 218–255.
16. Fossel. (2002). Op Cit.
17. Ehrlich, P. R. and Ehrlich, A. H. (1990). *The Population Explosion*, Simon and Schuster, New York.
18. Fossel. (1996). Op Cit, pp. 223–227.
19. Cohen, J. E. (1995). *How Many People Can the Earth Support?*, W. W. Norton and Company, New York.
20. Fossel. (1996). Op Cit, pp. 237–245.
21. Banks and Fossel. (1997). Op Cit.
22. Fossel. (1996). Op Cit, pp. 218–255.
23. Fossel, M. (2000). The role of cell senescence in human aging. *J. Anti-Aging Med.* **3**: 91–98.
24. Fossel. (2002). Op Cit.

Chapter 84

Advisability of Human Life Span Extension: Economic Considerations

Henry Rodriguez* and Dwayne A. Banks

Henry Rodriguez • Biotechnology Division, National Institute of Standards and Technology, 100 Bureau Drive, Mail Stop 8311, Gaithersburg, Maryland 20899
*Corresponding Author.
Tel: (301) 975-2578, E-mail: henry.rodriguez@nist.gov, Web: www.nist.gov
Dwayne A. Banks • Abt Associates Inc., Cambridge Massachusetts 02138

1. Introduction

Enormous gains in biomedical technology have occurred since the 1960s. This has led to great advances in treatment methodology, as well as enhanced diagnostic ability for physicians. Surgeons now routinely perform procedures thought impossible 40 years ago, such as organ transplantation. However, technological advancement has contributed to the rising costs of health care services in the following ways: (a) engendered increased labor specialization within the medical profession and, as a result, highly specialized professionals are able to command high rates for their services; (b) made possible by the treatment of maladies that only 20 years ago would have gone untreated; (c) medical technology (drugs, equipment and devices) is very expensive to purchase and maintain; and (d) the cost of medical research and development is passed along to consumers in the form of higher rates for services rendered. The high cost of treating the aged and the terminally ill is a result of this technological imperative; eventually, such life-sustaining technology forces society to make extremely difficult ethical decisions, in other words, the benefits to society of extending the human life span must be measured against the social cost incurred.

Currently we are incapable of altering the aging process in any meaningful way; however, it is important for society to understand that once such technology is adopted it is likely to be cost inflationary, distributed based upon ability to pay, lead to further specialization within the medical profession, have economy-wide effects, and force us to make tough ethical choices. Any assertions to the contrary are misleading and not supported by fact. It is, however, equally misleading to ignore the ongoing shift in our knowledge of cell and molecular biology — as we attempt to predict the future of health care and social policy in developed countries. The assumptions on which we base our predictions must be clearly understood.[1] Presently, there exists a reasonable degree of uncertainty in assuming that aging and age-related diseases will remain immutable in the coming decades. This will have profound implications for fiscal and social policy. It is evident, that any significant increases in the human life span will be accompanied by concomitant shifts in social spending on the aged. The magnitude or fiscal outcome of such increases cannot be accurately estimated; neither can their global distribution on the well being of populations. How social institutions accommodate impending changes in life-span extension and the age structure of the population will significantly affect the quality of life for everyone in the 21st century.

2. Aging, Technology, and Health Care Infrastructure

The provision of health care services is costly in all industrialized nations. Nations differ by their political, social, demographic, regulatory and legal structures; however, the underlying factors that govern the distribution and cost of health care services are the same: scarce national resources, changing demographic profiles, and changes in technology and consumers' expectations. These factors, coupled with an ever-aging population, have important implications for social planning in both developed and developing countries. While there exists some variation internationally in the health status of elderly cohorts,[2, 3] the social costs of increasing the maximum human life span will fundamentally depend upon the level of functional impairments and chronic disabilities among the aged.[4, 5] In fact, recent research has indicated that increases in the mean life span have been accompanied by a concomitant decline in the prevalence of age-related morbidity and disability.[6, 7] However, the magnitude of such a decline and its overall impact on health care cost relative to previous elderly cohorts will vary internationally by the structure of a nation's health care system, and the rules and regulations that govern the adoption and diffusion of medical technology. For example, in developed nations, the rate of technological diffusion and adoption has lead to a reduction in the risk of treating various age-related maladies, as well as overall reductions in recovery time.[8] Typically, the diffusion of new technologies in the economy is considered to be welfare improving due to reduced production costs or increased product differentiation (i.e. providing consumers with more choice of products or services). However, in the case of medical technologies, there are market failures that provide incentive for excess and inappropriate supply of services. Hence, the development of an effective regulatory strategy prior to the adoption of a new medical technology is a necessary condition for promoting the optimal application of a new technology within a treatment protocol.

Moreover, innovations in medical technology, whether therapeutic, diagnostic, or organizational for age-related conditions, need not coincide with the health care needs or promote the well being of those in developing countries. The immediate and future needs of the vast majority of these countries is the control of infectious diseases, dealing with the fatal consequences of drought, eradicating famine, preventing civil wars, economic stabilization, debt reduction, and the development of effective health planning infrastructures.[9] Therefore, when considering the international fiscal effects of increasing the maximum human life span and the impact of such innovations on the level of functional impairment among the aged, it is important to acknowledge that these advances are likely to benefit developed nations exclusively.

3. America's Elderly Population

America's elderly population is currently growing at a moderate pace. But over the next 30 years, the growth is expected to become rapid. So rapid, in fact, that by the middle of the next century, it might be completely inaccurate to think of us as a nation of the young since there could be more persons who are elderly (age 65 or over) than young (age 14 or younger).

The elderly population has grown rapidly throughout the country's history. From 1900 to 1960, the elderly increased 10-fold, while the population under age 65 was only 2.2 times larger.[10] Between 1960 and 1990, the elderly grew by 88%, compared to 34% for persons of age 65 and under (Fig. 1).

It is projected that during the period 1990–2010, the elderly growth rate will be lower than during any 20-year period since 1910, because of the low fertility of the 1930s[10] (Fig. 2). After this slow-growth period, an elderly population

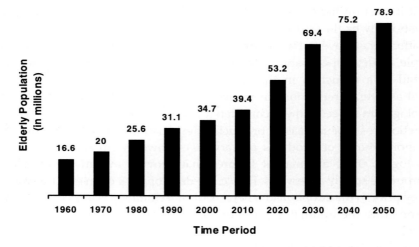

Fig. 1. United States elderly population (65 plus) from 1960 to 2050.

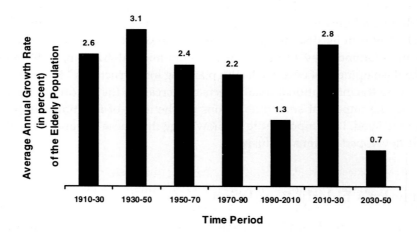

Fig. 2. United States average annual growth rate (in percent) of the elderly population: 1910–1930 to 2030–2050.

explosion between 2010 and 2030 is inevitable as the baby boom generation reaches age 65. The elderly, who comprised only 1 in every 25 Americans (3.1 million) in 1900, made up 1 in 8 Americans (33.2 million) in 1994. It is projected that by the year 2050, the elderly population in the United States will more than double its 1990 number of 31.1 million to 78.9 million.[10] By that year, as many as 1 in every 5 Americans could be elderly, with the majority of the growth occurring between 2010 and 2030, when the baby boom generation enters their elderly years. During this time period, the number of elderly will grow by an average of 2.8% annually.

Factors contributing to the elderly population growth increase include large decreases in early deaths caused by infectious diseases (resulting in a dramatic increase in the average life expectancy of humans due to improvements in sanitation, health care, housing, nutrition, the development of vaccines), and the discovery of antibiotics in the past century. These factors have also contributed to a sharp rise in the medium age of the United States, 20 years old in 1860 to 34 years old in 1994.[10]

The "oldest old", those aged 85 and older, are the most rapidly growing age group in the Unites States. The oldest old numbered 3.5 million in 1994, making them 10% of the elderly population and just over 1% of the total United States population.[11]

By 2020, the oldest old are projected to double to 6.5 million. The oldest old will again double to 13.6 million by 2040 as the survivors of the baby boom generation reach the oldest ages, and it is projected they could number as many as 18.2 million in 2050, making them 24% of the elderly American population and 5% of all Americans.

In 1990, of the 31 million people who were aged 65 and over, 37,306 were classified as being centenarians.[12] Between 1960 and 1994, those 85 and oldest rose 274% in contrast to the elderly population that rose 100% or the entire United States population that grew only 45 percent. Since the oldest old often have severe chronic health problems that demand special attention, the rapid growth of this population group will have significant social and economic implications for individuals, families, and governments.

4. Declining Earnings Power/Increase in Utilization of Health Care

In addition to the loss of family and friends, older persons face two potentially serious economic problems: (a) declining earnings power and (b) increased utilization of health care services. Declining earnings power is exacerbated by public and private policies that reduce the incentives of older persons to continue working, and reduce the incentives for employers to hire them. Increased utilization of health care services, something closely associated with the aging process, requires out-of-pocket payments for drugs, co-payments, deductibles, and premiums.

The two economic problems mentioned above are interrelated in the following ways. First, expenditures on health care, much like other goods and services, require that tradeoffs be made. For example, individuals and households face budget constraints. Therefore, when an elderly person pays out-of-pocket for drugs, co-payments, premiums and deductibles, they must forgo expenditures on other items. Hence, they must forgo their level of consumption of many necessity goods, such as optimal dietary supplements, proper heating, and air conditioning — to mention a few.[13] In addition, the likelihood of living in poverty appears to increase with age for both men and women, with women more likely to live below the poverty level at all ages.[14] Finally, the implication of declining purchasing power, among elderly cohorts, and the concomitant increases in their levels of health care consumption, raises important issues for the public provision of health care services. Faced with scarce and oftentimes declining, general revenues, governments must often make tough choices between the provision of health care services and other social services. The issues are further exacerbated by the particular economic attributes of health care services.

5. The Economic Attributes of Health Care

It has long been established that health care services possesses several attributes which sets them apart from other goods and services in the economy.[15] These attributes include: uncertainty in demand and prices; external effects; and principle-agent failures. It also represents a category of services whose allocation cannot be determined solely by market-based theories. Normative factors, which depend upon prevailing cultural and social norms, and social institutions interplay with these attributes in determining the final distribution of services. One of the most prominent attributes of health care services, are their level of uncertainty in demand. Uncertainty in demand and prices results from the fact that individuals cannot predict with certainty when the onset of an illness will occur, nor do they know with precision the actual costs of services once an illness does present itself. It is precisely for these reasons that individuals (through private or public means) choose to purchase health insurance, to offset the adverse financial consequences of demand and price uncertainty. In addition, an individual's decision to consume health care services may have positive welfare effects. Hence, the consumption of health care services, by the aged and others, may have what economist term "positive consumption externalities" on others.

External effects arise when an individual's level of consumption, of a particular good or service, has negative or positive welfare effects on others, and payment cannot be extracted from those who benefit. The consumption of health care services is best explained in terms of positive consumption externalities. In other words, the consumption of health care services by one person typically has welfare improving effects on another. This is most evident when one considers infectious diseases: inoculation by one individual prevents another from becoming ill. The same may be the case for chronic health maladies, which plague the aged. The aged are part of a larger social community of family members, friends, and others. Their ability to consume chronic health care services has positive welfare effects on one or more individuals in the society at large; hence, the occurrence of positive externalities in the consumption of chronic health care services. Moreover, knowledge for obtaining the optimal health care inputs and providers is difficult and quite expensive to obtain. Consequently, patients are dependent upon physicians and others as their fiduciary agents. These agents are responsible for selecting the optimal inputs into the production of improved health status for the patient. Therefore, tendency for the following market failures arise: the aged patient may purchase services at artificially high rates, as well as consume services in excess of optimal amounts. In an attempt to mitigate such effects, public and private insurers have implemented several utilization review mechanisms, as well as implemented changes in reimbursement policies.[16] Often the provision of health care services requires the input from many individually motivated actors (e.g. physician, hospitals and ancillary personnel); hence, the ability to calculate precisely the actual price of services, *a priori*, is precluded. It is for this reason that public and private financiers of health care services, such as Medicare and commercial insurers, have implemented Prospective Payment Systems (PPS), and Resource Based Relative Value Scale (RBRVS). Each has been specifically designed to offset the level of price uncertainty that prevails in the distribution of health care services to the aged. Finally, acknowledging the particular economic attributes of health care service is a necessary condition for designing optimal health care policy for the aged. Several options for policy design are likely, the most prominent of which is the choice between the public or private provision of services.

6. Public versus Private Provisions

Due to the inherent attributes of health care services, and various organizational forms capable of delivering these services, the decision by local or federal governments to internally produce or externally contract for service production is more complex than a simple dichotomous choice between internal versus external production.[17] Instead, governments face an array of options for external production, such as: contracting with non-profit firms, contracting with for-profit firms or contracting with other public agencies. Consistent with this conceptual framework is that each arrangement has relative production and transaction costs advantages. Governments, therefore, weigh these cost advantages when making their final choices. However, the final choice is constrained by their political and fiscal realities. Such organizational choice models rely upon three major assumptions: (a) that governments have an interest in cost containment in the delivery of health care services. This is due primary to citizens' demands for services and their resistance to tax increases, as well as the concurrent reductions in revenue sharing at the central government level; (b) that service delivery costs include both transaction and production costs. As a result, local governments consider both production and transaction costs efficiencies in making their organizational choices; and (c) that non-profit, for-profit and public firms represent three distinct governance structures. Each governance structure exhibits relative advantages and disadvantages for production and transaction costs economizing. Hence, governments must contend and weigh each of these issues when designing optimal policies for designing health care policies for the aged.

7. Mortality, Work, and Retirement

The relationship between mortality, work and retirement is a very complicated one indeed. Factors such as the extent of intergenerational transfers,[18] structure or prevalence of social security systems, as well as health differentials among population cohorts further confuses the issues. However, it remains evident that the continued decline in the levels of functional impairments among the elderly, along with increases in their active life expectancies[19, 20] has profound implications for work and retirement choices. Even though the effect of longevity on work and retirement choice remains unclear,[21] developed nations can expect years of productive life among elderly age-specific cohorts to increase. This need not imply, however, that increases in productive life years will be associated with increases in work years. In fact, the American population is aging rapidly, living longer, saving less, and leaving the work force at younger retirement ages.[22] This has profound implications for the future solvency of the nation's Social Security fund. Hence, to sustain current levels of well being, it is clear that several changes in both policy and behavior will have to occur: (a) individual private contributions to retirement accounts will have to increase, (b) longer life spans will have to be accompanied by correspondingly longer work years, (c) raising the age of qualification for benefits, or (d) increasing the social security payroll tax. The magnitude of the latter will be determined by several factors, the least of which is the willingness of current generations to subsidize future generations, as well as the ratio of workers to retirees.

In fact, in their study of the relationship between mortality decline and Social Security finances, Lee and Tuljapurkar estimated that each one-year increase in life expectancy would require a 3.6% increase in the Social Security payroll tax or a reduction of 3.6% in benefits received to sustain the system.[21] Moreover, the structure of public and private pension funds can in themselves affect the labor force participation rates of individuals. The retirement income provided by these funds, in addition to Social Security benefits, allows older workers to leave the labor force at younger ages and still support themselves during retirement years.[22] Hence, in estimating the overall impact of increasing the maximum human life span on the fiscal health of a nation, it is imperative that policy makers possess a clear understanding of the interplay between mortality, health status, work, retirement and the structure of retirement accounts.

8. Concluding Remarks

In the 21st century, the demographic realities of aging require that all countries take stock of what an aging population means to them. The question of how one goes about implementing the numerous options to ensure that the nation's elderly citizens are financially secure when needing medical services needs to be considered. While some might focus on improving access to curative and preventive health care services, through changes in reimbursement rates and technology, it is imperative, however, that the particular economic attributes of health care services be considered prior to evaluating policy options. Furthermore, policymakers, at all levels, must ensure that the resources employed, and the delivery systems designed are consistent with the political, institutional, cultural, and regulatory frameworks of their countries. It is only through such systematic planning that policy makers can assure that the optimal programs and policies are in place to satisfy the multi-dimensional needs of an aging population. In addition, the citizenry of these nations must acknowledge their role in the policy process.

It is critical for each person to understand the importance of comprehensive planning for his or her longevity, from their individual contributions to public and private pension funds to the officials whom they elect for designing policy on their behalf. In this paper we have attempted to illustrate the role and interplay of the policy process among various actors, individuals and their representatives, and the tough economic and ethical choice that they must be willing to make in order to assure optimal policy design. Thus far, governments have made great strides in reducing age-related disabilities; however, this has come at some costs to society. The costs have been in terms of resources allocated for age-related research, as well as the increased provision of resources consumed during retirement years. Hence, governments and their citizenry must weigh the various options presented in this chapter when assessing the benefits and cost of designing age-related health care policies.

References

1. Banks, D. A. and Fossel, M. (1997). Telomeres, cancer, and aging: altering the human life span. *J. Am. Med. Assoc.* **278**(16): 1345–1348.

2. Sallar, A. M., Hogg, R. S. and Schechter, M. T. (1996). Survival after age 80. *New England J. Med.* **334**: 537–538.
3. Manton, K. C. and James, V. W. (1995). Survival after the age of 80 in the United States, Sweden, France, England and Japan. *New England J. Med.* **333**: 1232–1235.
4. Lubitz, J., Beebe, J. and Backer, C. (1995). Longevity and medicare expenditures. *New England J. Med.* **332**: 999–1003.
5. Schneider, E. L. and Curalink, J. M. The aging of America: impact on health care costs. *J. Am. Med. Assoc.* **263**: 2335–2340.
6. Manton, K. G., Corder, L. and Stallard, E. (1997). Chronic disability trends in the elderly United States population: 1982–1994. *Proc. Natl. Acad. Sci. USA* **94**: 2593–2598.
7. (1997). *Seen Deadly Mythos: Uncovering the Facts About the High Cost of the Last Year of Life*, Alliance for Aging Research, Washington, DC.
8. Grossman, J. M. and Banks, D. A. (1998). Unrestricted entry and nonprice competition: the case of technological adoption in hospitals. *Int. J. Econ. Business* **5**: 223–245.
9. Lee, K. L. and Mills, A. (1983). *The Economics of Health in Developing Countries*, Oxford University Press, Oxford.
10. (1995). United States Department of Commerce, Bureau of the Census. *Sixty-Five Plus in the United States*.
11. (1995). United States Department of Commerce, Bureau of the Census. *Aging in the United States: Past, Present, and Future*.
12. (1990). United States Department of Health and Human Services, National Institute on Aging. United States Department of Commerce, Bureau of the Census. *Centenarians in the United States*.
13. Glied, S. (1997). *Chronic Condition: Why Health Reform Fails*, Harvard University Press, Cambridge, MA.
14. Rogers, Carolyn, C. (1999). *Changes in the Older Population and Implications for Rural Areas*, United States Department of Agriculture, Research Report Number 90.
15. Arrow, K. (1963). Uncertainty and the economics of medical care. *Am. Econ. Rev.* **53**: 941–973
16. Banks, D., Kunz, K. and Macdonald, T. (1994). *Health Care Reform*, Institute of Government Studies Press, Berkeley.
17. Banks, D. (1996). Transactions cost economics and its applications to health services research. *J. Health Service Res. Policy* **1**(4): 250–252.
18. Lillard, L. A. and Willis, R. J. (1997). Motives for intergenerational transfers: evidence from Malaysia. *Demography* **34**(1): 115–134.
19. Manton, K. G., Stallard, E. and Liu, K. (1993). Forecasts of active life expectancy: policy and fiscal implications. *J. Gerontol.* **48**: 11–26.
20. Sullivan D. F. (1971). A single index of mortality and morbidity. *HSMHA Health Report* **86**: 347–354.

21. Lee, R. and Tuljapurkar, S. (1997). Death and taxes: longer life, consumption, and social security. *Demography* **34**(1): 67–81.
22. Wise, D. A. (1997). Retirement against demographic trend: more older people living longer, working less, and saving less. *Demography* **34**(1): 83–95.

Index